ENGINEERING
Hacker ■ Burghardt ■ Fletcher ■ Gordon ■ Peruzzi ■ Prestopnik ■ Qaissaunee
&TECHNOLOGY

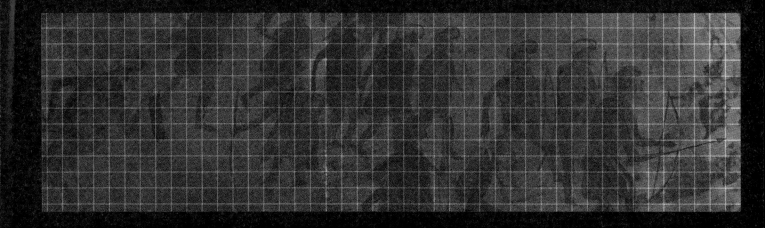

ENGINEERING

Hacker ■ Burghardt ■ Fletcher ■ Gordon ■ Peruzzi ■ Prestopnik ■ Qaissaunee

&TECHNOLOGY

This material is based upon work supported by the National Science Foundation under Grant No. DUE 0603403. Any opinions, findings, and conclusions or recommendations expressed in this material are those of the authors and do not necessarily reflect the views of the National Science Foundation.

DELMAR
CENGAGE Learning™

Australia • Brazil • Japan • Korea • Mexico • Singapore • Spain • United Kingdom • United States

Engineering and Technology
Hacker, Burghardt, Fletcher, Gordon,
Peruzzi, Prestopnik, and Qaissaunee

Vice President, Career and Professional
Editorial: Dave Garza

Director of Learning Solutions: Sandy Clark

Senior Acquisitions Editor: James DeVoe

Managing Editor: Larry Main

Product Manager: Mary Clyne

Editorial Assistant: Cris Savino

Vice President, Career and Professional
Marketing: Jennifer McAvey

Executive Marketing Manager: Deborah Yarnell

Marketing Manager: Jimmy Stephens

Marketing Coordinator: Mark Pierro

Production Director: Wendy Troeger

Production Manager: Mark Bernard

Content Project Manager: Mike Tubbert

Art Director: Bethany Casey

For product information and technology assistance, contact us at
Professional Group Cengage Learning Customer & Sales Support, 1-800-354-9706

For permission to use material from this text or product,
submit all requests online at **cengage.com/permissions.**
Further permissions questions can be e-mailed to
permissionrequest@cengage.com.

Library of Congress Control Number: 2008939730

ISBN-13: 978-1-4180-7389-3

ISBN-10: 1-4180-7389-X

Delmar
5 Maxwell Drive
Clifton Park, NY 12065-2919
USA

Cengage Learning is a leading provider of customized learning solutions with office locations around the globe, including Singapore, the United Kingdom, Australia, Mexico, Brazil and Japan. Locate your local office at: **international.cengage.com/region**

Cengage Learning products are represented in Canada by Nelson Education, Ltd.

For your lifelong learning solutions, visit **delmar.cengage.com**

Visit our corporate website at **cengage.com.**

Notice to the Reader
Publisher does not warrant or guarantee any of the products described herein or perform any independent analysis in connection with any of the product information contained herein. Publisher does not assume, and expressly disclaims, any obligation to obtain and include information other than that provided to it by the manufacturer. The reader is expressly warned to consider and adopt all safety precautions that might be indicated by the activities described herein and to avoid all potential hazards. By following the instructions contained herein, the reader willingly assumes all risks in connection with such instructions. The publisher makes no representations or warranties of any kind, including but not limited to, the warranties of fitness for particular purpose or merchantability, nor are any such representations implied with respect to the material set forth herein, and the publisher takes no responsibility with respect to such material. The publisher shall not be liable for any special, consequential, or exemplary damages resulting, in whole or part, from the readers' use of, or reliance upon, this material.

Printed in Canada
1 2 3 4 5 XX 11 10 09

TABLE OF CONTENTS

What do the followings items have in common: an MP3 player and a running shoe; a cell phone and a car; a kitchen utensil, a seed, and a toothbrush? You might have used all of these items before you even got to school today. Maybe you checked the weather on the Internet before you got dressed. Maybe you texted a friend for a ride to school. Maybe you ate a bowl of cereal for breakfast, then brushed your teeth.

All of these items affect the quality of your life, and all of them, even the seeds that produced the grains in your breakfast cereal, are examples of technological design. In the same way that engineers work continuously to create new cell phone features to keep you connected, they also use technology to create new breeds of grains and other plants to keep you fed.

You interact with technology all day, every day. But do you have a working knowledge of the technologies you and your family use? Is it enough to know when and how to use a device, or should you also understand how the device works? Should you learn more about the effects that device will have on the environment and on society?

Why This Book Matters

Engineering and technology permeate our daily lives. This book matters because modern nations depend on engineered innovations to keep their economies strong and their environments healthy. New technologies emerge continuously. Society's leaders and responsible citizens must make informed decisions about the impact that technologies, both emerging and existing, will have on society. Yet few people fully appreciate and understand the role played by engineering and technology in modern life.

Although technology can improve the quality of our lives, it can also create undesirable consequences. Many of today's major social issues involve a technological dimension. School is the ideal place to begin understanding technology and the role it plays in our culture. Whether technology is used to benefit or to afflict our society is our decision. Students can learn that people can and must control the development and application of technology so that it is used responsibly for the good of human kind with minimal negative consequences.

In this book, you'll learn how engineers transform their knowledge of math and science into practical applications. You'll explore how technological innovations have changed and continue to change our world. And you'll have the opportunity to develop and test your own design solutions to interesting technological problems through the process of informed design.

Through your study of engineering and technology, you will become more technologically literate. Technologically literate people recognize that technology is not magic, nor is it beyond their control. Technology can be understood and therefore designed, modified, and influenced by intelligent, voting citizens. Technological decisions, and public policies stemming from them, must not be left to a technological elite, but rather to a technologically literate citizenry.

How This Book Was Designed

Good textbooks are *engineered*. This textbook was designed by a team of authors, including teachers, engineers, scientists, and technologists. With sponsorship from the National Science Foundation, these national leaders in engineering and technology education worked together to answer a pressing question: "How can we help students develop the technological literacy they need to make informed decisions for tomorrow's society?"

Driven by the national Standards for Technological Literacy, this book addresses the most contemporary technological content, using engaging, pedagogically sound, informed-design activities. Woven throughout the text are passages that will acquaint students with the requirements, responsibilities, necessary personal attributes and attitudes, and educational pathways that will lead to success in the various technological areas.

Features of This Text

This text is rich in features that demonstrate the importance of developing technological literacy:

- "Engineering Quick Take" presents chapter ideas within an engineering context and provides age-appropriate mathematical and scientific analyses.

- "Technology and People" introduces readers to charismatic women and men who have made significant technological contributions.

- "Technology in the Real World" showcases interesting innovations and trends, fun facts, extreme engineering, and socio-technological impacts.

To the Student

- Design Activities are high-interest activities that place chapter ideas into engaging contexts. These activities lead students through the informed-design process, assisting them on their way to developing their own creative solutions to a wide variety of engineering design challenges.

- A companion Student Activity Guide contains Knowledge and Skill Builders (KSBs) that provide the underlying knowledge of concepts and skills students need to reach informed-design solutions, instead of merely engaging in unfocused trial-and-error problem solving.

DESIGN CHALLENGE 1:
Manufacturing an Antacid-Powered Rocket

- ### Problem Situation

We have learned the United States is competing in an increasingly global economy. As the number of scientists, engineers, and technologists in other countries such as China and India is growing, fewer students pursue these careers in this country. Getting young kids interested in careers in science, technology, engineering, and mathematics is a national imperative that many organizations have begun to address. For example, NASA (the National Aeronautics and Space Administration) has developed a number of exemplary materials for this very purpose. One such resource is *3-2-1 POP!*, an activity requiring students to build a small rocket with construction paper or heavy-weight stock paper.

More students might be able to try out the *3-2-1 POP* activity if they had easy access to the material required for the activity. In this Design Challenge, you will develop a means of mass-producing and marketing rocket kits that contain all the material required to build a rocket. You will use the *3-2-1 POP!* activity as a starting point.

- ### Your Challenge

Given the necessary materials, you will research rockets and propulsion systems, develop a prototype rocket, consider alternative materials and propulsion systems, and finally create the jigs, fixtures, and processes for the small-scale mass production of your chosen prototype. You will then design a packaging and marketing strategy that will excite younger students about careers in engineering and technology and encourage more students to pursue these careers.

- ### Safety Considerations

1. Use a great deal of care when working with electricity and power tools.
2. Wear eye protection at all times.

- ### Materials Needed

1. NASA's *3-2-1-POP!* activity
2. Construction paper
3. A variety of heavy-weight (60-110#) paper
4. Flexible plastic sheets
5. A variety of stock pieces of wood and aluminum (block, cylinders, cones, etc.)
6. A variety of fastener materials, including tape, glue, or additional means your instructor may provide
7. Access to a machine shop or hand tools
8. Scissors and an X-acto knife
9. Plastic 35-mm film canisters and other similar small containers with lids
10. Effervescing antacid tablets
11. Plastic packaging bags.
12. Self-adhesive labels
13. Markers
14. Stopwatch

- ### Clarify the Design Specifications and Constraints

Working as a team, you will create and demonstrate a small-scale mass production unit that produces rocket kits promoting careers in engineering and technology. The rocket should be safe to operate and use no hazardous materials. It should be

What Is Informed Design?

As scientists investigate phenomena to contribute to our understanding of the natural world, engineers and technologists design products, structures, and systems that contribute to life in the human-made world. Design teams need members who know how to go about the *process* of designing and who understand the mathematical, scientific, and technological ideas that support effective, functional design. Successful designers understand these underlying ideas well. Therefore, they approach the design process from a knowledgeable, *informed* perspective.

This book offers design challenges at the end of each chapter to help promote its *informed-design* approach to solving design problems. This unique approach prompts students to begin the solution-finding process by first analyzing design specifications and constraints. Once these are understood, students are invited to complete a series of KSBs. KSBs are short, focused, practical tasks that provide students with the important background knowledge and skills they need in order to develop detailed design solutions.

After completing the KSBs, students discuss and propose alternatives and then select an optimal design. This process fosters discussion among students and between students and instructor. Student designs are tested, evaluated, and modified. Finally, students communicate their approaches and design solutions through class presentations and final design reports.

Support for Teachers

A robust Instructor's Resource on CD provides teaching tools to help make classroom preparation efficient and effective:

- A complete instructor's guide that includes correlations to the ITEA standards, instructional outlines, answers to feedback questions, and background for implementing design activities
- Chapter-by-chapter PowerPoint presentations
- A computerized test bank that allows instructors to tailor chapter tests to their assessment needs
- A complete correlation to the ITEA Standards for Technological Literacy

Acknowledgments

The authors and publisher wish to acknowledge the following educators and subject-matter experts who contributed their valuable wisdom and perspective throughout the development process:

Richard Bright, Roosevelt High School, Des Moines, IA
Toss Cline, James Madison High School, Gainesville, VA
Thomas A. Frawley, Past President, New York State Technology Education Association
Nick Gilles, Spring Valley Middle School, Spring Valley, WI
Jason Hancock, Central Dauphin High School, Harrisburg, PA
David A. Janosz, Jr., Executive Director, Technology Association of New Jersey
Thomas Kubicki, Assistant Professor, State University of New York at Oswego
Thomas Liao, Distinguished Teaching Professor, Stony Brook University, Stony Brook, NY
Alto Jo Longware, Ausable Valley Central School District, Westport, NY
Mark Piotrowski, Lower Merion HS, Ardmore, PA
Richard Steven Price, Riverdale Technical Education Center, Riverdale, GA
Albert Rossner, Jackson Area Career Center, Jackson, MI
Tony Suba, Ben Davis High School, Indianapolis, IN
Kirk Woosley, Skyview High School, Billings, MT
Edward Zak, (High School) Past President, New York State Technology Education Association

In addition, the authors acknowledge the following instructors who provided invaluable assistance in field testing the Design Challenges:

Pat Boire, Bolton Central School District, Bolton Landing, NY
Chris DeHann, East Lansing High School, East Lansing, MI
Jenifer Lazare, Austin Independent School District, Austin, TX
Michael Norton, Elgin Independent School District, Elgin, TX
Robert Piotrowski, Keyport High School, Keyport, NJ
Kristy Rhodes, Oppenheim-Ephrata Central School, Oppenheim, NY
Lyma Robertson, Vander Cook Lake High School, Jackson, MI
Becky Thompson, Georgetown Independent School District, Georgetown, TX
Angela Wheeler, Austin Community College, Austin, TX
Lisa Wind, Richfield Springs Central Schools, Richfield Springs, NY

For their extraordinary development work and photo research, the authors and publisher acknowledge the contributions of Ann Shaffer and Mary Pat Shaffer, who have left an indelible impression on this text.

The authors would like to express our deepest appreciation to Mary Clyne, Product Manager for Engineering and Technology Education at Delmar Cengage Learning, whose managerial and publishing expertise gave us constant support during the development process, and whose good humor kept us on task and shepherded the project through its formative stages to its conclusion. Sincere thanks are offered to David Boelio and Jim DeVoe, who, as senior acquisitions editors, provided us with essential conceptual design support; and to Michael Tubbert, the Delmar/Cengage Learning content project manager who was our liaison to the editorial and production staffs. To our families, who supported us through the entire writing process, we express our sincerest appreciation.

We offer this work to all our friends and colleagues in engineering and technology education in the hope that it will contribute to the technological capability and career orientation of our students, as well as to the impact of engineering and technology education on teaching and learning.

About the Authors

Michael Hacker has served as a technology education classroom teacher, department supervisor, university teacher educator, and the New York State Education Department Supervisor for Technology Education. He has co-directed eight large-scale National Science Foundation (NSF) projects that advance K–14 science, technology, engineering, and mathematics (STEM) literacy. For more than forty years, technology education has been at the core of his professional life. He is a member of the prestigious International Technology Education Association (ITEA) Academy of Fellows, received the Epsilon Pi Tau Distinguished Service Citation; the ITEA Award of Distinction, and ITEA State Supervisor of the Year award; and the Institute of Electronics and Electrical Engineers Mathematics and Science Education award. He has authored or edited six books on technology education. He is currently codirector of the Center for Technological Literacy at Hofstra University.

David Burghardt is professor of engineering at Hofstra University. He is the author of ten texts in thermodynamics, diesel engines, and engineering fundamentals. He is a member of the National Academy of Engineering Committee on Understanding and Improving K–12 Engineering Education in the United States. He has codirected eight large-scale National Science Foundation (NSF) projects that advance K–14 STEM literacy. He is currently codirector of the Center for Technological Literacy at Hofstra University.

Linnea Fletcher is department chair of biotechnology at Austin Community College. She is a regional director for the NSF-funded Advanced Technological

Education center grant, Bio-link. She presently serves as a program officer at the National Science Foundation.

Anthony Gordon is director for information technology for Saginaw Public Schools. He consults widely on Science, Technology, Engineering, and Mathematic (STEM) education implementation. A native of England, he served as design and technology advisor in the Staffordshire Local Education Authority and is a recognized leader in design and technology in the United Kingdom. He has edited three books on math, science, and technology integration.

William Peruzzi, a science educator and former principal, presently serves as curriculum coordinator for a New York State Energy Research and Development Authority grant that has placed photovoltaic weather systems and data acquisition systems in fifty New York State schools.

Richard Prestopnik is professor of electrical and computer technology at Fulton Montgomery Community College in New York. He has authored four books on microprocessors and digital electronics and published an invention disclosure as a computer engineer with IBM.

Michael Qaissaunee is founding director of the Mid-Atlantic Institute for Telecommunications Technologies (MAITT). He also serves as coprincipal investigator for the National ICT Center, located in Springfield, Massachusetts. In 2007, he received the Global Wireless Education Consortium (GWEC) Wireless Educator of the Year award.

ENGINEERING
Hacker ▪ Burghardt ▪ Fletcher ▪ Gordon ▪ Peruzzi ▪ Prestopnik ▪ Qaissaunee
&TECHNOLOGY

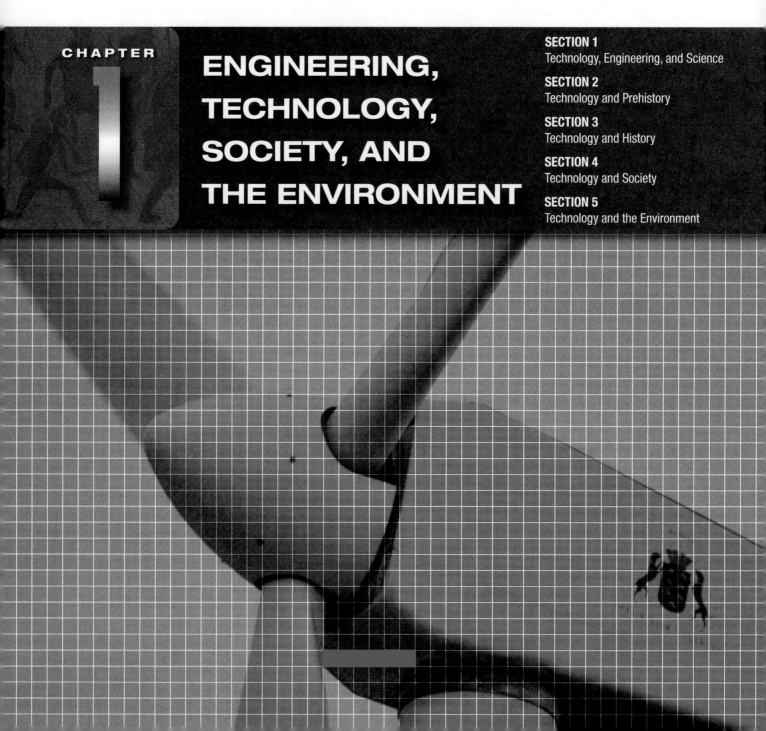

T

AKE A MOMENT to make a list of all the things you have used in the past twenty-four hours that were developed through technology. Your list might include complicated devices, such as a computer, a cell phone, a video-game controller or a car. Your list might also include less complicated devices, such as a bicycle or a stapler. But what about other, even simpler devices, such as a pencil or a toothbrush?

CHAPTER

1

ENGINEERING, TECHNOLOGY, SOCIETY, AND THE ENVIRONMENT

Figure 1.1 | You probably think of complicated devices, such as game controllers, as technology. But what about much simpler devices?

Figure 1.2 | Even a simple pencil is a form of technology.

We define technology as the means by which humans modify the world to address their needs and wants. That means that a toothbrush is a result of technology. So are a spoon, a notebook, and a running shoe. For that matter, the grains in the cereal you had for breakfast are the result of technology because humans selectively breed plants in order to create seeds that produce the best possible crops. Indeed, the seeds from which your cereal grains grew might be patented.

In this book, you will learn how engineers transform their knowledge of mathematics and sciences, such as physics, chemistry, and biology into practical applications. You will explore how technological innovations have changed and continue to change our world. As you will see, modern countries increasingly depend on engineered innovations to keep their economies robust and the environment healthy. New technologies are being adopted at a dizzying pace. Politicians and responsible citizens are expected to make informed decisions about the impact of technology on society and on each other. Yet, few people understand the role engineering and technology play in all this.

People interact with technology every day, but do you have a working knowledge of the technologies you and your family use? Is it enough to know when and how to use a device, or should you also know how the device works and the effects of the use of that device on society and the environment?

Because of the seamless nature of user-friendly technology, people typically have become accustomed to using devices without knowing what makes them operate, the implications of their use, or how they came to be. We tend to think of technology only in terms of its products—cell phones,

Figure 1.3 | Modern grains are a form of technology. They are selectively bred to create the best possible crops.

3

computers, air fresheners, amusement parks, or antibiotics. The knowledge and techniques used to create and market such items—research methods, engineering know-how, manufacturing expertise, technical and marketing skills—are also important to the quality of our lives.

One intent of this chapter is to narrow the gap between technological reality and public understanding. Another is to present a brief history of technology revealing the inevitable environmental trade-offs that occur when we adopt new technologies. If our environment and economy are to be healthy and robust, it is essential that decision-makers consider ethical factors as they grapple with the uses and misuses of technology.

SECTION 1: Technology, Engineering, and Science

Figure 1.4 | Science is both a body of knowledge about the natural world and the process of inquiry that generates this knowledge.

KEY IDEAS >

- The rate of technological development and diffusion is increasing rapidly.

- Technological progress promotes the advancement of science.

- Technological innovation often results when ideas, knowledge, or skills are shared within a technology, among technologies, or across other fields.

- Making decisions about the use of technology involves weighing the trade-offs between its positive and negative effects.

- Throughout history, humans have devised technologies to reduce the negative consequences of other technologies.

Figure 1.5 | Engineers devise practical applications for the knowledge generated by scientists.

You probably have a sense that technology, engineering, and science are related in some way. People typically use these terms interchangeably. However, these terms actually refer to different things. Before you can understand the role technology has played in human history, you need to make sure you understand these terms.

Defining Terms

Science is the systematic study of the natural world. We can think of science as having two main parts:

- A body of knowledge about the natural world
- A process of inquiry that generates such knowledge

Engineering is the practical application of the information acquired by science. Like science, engineering consists of two main parts:

- A body of knowledge about the design and construction of products
- A set of processes and techniques that are the methods for accomplishing the desired aim of creating the products

Generally speaking, science is about understanding the natural world, about discovering and explaining *what is*. Engineering is about humans creating *what has never been.* Think of present-day science and engineering as being tightly coupled. They each use the same tools, but their objectives differ. Science involves the analysis of newly observed or unexplained events or objects. The results of the analyses are then used to model, mathematically or graphically, the particular event or object. In contrast, engineers focus on designing solutions for specific problems. Engineers apply analyses and mathematical or graphical modeling as they design or modify socially useful products.

So how does the term "technology" fit into this? Technology, as we have already seen, is the means by which humans modify the world to address their needs and wants. In other words, technology is a combination of engineering and science. The term technology is used to refer to the:

- tangible artifacts of the human-designed world (e.g., bridges, automobiles, computers, satellites, medical imaging devices, drugs, genetically engineered plants);
- systems of which artifacts are a part (e.g., transportation, communications, health care, food production);
- the people, infrastructure, and processes required to design, manufacture, operate, and repair the artifacts. [NAE, 2002b]

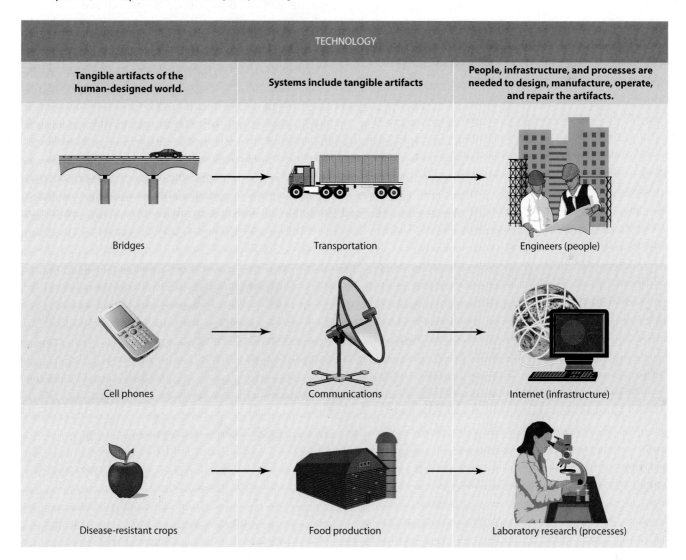

TECHNOLOGY		
Tangible artifacts of the human-designed world.	Systems include tangible artifacts	People, infrastructure, and processes are needed to design, manufacture, operate, and repair the artifacts.
Bridges	Transportation	Engineers (people)
Cell phones	Communications	Internet (infrastructure)
Disease-resistant crops	Food production	Laboratory research (processes)

Figure 1.6 | The term technology refers to artifacts designed by humans, the systems in which these artifacts operate, and the people who make and maintain the systems and artifacts.

5

Figure 1.7 | The media called the Apollo 11's trip to the moon a victory of science, but it is more accurate to call it a victory of technology.

You have probably heard people use the term "technology" to refer to specific things created by engineers—such as computers, or satellites. You might also have heard the term used to refer to scientific pursuits—such as the study of plant DNA (deoxyribonucleic acid) or the analysis of radio signals from deep space. But strictly speaking, technology refers to the many ways that humans have devised to resolve the problems we have faced. For example, we have used technology to acquire food, to travel from place to place, and to build ourselves homes. As you will learn next, we have also used technology to learn more about the world—in other words, to acquire more scientific knowledge.

The Relationship between Technology and Science

Throughout history, technological progress has promoted the advancement of science. Technology is and has been the basis for much scientific research. You cannot easily separate the achievements of technology from those of science. When the Apollo 11 spacecraft traveled to the moon, the media called the accomplishment a victory of science (Figure 1.7), but in fact the feat was a combination of science and engineering. Similarly, the development of disease-resistant plants should not be attributed to science alone. Science is integral to both of these landmark achievements, but it is more accurate to call both events technological developments—that is, a combination of science and engineering.

The rate of technological development and diffusion is increasing rapidly. Technological progress promotes scientific advancement. An example of scientific knowledge that would have remained unknown without a prior technological development is our understanding of human DNA. Thanks to the invention of gene-sequencing machines (Figure 1.8), modern scientists were able to unlock the secrets of human DNA. In a large-scale scientific/technological endeavor known as the Human Genome Project, decoded DNA information provided an enlightening new system of scientific knowledge. This knowledge has led to the design and implementation of health innovations that guide health-care

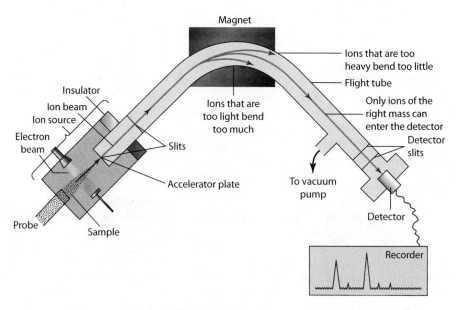

Figure 1.8 | This mass spectrometer, developed as part of the Human Genome Project, helped make the decoding of the human genome possible.

professionals during diagnostic and patient-healing processes. Spin-offs from the Human Genome Project include new understanding of the genetic factors responsible for conditions ranging from age-related blindness to obesity; genes representing these factors can now be detected through diagnostic techniques. Also, hundreds of biotechnology-based products that relate to the healing process are now undergoing or have undergone clinical testing. The ideas, knowledge, and skills resulting from the Human Genome Project spread swiftly to other fields, such as health care and diagnostic technologies.

Figure 1.9 | The inventing of the magnifying glass and the telescope allowed humans to see that the world was much more complicated than they had imagined.

You can probably think of other technological developments that allowed humans to share what they had learned about the world. A simple example is the magnifying glass, which allowed humans to observe things that they would otherwise not be able to see. Another example is the deep-water sailing ship, which allowed scientists, such as Charles Darwin to travel around the world, collecting specimens and observing and recording scientific data that helped him theorize.

Technological advances, such as the development of gene-sequencing machines, are made possible by innovation, which is the process by which preexisting products or ideas are transformed into new techniques and more useful or economical products. Technological innovation often results when ideas, knowledge, or skills are shared within a technology, among technologies, or across other fields. The rate of technological development and diffusion was slow during most of human history, but eventually the trial-and-error technological techniques that had been so common earlier were largely replaced (during the second industrial revolution) by an emphasis on research and the application of scientific understandings. In today's world, technological advancement is increasing rapidly due to the effects of research and the people and organizations that support change.

Solving Problems through Technology

Ultimately, people use technology to achieve the goal of solving human problems. In their pursuit of useful and practical solutions to vexing problems, such as hunger and disease, scientists and engineers do not always behave according to common expectations. Some scientists choose to become involved in designing practical applications of their discoveries. When they do so, they act like engineers. As engineers develop technologies or solve problems, they may need unknown information about the natural world. If they choose to obtain and explain that new knowledge themselves, through research, then they behave like scientists.

Technology Brings Benefits and Burdens

Making decisions about the development and use of new technologies involves weighing positive and negative effects. Technological advances involve trade-offs—giving up one thing in return for another. Be aware that while a new tool, product, or technique may benefit humankind, the same technology may, at the same time, burden society. Technology always influences history. Positive and negative consequences occur whenever a new product or process comes along. These consequences affect people and the environment. Negative consequences can be particularly harsh when they are unanticipated. For instance, radioactivity was discovered around the start of the twentieth century, and early in that century, glowing radioactive radium dials began to be sold that made watch dials easy to read during darkness. Everyone, was pleased, including the watchmakers. This innovation turned out to be a burden when those who were hired to brush the radium onto the watch faces began to die of tongue cancer from licking the brushes.

Throughout history, humans have devised technologies to reduce the negative consequences of other technologies. For example, recently, the pollution caused by gas-fueled automobiles has led to the invention of new energy sources for powering automobiles, including electricity. Can you think of other examples of new technologies that have been developed to counteract the effects of older technologies?

Figure 1.10 | The automobile has benefited humanity in many ways. Do you think the pollution it causes is an acceptable trade-off?

SECTION ONE FEEDBACK >

1. Use "science" and "technology" in a sentence.
2. Name the two main parts of engineering.
3. Name one technological development (other than those described in the chapter) that led to new scientific knowledge.
4. Describe one difference between science and engineering.
5. Which of these human activities is oldest—science, engineering, or technology?

KEY IDEAS >

- The evolution of civilization has been directly affected by, and has in turn affected, the development and use of tools and materials.

- The nature and development of technological knowledge and processes are functions of the setting.

Now that you have a better understanding of the nature of technology, you are ready to chart technology's progress from prehistoric times to the present day. As you will see, technology's history is largely the history of tools and power sources. This section and the next explore:

- The materials used during a particular period to create tools and other objects
- The main power sources used during a particular period
- The effects of innovation on the environment and human health throughout history

When discussing technology, it is helpful to refer to specific ages, or periods, in human history. Figure 1.11 shows the name, time range, and major innovation associated with each age. You will learn more about the Iron Age, the Bronze Age, and other ages of history throughout this section.

The Stone Age

Two million B.C.E. to about 3000 B.C.E. (B.C.E. means "before the common era," which sometimes is written as "B.C.," meaning "Before Christ.")

Stone Age cultures left no written records. The lack of written records led anthropologists—scientists who study human beings and their ancestors—to

AGE/ PERIOD	STONE AGE	BRONZE AGE	IRON AGE	MIDDLE AGES	WESTERN TECHNOLOGIES EMERGE	INDUSTRIAL REVOLUTION	20TH CENTURY	21ST CENTURY
Time Range	2,000,000 BCE to 3000 BCE	3000 BCE to 500 BCE	500 BCE to 500 CE	500 CE to 1400 CE	1400 CE to 1750 CE	1750 CE to 1900 CE	1901 CE to 2000 CE	2001 CE to 2100 CE
Major Innovation	Stone Tools, Use of Fire 	Smelting Metals 	Iron Tools 	Printing Press 	Energy from Coal 	Industrial Centers 	Airplanes 	?

Figure 1.11 | This time line shows technological developments throughout the ages.

Figure 1.12 | Stone Age humans were hunter-gatherers who obtained food by hunting, fishing, and foraging.

Just as mammals and birds today transform natural objects, such as twigs or stones into tools to help solve their problems (Figure 1.13), our ancestors made similar use of natural objects. Perhaps millions of years passed before someone purposely used a sharp stone to construct a tool. Much later, tool making was conducted on a regular basis. Later still, during the Stone Age, such tasks were carried out at specific sites by specialists.

characterize the time period and the people who lived then as "prehistoric." Can you imagine the difficulties involved in trying to recreate the construction of items made during times when people had no way to leave a written record of what they did?

Stone Age humans were hunter-gatherers; they obtained food by hunting, fishing, and foraging (Figure 1.12). Their main power source was their own muscles. The material transformed into tools during this age was primarily stone.

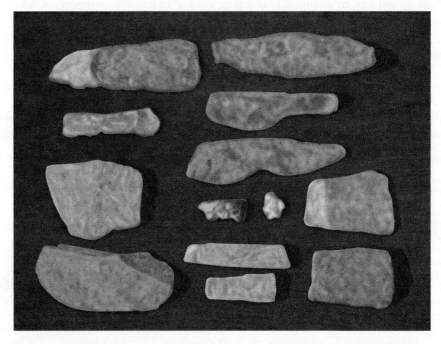

Figure 1.14 | Our Stone Age ancestors became experts at transforming stones into useful tools, such as these artifacts.

Figure 1.13 | A chimpanzee can sharpen a stick into a spear using its teeth. This is just one example of an animal making a tool to solve a problem.

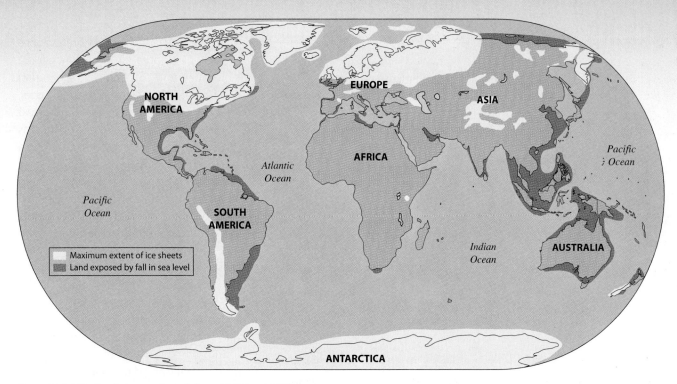

NORTH AMERICA

EUROPE

ASIA

AFRICA

SOUTH AMERICA

AUSTRALIA

ANTARCTICA

Atlantic Ocean

Pacific Ocean

Pacific Ocean

Indian Ocean

Maximum extent of ice sheets
Land exposed by fall in sea level

Figure 1.15 | The last ice age occurred about 10,000 to 15,000 years ago.

Figure 1.16 | Stonehenge, in southern England, is an example of prehistoric construction that suggests that its builders kept track of the relative positions of the Sun and the Moon through the strategic orientation of the structure itself.

Energy is the capacity to do **work**; work is done when energy is used to create a force that causes an object to be moved over a distance. **Power** is the rate at which work is done. Scientists typically measure work in units known as joules. Power is measured in units known as watts. If you do 100 joules of work in one second (using 100 joules of energy), the power expended is 100 watts.

Stone Age humans lived a nomadic life, traveling in small groups and surviving by gathering food through hunting, fishing, and foraging for edible plants. They learned how to acquire, use, and control fire. Taming fire was a technological development that contributed to the survival of early humans. Out-of-control fires would have been a major cause of periodic but temporary changes in the landscape.

Toward the end of the last ice age, about 10,000 to 15,000 years ago, some human communities made the transition to a more settled way of life. Evidence of this transition includes ancient tools, human remains, and prehistoric art.

The domestication of animals—breeding animals for a life in close association with and to the advantage of humans—marked a switch in power emphasis. The muscles of domesticated animals became an even more important source of power than human muscles.

As humans began to end their nomadic lifestyle, they had more time to develop agricultural innovations that resulted in reliable harvests and increased food supplies. Over time, they learned how to manufacture special farming tools. These tools, which included sickles and grinding stones, were used for tending, harvesting, and processing crops. Just as the development and use of tools and materials evolved, so did civilization evolve, in this case from a nomadic life to an agricultural life.

Manufacturing refers to the process of making raw materials into products by hand or by machinery. **Materials** are the substances of which something is composed or can be made. **Raw materials** are crude or processed materials that can be converted, by manufacture, further processing, or combination into useful products.

ANCESTOR	FARM ANIMAL
Jungle Fowl	Chickens
Wild Boar	Pigs
Tarpan	Horses
Auroch	Cattle
Mouflon	Sheep
Capra Aegagrus	Goats

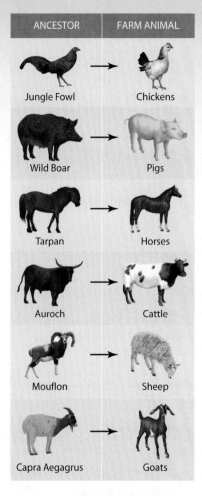

Figure 1.17 | Our domesticated farm animals had wild ancestors, but were bred to live with and help humans.

These technological changes took place slowly over long periods of time. In response to the need for food and protection, people became adept at devising tools featuring points or barbs. The stone-headed spear, arrow, and the harpoon were in widespread use (Figure 1.19) before 3000 B.C.E. Humans first used stone tips some time between 21,000 to 17,000 years ago.

The study of artifacts—objects created by humans for practical purposes—left by ancient peoples, as well as observations of people from current civilizations who continue to use Stone Age technologies, reveals much about the culture of early human civilizations. Culture refers to the customary beliefs, social forms, and material traits of human groups.

Since this early period of human civilization, the evolution of civilization has been continually and directly affected by the development and use of tools and materials. The nature and development of technological knowledge and processes are functions of the setting. Typically a new setting—a necessary combination of circumstances—would spur a new development. For example, the advent of an ice age might spur people to develop a technique for stitching together furs to create warm coverings for their feet. Can you see that the development of this technological knowledge and process is a function of the setting?

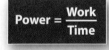

$$\text{Power} = \frac{\text{Work}}{\text{Time}}$$

Figure 1.18 | You can use this formula to calculate the amount of power expended when a human or a machine performs a specific job.

Figure 1.19 | Stone tips for spears, harpoons, and arrows were already in wide use before 3000 B.C.E

It is interesting to note that the Yanomamo tribe, native South Americans from the Amazon basin, continued to use Stone Age technologies until the 1950s (Figure 1.20). Even today, the Yanomamo have kept to their original ways of living, even though they now save time by using imported metal tools. Perhaps they never experienced the appropriate setting thought necessary to drive them to develop their own metallurgical technologies.

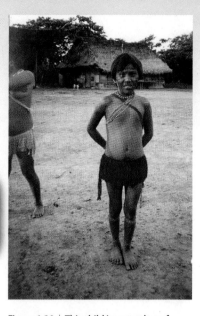

Figure 1.20 | This child is a member of the South American Yanomamo tribe, which, until recently, relied on Stone Age technologies. For example, their axes, machetes, clay cooking pots, fishhooks, and fishing line all were made from nonmetals.

SECTION TWO FEEDBACK >

1. Name an important communication behavior that Stone Age humans lacked.
2. How have prehistoric artifacts helped us to understand prehistoric cultures?

SECTION 3: Technology and History

KEY IDEAS >

- Throughout history, technology has been a powerful force in reshaping our social, cultural, political, and economic landscapes.

- The Iron Age was defined by the use of iron and steel as the primary materials for tools.

- The Middle Ages saw the development of many technological devices that produced long-lasting effects on technology and society.

- Early in the history of technology, the development of many tools and machines was based not on scientific knowledge but on technological know-how.

- Selecting resources involves trade-offs between competing values, such as availability, cost, desirability, and waste.

- Inventions and innovations are now the results of specific, goal-directed research.

- The Renaissance, a time of rebirth of the arts and humanities, was also an important development in the history of technology.

- The Industrial Revolution saw the development of continuous manufacturing, sophisticated transportation and communication systems, advanced construction practices, and improved education and leisure time.

- Energy resources can be renewable or nonrenewable.

- New technologies create new processes.

Throughout history, technology has been a powerful force in reshaping the social, cultural, political, and economic landscapes. In this section, you will examine some examples of how history has been shaped by emerging technologies.

The Bronze Age: 3000 to 500 B.C.E. During the Bronze Age, humans developed and mastered a new and extremely important technique—smelting. Smelting is

13

Figure 1.21 | In the Bronze Age, humans learned to make alloys that had properties different from any of their individual ingredients. These alloys were then used to make strong, effective tools, such as this axhead.

the extraction of a metal from its source (typically referred to as an ore). In the early part of the Bronze Age, humans learned how to smelt copper to make objects from the molten metal (Figure 1.21). Later, they learned to smelt tin and combine it with copper to create an alloy (a combination of metals) known as bronze. Bronze was an improvement over copper because bronze is more durable. Also, because bronze melts at a lower temperature, it was easier to work.

The Bronze Age began in an area of the Middle East known as the Fertile Crescent, which consisted of parts of present day Pakistan, Iraq, and Syria. Thanks to rivers that overflowed seasonally, the Fertile Crescent had good soil in which crops flourished. The area also benefitted from the abundant sunshine of the Mediterranean climate. The long, dry season with a short period of rain made the Fertile Crescent suitable for the first attempts at domesticating plants. The inhabitants found that small plants with large seeds, such as wheat and barley, were especially suitable for domestication because they were easy to harvest and store and because wild forms of the plants were widely available. The Fertile Crescent comprised a large area of varied geographical settings and altitudes that other regions lacked. These special situations made an agricultural life more profitable for former hunter-gatherers. The great variety of ecosystems meant a wider variety of plants to domesticate. This, in turn, made farming more profitable than nomadic life.

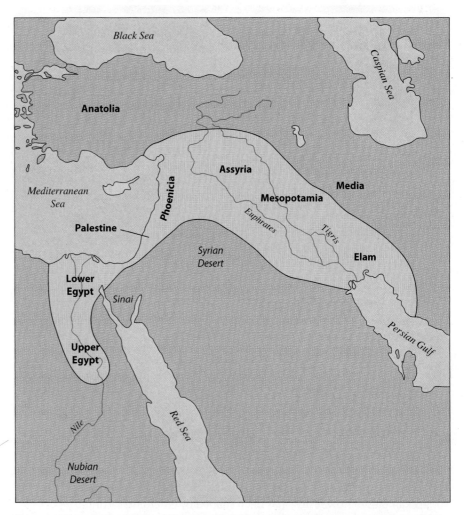

Figure 1.22 | Fertile Crescent Map: The Bronze Age began in the Fertile Crescent, an area made up of parts of present-day Pakistan, Iraq, Israel, Egypt, and Syria, where crops thrived and complex societies developed.

A thriving agrarian, or farming, culture arose in the Fertile Crescent. The plentiful food and generally favorable conditions led to an increase in the human population.

As human society expanded, it naturally became more complicated. One important example of this increasing complexity was the division of labor, through which work was broken down into its component tasks and distributed among different persons or groups, thereby increasing efficiency, production, and profits. Division of labor, in turn, created a need for specialized tools made from copper and then bronze. This demand for tools, in turn, resulted in expanded trade routes that fostered economic growth throughout the region. Thus, the development of smelting technologies allowed society to expand and become even more complicated. However, the consequences of smelting technologies were burdensome as well as beneficial. Smelting ores required a massive amount of fuel, entailing the destruction of up to four acres of forest per year in the Fertile Crescent. The denuded forests and smoke-filled air likely became a health and environmental problem for the areas now occupied by Greece, Turkey, Cyprus, and Lebanon.

Figure 1.23 | In the Bronze Age, humans learned to smelt ores in forges in order to create stronger tools.

The Iron Age: 500 b.c.e. to 500 c.e. (c.e. means "common era.") The Iron Age was defined by the use of iron and steel as the primary materials for tools. The first step in this development was learning how to smelt iron ore. This technology likely was developed by metallurgists from Asia Minor (Figure 1.24), located east of present-day Greece between the Mediterranean and the Black seas. This region was an important center of early civilization. Extracting iron from its ore was a major technological achievement because it allowed people to manufacture tools that were stronger and less expensive than bronze tools.

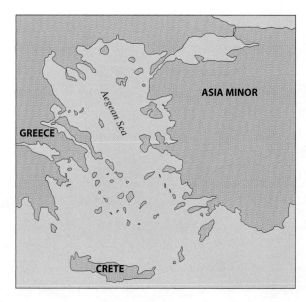

Figure 1.24 | The technique of iron smelting is believed to have been developed by metallurgists in Asia Minor, which is the region that corresponds roughly to modern-day Turkey.

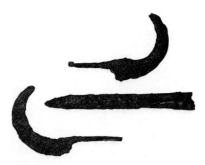

Figure 1.25 | Iron farm tools produced during the Iron Age were stronger and less expensive to make than bronze tools.

Some civilizations did not progress from stone, to bronze, and then to iron. For example, some regions lack deposits that contain the tin needed to make the alloy bronze. Over time, those civilizations that skipped the development of bronze progressed directly from copper to iron. This is another example of how a civilization's setting can affect the development of technological knowledge and processes.

Figure 1.26 | This farmer, working land in Madagascar, Africa, is taking advantage of technology developed in the Iron Age.

Among other things, Iron Age technology produced the iron-tipped plowshare, which permitted people to cultivate heavier soils and to plow more deeply, which in turn resulted in higher profits and increased food supplies. Over time, traders carried iron-smelting technology, along with iron products, from Asia Minor to other regions.

As civilizations advanced technologically in China, Egypt, Greece, Persia, and Rome, they became more dependent on water supplies. Water powered many of their industries, providing the force that did the work in enterprises, such as

Figure 1.27 | Peat cut from bogs was once a major source of fuel.

grain-grinding mills and iron-smelting facilities. Wood and peat fires then served as energy sources for forging metals (manufacturing tools by heating and hammering metal), heating homes, and bathing.

During the Iron Age, tooling—shaping, forming, or finishing with a tool—led to the practice of small-scale metal grinding. Since ductile iron was not discovered until the mid-twentieth century, workers had to grind iron objects that had been cast, and the inevitable breakages resulted in considerable waste of materials. The accumulation of discarded material from such techniques likely created one of the first solid-waste disposal problems for humans.

Figure 1.28 | In the Middle Ages, horses provided the muscle power required to grow crops.

The Middle Ages: 500–1500 c.e. The Middle Ages saw the development of many technological devices that produced long-lasting effects on technology and society. In the early Middle Ages, Europe was in turmoil due to repeated invasions following the collapse of the Western Roman Empire. As a result, industrial activity and population growth decreased. People felt unprotected and were afraid to travel or carry goods great distances by land. Manufacturing for export and trade purposes dropped off sharply, as did profits.

Horses became the main muscle power source for agriculture, and sailing ships were used to take advantage of wind power, eventually replacing the power that had been provided previously by groups of humans who sat on benches and rowed using oars.

Many technological devices developed during the Middle Ages produced long-lasting effects on technology and society. They included:

- a three-field system of crop rotation that involved planting legumes to replenish the soil with nitrogen and letting a third of each field lie fallow (rest) every year
- mills for processing cloth, brewing beer, and crushing pulp for paper manufacture

Early Middle Age technologies had a somewhat limited impact on the environment because industrial activity and population levels generally were in decline. Crops failed for several years following cooler and wetter summers and earlier autumn storms. Widespread famine resulted because people lacked technologies, such as refrigeration that would have allowed them to stockpile foodstuffs. Bubonic plague, commonly called the Black Death, was one of the worst natural disasters in history—it killed off a third of the European population in the mid-1300s. (Bubonic plague is caused by bacteria carried in fleas that leap from rats to humans and transmit bacteria through their bites.)

It was 400 years before Europe's population returned to its previous level, but toward the end of the Middle Ages, growing human populations became concentrated within cities. There, emerging technologies increasingly fouled the air and water, and untreated human and animal wastes contaminated water sources, resulting in fatalities from diseases. Incomplete or inaccurate beliefs about the natural world also contributed to the dangers of living during this time. For example, people believed that bad-smelling air was the cause of fatal diseases, and even most physicians failed to realize that washing hands was important.

Figure 1.29 | In the first outbreak of the bubonic plague, in the 1300s, 25 million people died in five years. By the 1600s, people still had no idea how to defend themselves from the plague. Desperate doctors devised plague robes like this one. They were covered with a smooth, waxy substance that was supposed to prevent the contagion from sticking to them; the mask's beak was stuffed with perfumed substances.

Western Technologies Emerge 1400–1750 c.e. The Renaissance, a time of rebirth of the arts and humanities, was also an important development in the history of technology. This period extended approximately from the 1400s to the middle of the 1600s. During the Renaissance, the arts and humanities flourished again after a long interruption during the Middle Ages. Important innovations included the printing press, clocks, gun powder, eyeglasses, flush toilets, microscopes, and submarines.

Figure 1.30 | Coal's introduction as a major power source increased industrial output, but it also resulted in severe air pollution.

Coal is just one type of **fossil fuel**—that is, a substance that was formed below the earth's surface long ago from decaying plant and animal remains. Petroleum and natural gas are also fossil fuels.

Beginning about 1500 C.E., Western Europe entered a period of colonial expansion. That is, countries, such as England, Holland, Spain, and Portugal expanded their empires to control resources found in what became their colonies. As European nations expanded into the Indian Ocean region and the Americas, goods from those areas were brought back to Europe. Selecting resources involves trade-offs between competing values, such as availability, cost, desirability, and waste. For example, leaders of European nations faced many trade-offs when they decided to import resources from their colonies. They decided that:

- resources were more available in their colonies than at home;
- the low cost of colonial resources would ensure sufficient profit to offset the time to transport the raw materials and the risk of events, such as pirate raids.

Manufacturers and merchants appeared to value gaining new resources more than the negative impacts that befell the colonies, such as slavery and environmental degradation. An unexpected consequence of these decisions was that these new goods and the profits they brought fostered new social habits and the development of innovative manufacturing techniques for processing the new resources.

Increased production required increased energy use. Coal replaced wood as the power source for smelting. The use of coal improved industrial output, but it worsened air pollution problems. The benefits of expansion to new colonies improved standards of living for Western Europeans. But, at the same time, the increased mining and deforestation of colonies, as well as burgeoning industrial and manufacturing processes at home worsened environmental problems.

The Industrial Revolution: 1750–1900 C.E. The Industrial Revolution saw the development of:

- continuous manufacturing;
- sophisticated transportation and communication systems;
- advanced construction practices;
- improved education and leisure time.

This important period in human development was not a sudden, local change; it was a *thorough* change that affected the entire world. Agriculture, economic policies, and social structures were never the same again.

Each major change was interrelated; increased activity for one brought increased activity for the others. For example, improved yields in the agricultural sector were driven by the following changes:

- The transition from a system in which fields were held in common by a village to a system of private land holdings
- Improvements in farming techniques, including improved breeding of livestock, control of insects, irrigation and farming methods, and development of new crops
- The use of horsepower in the fields to replace oxen, the previous main source of power

Figure 1.31 | In this advertisement for a glass factory, you can see several smokestacks, evidence of a large operation that produced far more output than the much smaller cottage industries of earlier times.

The increased food supply enabled people to move from farms to urban areas, where they found work in various industries. An industry is a distinct group that provides systemic labor for the purpose of creating something of value. More raw materials derived from

Figure 1.32 | In eighteenth-century America, carding, spinning, and weaving to make cloth were family affairs, with all three tasks carried out in a single household.

colonial expansion boosted the economy and created opportunities for innovators to improve industrial organization. Rates of production and overall efficiency improved, resulting in higher profits. Food became more widely available to more people.

Prior to 1760, textiles manufacturing was a cottage industry—that is, an industry that relied on family units working at home, using their own equipment (Figure 1.32). In a typical family textile business, the children might be responsible for untangling the cotton or wool fibers, the mother for spinning the fibers into thread on a spinning wheel, and the father for weaving the thread into cloth.

The viability of home textile production was challenged in the mid-1760s by the introduction of an automated spinning device called the spinning jenny. The spinning jenny could spin threads simultaneously much faster than human hands could. Unlike family workers in cottage industries, who relied on their own muscle power to complete their work, textile factories (featuring devices, such as spinning jennys) were located on rivers so that they could draw on the relentless flow of water to operate machines. Factories also provided the advantage of allowing manufacturers to produce finished goods in a single location, rather than subcontracting the work to various family businesses in scattered locations.

Power sources continued to change as innovations continued to emerge. The steam engine (Figure 1.34) enabled factory machinery to work more efficiently, and locomotives moved raw materials and products quickly from place to place. In the 1880s, new technologies allowed people to derive electrical power from new sources, such as coal-driven steam. Electrical lighting changed how people worked and conducted business. Labor was no longer limited by how long natural daylight was available.

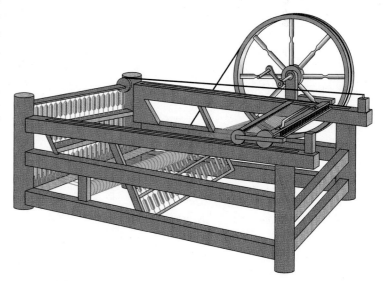

Figure 1.33 | The spinning jenny, an automated spinning device that could spin multiple threads at one time, threatened the viability of home textile industries.

19

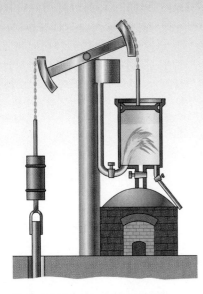

Figure 1.35 | The first internal combustion engines, precursors to modern automotive engines like this one, were introduced in the late 1880s.

Figure 1.34 | Innovations, such as this early steam engine gave factory machinery the power to work more efficiently. Steam engines were also adapted for use on locomotives, allowing people to move raw materials and products over land at speeds that had never been seen before.

In the mid to late 1800s, internal combustion engines were introduced (Figure 1.35). These engines obtained mechanical energy directly from the expenditure of chemical energy from fuel burned in a combustion chamber. The engines typically ran on gasoline derived from petroleum, a fossil fuel. These new engines helped reduce the cost of transporting people and products over land and water.

A by-product of the industrial revolution was environmental pollution levels that were similar to the levels we experience today. Increased population of humans and domesticated animals, filthy and unregulated factory towns, and periodic outbreaks of fatal diseases were persistent problems of the late 1880s. Great factories consumed coal in huge quantities, resulting in health-threatening air pollution and chemical discharges. Increasing amounts of untreated waste in factory towns left over from industrial processes, as well as waste from humans and domestic animals, greatly affected the environment and human health.

A so-called Second Industrial Revolution took place at the end of the nineteenth century. It featured a lasting change in the pattern of innovation. Earlier, the

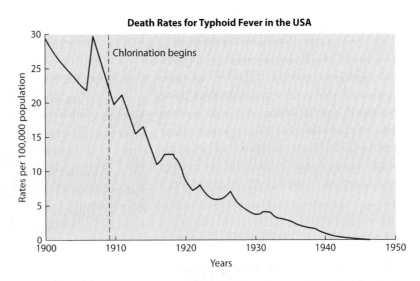

Figure 1.36 | Typhoid fever is an extremely debilitating and sometimes fatal intestinal disease that is caused by ingesting contaminated food or water. Throughout the nineteenth century, it posed a constant menace to public health. In the twentieth century, the introduction of chlorinated water vastly reduced the threat. Source: U.S.; Centers for Disease Control and Prevention, Summary of Notifiable Diseases, 1997.

development of many tools and machines was based not on scientific knowledge but on technological know-how. The Second Industrial Revolution instead was characterized by a switch to an emphasis on the results of research. This switch contributed to the rapid development of inventions and innovations having to do with petroleum, chemicals, electricity, and steel. For the first time in history, technological innovations most often were achieved by those engineers who had learned to apply structured, goal-directed research methods that involved a blend of scientific and technological knowledge and techniques. For instance, after the British scientist Michael Faraday conducted research that demonstrated how to produce electrical current, later inventors, such as Thomas Edison in the 1880s, developed products, such as the light bulb and electricity generating facilities that improved life by supplying people with electricity to provide light and power machines.

Twentieth-Century Technology 1901–2000 c.e. New technologies create new processes. During any ten-year period following World War I, more technological advances occurred than the sum total of new technological developments *in any previous century*. The innovations produced global benefits and burdens. Enlightened scientific methods and informed engineering design led to the rapid development of new technologies and the improvement of existing technologies. The problem of automobiles that were too expensive for the masses drove Henry Ford to develop a new set of processes. He established the assembly line, making the automobile affordable to the common person. In a similar way, the invention of propellers led to new processes that soon had airplanes soaring overhead. These new technologies served as a catapult for new development. Changes in power technologies resulted in jets, rockets, and then space shuttles replacing each other as the fastest, most powerful vehicles of flight.

Figure 1.37 | Henry Ford's introduction of an affordable automobile transformed the American landscape. This early highway, near Englewood, New Jersey, was bordered by street lamps and a pedestrian walkway.

Figure 1.38 | Before the twentieth century, clothing was made from natural fibers, such as wool and linen. Modern clothes can be made from a host of artificial fibers, such as nylon and rayon.

Iron was alloyed with other metals and materials to produce stronger construction products and appliances. Artificial fibers became commercially important, and plastics were used to great effect in the metallurgy and ceramic industries.

The twentieth century's first fundamental power innovation occurred in 1945, when scientists and engineers of the Manhattan Project (a joint effort of the U.S., Great Britain, and Canada to develop nuclear weapons during World War II) succeeded in releasing energy from the uranium atom's nucleus by splitting it into two smaller fragments, accompanied by an enormous release of force.

Other than that important event, the century's progress in energy-related matters consisted of improvements to existing systems. The internal-combustion engine was continuously improved, addressing the needs of road- and water-vehicles and airplanes. Gas turbines were developed that compressed and burned air and fuel in a combustion chamber. The exhaust jet released by this process was used to propel engines forward. The fuel for turbines—kerosene or paraffin—was refined from petroleum, which was in turn manufactured from crude oil. As a result, industrialized nations came to depend on the producers of crude oil. Petroleum became and continues to be a raw material of immense economic value, as well as the focus of international, political, and military conflict.

New technologies have traditionally been considered a solution to the problems created by old technologies. The twentieth century's vast increase in fossil fuel consumption depleted natural resources and produced extreme local and global climate changes, such as air pollution and higher average temperatures worldwide. Decreases in natural resources were offset for awhile by advances in drilling technologies. But, toward the end of the twentieth century, humans came to realize that devising newer technologies to reduce the impact of older ones was not enough to solve recurring energy crises. Instead, they began to focus on the differences between types of energy sources.

Figure 1.39 | Turbine engines generate enough power to propel a jet forward.

Figure 1.40 | Solar energy is an endlessly renewable source of power.

Energy sources can be renewable or nonrenewable. A renewable energy source is one that can be replenished naturally in a short period of time, such as wind, solar radiation, flowing water, geothermal sources, and biomass. A nonrenewable energy source is considered nonrenewable because it can not be replenished (made again) in a short period of time. Most are fossil fuels that come out of the ground as liquids (crude oil), gasses (propane and other natural gasses), and solids (coal), but there are some nonrenewable energy sources that are *not* fossil fuels, such as uranium ores. Nations began to use increasing amounts of renewable energies, such as solar and wind power, in an attempt to decrease dependence on fossil fuels. This quest for solutions continues today.

Inventions and innovations are now the results of specific, goal-directed research. The economic costs of growth are not usually included when "progress" is calculated. As emerging nations struggle to improve standards of living, they cut and clear forests to provide land for agriculture. In places, such as the Amazon

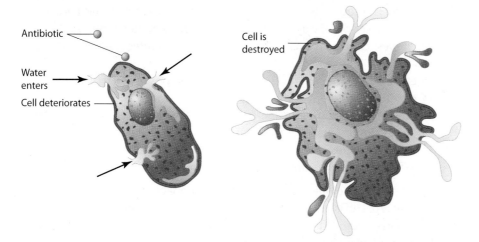

Figure 1.41 | The loss of biodiversity in the Amazon Basin of South America could mean the loss of important disease-fighting substances, such as antibiotics. This diagram shows an antibiotic disrupting the cell membrane of a germ.

23

Basin of South America, deforestation has led to major losses in biodiversity. **Biodiversity** refers to the number of species present in an area. There are countless more species in forests than there are on farms. From research on rainforest species of plants, animals, and microorganisms, people have discovered valuable disease-fighting substances, such as antibiotics. As habitats are destroyed and species die off, it becomes clear that the unanticipated costs of decreasing biodiversity may far exceed the economic value of new farmlands. It is a cost which present and future generations will pay.

Twenty-First Century Technology: A Glimpse into Our Future What new technologies will arise in this century, the twenty-first century? Which existing technologies will evolve through a series of refinements to become a new invention? Will some new technologies become profitable, yet provide minimal damage to the environment?

Some futurists, people who study and predict the future on the basis of current trends, believe that change is the one sure constant. They argue that:

- All technologies end, so every technology that is in use today will someday be replaced.
- Employment in industries will be surpassed by employment in occupations that provide services.
- Information technologies, which emphasize the processing and exchange of information, are rapidly becoming more important than industrial and manufacturing technologies.

Some futurists predict that in the future we will make use of technologies such as:

- Biosensors: Handheld devices that someday will be used to monitor postoperative and trauma patients for signs of infection in their bloodstreams.
- Cars that drive themselves: Such cars might reach speeds of up to 60 mph guided by lasers, a video camera, and a computer that recognizes signs and detects obstacles.
- Extreme ultraviolet microprocessors: X-ray microscopes that will be used to scan cells much as we currently use X-rays to create computerized tomography (CT) scans of bones and internal organs; this new technology will help researchers learn how organisms function at the cellular level.
- quantum cryptography: A system designed to harness subatomic particles to create a hacker-proof way to communicate over computer networks.
- nanotubes: Extremely small, tube-like carbon structures that will some day serve as:
  - extremely strong reinforcement fibers for composite materials;
  - conductors in flat panel displays;
  - tools for scanning tunneling (a powerful technique for viewing surfaces at the atomic level), magnetic resonance (medical imaging that provides detailed images of the body in any plane);
  - extremely small manipulators and tweezers.

As these new technologies emerge, others will probably fall into disuse or die off. Those that may die off include handwritten checks, the space shuttle, sign language, the fax machine, traditional AM-FM radio, broadcast television, and communication structures based on wiring.

What can you predict as additional examples of emerging and/or dying technologies?

Design a Video Game

In 1958, William Higinbotham, the head of the instrumentation division of Brookhaven National Laboratory (BNL) was preparing for the annual BNL Visitors Day. (Brookhaven is a government facility operated for the United States Department of Energy, and it specializes in nuclear physics research.) Thousands of people would be arriving to tour the labs and view exhibits set up in the gymnasium. He wanted the visitors to be impressed by what his specialized research and development (R&D) division could design and build. But he knew from past visitor days what people really liked; it was not fixed, unmovable displays. He came up with a creative idea for a hands-on display—a video tennis game.

The game his BNL engineering team developed may have been the very first video game ever played. From the crowd reaction, Higinbotham knew the display had been successful. He soon needed to use the display's equipment elsewhere, and the display was taken apart. Later he said that, had he realized how significant this innovation would turn out to be, he would have taken out a patent, one that would have been owned by the U.S. government.

His game had some features similar to those of video games developed later. His video, showed the edge of a tennis court, with the edge of the net perpendicular to it. This all was simulated on a screen through a vertical, side view. Each player had a knob to control angles, as well as a button to send the "ball" toward the opposite side of the court.

If you would like to see the game in action, go to BNL's Web site, http://www.osti.gov/accomplishments/videogame.html. You will find a video of the BNL's 1958 "Tennis for Two" video game.

To get a sense of what it must have been like to create the first video game, examine the following set of steps that describe the sequence of tasks the engineers probably followed. These steps are *not* listed here in the correct order. They are purposely jumbled. Use your engineering intuition to reassemble the steps for the project in the order they most likely occurred.

Jumbled Steps:

1. We met as a group, engineers and technicians together, and after much debate decided to develop one of the solutions.
2. We collected and analyzed performance data concerning how the model performed. We compared the model's performance to their expectations for the display. We verified that some performance problems remained and judged which could be resolved in a timely way.
3. We each gathered as much information as we could on the topic by reading journal articles and talking to colleagues. We presented what we learned to each other.
4. There was a big crowd on visitor's day. By word of mouth, adults and children were soon lined up to play our video game. It was a good example of how well our instrumentation laboratory designs and builds things.
5. We thought about the particulars of the problem we had described, including the restrictions that defined what we might do. Each of us then generated one or more schemes to solve the problem. The plans included our ideas, sketches, and descriptions of the work to be carried out.

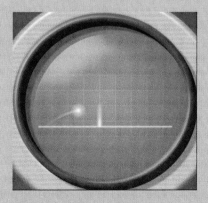

Figure 1.42 | The world's first video game, named "Tennis for Two," was played on the 5-inch-wide screen of an oscilloscope. An oscilloscope is a device that is commonly used to measure the voltage and frequency of an electric signal.

6. We revised the appearance and some functions of the display based upon the evidence we had gathered.

7. It became clear that
- We only had three weeks to design and construct the display.
- The audience would be children and adults.
- The display must be hands-on.
- The display must be visible by participants and by small groups of nearby spectators.

8. We worked together to make a model. It turned out to be a refinement of the original plan. The model helped us deepen and elaborate our ideas.

SECTION THREE FEEDBACK >

1. Why is there a question mark under "Major Technology" under "Twenty-first Century" in the table labeled Figure 1.11?
2. Why are bronze tools preferable to copper tools?
3. What did colonial expansion provide to Western Europeans that helped them develop new technologies?
4. During which time period in human history did technological innovations occur at the greatest rate?
5. Why are fossil fuels considered to be nonrenewable resources?

SECTION 4: Technology and Society

KEY IDEAS >

- The main incentive for developing new technologies is the "profit motive," or the potential for making a profit.

- The decision to develop a technology is influenced by societal opinions and demands, in addition to corporate cultures.

- Technological problems must be researched before they can be solved.

- Research and development is a specific problem-solving approach that is used intensively in business and industry to prepare devices and systems for the marketplace.

- Technological ideas are sometimes protected through the process of patenting.

Society's relationships with its tools and techniques greatly influence the extent to which societies can control their surroundings. A society is a broad grouping of people having common traditions, institutions, and collective activities and interests.

This section explains the relationships between technology and society, with a focus on two major components of society—culture and economics.

Humans require technological assistance to survive in the natural world. When compared to predatory animals, humans lack:

- Strong muscles
- Keen eyesight
- Claws
- The ability to fly on their own

Humans are able to compensate for their lack of physical attributes through survival-promoting mental attributes, such as:

- Consciousness of self
- Inclination to work cooperatively with others
- Ability to reason and plan
- Motivation to create and use technologies

Without such survival-promoting attributes and the products and techniques of technology, humans very likely long ago would have become extinct.

Figure 1.43 | Humans lack many of the physical attributes of this sharp-toothed predator. We have been able to compensate with mental attributes, such as self-consciousness and the ability to reason and make plans.

Culture refers to the learned, socially transmitted behaviors that develop due to social interaction. Cultural behaviors are based on beliefs, values, ideas, norms, and technologies. Norms are those actions that are considered necessary and that are seen as binding upon members of a group. Norms serve to guide, control, or regulate what society considers proper and acceptable behavior. Culture influences patterns of social interaction—the ways people respond to each other.

Technology influences culture. For example, after the American Civil War, the invention of barbed wire led to the fencing in of large sections of grazing lands in the American West. From then on, cattlemen and farmers did not get along very well. The cattlemen resented the barriers erected by the farmers, and the farmers resented the damage done by herds of grazing cattle when the fencing failed to keep them out.

Culture also influences technology, as we can see in the history of typewriter and computer keyboards. In early typewriters, the keyboard was designed to allow typists to type as fast as possible. The problem with this design was that adjacent

Figure 1.44 | The keyboards on early typewriters were designed to allow the typist to type as fast as possible.

Figure 1.45 | On early keyboards, the type bars jammed easily if the typist pressed the keys too fast. So, keyboards were redesigned to intentionally slow the typist and prevent jams.

Figure 1.46 | Jammed type bars are no longer an issue for modern keyboards. Nevertheless, current keyboards retain the deliberately inefficient design of older keyboards because people are accustomed to them.

A keyboard with the letters QWERTY arranged in one row in the upper left is known as a QWERTY keyboard.

and commonly used pairs of letters jammed when pressed down in rapid succession. The raised type bars, shown in Figure 1.45, stuck together and typing stopped.

To resolve the problem, engineers purposely designed an *inefficient* keyboard that purposely slowed rates of typing. Commonly used pairs of letters were spread across rows, and frequently used letters were concentrated more on the left side to force increased use of the left hand. The result was the keyboard design you probably have seen on numerous keyboards, with the letters QWERTY arranged in one row on the upper left side. This new design achieved its aim to slow down typing speeds and avoid jammed type bars. A "sticky" problem was resolved.

Over time, however, typewriters were improved, and the issue of jamming type bars was less of a concern. By 1932, engineers knew that keyboards laid out more efficiently would allow typists to work at twice the speed with much less effort. Still, the QWERTY keyboard remained the standard. Even with the introduction of computers, for which jamming typebars were not an issue, the QWERTY keyboard continued to be used. Why was the inefficient QWERTY keyboard never replaced by a more efficient arrangement of keys?

The reason is that QWERTY keyboard was thoroughly entrenched in society; people did not want to relearn a different arrangement of keys. The QWERTY keyboard is an example of a socially transmitted behavioral practice that people are reluctant to give up.

Economics and Technology

Just as culture and technology affect each other, so do economics and technology. Economics is a field of study that involves the description and analysis of the production, distribution, and consumption of goods and services. The decisions that individuals and companies make concerning whether to continue existing technologies or to develop, adopt, and use new technologies are largely determined by economic considerations.

Technological changes typically provide economic benefits for society as a whole, while imposing losses on some groups. This is particularly true when the demand for people's skills and resources have declined. From 1811 to 1816, a group of people known as the Luddites led machine-wrecking riots in northern England as a revolt against the unemployment caused by the introduction of machines during the Industrial Revolution.

You have probably noticed that the one constant about technologies is that they undergo change. Some technologies are used extensively for a while and then become obsolete; for example, the cottage industry for cloth making was replaced by the spinning jenny. This happens when replacement technologies are more effective, or economical; has anyone seen floppy disks for sale lately? Some technologies survive longer and coexist with new technologies. For example, some people compose using a keyboard for long tasks and a pencil for short tasks.

Some technologies become popular (are profitable), drop out of use (become worthless), and then become popular again (become profitable again). For example, wood stoves heated homes in the nineteenth and the early part of the twentieth centuries. Their use plummeted when electricity began to control oil- and gas-fueled stoves. Following the early 1970s oil crisis, wood stoves came back into fashion for awhile because of the scarcity of affordable oil.

Figure 1.47 | The oil crisis of the 1970s spurred a new interest in wood stoves.

Technology and Opportunity Costs The concept of opportunity cost is fundamental in economics. Opportunity costs are a combination of a direct outlay of funds, plus costs that do not involve an outlay of funds. This second part is not easy to comprehend; think of it as lost opportunities for financial benefit. Do you see that when you lose an opportunity to make money, this costs you?

To clarify this concept, suppose you make the following list of possible activities for a Friday evening: go to a movie, play basketball, visit your grandparents, go to a dance. Then suppose you choose to play basketball. The activities you missed by choosing basketball are lost opportunities, the second part of opportunity cost. Now apply the concept to money. Say you have $50 in your pocket; what do you do with it? What opportunities do you lose once you carry through on a decision to do something with the money? The opportunity cost of buying a new pair of sneakers would be the money you spent on the sneakers, plus the lost opportunity to do something else that might help you financially—perhaps buy a shovel and get paid back for clearing snow from driveways; perhaps deposit the $50 in the bank where it would earn interest.

When two technologies provide the same service, such as providing heat or transportation, the wise consumer—that is, the individual who purchases the service—would select the service with the lowest opportunity cost. For example, suppose you are trying to decide whether to travel by car or airplane to a vacation destination. You would probably start by considering the amount of money you would have to pay for each form of travel. For a car trip, you would have to pay for gasoline, car maintenance, and food and lodging while you are on the road. For an airplane trip, you would have to pay for the plane ticket and possibly one or two meals. Then you would also consider how each option might block opportunities to earn income. Travelling by airplane rather than by automobile results in a shorter vacation, which means you might miss fewer days of work, or, in other words, lose fewer opportunities to earn money. Economic considerations also affect the adoption of innovative technologies, such as a new consumer product. Whether or not consumers adopt a new technology often depends on whether or not the product provides its users with economic benefits that exceed opportunity costs. For instance, a person might purchase a small truck rather than a sedan if they otherwise would have to give up a part-time job that involved transporting equipment or materials.

Incentives for New Technologies Economists spend a great deal of time studying how incentives affect the choices made by businesses. Economic incentives are rewards that help drive innovations, by making it easier for companies to take risks and for other investors to back a project. Incentives also reward and motivate the inventors and innovators. In the modern marketplace, the main incentive for developing new technologies is the profit motive (the potential for making a profit).

When society faces a serious problem, this need for incentives has been known to delay solving that problem. After all, technological problems must be researched before they can be solved. But companies and investors compare potential profit against potential costs before making a commitment to expensive research that may or may not result in innovation. They may not be willing to invest in a solution if the economic incentives are not sufficient. Technological breakthrough is fostered when society chooses to spend money on research. As you will see, society provides incentives because of expected benefits that are associated with solving important problems. Monetary incentives, however, do not guarantee solutions to specific problems. Despite large yearly allocations of funds by government and foundations for cancer research, cancer continues to be a major cause of disability and death.

Unfortunately for investors, the benefits of innovation are likely to show up much later than the costs, and neither costs nor revenues can be predicted accurately. Accordingly, innovative projects tend to be seen as risky investments. Sometimes what we need the most is most difficult to achieve.

Technology and Corporate Culture The decision to develop a technology is influenced by societal opinions and demands, in addition to corporate culture. Corporate culture includes the values and norms that are shared by people and groups in an organization. Values and norms control the way business executives interact with each other and with outside investors.

Some companies are known for aggressive cultures; they tolerate risk well. The culture in many other companies is risk aversive. Managers in companies having risk aversive cultures tend to decline risks that are statistically even. They do not risk the loss of a dollar to gain a dollar. Before a risky new project is taken on, the potential gain needs to be sufficiently greater than the potential loss.

Research and Development Three stages in the development of a new technology are:

- Research and Development (R&D)—This is the specific problem-solving approach that is used intensively in business and industry to prepare devices and systems for the marketplace.
- Innovation—Innovation is the first application of the new invention. The person or company that first uses an innovation is known as the innovator.
- Diffusion—This term designates the adoption of the new invention by users other than the innovator.

Development of a new computer-based diagnostic tool in medicine might involve R&D carried out by physical scientists, physicians, computer

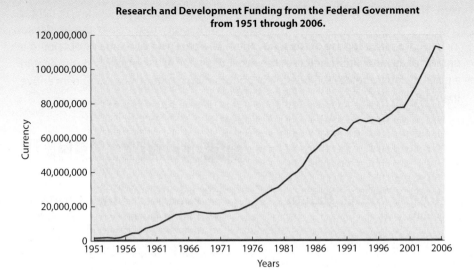

Research and Development Funding from the Federal Government from 1951 through 2006.

Figure 1.48 | This chart, based on data from the National Science Foundation, shows a trend toward a steady increase in R&D allocations from the federal government since 1953.

programmers, and biomedical engineers. The innovator would be the first hospital to use the innovation. Diffusion would occur when other hospitals adopt the successful innovation.

When a for-profit company considers R&D ventures, it always asks this question: "Will the proposed project increase or decrease the company's profits?" Given the risks involved, economic incentives typically help drive progress by ensuring that R&D takes place. The main providers of economic incentives for technological innovations are the federal and state governments, universities, and charitable foundations. Government subsidies for R&D are monies that Congress earmarks for research, some of which goes directly to universities or private companies that then undertake important projects.

The federal government provides the lion's share of incentives for technological research, and it monitors industries to protect the public. Government agencies that conduct this monitoring include:

- The Office of Safety and Health Administration (OSHA), which is responsible for the safety of employees at businesses and work sites
- The Environmental Protection Agency (EPA), which has responsibility for clean air and water and for appropriate disposal of solid waste
- The Federal Drug Administration (FDA), whose responsibilities for public welfare include safe food and the safety of new medications

Patenting New Technology

Technological ideas are sometimes protected through the process of patenting. Patents can also influence the development of new technologies. A U.S. patent provides the successful applicant with exclusive rights to the use of an innovation in the United States for a period of 17 years from the day on which the patent is

31

filed. The process of patenting thereby protects technological ideas. The patent creates monopoly rights over a process or commodity that allows the holder to increase the price for use of the innovation. The potential outcome of obtaining a patent for an innovation increases the expected value of a company's investment in a project, and therefore the possibility of obtaining a patent serves as an incentive to undertake R&D.

US007370016B1

(12) **United States Patent**
Hunter et al.

(10) Patent No.: **US 7,370,016 B1**
(45) **Date of Patent:** **May 6, 2008**

(54) **MUSIC DISTRIBUTION SYSTEMS**

(75) Inventors: **Charles Eric Hunter**, Hilton Head Island, SC (US); **Bernard L. Ballou, Jr.**, Raleigh, NC (US); **Kelly C. Sparks**, Raleigh, NC (US); **John H. Hebrank**, Durham, NC (US)

(73) Assignee: **Ochoa Optics LLC**, Las Vegas, NV (US)

(*) Notice: Subject to any disclaimer, the term of this patent is extended or adjusted under 35 U.S.C. 154(b) by 491 days.

(21) Appl. No.: **09/684,442**

(22) Filed: **Oct. 6, 2000**

Related U.S. Application Data

(63) Continuation-in-part of application No. 09/502,069, filed on Feb. 10, 2000, now Pat. No. 6,647,417, and a continuation-in-part of application No. 09/493,854, filed on Jan. 28, 2000, now abandoned, and a continuation-in-part of application No. 09/487,978, filed on Jan. 20, 2000, now Pat. No. 6,952,685, and a continuation-in-part of application No. 09/476,078, filed on Dec. 30, 1999, and a continuation-in-part of application No. 09/436,281, filed on Nov. 8, 1999, now abandoned, and a continuation-in-part of application No. 09/385,671, filed on Aug. 27, 1999.

(51) Int. Cl.
G06Q 99/00 (2006.01)

(52) **U.S. Cl.** 705/57; 705/58; 705/40

(58) **Field of Classification Search** 705/57, 705/58, 72, 67, 40; 380/200 204; 340/825.35; 369/30
See application file for complete search history.

(56) **References Cited**

U.S. PATENT DOCUMENTS

3,373,517 A   3/1968 Halperin

3,376,465 A   4/1968 Corpew
3,848,193 A   11/1974 Martin et al. 325/53
3,941,926 A   3/1976 Slobodzian et al. 178/7.3
3,983,317 A   9/1976 Glorioso 178/6.6

(Continued)

FOREIGN PATENT DOCUMENTS

EP   0 683 943 B1   11/1993

(Continued)

OTHER PUBLICATIONS

"Wink Television Press Room," http://www.wink.com/contents/PressReleases.shtml, downloaded and printed on May 14, 2002.

(Continued)

Primary Examiner—Pierre Eddy Elisca
(74) *Attorney, Agent, or Firm*—Woodcock Washburn LLP

(57) **ABSTRACT**

Music is blanket transmitted (for example, via satellite downlink transmission) to each customer's user station where selected music files are recorded. Customers preselect from a list of available music in advance using an interactive screen selector, and pay only for music that they choose to playback for their enjoyment. An antipiracy "ID tag" is woven into the recorded music so that any illegal copies therefrom may be traced to the purchase transaction. Music is transmitted on a fixed schedule or through an active scheduling process that monitors music requests from all or a subset of satellite receivers and adjust scheduling according to demand for various CD's. Receivers store selections that are likely to be preferred by a specific customer. In those instances where weather conditions, motion of atmospheric layers or dish obstructions result in data loss, the system downloads the next transmission of the requested CD and uses both transmissions to produce a "good copy". In conjunction with the blanket transmission of more popular music, an automated CD manufacturing facility may be provided to manufacture CD's that are not frequently requested and distribute them by ground transportation.

30 Claims, 15 Drawing Sheets

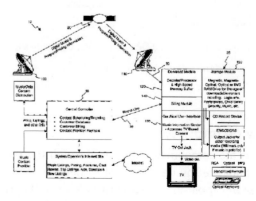

Figure 1.49 | This is a patent for a music device.

The Downside of Patents

▷ Patents can sometimes result in the following negative outcomes:

- Large monetary rewards go to the first company to obtain a patent for a new invention. As a result, multiple companies might spend money in a race to be the first to succeed in developing an invention. Such races are redundant and can be wasteful.

- Companies might develop multiple products that perform practically identical functions, but by different mechanisms. Competitors' products are also eligible to receive a patent, provided the mechanism by which the variation works differs from the original. Such practices are wasteful from a social perspective, but were a common ploy in the pharmaceutical industry during the mid-twentieth century until the formulas for chemically based drugs became patentable, protecting the R&D efforts of companies.

- Rates of technological diffusion become slower than is socially optimal. The diffusion of ballpoint pens is an example. The cost for ballpoint pens in the 1950s averaged about $5 each. When the original patent elapsed, competing companies could and did develop their versions of the ballpoint pen. Within five years, competition drove prices down to pennies. The public had waited over two decades for the price of a ballpoint pen to drop! The public did, however, decline to adopt the innovation until the price was right.

Figure 1.50 | The public preferred to use fountain pens until the patent for the ballpoint pen had expired, and the price of ballpoint pens dropped down to pennies.

TECHNOLOGY AND PEOPLE:
Ibn Ismail al-Razzaz al-Jazari

Let's travel back in time to the Middle Ages (the heyday of the Fertile Crescent region—modern Iraq and northeastern Syria). There we discover a very interesting person now called by many now call "the father of modern-day engineering." His name is Ibn Ismail al-Razzaz al-Jazari. He was an Arab Muslim scholar, inventor, and mechanical engineer who lived from 1,136 C.E. to 1,206 C.E. He was named after the area where he was born, al-Jazira, the traditional Arabic name for the Fertile Crescent.

Al-Jazari, while serving as chief engineer of an Artukid palace, the ruling Turkish dynasty of the time, developed several innovative mechanical and hydraulic devices and wrote a book, including construction drawings, called *The Book of Knowledge of Ingenious Mechanical Devices*. His invention of the crankshaft/crank mechanism is considered the most important mechanical invention after the wheel. Modified forms of this device are still used as components of steam engines and internal combustion engines.

Al-Jazari also developed what may have been the first entertainment industry. In 1206 c.e., he created an early precursor of a programmable humanoid robot. His automaton was a boat, featuring four automatic musicians that floated on a lake to entertain guests at royal parties. His programmable drum machine had pegs (cams) that bumped into tiny levers, creating percussion.

He was indeed a mechanical genius.

Al-Jazari's book and drawings showed others how the following devices worked:

- Crankshaft, crank mechanism, connecting rod
- Programmable automaton and humanoid robot
- Reciprocating piston engine, suction pipe, suction pump, double-acting pump
- Combination lock
- Cam, camshaft, and segmental gear
- Mechanical clocks driven by water and weights

SECTION FOUR FEEDBACK >

1. Name an economic factor to consider while developing an innovation, other than cost.
2. Name the socially transmitted behavioral practice that serves to control what society considers proper and acceptable behavior.
3. What process is the last stage in the development of a technological innovation?
4. How long does a U.S. patent last?

SECTION 5: Technology and the Environment

KEY IDEAS >

- Humans can devise technologies to conserve water, soil, and energy through appropriate techniques.

- Ethical considerations are important in the development, selection, and use of technologies.

- Implementing technology involves decisions that weigh trade-offs between predicted positive and negative effects on the environment.

- With the aid of technology, various aspects of the environment can be monitored to provide information for decision making.

- The alignment of technological processes with natural processes maximizes performance and reduces negative impacts on the environment.

Figure 1.51 | Too many animals in a pasture will eventually ruin the pasture. In an earlier age, when animals grazed on land held in common by a village, a ruined pasture might precipitate an economic crisis.

In 1969, an early environmentalist, Garrett Hardin, wrote in an essay, "The Tragedy of the Commons," that "Ruin is the destination toward which all . . . rush, each pursuing . . . [his or her] own best interest in a society that believes in the freedom of the commons."

Hardin was referring to an earlier age when herders raised cattle on a *common* pasture—land that was considered available for the use of all. Herders raising cattle on a commons would continue to increase their own herd's size without considering the effect this had on the shared pasture. Since the grass was owned by all and freely available to all, a herder could increase the size of his or her herd without worrying about the cost of feeding the cattle.

The problem, of course, is that having too many animals grazing in a pasture will eventually ruin the pasture. The overgrazing, caused by a few herders increasing their herds, would affect all the herders, even those who did not increase the size of their herds. Everyone suffered from the resulting economic crisis.

According to Hardin, in the modern world, humans have behaved like these early herders. Industrialists, manufacturers, and consumers have considered global ecosystems as limitless common pasture and treated them as such. As a result, the planet is showing signs of environmental crises—global warming, depletion of the ozone layer, loss of biodiversity, and pollution of air, water, and soil. Humans *can* devise technologies to conserve water, soil, and energy through appropriate techniques. Many people and companies are hard at work on this problem, but science and technology alone will never solve these problems. Difficult political decisions need to be made, and individuals and groups need to make lifestyle sacrifices. Decisions about implementing technologies involve the weighing of trade-offs between predicted positive and negative effects on the environment.

A Reorientation of Values

The resolution of environmental crises calls for a reorientation of human values. Ethical considerations are important in the development, selection, and use of technologies. Stewardship is the careful and responsible management of the environment. Vigilant and ethical stewardship should be practiced by individuals, organizations, and industries. Environmentally literate citizens are needed

Figure 1.52 | Modern technologies have contributed to massive environmental pollution.

who comprehend the scientific, technological, economical, legal, and moral issues embodied in environmental problems, and who will vote wisely on proposed solutions to the burdens that the misuses of technology have spawned. With the aid of technology, various aspects of the environment can be monitored to provide information for decision making.

If environmental crises are to be remedied, decision-makers, such as politicians, must strive to promote the following:

- The gathering of pertinent and sufficient information, before remedial actions are applied.
- Development of technologies that monitor the environment, providing decision-makers with valid, reliable, and timely information to help them take appropriate actions.
- The alignment of technological processes with natural processes to maximize performance and reduce negative environmental impacts.
- Consideration of ethics in the development, selection, and use of technologies in order to remediate environments and take into consideration the earth's limited resources.
- Technological compromises resulting in political solutions that enhance or maintain a sustainable world—a world in which resources are used in ways that they are not depleted or permanently damaged.
- Education about the relationship between technology and the environment; an education that helps individuals realize that personal lifestyles influence the environment and human health.
- Changes in ingrained practices concerning, for example, the consumption and disposal of products—a necessity if a sustainable world is to be achieved and maintained.

TECHNOLOGY IN THE REAL WORLD: Benefits and Burdens of Nanotechnology

Nanotechnology deals with extremely tiny structures at the atomic and molecular levels that are manipulated using powerful scanning tunneling microscopes to create a vast array of new materials and devices. These new objects are made in two main ways: by reducing "building blocks" or (patches of molecules), to sizes as small as a nanometer—about one-hundred-thousandth the width of a human hair—or by manipulating individual atoms and molecules to form shapes and structures, much as if they were tiny Lego pieces.

In 1959, the physicist Richard Feynman was the first to suggest the possibility of manipulating materials at the nanoscale level, which means measuring in nanometers (one billionth of a meter). He realized that through nanotechnology an image of the whole Encyclopedia Britannica could be placed on the head of a pin. In 1974, Norio Taniguchi of the University of Tokyo coined the term "nanotechnology" to signify machines having nanoscale mechanical tolerance. The microelectronic industry at that time was encouraging miniaturization, since it was in fierce competition to develop smaller, and thus faster, integrated circuits (entire circuits made with a single crystal). Since then, R&D has driven this fast-growing set of technologies, and over 300 new products have been diffused into the global marketplace.

Approaches to Nanoscale Manufacturing

▷ There are two general approaches in nanoscale manufacturing: Top-down manufacturing (TDM) and bottom-up manufacturing (BUM). In TDM, we start with a bulk material and then break it into smaller pieces using mechanical, chemical, or other forms of energy. In BUM, we synthesize building blocks, from materials at the atomic or molecular level through physical and/or chemical processes. This allows the building blocks to continue to self-assemble into macroscale (bulk) materials and devices. Both approaches can be carried out in settings that are either gaseous, liquid, supercritical fluids, solid states, or in a vacuum. Self-assembly is a reversible process in which random preexisting building blocks assemble themselves into ordered patterns.

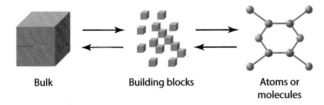

Bulk   Building blocks   Atoms or molecules

Figure 1.53 | In one approach to nanoscale manufacturing, known as bottom-up manufacturing (BUM), tiny building blocks assemble themselves into larger, bulk materials. In top-down manufacturing (TDM), materials at the atomic or molecular level are synthesized into building blocks.

The materials used in nanotechnology have properties not found in conventional bulk forms of the same materials. Some nanoparticles, such as that of silver, have greatly increased chemical reactivity resulting from special properties due to their small size. These highly reactive, tiny particles can penetrate our skin, travel rapidly within our bodies, and react with cell structures. This increased chemical and biological reactivity has allowed engineered nanoparticles to achieve specific purposes—innovative healing medications, for example, or construction materials that contain carbon that is 100 times stronger than steel. Some other interesting properties are seen in aluminum (which turns highly explosive) and gold (which melts at room temperature and is ruby red in color).

R&D has led to other uses for nanoparticles, including:

- microscopic tubes, spheres, wires, and films for specific tasks, such as generating electricity or transporting drugs in the body
- "smart" anticancer treatments that, among other things, deliver drugs *only* to tumors
- storage techniques that place information equal to that in the Library of Congress into a device the size of a sugar cube
- nanoparticles of silver that inhibit the growth of bacteria

The special properties of the nanosized element carbon make it worth exploring in greater detail. Carbon nanomaterials that are spherical in shape are known as fullerenes. They consist of repeating hexagonal and pentagonal rings of carbon atoms.

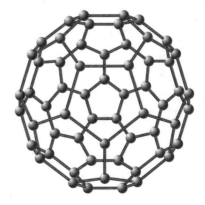

Figure 1.54 | A buckyball is a special type of fullerene that is extremely stable.

A special kind of fullerene, called a buckyball, is extremely stable. Buckyballs are fullerene molecules that can trap other molecules inside their carbon structures. Manufacturers now use fullerenes in products, such as cosmetics, fuel cells, and anti-aging facial creams. Another carbon nanomaterial is the tube-shaped structure known as a nanotube. These extremely strong conductive tubes are sealed in composite materials and used in automobile body parts, electronic equipment, and sports gear.

The U.S. government recognizes the tremendous potential for nanotechnology, and has provided nanotechnology R&D projects with billions of dollars as economic incentives. According to reports from the National Science Foundation, about $2.6 trillion worth of worldwide goods are expected to involve nanotechnology by the year 2014, up from $32 billion in 2005 and $50 billion in 2006.

Potential Negative Consequences of Nanotechnology

The tiny particles used in nanotechnology could potentially pose a looming problem. Conventionally sized materials are sometimes toxic when they are reduced to nanoscale. This may be due to the special physical forces that act upon particles at smaller scales. It may be due, as well, to the area to volume ratios that exist for nanoparticles. Generally, nanoparticles tend to react more readily with human tissues and other substances than do conventional forms of the same materials.

Health Concerns Unlike conventional carbon, buckyballs are partially water soluble, and have been shown to cause brain damage in fish. Some experts worry that buckyballs and other nanoparticles could have the same effect on humans, making people more prone to nervous-system diseases later in life.

Despite these concerns, one-third of the nanotechnology products currently on the market are intended to be eaten or applied directly to the skin. Because nanoparticles are about the same size as small components of human cells, some experts worry that they might damage our cells. Although cautions should be taken, these concerns can be placed in context by noting that humans have always been exposed to some nanoparticles. They occur through industrial pollution and natural sources, such as air photochemistry and forest fires.

Some typically harmless materials have been shown to be toxic when nanosized, and different risks are posed by nanoparticles compared to the same materials at conventional sizes. Determining the potential risk of manufactured nanoparticles has been difficult due to the lack of information about:

- the extent to which nanoparticles will be used in products
- the likelihood of such particles being released into the environment in a form or quantity that could harm humans or the environment

Engineers expect that exposure from composite materials containing nanoparticles and nanotubes will be low, since they typically make up a very small fraction of a given product. Such an assumption should be tested for each new product. The design process for products containing nanoparticles should consider risks from the release to the environment of these very tiny components. New products should be rigorously tested, assessed, regulated, and reported before any large-scale commercial applications are made available to the public.

Environmental Concerns and Nanotechnology Regulation The Environmental Protection Agency (EPA) has recently begun to consider regulating consumer items that contain nanoparticles of silver. The potential effect of the use of silver-embedded textiles—bandages and shoe liners embedded with nanoparticles of bacteria-killing silver—is large, but regulation would hold off widespread use until toxicity questions are addressed. For instance, what impact might the high reactivity of silver nanoparticles have on plants, animals, and other ecosystem inhabitants? Regulation of nanotechnology should be automatic until appropriate research has been undertaken to determine the potential benefits and risks of its products.

Many people think that insufficient funds are being devoted to environmental concerns raised by nanotechnology and to the safety of nanotechnology workers and the general public. Consider this: During a six-year period, of 80,000 journal articles on toxicology (a science that deals with poisons and their effects), only 0.6% even mentioned nanoparticles. Also, consider this: In 2006, the U.S. government spent 1.06 billion dollars on nanotechnology R&D, while spending $5 million (EPA) and $175 million (NIH and NIOSH), on environmental and health-related research.

We only have to look to the recent past to find examples of other materials, such as radium, DDT, asbestos, and chlorofluorcarbons, whose unintended consequences created health and environmental crises. Increased government regulation now could avoid similar problems as nanotechnology emerges.

Here are a few issues related to the regulation of the nanotechnology industry:

Some manufacturers have not been labeling their nanotechnology products. The government could make ingredient labeling mandatory for nanotechnology products and techniques.

- Funding and economic incentives for the study of nanotechnology risks have been meager. The government and research institutions could increase allocations to promote nanotechnology health and safety research. For instance, programs could be set up to monitor the health of nanotechnology employees and effects of nanotechnology on the environment to provide timely, reliable information to decision-makers.
- Manufacturers and industries that work with nanotechnology have been under-regulated. Regulatory agencies and organizations, such as the FDA, NIOSH, EPA and the Consumer Product Safety Commission need to provide greater oversight of nanotechnology development.

What Can *You* Do?

As a private citizen, consumer, and student, you can:

- Demand that labels indicate the presence of nanotechnology ingredients to help you make informed choices about which products to purchase.
- Visit Web resources to learn more about nanotechnology and to identify benefits and burdens as they unfold.
- Speak up, sharing the insights you gain with family, friends, and classmates.

SECTION FIVE FEEDBACK >

1. How does a "common pasture" field differ from "private farmland"?
2. Describe an action that you think people would agree represents stewardship of the environmnt.
3. Describe what a "sustainable world" would be like.
4. Clip and bring to class (or, find and print out) an article from a newspaper that you think could be labeled "environmental crisis." Tell why you believe this to be so.

Matching Your Interests and Abilities with Career Opportunities: Environmental Scientist

Environmental scientists conduct research or investigations to identify and eliminate pollutant sources or other hazards that affect the environment. They apply scientific and technological knowledge as they collect, synthesize, study, report, and take action based on data derived from air, food, soil, and water measurements or observations.

Nature of the Industry

There are more than 80,000 employed environmental scientists, not counting those identified as teachers who work in colleges and universities. Most environmental scientists are employed in state and local governments. You will also find environmental scientists working as managers; serving as scientific and technical consultants; offering architectural, engineering and related services; and working in the Federal government. Some are self-employed.

In their work, environmental scientists:

* Conduct environmental audits and inspections and investigations of violations
* Evaluate problems to determine appropriate regulatory actions, or to provide advice on the prosecution of regulatory cases
* Communicate scientific and technical information through oral briefings, written documents, workshops, conferences, and public hearings
* Review and implement environmental technical standards, guidelines, policies, and formal regulations
* Provide guidance and oversight to environmentalists, industry, and the public
* Offer advice on proper standards and regulations
* Analyze data, determining validity, quality, and scientific significance, and interpreting relationships among human activities and environmental effects

Working Conditions

Entry-level environmental scientists spend a fair amount of time in the field. More experienced workers devote more time to office or laboratory work. Field trips that require stamina are sometimes necessary. These trips may involve work in both warm and cold climates, in all kinds of weather. Environmental scientists might have to dig or chip with a hammer, scoop with a net, get soaking wet, and carry equipment in a backpack.

Environmental scientists in laboratories conduct tests and experiments and write reports; those in Federal government research positions, in colleges and universities, or in consulting positions more typically design programs and write grant proposals to continue their research projects.

Training and Advancement

A bachelor's degree is adequate for a few entry-level positions, but environmental scientists increasingly need a master's degree. A doctoral degree is necessary for college teaching and most high-level research positions.

Many in the field earn degrees in life sciences, physical sciences, or engineering, and then, through further education or research interests and work experience, apply their education to environmental areas. Others earn a degree in environmental science.

Computer skills are essential for prospective environmental scientists, as are excellent interpersonal skills for teamwork with other scientists, engineers, and technicians. Strong oral and written communication skills also are essential. Careers tend to begin with field exploration, or, occasionally, in laboratories or offices in the roles of research assistants or technicians. More difficult assignments are taken on as experience is gained. Eventually, there may be promotion to project leader, program manager, or some other management or research position.

Outlook for Environmental Scientists

The growth in job opportunities is expected to keep pace with that for all occupations through the year 2014. Growth should be strongest at private-sector consulting firms. Demand for environmental scientists will be spurred largely by public policies that oblige companies to comply with complex environmental laws and regulations. Opportunities also will open up due to a general awareness of the need to monitor the quality of the environment, to interpret the impact of human actions on land and water ecosystems, and to develop strategies for restoring ecosystems.

Earnings

Median annual earnings of environmental scientists were $56,100 in May 2006. The middle 50% earned between $42,840 and $74,480. The lowest 10% earned less than $34,590, and the highest 10% earned more than $94,670.

According to the National Association of Colleges and Employers, beginning salary offers in July 2005 for graduates with bachelor's degrees in environmental science averaged $31,366 a year.

[Adapted from Bureau of Labor Statistics, U.S. Department of Labor, Occupational Outlook Handbook, 2008–09 Edition, visited April, 2008, http://www.bls.gov/oco/]

Summary >

Technology is defined as the process by which humans modify the world to address their needs and wants. Informed engineering design is a critical set of technological techniques that involves identifying a problem and applying criteria and constraints to solve that problem. Engineering design does not result in a unique or correct solution, but instead leads to a best or optimum solution.

Present-day technology is a blend of engineering and science. Technology is closely associated with innovation. Innovations result when preexisting products or ideas are transformed into new and more useful or economical products or techniques.

Different materials and power sources have been used during various time periods throughout history. Early in the history of technology, innovation was driven more by technological know-how and trial and error than by scientific knowledge. Technological advances generate trade-offs, the giving up of one thing in return for another. Whenever a new product or process comes along, both positive and negative consequences arise.

As history progressed, technological innovations affected the environment and human health. The Middle Ages saw the development of many technological devices that produced long-lasting effects on technology and society, including the printing press, developed during the Renaissance that greatly expanded communications. During the Industrial Revolution, massive industrial centers were derived from such things as advanced construction practices and sophisticated transportation and communication systems.

As the use of energy for power purposes changed—energy from muscles, to energy from wood and charcoal, to energy from fossil fuels, such as peat and coal, to electrical energy, to atomic energy—each change resulted in greater degrees of damage to the environment. In the twenty-first century, change will continue to be a recurrent theme for technology.

Two components of society, culture and economics, have affected technology greatly. The relationship that society has with its tools and techniques greatly influences the extent to which societies can control their surroundings and thereby prosper.

Culture influences patterns of social interaction and norms serve to guide and regulate what society considers proper and acceptable behavior. Culture influences technology, and the opposite is also true, technology influences culture.

The decisions individuals and companies make about whether to continue to use existing technologies, or to develop, adopt, and use new technologies are usually decided through economic considerations. A characteristic of some technologies is that they are used extensively for a period of time, fall out of use, and then come into fashion again.

Opportunity cost is fundamental in economics; such costs are a combination of money costs and costs that interfere with opportunities to earn additional income, or otherwise prevent one from benefiting financially. When alternative technologies provide the same basic service, those who need the service tend to select the technique or product that has the *lowest* opportunity cost.

Society provides economic incentives for the development of innovations because of the expected benefits that are associated with solving important problems. Corporate culture in private for-profit companies strongly influences decisions on whether to develop technologies.

In the modern world, we identify three stages in the development of a new technology: research and development (R&D), innovation, and diffusion. The riskiest stage, R&D, is most in need of economic incentives. The federal government regulates industry and is also the greatest provider of incentives for technological R&D. A U.S. patent provides exclusive rights to the use of an innovation in the United States for a period of 17 years, and it creates monopoly rights that allow the holder to increase the price for its use. In effect, this protects the technological idea or innovation.

Decisions regarding implementation of new technologies should involve the weighing of trade-offs between predicted positive and negative outcomes for the environment. Although the medieval practice of having common fields for the use of all has been replaced by private ownership of farms, there is a growing social awareness that humans have continued to behave like herders on the commons. Accordingly, the planet is showing signs of environmental crisis—global warming, ozone depletion, loss of biodiversity, and pollution of air, water, and soil.

The development of technologies that appropriately remediate the environment requires informed engineering practices that take into consideration the earth's limited resources. Education is an important vehicle for getting individuals to realize that personal lifestyles influence health and environmental conditions. Changes of certain ingrained practices (consumption and disposal of products) are necessary, if a sustainable world is to be achieved and maintained.

Ethical considerations are important in the development, selection, and use of technologies. There is no such thing as a technological fix for the environment. People need to realize that other actions are needed besides the devising of new technologies to reduce the negative consequences of previous technologies. (Yes, adding scrubbers to the chimneys of coal-burning facilities does lower pollution amounts, but convincing society to alter lifestyles may be a better long-term solution.)

FEEDBACK

1. Why did hunter-gatherers live a nomadic life?

2. Explain why some early civilizations never adopted Bronze Age technologies, skipping instead from the Stone Age to the Iron Age.

3. Choose two different pre-Industrial Revolution technologies and for each, describe the burdens on society that resulted from them.

4. Why did the human population in Europe decrease during the early Middle Ages?

5. Describe what the term "revolution," as used in the phrase "Industrial Revolution," means.

6. What effect does a lack of biodiversity have on humans?

7. The twenty-first century section of this chapter lists technologies that some futurists believe might die out in this century. Research two such technologies, then explain why they might die out and what would probably replace them.

8. Describe a societal norm and relate how it has influenced your actions.

9. Describe one example of how culture influences technology, and one example of how technology influences culture. Use books, journals, or the Web to conduct library research concerning each of your examples.

10. Imagine and then describe a purchase you might make for which you would consider "opportunity cost." Explain why you would choose to consider opportunity cost in that situation.

11. Why is it that for-profit companies rely on sources, such as the federal government for economic incentives and subsidies, rather than paying for their own R&D projects?

12. Compare cows grazing on a common field to one of the following:

    a. coal-burning energy facilities releasing combustion products out of chimneys that lack scrubbers

    b. industries releasing untreated toxic byproducts into local streams

13. We know that humans often devise technologies to reduce the negative consequences of other technologies. Do you believe this approach will ensure that our environment remains operational? Explain why or why not.

14. Describe a change in lifestyle that you would be willing to make to help promote a healthier environment.

43

DESIGN CHALLENGE 1:
Using Energy Wisely

• The Problem Situation

As more companies, industries, and consumers purchase and use electrical appliances, the total amount of energy used has increased tremendously. Facilities that convert fossil fuels—coal, oil, and natural gas—and renewable sources of energy into electrical energy are sometimes unable to produce sufficient energy to address demands. Brownouts and blackouts result. Brownouts are a drop in voltage in an electrical power supply, so named because they typically cause lights to dim. Blackouts are large-scale disruptions in electric power supply. Both of these events tend to take place during peak electrical-use seasons—winter for heating; summer for cooling. What might consumers do to help avoid such annoying and costly events?

• Your Challenge

After inventorying your household for:

- types of electric appliances used
- how many watts each appliance uses
- how many hours each appliance is used in a week

design an energy-saving plan for a household that will reduce that household's use of electrical energy. Please go to the *Student Activity Guide* and follow the instructions to restate the design challenge in your own words.

• Materials Needed

1. pencil and paper
2. electric appliance list
3. 1 copy of a recent electricity bill for your household

• Clarify the Design Specifications and Constraints

To solve this problem, your design must address appropriate specifications and constraints. In the *Student Activity Guide,* describe the problem clearly and fully, noting constraints and specifications. Constraints are limits imposed upon the solution. Specifications are the performance requirements the solution must address.

• Research and Investigate

To better complete the design challenge, you need to gather information that will help you build a knowledge base. Complete the following steps:

1. Search for and discuss solutions that currently exist to solve this or similar problems. Identify problems, issues, and questions that relate to addressing this design challenge.
2. In the *Student Activity Guide,* complete knowledge and Skill Builder 1 (KSB 1): How Has the Use of Electrical Appliances Changed over Time?
3. In the Guide, complete KSB 2: Calculating the Amount of Electric Energy Used in Households.

• Generate Alternative Designs

In the *Student Activity Guide,* describe at least two possible solutions to the problem. Do not stop when you have a single solution that might work. Continue by approaching the challenge in new ways. Describe the alternative solutions you develop.

• Choose and Justify the Optimal Solution

In the *Student Activity Guide*, explain why you selected the solution you did, and why it is the best or the optimal choice.

• Communicate Your Achievements

Defend your selection of an alternative solution: Why is this solution the optimal choice? Use engineering, mathematical, and scientific data and employ analysis techniques to justify why the proposed solution is the best one for addressing the design specifications.

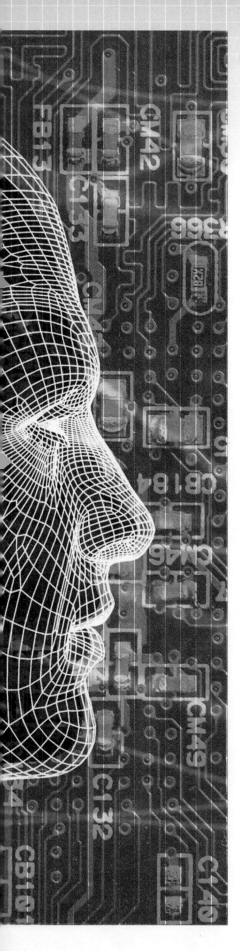

DESIGN CHALLENGE 2:
Separating Plastics

● The Problem Situation

To maintain a sustainable world, industries, organizations, and consumers must become committed stewards of the earth's resources. A socially transmitted set of behavioral practices that may become a norm is the "reduction in use, recycling, and reuse" of solid wastes. The goal is to keep as high a percentage of discarded items out of the solid waste stream as possible. Discarded resources fill landfills prematurely, thereby wasting land; discarded resources litter streets and are time consuming and expensive to recover. One challenge in setting up a workable recycling program is distinguishing among items that can and cannot be recycled. You probably already have noticed one technique that simplifies the process of sorting recyclable items—stamping plastic products with a code that indicates the type of plastic from which they are made. But what about plastic items that have been damaged, so that this code is no longer readable? How might you distinguish between types of plastics without this code?

● Your Challenge

Given samples of plastics that lack the recycling symbol, design a technique based on the physical properties of plastic that will sort the various kinds of plastics. Use the *Student Activity Guide* to complete this activity.

● Safety Considerations

1. Wear safety goggles to avoid splashing isopropyl alcohol in your eyes.
2. Take care not to spill the isopropyl alcohol.

● Materials Needed

1. Sealable plastic bags containing chips of plastic
2. Water/sugar solution
3. Water/isopropyl (rubbing) alcohol solution
4. Dishwashing liquid (optional)
5. Safety goggles
6. Petri dishes or other transparent containers to hold one chip at a time

● Clarify the Design Specifications and Constraints

To solve the problem, your design must address the appropriate specifications and constraints. In the *Student Activity Guide*, describe the problem clearly and fully, noting constraints and specifications. Constraints are limits imposed upon the solution. Specifications are the performance requirements the solution must address.

● Research and Investigate

1. To better complete the design challenge, you need to gather information that will help you build a knowledge base. Search for and discuss solutions that presently exist to solve this or similar problems. Identify problems, issues, and questions that relate to addressing this design challenge.
2. In the *Student Activity Guide*, complete Knowledge and Skill Builder 1 (KSB 1): Which Material Does Each Plastic Recycling Code Stand For?
3. In the *Student Activity Guide*, complete KSB 2: What Are the Unique Properties of Different Kinds of Plastics?

- ## Generate Alternative Designs

In the *Student Activity Guide*, describe at least two possible solutions to the problem. Do not stop when you have a single solution that might work. Continue by approaching the challenge in new ways. Describe the alternative solutions you develop.

- ## Choose and Justify the Optimal Solution

In the *Student Activity Guide*, explain why you selected the solution you did, and why it is the best or the optimal choice.

- ## Communicate Your Achievements

Defend your selection of an alternative solution: Why is this solution the optimal choice? Use engineering, mathematical, and scientific data and employ analysis techniques to justify why the proposed solution is the best one for addressing the design specifications.

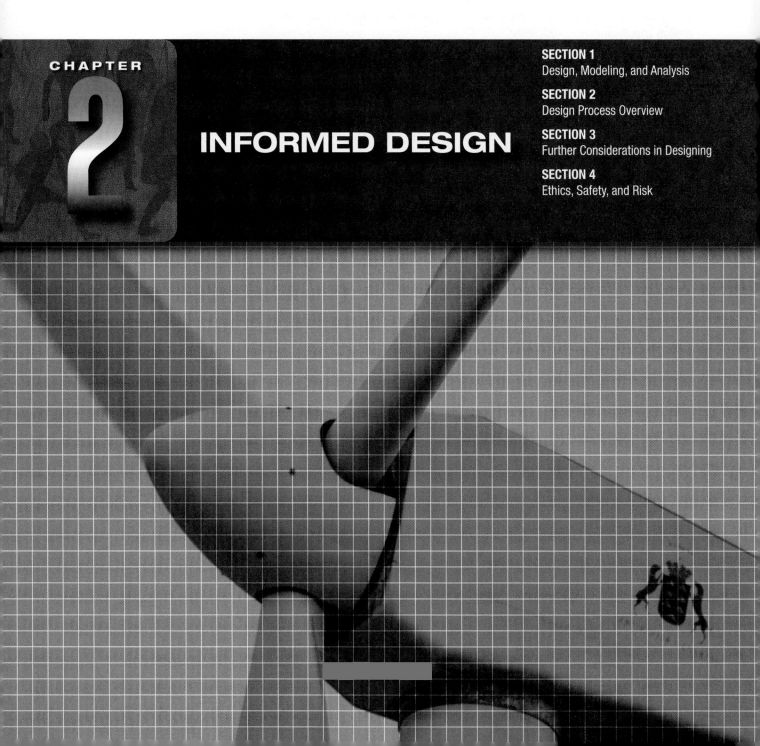

O **NE OF THE** wonderful aspects of engineering technology is that it is a very creative profession. Engineers make products and processes that solve a human need. Furthermore, engineers solve problems creatively, seeking solutions that did not exist before. Creating designs and analytically modeling them to predict how they will perform requires a blend of talents. For example, imagine that you are

CHAPTER

2

INFORMED DESIGN

SECTION 1
Design, Modeling, and Analysis

SECTION 2
Design Process Overview

SECTION 3
Further Considerations in Designing

SECTION 4
Ethics, Safety, and Risk

given the challenge of devising a new birdhouse, one that can be attached to a fence or building using natural resources. You need to learn about the types of nests the birds like to build and how to position the birdhouse so that squirrels and other animals cannot get to the food. You also need to know about types of birds you want to attract—large birds, small ones, or all types. Figure 2.1 illustrates one solution that prior generations used. Basically, a pottery urn was turned on its side and held by a nail. It used a twig for the bird's standing perch so it could hop into the house. The early Americans wanted insect-eating birds nearby, and this birdhouse was one solution.

SECTION 1: Design, Modeling, and Analysis

KEY IDEAS >

- Design and analysis are used in engineering design.

- Engineering design is influenced by visualizing and thinking abstractly to develop creativity.

- Regulations and codes both help and hinder technology.

- Most technological development has been evolutionary, the result of a series of refinements to the basic invention.

- Visual thinking is central to the creative process.

Figure 2.1 | **This colonial birdhouse is made from pottery, formed from clay, a natural resource. Colonists wanted nearby birds to eat pests and so created appealing homes for the birds.**

Consider the following situation. A deep stream that is five feet wide needs a bridge so people can walk across it. You find a plank that is two inches thick, twelve inches wide, and eight feet long and place one end on each bank, crossing the stream (see Figure 2.2). No engineering design was needed in this case; from your experience, you decided that the plank should do the job. In fact, you are so pleased that you decide to go into business making plank bridges. But now people who might buy the bridges have questions about how great a load (weight) the bridge can support. Will it support six people? What size are the people, you ask, and so it goes. From an engineering view, what exists in this case is a beam supported on two ends. In this simplest view, the beam will not deflect, not bend.

To be able to respond to potential customers, you need to analyze this beam. So you create a representation, a physical model, of the actual beam, and this can, in turn, be represented by mathematical equations. This requires that you understand the forces throughout the beam caused by the weight of people standing on it or walking across it. You will also research the properties of the wood, including how strong it is, and then combine this information with the mathematical analysis to see if the wooden plank would break when the forces (the load) are applied.

Plank bridge

Figure 2.2 | **No engineering design is needed to create a simple plank bridge.**

At this point, you are analyzing what is assumed to be. You are not analyzing the wooden plank that exists; you have developed a physical model of the plank and represented that model mathematically, including the assumption of certain properties for the wood itself. Based on your analysis, you feel certain that the plank can hold 750 pounds but, to be on the safe side, you guarantee it will support 500 pounds. What you have inadvertently introduced is a factor of safety, which all good designs have and which accounts for unexpected load conditions and property variations. We are using the word *design* a little casually in this case, as you did not really design the plank but assumed it would be sufficient. Another consideration is that, although the model will support 750 pounds, the actual plank might not. Your analysis assumed that the wood had certain properties and could be modeled in a certain mathematical fashion. Perhaps, however, a knot at the center of the plank causes it to break under a load of 500 pounds. More sophisticated analysis can then be used to model a plank with a knot.

An equally important aspect is that a satisfactory design analysis should be carried out during the actual building construction. Often, a disaster we read about is caused by a difference between design and construction. If differences occur in construction or in materials used, then more analysis is needed to make sure that the revised design meets the load requirements.

A case in which changes tragically affected a project was the construction of elevated walkways in the Hyatt Regency Hotel in Kansas City, Missouri. Two elevated walkways, shown in Figure 2.3, spanned the lobby. In July 1981, a dance contest was in progress. More than 1,000 people were gathered on the walkways, watching the participants below. Some of the observers above were dancing, as well. The walkways were not constructed as designed. Moreover, they were bearing loads that were twice that of the original specifications. The walkway supports failed; 111 people were killed, and another 188 were injured. Several factors led to the walkway failure, including the dynamic load caused by the dancing, a welded seam that supported the load, and the possible omission of some load-bearing washers. The constructed walkway had a different design from the original, and unfortunately that design was far worse.

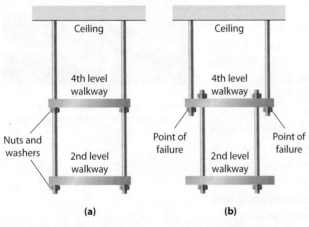

Figure 2.3 | The diagrams show the walkway of the Kansas City Hyatt Regency Hotel, both as designed (a) and as constructed (b).

Focus again on the bridge you are building. Imagine that a competitor enters the picture, as someone else has decided that you are not the only person who can lay planks across a stream. To be competitive and create a better product, you want to design something new. You could use cost-saving tactics first, such as shortening the plank to reduce support on each bank. You could also reduce the plank's thickness or even reduce the factor of safety. Using your knowledge of analysis and design, you create a new bridge design, which is modeled and analyzed to see if it meets the load requirements and the factor of safety you want. There is a significant difference in design between creation and analysis—both processes are necessary. Figure 2.4 shows an example that is far more complex than your plank. Design and analysis are used in engineering design.

Let's go back to the competitive situation that exists between you and others in the same field. Being an ethical person, you selected a very conservative factor of safety. Your competitor chose a smaller factor of safety and feels that it is adequate. That means your competitor's bridge, if it has a similar design, will cost less than yours and hence have a greater chance of being selected. Municipalities and states have recognized this competitive problem and have established building codes, which define the minimum requirements for construction. Very often the contract specifications from the company requesting the design will detail the design requirements in addition to statutory requirements. This means that everyone is operating from the same vantage point. As we often find in the technological world, there are problems associated with codes and regulations. These requirements are based on existing designs and materials, and so design innovation can be restricted by these codes. Regulations and codes both help and hinder technology.

This is particularly true when new materials allow for designs that were not previously possible. Most technological development has been evolutionary, the result of a series of refinements to the basic invention.

Figure 2.4 | Engineering designer Othmar Ammann used his knowledge of bridge design and traffic flow to create a dramatically new design—the George Washington Bridge connecting New York City to New Jersey.

Designing

Designing is a process that people use to plan and produce a desired result. The result may be a product, process, or system that meets a specified human need or solves a particular problem (see Figure 2.5). In engineering, we are interested in technological designs that require the blending of creativity and technical know-how. Design and creativity go hand in hand; enhancing your creativity will improve your designs and deepen your technical knowledge. Engineering design is influenced by creative abstract thinking and visualizing.

Creativity

Creative people are willing to take risks. Although the knowledge gained from success is very appealing, the knowledge gained from failing, learning, and then succeeding is more long lasting. In learning a new skill, a person must experience

Figure 2.5 | Designing turnpike connections requires multilevel visual understanding.

Figure 2.6 | Mark Zuckerberg is the founder of Facebook.

Figure 2.7 | Three blocks can be stacked to create a simple arch.

Figure 2.8 | Dr. Barbara McClintock investigated corn and found that genes for physical traits were carried in chromosomes. She won the Nobel Prize for her work in 1983.

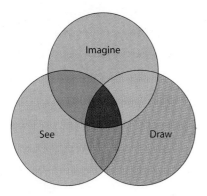

Figure 2.9 | A Venn diagram illustrates the connections between seeing, imagining, and drawing.

periods of awkwardness and inability. Mistakes are made in the process of learning. Top athletes do not start at the top; they learn from the baskets they miss and from their strikeouts. Facebook founder Mark Zuckerberg, shown in Figure 2.6, believed so much in his dream of creating an online community that he borrowed funds and stopped attending college (in general, not something most people should do).

Seeing connections where none were perceived before is another important attribute. Thinking visually can allow people to see interconnections. Imagine placing three blocks together, as shown in Figure 2.7. How might this figure be described? Did your list include a table, a pleasing form, a chair, an inverted "u," a hole in space, a bridge, or the Greek letter *pi*? Visual thinking is an enabling intelligence, empowering you to see in different ways and to make connections between varied perspectives. Visual thinking is central to the creative process.

Certainly, you need to be knowledgeable about your subject for your creative thoughts to have foundation. Although gaining knowledge can bring the accompanying danger of strengthening a fixed view and thereby closing off other alternatives, it also strengthens intuition. Increased knowledge can also help you understand why some of your past efforts were successes and some were mistakes.

Determination is the fourth ingredient common to creative people. As Thomas Edison said, creative leaps are the results of perspiration as much as inspiration. The determination to make creative change requires dealing with periods of uncertainty. In the process, a resiliency, not a hardening, emerges. Resiliency connotes strength and flexibility, whereas hardness connotes strength and rigidity. Dr. Barbara McClintock (see Figure 2.8) exhibited these traits in her research; as a scientist in the 1940s and 1950s, she encountered resistance from the scientific community for her theories and experiments about how genetic information is transferred from one generation to the next. But, in 1983, she was awarded the Nobel Prize for her pioneering work.

Visual Thinking

Visual thinking is central to the creative process. It is aided by quick sketching, which permits ideas to become further refined. We think visually all the time. The English language contains many words that link vision and thought, such as insight, foresight, and hindsight. Indeed, the etymological root of the word *idea* comes from the Greek word *idein*, which means "to see." Thinking is a pervasive, yet elusive, concept. Thinking, as manifested by the brain's electrical activity, occurs even while we sleep. It is a complex activity that involves the entire body—physiologists have shown that muscle tone affects thought. Neurologists link the nervous system and psychologists link feelings and emotions to thought processes. The process is visual, verbal, and mathematical, and it is affected by anything that affects our bodies. Of course, not all thinking is fruitful or purposeful; most of the time it produces little of value.

Visual thinking pervades all our activities, from the profound to the mundane: from Einstein viewing himself traveling at the speed of light in his research for the theory of relativity, to an athlete picturing herself winning a medal, to a walk through a crowd of people. How can we use our visual thinking in a way that will let us purposefully create? We use three types of activities: seeing, imagining, and sketching. Figure 2.9 illustrates these overlapping circles in a Venn diagram, with one activity reinforcing the other. Seeing is what helps us sketch, and sketching helps us see more clearly, which can stimulate our imagination and provide a greater desire to sketch. The act of sketching helps you quickly capture images in your mind; the thought and memory of these images thus stimulates your imagination.

SECTION ONE FEEDBACK >

1. What are the differences between designing a birdhouse and making one?
2. Can you design a birdhouse without analyzing it?
3. How are seeing, imagining, and sketching related?

SECTION 2: Design Process Overview

KEY IDEAS >

- Design problems are often not presented in clearly defined terms.

- The design process is iterative, in that it can go back and repeat steps several times, and it includes several steps, such as brainstorming and optimization.

- Technological problems require research before they are solved.

- Requirements involve identifying the specifications and constraints of a product or system and determining how they affect the final design and development.

- The process of engineering design takes into account a number of factors.

- Designs are critiqued and then refined and improved.

- Established design principles guide the design process. Optimization and trade-offs are important elements of the process, as design requirements may compete with each other.

- A prototype is a working model used for testing the design.

- Research and development are part of the design problem-solving method; they are also used in business for researching marketplace requirements.

The design process starts with a problem that requires a solution. The problem may be narrowly defined, such as developing a sports bag, or even more loosely defined, such as improving the water quality in a pond. Before creating solutions, an engineer researches and investigates the topic.

Researching the Problem

Research can be technical or nontechnical. The type of research required depends on the problem.

Often, design problems are not presented in clearly defined terms. In that case, research can help define the problem more clearly. Research can also help the engineer understand the context that created the problem. For instance, what factors contribute to a pond's low water quality—decreased water supply because of urban development or increased nitrates from lawn fertilizers? Regarding the sports bag, were there problems with previous versions? Did customers complain about the straps? Were the bags too large or too small, as shown in Figure 2.10.

Figure 2.10 | Sometimes a backpack is a bit too big.

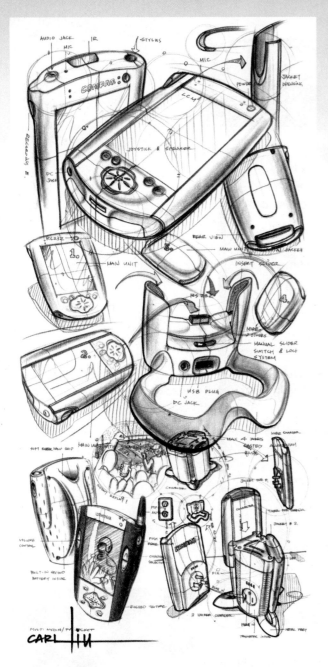

Figure 2.11 | Ideation sketches are quick sketches to capture the essential elements of a design and prompt further thinking about the design.

In the case of the sports bag, designers may need to examine other types of sports bags and learn about their characteristics, both good and bad. This examination leads to the specifications and constraints area of the problem statement, in which the output requirements are noted. These requirements could include the bag's capacity, size, use of detachable parts, and availability in a variety of colors. Constraints may be imposed to limit the variety of possible solutions; for example, perhaps the material must be nylon, and the cost must not exceed $10. Requirements involve identifying the specifications and constraints of a product or system and determining how they affect the final design and development.

Brainstorming a Solution

At this point, the design engineer more thoroughly understands the problem in two ways: technically, through research and investigation, and philosophically, through the design requirements. Now the designer's creative side is freed to brainstorm. Several quick sketches can lead to different approaches to solving the problem, as shown in Figure 2.11.

This is perhaps the most challenging part of the design process. We often seize on one idea, judge it to be satisfactory, and then stop thinking creatively and stop searching for the best design. In the creative sphere, other designs can help validate the best solution by providing alternatives to check against. In this process, positive and negative features of the designs are examined, and in this way the best design is determined.

Analyzing the Design

Before the design is constructed, it needs to be analyzed. Will it perform the required functions and meet the design specifications? Engineering analysis involves creating the mathematical equations that represent the system under consideration and the solution of these equations. It is one of the important components of the engineering design process. These equations are a mathematical model of the system.

The analysis of the model includes use of specifications and constraints that have been developed during the problem definition phase. For example, one aspect of analysis might be to calculate the bag's volume and evaluate the pressure you feel on your shoulders when carrying it. The questions you answer here include whether the device will meet the functional requirements. Perhaps an original specification is too demanding and needs to be revisited.

Constructing, Testing, and Evaluating a Prototype

Once the analysis is completed and any modifications are made to the design as a result, the construction of a prototype begins. A prototype is a working model that is used for testing a design. During construction of the prototype, changes invariably are made to conceptual design. You should recognize that such changes are fine—they are part of the creative process. Of course, you must make certain that the prototype solves the problem, and this is where testing and evaluation occurs. The testing should be conducted in a scientifically correct manner so that

Statement: A wheelbarrow carries a 200 lb_f load. The wheelbarrow's handle is 5 feet long and held at an angle of 20° to the horizontal. The load's center of gravity is 2 feet from the wheel's axis. Determine the lifting force and the force on the tire.

<u>SOLUTION</u>

Given: A wheelbarrow has a known load acting at its center of gravity. The length & angle of the handle are known.

Find: The lifting force and the force on the tire.

Sketch & Given Data:

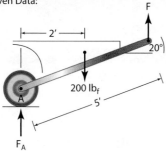

Assumptions: The wheelbarrow is in static equilibrium.

Analysis: From the $\Sigma M_A = 0$

$$\Sigma M_A = 0 = -(200\ lb_f)(2\ ft) + (F\ lb_f)(5 \cos 20°\ ft)$$

$$F = \frac{400}{(5 \cos 20°)} = \underline{85.1\ lb_f}$$

$$\Sigma F_y = 0 = +F_A - 200 + 85.1$$

$$F_A = 114.9\ lb_f$$

Figure 2.12 | Mathematical analysis of a wheelbarrow allows people to predict the amount of force required to lift the load.

the data is valid and reliable. Thus, the process of engineering design takes into account a number of factors. We will investigate other factors toward the end of the chapter, taking safety and environmental concerns into consideration.

The Design Process

Figure 2.13 illustrates the informed engineering design process, which is guided by established design principles. The process is nonlinear, as is typical of creative processes, and in contrast with the linearity and logic of mathematical analysis and scientific inquiry. Notice the inner arrows in the figure; these indicate that at any time in the process, it is fine to go back and add information (perhaps a specification) or eliminate a constraint. The design process is iterative, and includes several steps such as brainstorming and

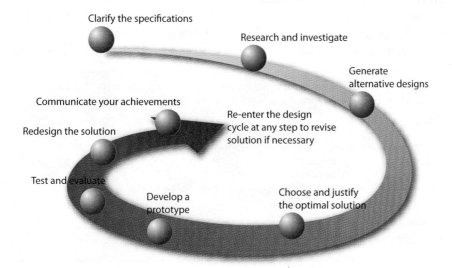

Figure 2.13 | The informed engineering design process is nonlinear, but it is guided by established design principles.

optimization. In the informed engineering design process, there are guided research and investigation activities for students that model what engineers do in practice. These guided activities are called Knowledge and Skill Builders (KSBs). Engineers design from a knowledgeable and skilled background to create new products or processes.

Design Cycle Phases

Now we will examine the phases of the design cycle in more detail. Table 2.1 lists the phases.

Table 2.1 | Phases of Informed Engineering Design

PHASE	DESCRIPTION
Phase 1	Describe the design problem clearly and fully.
Phase 2	Research and investigate the problem.
Phase 3	Generate alternative designs.
Phase 4	Choose and justify your optimal design.
Phase 5	Develop a prototype.
Phase 6	Test and evaluate the design solution.
Phase 7	Redesign the solution with modifications.
Phase 8	Communicate your achievements.

Phase 1: Describe the Design Problem Clearly and Fully People with arthritis in their fingers have a hard time gripping small objects. They need an easy way to carry out such tasks as unlocking a door or hunting for the right object in a drawer. The specifications are the performance requirements or output requirements that the solution must fulfill. For instance, the design specifications for toothpaste might include that it cleans plaque from teeth, tastes good, and can be squeezed easily out of a tube. The design specifications for a sports bag might require its material to be both colorful and waterproof, as shown in Figure 2.14. A design specification for a certain type of car might be that it can accelerate from 0 to 60 mph in under 10 seconds. Also, design specifications often include safety considerations. For example, a passenger elevator might require a safety factor 10 times greater than the load it is expected to carry, or the front of a car might have to be completely undamaged after a crash at 5 mph.

The constraints are limits imposed upon the solution. Constraints are often related to resources, such as what kind of materials the designer is able to use, how much money a finished product can cost, or how much time can be devoted

Design Challenge: Design and construct a sports bag that is durable and can be used for students engaged in a variety of sports activities—volleyball, tennis, baseball, and soccer.

Specifications:
- The sports bag should be large enough to hold clothing and sports equipment associated with each sport; however it does not need to be long enough to hold a baseball bat or tennis racket within the bag.
- The material should be colorful and waterproof.
- There should be straps that allow for over-the-shoulder carrying as well as carrying by handles.

Constraints:
- The sports bag should not cost more than $25 retail.
- The straps should be adjustable.

Figure 2.14 | The specifications for a product (here, a sports bag) are the performance requirements or output requirements that the solution must fulfill.

Figure 2.15 | You need to do thorough research in order to identify problems, issues, and questions that relate to a design challenge. In addition to searching the Internet and conducting interviews, you can do research at a library.

to producing it. Other limitations can involve the availability of certain kinds of workers or the need to limit negative effects of the design on the environment.

Phase 2: Research and Investigate the Problem Virtually all technological problems are researched before they are solved. In this phase, you search for and discuss solutions that presently exist for the problem at hand or similar problems. In the process, you will learn a lot about good and bad solutions.

In the course of your research, you can identify problems, issues, and questions that relate to addressing the design challenge. You can gather information from other people, from library research (see Figure 2.15), from searching the Internet, from catalogues, and through visits to stores that sell products like the one you are designing. In addition, you normally will need to make measurements, collect data about the materials you use and about the performance of the design, and rate alternatives against one another. You will need to use mathematics in doing this work.

You may also perform scientific investigations and conduct experiments to help you learn how certain choices affect a design's performance. For example, if a designer wants to create a food cooler for camping trips, an experiment might be set up to determine how well different materials keep food cold. The results from this investigation would be used to help make design decisions about the cooler. You may want to investigate current sports bags, thinking about their good and bad features. If you put a certain amount of weight in the bags (for instance, 20 pounds), are they comfortable to carry? What could be done to make them more comfortable?

When doing your research, think about the design criteria. Thinking about your design requirements can help you identify questions that you need to answer in order to come up with a solution. If you are trying to design a product that people will use, you may want to collect data from the library about people's heights, weights, lengths of reach, or other design factors that will make the product easier for certain age groups to use.

Companies also perform market research to determine if customers will like a new product. For example, companies may ask potential buyers to fill out a questionnaire to find out their preferences. Suppose a company wants to develop toothpaste for

Figure 2.16 | Companies conduct market research to find out customer preferences, such as what they like or don't like about their current toothpaste.

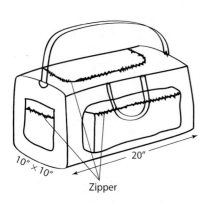

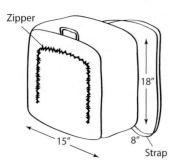

Figure 2.17 | Brainstorming can lead to multiple ideas, as shown in these sketches of possible sports bag designs.

teenagers. The questionnaire might ask what teenagers like or dislike about the toothpaste they're using now. Do they like its taste? Do they like the way it feels in their mouths? What kind of dispenser do they prefer to use, a tube or a pump? What colors do they prefer for the toothpaste, tube, and box (Figure 2.16)? The company will use the results of this market research to design the product so it appeals to the greatest number of people. Research and development are part of the design problem-solving method; they are also used in business for researching marketplace requirements.

Phase 3: Generate Alternative Designs Don't stop when you have one solution that might work. Continue by approaching the challenge in new ways. Describe the alternative solutions you develop.

As a result of your research and the knowledge you have gained, you may have one or more possible solutions to the problem. The ideas can be totally different, or they can be improvements to your first idea. There is almost always more than one solution to a technological design problem, and good designers are rarely satisfied with the first idea that pops into their minds. You can suggest several ideas—each one of which might do the job—but try hard to think about better ways (see Figure 2.17).

Another way of coming up with alternative designs is called brainstorming. During brainstorming, each person in a group can suggest ideas. One person writes all the ideas down, and no one is allowed to laugh at or criticize ideas, no matter how foolish or unusual they might seem. The brainstorming process is used to help people think more creatively. People feel free to share any wild ideas. Sometimes one person's wild idea will open up someone else's mind to a totally new approach. An advantage of brainstorming is that your mind begins to make unusual connections.

Sometimes an idea may just pop into your head. Usually these sudden ideas are followed by the "Aha!" response ("Aha! I've got it!"). This insight comes from being thorough in researching the problem and from being creative in thinking about the problem from many different angles. Even when you are not consciously thinking about the problem, your brain may still be working on it. After many ideas have been proposed, the group reviews them all. The best ideas are developed further.

Phase 4: Choose and Justify Your Optimal Design For the alternatives that remain, you should list their strengths and weaknesses in relation to the design criteria. To make your decisions, you might have to do more research, testing each alternative and gathering data about its performance. These test results must be recorded and compared with results from other tests, so that fair and accurate decisions can be made about which solution is best. Sometimes, the testing will suggest that if you change one alternative slightly, or combine two or more alternatives, you will end up with a better solution. The design improvements can lead to better performance, increased safety, and lower cost.

The process of improving each alternative or improving each part of a design is called optimization. Often, different alternatives will be better in different ways. For example, one material may be stronger, but a second material may cost less. When people choose the best solution, they normally make trade-offs. That is, they give up one desirable thing for another. In such cases, we must decide which criteria are the most important, and then arrive at the best overall solution to the problem. In other words, we can optimize the alternatives and then make trade-offs to determine the best possible solution. Sometimes design requirements compete with each other. The idea is to decide on a design that best meets the specifications, fits within the constraints,

Figure 2.18 | How do you decide which television to purchase?

Figure 2.19 | A prototype provides a working model for testing the sports bag design.

and has the least number of negative characteristics. Established design principles guide the design process. Optimization and trade-offs are important elements of the process, as design requirements may compete with each other (Figure 2.18).

Defend your selection of the best alternative solution: Why is it the optimal choice? Using information you have gained from research and investigation, justify why the proposed solution is the best one for addressing the design specifications. The chosen alternative will be the basis of your preliminary design. Each of the alternatives must be examined to see if it meets the design criteria, specifications, and constraints that were defined in the first step. Usually, you do not use alternatives that do not sufficiently address the specifications and constraints.

Phase 5: Develop a Prototype In this phase, you make a model of the solution, identify possible modifications that would lead to refinement of the design, and then carry out these modifications. Most often, a model of the solution is made first; when this model is a full-scale version of the solution, it is a prototype—a working model for testing the design. For example, Figure 2.19 shows a prototype for a sports bag.

Models of the proposed solution are important if the solution is complex and costly, if a large quantity of the final products must be manufactured, or if the proposed solution presents risks to people and the environment. In all of these situations you would want to minimize the potential for costly errors. For example, a model of a nuclear power plant would be built and tested to discover and eliminate any problems that were not apparent during the analysis phase, before the actual plant was built. One kind of

Figure 2.20 | This small model of an F/A 18 EF airplane was tested in a wind tunnel before the first actual plane was built.

model might be a mathematical model that simulates how a nuclear reaction occurs and how it can be controlled. Another model could be a smaller-scale version (or even a larger version of very small objects) of the proposed solution to the problem. For example, a small model of a new airplane would be built to test in a wind tunnel before the first actual plane was built (see Figure 2.20). A

59

large-scale version of a tiny integrated circuit might be built so that all the parts and connections could be clearly seen. Skilled craftspeople and technicians are often employed to make models before full-scale construction is started. As you construct your model or prototype, you will probably need to make adjustments to your original design. Be certain to note these changes and why you are making them. If the changes are significant, you must make certain that your design is still the best alternative.

TECHNOLOGY IN THE REAL WORLD:
Apple versus Microsoft

When you are designing something for yourself, you have only yourself to answer to and to satisfy. However, in the commercial world, a variety of constituencies and needs must be satisfied—the end user (the customer), the technological limitations (engineering), and business (profitability and fit with other products). The axiom known as Keep It Simple (KIS) is often championed, but not always applied. There is a tension between what marketing people want, what engineering people think they can do, and how people like to use the product or what they want from it.

One challenge that engineers and designers face is to keep some features out, as some people mistakenly think that adding more features always adds more market appeal. Often a comparison is made between Apple and Microsoft regarding simplicity and elegance versus multiplicity of features. Apple has concerned itself with everything from the computer hardware, how the keys feel when you push them (Figure 2.21), to the software applications. Microsoft, on the other hand, has dealt mainly with software. The menu bar in Microsoft Word (see Figure 2.22) has many features that typical users do not need; many users do not understand the iconic representations of these features on the menu bar. The design seems complicated for the typical user.

Figure 2.21 | Here Steven Jobs introduces the keyboard for the new iMac. The resistance on the keys of an Apple iMac increases with finger pressure.

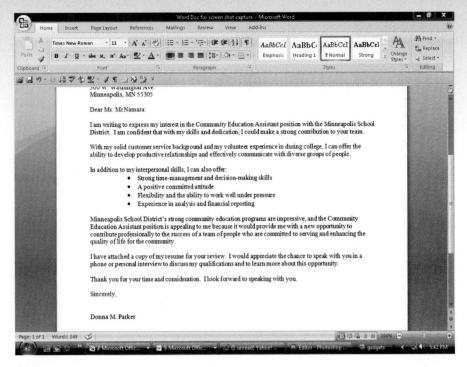

300 W. Washington Ave.
Minneapolis, MN 55305

Dear Ms. McNamara:

I am writing to express my interest in the Community Education Assistant position with the Minneapolis School District. I am confident that with my skills and dedication, I could make a strong contribution to your team.

With my solid customer service background and my volunteer experience in during college, I can offer the ability to develop productive relationships and effectively communicate with diverse groups of people.

In addition to my interpersonal skills, I can also offer:
- Strong time-management and decision-making skills
- A positive committed attitude
- Flexibility and the ability to work well under pressure
- Experience in analysis and financial reporting

Minneapolis School District's strong community education programs are impressive, and the Community Education Assistant position is appealing to me because it would provide me with a new opportunity to contribute professionally to the success of a team of people who are committed to serving and enhancing the quality of life for the community.

I have attached a copy of my resume for your review. I would appreciate the chance to speak with you in a phone or personal interview to discuss my qualifications and to learn more about this opportunity.

Thank you for your time and consideration. I look forward to speaking with you.

Sincerely,

Donna M. Parker

Figure 2.22 | The design of the Microsoft Word menu bar might seem complicated to some users.

One goal of designers is to create devices that are minimally complicated; every function should be needed, and each function should be intuitively easy to understand. This is a great challenge, as technologically able designers want to use the various features of their design programs. The challenge of differentiating your product with simplicity is demanding. For example, a technology such as keyboard resistance may not be obvious to a user, but the technological creativity required to create this feature is significant.

The iMac is an example of design simplicity, sometimes called elegance. Unlike other desktop computers, the iMac contains no cooling fans. No fans mean no noise and one less thing to fail. In the iMac, the computer and monitor are cooled by natural convection—air passes through the computer in the same way that air passes over a radiator, convecting heat from the surface. Other computers rely on small fans within the casing to blow air over the circuit boards and keep them cool.

Phase 6: Test and Evaluate the Design Solution Develop a test, or several tests, to assess the performance of the design solution. Test the design solution, collect performance data, and analyze the data to show how well the design satisfies constraints and specifications. A form that might facilitate such assessments is shown in Figure 2.23.

Observing (monitoring) the results of the tests may suggest how you can improve the design or the construction of the solution. The testing is done scientifically by changing only one factor at a time. You should note the factors that affect the performance of your design; such factors are called variables.

All the information from your testing is gathered and analyzed to see if the design solves the problem. You will have learned a great deal about the problem and its solution as you construct and test the prototype. In the case of the sports bag, for example, perhaps only one person believed the straps should be wider, but six people wanted the top opening enlarged and four wanted a personal storage

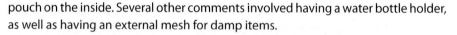

Ten people were interviewed about their opinions of the prototype sports bag and to ascertain that the bag met with specifications. The bag is 25 inches long and 13 inches square at each end. The following questions were asked:

1) Is the bag large enough for you to consider using as a sports bag?
2) Are the handles satisfactory?
3) Do you like the having the option of carrying the bag with handles and with an over the shoulder strap?
4) Is the top opening large enough for your sports equipment?
5) Is waterproof material important in a sports bag?
6) Are internal compartments useful to you?
7) Are there any features missing that you think are important?

Question	Yes	No	Comments
1	10		
2	9	1	Wider shoulder strap
3	10		
4	4	6	Max out the opening
5	10		
6	6	4	Prefer personal items on inside
7			Water bottle holder; external mesh for damp items

Figure 2.23 | The information from product testing is gathered and analyzed to assess the sports bag design.

Figure 2.24 | The prototype will be modified to include a bottle holder, mesh side bag, internal compartment for personal items, and larger top opening.

Figure 2.25 | After any needed changes have been made, production of the full-scale design can begin.

pouch on the inside. Several other comments involved having a water bottle holder, as well as having an external mesh for damp items.

Phase 7: Redesign the Solution with Modifications In the redesign phase, you critically examine your design and note how other designs perform to see how you can improve your own. In Figure 2.24, the decision was made to eliminate the external side pouches for personal items, and to replace them with an external mesh that opened at the top. A zippered enclosure on one end of the bag was replaced with a mesh bottle holder. The zippered enclosure on the other end was retained.

Identify the variables that affect the performance of your design. Think about scientific concepts that underlie these variables. Indicate how you will use scientific concepts and mathematical modeling to further improve the performance of your design. For example, perhaps in the design of a warning light, you found in an impact test that if you used a regular light bulb, it would be easily damaged. However, if you replaced the bulb with a light emitting diode (LED), the visibility would be just as good, and the design would be more rugged.

After any needed changes have been made, production of the full-scale design can begin. Perhaps a large building will be constructed, or a product will be mass-produced, as shown in Figure 2.25. Feedback should be obtained and used over the life of the product or solution to make sure that it continues to meet the design criteria. Designs are critiqued and redefined, then improved during development and during the product's life. If necessary, additional changes are made to the product during its life. Changes occur when there are product recalls; for example, the windshield-wiper motors on a car might have to be redesigned if information indicates that the original design can create a safety hazard.

Phase 8: Communicate Your Achievements Complete a design portfolio or design report that documents the previously mentioned steps. Make a presentation justifying your design solution. Present your final design along with a summary of how it meets the specifications and constraints to solve the design challenge. Describe the design process that you went through and which phases were most important in creating your final solution.

1. Describe the differences between specifications and constraints.
2. Do you need to follow the design process sequentially? Why or why not?
3. Why is technical knowledge and skill important in designing?
4. What does optimization mean, and how is it related to the design process?

SECTION 3: Further Considerations in Designing

KEY IDEAS >

- Modeling helps engineers test for different conditions to which the product or process might be subjected.

- Teams and teamwork are very important during design projects.

- Many new designs are ecological; they fit with the natural environment and require a multidisciplinary team approach.

- Design includes factors such as safety, environmental considerations, maintenance, and repair.

- Design solutions change over time.

In design, balance is the key to success. Balance must exist between theory and practice; in other words, balance must exist between modeling and analysis skills and between hardware implementation and measurement skills.

Modeling

Modeling helps engineers test for the different conditions to which the product or process might be subjected. For instance, suppose we want to build a simple series circuit that contains a battery, a switch, and a light bulb. A flashlight operates on a simple series circuit, as illustrated in Figure 2.26. A physical model of this circuit is illustrated in Figure 2.27. In this figure, the battery is replaced by a symbol for a voltage source (battery), the switch is symbolically shown, and the light bulb has been replaced by a symbol for resistance. (A light bulb is essentially a resistor that heats up to a high temperature and radiates.)

What happens when the switch closes? Electrons flow from the battery's anode through the circuit, the bulb lights up, and the electrons return to the battery at the cathode. Using an ammeter, we can measure the current flow. Let's assume that you want to change the voltage, add another battery, and determine what happens.

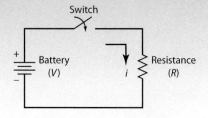

Figure 2.26 | A flashlight operates on a simple battery, switch, and light bulb series circuit.

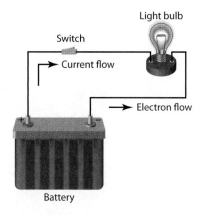

Figure 2.27 | A physical model shows a representation of a battery, switch, and light bulb circuit.

Let's also assume that you want to do this for a variety of battery combinations. This can be time-consuming, which is why engineers use physical models that can be mathematically modeled. Using Ohm's Law, the voltage drop across a resistor is equal to the current flow through the resistor times the resistance ($V = i \times R$). From this, we can model the current flow as $i = V/R$. In this instance, the voltage, V, is the battery voltage and R is the resistance measured in ohms. We can calculate the current for various voltages, and we are not required to set up a series circuit and measure the current each time.

It is important to test the mathematical model against the experimental model to make sure the results are the same. Otherwise, using the mathematical model could lead us to make erroneous conclusions. For instance, even in this simple electric circuit, we are neglecting resistance in the wire and in the switch; we are also assuming that the battery voltage remains constant. These are reasonable assumptions in this situation, as copper wire has very little resistance, and the resistance of the switch is also very small. However, differences will often exist between the physical system, the physical model, and the mathematical model.

Figure 2.28 illustrates how physical (designed) systems, physical models, and mathematical modeling are related. A physical model is an imaginary physical system, which resembles an actual system in its salient features but is simpler and thereby more amenable to analytical studies. Mathematical models are made from the physical model; this idea is at the heart of engineering analysis. Technological judgment is needed here to make astute approximations; such approximations are made at the onset of an investigation, and they are the crux of engineering analysis.

Engineers employ the methodology in Figure 2.28 when they develop design solutions. They investigate alternative solutions to determine the best solution. They also employ the methodology to determine the performance characteristics of an actual physical system. However, engineers also need to check the accuracy of the modeling, so tests are performed on the actual system, as illustrated in Figure 2.29 and Figure 2.30.

The results are compared to the modeled system to check the accuracy of the model and provide direction for model improvement as necessary. This concept is illustrated in Figure 2.31.

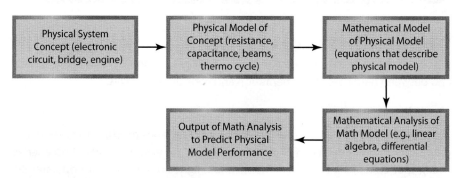

Figure 2.28 | This diagram indicates the relationships between physical systems, physical models, and mathematical models.

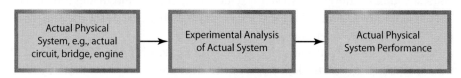

Figure 2.29 | Testing of an actual physical system, coupled with analysis of experimental data, yields performance indicators of the actual system.

Figure 2.30 | A reusable rocket engine is tested in a NASA test facility.

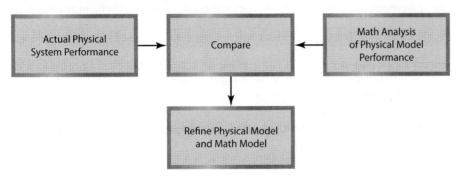

Figure 2.31 | Comparing modeled and actual system performance allows you to refine the physical and mathematical models.

Engineering design is not trial-and-error **gadgeteering**. Engineers use their knowledge of science and engineering science to understand what is happening physically, their use of mathematics to create models for analysis, and their understanding of prior technological solutions so they can innovate. Then they create design solutions. This approach is in contrast to the process used by inventors, who may gadgeteer until they arrive at a workable solution that they can patent or manufacture.

The use of modeling, with its inherent predictive analysis, is an important characteristic of engineering. The importance of modeling and analysis in the design process has never been more important than it is today. These design concepts can no longer be evaluated by the build-and-test approach because it is too costly and time-consuming.

A variety of engineering models may be developed based on the particular need. Engineers ask: "Why am I modeling the physical system, and what is the range of operation for which I want my model to be valid?" If the need is system-design iteration, then a "design model" is needed. Iterations can then be performed, using, as a starting point, the results of the work performed with the design model. Models only need to be valid for the particular range of operation of interest. Thus, for the simple series circuit we discussed earlier, we can test the system for a wide variety of battery voltages and resistances.

Ecological Design

A trend in engineering and other design professions, such as architecture, is to seek designs that are compatible with the natural environment in their development, operation, and disposal. This trend springs from the growing suspicion that the affluence we currently enjoy is, in part, borrowed from the future. We are depleting natural resources, minerals, and energy at an irreplaceable rate. In the process, we could be creating an unsustainable world. Some people consider that we are living in two worlds. The first is the natural world that has evolved over 4 billion years. The second is the human-made world of artifacts, farms, roads, and cities designed over the last few millennia. The condition that threatens both worlds—unsustainability—results from a lack of integration between them. Many new designs are ecological; they fit with the natural environment and require a multidisciplinary team approach.

Ecological design is simply the effective adaptation to and integration with nature's processes, as shown in Figure 2.32. It tests solutions with a careful accounting for their full environmental impacts. For example, some sewage treatment plants use man-made marshes to simultaneously purify water, reclaim nutrients, and provide habitat. These marshes also become agricultural systems that mimic natural ecosystems and merge with their surrounding landscapes. New types of industrial systems use the waste stream from one process as useful input to the next, thus minimizing pollution. Multidisciplinary teams are required to solve these complex problems.

Engineers with an ecological focus have observed that the traditional model of industrial activity—in which individual manufacturing processes take in raw materials and generate products to be sold, plus waste to be disposed of—should be transformed into a more integrated model: an industrial ecosystem. In such a system, the consumption of energy and materials is optimized, waste generation is minimized, and effluents of one process—whether they are spent catalysts from petroleum refining, fly and bottom ash from electric power generation, or discarded plastic containers from consumer products—serve as raw materials for another process.

Figure 2.32 | Plants can clean the waste from water through a series of engineered ecosystems.

Engineering traditionally concerns itself with safety and efficiency; ecological design asks that we do more. We have already made dramatic progress in many areas by substituting design intelligence for the extravagant use of energy and materials. Computing power that, fifty years ago, would fill a small house with vacuum tubes and wires can now be held in the palm of your hand. The old steel mills whose blast furnaces, slag heaps, and towering smokestacks dominated the industrial landscape have been replaced with efficient scaled-down facilities and processes. Many products and processes have been miniaturized, dramatically reducing the energy and materials required to fabricate and operate them. For instance, some manufacturing now uses molecular nanotechnology. If we think about products at their most fundamental level as being atoms, then the properties of these materials depend on how the atoms are arranged.

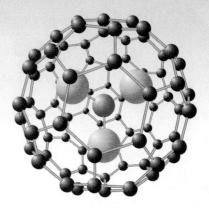

Figure 2.33 | A nanomanufactured spherical molecule, trimetasphere, has 80 carbon atoms that can contain metallic atom ions.

In Figure 2.33, the nanomanufactured molecule was designed to be used in MRI scans to provide better contrast and perhaps attack cancer cells in the future. The possibilities of nanomanufacturing and nanodesign are seemingly endless. If coal atoms are rearranged, diamonds can result. Similarly, by rearranging the atoms of sand and adding a few trace elements, computer chips are made. Today's manufacturing methods are very crude when viewed from a molecular level— casting, grinding, milling, and even lithography move atoms in huge groups. Molecular nanotechnology or manufacturing allows positioning of every atom in its correct place, yielding reduced costs in terms of materials and energy.

Certainly, society plays a pivotal role in creating a climate for change. For instance, in Germany, manufacturers are now required by law to either return and recycle old packaging or pay a steep tax. This law transformed the German packaging industry. Questions now include: How can durability and reuse be incorporated into packaging design? How can easy disassembly of packaging components facilitate recycling? These questions have triggered extraordinary innovations in reusable or recyclable packaging with corresponding environmental benefits, including decreased waste and decreased use of virgin materials. Traditional design does not ask these questions; instead, it emphasizes the importance of cost or convenience with minimal or no environmental considerations. Design solutions change over time. New technologies and new laws and regulations create different requirements for design. Design includes factors such as safety, environmental considerations, maintenance, and repair.

Figure 2.34 | The Norwegian coast is very dramatic, with deep fiords and rugged coastlines rising steeply to hilltops. To hilltops hence there can be great ecological diversity in a short distance.

In a sense, evolution is nature's ongoing design process. The wonderful thing about this process is that it happens continuously throughout the biosphere. A typical organism has undergone at least a million years of intensive "research and development." A few years ago, two Norwegian researchers set out to determine the bacterial diversity of a small amount of beech forest soil and of shallow coastal sediment (Figure 2.34). They found more than 4,000 species in each sample, which more than equaled the number listed in the standard catalog of bacterial diversity. Even more remarkable, the species present in the two samples were almost completely distinct— nature's design process yielded different solutions under different constraints.

In sustainable design of products and processes, three strategies are employed, with an eye toward ecological awareness, to address environmental effects: conservation, regeneration, and stewardship. Conservation slows the rate at which things are getting worse by allowing scarce resources to be stretched further. Typical measures are recycling, adding insulation, and designing fuel-efficient cars. Regeneration is the expansion of natural resources by active restoration of degraded ecosystems and communities. Stewardship is the practical quality of care in relation to other living creatures and the landscape. Stewardship requires continual reinvestment, observation, and design innovation. Consider the challenge

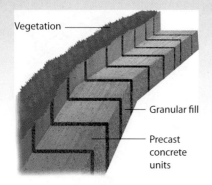

Vegetation

Granular fill

Precast concrete units

Figure 2.35 | A greenwall is a compromise between solely ecological design and traditional retaining-wall design.

of how to control erosion on a steep hillside. A conventional design could use a thick concrete retaining wall to hold the earth.

In ecological design, the same goal is accomplished with natural processes, such as seeding the hillside with willow branches that sprout and develop articulated roots, holding the soil in place. However, the willows must be tended until the roots develop; knowledge of trees is required to distinguish between willows with deep root systems and other species with shallow root systems. Notice that the ecological solution uses very little energy and matter, but does require stewardship, while the traditional design method uses much more energy and matter, but requires little stewardship. As a society, we have expected engineers to bend an inert world into shape. An alternative is to catalyze the self-designing potentials of nature, allowing useful properties to emerge rather than deliberately imposing them. Figure 2.35 shows how these two approaches can come together in a compromise.

As you have seen, engineering design includes a number of factors. These factors often compete with each other, but all are important.

ENGINEERING QUICK TAKE

Technical Decision Analysis

It is comparatively easy to select the best design from many if one of them clearly stands out as the best choice on several fronts. Nevertheless, competing designs will offer distinct features and have different advantages and disadvantages. Again, an inherent characteristic of the design process is its trade-offs, including cost versus material, material versus reliability, and reliability versus performance. This approach of weighing many different factors contrasts with using a single evaluation criterion, which often is cost.

Multi-Criteria Decision Analysis

The multi-criteria decision analysis technique allows us to balance criteria in a rational fashion and arrive at a best product. For example, in the next year or so, you may be considering a university to attend. How might you decide, in a rational fashion, among several universities? The multi-criteria decision analysis technique provides you with a methodology used by engineers in deciding between competing designs. Table 2.2 lists some typical criteria and their weighting factors that you might use in selecting a college. Not all criteria are equally important, so the relative importance of each criterion is denoted by the weighting factors. In general, determining the criteria and weights is not an easy task.

Table 2.2 | Criteria and Weighting Factors for Selecting a College

CRITERION	WEIGHTING FACTOR (1–10)
1. Cost	9
2. Class size	6
3. Compatible student body	7
4. Major desired	8
5. Reputation	7
6. Nearness to home	5

Figure 2.36 | This Google Earth image of Iowa State University was taken at an elevation of 2.5 km. You can use Google Earth to measure the distance across the campus.

Figure 2.37 | Bryn Mawr is a small liberal arts college.

Selecting Criteria

The criteria in Table 2.2 were used in selecting a university: Cost is a very important criterion and deserves a top weighting factor for most people. Nearness to home may not be as important to you, so it was given a lower weighting factor. One of the keys to success at a university is how you relate to your fellow students; if you find a compatible group of peers, you will be much more likely to have a successful college career. The size of the university may not be important to you. Whether the college or university has 2,000 students or 20,000 may matter less than whether it has the major you desire or enjoys a stellar reputation (see Figures 2.36 and 2.37). You might assign different weights to these factors, depending on which of the factors matters most to you.

Rating Criteria

At this point, you can select alternative schools and rate them with a score of 1 to 10 based on how well they perform in each of the six categories. The score is multiplied by the weighting factor, and the total for each school is determined. The highest total score indicates the best school based on these criteria. For example, you can contrast a large public university (LPU) with a medium-size private university (MPU). Table 2.3 indicates the scoring that might occur. In this example, the tuition cost of the public university is as inexpensive as possible, except perhaps for the community college; this fact yields a high score. The class sizes and number of students enrolled are much smaller at the private university; hence, its higher rating. The student-body characteristics of the private university were more appealing to this hypothetical student. On the other hand, the LPU did have the exact major the student desired (for example, ceramic engineering), whereas the MPU did not (for instance, materials science). The reputations of the faculty and graduates from both institutions were very good, with the edge going to the LPU because it has more research and publications. The MPU was nearer to home, which was a plus in this instance. You can see from the totals that the two universities are virtually tied in rating, indicating the difficulty in making a decision between two similar options.

Section 3 ▲ Further Considerations in Designing

It is possible for bias to enter the scoring, but there are ways to minimize this effect through benchmarking, which will be discussed in the following section.

Table 2.3 | Weighted Comparison of Two Universities

CRITERION	WEIGHT (1–10)	PRIVATE UNIVERSITY RATING	PUBLIC UNIVERSITY RATING	PRIVATE UNIVERSITY TOTAL	PUBLIC UNIVERSITY TOTAL
Cost	9	5	10	45	90
Size	6	10	5	60	30
Student body	7	8	6	56	42
Major	8	7	9	56	72
Reputation	7	7	8	49	56
Closeness to home	5	8	5	50	25
			Total	316	315

Benchmarking Criteria

Benchmarking entails establishing what is the best for a given criterion, quantifying it, and then comparing a given product or process to the benchmark. For instance, in considering four-year undergraduate schools, the cost at the LPU is $10,000 per year while the MPU costs $20,000 annually. As noted earlier, the LPU has the lowest possible cost, and so its tuition becomes the benchmark value. To determine the rating of the universities, divide the benchmark value by the university's cost, and then multiply by 10 to convert the number to a value between 1 and 10. For instance:

$$\text{LPU cost rating} = \frac{\$10,000}{\$10,000} = 1.0 \qquad (1.0)(10) = 10$$

$$\text{MPU cost rating} = \frac{\$10,000}{\$20,000} = 0.5 \qquad (0.5)(10) = 5$$

Figure 2.38 | Students often work in teams in laboratory courses; even at large universities, the laboratory sections are comparatively small.

Figure 2.39 | Unlike a typical high school class, students in a college course might fill a large lecture hall.

The average class size in the MPU is 25 students, which again is viewed as the lowest possible value and hence the benchmark value. The LPU has an average class size of 50. Class size can be an important criterion when selecting a college or university, as shown in Figures 2.38 and 2.39.

$$\text{LPU class size rating} = \frac{25}{50} = 0.5 \qquad (0.5)(10) = 5$$

$$\text{MPU class size rating} = \frac{25}{25} = 1.0 \qquad (1.0)(10) = 10$$

The analysis is very adaptable to spreadsheets, so adding more universities for comparison is not a chore. Determining the benchmark values requires research because the quality of your multi-criteria decision analysis depends on the quality of the data you use.

Teamwork

Very often, your education has focused on individual development, building competencies and understanding in a variety of subject areas, being individually accountable, and competing with your fellow students. The practice of engineering requires teamwork; you will work in teams on projects, and you will have both individual and group responsibilities (see Figure 2.40). Your advancement is based in large measure on whether your group achieves its goals. Teams and teamwork are very important during design projects.

Effective teams have many attributes, including respect for one another, the ability to listen carefully to others' opinions, members taking responsibility for themselves and for the group, participation by all, and a common goal or purpose. Unfortunately, groups are often not organized as teams, but simply as people put together to work who may have no interest in assisting one another. Indeed, the reward system is often based on individual accomplishment, so, in

Figure 2.40 | Teamwork is necessary in this Habitat for Humanity home construction project.

Figure 2.41 | Outward Bound participants need to work as teams to survive challenges that a single person would be hard pressed to solve alone.

effect, team members are competing against each other. This system will not yield a positive outcome from a team perspective. A positive, cooperative group or team requires individual and group goals, a common sense of purpose and support of one another's achievements, and help with one another's problems (see Figure 2.41). This requires time together and positive communication with one another.

Learning to work with others is a very important skill, and it is one that can be developed. Perhaps the most important attribute is your communicative ability—the ability to listen as well as to speak. Listen to other people's ideas; try to understand why they think as they do without critically examining their words in the hopes of refuting them to promote your own ideas. In addition, you will need to express yourself clearly and thoughtfully, respecting others' opinions, even when yours might be at odds with theirs. This open-minded attitude facilitates consensus. This is especially important on multidisciplinary teams in which people have different perspectives that need to be valued.

Seth Rosenberg and Eyal Angel are two mechanical engineering students who decided to design and build their own Formula One-style racing car—from scratch, not from a kit. This was a challenging undertaking, but the two followed the engineering design process. They knew the type of car they wanted: one that was similar to Formula One racecars, which are lightweight and very fast. They used the Ariel Atom 2, a British racecar, as their benchmark model.

Automobiles are made of systems that in turn are made from components, and everything has to work together. Seth and Angel, as he prefers to be called, did not take any courses in automotive design, but they were confident that their education in engineering courses would enable them to learn more. And learn more they did: During the research and investigation phase, they developed their own knowledge and skills involving suspension systems, exhaust systems, braking systems, and chassis design.

When designing the chassis, Seth and Angel used two computer-aided design (CAD) systems to help them visualize and analyze the car's structure. Seth is a qualified welder and machinist, having worked in his grandfather's machine shop since he was in middle school. These skills were essential, as he was the lead fabricator in welding the tubular steel together. Seth and Angel needed to know the size requirements for the various components before constructing the frame.

Their final design had a 2.4-liter, turbocharged rear engine, rack and pinion steering, and a five-speed manual transmission, fitted together to create a 1,200-pound car. The engine and transmission were from a wrecked 2004 Dodge Neon purchased on eBay. The Neon is a sport compact car, and the stock drive shafts that connect the wheels to the transmission were not long enough for their Formula One. Therefore, Seth and Angel had to design a new length for the parts and order longer driveshafts from a custom driveshaft shop.

Figure 2.42 │ Seth Rosenberg and Eyal Angel designed their own Formula One-style automobile.

The cooling system posed its own set of design challenges. Seth and Angel wanted the radiator to be in the front of the car so it could get good air flow, but the engine was in the rear. The rack and pinion steering system was also custom designed, from the two universal joints that connected the steering wheel to the pinion, to the A frame that holds the wheels and the shock absorbers. The project required a tremendous amount of visual thinking, imagining how parts would move and interconnect, and fundamental analysis skills to make sure the components were strong enough.

Seth and Angel are delighted with their project, which they consider the dream of a lifetime come true. They hope to make their model "street legal" and to attract investors who would like a racecar of their own.

Figure 2.43 | Making a float for a parade requires teamwork, using the abilities of many different people.

A positive group will include shared leadership roles and shared responsibility, so everyone is empowered. There will be individual and group accountability, dividing tasks and supporting one another in accomplishing them, and a collective work product. For instance, you might have the opportunity to join with others in creating a float for a parade. This can be a terrific design project, a great chance to collaborate with people of many backgrounds, to develop teamwork, to have fun.

Both teams and working groups have tasks to accomplish, and the methods by which the goals are achieved are different. These divergent paths can often produce results of dissimilar quality. For instance, a working group should have a strong, clearly focused leader. This person can clearly articulate the group's goals and tasks and assign different tasks for members to accomplish. On the team, the leadership role is more diffuse; there are shared leadership responsibilities in recognition of the talents that team members possess. In designing a parade float, as shown in Figure 2.43, people with an art background and heightened aesthetic sensibilities are essential, as well as people with technical abilities to create structural supports or moving parts. Their contributions need to be respected and other team members must listen to their opinions.

A focused leader model does not enfranchise these various opinions. The need for individual accountability and mutual accountability extends throughout the team structure. A working group has individual accountability; a team does too, but its members listen to and help one another to achieve their goals.

How can this spirit of positive collaboration be accomplished? A normal method is to have meetings in which open-ended discussion occurs and various voices are heard and respected, resulting in group problem solving in which all members are invested in the outcome. A group that incorporates leaders who delegate tasks to other team or group members can result in the same investment. A spirit of collaboration requires that we develop a respect for people with diverse backgrounds that are different from our own. These different perspectives will broaden our abilities and knowledge while we become more proficient engineers.

Building a consensus does not mean that everyone has to agree to every element of a group decision. Indeed, you may not agree with a decision, but you support it because it was arrived at through open discussion. In the open discussion, your point of view and those of others were listened to and understood. A collaborative decision evolved, based on an understanding of all points of view. This is difficult to achieve; it requires good communication skills, both in articulating points of view and listening to and understanding others.

As you can see, an important feature of effective teams is communication. Good communication skills are an essential attribute of successful engineers and engineering managers.

SECTION THREE FEEDBACK >

1. Describe an instance in which teamwork helped you.
2. What are the differences between the physical system and the mathematical model of the system?
3. Think about a container that is thrown away. How might it be redesigned so it could be recycled or reused?

SECTION 4: Ethics, Safety, and Risk

KEY IDEAS >

- Products that are dangerous to use should fail safely.

- Public safety is of paramount importance when designing products and processes.

- Risk assessment tries to determine the probability of a negative event's occurrence.

- In developing products, applicable regulations, codes, and standards are identified, and they establish the products' basic safety requirements.

- Manufacturers cannot move too far ahead of what society expects in terms of product quality and still stay in business.

Engineers often would like to ignore moral dilemmas and would rather deal with issues related to problem solving in the physical world. These dilemmas are problems to which there is no right solution, but they indicate the awareness that engineers must develop if they are to participate fully in society. Being attuned to issues other than those in technical areas is part of the obligation of being a professional engineer. Not all problems are technological and not all problems can be solved technologically. Some problems have political implications, while other problems offer no complete solution, but only a way to minimize poor solutions. These problems are aspects of life that engineers must constantly recognize.

For example, many communities need affordable housing for people and families with low, medium, and high incomes. As with all design decisions, there are trade-offs in terms of advantages and disadvantage to each possible solution to this problem (see Figure 2.44). For instance, housing development increases housing density, which theoretically should lower the cost of homes and make them more affordable to families. However, there are other impacts on transportation, schooling, and utilities that mitigate against high-density housing and often cause a net tax increase for the entire community involved.

Safety

Engineers most often work as part of a design team, making decisions about a product or service that affects the whole, and they are affected by the works of others. Engineers are increasingly aware of product safety when designing new or improved products. Public safety is of paramount importance when designing products and processes. Society holds a manufacturer responsible for a product's

Figure 2.44 | A housing development reflects a trade-off between density, the need for housing, affordability, and societal impact.

75

performance, even if that product is used in a manner unintended by the manufacturer, and even if intermediaries have sold the product. This does not mean that a product cannot wear out, but it should do so in a safe fashion. Engineers cannot foresee every circumstance in which a product may be used, so they must be diligent in assessing how it fails. Even though the company for which an engineer works, not the individual engineer, might be held liable when a product fails, an engineer's responsibility to society and to the company requires attentiveness and care. Even products that can be dangerous to use should fail safely. When products fail safely, they cease to function, but do not cause injury in failing.

Product quality involves gray areas. A company must be able to make its products at a cost that is low enough to be competitive with others. If a product has the highest quality, but it consequently has a high and uncompetitive price, the company may not be able to remain profitable and may be forced out of business.

Many household products require built-in safety features. Today, for example, lawn mowers for homeowners have two handles that need to be squeezed together, as shown in Figure 2.45. One is the handle needed to push and direct the mower, and the other is an automatic cutoff. If the cutoff handle is not held simultaneously with the other handle, the mower will not run. If you let go of the cutoff handle, the mower will stop. Several decades ago a cutoff handle was seldom found on lawn mowers. However, people would injure themselves by trying to pick up mowers while they were still running, losing fingers and hands in the process. Modern regulations exist to try to prevent such problems.

Figure 2.45 | Notice the lawn mower has two handles that must be grasped for the engine to operate. Releasing the lower handle stops the engine.

Manufacturers cannot move too far beyond society's expectations in terms of product quality and remain in business. With new technologies and materials available, an engineer can often redesign and improve product quality without increasing cost. In today's world marketplace, engineers compete with other engineers from all nations. The need to maintain or increase product quality while decreasing product cost is ever more important.

Because companies become responsible for products they manufacture that are later misused, engineers must design high-quality products that meet standards of excellence. A company may be held liable for a variety of reasons:

1. The product's design is defective, so it cannot be safely used for its intended purpose.
2. The manufacturing process is flawed; for instance, the testing and inspection are deficient.
3. The product labeling is inadequate regarding warnings and proper use.
4. The product packaging allows safety-related damage during shipping, or the product parts may be separated, allowing sale in dangerous form.

Certainly this seems like a formidable list of requirements, but these requirements do not constitute an impossible challenge for a company to overcome. Quality control and quality assurance have become significant in all phases of manufacturing, from design to distribution.

Risk and Risk Assessment

Ideally, all products and projects would be 100 percent risk free. That ideal is unattainable, but a design should nevertheless present minimal risk to people and the environment.

Risk assessment is a combination of art and science. Its purpose is to try to determine the probability of some negative event's occurrence. While the science of making the calculations is quite easy, the art of assessing the validity of the calculations is not (see Figure 2.46). For instance, approximately 24,000 people died in automobile accidents in 1987, out of a total of 2.5 trillion miles driven. Thus, the risk (probability) of an accident occurring per mile was:

$$\text{Probability/mile} = \frac{24,000 \text{ deaths}}{2.5 \times 10^{12}} = 9 \times 10^{-9} \text{ deaths/mile}$$

If you were to take a-mile trip, the probability that you would have a fatal accident is 9×1^{-7}, or about one chance in a million. If you were not careful in selecting the data for the previous calculation, you might have included pedestrians killed by automobiles in the total number of deaths, which in 1987 was approximately 49,000 people. Including pedestrians would therefore have resulted in calculation of a higher probability of a fatal accident. Additional methodology could be used to further refine the data with other considerations, such as whether you were the driver or passenger, wore a seat belt, drove in a certain state, or drove in certain weather conditions. In any case, data is available, and the calculations are straightforward.

In the area of biological sciences, data is not as available, which increases the uncertainty of a risk assessment. For instance, peanut butter (see Figure 2.47) contains very low levels of a chemical called aflatoxin B. This chemical is produced by a fungus that attacks peanuts when they are not carefully dried and stored, and it has been shown to cause liver cancer. One study found that consuming a 10-ounce jar of peanut butter would increase the risk of death due to liver cancer and aflatoxin B by about one in a million (similar to the risk factor of death when driving 100 miles).

But how certain are the numbers used in the study? Much more interpretation is necessary. Simply finding the number of deaths due to liver cancer per year and dividing by the total consumption of peanut butter does not provide the answer, for many reasons. For example, there are many other causes of liver cancer, such as excessive alcohol consumption. In addition, other grains and nuts that are part of the food chain contain aflatoxin B. To further complicate the data, a person does not die immediately after consuming aflatoxin B; it can take up to 30 years, during which other events can occur that may aggravate liver cancer. One methodology used in such research is epidemiological study, in which statistics are used in attempts to find correlations between groups of people. It is easy, however, to draw erroneous conclusions using this methodology.

Biological studies rely on the use of animals, and the results of these studies must be extrapolated to humans, as humans are not used for initial testing. Such extrapolation introduces another type of error. For instance, in one experiment to test the lethality, a lethal dose, of a certain chemical, a guinea pig required one microgram/kg of its body weight, a hamster required 5,000 micrograms/kg of its body weight, a male rat 22 micrograms, and a female rat 45 micrograms. Which test animal's data would be best to use if the results were extrapolated to humans? This is a very important question to answer. To further complicate the situation, animal tests usually use high dose levels to determine if a chemical is toxic. As you can see from the animal studies, there is not a straight-line connection from the weight of one animal to that of another. It is very difficult to extrapolate what level is dangerous to a human from small-animal testing. This further contributes to uncertainty in describing risk. Some forms of testing for environmental hazards face the same uncertainties.

Figure 2.46 | Congested traffic affects many urban areas; a minor benefit is that cars stuck in traffic must travel so slowly that most accidents are not fatal.

Figure 2.47 | How much, if at all, does peanut butter contribute to liver cancer?

Figure 2.48 | Skiing can be fun, but it can also be risky.

Another question arises: How does one assess a new technology's risk where no negative events yet exist? The fact that a negative event has not yet happened does not mean it can never occur.

Also, in a study of pollution risks, the public may fear that not all factors are included in the risk assessment or that some risks may be hidden. Studies have shown that we are much more accepting of a known and uncontrollable risk (such as the risks of skiing, as shown in Figure 2.48) than a hidden (see Figure 2.49) or unknown risk such as those posed by air or water pollution, in which not all factors can be quantified.

The perception of risk includes many factors, such as the following:

1. Imposed risks seem greater than those that are voluntary. For example, some people will risk skiing more readily than they will eat foods that contain preservatives.
2. Unfairly shared risks are viewed negatively, as the person receives no benefit from the risk.
3. Risks that are personally controllable are easier to accept.
4. Natural risks are less threatening than risks created by humans. Thus, naturally occurring radon in soil is less objectionable to most people than radon found in radioactive mine debris.
5. Risks from catastrophes, whether natural (earthquakes) or man-made (Bhopal), are very frightening to people.
6. Risks associated with advanced technologies, such as hormones that increase milk production in cows, are far less acceptable than risks from known technologies, such as car and train crashes.

In developing a product, the applicable regulations, codes, and standards are identified, and these serve to establish the basic safety requirements. The product concept is weighed, with safety requirements on one side, and the schedule, feasibility, and costs on the other side. The designer must examine safer alternatives if they exist, and if a product liability lawsuit is a possibility. The engineer must

Figure 2.49 | Machines like this were used in the 1940s and 1950s to X-ray people's feet and determine how well their shoes fit. The practice stopped when people learned that the X-rays were giving off dangerous amounts of radiation.

consider the user, as well as user error and safety, in the design of any product or project. This becomes very difficult for a large project, as the complexities of the systems create a staggering array of possibilities for misuse.

SECTION FOUR FEEDBACK >

1. Are electric toasters dangerous? How can their safety be improved?
2. Why do you think risks associated with advanced technologies are far less acceptable to people than those from known technologies?

CAREERS IN TECHNOLOGY

Matching Your Interests and Abilities with Career Opportunities: Automotive Service Technician

You know that automobiles are technologically sophisticated machines, from computer-controlled engines to hydraulic and pneumatic systems. With more and more cars and trucks in service, the demand for people who can repair and maintain them is strong. People with good diagnostic and problem-solving skills who can understand these complex system interactions are in demand. Potential automotive service technicians also need to possess a good knowledge of electronics and must have good mechanical aptitude. Because automotive technology is continually changing, service technicians must continually upgrade their knowledge and skills.

Nature of the Industry

Today's vehicles are complex systems that integrate computer-controlled electronics to run the vehicles and measure their performance while on the road. For instance, if a car's engine starts to knock because of gasoline detonation, the engine's timing will automatically be retarded, or delayed, to reduce the knocking. Technicians must have a broad-based knowledge of the systems and how they interrelate, as well as the ability and training to operate electronic diagnostic equipment. Most often, information about a vehicle is stored in computer files that are automatically updated by the manufacturer.

Working Conditions

Most service technicians work at least 40 hours per week (including some evenings and weekends to accommodate customer needs). Usually, the work environment is well-lighted and well-ventilated, though it can be drafty and occasionally noisy. The work requires handling dirty parts and sometimes working in awkward positions. This can result in minor cuts and bruises, but seldom serious accidents.

Training and Advancement

People who want to be service technicians should complete formal training in a high school, postsecondary vocational school, or community college. Some people still learn a trade in an apprenticeship from experienced workers; however, because of the increasing complexity of automotive systems, this is not the best training experience. High school programs vary in complexity and scope; some are extensive and keep up with current technologies, but others do not. Postsecondary training programs also vary a great deal, but most combine classroom and hands-on laboratory instruction. The training may vary from six months to a year, and a graduate receives a certificate of completion. Community college programs typically are two years long; the graduate receives an associate degree or certificates related to the training.

Training in electronics is essential because electronic malfunctions, including those related to the computer systems in vehicles, are involved with the majority of problems in automobile malfunctioning. Typically, technicians purchase their own tools, which eventually can amount to an investment of thousands of dollars. The service station or garage provides the diagnostic and engine-analyzing equipment.

This career path does not necessarily lead to becoming a service technician. Candidates who have good communication and interpersonal skills and who work well with customers can become service managers.

Outlook

The area of automotive repair will experience continued growth, consistent with the growth of the economy, through the year 2014. Because service stations will continue to offer fewer repair services, employment will primarily be in automobile dealerships and independent repair facilities.

[Bureau of Labor Statistics, U.S. Department of Labor, Occupational Outlook Handbook, 2008–09 Edition, accessed January 7, 2008, http://www.bls.gov/oco/]

Summary >

Engineers are creative people who imagine and design new technologies that solve problems. In the process, they often improve our standard of living. In their work, engineers bring together knowledge from science, mathematics, social sciences, and humanities.

Engineers use the design process to creatively apply their knowledge to design and develop new products and processes. Design, modeling, and analysis are key concepts in the design process. The design process has several stages that are iterative and nonlinear. These stages include

- determining the problem with specifications and constraints,
- conducting research and investigation,
- generating alternative solutions,
- deciding on the best solution,
- developing prototype,
- testing, evaluating, redesigning, and
- communicating information about the design.

The informed design process models the engineering design process with guided research and investigation.

Engineers work on teams. They use modeling in developing new designs. Modeling and analysis to predict system and product behavior has replaced the build-and-test method for evaluating design concepts. This approach can make a company very efficient in the process of designing new products.

The projects that engineers work on and the designs that engineers create all require documentation. Your written communication should persuade others that you are correct, that your reasoning is sound, and that your conclusions are convincing.

Engineers are very concerned about creating products that are safe to use. Ethics is the application of a moral philosophy to standards of behavior. When you are a practicing engineer and a contributing member of society, you sometimes must face dilemmas that have no one right answer. Society holds a manufacturer responsible for a product even if it is used in an unintended manner or if intermediaries have sold the product. This does not mean that a product cannot wear out, but it should do so in a safe manner. Design engineers must be diligent in assessing how a product fails, and they must design products to fail safely.

We would like all products and projects to be risk free, but this is not possible. A design should present minimal risk to people and the environment. Risk assessment is a combination of art and science used to help determine the probability of a negative event occurring.

FEEDBACK

1. The purpose of this problem is to demonstrate sketching and creativity. Take a large piece of drawing paper and draw approximately 30 freehand circles that are about 1 inch in diameter and 2 inches apart. Then, in a timed exercise of 5 minutes, sketch details on each circle that define it. Examples would be a flower blossom or a baseball.

2. In the game called Pictionary, you must sometimes sketch concepts. Make an abstract sketch of each of the following things or concepts: animal, tree, shut, penetrate, collapse, turbulent, sharp, decayed. The idea of this exercise is for you to communicate ideas to yourself, not to others.

3. Select a consumer product and determine the best or optimal brand. Include the characteristics you believe are important, and why you believe they are important. Select four criteria you would use in testing the product.

4. Consider a consumer product that you purchased or examined in a store. How might the packaging be redesigned for reuse? For recycling? How might the product have different design features so that it could be recycled into component parts?

5. Pick a popular consumer product (such as soda, lipstick, pickles, or cookies) and determine the best variety. Develop criteria for testing and evaluating the product, perform the tests, and prepare a report justifying your selection.

6. What is the role of ecological design in the design process?

7. Investigate the following products, noting any design deficiencies that should be addressed or features that you believe should be added: vacuum cleaner, clothes dryer, snow blower, popcorn popper, toaster, coffee maker, telephone answering machine.

8. Consider a procedure such as changing a car's tire or replacing a bicycle tire. List the necessary steps and tools needed for each step. Devise alternative sequences for performing the same task.

9. Expand the following problems with more complete definitions:
   a. Design a lamp.
   b. Design a car jack.
   c. Design a hair dryer.
   d. Design an apartment building.

10. Develop additional definitions for the following design problems:

    a. a reserved-seat system for commuter trains

    b. a manufacturing assembly line for making picture frames

    c. a sorting machine for plastic bricks of various colors

11. Develop a list of ideas for solving the following problems:

    a. finding a dropped contact lens

    b. locking all of a house's exterior doors from one location

    c. creating a wake-up alarm for sleepy drivers

    d. collecting golf balls on a driving range

    e. filling a beverage bottle on the ground from a second-story window

12. The following products have inherent dangers while being used. Identify the danger and develop ideas to overcome or limit it:

    a. electric toaster

    b. electric hedge trimmer

    c. chain saw

    d. incandescent light bulb

13. Satellites are used for many types of communication, including telephone and television. Describe how they are used; include the satellite system's age and any vulnerability inherent in it.

14. The global positioning system (GPS) uses geosynchronous satellites that pinpoint locations on earth using comparatively inexpensive equipment. Discuss how the system operates and then discuss the implications of having this information freely available.

15. Thomas Edison was a dynamic inventor, engineer, and businessman. Investigate his life; note the hours he worked when seeking a solution, his use of teams to develop ideas into products, and his business acumen in searching for products that had market potential.

16. Alexander Graham Bell created the telephone, but he originally wanted to improve communication for deaf people because he had a deaf child. Investigate the telephone's development; note the integration of other people's ideas into a completely new invention.

17. Develop a list of criteria that you would use in comparing different brands of the following products:

    a. bicycle

    b. backpack

    c. laptop computer

    d. tennis racket

18. Investigate "cookies," which are applets that are downloaded by Web sites onto computer hard drives. Is this a wise idea? Are there ways to control the downloading of cookies, small text files that contains Web site addresses that are installed on your computer by software from the website?

19. Privacy and the Internet are major policy concerns that are currently being addressed by governments worldwide. Many European nations hold different beliefs from those held in the United States about what information should be made available over the Internet. This is particularly true in regard to financial information, such as credit cards. Investigate these differences between the United States and Europe, noting the different views of what protections governments should provide.

20. Construct a list of advantages and disadvantages between three modes of transportation: automobile, train, and airplane. Some of the factors to consider are cost, accessibility, safety, luggage storage, and sleeping. The advantages and

disadvantages may change depending on the distance traveled (for example, 10 miles, 500 miles, or 3,000 miles).

21. Cellular phones are becoming the primary means for telephone communication in developing countries. Hardwiring (copper or fiber) does not have to be in place to use cell phones. Discuss how cell phones operate, including the need for cell towers, roaming, and satellite connectivity. Include the issue of privacy in your discussion. It is very easy to listen to other people's cellular conversations, but is it legal?

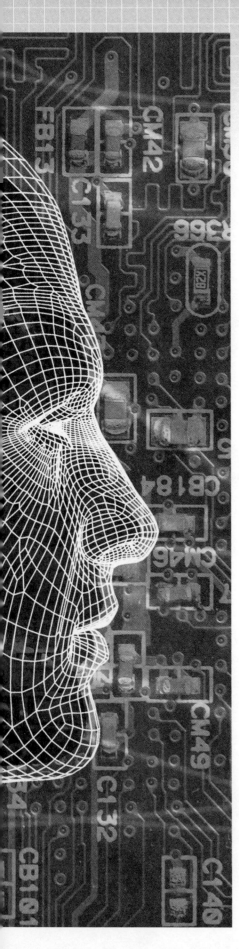

DESIGN CHALLENGE 1:
Creating a Dehydrator

● Problem Situation

An elementary-school group is planning a weekend hike during which participants must carry everything, including their food and clothing. The group is concerned about the heavy load each person will have to carry. Unable to think of ways to substantially reduce the weight of their backpacks, the group has turned to you for advice.

● Your Challenge

As part of a group, you will design, construct, and test a dehydrator that dries fruit as efficiently as possible while still maintaining its quality.

Refer to this activity in the student workbook. Complete the first section of the Informed Design Folio (IDF), stating the design challenge in your own words.

● Safety Considerations

1. Never cut toward your body or point any sharp item at your body or at anyone else.
2. Do not eat the apples.
3. Do not put your fingers near a rotating fan blade.
4. Wear protective glasses at all times.
5. Be careful with the ends of the wire mesh; they can be sharp.

● Materials Needed

1. Six apples
2. Knife for slicing apples
3. Four wooden strips, each with dimensions of 60 cm × 1 cm × 1 cm
4. Wire cutter
5. Bell wire
6. Several kinds of wire mesh, including plastic-coated
7. 3-volt electric motor
8. Two battery holders
9. Fan blade
10. Two C batteries
11. Aluminum foil, 10 square feet
12. One 100W–120W floodlight or heating lamp
13. Cardboard and/or foam board, 10 square feet
14. Duct tape
15. Graph paper or Microsoft Excel spreadsheet application
16. Scale for weighing apples and wire mesh

● Clarify the Design Specifications and Constraints

To satisfy the problem, your design must meet the following specifications and constraints.

Specifications

Dehydrate the maximum weight of apples for a single dehydrating session. Dehydrating time will be set by the instructor. Teams must estimate their design's efficiency based on the data. Efficiency, or e (grams/min), is defined as the time it takes to achieve a specified weight loss.

$$e = (\text{initial weight} - \text{final weight})/\text{time}$$

Constraints

1. You may only use approved materials.
2. The drying surface must have an area no greater than 196 square inches.
3. You must take all apple slices from a single apple, discarding the initial slices that are covered by skin and the core, before taking slices from another apple.
4. The drying tray must be removed to facilitate weighing.
5. Refer to the IDF and state the other specifications and constraints. Include any others that your team or your teacher included.

● Research and Investigate

1. Refer to the Student Workbook and complete KSB 1, Dehydration Techniques.
2. Refer to the Student Workbook and complete KSB 2, Humidity and Evaporation.
3. Refer to the Student Workbook and complete KSB 3, Data Analysis.
4. Refer to the Student Workbook and complete KSB 4, Geometric Modeling and Analysis.

● Generate Alternative Designs

Refer to the Student Workbook and describe modifications to the standard design solutions, indicating with sketches what you consider to be each solution's strengths and weaknesses. Your sketches should show sufficient detail, and they should include justifications for your different solutions.

● Choose and Justify the Optimal Solution

Refer to the Student Workbook. Explain why you selected a particular solution and why it was the best.

● Display Your Prototypes

Construct the fruit dehydrator. Include drawings of the various elements used to construct it. Also include a sketch of the final design or take a photograph of your model and put it in the Student Workbook.

In any technological activity, you will use seven resources: people, information, tools/machines, materials, capital, energy, and time. In the Student Workbook, indicate which resources were most important in this activity. How did you make trade-offs between them?

● Test and Evaluate

Did your new design meet the initial specifications and constraints? Indicate the tests you performed, including the experiments you performed to verify the design. What was your dehydrator's efficiency? Include the time-versus-weight data.

● Redesign the Solution

Respond to the questions in the Student Workbook about how you would redesign your solution based on knowledge and information that you gained during the activity.

● Communicate Your Achievements

In the Student Workbook, describe the plan you will use to present your solution to your class and show what handouts or graphs you will use (You may include PowerPoint slides).

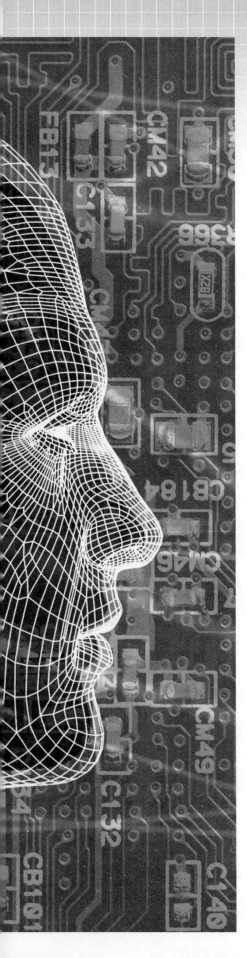

DESIGN CHALLENGE 2:
Creating a Mechanical Pinball Machine

● Problem Situation

It is a rainy day, and you are visiting a friend's home. Rather than playing a pinball video game on the computer, you tell him it is possible for you to build your own pinball machine. He finds this hard to believe.

● Your Challenge

You and your team members are to create a mechanical pinball machine, without electronics, that is fun to play.

Refer to this activity in the student workbook. Complete the first section of the Informed Design Folio (IDF), stating the design challenge in your own words.

● Safety Considerations

1. Never cut toward your body or point any sharp item at your body or at anyone else.
2. Wear protective glasses at all times.
3. Be careful handling the edges of the Plexiglas and metal strips.
4. Glue guns are hot; do not touch the tip of the gun.
5. Do not overextend rubber bands or snap them.

● Materials Needed

1. One 24" × 24" piece of plywood
2. Push pins, nails
3. One 24" × 24" piece of foam board
4. Glue
5. 6 sticks, 1 centimeter square and 60 cm long.
6. Rubber bands of different widths (1/8", ¼", and 3/8")
7. Ball bearings with a diameter of 5/8"
8. Plexiglas strips and metal strips

● Clarify the Design Specifications and Constraints

To satisfy the problem, your design must meet the following specifications and constraints.

Specifications

1. The board must be tilted at a minimum of 10 degrees to the horizontal.
2. The ends of the flippers must be no closer together than twice the ball diameter.
3. The ball must travel at least twice the board length in the maze you create.

Constraints

Refer to the IDF and state the other specifications and constraints. Include any others that your team or your teacher included.

● Research and Investigate

1. Refer to the IDF and complete Knowledge and Skill Builder 1, Checking Out Toys.
2. Refer to the IDF and complete Knowledge and Skill Builder 2, Angles and Bumpers.

3. Refer to the IDF and complete Knowledge and Skill Builder 3, Creating a Maze.

4. Refer to the IDF and complete Knowledge and Skill Builder 4, Designing Flippers.

• Generate Alternative Designs

Refer to the IDF and describe four of your possible solutions to the problem.

• Choose and Justify the Optimal Solution

Refer to the IDF. Explain why you selected a particular solution and why it was the best.

• Display Your Prototypes

Construct your solution. Put a photograph or sketch of your final design in the IDF.

In any technological activity, you will use seven resources: people, information, tools/machines, materials, capital, energy, and time. In the IDF, indicate which resources were most important in this activity. How did you make trade-offs between them?

• Test and Evaluate

How will you test and evaluate your design? In the IDF, describe the testing procedure. Justify how the results will show that the design solves the problem and meets the specifications and constraints.

• Redesign the Solution

Respond to the questions in the IDF about how you would redesign your solution based on knowledge and information that you gained during the activity.

• Communicate Your Achievements

In the IDF, describe the plan that you will use to present your solution to your class and show what handouts or graphs you will use. (You may include PowerPoint slides.)

N A WORLD driven by technology, it is imperative that all citizens develop some degree of technological capability. Whether or not you plan to enter a scientific or technical career, having a technological knowledge base will help demystify the human-made world in which you live. In addition, it will help you better adapt the natural environment to your needs.

To be a fully integrated member of our highly technological society, you should understand how technological products, systems, and

CHAPTER

3

ABILITIES FOR A TECHNOLOGICAL WORLD

SECTION 1
Applying the Design Process

SECTION 2
Using and Maintaining Technological Products and Systems

SECTION 3
Assessing the Impact of Products and Ideas

SECTION 4
Management of Technological Endeavors

Figure 3.1 | These students are using their technological know-how to collaborate on a video-editing project. The more you know about technology, the more comfortable you will be as you tackle the challenges you will face throughout your life.

environments are designed; be able to properly use and maintain technical products and systems; have the ability to assess technological benefits and risks; and be able to effectively manage technological endeavors. This chapter will discuss the abilities you need in order to be a savvy consumer, a good citizen, an efficient and creative worker, and an efficient manager in our highly technological society.

SECTION 1: Applying the Design Process

KEY IDEAS >

- Once a design problem has been identified, it must be analyzed to determine if potential benefits justify investing time and other resources in trying to search for solutions.

- Design specifications and constraints will affect the design

concept, its implementation, its marketability, and its commercial profitability.

- Designers and engineers refine a design by using prototyping and modeling to ensure quality, efficiency, and productivity of the final product.

- Modeling possible design solutions allow the designer

to try out "What if?" scenarios before actually implementing and producing the design.

- The design solution that you choose should be evaluated, and then the final design should be communicated to interested people using techniques such as 3-D modeling, and verbal, graphic, quantitative, virtual, and written means.

- Design solutions are evaluated at various intervals during the design process using a variety of models, including conceptual, physical, and mathematical models, in order to check for proper design functioning and to note areas where improvements are needed. Quality control is a planned process that ensures that a product, service, or system meets established criteria.

- Quality control measures ensure that products and systems meet customer requirements and industry standards.

- Research and development is a specific problem-solving approach that is used intensively in business and industry to prepare devices and systems for the marketplace.

Design is an iterative process. When products or systems are designed for commercial use, they must be reliable, cost effective, and marketable. Developing such a robust design often requires numerous cycles (or iterations) of clarifying design specifications (sometimes referred to as design criteria) and constraints (limitations). Design also involves research, generating alternatives, modeling, testing, and redesign.

Complex design problems require a great number of iterations. Think about the complexities involved in designing a spacecraft. Lockheed Martin Corporation is designing the nation's next spacecraft for the National Aeronautics and Space Administration (NASA). Named Orion, after one of the brightest and most recognizable constellations, it will replace the space shuttle and be used to service the International Space Station and carry astronauts to the Moon and beyond.

According to NASA, industry teams spent 13 months refining concepts, analyzing requirements, and sketching designs for Orion. The new capsule's updated computers, electronics, life support, propulsion, and heat protection systems represent a marked improvement over older systems. Unlike the space shuttle, Orion will be positioned on top of its booster rocket to protect it from ice, foam, and other launch-system debris during ascent. An abort system will be responsible for separating the capsule and crew from the booster in an emergency. NASA recently completed a series of tests that will aid in the design and development of a parachute recovery system for the Orion crew capsule. And these are only some of the many features that will ultimately be incorporated into this exceedingly complex spacecraft.

Measuring 16.5 feet in diameter, Orion (shown in Figure 3.2) will have more than 2.5 times the interior volume of the three-seat Apollo capsules that carried astronaut crews in the late 1960s and early 1970s. Those early missions lasted only several hours to several days. By contrast, Orion missions might last several weeks.

You can imagine how many design challenges the engineers at the Lockheed Martin Corporation continue to face in building this new spacecraft, and how many design iterations will ultimately be required.

Applying the Design Process to Roller Skates

Just as a complex design problem such as the Orion spacecraft requires a great number of iterations, less complicated designs also benefit from multiple iterations.

Figure 3.2 | The spacecraft Orion is currently being developed by NASA and the Lockheed Martin Corporation. The new spacecraft design requires hundreds of thousands of design decisions. Here we see what Orion will look like approaching the International Space Station.

An interesting example of how the design process is applied can be seen in the development of a new kind of roller skate—LandRollers.

Skaters call rough ground "gatorback" since it is as rough as an alligator's back. If you are skating and come across gatorback, what do you do? One thing you could do is take off your skates and walk, but that's not much fun. Rough ground is much easier to skate over if you use skates designed for the purpose.

How would you design roller skates that would roll effectively over rough terrain? How might these roller skates be different from conventional in-line skates? This was exactly the problem facing inventors Bert Lovitt and Warren Winslow.

Winslow was an avid skater and skier. He wanted a wheeled ski that would allow his dogs to pull him around his Illinois farm. He decided to take a pair of skis, shorten them, and add large wheels at the ends of each ski. The result was similar to a scooter, but with one "ski" for each foot.

Once a design problem has been identified it must be analyzed to determine if potential benefits justify investing the time and other resources needed to search for solutions. Lovitt and Winslow had identified the design problem, and they believed that their innovation could be a winner in the marketplace.

Lovitt, from California, disliked the "scooter for each foot" idea. He advocated a more elegant design, one that could ultimately be patented. Thus began the process of iterative design. Lovitt suggested that they take a pair of large wheels, with the hubs removed, and add a ski boot on rollers inside the rim of each wheel, as shown in Figure 3.4.

The problem with this design is that the skater couldn't insert his or her foot into the boot without hitting the metal rim. This required a radical innovation: attaching the metal wheel rims at an angle, as shown in Figure 3.5.

Lovitt and Winslow built a prototype. The first prototype worked; the boots rolled freely inside the large wheels. But skating using the prototype was difficult. The skates were extremely unstable. To make the skates easier for the average person to manage, Lovitt and Winslow decided to add a small wheel in the back of each skate for balance (see Figure 3.6).

This was a definite improvement. However, it turned out that the small wheels added for balance absorbed most of the skater's weight but didn't allow great maneuverability. To improve weight distribution and therefore make the skates even more stable and maneuverable, Lovitt and Winslow redesigned the skate yet again, this time moving the small balance wheels from the back to the front, as shown in Figure 3.7.

Figure 3.3 | Skaters struggle on rough terrain they call "gatorback" because it is rough as an alligator's back. Gatorback poses a technological challenge for Rollerblade designers.

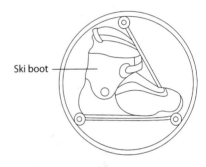

Ski boot

Figure 3.4 | One of Lovitt and Winslow's early attempts at a Rollerblade that could roll over rough terrain consisted of a ski boot on rollers positioned inside a metal wheel rim.

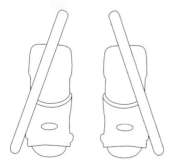

Figure 3.5 | An important design innovation was to angle the wheels. Here you see a top view of the boots, with the metal wheel angled from the heel of each boot to the outer edge. This allowed the skater to slip his or her foot into the top of the boot, without hitting the wheel.

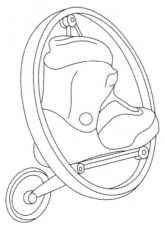

Figure 3.6 | The LandRoller designers added a small wheel in the back of each skate for balance.

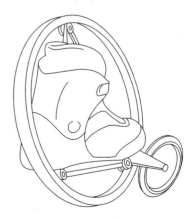

Figure 3.7 | In yet another version of the LandRoller skate, the small balance wheels were moved from the back to the front to improve weight distribution.

Figure 3.8 | iteration involved using two 8-inch, air-filled tires on each skate.

After a few more models, they decided to replace the rear hub-less wheels with spoked wheels from a children's bicycle. Then a friend convinced Lovitt and Winslow that the skates were too large and heavy to be practical. They decided to make the wheels smaller and shorten the wheelbase. They built a new pair with two 8-inch air-filled tires on each skate.

Now they were getting somewhere. They had a skate that could manage easily over bumpy roads, grassy slopes, and pavement that was cracked or smooth. The skates were still too heavy and unwieldy, though. The large front wheel made it hard for the skater to move freely.

An unexpected breakthrough occurred one day when Winslow happened to come across a snowmobile. The snowmobile treads, which wrapped around several sets of suspension wheels, gave him an idea. The wheels varied in size, with some seven inches in diameter and others just over five inches in diameter. The team tried out a few sets of snowmobile wheels on their skates. They created a new pair of frames and angled the new wheels as in the earlier iterations. It worked! The improved wheel configuration looked great and coasted along over rough and smooth terrain with ease. The many iterations of the design allowed the team to solve a variety of problems and produce a new and exciting product. To learn more about the LandRoller story and to see the final product, visit the company's Web site at *www.landroller.com*.

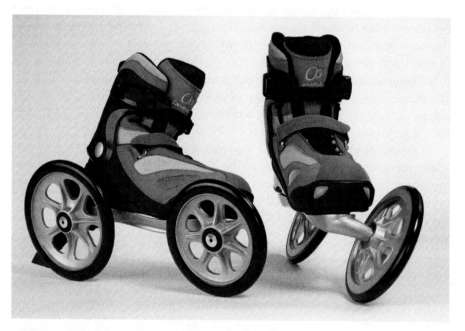

Figure 3.9 | After years of design and development, the LandRoller is a success.

Designing for Commercial Success

The design process is affected by the economic system in which the product is being developed. Capitalism (the dominant economic system in the United States, Canada, Mexico, and most European countries) is an economic system in which the means of production are mostly privately owned, and capital is invested in the production, distribution, and other trade of goods and services, for profit. In capitalist societies, most designers and engineers work for companies that thrive on marketplace success. For this reason, once a design problem has been identified, it must be analyzed to determine whether potential benefits justify investing time and other resources in searching for solutions.

Certainly, many design professionals engaged in design also orient their work to benefit the common good, but most design activity is focused on bringing ideas to fruition that will be financially profitable. Normally, if a design idea won't make money, it will be rejected. In order to bring ideas to market, research and development (R & D) is done to increase the knowledge base that can be applied to product development. Research and development is a specific problem-solving approach that is used intensively in business and industry to prepare devices and systems for the marketplace.

Design Specifications and Constraints

Design specifications are the specific goals, or performance requirements, that a project must achieve in order to be successful. Consider a design for an office building: A large corporation wishes to establish a national headquarters in a large city. One of the design specifications that the building must meet is to be able to provide sufficient office space for 1,000 employees. (Figure 3.10)

In art and in architecture, there is an axiom that states that "form follows function." If form indeed follows function, we can see how design specifications affect the design concept itself. For example, a building's shape will reflect its intended purpose. If a design criterion states that a building is to house a thousand offices in a city where land costs are very high, then the form (a skyscraper, for example) is dictated by the functional requirements.

The design specifications and constraints affect the design concept's implementation, marketability, and commercial profitability. In the example of LandRoller skates, design specifications included:

- The skates must be useable over rough terrain.
- They must be lightweight and comfortable.
- They must be durable.
- They must be attractive and marketable to young people of both genders.
- They must be easy to learn to use.

Design constraints are the limitations that are imposed on the designer and the design. Constraints ensure that the design meets its performance requirements. These constraints often limit resources and define parameters such as size, shape, cost, and production time. If you were developing some skates of your own, you might have to work within these design constraints:

- Development costs must be limited to $75,000.
- Patent costs must be limited to $9,000.
- Marketing costs must be limited to $50,000.
- Materials costs must be limited to $30 per pair.
- Construction materials must be readily available.
- The coefficient of rolling friction for a skate wheel on an asphalt surface must be below 0.015.
- The upper portion of the skates must be made of recyclable materials.

In today's society, designers and engineers must concern themselves with constraints that are related to societal and quality-of-life issues. These realistic constraints include economic, environmental, social, political, ethical, health and safety, manufacturability, and sustainability limitations. Such constraints require designers and engineers to make trade-offs that mean giving up some benefits to reach an overall, acceptable solution.

Figure 3.10 | In an urban setting, where land is extremely expensive, a skyscraper is a good way to satisfy design specifications that require a building to house a thousand offices.

Section 1 ◬ Applying the Design Process

Figure 3.11 | The need to prevent air and water pollution is an important design constraint for modern manufacturers.

Realistic Constraints

▷ Designers and engineers must carefully consider a variety of realistic constraints when designing products and systems. Realistic constraints that help ensure that a product satisfies modern societal and environmental demands include the following:

Economic:

- Consider the prices of related or similar products on the market when setting a price and predicting profits.
- Consider the potential impact to the local and U.S. economy.
- Keep maintenance costs low enough to allow for a profit.

Environmental:

- Limit vibration-induced noise to workers, product users, and the public such as large power transformers, road lamps.
- Limit air pollution (by using electric or hybrid engines) and water pollution, such as toxic waste into a river (see Figure 3.11).
- Limit use of non-biodegradable bags and computer cases made from plastic to protect the landscape.
- Control temperature of exhaust gasses to limit global warming.
- Establish procedures for manufacturing waste collection and processing.
- Use energy-saving devices.

Social:

- Avoid designs that favor only certain people or groups.
- Consider how the design will balance worker and union demands versus employer needs.
- Follow current government codes to protect society.

Political:

- Avoid producing and marketing products (e.g., computer games, messages on clothes) that promote racial or gender stereotypes.
- Avoid producing products for customers who might use them to negatively affect domestic security in the nation where the products are produced.

Ethical:

- Avoid producing products that can hurt people, either physically or psychologically.
- Pay the necessary fees, in order to make use of products, designs, or concepts that are protected by patents.
- Be wary of products that use materials that look better than those of their competitors, but that are toxic.
- Avoid skimping on the design process (that is, avoid under-designing), in order to create cheap but potentially dangerous products.
- Weigh the ethical concerns involved in designing products that facilitate the secret surveillance of someone's private life.

Health and Safety:

- Consider the safety of workers, consumers, and the general public
- Be aware that noisy products can cause hearing loss.
- Avoid using materials and environments that are hazardous for workers.

- Consider the impact of products that require the use of radioactive materials.
- Require special safety requirements for products for infants and children.
- Design control systems with acceptable stability margins for machinery where safety is of concern.
- Consider actual environmental factors (extreme working temperature, corrosive fluid, abrasive air, severe radiation in space, etc.).

Sustainability:

- Determine what conditions would threaten the survival of the business.
- Ensure that the product has a well-defined life span under the assumed normal operation conditions.
- Ensure that all parts have a similar designed life span.
- Ensure the reliability and durability of the product's supposed function.

Legal:

- Be aware of legal issues concerning products protected by patents.
- Follow all city, state, and federal codes related to your product.
- Ensure that products that are designed to secretly collect information are legal.
- Check to make sure that federal, state, and local laws do not limit the use of your product. For example, if you were designing a radar-detection device for cars, you would find that your product was illegal in some states.

Codes and Standards

Codes and standards have been developed by government agencies and industries to ensure that products, structures, and environments meet safety and operational requirements. Thousands of standards have been set for things as diverse as water quality, head and face protection for use in ice hockey, clothing for protection against heat and flame, machine safety, vehicle emissions, food labeling, electronic equipment, aircraft, personal care items, and children's booster seats. Designing to meet these standards may require that additional trade-offs be made. For example, to meet safety standards, automobiles must include seat belts, airbags, and front and rear bumpers that add to the vehicle's equipment and weight, thereby increasing its cost and reducing its fuel economy.

Prototypes and Modeling

A prototype is a model of a new product or system. Sometimes the prototype is a working model, sometimes it is a scale model. Because there are often uncertainties about how a completed design will perform, prototypes are built to demonstrate how the design functions before a product is produced in large quantities, or before a costly system is implemented. Prototyping is a step along the way to developing an optimal design. Modeling possible design solutions allows the designer to try out "What if?" scenarios before actually implementing and producing the design. Designers and engineers refine a design by using prototyping and modeling to ensure the quality, efficiency, and productivity of the final product.

Prototypes are produced during the design of most technological systems, including software programs, electronic circuits and components, household and personal care products, architectural layouts, and automobiles.

Figure 3.12 | The Ford Explorer America concept vehicle serves as a platform for Ford's new family of engines designed to provide the power of larger engines while achieving 20–30% gains in fuel economy.

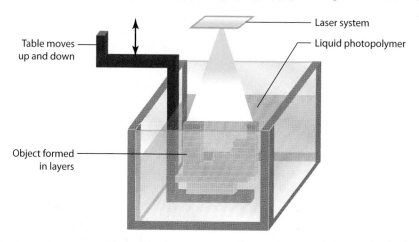

Laser system

Table moves up and down

Liquid photopolymer

Object formed in layers

Figure 3.13 | A stereolithography machine creates three-dimensional models based on graphics stored on a computer.

Figure 3.14 | This stereolithography machine, manufactured by 3D Systems Corporation, is used to create three-dimensional models.

Sometimes, manufacturers need to do rapid prototyping in order to develop a model of an intended product quickly. Often, the model might not be fully functional but would provide a good example of shape and size.

One of the technologies used to do rapid prototyping is stereolithography. This is a process that builds a three-dimensional model from a 3-D graphic stored in a computer. The process makes use of a special liquid plastic, or resin, that hardens when exposed to an ultraviolet laser beam. In a stereolithography machine, shown in Figure 3.13 and Figure 3.14, a laser draws the outline of a model on a pool of resin. The outline solidifies, and the model is lowered by several thousandths of an inch into a vat of liquid plastic. The laser then solidifies the next microscopically thin layer on top of the preceding one. In this way, a model is formed, layer by layer.

Three-dimensional models of some complexity (such as the one shown in Figure 3.15) can be formed by this process in a matter of hours. The machinery can be expensive for a small manufacturer. Instead of purchasing their own machines, companies can send their designs over the Internet to a stereolithography company, which will ship back a model within a day or two. The objects generated from the resins do not have the physical, mechanical, or thermal properties typically required of end products. They are useful, however, in displaying how a prototype would look (see Figure 3.15 for an example).

Communicating and Evaluating Design Solutions

The design solution that you choose should be evaluated, and then the final design should be communicated to interested people using techniques such as 3-D modeling and verbal, graphic, quantitative, virtual, and written means. How would you best communicate a proposed design solution to a group of interested people? You might describe your idea orally or in writing; another way is to draw a picture or a sketch (is it true that a picture is worth a thousand words?). Depending upon the complexity of the design and the intended audience, one method might prove more effective than another.

Suppose you wish to design a personal Web page and have decided what information you want to present. Before you begin to lay out your page, you most likely will want to get ideas by visiting other Web pages that present similar information. Once you have a pretty good idea of what your Web page should contain, you would create a storyboard (a sequence of illustrations) that would represent the flow of ideas that you want to present. In this case, you have chosen to communicate your design ideas graphically (see Figure 3.16).

Figure 3.15 | This model of an automobile tire was made by a stereolithography machine.

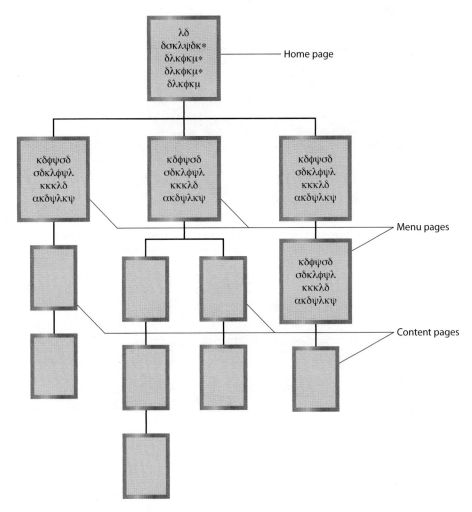

Figure 3.16 | In the context of Web page design, a storyboard is a diagram that shows the flow of pages within a Web site.

97

Computer-Aided Design

Very accurate and reliable product designs can be developed and communicated using computer-aided design (CAD) software. These programs are used to design products, systems, and structures. CAD software is used by product designers to establish part geometry (the dimensions, shapes, and tolerances of parts). The software often includes mathematical algorithms that allow the designer to analyze the strength or operating characteristics of components. CAD drawings can be very sophisticated and complex. They can communicate the details of a design quite clearly.

A CAD system includes a computer, keyboard, monitor, and a printer or plotter. The CAD system can convert the graphic designs to digital data that can be output to rapid prototyping devices like the stereolithography machines described earlier in this chapter. The resulting three-dimensional model (such as the one shown in Figures 3.17 through 3.20) gives potential customers a very clear idea of what the final product will look like.

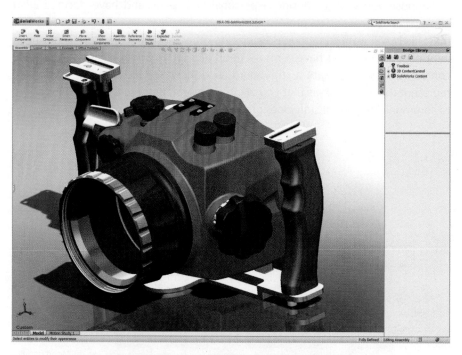

Figure 3.17 | Very accurate and reliable product designs can be developed using the two-dimensional and three-dimensional capabilities of computer-aided design (CAD) programs.

Figure 3.18 | This 3-D model of a chaise lounge was created by a CAD-CAM system.

Figure 3.19 | This looks like a real car, but it is really a 3-D model created by a CAD-CAM system.

A CAD system can also output its data to computer numerical controlled (CNC) industrial machines that convert the data to a set of instructions that control manufacturing tools. In this way, designs are communicated not only to people, but directly to machines. Such a system, which is called a CAD-CAM system (computer-aided design, computer-aided manufacturing) can produce actual finished products in addition to prototypes.

Figure 3.20 | **This model of a hand drill was created by a CAD-CAM system.**

Evaluating a Design

▷ Design solutions are evaluated at various intervals during the design process using a variety of models, including conceptual, physical, and mathematical models, in order to check for proper design functioning and to note areas where improvements are needed. In addition to studying various models, designers and engineers discuss their ideas with technical and production staff and with potential customers to refine their thinking and to ensure that designs can actually be produced to specifications. Very often, quantitative data is also used to describe the design so that its capabilities are clearly communicated. For example, a bridge design includes a vast amount of mathematical data that describes how the various components will stand up under different loads.

Ensuring Quality

Quality control is a planned process that ensures that a product, service, or system meets established criteria. Quality-control measures ensure that products, systems, and services conform to design specifications and tolerances and that they meet customer requirements and industry standards.

The maintenance of quality is therefore a critical area of concern, not only for producers of goods and services, but for national governments. A nation's strength and the well-being of its citizens are determined to a large degree by its economic competitiveness. Manufacturing is the area in which international economic competition is most intense. It plays the dominant role in world trade. Since great international competition exists, products and systems with the best design and highest quality will be in greatest demand.

Quality standards and quality control procedures are not confined to the manufacturing sector. Such standards are well established in many industries, including administrative services, banking, consulting, insurance, information and communications systems, computer software design, retailing, and transportation.

Quality is ensured in several ways. A system or product is only as good as its design. Quality levels are set by producers who can establish high or low quality standards. Setting and meeting high standards (doing it right the first time) will go a long way toward ensuring quality.

Employees at various levels also influence quality. Senior management defines quality standards and develops goals, budgets, and strategic plans. Engineers design process control systems. Machine operators monitor and control the machinery. In modern manufacturing plants, machine operations are automated. The machines include monitoring and control systems so that they can automatically correct problems when products or systems fall outside of established quality parameters.

In manufacturing, quality is maintained by carefully controlling the processes of production. Monitoring the raw materials, assemblies, products and components; services related to production; and management, production, and inspection processes are all part of quality control.

99

Figure 3.21 | This IBM engineer is inspecting a semiconductor wafer.

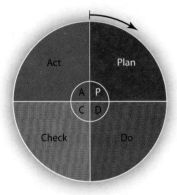

Figure 3.22 | William Edwards Deming worked in Japan to implement quality improvement (kaizen). Kaizen involves components related to maintenance and operating standards. Improvement is focused on making current standards even higher.

The goal of quality control should be to identify steps in the processes that have the most impact on outcomes. By establishing effective control systems at the early stages of production, companies eliminate the rework, scrap, re-inspection, and delays that occur when controls are added too late in the process.

Statistical Process Control Statistical process control (SPC) is used to assure that quality is maintained at a high level and reduces the likelihood of unexpected failure. Statistical process control in manufacturing operations involves random sampling and testing of a fraction of the output. Variances from critical tolerances are tracked, and processes are corrected before bad parts are inadvertently produced.

Total Quality Management Total Quality Management (TQM) is an approach toward establishing quality in an organization based on the cooperation of all workers at all levels. The goal is to ensure customer satisfaction. The method was developed by an American statistician, Dr. William Edwards Deming, while working in Japan in the 1950s. Deming's work helped Japan establish an admirable reputation for manufacturing high-quality, innovative products. One of the core principles of Japanese TQM is called *Kaizen* (see Figure 3.22). In Japanese, Kaizen means *improvement* and focuses on continuous improvement of technological processes to make these processes *visible*, *repeatable*, and *measurable*.

SECTION ONE FEEDBACK >

1. Explain why LandRoller skates took over eight years to bring to market and the new, hugely complex Orion spacecraft took 13 months of development.

2. Using LandRoller skates as an example, explain how the design specifications and constraints affected the design concept's implementation, marketability, and commercial profitability.

3. Auto sales data show that while Japan's Big-3 (Toyota, Honda, and Nissan) continue to experience strong U.S. sales, the U.S. Big-3 (DaimlerChrysler, General Motors, and Ford) saw their combined market share fall from 70% to 57% in the last five years. What, in your view, is the reason for this decline? Do research to back up your position.

SECTION 2: Using and Maintaining Technological Products and Systems

KEY IDEAS >

- A variety of techniques including flow charts, symbols, spreadsheets, graphs, and time lines can be used to document processes and procedures and communicate them to different audiences.

- Engineers and technicians rely on technical knowledge and computer-based and electronic tools to maintain technological systems.

- Monitoring the operation of a system and making adjustments as needed help ensure precision and safe and proper functioning.

We are surrounded by examples of systems. Some are natural systems, like living organisms. Some are technological systems. Examples range from iPods to automated machines, to highway systems and nationwide telephone systems. A system is the means by which we achieve a desired result. All technological systems have certain common elements. These include inputs, processes, and outputs. Natural systems (like living organisms) take in nutrients, process them, and produce outputs like waste products and carbon dioxide.

Technological systems combine resources like materials and energy to produce outputs. Both natural systems and technological systems can be controlled by feedback that keeps the system operating as intended. An example of feedback in a natural system is how we humans perspire when we get too hot; the perspiration then evaporates, cooling us back down. Technological systems are not only controlled by feedback, but also by on-off or proportional control systems (discussed later in this chapter).

Properly using and maintaining technological products and systems ensures that they function as intended and that they have the longest possible life cycle. Products and systems can be extremely simple (such as a toy race car with only a few moving parts), or can be exceedingly complex (like an automobile or airplane), where many interrelated components work together.

Most systems are made up of subsystems: For example, the nation's transportation system is made up of subsystems that include not only the vehicular systems and highways systems, but also the support subsystems, such as communication, lighting, and power. Even small systems like the child's race car are made up of subsystems. In the race car, the wheels and axles comprise one subsystem; the body comprises another; the propulsion system (perhaps a CO_2 cartridge) comprises yet another. When we maintain systems or troubleshoot system problems, it is often helpful to first identify how the subsystems interrelate, what their functions are, and then determine within which subsystem the problem resides.

Documenting and Communicating Ideas

Computers and calculators are used to access, retrieve, organize, process, maintain, interpret, and evaluate data and information in order to communicate. A variety of techniques including flowcharts, symbols, spreadsheets, graphs, and time lines are used to document system processes and procedures and communicate them to different audiences. You can use these tools to assure that technological systems function correctly.

Flowcharts are diagrams that depict the major steps in a process (see Figure 3.23). They are used to show how one step leads to another. Computer-program flowcharts show the sequence of instructions in a computer program. In manufacturing, flowcharts show the sequence of steps in product design or production.

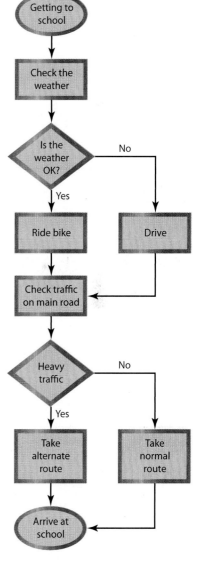

Figure 3.23 | This flowchart illustrates the decision-making process involved in choosing the best route for driving to school depending upon the weather. Charts such as these are used to plan the sequence of production processes.

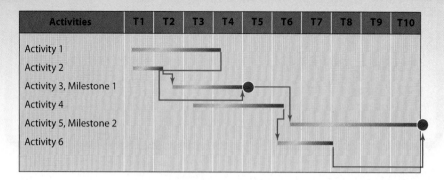

Activities	T1	T2	T3	T4	T5	T6	T7	T8	T9	T10
Activity 1										
Activity 2										
Activity 3, Milestone 1										
Activity 4										
Activity 5, Milestone 2										
Activity 6										

Figure 3.24 | This Gantt chart lists each activity and the time periods during which the activity will be completed. The time periods in a Gantt chart could be months, seasons, years, or any other period you specify.

Gantt and PERT Charts Gantt charts and PERT charts are popular management tools that give project managers graphical illustrations of progress toward meeting deadlines and goals. A Gantt chart focuses on the sequence of tasks and shows them along a horizontal axis on an x-y chart. The length of the bar corresponds to the length of the task. Among other things, you can use a Gantt chart to store information about tasks, such as the names of the individuals assigned to them. You can create a Gantt chart on graph paper. However, most people create Gantt charts, like the one shown in Figure 3.24, using project management computer software. The first Gantt chart was developed by in 1917 by an American engineer named Henry Gantt.

A PERT chart is a tool used to schedule, organize, and coordinate project-related tasks. (PERT is short for "Program Evaluation Review Technique.") This type of chart shows how one task flows into another. It is an effective way of portraying task dependency. PERT charts allow managers to see whether a task is falling behind schedule and to determine how that delay might affect the entire project. PERT charts were developed by the U.S. Navy to manage the Polaris submarine missile program in the 1950s (see Figure 3.25 for an example of a PERT chart).

Symbols Symbols are shortcuts that can be used to graphically represent processes (such as those on a flowchart), features (such as symbols on a map that represent

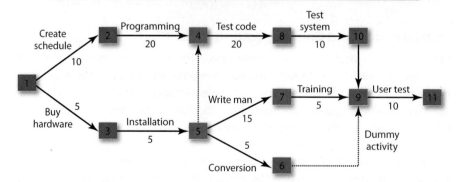

Figure 3.25 | This is an example of a PERT chart that is used to schedule, organize, and coordinate tasks within a project.

campgrounds or county borders), components (such as those in electrical schematic drawings), company names (such as the symbols used to represent stocks), and mathematical ideas (such as greater than [>], less than [<], equal to [=], etc.). Many symbols, such as those used in schematic diagrams, road signs, and maps, can be understood universally (see Figures 3.26 through 3.29 for examples of symbols and their uses).

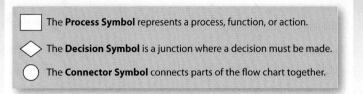

The **Process Symbol** represents a process, function, or action.

The **Decision Symbol** is a junction where a decision must be made.

The **Connector Symbol** connects parts of the flow chart together.

Figure 3.26 | These symbols are used in a flow chart.

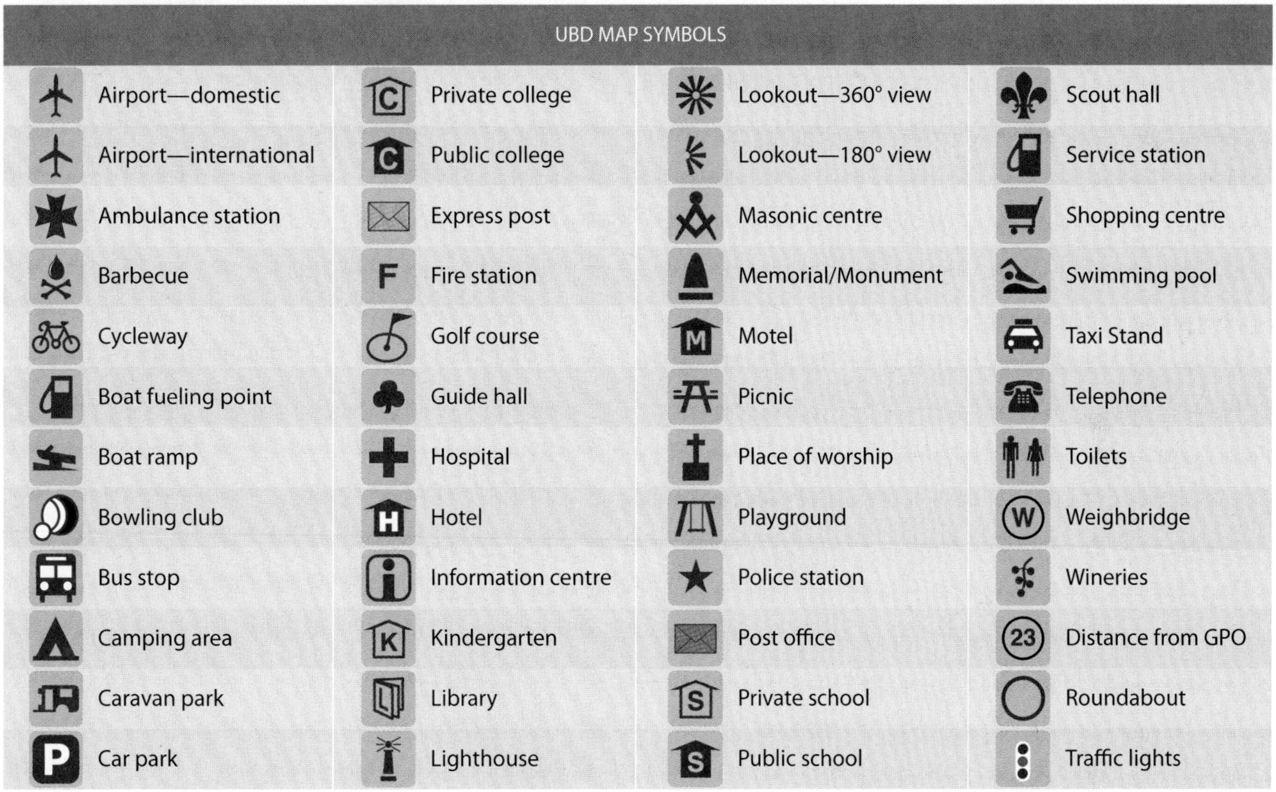

UBD MAP SYMBOLS						
✈	Airport—domestic	C	Private college	※	Lookout—360° view	Scout hall
✈	Airport—international	C	Public college	←	Lookout—180° view	Service station
✚	Ambulance station	✉	Express post	⚒	Masonic centre	Shopping centre
	Barbecue	F	Fire station		Memorial/Monument	Swimming pool
	Cycleway		Golf course	M	Motel	Taxi Stand
	Boat fueling point	♣	Guide hall		Picnic	Telephone
	Boat ramp	✚	Hospital		Place of worship	Toilets
	Bowling club	H	Hotel		Playground	W Weighbridge
	Bus stop	i	Information centre	★	Police station	Wineries
	Camping area	K	Kindergarten	✉	Post office	23 Distance from GPO
	Caravan park		Library	S	Private school	Roundabout
P	Car park		Lighthouse	S	Public school	Traffic lights

Figure 3.27 | These are Universal Building Design symbols used internationally on maps and sign posts.

Spreadsheets

Spreadsheets are software programs that computerize record keeping and budgeting. A spreadsheet is a grid of boxes (known as cells) that identify columns by letters and rows by numbers, so a cell can be designated by a letter and a number. Figure 3.30 shows some data stored in one of the most commonly used spreadsheet programs, Microsoft Excel.

You can place a separate numerical value in each cell and then manipulate the values mathematically using formulas. Because the spreadsheet recalculates the results of the formulas each time you change a value, you can use a spreadsheet to answer "What if?" questions. For example, in Figure 3.31, the data in column B shows the minimum-wage salaries for workers in several states. Column C shows the salaries for each state if the minimum wage were raised by 10%. Column D represents the minimum wage for each state if the minimum-wage rate were raised by 12%. In other words, the spreadsheet answers the following "What if?" questions:

- What if the minimum wage were raised by 10%?
- What if the minimum wage were raised by 12%?

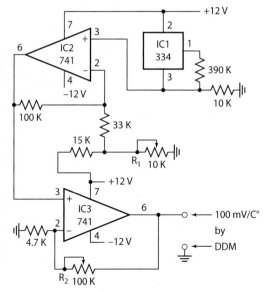

Figure 3.28 | This drawing uses symbols in a schematic diagram of an electronic thermometer circuit.

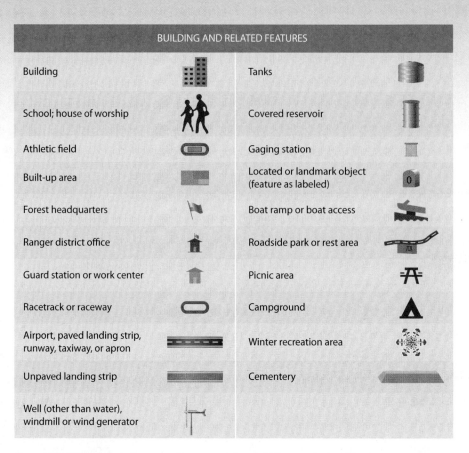

BUILDING AND RELATED FEATURES		
Building	Tanks	
School; house of worship	Covered reservoir	
Athletic field	Gaging station	
Built-up area	Located or landmark object (feature as labeled)	
Forest headquarters	Boat ramp or boat access	
Ranger district office	Roadside park or rest area	
Guard station or work center	Picnic area	
Racetrack or raceway	Campground	
Airport, paved landing strip, runway, taxiway, or apron	Winter recreation area	
Unpaved landing strip	Cementery	
Well (other than water), windmill or wind generator		

Figure 3.29 │ These symbols are used in maps to show buildings and related features.

	A	B	C	D
1				
2				
3				
4				
5				

Figure 3.30 │ A spreadsheet is a matrix of boxes (cells) that identify columns by letters and rows by numbers, so a cell can be designated by a letter and a number. For example, the shaded cell shown in Figure 3.30 would be designated as B3.

	A	B	C	D
	State	**Min wage**	**With 10% increase**	**With 12% increase**
1				
2	Colorado	5.15	5.665	5.768
3	Florida	6.40	7.04	7.168
4	Missouri	5.15	5.665	5.768
5	New York	7.15	7.865	8.998
6	Washington	7.63	8.393	8.5456

Figure 3.31 │ This spreadsheet displays a "What if?" scenario (what if the wages were increased by various percentages)?

Using Spreadsheets

Spreadsheets use mathematical formulas to do their calculations. What makes spreadsheets so useful is that the formulas do not simply manipulate specific numbers. For example, they do not simply add .345 and 45.2. Instead, you can use spreadsheet formulas that manipulate the *contents* of the various cells. For example, you could create a formula that adds the contents of cell B2 to the contents of cell C2. Then, if you change the values in these cells, the spreadsheet program automatically recalculates the formulas to reflect the new values. In the example shown in Figure 3.31, the increased minimum wage in cell C2 is calculated using a formula that multiplies the contents of cell B2 (which contains the current minimum wage for Colorado) times 1.10. (This is correct because, if the minimum wage in Colorado were to rise 10%, it would be 1.10 times the present wage.) Cell D2 would contain a similar formula to calculate the Colorado minimum wage after a 12% increase.

Now suppose that you discover that you were misinformed about the current minimum wage in Colorado, and that you had entered the wrong value in cell B2. You can fix this mistake by entering the correct value in cell B2. The spreadsheet then automatically recalculates the values for the minimum wage increased by 10% and 12%. As you can see, spreadsheets allow us to perform repetitive calculations very efficiently.

Consider the following problem that a rancher must face in raising cattle. The rancher needs to know the best combination of grain and hay to produce satisfactory weight gain in a steer. The spreadsheet in Figure 3.32 shows combinations of grain and hay that could be used as feed to produce satisfactory weight gain in a steer.

The spreadsheet in Figure 3.32 shows costs of $0.06 / pound for hay and $0.15 / pound for grain. The equation to create the total cost for hay is created by entering in cell D3 the instruction =B3*0.06. The equal sign tells the Excel program that what follows is an equation that multiples the value in cell B3 by 0.06. Similarly, the equation in cell E3 is =C3*0.15 and the equation for the total cost in cell F3 is = D3+E3.

	A	B	C	D	E
1	Pounds of Hay	Pounds of Grain	Hay Cost	Grain Cost	Total Cost
2	1000	1316	60	197.4	257.4
3	1100	1259	66	188.85	253.85
4	1200	1208	72	181.2	253.2
5	1300	1162	78	173.3	252.3
6	1400	1120	84	168	252
7	1500	1081	90	162.15	252.15
8	1600	1046	96	156.9	252.9
9	1700	1014	102	152.1	253.1
10	1800	984	108	147.6	255.6
11	1900	957	114	143.55	257.55

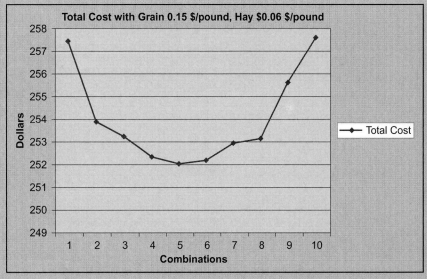

Figure 3.32 | A spreadsheet showing combinations of grain and hay that could be used as feed to produce satisfactory weight gain in a steer.

The total cost may be plotted using the Excel line graph function, which indicates the least cost is for combination five. When there are many combinations to view, it is often easier to visualize the results in graphic form.

What happens if the price of grain increases? Will there be a new optimum point? A quick change of the data provides the answer. In this case, the table with grain at 0.18 $/pound is created. Visual inspection shows that the least cost occurs at combination seven. The graph, shown in Figure 3.33, indicates the same.

	A	B	C	D	E
1	**Pounds of Hay**	**Pounds of Grain**	**Hay Cost**	**Grain Cost**	**Total Cost**
2	1000	1316	60	236.88	296.88
3	1100	1259	66	226.62	292.62
4	1200	1208	72	217.44	289.44
5	1300	1162	78	209.16	287.16
6	1400	1120	84	201.6	285.6
7	1500	1081	90	193.58	283.58
8	1600	1046	96	188.28	283.28
9	1700	1014	102	182.52	283.52
10	1800	984	108	177.12	285.12
11	1900	957	114	172.26	286.26

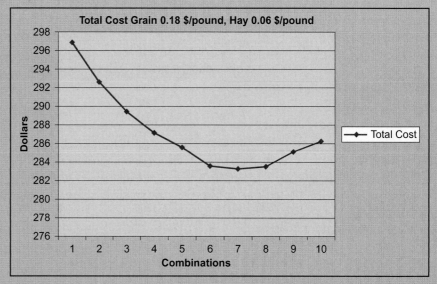

Figure 3.33 | This revised spreadsheet and graph show that the least cost occurs at combination seven.

Graphs

A graph is a pictorial representation of a relationship between variables. The graph allows us to interpret information more easily than reading about it or seeing a table of data. Graphs are often used to show trends and to compare sets of data (see Figures 3.34 through 3.36 for examples of different graphs).

A line graph is drawn by connecting points, representing data, with a line. A bar graph uses columns whose heights represent values of the data being displayed. A pie chart is used to compare categories within the data set to the whole.

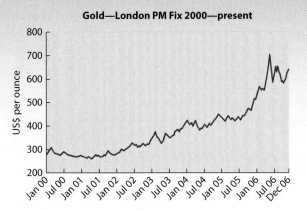

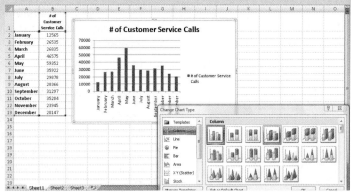

Figure 3.34 | This line graph shows how the price of gold has changed from 2000 to 2006. Can you see a trend?

Figure 3.35 | Graphs and charts can be drawn using programs such as Microsoft Excel.

Time Lines

A time line is a historical list of events that lead up to a particular result. Often time lines display technological events or inventions. Time lines are also used as a project-planning tool to keep progress on track. For example, the time line shown in Figure 3.37 shows technological developments from 1752 to 1863.

Flowcharts, symbols, spreadsheets, graphs, and time lines can be used to document processes and procedures and communicate them to different audiences.

Maintaining and Controlling Technological Systems

Engineers and technicians rely on technical knowledge and computer-based and electronic tools to maintain technological systems. The tools can be used to control technical processes, to monitor the operation of systems, to troubleshoot, and to perform repairs. Electronic tools, such as meters and oscilloscopes, can monitor and display voltage and current levels. Electrical current sensors measure AC and DC current levels; electronic sensors (such as photocells and microphones) detect a change in the environment and produce an electrical signal in response. Mechanical sensors can monitor

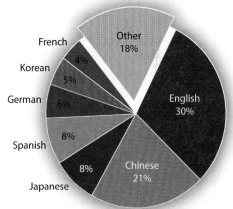

Figure 3.36 | This pie chart shows the languages people use when they are online using the Internet. (Source: Report of the Internationalized Domain Names Working Group, ICANN.)

Figure 3.37 | This time line shows technological developments from 1752–1863.

temperature, pressure, sound and light levels, flow of liquids and gasses, chemicals (like oxygen sensors), radiation, and motion. Sensors are transducers that detect a change in one type of energy and convert that change into another form of energy, usually electrical current or mechanical motion.

System Control

Controlling technological systems is largely a matter of using feedback from sensors to activate a control system. Feedback is a sample of a system's output that is detected by a sensor. Feedback serves as a measure of whether the system is operating within desired parameters. Systems that rely on feedback are called closed-loop systems. They are able to adjust their output if external conditions change.

In a home heating control system, the feedback comes from a measurement of the actual temperature. If the room temperature is too high, a controller (a thermostat) turns off the furnace. If the room temperature is too low, the thermostat turns the furnace on. Another example of feedback control is in a modern flush toilet. After flushing, the water level in the toilet tank decreases. A float falls with the decreasing level, and as it falls it causes a valve to open, allowing water to flow back into the tank. As the tank fills, and the water level rises, the float rises as well and closes the valve to maintain the proper water level (see Figure 3.38).

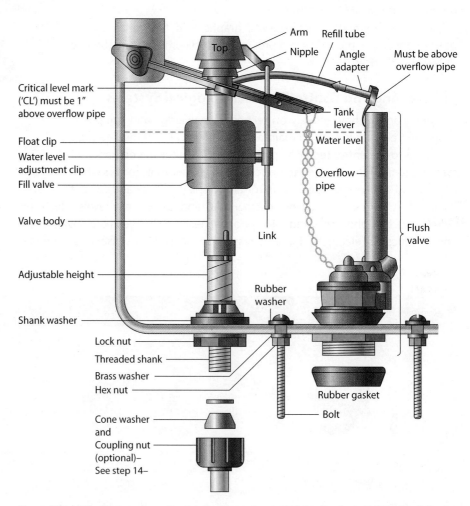

Figure 3.38 | This diagram shows the details of a modern toilet's feedback-control system. A float activates a valve to turn the flow of water on or off.

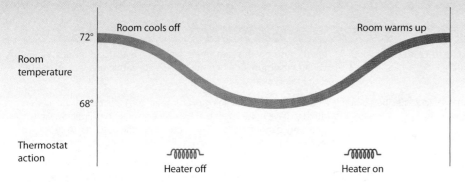

Figure 3.39 | A home heating system thermostat is an example of a bang-bang controller.

On-Off Control

The simplest type of on-off control is a switch that enables a person to make or break an electrical circuit by hand. A thermostat is another example of an on-off control. When the temperature in the home drops too low, the furnace comes on. When it gets too hot, the furnace turns off. This on-or-off control is called *"bang-bang"* control. When it gets too cold, *bang*, the furnace turns on. As soon as the temperature rises above the set point, *bang,* the furnace turns off (see Figure 3.39).

On-off controls have drawbacks. If a thermostat waits until it senses the desired temperature before it turns off the furnace, there could be some residual heating and the temperature may overshoot the desired room temperature a bit. Normally, in the typical residential building, this is not a huge problem. However, in a system designed to control an elevator, an overshoot would be troublesome indeed.

On-off control can be used to set a sequence of events in motion. For example, the on-off controller in a washing machine operates the wash, rinse, and spin cycles. Timers are another example of on-off controllers. These often have a number of programmable set times that can be used, for example, to turn outside lights on or off at different times of the day or night.

Proportional Control

Another type of controller adjusts the system incrementally. This controller is called a proportional controller. Suppose that instead of just switching the furnace on and off, we used the temperature reading to adjust the gas valve on the furnace. If it gets very cold in the room, the gas valve opens more and the furnace produces more heat. If it gets warmer, the gas valve closes a bit; if it gets very hot, the valve closes even further. The controller allows an amount of gas to flow into the furnace that is proportional to the temperature of the room. Using this method, the control system ensures that the *adjustment is proportional to the difference between the desired result and the actual result.*

Regardless of whether a system uses an on-off or a proportional controller, designing the system so that its operation is monitored and adjusted as needed helps ensure precision and safe and proper functioning. Complex systems have many layers of controls and feedback loops to provide information.

System Limits

Systems should be operated so that they function in the way that they were designed (see Table 3.1). Every system has intrinsic performance limits that are a function of the system design. Exceeding these limits has the potential

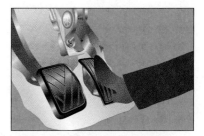

Figure 3.40 | A car's speed can be controlled by adjusting the gas pedal based on the driver's observation of the difference between the desired speed and the actual speed. "Bang bang" control refers to situation in which the gas pedal is used like a switch: either it is fully on or fully off. This would control the speed, but it would also result in jerky motion.

Figure 3.41 | Proportional control refers to a situation in which the gas pedal is pressed in proportion to the difference between the desired speed and the actual speed, allowing for a smoothly controlled level of speed.

Table 3.1 | The Limits of a Variety of Systems

SYSTEM	SYSTEM LIMITS	DANGER OF EXCEEDING SYSTEM LIMITS
Infant Safety Seat	For children weighing up to 22 lbs (10 kg) and up to 29 in. (74 cm) tall.	Infants outgrow this seat when they are heavier than the seat's weight maximum or when their heads are within one inch of the top of the safety seat.
Children's Toys	Noise level should not exceed 85 decibels when held close to the ear.	A mechanical toy machine gun and a cap gun were tested and all were found to have noise levels that exceed recommended limits, making them very dangerous toys. These guns have the potential to cause serious damage to hearing and could cause instant hearing loss.
Highways	Speed limits are set for highways and secondary road systems.	The speed limit is the absolute maximum and does not mean it is safe to drive at that speed irrespective of conditions. Driving at speeds too fast for the road and traffic conditions can be dangerous. Since emissions of nitrous oxide increase as vehicle speeds increase above about 48 mph, speed-limit changes may have important environmental consequences.
Cell Phone	Specific absorption rate (SAR) is a measure of the rate of radio energy absorption in body tissue, and the SAR limit recommended by the International Commission of Non-Ionizing Radiation Protection is 2 watts /kg.	Most phones emit radio signals at SAR levels of between 0.5 and 1 w/kg. However, radio waves from mobile phones harm body cells and damage DNA in laboratory conditions, according to a study funded by the European Union.

Figure 3.42 | Every time you use a cell phone, you are using a device with an embedded computer.

Figure 3.43 | This programmable digital controller monitors and controls temperature, pressure, flow, level, and other process variables in applications such as furnaces, ovens, environmental chambers, packaging machinery, and plastic processing machines.

to reduce performance, create safety hazards, and damage the environment. For example, the physical design of a computer and the network to which it is connected impose limits on the amount of data the computer can process. Only by redesigning the computer system (adding memory, using a faster processor, using a network with a higher bandwidth) can these limits be changed. An example that you may be familiar with is that a computer can open only a certain number of programs without slowing down; exceeding the limit can severely reduce the computer's performance.

Table 3.1 lists the limits for a number of systems and explains the dangers of exceeding those limits.

Computer Control

Many systems are controlled by computers. When small computers are built in to the devices they control, they are called embedded computers. Embedded computers are typically very small computers, known as microprocessors. They are used to control a wide range of systems, including cell phones, industrial robots, MP3 players, gasoline pumps, airplanes, digital cameras, appliances, and toys (see Figure 3.42).

Analog and Digital Controllers

Controllers can be analog or digital control devices. Analog controllers such as variable resistors (rheostats or potentiometers) control continuously varying quantities like voltage and current. Digital controllers are based on the principle that digital signals have only a finite set of states (on or off) and thus, error correction is easier. They work using only two voltage levels: low or high, with low being near zero volts and high being dependent on the supply voltage that is used. The high and low voltage levels used in monitoring signals comprise a logic system that tells a device when to start and when to stop. Digital controllers are normally more accurate and reliable than analog control devices (see Figure 3.43).

A great advantage of using digital controllers is that they are programmable and that changes can be made through software. There is no need to change a hardware device.

1. What might be the best graphic project planning tool for monitoring progress on a project?
2. What type of software program would best be used to show various budget scenarios?
3. Give an example of an electronic tool that could be used to monitor circuit conditions.
4. The human body shivers to warm up and perspires to cool down. Is this an example of a bang-bang control system or a proportional control system? Explain your answer.

SECTION 3: Assessing the Impact of Products and Ideas

KEY IDEAS >

- When collecting information about the impact of technological products and systems, it is important to evaluate the information to make sure it is of high quality.

- To draw accurate conclusions about the effect of technology on the individual, society, and the environment, data must be collected and synthesized and trends must be analyzed.

- Forecasting techniques can be designed and used to evaluate the results of technologies that could potentially affect the environment and alter natural systems.

- Assessment techniques such as trend analysis and experimentation can help inform decisions about the future development of technology.

- Once a malfunctioning system is diagnosed, it can be repaired using appropriate tools, materials, machines, and knowledge.

When a technology is implemented, there are always consequences. We expect that the consequences will be desirable, but sometimes there are unexpected and occasionally undesirable consequences, as well. Think about television as an example. When commercial television was introduced in the 1940s, the inventors did not expect that the average U.S. citizen in 2008, would, between the ages of 2 and 65, have spent almost nine years of 24-hour days watching TV. Nor did the inventors of the automobile expect that much of the beautiful North American landscape would be paved over by roads.

Technology is neither good nor evil. Its impact depends upon how it is used. The famous early television broadcaster Edward R. Murrow said (about television, see Figure 3.44).

"This instrument can teach, it can illuminate; yes, and it can even inspire. But it can do so only to the extent that humans are determined to use it to those ends. Otherwise, it is nothing but wires and lights in a box."

(EDWARD R. MURROW
RTNDA Convention, Chicago, October 15, 1958)

Ensuring the Quality of Information

When collecting information about the impact of technological products and systems, it is important to evaluate the information to make sure it is of high quality. For example, how do you know for sure that by the age of 65, the average U.S. citizen will have spent nearly nine years watching television? You read it in this book, and books are normally reliable sources of information; but if you were part of a policy-making group and were called upon to make decisions about teaching young children good television-watching habits, wouldn't you want to make sure that the information you are guided by is entirely correct?

Sometimes, incorrect information about technology is passed along either out of ignorance or because people have a particular point of view they want to advance. Out of ignorance, a TV owner might blame television-reception interference on a neighbor's ham radio equipment. Although it is possible that the ham radio transmitter is the culprit, it is more likely that the ham equipment is operating properly and that something else is causing the interference, such as a faulty power line or a malfunctioning household appliance. Sometimes the design of the TV itself is the problem. In that case, the interference can be easily remedied by attaching an inexpensive filter to the TV set. Before faulting the ham radio operator, it would be important for the TV owner to make sure that the information about the cause of interference is correct. Once the malfunctioning system is diagnosed, it can be repaired using appropriate tools, materials, machines, and knowledge.

Sometimes people use information about technology incorrectly just to advance a particular point of view. A good example of this comes from the time of the Industrial Revolution. During that time, a group of people called Luddites (named after a probably fictitious leader, Ned Ludd) who worked in the textile mills in England, claimed that knitting machines would threaten workers' jobs. They passed that information along to other workers and the Luddites roamed the countryside smashing the machines (see Figure 3.45). In truth, even though some jobs were outmoded, the machines led to progress and new trends, and ultimately to the creation of many more new jobs. This case also illustrates how information that is collected to assess the impact of technology must be evaluated with respect to its quality. To draw accurate conclusions about technology's effect on the individual, society, and the environment, people must collect and synthesize data and analyze trends.

Drawing Conclusions about the Effects of Technology

A compelling reason to become technologically literate is that it allows you to draw conclusions about the effects of technology. This then allows you to take a position about supporting or restricting the development of the technology in question. For example, how would you determine whether genetic modification of crops is of benefit to society? A technologically literate person would understand that if foods are genetically modified, they include a new protein. They become herbicide resistant. This means

Figure 3.44 | The famous early television broadcaster, Edward R. Murrow, understood that television could teach and inspire, but only if humans decided to put it to good use.

Figure 3.45 | In nineteenth-century England, the Luddites were a group of people who were convinced that industrial machinery would spell the end of most manufacturing jobs. To protest this new technology, they rioted and destroyed machines.

Chapter 3 △ Abilities for a Technological World

112

that weeds can be sprayed from airplanes and exterminated, while the genetically modified foods can survive the spraying. However, questions still need to be asked: How does genetically modified food affect humans? What about organic foods? Are the health benefits of organic foods worth the extra cost, and are they really healthier? Some say that the manure used in organic compost may be more harmful than the health risks from pesticide residues. Are there inherent risks that outweigh the benefits? How do you decide? Only after studying the technology and attaining some degree of technological literacy can you make an informed judgment.

Today, the term *Luddite* refers to a person who is opposed to or afraid of new technology.

From the financial point of view, drawing conclusions about the benefits or risks of a technology can help people and companies decide on business and investment strategies. In order to draw conclusions about the effects of technology on the individual, society, or the environment, data must be collected and synthesized, and trends must be analyzed. Historical events (such as the Luddite incidents), global trends (in the case of the Luddites, mechanization), and economic factors (the possibility of creating a greater number of new jobs) should be taken into account. Part of the assessment about whether a technology should be further developed or halted should be based on whether the benefits outweigh the risks, and if there are risks, whether they can be well managed so as to cause only an acceptable level of negative impact. Assessing a product or system can prove to us that it is dangerous, but it cannot prove to us with certainty that it is safe.

TECHNOLOGY IN THE REAL WORLD:
Drawing Conclusions about the Mechanization of Farming

Agriculture contributes positively to our foreign-trade balance (the difference between the value of our exports and the value of our imports), and it remains one of the nation's larger industries in terms of total employment. Technology continues to enable us to produce more agricultural products with fewer workers and fewer farms. Proponents of large, highly mechanized farms argue that the classic family farm is obsolete in that it is no longer able to compete with large fully mechanized farms. Others disagree, claiming that family farms are the bedrock of American society and need to be protected.

Cattle farms today use highly mechanized feedlots to feed large numbers of animals and measure their weight until the desired weight is reached. Computers control the release of feed, formulated in laboratories, and dispense it mechanically in the exact amount needed by each animal. The animals are fed special growth hormones and provided with medications to prevent disease. These technological approaches allow modern farmers to get cattle to market in about half the time (13 to 16 months) that it took several decades ago. Sensors, placed in the animals' skin, are able to detect health problems. These sensors make it possible to alert the farmer that animals might be sick, thereby reducing the potential for contaminated meat.

Modern dairy farms are as automated as modern robotic factories; in fact robotic milking machines are commonly employed. Cows are taught to gather at automated milking machines that use lasers to guide the milking mechanisms to their udders (see Figure 3.47).

Figure 3.46 | A silage machine creates a huge 50-foot swath, chops it, and blows it into trucks driving alongside.

Although some of the work on large ranches such as branding and herding is reminiscent of western movies, technology has changed the way the farming is done. Branding cattle and vaccinating them are processes that are most often mechanized today. Ranchers use motorized vehicles rather than horses to corral and herd cattle, and they make use of global positioning systems (GPS) to pinpoint areas of land where fertilizer, herbicides, and water need to be added.

Even waste disposal methods on farms are highly technological. Modern dairy farms produce large amounts of animal waste. In many dairies, cows are housed in stalls where waste is removed by flushing water across the floors. The water and the solids are separated by machines. The solids are commonly used as soil conditioners. The liquid is used as an inexpensive fertilizer that is pumped to surrounding crop fields. Although this type of waste system is common, little is known about the bacteria that inhabit them. Furthermore, some believe that waste-water holding lagoons may be environmental reservoirs for disease-causing bacteria.

The financial and societal positive and negative consequences of mechanized farming must be considered when assessing its impacts on our culture.

Forecasting techniques

When implementing new technologies, it is important to ensure that they do minimal harm to the environment and living systems. Sometimes, harm to the environment is accepted in light of the benefit to humans; as a society, we accept the proliferation of automobiles, even though we are aware that auto exhaust contributes to air pollution and global climate change. As a society, we make trade-offs, weighing benefits against risks. We should carefully think through these risk-benefit trade-offs before implementing technologies on a wide scale.

Forecasting techniques can be designed and used to evaluate the results of technologies that could potentially affect the environment and alter natural systems. Natural systems affected by technology include lakes (building homes around the shore), rain forests (cutting them down for the wood), or land (strip mining for coal).

One useful forecasting technique is a cross-impact analysis. In this form of analysis, we identify several possible futures. We then try to predict the potential impact of each of these futures if they occurred in combination.

Figure 3.47 | A modern dairy farm uses automated milking machines like this one to milk many cows at one time.

Table 3.2 | This Cross-Impact Matrix Describes the Potential Effects of Several Proposed Technologies

	GENETICALLY ENGINEERED HUMANS	CITIES ON THE MOON	LIFE ON OTHER PLANETS	CITIES UNDER THE SEA
AIR POLLUTION	• Humans will be able to breathe sulfur dioxide.	• Moon societies will safeguard against the kind of pollution that occurred on Earth.	• People will vacation on non-polluted planets.	• Polluted air will be used for fuel by sea cities.
RISING ENERGY COSTS	• Humans will be able to survive on fewer calories.	• Energy will be beamed to Earth from a moon base.	• Energy sources will be exported from other planets to Earth.	• There will be new uses of sea water to provide energy.
OVERPOPULATION	• Child-bearing will be forbidden. • Only genetically engineered humans needed for specific purposes will be authorized. • People will debate whether we should alter human life.	• Societies will move to the Moon.	• Marriages will occur between earthlings and humanoids from other planets. • Attempts to transfer populations will be met by resistance. Interstellar wars will be fought.	• People will work in new industries, like mining sea beds.
LONGER LIFE SPANS	People will be altered several times during their lifetimes as conditions change on earth.	• People will commute to summer homes on the Moon. • New travel agencies will specialize in Moon/Earth travel.	• People will spend more time traveling to distant planets.	• People will live part of their lives on land and part in sea cities to learn about all forms of life.

A cross-impact matrix is a type of table that we can use when conducting a cross-impact analysis. We can use a cross-impact matrix to generate ideas about the possible effects of a variety of future technologies if the technologies occurred in combination. Table 3.2 shows an example of a cross-impact matrix.

Another forecasting technique involves the use of a futures wheel. A futures wheel is a diagram with one scenario at the center that might lead to other scenarios. These in turn, spawn other possible outcomes. These new outcomes are called second-order impacts. These may spawn yet other outcomes (third-order impacts) and so on.

Trend Analysis

Assessment techniques such as trend analysis and experimentation can help inform decisions about the future development of technology. By analyzing trends, entrepreneurs can identify new product ideas and opportunities to reach prospective customers. Paying too much attention to the present can lead to missed opportunities in the future. An analysis of the do-it-yourself trend led to the establishment and success of home improvement "super stores," such as Home Depot and Lowe's. The concern with obesity has led to a trend toward healthier eating and has opened opportunities for companies that produce organic foods, and foods with reduced transfats. Since there is a growing population of baby boomers who are nearing retirement age and will have more time to prepare meals, the healthy eating trend has led to a demand for more healthy and upscale foods.

Trend assessment is an evaluative technique based on iterative procedures that involve analyzing trade-offs, estimating risks, and choosing a best course of action. Consider trends in food. Today, there are trends toward healthier foods. Looking at the futures wheel in Figure 3.48, which of the possible outcomes do you think

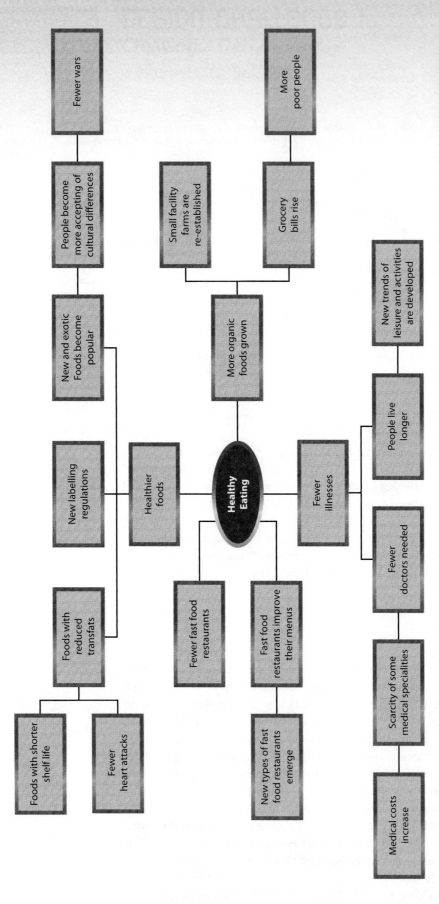

Figure 3.48 | This futures wheel shows a scenario in which people become more concerned with healthier eating. The center scenario, "Healthy Eating" spawns the "More Organic Foods Grown" scenario. These two scenarios, in turn, lead to other, related scenarios.

might be realistic, and what might be some of the trade-offs we would have to make between benefits and risks? As a society, we must have the technical knowledge to do these analyses so that we can advocate for and pass legislation that would provide the best course of action.

SECTION THREE FEEDBACK >

1. Some consumers in the United States worry that the extended shelf life of UHT milk (milk that is processed under ultra high temperatures, about 275° F) may not be healthy because such milk is delivered and stored without refrigeration. Collect information about UHT milk and explain whether you favor its use in your own household.

2. Figure 3.49 illustrates how environmental pollution, while worrisome, contributes far less to cancer than poor diet and smoking. Figure 3.50 illustrates trends in fast-food consumption. What do these two trends, taken in combination, indicate to you? What actions can you take to remedy these situations?

3. Develop a futures wheel. At the center, list laws that eliminate smoking. Project possible second- and third-order impacts of this legislation.

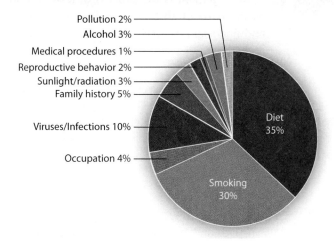

Figure 3.49 | The figure shows how environmental pollution contributes only minimally to cancer compared to other known causes. (Source: Doll and Peto (1981), Journal of the National Cancer Institute.)

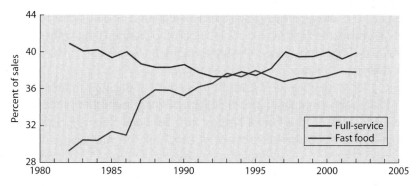

Figure 3.50 | This chart illustrates trends in fast-food consumption. (Source: USDA. Economic research service.)

KEY IDEAS >

- Management is the process of leading and directing all the parts of an enterprise by organizing the human and financial resources and time in the most profitable and beneficial way.

- Good leaders motivate, inspire, and enable their employees to carry out the mission and goals of an organization in effective ways.

- A project can be modeled using a systems diagram with desired results; processes using resources to attain outcomes; and feedback loops involving monitoring and control subsystems.

GOOD MANAGEMENT INVOLVES			
Leadership and motivation of employees	Project planning and organization	Project implementation	Monitoring and controlling operations

Figure 3.51 | Elements of good management.

Management is the process of leading and directing all the parts of an enterprise by organizing the human and financial resources and time in the most profitable and beneficial way. Good management requires leadership, teamwork, and clear identification of roles and responsibilities. Management is responsible for scheduling and project tracking, project financing, and oversight of human resources. Broadly stated, management involves leadership and motivation of employees; project planning and organization; project implementation; and monitoring and controlling operations (see Figure 3.51).

Leadership and Motivation

Good leaders motivate, inspire, and enable their employees to carry out the mission and goals of an organization in effective ways. Good leaders recognize that employees are the most crucial resource. Leadership does not always have to come from a person in authority. It can just as well come from someone with special skills or expertise; it can come from a trendsetter; it can come from people who are particularly respected for their ideas. The challenge that leaders face is to do three things well: to get their staff to accomplish goals and planned tasks with excellence; to develop the skills and abilities of people whom they supervise; and to take good care of their subordinates.

Often, becoming a leader is a result of a personal decision to assume a lead role. We have seen examples of this in politics, business, education, entertainment, religion, and even in the home. Names such as Oprah Winfrey, the television personality; Michael Bloomberg, mayor of New York City and owner of a media empire; and Sergey Brin, the co-founder of Google (see Figure 3.52), come to mind as examples of people who have risen to leadership and prosperity even though they came from humble beginnings.

Figure 3.52 | Sergey Brin, co-founder of Google, was born in August 1973 in Moscow, Russia. He and his family emigrated to the United States to escape religious persecution (they are Jewish) in 1979.

TECHNOLOGY AND PEOPLE:
A Profile of Leadership: Russell Simmons

Russell Simmons was born in New York City in 1957 and attended the City College of New York (see Figure 3.53). He is an entrepreneur, the co-founder of the hip-hop label Def Jam, and founder of the Russell Simmons Music Group. His business interests include a clothing company called Phat Farm, a movie production house, television shows, and an advertising agency. His Rush Communications company includes a company that produces Legacy and Arthur Ashe sneakers.

Simmons brought an award-winning live show called "Def Poetry" to Broadway, but his leadership extends beyond his music. He has made important contributions to the civil rights movement and participated in get-out-the-vote efforts in the 2004 U.S. presidential election. He has taken public positions against cruelty to animals in factory farms and slaughterhouses. Currently, he is working on a project to fight anti-Semitism.

Russell Simmons is a person who has used his talents and his entrepreneurial ideas not only to amass personal wealth, but to improve society—the mark of a true leader.

Figure 3.53 | Russell Simmons is an entrepreneur who co-founded the hip-hop label Def Jam, then went on to develop business and philanthropic interests in many areas.

Planning and Organization

Project planning involves making decisions about what should happen to ensure that a project succeeds. During the planning phase, the people responsible for a project identify goals, strategic plans, and time lines. They also create a budget for the project in a spreadsheet. In addition, they create a management plan that spells out roles and responsibilities of all personnel and they decide which methods they will use to measure and evaluate the success of the project. The planning process continues throughout the project's duration as the team assesses its progress toward meeting goals—and then revises strategies and plans as necessary.

Project organization involves setting up procedures that make the best use of resources—human, financial, material, energy, equipment, information, and time—so that the plans that have been developed can be implemented successfully. In order to make the best possible decisions about these procedures, it's important to include people in the decision-making process (the stakeholders) who will be involved in carrying out the procedures. In particular, it's important to ensure that the stakeholders help to define what needs to be accomplished and how. Organizational charts are helpful in displaying how the various responsibilities, agencies, offices, and lines of authority are related (see Figure 3.54).

Project Implementation

Project Implementation is the process of carrying out plans in ways that maximize success and preserve the core values of the organization or business. Organizational values may include honesty and integrity, excellent customer service, high performance, openness and teamwork, and respect for diversity. Without clear organizational values, individuals within the organization may pursue courses of action that may lead to results that the organization doesn't wish to support.

Implementation strategies involve ensuring that plans and policies are carried out, time lines are met, and that human resource management and financial

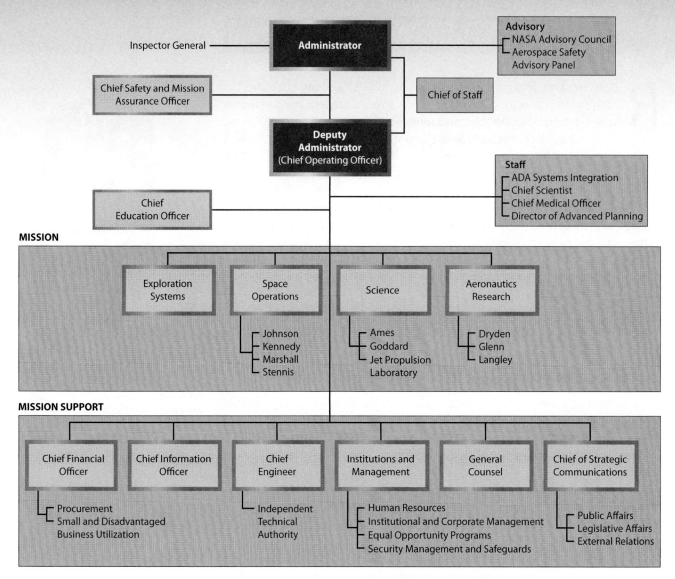

Figure 3.54 | This is an example of an organizational chart used by the National Aeronautics and Space Administration (NASA).

management strategies are put into place and monitored. An important element in properly implementing a project is collecting data that contributes to making good management decisions and is linked to project goals. To assess overall performance, input, output, and outcome data are collected. Input data provide a measure of the resource inputs (people, information, equipment, materials, energy, capital, and time). Output data provide a measure of what it is you've actually produced and help you determine if you are making progress toward attaining your goals. Outcome data allow you to assess to what extent you've actually reached your goals.

Monitoring and Controlling

A project can be modeled as a system with desired results; processes using resources to attain outcomes; and feedback loops involving monitoring and control subsystems. The *desired results* are the planned goals; the *process* uses resources to accomplish project implementation strategies; and you know from your previous study of systems that to adjust for changing conditions, *feedback* allows the system to be properly monitored and controlled (see Figure 3.55).

In a project, whatever the goals, checking progress against plans allows for making modifications that help ensure that goals are reached. Feedback should be gathered about time management; people's performance; financial expenses, profits, and losses; use and costs of energy, material, and equipment resources; and the quality of end results (products or services). In a system that is based upon people working together to achieve a business organization's goals, management personnel are ultimately responsible for performance. In the past, it was assumed that top management would assume the entire responsibility for deciding on monitoring and control methods. Now, in many modern organizations and businesses, other staff members are increasingly involved in decision making.

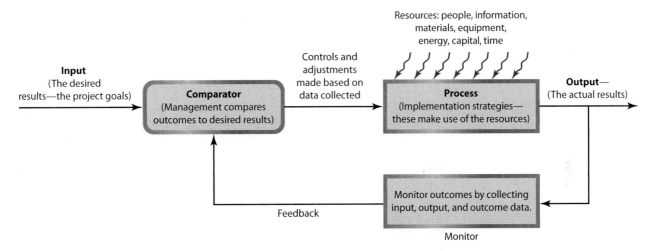

Figure 3.55 | A project can be thought of as an input-process-output feedback control system. The input to the system is the desired result; the process consists of the implementation strategies that make use of resources; the output is the actual result. The output (the project outcomes) may not be what was intended. Data are collected (feedback) and a comparison is made, usually by management, to determine if the outcomes are desired. If not, adjustments are made to the process. The adjustments may mean changing resources or changing procedures.

SECTION FOUR FEEDBACK >

1. Develop a management plan for a project that will develop a strategy game to be played on a game console or a computer. The management plan should include:
   - A clear statement of the goal, or goals, of the project (not the game goals, but the project goals)
   - A time line that illustrates graphically what the important milestones would be during the development of the game
   - A management plan that identifies the roles and responsibilities of project managers, technical consultants, and other workers
   - A project budget
   - Identification of the data that you would collect that would ensure that project goals would be met
2. Select a project of your own choosing. It might be a project that results in a product, a service, or an environment (like a park or garden). Draw a systems diagram of the project that indicates the desired results, the specific resources used, the data that would need to be collected to ensure project success, and the ways in which management might implement control mechanisms (technological, financial, procedural) to ensure success.

Matching Your Interests and Abilities with Career Opportunities: Computer Professional

You are probably quite aware that computer skills are basic abilities people need to succeed in our technological world. One of the fastest growing career clusters requires the ability to design computer systems and provide related technical services.

Significant Points

* The computer systems design and related services industry is expected to experience rapid growth, adding 453,000 jobs by 2013.
* Professional workers enjoy the best prospects, reflecting continuing demand for higher level skills needed to keep up with changes in technology.
* Computer systems analysts, computer software engineers, and computer programmers account for about 60% of all employees in this industry.

Nature of the Industry

Services include custom computer programming services, computer systems design services, and computer facilities management services.

Working Conditions

Most workers in this industry work in clean, quiet offices. Those in facilities management and maintenance may work in computer operations centers. These are facilities that house powerful computers that provide the information technology services needed by a company, a university, or other such entity.

Occupations in the Industry

Programmers write, test, and maintain the detailed instructions, called program code, that computers must follow to perform their functions.

Computer engineers design, develop, test, and evaluate computer hardware and related equipment, software programs, and systems.

Systems analysts study business, scientific, or engineering data-processing problems and design new flows of information.

Database administrators determine ways to organize and store data and work using database management systems software.

Computer support specialists provide technical assistance, support, and advice to customers and users. This is actually a group of occupations that includes workers with a variety of titles, such as "technical support specialist" and "help-desk technician." Other skilled computer workers are involved in the design, testing, and evaluation of network systems, such as local area networks (LANs), wide area networks (WANs), Internet, and other data communications systems.

Web developers are responsible for designing and creating Web sites. Webmasters are responsible for the technical aspects of a Web site, including day-to-day performance issues, such as speed of access, and for approving site content.

Network or computer systems administrators install, configure, and support an organization's LAN (local area network), WAN (wide area network), network segment, or Internet functions. They maintain network hardware and software, analyze problems, and monitor the network to ensure availability to system users.

Training and Advancement

Many employers seek applicants who have at least a bachelor's degree in computer science, information science, or management information systems (MIS).

Outlook

This is one of the 25 fastest growing industries in the United States. Wage-and-salary employment is expected to grow 40% by the year 2014, compared with only 14% growth projected for the entire economy.

Earnings

Non-supervisory workers in the industry averaged earnings that were almost twice as high as the average for all industries.

[Bureau of Labor Statistics, U.S. Department of Labor, Occupational Outlook Handbook, 2007–08 Edition, visited November 2007, http://www.bls.gov/oco/]

Summary >

To be a fully integrated member of our highly technological society, a person needs to develop abilities that provide technological literacy. These abilities include having a familiarity with how technological products, systems, and environments are designed; being able to properly use and maintain technical products and systems; having the ability to assess technological benefits and risks; and being able to effectively manage technological endeavors.

Good designers know that once a design problem has been identified, it must be analyzed to determine if potential benefits justify investing time and other resources in trying to search for solutions. The design specifications and constraints will affect the design concept, its implementation, marketability, and its commercial profitability.

Developing prototypes and using modeling techniques help ensure the quality, efficiency, functionality, and dependability of the final product. Modeling possible design solutions allow the designer to try out "What-if?" scenarios before actually implementing and producing the design.

Design solutions are evaluated and communicated using three-dimensional models, and verbal, graphic, quantitative, virtual, and written means. Quality control measures ensure that products and systems meet customer requirements and industry standards.

A variety of techniques, including flowcharts, symbols, spreadsheets, graphs, and time lines can be used to document processes and procedures and communicate them to different audiences. Computer-based and electronic tools are used along with technical knowledge to maintain the proper operation of technological systems. Monitoring the operation of a system and making adjustments as needed help ensure safe and proper functioning.

Systems should be operated so that they function in the way they were designed. Computer systems with appropriate software are used to process information needed to ensure that technological systems are kept in good working order.

123

Information that is collected to assess the impact of technological products and systems must be evaluated with respect to its quality. To draw conclusions about the effect of technology on the individual, society, and the environment, data must be collected and synthesized and trends must be analyzed.

Forecasting techniques can be designed and used to evaluate the results of altering natural systems. Assessment techniques such as trend analysis and experimentation can help inform decisions about the future development of technology.

To ensure smooth function of a technological project, appropriate management techniques must be employed. Management is the process of planning, organizing, and controlling work. Effective project management involves scheduling and project tracking, financing the project, and building a team-based personnel structure.

FEEDBACK

1. Do you think design ideas that aren't likely to become commercial successes should be rejected? Explain your position.

2. Develop a flowchart showing the steps required to set up and test a newly purchased computer system that includes the computer, a mouse, keyboard, speakers, and an Internet connection.

3. Design a Gantt chart that might be used in planning an evening event for 50 people that includes music and food.

4. Develop a spreadsheet that shows itemized costs of food purchased for your family for one week. How might you save on food costs?

5. Give an example of how electronic tools are used to maintain the proper operation of technological systems. What knowledge is required to use these effectively?

6. How does feedback allow proper adjustments to be made to a system to ensure proper functioning?

7. An "amusement ride" is defined as a system that moves people through a fixed course within a defined area for the purpose of amusement. Choose an example of such a system and describe the consequences that might exist if it were to be operated in a way other than the way it was designed to function.

8. Give an example of a technological system that is computer controlled and explain how the controller keeps the system operating properly.

9. Cite an example of how misinformation about the impact of a technological product or system can be used to advance a particular point of view.

10. Read an article about the Luddites. Explain the issues they were facing and take a position that either supports or argues against their concerns and methods.

11. Trends might lead toward foods with added vitamins and minerals, organic foods, more relaxed dining, foods with less fat and fewer calories, new packaged food products that are easy to prepare yet provide good nutrition, more nutritious fast food, nutrition food bars, and foods from foreign cultures. Make a table showing five possible food trends. In one column, identify possible benefits; in another column, identify possible risks. In a summary statement, discuss what you believe to be the most logical approach a community could use to advocate for healthy nutrition.

12. Some managers are autocratic. Define the term "autocratic" and take a position supporting or rejecting autocratic management. Clearly explain your position.

DESIGN CHALLENGE 1:
Pelicans and Green Men

• Problem Situation

A pedestrian crosswalk is a section of a street in which pedestrians have the right of way. Such crossings are often marked by signs, painted lines, flashing lights, or traffic symbols. In England, pedestrian crosswalks are called "pelican crossings," but such crossings exist worldwide. "Pelican" is actually slang for a **PE**destrian **LI**ght **CON**trolled crossing, or pelican. Often, crosswalks are controlled by traffic signals with lights or images of red or green men. In this activity, you will design and produce a model of a pelican crossing.

• Your Challenge

You are asked to design and construct a model of a pedestrian crosswalk—a pelican crossing—that is marked by symbols and/or signs and uses a set of traffic lights with a system to control them.

• Safety Considerations

1. Wear eye protection at all times. Be particularly careful about using cutting tools, spraying paint, and applying finishes.
2. Make sure that all machines are used with all guards in place.
3. If you are using a large plywood board, get help handling it.
4. Be careful with electrical and electronic devices. Make connections first, and then turn on power. Be sure there are no short circuits.
5. Only use low-voltage sources of electricity for lights, mechanical devices, and control systems.
6. Be sure that the combination of bulbs, relays, and other devices that are connected to a power supply do not exceed the current capacity of the supply.

• Materials Needed

1. ½-inch thick plywood board measuring at least 2 feet by 4 feet
2. Paints of various colors, including white paint
3. Paint brushes and rollers
4. Low-voltage power supply
5. Low-voltage light bulbs or LEDs
6. Low-voltage relays
7. Timers
8. Assorted hardware
9. Assorted pieces of wood, metal, or plastic for making model signs
10. Computer with Internet access

• Clarify the Design Specifications and Constraints

Your model must be made on a sheet of ½" thick plywood measuring at least 2′ x 4′. In your design, the pelican crosswalk must include at least two attention-getting symbols, a pattern of painted lines, and a traffic light with at least two lights of different colors. The lights must be activated by a control system that turns the lights on and off and keeps the lights on and off for an appropriate period of time. Include at least one enhancement that makes crossing easier or safer for a person with a disability.

125

• Research and Investigate

To complete the design challenge, you need to first gather information to help you build a knowledge base.

1. In your guide, complete the Knowledge and Skill Builder I: Rules governing crosswalks.
2. In your guide, complete the Knowledge and Skill Builder II: Crosswalk design.
3. In your guide, complete the Knowledge and Skill Builder III: Signs and symbols.
4. In your guide, complete the Knowledge and Skill Builder IV: Investigating times of red, green, and yellow lights.
5. In your guide, complete the Knowledge and Skill Builder V: Enhanced crosswalk features to assist persons with disabilities.

• Generate Alternative Designs

In your student activity guide, describe two of your possible alternative approaches to designing the pelican crosswalk. Discuss the decisions you made in a) choosing signs and symbols, b) choosing the length and width of the crosswalk, c) designing the control system for the lights, and d) enhancing the crosswalk with features to help persons with disabilities. Attach drawings if helpful and use additional sheets of paper if necessary.

• Choose and Justify the Optimal Solution

What decisions did you reach about the design of the signs and symbols, the length and width of the crosswalk, the control system for the lights, and enhancements to aid persons with disabilities?

• Display Your Prototypes

Construct a model of your pelican crosswalk on a ½-inch thick plywood board measuring at least 2 feet by 4 feet. Include descriptions and either photographs or drawings in your guide.

• Test and Evaluate

Explain whether or not your pelican design met the specifications and constraints. What tests did you conduct to determine this? Troubleshoot, analyze, and maintain your pelican system to ensure safe and proper functioning.

• Redesign the Solution

What problems did you face that would cause you to redesign signs and symbols, the size of the crosswalk, the control system for the lights, or enhancements to aid persons with disabilities? What changes would you recommend in your new designs? What additional trade-offs would you need to make?

• Communicate Your Achievements

Describe the plan you will use to present your solution to your class. (Include a media-based presentation.)

DESIGN CHALLENGE 2:
Finding Money

• Problem Situation

Sometimes leaders in technology have excellent ideas, but they lack the funds to implement those ideas. In many cases, these leaders seek funding from government agencies or foundations to finance their ideas. In order to do so, the individuals must submit a proposal to the funding agency. The proposal is a written document that explains the project clearly, presents a compelling case that the proposed ideas can work, and provides a management plan and a budget that will convince the funding agency that the people making the proposal have thoughtfully considered all aspects of the proposed project.

• Your Challenge

Develop a proposal that would generate enough funding for your class to engage in an engineering design project that would be of benefit to people living in the community. Assume you are directing your proposal to an educational or community group that has funds to distribute to a deserving applicant.

• Safety Considerations

Ensure that you use computer hardware and software only as instructed by your teacher.

• Materials Needed

1. Computer with Internet access
2. Software (such as Microsoft Word, Excel, or project management software) for making flowcharts, Gantt charts, Pert charts, and time lines
3. Spreadsheet software (such as Microsoft Excel) for budget development

• Clarify the Design Specifications and Constraints

Pick a project that, when completed, will satisfy a real need. The project must be completed within one school year. Your proposal must include the following sections:

- Project Title
- Need for the Project
- Project Goals
- Project Overview
- Work Plan and Time Line (in chart form)
- Plan to Evaluate Your Project
- Role of Each Project Team Member
- Budget (In Excel or similar software)
- Letters of Support from School or Community Personnel

You must limit your proposal to 15 pages, single spaced, using a font size of between 10 and 12 points, with one-inch page borders.

• Research and Investigate

To complete the design challenge, you first need to gather information to help you build a knowledge base.

1. In your guide, complete the Knowledge and Skill Builder I: Identifying a community need.
2. In your guide, complete the Knowledge and Skill Builder II: Meeting with potential funders.

3. In your guide, complete the Knowledge and Skill Builder III. Proposal preparation.
4. In your guide, complete the Knowledge and Skill Builder IV: Sources of funding.
5. In your guide, complete the Knowledge and Skill Builder V: Using spreadsheet software.
6. In your guide, complete the Knowledge and Skill Builder VI: Making charts and time lines.
7. In your guide, complete the Knowledge and Skill Builder VII: Presenting your proposal.

● Generate Alternative Designs

Describe at least two possible project ideas. Discuss the level of funding required to implement each idea. Discuss the decisions you made in choosing the project you wish to pursue.

● Choose and Justify the Optimal Solution

Explain what prompted you to pick the particular project you chose. Explain your decisions regarding: 1) the design of the project, 2) the level of funding needed, 3) the project management, and 4) the people or organizations you will approach for funding.

● Display Your Prototypes

Display your completed proposal on a tri-fold board. Label each of the sections. Include any drawings that would help explain your proposal ideas. Include letters of support and commitment that you have received from interested parties.

● Test and Evaluate

Explain whether your proposal was successful in obtaining needed funding. If you were not successful in obtaining funding this time, what would you have to do to enhance your funding opportunities?

● Redesign the Solution

What problems did you face that made it difficult for you to frame an effective proposal? What changes would you make if you were to rewrite the proposal and resubmit it? What changes would you recommend to the engineering design project? What additional trade-offs would you have to make in recasting your proposal and project?

● Communicate Your Achievements

Describe the plan you will use to present your solution to your class. (Include a media-based presentation.)

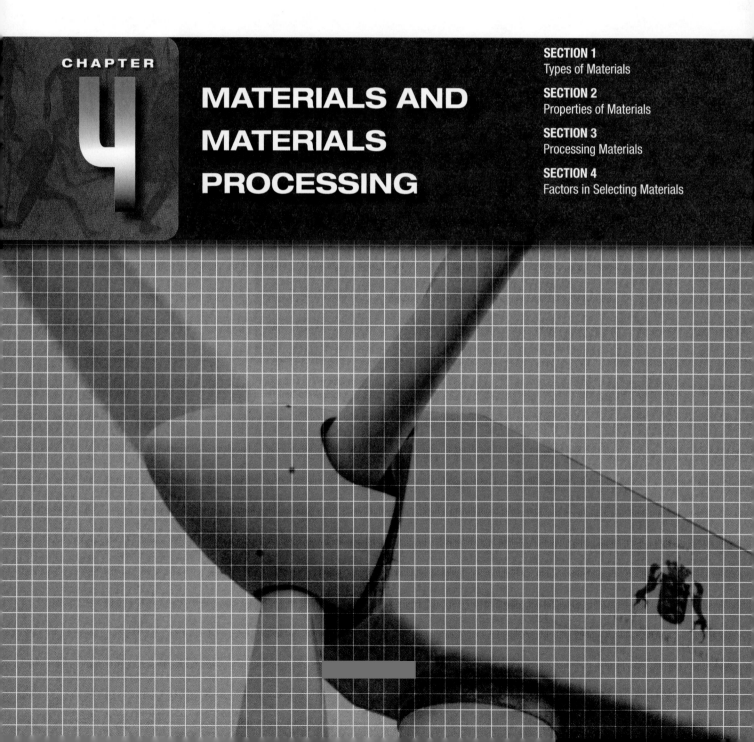

T HE TERM MATERIALS refers to the substances from which things are made. Materials are the lifeblood of both the natural and built worlds. Whether in living organisms or in manufactured projects; whether in the smallest living cells or in the most massive constructions, atoms combine to form materials with special properties that define the characteristics of the matter they make up. If the bonds between atoms are very strong, the material is probably a metal. Because of the strength of the atomic bonds, metals tend to have high melting points, since it takes a great amount of heat energy to break down those strong bonds. In most metals, the electrons in the outermost atomic orbits are loosely held by the nucleus. In that case, the electrons can become dislodged by a force (like an electrical voltage). Hence, those metals are good conductors of electricity. In the human-made world, we choose materials carefully based on their properties, which determine how products and systems function and behave.

The study of materials is fascinating, taking you from the ordinary to the extraordinary. Some materials are quite familiar. Wood and some types of metal are used to build furniture for homes and offices. Plastic is formed into automobile parts, and clay is used for stoneware and porcelain dishware. Other materials are probably less familiar to you. For example, graphite, fiberglass, and even titanium are used to produce very strong tennis rackets. Even more extraordinary are nanoscale materials that are made from molecular components that are as small as one billionth of a meter (10^{-9} meters) in size and that assemble themselves chemically. Carbon nanotubes are single molecules of carbon that have a diameter of one to three atoms. A material made from carbon nanotubes is about 100 times stronger than steel, but weighs about six times less. This chapter will introduce you to the types and properties of materials, how they are processed, and how people decide to select materials for a particular application.

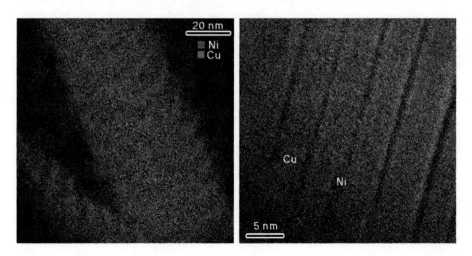

Figure 4.1 | Nanoscale wires of copper and nickel.

Figure 4.2 | The Young Endeavor sailing ship.

HEY IDEAS >

- Materials have different origins and may be classified as natural, synthetic, or mixed.

- The most common materials from which products are made are wood, metal, plastics, ceramics, and composites.

- Recyclable materials allow what otherwise would be considered waste to be reprocessed into useable products.

Materials have different origins and may be classified as natural, synthetic, or mixed. Natural materials are those that occur in nature. Some natural materials are organic materials that originate from a living organism and have carbon-based structures. These are capable of decay. Organic materials include wood and natural fibers (like cotton, hemp, and wool). Some natural materials are inorganic. These substances are minerals and are not characterized by carbon-based structures. Inorganic materials include stone, clay, and metals.

Synthetic materials are human-made. Many substances that occur naturally can be produced synthetically. This is often the case with pharmaceutical products, which might have exactly the same chemistry as the natural materials they replace. For example, most synthetically made vitamins are identical to those extracted from natural sources. The advantage of synthetic vitamins is that they are less expensive to produce. Some other common synthetic materials are rubber, plastics, Teflon, Nylon, and Kevlar.

Some synthetic materials are designed to have entirely new physical properties. Nylon was a breakthrough product that revolutionized the clothing industry because it is strong, light-weight, easy to wash, and quick to dry. Kevlar is one of a number of aramids (aromatic polyamide fibers) that are very strong and heat-resistant. (You'll learn more about aramids in the Engineering Quick Take later in this chapter.) Aramids are used in applications as wide-ranging as bicycle frames, tennis rackets, and bulletproof vests.

Mixed materials are a combination of natural and synthetic materials. Examples of mixed materials are plywood, paper, and wool-polyester blends of fabrics. Plywood is an engineered wood that is created by cutting thin sheets of wood (called plies) to precise specifications, and then laying them on top of one another (laminating them) at alternating 90° angles. The plies are glued together using synthetic adhesives.

Paper fibers are typically composed of wood fibers (also called cellulose) but synthetic plastic fibers are often added to improve the paper quality. Fabrics are often made from a blend of cotton or wool and synthetic materials. The mix of materials gives the fabric the beauty of the natural material but adds strength, plus shrinkage and stain resistance.

The most common materials from which products are made are wood, metal, plastics, ceramics, and composites (combinations of materials). In the remainder of this section, we'll examine these materials in more detail.

Synthetic Aramid Fibers

Figure 4.3 | Atomic structure of an amide.

The name "aramid" comes from the term aromatic polyamide. Amides are members of a group of chemical compounds containing carbon-oxygen bonds and nitrogen (see Figure 4.3). The term polyamide refers to many amide molecules connected into a single unit.

If something is aromatic, it generally means that the substance has an odor. Originally, materials such as wintergreen leaves, vanilla beans, cinnamon bark, and anise seeds were called aromatic because of their pleasant aromas. In engineering terms, aromatic refers to substances that have a very stable atomic structure, often containing six carbon atoms (see Figure 4.4).

Figure 4.4 | An aromatic chemical bond.

Aramid fibers are synthetic materials with very strong atomic bonds that are used in aerospace and military application. They are often used as substitutes for asbestos, which has proven to be carcinogenic (cancer-producing). Kevlar is an example of an aramid fiber (see Figure 4.5).

Aramids have a unique set of properties. These include:

- High strength-to-weight properties
- Good impact resistance
- High stiffness
- Resistant to chemicals
- Nonconductive
- Flame- and heat-resistant

Some applications include:

- Flame- and heat-resistant clothing
- Protective clothing like gloves and chaps
- Body armor
- Sports equipment
- Brake pads
- Rope
- Boat sails and hulls
- Reinforced concrete

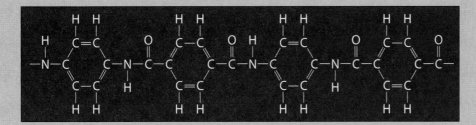

Figure 4.5 | Atomic structure of Kevlar.

Wood

Since before recorded history, wood has played an important role in human technological activity. Historians tell us that there is evidence that early humans (hominids) used wood for cooking and heating some 1.5 million years ago. Wooden boats replaced boats made from papyrus in Egypt about 4000 B.C.E. Egyptian tombs

133

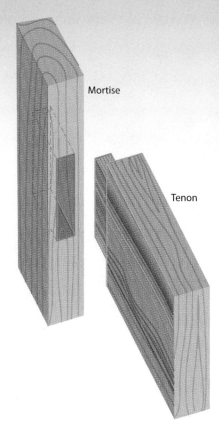

Mortise

Tenon

Figure 4.6 | A mortise and tenon joint.

Figure 4.7 | Softwood structure.

give a wealth of information about the sophisticated woodworking technology that was used during the period of 3300–3000 B.C.E. Wooden objects using mortise and tenon joints were found. A tenon is made by shaping the end of a piece of wood so that it fits into a hole cut into a second piece. This type of wood joinery does not require fasteners (see Figure 4.6). For many centuries, wood has been used as a primary construction and manufacturing material to build structures, tools, furniture, and weapons.

Today, wood is used in tools, paper, buildings, bridges, guardrails, railroad ties, posts, poles, mulches, furniture, packaging, and thousands of other products. Different types of trees produce wood with different properties.

Wood is a renewable material, since it comes from trees that can be replanted once harvested. Wood is recyclable and biodegradable, therefore it is an environmentally friendly resource. Recyclable materials allow what otherwise would be considered waste to be reprocessed into useable products. Recycling includes collecting recyclable materials, sorting and processing them, and using them to produce new products.

Types of Wood Wood is categorized as hardwood or softwood. These are terms that refer to the cells that carry water in the living tree. They do not refer to the actual hardness or softness of the wood itself. Some hardwoods are softer than some softwoods. For example, basswood and poplar are soft hardwoods, while fir and yew are hard softwoods. Balsa is one of the lightest, least dense woods, but it is considered a hardwood.

Most softwood trees are conifers (or evergreens), although there are exceptions, such as cypress trees. Softwood trees have needlelike leaves, and they produce conifer cones (such as pine cones). In softwoods, the wood is made up of long, narrow cells, called tracheids, with hollow centers (see Figure 4.7).

Hardwood trees are broad-leaved trees, such as oak, ash, beech, teak, and mahogany. Hardwoods have tubular cells that are shorter in length (normally about one millimeter long) than softwood cells and have thicker walls (see Figure 4.8). Hardwood trees are deciduous trees; that is, they shed their leaves each year.

Structure of Wood Wood (and the timber that is made from it) is primarily made up of cells. Most of the cells in the woody part of a tree are dead. However, some living cells, called xylem (pronounced zy-lem) conduct water and nutrients through the wood to the leaves of the tree. The outside of a tree is composed of bark that has an inner portion and an outer portion. The outer portion of the bark is composed of nonliving cork material that protects the tree from damage by weather and insects. Only the inner portion of the bark is alive. This inner layer is called the phloem and is made up of cells that form thin tubes. The walls of the tubes are made of cellulose. These tubes carry the sugars and other materials made in the leaves to the other living parts of the tree (see Figure 4.9).

The oldest portion of the tree is the heartwood. It is the nonliving center of the tree. It is the hardest and most dense part of the tree. The outer portion, just under the bark, is called the sapwood. This living, younger wood is softer than the heartwood.

If you cut a tree horizontally, so that you see a cross section of the trunk, you will notice that a series of rings called growth rings or annual rings can be seen. Each ring shows one year of age of the tree. The newest rings (representing the newest growth) are closest to the bark. The rings are formed because a tree grows at different rates during the year and the density of the wood changes during each growing cycle.

Figure 4.8 | Hardwood structure.

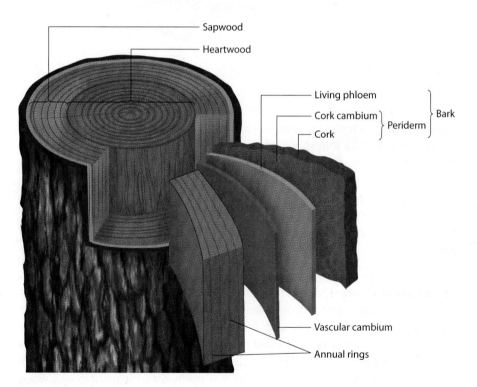

Figure 4.9 | Tree layers.

Uses of Wood When trees are processed into industrial materials, they are cut into planks called timber or lumber. Depending upon the wood, these are further processed into planks and boards (used for boat building, flooring, furniture, and shelving); sheets (used as veneers and for making plywood); and dimensional lumber (such as the 2 × 4s and 2 × 6s used in framing buildings).

Engineered woods are made by combining particles and fibers of wood with adhesives. These human-made composites can be designed (engineered) to meet

Figure 4.10 | Wood veneer.

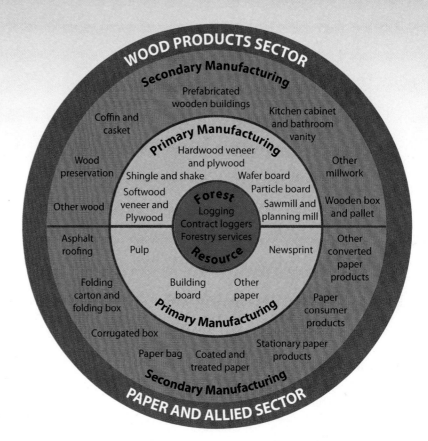

Figure 4.11 | Wood products.

specific requirements. Wood chips are used to produce a variety of engineered woods, such as chipboard and hardboard. Other engineered woods include plywood (made from thin sheets of wood that are laminated and glued together in a hot press); waferboard (sometimes referred to as oriented strand board, or OSB) that is made from wood ground into thin strands, mixed with wax and glue, layered, and pressed in a hot press; and microlam (similar to plywood, but using many layers of thin wood glued together, strong enough to be used for beams). These engineered woods are not only strong, but cost effective. They are made of low-value wood resources (such as wood chips), but they are used to produce high-value products, such as furniture and building materials.

Properties of Wood The thickness of a wood's cell walls determines its density. Cells with thin walls and a large hollow center (such as balsa wood) are lighter. Wood with thick-walled cells and a narrower, hollow center are denser. The strength of a particular type of wood is related to its density. Thus, balsa, with a density of 170 kg/m^3, is very light and suitable for building model airplanes, but it isn't strong enough to be used in furniture making. By contrast, mahogany, a wood with medium density, is suitable for all types of furniture. Rosewood, because of its high density (about 750 kg/m^3) and beautiful color, is used to build expensive furniture and high-end musical instruments. The hardest wood is lignum vitae. It is so dense (over 1,200 kg/m^3) that it will sink in water (water has a density of 1,000 kg/m^3). Lignum vitae is used to make mortars and pestles that are used for grinding medicines and spices.

Table 4.1 shows the densities of common woods. Water has a density of 1,000 kg/m^3 or 62.4 lbs/cu.ft. (pounds per cubic foot). Can you tell from reading the table which of these woods will sink?

Table 4.1 | Wood Densities

TYPE OF WOOD	DENSITY (kg/m³)	TYPE OF WOOD	DENSITY (kg/m³)
Apple	660–830	Larch	590
Ash, black	540	Lignum vitae	1280–1370
Ash, white	670	Mahogany–Honduras	545
Aspen	420	Mahogany–African	495–850
Balsa	170	Maple	755
Bamboo	300–400	Oak	700
British birch	670	Pine–Oregon	530
Cedar, red	380	Pine–Red	370–660
Cypress	510	Redwood–American	450
Douglas fir	530	Rosewood–East Indian	750
Ebony	960–1120	Spruce–Sitka	450
Elm–English	600	Teak	630–720
Elm–Rock	815	Willow	420

A wood's color is a function of the chemicals in its cell walls. Heartwood is generally darker than sapwood. The term grain refers to the way the wood fibers grow. Sometimes, the grain is very straight, a product of a straight-growing tree. When the tree trunk twists as it grows, the grain sometimes looks as though it swirls or spirals. Wood grain adds great beauty to wood products, and wood is very often chosen for its grain. For example, the tops of fine guitars are often made of carefully selected, close-grained spruce. When woodworkers cut or shape wood, they always take the grain direction into consideration (see Figure 4.13).

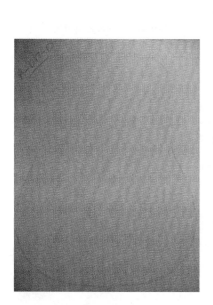

Figure 4.12 | Guitar top.

Figure 4.13 | Wood grain images.

137

Because wood cells contain water, before wood is processed into products, it first is dried to make it dimensionally stable and to eliminate shrinkage, checking, and warping. This is especially important when wood is to be made into furniture and musical instruments. Wood can be dried in air or in a kiln (a drying oven) (see Figures 4.14 and 4.15).

Figure 4.14 | Air-drying lumber.

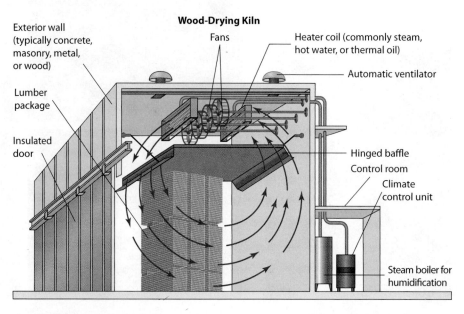

Figure 4.15 | Kiln-drying lumber.

Metal

Metals are categorized as either ferrous or nonferrous. Ferrous metals contain iron. (In Latin, the word *ferrum* means "containing iron.") Some examples of ferrous metals are cast iron and steel. Nonferrous metals do not contain iron. Examples of nonferrous metals include aluminum (the most abundant metal in the earth), cobalt, copper, lead, tin, tungsten, and zinc. The precious metals (gold, silver, and platinum) are also nonferrous.

Ferrous metals are usually, but not always, magnetic. For example, certain types of steel that are made with an alloy (a mixture of two or more metals) of manganese and nickel or chromium are ferrous metals, but they are not magnetic.

Nonferrous metals are usually nonmagnetic, but there are exceptions. (For example cobalt and nickel are nonferrous and magnetic.)

Many steels are alloys that include iron and a small percentage of carbon. As the carbon percentage increases, so does the strength of the steel. Carbon steels are categorized as low-carbon steel (less than 0.3% carbon), medium carbon steel (0.3 to 0.6% carbon), and high carbon steel (more than 0.6% carbon). Low carbon steel is generally used for inexpensive products such as nuts and bolts, sheets, and tubes. Medium carbon steel is used for automotive applications. High carbon steel is used for cutting tools, springs, and cutlery.

Specialty steels are steels that are alloyed with metals such as nickel and cobalt, titanium, and tungsten. They are used in conditions under which other metals would corrode or degrade in response to high temperature.

Structure of Metal As with all materials, the atomic structure of metals determines their properties. Metals are crystalline structures (see Figure 4.17). Their atoms are packed tightly together in rows with repeating, orderly patterns. They therefore have high densities. Table 4.2 shows the densities of common metals.

Metals are actually formed from many small crystals that arrange themselves in different orientations. Thus, metals are truly polycrystals, materials made from many crystals (see Figure 4.18). The crystalline structure of metals gives rise to the way the material behaves.

Because the atoms in metals are packed tightly together, it takes a great deal of energy to break the bonds. Metals, therefore, are strong. In most

Magnetic properties are important; for example, titanium is used to make screws that hold body parts together because titanium is not magnetic and therefore does not interfere with MRI (magnetic resonance imaging) the way steel parts would. As an interesting note, when titanium is deformed under stress, it acts in a similar way to bone, so that it works well in bone and joint repair (see Figure 4.16).

Figure 4.16 | Titanium is commonly used in orthopedic implants such as joint replacements and bone pins, plates, and screws. The photo shows parts of a hip replacement made of a titanium alloy.

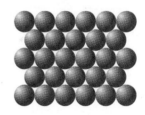

Figure 4.17 | Crystalline structure.

400 μm

Figure 4.18 | Polycrystalline structure of metal.

Table 4.2 | Metal Densities

METAL OR ALLOY DENSITY (kg/m³)			
Aluminum	2,643	Nickel silver	8,400–8,900
Brass	8,550	Platinum	21,400
Bronze (8–14% Tin)	7,400–8,900	Silver	10,490
Cast iron	6,800–7,800	Solder 50/50 (lead and tin)	8,885
Cobalt	8,746	Stainless Steel	7,480–8,000
Copper	8,930	Steel	7,850
Gold	19,320	Tin	7,280
Iron	7,850	Titanium	4,500
Lead	11,340	Tungsten	19,600
Magnesium	1,738	Uranium	18,900
Mercury	13,593	Zinc	7,135
Nickel	8,800		

metals, the electrons in the outermost shell of the atom are loosely bound to the nucleus. They can become free electrons easily and can travel from atom to atom.

Uses of Metal   Because of the atomic structure of metal, particularly the loosely bound electrons in the outermost atomic shell, metals are good conductors of electricity and heat. Therefore, metals are used to make pots and pans and electrical cables and wires. Since metals have high densities, they are used where strength is important, such as in the construction of bridges and building structures. Some metals (such as aluminum) are lighter and are used where weight must be reduced, such as in airplanes and fuel-efficient automobiles.

Types and Uses of Common Metals

▷   **Iron** is the most common and least expensive metal to produce. In steel making, it is alloyed with carbon. Iron products include bridges, cars, steel beams, rails, and cutlery. **Aluminium** is soft and lightweight and is the most commonly used metal after iron. It is used in vehicles, packaging, building construction (aluminum siding), and for pots and pans. **Cobalt** is a tough, lustrous, silvery-gray element. It is actually harder than iron and is used to make steel alloys used for magnets, razor blades, metal cutting, and surgical steel. It is also used to make paints and provides the blue color in glass and enamels. **Copper** is used to make electrical wires because of its high conductivity. It is a reddish, metallic element that is often used to make inexpensive metal jewelry.

Gold is a precious metal, much of which is kept by countries in reserve as a form of wealth. Coins and jewelry are made from gold. Gold is so dense ($19,300$ kg/m^3) that U.S. \$10M dollars worth of gold takes up less than one cubic foot of volume. Jewelry and dishes made from less expensive metals are often plated in gold to add to their appeal. Gold is also used to plate electronic parts because it is a very good conductor of electricity and does not oxidize as easily as silver or copper. Because it is such an inactive metal and doesn't have any taste, it is used in dentistry and for plating the inner surfaces of expensive drinking goblets and wine glasses.

Lead is the heaviest and softest of the common metals. Lead resists corrosion and is used in storage batteries, paint, glass, solder, and explosives. **Silver** is a white, metallic element that has the highest thermal and electric conductivity of any substance. Silver is used to make high melting point solder, jewelry, coins, silverware, hollowware (cups, chalices, and bowls), and electronic components. Internationally, about 70% of silver produced is used for coins. **Nickel** is a hard silvery-white metal. It is used in magnets, coins, and stainless steel. It produces a green color in glass. It is also widely used to make many other alloys, such as nickel brasses and bronzes, and alloys with copper, chromium, aluminum, lead, cobalt, silver, and gold. **Zinc** is a lustrous, bluish-white metallic element used in many alloys like brass (copper and zinc) and solder (zinc and tin) and is used for coating (galvanizing) iron and steel to protect them from rusting, electrical fuses, and casting metal parts. Small quantities are also used in some medicines and chemicals.

Properties of Metal The most important properties of metals include density, toughness, impact resistance, high strength, and the ability to be deformed without breaking. Metals generally conduct heat and electricity well. Metals are opaque, solid (except mercury, which only becomes solid at about −38 degrees F), dense, and have a luster. Many are very strong, yet they can be formed into a variety of shapes. As mentioned earlier, some metals are magnetic. At a temperature of absolute zero (273.15° C and −459.67° F), certain metals and metallic compounds become superconductors (where their electrical resistance drops to zero).

Plastic

Plastic has become one of the industrial world's most commonly used materials. The United States produces over 100 billion pounds of plastics yearly. It is an extremely versatile product that can be made into films, fibers, and solids and inexpensively molded into a huge variety of products and shapes. The term "plastic" comes from the fact that the material has high plasticity—that is, it can be formed into shapes without breaking. Because it does not degrade and decompose naturally, plastic disposal raises environmental concerns. One third of all plastics is used in packaging. Since packages are used for only a short period of time before they are discarded, plastic packaging comprises a large amount of environmental waste.

Structure of Plastics Plastics are polymers. The word polymer comes from the Greek word *polumeres* and means "many parts." Polymers are chemical compounds that include long chains of monomers. A monomer is a small molecule (one part) that can become part of a larger molecule.

There are naturally occurring and synthetic polymers. Naturally occurring polymers include lignum in wood, proteins, starches, and cellulose. Plastics are synthetic polymers.

Polymers are built from carbon atoms bonded to each other and to atoms of other elements, primarily hydrogen. Polymers are light, resist corrosion, and can be economically made into a wide variety of products. They resist the flow of electricity and heat and are therefore good insulators. Although most polymers do not have high strength if bent, they can be made stronger by adding other materials, such as glass fibers.

As an example of how a polymer is formed, consider the plastic polyethylene. Polyethylene is made up of many ethylene molecules. The chemical formula for ethylene is C_2H_4 (see Figure 4.19). The formula means that two carbon atoms are bonded to four hydrogen atoms. Polyethylene is a long chain of ethylene molecules (see Figure 4.20).

Figure 4.19 | An ethylene molecule.

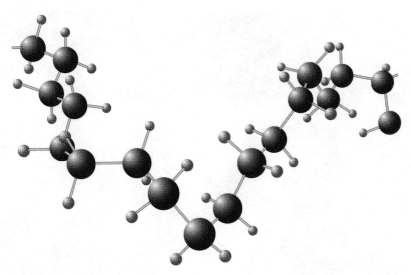

Figure 4.20 | Polyethylene.

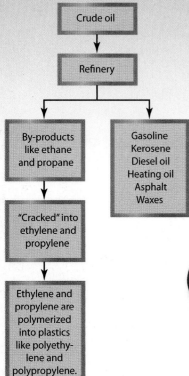

Figure 4.21 | Flowchart of crude oil to plastic.

Most plastic is made from petroleum. About 10% of the oil produced in the United States, or about two million barrels of oil per day, is used to produce plastic.

When crude oil comes out of the ground it must be processed (refined) to make it useable (see Figure 4.21). The processing is done in huge oil refineries which remove non-hydrocarbon substances from the oil and separate the crude oil into oils of different viscosities (thicknesses). The lighter oils are used for gasoline and kerosene; the heavier oils are used to make waxes and lubricating oils.

Refinery by-products of oil (and natural gas) are processed into ethane and propane, which are used as industrial feedstocks (raw materials) for plastic production. The complex molecules of ethane and propane are broken (cracked) into simpler ethylene and propylene molecules. In a process called polymerization, thousands of ethylene and propylene molecules join together with oxygen under heat and pressure to make plastics like polyethylene (which is what plastic cling wrap is made from) and polypropylene (used to make plastic containers and rope).

TECHNOLOGY AND PEOPLE:
Cassandra Fraser

Dr. Cassandra Fraser is a chemist at the University of Virginia who specializes in building polymers, large molecules consisting of repeated chemical units that are strung together in a line or that radiate from a central core (see Figure 4.22). Polymers are the stuff of everyday life—plastic bags, compact disks, and Frisbees are all made of polymers—but they are also the basis of human life. Proteins, the building blocks of living organisms, are essentially polymerized amino acids, and DNA is the macromolecule containing our genetic code.

Fraser is conducting pioneering research on advanced biomaterials that use metals as the hub of polymer chains. "Metals offer many different possibilities for molecular structure and bonding," Fraser notes. "You can, for instance, design a metal-centered polymer in which particular segments come off in response to events in the environment." Using a library of polymeric chains and the metals found in the periodic table, Fraser has devised a method for producing macromolecules that can glow (fluoresce) or alter their structure and color in response to changes in their surroundings. These qualities make them valuable as vehicles for targeted delivery and triggered release of drugs, and as probes in biological research, as well.

Figure 4.22 | Cassandra Fraser.

Types of Plastic Plastic materials are categorized as thermoplastic or thermoset. A thermoplastic material can be softened by heat, and then hardened again by cooling. Thermoplastic materials can thus be repeatedly heated, softened, and shaped. This gives them the important advantage of being recyclable. Thermoset plastics include bakelite and polyurethane. Once they are heated and molded, they cannot revert back to their original form. The most common plastics are thermoplastics and include polyethylene, polypropylene, polyvinyl chloride (PVC), polystyrene, and polymethyl methacrylate (acrylic plastic, often called Plexiglas). Nylon is a common polymer that is used both as a fiber and as a solid thermoplastic.

Uses of Plastics Plastics play an increasingly important role in our lives. Plastics are economical, easy to maintain, lightweight, stand up to hard use over time, resist corrosion, and can be made to look very attractive. The material has an enormous number of uses in such applications as automobiles, consumer electronics and computers, household appliances, furniture, industrial parts, packaging, and toys. Plastic is used in the construction industry as insulation and vinyl siding.

Figure 4.23 | Polyethylene bags.

Polyethylene is used to make soft drink bottles, plastic bags, cellophane wrap, and squeeze bottles (see Figure 4.23). Polypropylene has a higher melting point than polyethylene and is therefore often used for dishwasher-safe food containers (see Figure 4.24). PVC is used to make water pipes for residential and commercial construction (see Figure 4.25). Polystyrene, in its solid form, is a colorless hard plastic used for producing small plastic parts like cutlery, CD cases, and models; but it is more often mixed with a gas to form expanded polystyrene, known by its trade name, Styrofoam® (see Figure 4.26). Styrofoam® is polystyrene that is injected with a blowing agent (typically a gas), causing gas-filled pockets to occur within the plastic. Gasses, like air, do not transfer heat very well because the molecules are loosely packed. Therefore, Styrofoam® is an excellent insulation material and is used extensively in the building trades. Polyurethane is available in different hardnesses and can be used in products as diverse as pencil erasers and bowling balls. As a liquid resin, polyurethane is used as a hard, tough wood finish (see Figure 4.27).

Figure 4.24 | Polypropylene products.

Figure 4.25 | PVC pipe.

Figure 4.26 | Expanded polystyrene.

Figure 4.27 | Polyurethane.

Properties of Plastic Plastics are generally good heat and electrical insulators; they are strong, lighter than many materials of comparable strength, and can be formed and shaped easily. Some plastics (like low density polyethylene) are extremely strong, yet are very flexible. Under load or heat, all plastics will creep (deform).

Plastics have densities that are less than metals, greater than most woods, and similar to crude oil. Polyethylene, for example has a density of 920 kg/m^3

and polyurethane has a density of 1,000 kg/m³. Pine wood has a density of 650 kg/m³, and aluminum has a density of 2643 kg/m³. Crude oil has a density of about 900 kg/m³.

Plastic can be made transparent, translucent, or opaque; some plastics have very high resistance to chemicals, and one, polytetrafluoroethylene (better known as Teflon®), is virtually insoluble in all known solvents.

Ceramics

Ceramic materials are made from clay and other nonmetallic inorganic materials such as sand. Inorganic materials come from nonliving substances (minerals). The word *ceramics* comes from the Greek word *keramos*, which means pottery. The global production of ceramic products is about 150 million tons yearly. China is the world's largest producer, contributing about two thirds of the global supply.

Structure of Ceramics Ceramic materials are normally crystalline substances, which means they have an atomic structure that forms a 3-D geometric shape like a cubic, rectangular, or hexagonal prism. They are compounds formed by combinations of metallic and nonmetallic elements such silicon and oxygen (silicon dioxide, the principal component of sand); aluminum and oxygen (alumina); and calcium and oxygen (quicklime).

Silicon is an example of a crystalline ceramic material (see Figure 4.28). One silicon atom bonds to four other silicon atoms forming a regular pattern. Integrated circuit chips are made primarily of silicon.

Glass, on the other hand is a noncrystalline ceramic. When it cools from a molten state, it cools so quickly that it becomes solid before crystals can form. The atoms are also not as closely packed as they would be in a crystalline substance, meaning that glass is less dense than a comparable ceramic crystal (see Figure 4.29).

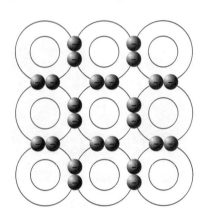

Figure 4.28 | Silicon crystal.

It is generally accepted that the longest word listed in major English language dictionaries is the word pneumonoultramicroscopicsilicovolcanokoniosis. It is a 45-letter word that refers to the black lung disease that affects miners when they breathe in fine particles of silicon.

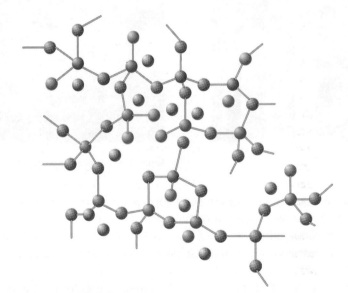

Figure 4.29 | Atomic structure of glass.

Uses of Ceramic Materials Ceramic materials are used to make bricks, dishes, bathroom and kitchen fixtures, electrical insulators, glass, sandpaper and other abrasives, and concrete. Glass can be produced in many varieties from inexpensive machine-blown glass used in soda bottles, to sheet glass used for plate glass windows, to high quality ground glass used for optical lenses. A class of ceramics called *advanced ceramics* can withstand very high temperatures and resists corrosion. These ceramics are used to make automobile engine parts and catalytic converters, cutting tools, and special magnets.

Properties of Ceramic Materials Ceramic materials have high melting points; high strength and hardness; are resistant to wear and corrosion; are very brittle; can be transparent, translucent, or opaque; and have fairly low densities. Clay has a density of about 1,300 kg/m^3 as compared to steel, which has a density of 7,850 kg/m^3.

Figure 4.30 | Examples of ceramics at a Chinese exposition.

Figure 4.31 | Collage of ceramic products.

Interestingly, ceramic materials can be produced with special properties. Although most ceramics are excellent electrical insulators, some can be electrically polarized and are used to make random access memory (RAM) chips for computer memory. Ceramics can even be magnetic—special ceramics are used as ferrite cores in electronic circuit coils. Some ceramic materials generate an electrical voltage when subjected to mechanical pressure. Called piezoelectric, these materials are commonly used as fire lighters that generate the sparks necessary to ignite gas barbecue grills.

Figure 4.32 | Chevrolet corvette.

Composite Materials Composite materials are made from a combination of two or more materials with different properties. The materials making up the composite do not fully blend as they do in an alloy; materials in a composite retain their separate identities. The advantage of producing and using composites is that the combination of materials produces a new material with improved properties. They are often stronger, stiffer, and lighter than the materials they replace. Composites can be more expensive to produce, however, and their cost is a factor in determining their use.

Composite materials are currently being engineered on a large scale, although some examples of composite materials are found in nature. Our bones and teeth are composites; they contain mineral crystals and organic molecules. Wood is also a composite material; it is made of cellulose fibers held together by a lignin, an organic polymer. Early examples of engineered composites are bricks made of mud and straw. A brick made of mud can break if it is under tension (pulled upon), but it is fairly strong when it is under compression. Adding straw, which is strong when pulled upon, makes a composite brick that has reasonably good compressive strength and tensile strength.

Fiberglass was the first modern composite, developed in the late 1930s. Fiberglass uses a polymer material that in reinforced with extremely fine glass fibers. It is a very commonly used composite material used to strengthen molded plastic parts in airplanes, cars, and boats. It is also used to make surfboards, and it is widely used as insulation in buildings.

Figure 4.33 | Glass fibers are used with a wide variety of resins to make automotive body parts, pick-up truck caps, bathtub and shower units, boats, and swimming pools. Glass fiber composites have high strength to weight ratios and impact resistance, resist corrosion, and can have very smooth surfaces.

Figure 4.34 | Boeing 787 Dreamline.

787 Construction Materials

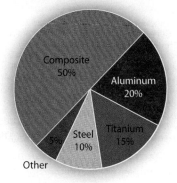

Figure 4.35 | Composite materials are used extensively in the Boeing 787.

An example of effective composite use can be seen in the new Boeing 787 Dreamliner, an airplane that can carry up to 300 people (see Figure 4.34). Safety, production costs, and fuel savings outweigh the cost of using composites, and more than 50% of the airframe is comprised of carbon fiber composites (see Figure 4.35). The composite material is lightweight, flexible, and is stronger than the aluminum it replaces. Typically, about 50,000 rivets and fasteners would be needed for an aircraft like the 787. Using composites allows the frame sections that used to be connected together by fasteners to be joined together end-to-end. This reduces the number of fasteners by 80%.

Composites are formed by using materials of two types: one type is called the matrix material; the other type is called the reinforcement material. The matrix material is a binder that envelops the stronger reinforcement material. In a mud brick, the mud is the matrix material and the straw is the reinforcement material. In concrete (made of cement and stones), the cement is the matrix, the stones are the reinforcement. In wood, the cellulose is the matrix and the lignin is the reinforcement. In fiberglass, the matrix is a polymer and the glass fibers are the reinforcement.

Materials engineers can design composites with enhanced properties by selecting the reinforcement and matrix materials with those new properties in mind. For example, when thermosetting plastics are used as a matrix, they contribute high durability and chemical resistance. When ceramic materials are used as the matrix, the new composite material can withstand high temperatures. Since carbon fibers are very strong, they are used as reinforcement materials in airplanes, tennis rackets, and golf clubs.

Composites are categorized into three matrix types: Polymer matrix composites, metallic matrix composites, and ceramic matrix composites. Polymer matrix composites consist of strong fibers embedded in a resilient plastic that holds them in place. These are used in lightweight bicycles and boats that need to resist corrosion from salt water. Metallic matrix composites have very high temperature limits and are tough and strong. They are used on the skin of hypersonic aircraft that fly at up to five times the speed of sound (Mach 5, or 3,000 mph). Ceramic matrix composites are used in advanced engines and allow the engines to operate at a higher temperature so that cooling fluids can eventually be eliminated. The major difficulties associated with this promising composite material are that it is both brittle and expensive to produce.

Over recent decades, many new composites have been developed, some with very valuable properties. By carefully choosing the reinforcement, the matrix, and the manufacturing process that brings them together, engineers can tailor the properties to meet specific requirements. They can, for example, make the composite sheet very strong in one direction by aligning the fibers that way, but weaker in another direction where strength is not so important. They can also select properties such as resistance to heat, chemicals, and weathering by choosing an appropriate matrix material.

Figure 4.36 | Mine hunter ship with composite hull.

SECTION ONE FEEDBACK >

1. Explain the difference between organic and inorganic materials and give two examples of each type.
2. How would you typically distinguish between hardwood and softwood trees in a forest?
3. Describe the molecular structure of a polymer.
4. What is the difference between thermoplastics and thermoset plastics?
5. Give an example of an application in which a ceramic material is preferred to metal. Next, explain why the ceramic metal is a better choice.

KEY IDEAS >

- Materials are chosen based on properties that make them appropriate for particular applications.

- The properties of solid materials can be grouped into the following categories: Physical, mechanical, electrical, magnetic, thermal, optical, and acoustic.

- Mechanical properties include a material's strength and other properties, such as its hardness, ductility, elasticity, plasticity, brittleness, toughness, and malleability.

Figure 4.37 | Forging metal.

Materials are chosen based on their properties, which make them appropriate for particular applications. For example, copper, which conducts electricity easily, might be appropriate for electrical wires, but not for making twist drills, because it is so soft. A spring that reverts to its original length after being stretched would not be made from the same material as a metal rod that would be stretched into wire.

Natural materials, those found in nature, have been used for millennia for tools and building materials based upon their inherent properties and availability. As life became more technologically based and materials were needed for more complex products, synthetic materials with improved properties were developed. These were engineered to meet specific needs.

The properties of materials can be changed by treating them in various ways. For example, adding carbon to steel can increase its strength, and heating metal can soften it. In this section, you'll learn how the properties of materials influence their appropriateness for specific technological applications.

Categorizing Properties of Materials

The properties of solid materials can be grouped into the following categories: Physical, mechanical, electrical, magnetic, thermal, optical, and acoustic. The following discussion describes these properties in detail.

Physical Properties Physical properties refer to the density and to the freezing, melting, and boiling points of materials. Density is a measure of how tightly the atoms of the material are packed together. Metals are generally more dense than polymers; polymers are generally more dense than ceramics; ceramics are generally more dense than woods.

Density is a measure of how compact a substance is and relates to how much of the mass of the material can be packed into a particular volume. The symbol used to express density is the Greek letter Rho (ρ). The formula for density looks like this:

$$\text{density} = \text{mass/volume} \ (\rho = m/v)$$

TECHNOLOGY IN THE REAL WORLD:
Freezing, Melting, and Boiling Points

All materials' freezing, melting, and boiling points exist on a temperature continuum (see Figure 4.38). When most solids (with the exception of polymers) are heated to their melting points, the change from solid to liquid is abrupt and rapid. When solder (an alloy that is used to join metal parts) reaches its melting point, for example, it flows suddenly. Likewise, when molten solder is allowed to cool, it suddenly solidifies upon reaching its freezing point. If the liquid is allowed to cool, it becomes solid again (it reaches its freezing point).

Freezing results in a release of heat known as the heat of fusion. For most materials, the melting point temperature and the freezing point temperatures are the same. At that point, an increase in temperature will cause the material to melt; a decrease will cause it to freeze. For example, if water is at its freezing point (at that point it will be ice) and gets warmer, it becomes liquid water. If ice is at its melting point (at that point it will be water) and gets cooler, it becomes solid ice. When a liquid is heated to its boiling point, it turns to a gas.

Polymers don't transition abruptly from solid to liquid when they are heated. When a polymer melts, there is usually a sizeable temperature range over which its viscosity changes from high to low as it changes from solid to liquid.

Glass behaves in a different manner than do most solids. When glass is heated to high temperatures, it gradually becomes softer, achieving an increasingly lower viscosity (thickness), but there is no abrupt change of state from solid to liquid. When molten glass is cooled, it gradually becomes more viscous until it appears solid.

Upon reaching the solid state, glass continues to become more viscous, but at a much slower rate. When glass has been cooled below its set point, it still remains theoretically "liquid." For this reason, glass is often referred to as an undercooled or supercooled liquid.

In addition, most liquids form crystalline structures when cooled. That is, their molecules align in predictable, rigid networks. Glass molecules, however, form random networks and do not crystallize. The fact that molten glass cools into a noncrystalline solid is a distinguishing characteristic of the material.

Figure 4.38 | Solid to liquid to gas continuum.

Mechanical Properties Mechanical properties are those that influence a material's ability to endure and withstand applied forces. These properties define the strength of a material, and govern how easily it can be formed into a shape. These properties include a material's strength, as well as other properties such as its hardness, ductility, elasticity, plasticity, brittleness, toughness, and malleability.

Strength of Materials When we talk about the strength of a material, we are referring to the amount of stress it can withstand. The compressive strength of a material is the maximum stress a material can withstand when a load is placed upon it. Compressive strength is found by dividing the load by the cross sectional area of the material. An example of a compressive force is when a column that holds up a bridge is subjected to the heavy load of the bridge and the traffic. Compressive forces tend to shorten an object.

Tensile forces tend to lengthen an object. They stretch a material (as when you stretch a piece of rubber). The tensile strength of a material is the maximum amount of tension (stretching) that the material can withstand before it fails. We can measure tensile strength in two ways: tensile yield strength and the ultimate tensile strength. The tensile yield strength is a measure of the stress at which the material becomes permanently deformed. The ultimate tensile strength is a measure of the stress at which the material will fail.

Measuring Tensile and Compressive Strength

▷ Tensile and compressive strength are measured in pounds/square inch (psi) in the English system, and in newtons/square meter in the SI system. One newton/square meter is called one pascal, after the seventeenth century French scientist, Blaise Pascal, who contributed a great deal to the understanding of pressure and vacuum. A more common measure of strength is the megapascal (MPa) which is 10^6 pascals or about 145 pounds/square inch. Ceramic materials like concrete can have large compressive strength but often have low tensile strength. Steels and certain synthetic polymers have very high tensile strengths. Table 4.3 compares the densities, tensile and compressive strength, and stiffness (Young's modulus) of many common materials.

The shear strength of a material is the maximum stress (measured in psi or MPa) required to shear a piece of material so that sheared parts are totally separated from one another. Shear forces tend to cause one part of the sheared object to slide over the other part (as when you shear a bolt or rivet with a cold chisel, as shown in Figure 4.39); if shear forces are applied to a rectangular block, they generally cause the block to deform into a parallelogram.

Flexural strength refers to the amount of stress it takes to bend a material to the point of failure. The flexural strength, also measured in psi or Mpa, actually refers to the stress on the outermost surface of the material being flexed. Flexural forces occur when a material (like a sheet of plastic) is bent. For materials that bend but do not break, the flexural strength, in engineering terms, is reached when the outermost surface is 5% deformed.

A material that won't break easily but will bend has a lower flexural strength than a material that is strong but brittle. When a material is bent, tensile stress occurs on

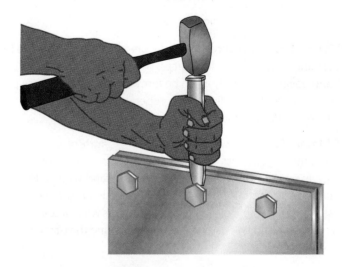

Figure 4.39 | Cold chisel shearing a bolt.

Table 4.3 | Tensile and Compressive Strength Table

MATERIAL	DENSITY (g/cm³)	TENSILE YIELD STRENGTH (MPa)	ULTIMATE TENSILE STRENGTH (MPa)	COMPRESSIVE YIELD STRENGTH (MPa)	YOUNG'S MODULUS (ELASTICITY) GPa
METALS					
Aluminum (pure, annealed)	2.7	N/A	50	N/A	70
Aluminum Alloy 2014-T6	2.8	414	483	470	72.4
Brass	8.498	217	403	N/A	96–110
Bronze	8.3	195	490	245	96–120
Copper 99% Alloy	8.92	33	210	35	110
Iron, cast (ASTM 20)	7.75	151	270	572	152.3
Iron, wrought	7.75	355	400	200	190
High carbon steel (SAE/AISI 1095)	8.03	455	827		195
AISI Type S20910 Stainless Steel, high strength, typical	7.89	1000	1105	1000	200
Titanium, pure	4.5	140	220	145	116
Titanium, iron and aluminum alloy	4.65	1240	1430	1280	110
Tungsten	19.3	750	980	1100	406
CERAMICS					
Aluminum Oxide	3.96	N/A	300	3000	390
Concrete	2.24–2.48	N/A	1.4–7	14–70	18–30
Fiberglass (E-glass)	2.6	N/A	3400	N/A	22–80
Glass	2.53	N/A	50	50	70
Marble	2.8	N/A	14	155	50–100
Quartz	2.5–2.7	N/A	48	1100	75
POLYMERS					
Acrylic (Plexiglas)	1.19	124	80	110	3.1
Kevlar 49	1.44	3000	3600	36	60
Nylon	2–4	45	75	50	2–4
Polypropylene	0.91	12–43	19.7–80	40	1.5–2
Polystyrene	1.05	37	42	70	3.0–3.4
WOODS					
Balsa	.16	N/A	1	1	4.1
Oak	.56	N/A	4.5	8.6	11
Walnut	.55	N/A	4.8	7	9.8
White Pine	.35	N/A	2.1	3.5	8.2
OTHER SELECTED MATERIALS					
Carbon Fiber	1.75–2	3,500	5,650	N/A	230–525
Carbon Nanotube	1.34	~150,000	~200,000	~150,000	>1,000
Diamond	3.5	N/A	N/A	8,680–16,530	1,100
Natural Rubber (high-quality plantation (Hevea) rubber)	.93	20	27	N/A	.01–0.1
Spider silk	1.3	1150	1200	N/A	1–10

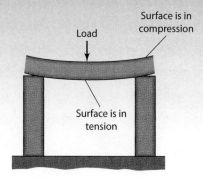

Figure 4.40 | Flexural forces causing surfaces to be in compression and tension.

one surface, and compressive stress occurs on the other. Polymers have low flexural strengths, steels have high flexural strengths.

Torsional strength is the stress that a material can withstand when twisted (like when you twist a piece of licorice). Torsional forces are applied on a cross section of a material.

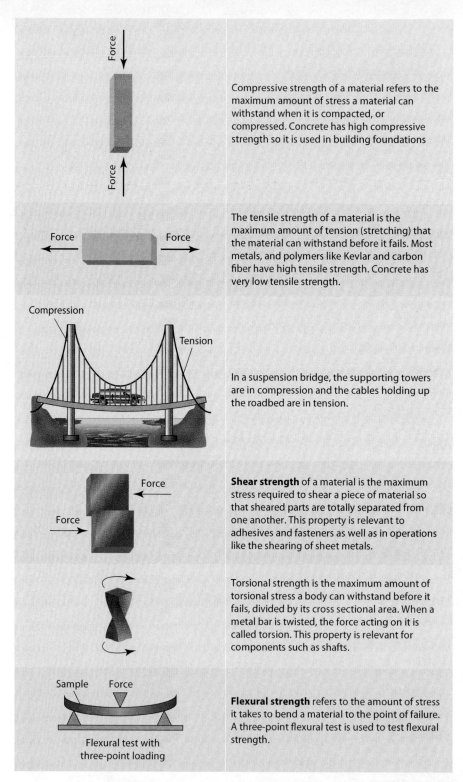

Compressive strength of a material refers to the maximum amount of stress a material can withstand when it is compacted, or compressed. Concrete has high compressive strength so it is used in building foundations

The tensile strength of a material is the maximum amount of tension (stretching) that the material can withstand before it fails. Most metals, and polymers like Kevlar and carbon fiber have high tensile strength. Concrete has very low tensile strength.

In a suspension bridge, the supporting towers are in compression and the cables holding up the roadbed are in tension.

Shear strength of a material is the maximum stress required to shear a piece of material so that sheared parts are totally separated from one another. This property is relevant to adhesives and fasteners as well as in operations like the shearing of sheet metals.

Torsional strength is the maximum amount of torsional stress a body can withstand before it fails, divided by its cross sectional area. When a metal bar is twisted, the force acting on it is called torsion. This property is relevant for components such as shafts.

Flexural strength refers to the amount of stress it takes to bend a material to the point of failure. A three-point flexural test is used to test flexural strength.

Figure 4.41 | Full page of compressive, tensile, shear, torsional, and flexural strength diagrams.

Other Mechanical Properties of Materials The physical and mechanical properties of materials determine how they behave under stress, and, therefore, how they can be formed into various shapes. For example, some materials (like rubber) will yield under even a small tensile stress and will stretch a great deal. Others (like glass) won't stretch very much at all and will break when stressed.

Hardness determines the ability of a material to withstand scratching or penetration. A tool used for cutting must always be harder than the material to be cut. Some very hard materials are diamond and silicon carbide. Steel cutting tools are sometimes fitted with diamond or carbide tips to make them more effective.

Ductility is a measure of how much a material will yield before breaking. Metals such as copper and brass can be stretched (drawn) into wires because they are ductile. Ductility is often expressed as a percentage of elongation. A ductile material, like a piece of soft clay, can be stretched without having to apply high tensile forces. A material with high tensile strength requires a large tensile stress to stretch it; therefore materials with high tensile strength are not as ductile as materials with low tensile strength.

A property of materials closely related to ductility is the material's elasticity. As ductility refers to how a material can be stretched without breaking, its elasticity refers to how much it can be stretched without being permanently deformed. Elasticity is the ability a material has to stretch under load but then return to its original shape and length. A rubber band is very elastic, as it can stretch up to 500% of its original length, but the deformation is temporary, because the rubber band returns to its original length when the load is removed. Elastic materials can be stretched only so much before they reach a point where they won't go back to their original length; at that point they stay stretched. This point is called the elastic limit. The elastic limit is the maximum stress to which a material could be subjected and still be able to return to its original shape when the load is released. Different materials have different elasticities and therefore are stiffer or less stiff. For example, steel is elastic, but not as elastic as nylon.

Young's Modulus

▷ A simple measure of elastic strength (the stiffness of a material) is Young's modulus

(E). It can be found by dividing the stress applied to a material by the amount the

material moves (called the strain) when subjected to that stress. Stress is the force that

is applied over a particular cross sectional area of the material; the strain is the fractional

change in length that results. Therefore:

$$E \equiv \frac{\text{tensile stress}}{\text{tensile strain}} = \frac{F/A}{\Delta L/L_0}$$

where

E is the Young's modulus (modulus of elasticity) measured in pascals;

F is the force applied to the object;

A is the cross-sectional area through which the force is applied;

ΔL is the amount by which the length of the object changes;

L_0 is the original length of the object.

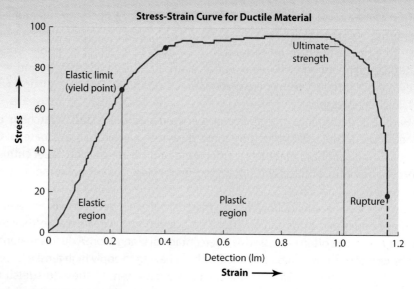

Stress-Strain Curve for Ductile Material

Stress

100

80 — Elastic limit (yield point)

Ultimate strength

60

40

20 — Elastic region

Plastic region

Rupture

0

0   0.2   0.4   0.6   0.8   1.0   1.2

Detection (lm)

Strain ⟶

Figure 4.42 | Stress-strain curve for ductile materials.

Once a ductile material has reached its elastic limit, it can continue to deform, however, because of the property called plasticity. Plasticity is the property in a material which permits permanent deformation to occur before it ruptures. Plasticity allows the material to continue to deform under load until it reaches the breaking point. Materials commonly called plastics (like Plexiglas) have the quality of plasticity. Putty and plasticine (modeling clay) have high plasticity.

Yield point is an engineering term that refers to the amount of stress applied to a material when it starts to deform plastically. Before the yield point is reached, the material will be in an elastic range and will not be permanently deformed. Once the yield point is reached, the deformation will be permanent (see Figure 4.42).

Brittleness determines how easily a material under stress will fracture without significant deformation. Glass, cast iron, and ceramic materials (like concrete) are brittle. Generally, brittle materials are high in compressive strength but low in tensile strength. For example, cast iron will elongate only slightly under tensile stress. Once the yield point is reached, cast iron can fail suddenly, demonstrating its brittleness.

Toughness is a measure of how a material can survive a sudden impact without fracturing. A material that is tough will require a lot of energy to break it. A material can be strong (can withstand a static load) but not tough (cannot withstand an impact). Ceramic materials are strong (in compression) but not tough. Rubber is tough, but not strong; stainless steel is strong and tough.

Malleability is the property that determines how well a material can be hammered and pressed into shape. Malleable materials behave oppositely to brittle materials. A malleable material, like copper, can be formed by using methods that are unsuitable for forming brittle materials. For example, U.S. pennies were once made of mostly copper but to reduce manufacturing costs, pennies minted in 1983 or later are made of 97.5% zinc, with a thin copper coating. Copper and zinc are both malleable metals. To manufacture the pennies, metal blanks are stamped with designs and inscriptions by presses that exert 35 tons of pressure on the coin. More brittle metals like cast iron must be melted and poured into a mold to be used to make products.

Electrical, Magnetic, Thermal, Optical, and Acoustic Properties of Materials

Besides mechanical properties, materials also have unique properties that cause them to react differently to electricity, magnetism, heat, light, and sound.

Electrical Properties Electrical properties include conductivity (signified by σ, the Greek letter sigma); and resistivity (signified by ρ, the Greek letter rho). Note that the symbol *rho* is used to represent both density and resistivity. This is a potential source of confusion, but the meaning of this symbol is usually clear because of the context within which the symbol is used.

Conductivity is a measure of how a particular material allows electrons to flow through it; resistivity is a measure of how a material restricts the flow of electrons. Conductivity and resistivity are inversely related, that is, the higher the conductivity, the lower the resistivity or: σ = 1/ρ. In practical terms, this means that materials with low resistivity are better conductors.

Resistivity is not exactly the same as electrical resistance. Resistivity is a function of the material from which something is made and the temperature of the material. Normally (but not always), if a material is heated, its resistivity increases. When we speak of resistivity, we must speak of it at a particular temperature. Copper is copper; a one-mile length of 18 gauge copper wire (0.0403 inches in diameter) has the same resistivity as a one-inch length provided they are both at the same temperature.

Resistance, on the other hand is a measure of the opposition to current flow of a specific piece of a material. Resistance is measured in ohms (Ω, the Greek letter Omega), after the German Physicist Georg Simon Ohm. One thousand feet of #18 gauge copper wire has a resistance of 6.385 ohms.

The relationship between resistivity and resistance is expressed by the equation:

$$\rho = R\frac{A}{\ell}$$

where

ρ is the resistivity (measured in ohm metres, Ωm);
R is the resistance of the material (measured in ohms, Ω);
ℓ is the length of the specimen (measured in meters, m);
A is the cross-sectional area of the specimen (measured in square metres, m²).

Different materials have different resistivities. Materials with lower resistivities have less electrical resistance and are therefore better electrical conductors. The material with the lowest resistivity is silver. Since plastics and rubber have high resistivities, they are widely used as electrical insulators. Nichrome wire has high resistivity and its opposition to electron flow causes heat. That is why nichrome wire is used in toasters. Table 4.4 lists resistivity values for various materials at temperatures of 20° C.

Table 4.4 | Table of Resistivities

TYPE OF MATERIAL	RESISTIVITY IN OHM-METERS
Silver	1.59×10^{-8}
Copper	1.7×10^{-8}
Gold	2.4×10^{-8}
Aluminum	2.8×10^{-8}
Tungsten	4.6×10^{-8}
Nichrome	150×10^{-8}
Polystyrene	$10^{7} - 10^{11}$
Polyethylene	$10^{8} - 10^{9}$
Rubber	10^{13}

Magnetic Properties All materials exhibit some magnetic effects because the electrons in atoms of matter spin and create tiny magnetic fields. Magnetic properties are classified as diamagnetic, paramagnetic, or ferromagnetic. Diamagnetic materials exhibit extremely weak magnetism that exists only in the presence of a magnetic field. Most often in materials, electrons are paired and the individual electrons spin in opposite directions. In an electron pair, the spins cancel each other's magnetic fields. Diamagnetic materials have unpaired electrons and therefore a weak magnetic field will exist. These materials will very slightly repel an external magnetic field. Practically, diamagnetic materials are so weak that we can regard them as non-magnetic. Examples are water, oil, and gold.

Paramagnetic materials (like aluminum, barium, calcium, sodium, and iron oxide) will become weakly magnetized in a magnetic field. However they do not retain their magnetism when the magnetic field is removed.

The materials that exhibit the strongest magnetic properties are ferromagnetic materials. In ferromagnetic materials, the unpaired electrons organize themselves into domains with strong magnetic fields. An external magnetic field causes the domains to align and create a magnetized material. Ferromagnetic materials include iron, nickel, and cobalt.

All ferromagnetic materials will lose their magnetism when heated. The temperature at which a ferromagnetic material loses its magnetism is called the Curie point. For iron, the Curie point is about 1,400° F.

Magnetic materials that retain their magnetism have high retentivity. This is desirable for permanent magnets that are used for electric motors, loudspeakers, hard drives, magnetic recording tape, and for separation of magnetic minerals from nonmagnetic particles.

Electromagnets use soft iron for the core around which wire is coiled. Soft iron has very low retentivity; this means that, when the electric current (the electric field) is switched off, the device no longer acts as a magnet. Automobile salvage yards use this magnetic property in the large electromagnets on cranes used to pick up heavy autos. When the operator turns off the electric current, the automobile drops into the desired place. Good magnets are made from materials that easily conduct magnetic lines of force. This ability is called permeability and is similar to conductance in electrical circuits.

Thermal Properties Thermal properties refer to a material's ability to conduct heat, and to expand when heated. The ability to conduct heat is referred to as thermal conductivity. The ability to expand is referred to as the material's coefficient of thermal expansion. Most materials expand when heated; some, like water, expand when cooled (water turns to ice) and therefore have a negative temperature coefficient.

If you examine the thermal expansion coefficients in Table 4.5, you will find that steel expands more than glass. That is why if you have a steel lid on a jar and run it under hot water, the lid expands more than the jar, making it easier for you to loosen the top. Additionally, the thermal conductivity of the steel is much greater than that of glass, so the top gets warmer more quickly. In constructing systems with tightly fitted parts, it is important that the materials of each part are the same. Otherwise, the coefficients of thermal expansion would be different, and the parts would not fit together properly. In designing bridges, steel structures, and pipelines to be used in environments where temperatures vary considerably, the thermal properties of the materials are taken into account by engineers.

All materials have a temperature limit. The temperature limit (not to be confused with the melting point) is the highest or lowest temperatures at which the material would retain its properties. For example, most metals have upper temperature

Advanced magnetic materials are being used in an innovative experimental way for cooling. All atoms vibrate and create some measure of heat, but an external magnetic field can align the atoms and reduce the heat buildup. The strong magnetic field cools an alloy of gadolinium, silicon, and germanium that show a very high response to a magnetic field. The material cools water flowing around it. The water replaces the fluorocarbons used in typical refrigerators making this new process one that may have positive environmental consequences.

Table 4.5 | Thermal Properties of Some Common Materials

MATERIAL	THERMAL CONDUCTIVITY (W/m · K) AT 25° C	THERMAL EXPANSION COEFFICIENT (m/m · K × 10^{-6})
Air	.024	3.67
Aluminum	237	23.0
Brass	109	19.1–21.2
Brick	0.69	4.00–7.00
Concrete	0.42	7.00–14.0
Copper	410	16.6–17.6
Diamond	900–2320	1
Glass	1.05	4.00
Gold	317	14
Iron (Cast)	55	9.90–12.0
Magnesium	156	24.2
Nickel ,	90.7	13.0
Nylon	0.25	74.0–100
Polyethylene	0.42	180–200
Polypropylene	0.1–0.22	200
Polystyrene (expanded)	0.03	50–70
Polyvinylchloride (PVC)	0.19	70
Platinum	71.6	9
Rubber	0.16	77
Silicon	148	3
Silver	429	19.2
Steel	46	10.0–18.0
Tungsten	174	4.30
Wood	0.04 (balsa)–0.17 (oak)	1.1
Zinc	116	30.2

limits between 400° C and 800° C. The strength of these metals drops quickly as they approach the temperature limit.

Materials have different thermal properties. Materials with higher thermal expansion coefficients expand most when heated. Materials with higher thermal conductivities are good conductors of heat. Those with low thermal conductivities are good heat insulators. From the table above, it is apparent that among the materials listed, air is the best insulator and expanded polystyrene is also very good because it is filled with air pockets. They are better insulators than glass or wood. It appears from Table 4.5 that most metals are good conductors of heat, but diamond is much better. The materials in Table 4.5 that expand the most with heat are plastics; the ones that expand the least are diamond and wood.

Optical Properties For some applications, materials are chosen on the basis of their optical properties, primarily their ability to reflect, absorb, refract, or transmit light. Some materials, like bright white glossy photography paper, can reflect 90% or more of the light that strikes it (incident light). Some materials, such as black rubber, reflect only a small percentage of incident light (about 2%).

Mirrors make use of the properties of reflectance of metals like aluminum or silver. Very thin coatings of these metals are deposited on another material, generally glass, with a uniform surface. Aluminum is a common coating because it is inexpensive and provides good reflectivity. Silver is even more reflective, but more expensive. In cases where the quality of the mirror must be exceptionally high, for example in high-powered telescopes, liquid metals are heated to the point where they vaporize. The vapor condenses on a sheet of glass, providing an extremely high reflective surface.

A material can absorb light of different wavelengths. In an absorbent material, light energy at a particular wavelength is absorbed (converted into heat energy) so that the amount of light reflected by the material is less than the light entering the material. The opposite of *absorption* is *transmittance*, which refers to the amount of light at a particular wavelength that can pass through a material.

Light is refracted (bent) as it travels from one medium to another. The *refraction* takes place at the boundary between the materials. The refraction occurs because light travels faster in some materials than in others. It travels fastest in a vacuum. The ratio of the speed of light (at a particular wavelength) in a vacuum to the speed of light in any other material is called the *refractive index*. If the refractive index is high, it means that the material refracts light more than if the index is low. Eyeglasses are a good example. You often have a choice of buying plastic (polycarbonate) lenses or glass lenses. Most glass lenses have a refractive index of about 1.5–1.6. Plastic lenses have a refractive index of about 1.74, meaning that they can be thinner and lighter in weight than glass lenses.

Acoustic Properties *Acoustic properties* relate to how the velocity of sound changes through various materials. Sound absorption, like light absorption, relates to the sound energy lost as a particular sound wave moves through a material.

The speed at which sound can travel through a material depends upon the material's stiffness and density. Sound travels through stiff materials, like aluminum and glass, faster than it travels through flexible materials like cork or rubber. Materials that slow the speed of sound are used as insulating materials. Engineers calculate the speed of sound in a solid material by using this formula:

$$\overline{v}_{solids} = \sqrt{\frac{E}{\rho}}$$

where v is the velocity of sound in the material, E is Young's Modulus for the material and ρ is its density. Some examples of the speed of sound in some common materials are shown in Table 4.6. Note that the speed of sound increases as the temperature of the material increases. The speeds listed in Table 4.6 presume the various materials are at 25° C.

Table 4.6 | Speed of Sound in Some Common Materials

MATERIAL AT 25° CENTIGRADE	SPEED OF SOUND m/s
Air	330
Cork	500
Water	1,493
Rubber	1,600
Polyethylene	1,900
Nylon	2,300
Oak Wood	3,860
Aluminum	5,100
Steel	5,900
Window Glass	6,800
Diamond	12,000

Materials Science and Engineering

The fields of material science and engineering are bringing about advancements in virtually all areas of technology. In the field of transportation technology, for example, high-tech materials find applications in polymers for car parts and bodies, new hybrid power plants using fuel cells, and tires with improved characteristics.

Materials Science

Materials science primarily investigates the properties of materials and how the atomic structure of a material causes it to behave in the way it does. Materials scientists work in areas that include biomaterials, composites, polymers, ceramics, electronics, metallurgy, photonics, and surface science.

Materials scientists engage in a process called *synthesis,* which can be thought of as the preparation of a new material. During synthesis, design criteria and specifications for the new material are set, basic materials are chosen, and the materials are combined chemically to form a microstructure that is carefully controlled. Once the materials are synthesized, the new material is *characterized* to accurately identify its properties. Characterization helps materials scientists to understand the way the atomic structure contributes to the material's behavior, to suggest the best materials for a particular use, and to improve its performance under specified conditions. After a material has been synthesized and characterized, it is ready to be tested for ultimate use in manufactured products.

Materials Engineering

Materials engineering is concerned with how materials can be designed, put into use, and tested. Materials engineers ensure that engineering materials are fabricated to meet cost constraints, proper compositions and tolerances, and quality requirements.

Materials engineers oversee the processes involved in producing materials. These can be divided into two main categories: primary fabrication and secondary fabrication. Primary fabrication converts raw materials from their mineral or organic origins into industrial materials that would be further processed into products. It is during primary fabrication that most of the characteristics of the microstructure of the material are determined.

The process of converting iron ore into pig iron (raw iron) is an example of primary fabrication. The process of transforming the pig iron into finished products using processes such as rolling, casting, and forging is known as secondary fabrication. Secondary fabrication can have some effect on a material's basic properties, but normally has less of an effect than primary fabrication does. For example, the process of rolling iron plates can harden the iron's surface.

There is no clear line separating the work of materials scientists and materials engineers. In fact, in the high-technology worlds of new composites and nanoscale materials, interdisciplinary teams—often involving materials scientists, chemists, physicists, and engineers—collaborate to better understand microstructures of materials and develop applications for them. In the coming years, materials scientists and engineers will increasingly be relied upon to develop and find uses for materials hitherto unknown. New materials will have to be environmentally friendly, inexpensive to produce, and provide high performance and reliability.

The fields of material science and engineering are sometimes considered one field, which is referred to using the acronym MSE.

1. Give one example each of materials that would be used for a) electrical conductors, b) magnets, and c) acoustic insulation.
2. Explain the difference between tensile yield strength and ultimate tensile strength.
3. Are elastic materials always ductile? Explain your answer.
4. Using the formula $\rho = R\frac{A}{\ell}$ determine the resistance of a 1 mile length of 12-gauge copper wire. Note: One mile = 1,609 meters; the diameter of 12-gauge wire = 0.2117 cm. Note: find the resistivity of copper from Table 4.5.

SECTION 3: Processing Materials

KEY IDEAS >

- The term "processing" means transforming basic (raw) materials into industrial materials, and then into finished products.

- We can categorize processing methods according to whether they produce a mass change, phase change, or structure change. We can also divide processing methods into those that cause deformation of the material (that is, change its shape) and those that consolidate (or combine) materials. New developments in materials processing will be driven by very highly technical processes and the use of hitherto unknown materials.

Figure 4.43 | Coils of steel.

The term processing means transforming basic (raw) materials into industrial materials, and then into finished products. When processing materials, manufacturers must address a host of issues, such as cost, quality, and performance. They must also consider the environmental impact of each processing method. For example, they must consider the energy required to convert a material to a finished product, as well as the disposability of the product after it has fulfilled its intended life cycle.

Material Processing Methods

Raw materials are the starting point. Examples of raw materials are metal ores, clay, natural rubber, and petroleum. These raw materials are processed into industrial material (such as iron and various types of plastics), which are then further processed into end products .

Materials can be processed while they are in solid, liquid, or gaseous states. In the solid state, materials are formed by using forces that place tensile, compressive, shear, flexural, or torsional stresses upon them.

Solid materials can be hard (for example, sheet metal); soft (for example, clay), or powders (for example, the powders from which pills are formed). In the liquid state, materials are poured into molds in a process known as casting. It is less common to process materials in their gaseous state; however, some processes,

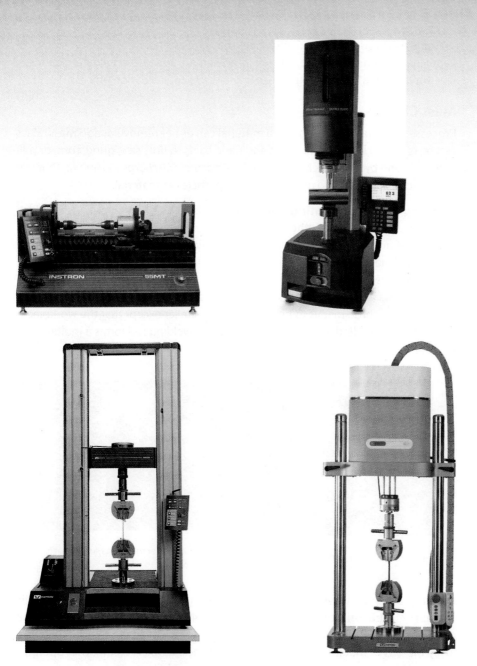

Figure 4.44 | A variety of materials testing machines.

notably *chemical vapor deposition*, vaporize a substance into a gas and deposit a thin film of high purity material onto a surface. One example of the use of these thin films is in the electronics industry, where materials such as silicon are deposited onto semiconductor wafers and then etched to become integrated circuits (ICs).

Categorizing Materials Processing Methods

Manufacturers make use of many different types of processing methods. These methods can be categorized in different ways. We can categorize processing methods according to whether they produce a mass change, phase change, or structure change. (You'll learn what these terms mean shortly.) We can also divide processing methods into those that cause deformation of the material (that is, change in shape), and those that consolidate (or combine) materials.

161

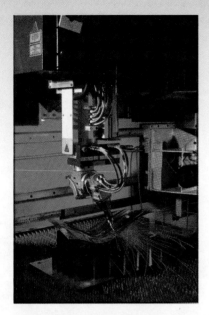

Figure 4.45 | Laser profile cutting.

Some materials processes fit into more than one category. For example, chemical vapor deposition is a phase change process and also a consolidation process. Electroplating is a mass change process and also a consolidation process.

Mass Change

Processes that *change the mass* of a material can either remove or add mass. Mass can be removed by using machine and hand tools to drill, saw, grind, turn, or mill materials; and by using techniques such as *electrical discharge machining* that use high electrical currents to melt away small particles of material.

Cladding and Electroplating Mass can be added by cladding processes (which bond different metals together by pressing them together under high pressure; and by electroplating, which is an electrochemical process that deposits a adds a thin layer of metal onto a *substrate* (underlying material) (see Figure 4.46). Soldering and brazing use heat to join metals by adding a melted filler material (made of a metal that has a lower melting point than the base metals, called solder or braze) that causes the base metals to bond. Welding is similar except that the workpieces themselves are melted at the weld point and the welding rod forms a molten pool of material that joins the parts.

Figure 4.46 | Laser cladding.

Figure 4.47 | Electrical discharge machining.

Figure 4.48 | Drilling.

Drilling Drilling involves using a rotating bit, held in the chuck of a hand drill, portable electric drill, or drill press (see Figure 4.48). The drill bit has a tip made of hardened steel or silicon carbide. The flutes (or grooves) on the drill bit carry away the bits of material that are being cut. A *jackhammer* is a drill that operates pneumatically (uses compressed air) and is used to drill into rock or pavement. It works by hammering a chisel-like tip into the material that the operator wants to break into pieces.

Sawing Sawing is a mass-change process that uses a saw blade with sharp teeth. The blade removes material the size of the saw cut from a larger piece and thus separates material into parts and shapes. Sawing is done with hand saws using straight or curved blades; band saws using an endless steel band having a continuous series of teeth that runs over wheels or pulleys; and circular saws that are machines that use a steel blade that has teeth on its perimeter and is mounted on a rotating shaft.

Figure 4.49 | Bench grinder with pedestal.

Grinding Grinding uses abrasive materials embedded into a solid circular wheel or belt to remove small particles of material. Sanding is similar but uses abrasive materials bonded to a flat sheet of cloth or paper. Grinding and sanding are processes use to smooth the surface of a material. When very tiny abrasive particles are used, the effect is to smooth the surface so much that it becomes polished.

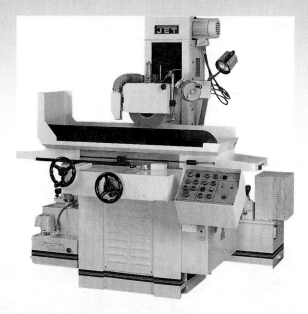

Figure 4.50 | Surface grinder.

Figure 4.51 | Woodworking lathe.

Surface grinders are used for various grinding processes, including truing flat metal stock, squaring material within tenths, sharpening cutting tools, and for sharpening punch and die sets (see Figure 4.50).

Turning Turning is done on a lathe (see Figures 4.51 and 4.52). The workpiece is rotated so that its surface is cut by a stationary cutting tool. The tool can taper the workpiece, reduce its diameter uniformly, or cut grooves into it.

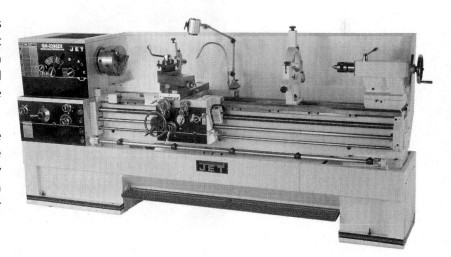

Phase Change

Figure 4.52 | Metalworking lathe.

Processes that *change the phase* of a material make solid parts from liquid or gaseous materials. Some examples of phase-change processes are metal casting, injection molding, and blow molding of plastics. In another interesting phase change process, known as infiltration, liquid metal fills the microscopic pores in the surface of a mold and becomes part of the mold surface when it cools. This process is used in the production of automobile engine parts. *Chemical vapor deposition,* discussed earlier, is another phase-change process. A fascinating example of a phase change process is seen in Chapter 6 in the discussion of zero-energy houses. In this case, a wax (solid) is encapsulated in plastic and used as an energy storage device in a wall. When the temperature is warm (above 75° F) the wax changes phase (melts) and stores the heat energy. When the temperature outside falls, the material releases the stored heat energy and heats the house.

Casting

Molds for casting can be made from ceramic materials, polymers, metals, plaster, rubber, wood, or composites. The mold is an exact negative copy of the object that is to be cast. Often, solid materials (like iron or thermoplastics) must be melted before being cast. Sometimes, as in the case of plaster, slip (liquid clay), and concrete, the materials are mixed cold and poured into molds or forms, where they harden.

Some molds are designed to be used only once. These molds are called expendable molds and are often made of sand, plaster, wax, or clay. After the casting is made, the mold is broken apart to remove the finished product (some plaster molds used for casting clay objects from liquid clay can be used to make multiple castings).

Some molds can be reused many times. These molds are called "nonexpendable molds." Some examples are steel molds used to make parts out of nonferrous metals; molds used for injection molding of plastic; and molds used for pressing and blowing glass (see Figures 4.53 and 4.54).

Figure 4.53 | Pouring iron into a mold to form an ingot.

Figure 4.54 | Glass making.

There are some very exciting new developments in material processing. One is to process materials into products that require very little finishing or post processing. An example is casting products with complex geometries into their final shape. Normally the surfaces of these castings are not finished to perfection and require further heating, smoothing, or machining. The method, referred to as net shaping, reduces the amount of materials used by up to 50% and saves energy needed in the production process because the casting is more perfect and does not require the need for further machining or heating.

Another emerging idea is to use microfactories to process materials. These are small, desktop factories that match the size of the production system to the size of the parts being made. They are linked to engineers, suppliers, and customers over the Internet. They can be easily reconfigured to produce different parts as needed. A *fabber* is a type of microfactory that builds objects by adding plastic materials, layer by layer, similar to the way a layer of ink is printed on paper. New developments in materials processing will be driven by very highly technical processes and the use of new materials.

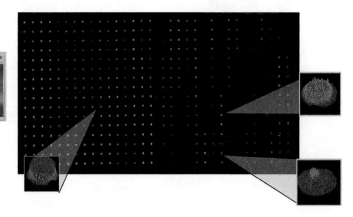

Figure 4.55 | Printing polymers.

Figure 4.56 | This is an assembled Fab@Home Model 1 with a dual syringe deposition system. As of 2008, it sells for about US $3000. It is 18.5 inches wide, 16 inches deep and 18 inches tall. It can be used with a variety of materials that are soft or fluid enough to push through a syringe, yet solid enough to form layers. It can build objects up to 8" x 8" x 6" tall. Model 1 systems have been used by individuals all over the world to build products ranging from plastic parts to complete flashlights, and even 3D chocolate sculptures and cake decorations.

Figure 4.57 | Flashlight "printed" with a fabber.

Pressing Pressing of molten materials is similar to casting, except that after the molten material is poured into a mold, a plunger with the shape of the inside of the object to be produced is lowered into the mold. The material squeezes around the plunger. It takes the form of the mold on the outside and the form of the plunger on the inside. Pressed glassware is made by pressing molten glass into a mold to create the shape and surface design desired.

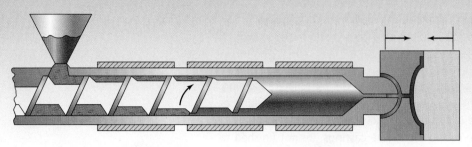

Injection Molding Injection molding is a common industrial process used to manufacture plastic parts to a high degree of accuracy. Plastic pellets of many different colors and types (such as polystyrene, polypropylene, and polyethylene) can be used. The pellets are placed into a hopper and fed into a heated cylinder. A hydraulic plunger or a mechanical screw forces the softened plastic into a mold that is clamped under pressures that can be as high as 1,000 tons or more. Molds can have multiple cavities, allowing many parts to be cast at the same time. The advantage of using a mechanical screw is that it can be controlled to inject a very small amount of the melted plastic and therefore is suitable for making small parts (see Figure 4.58). The sections of the mold are separated once they cool and the plastic parts are then removed. Household items, plastic gears, toys, and LEGO parts are among those that are injection molded.

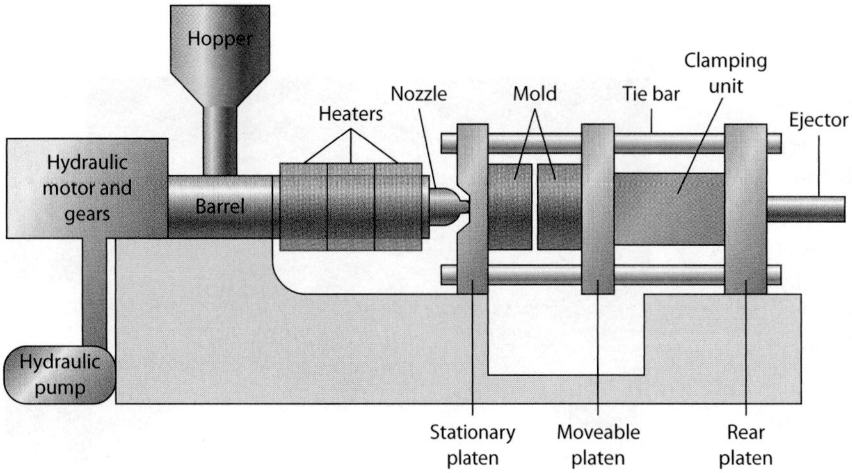

Figure 4.58 | Injection molding using a mechanical screw.

Figure 4.59 | Injection molding machine schematic.

Figure 4.60 | LEGO parts are made using an injection molding process.

Blow Molding and Vacuum Forming In blow molding, a gob of molten plastic (called a *parison*) is enclosed in a metal mold that clamps around it (see Figure 4.61). The parison is inflated by air pressure and takes the shape of the mold. Hollow plastic bottles and containers are made this way. Low air pressure of about 100 psi is used, and therefore the mold clamping pressures are much lower than in injection molding, making this a very cost-effective process.

Vacuum forming is a method of forming a sheet of plastic, usually polystyrene or acrylic, around a model or mold (see Figure 4.62). The plastic sheet is heated until it softens, draped over the model, and a vacuum system is used to pull the sheet tightly down. The process is used to make plastic "blister packaging" that hangs on hooks in retail displays, as well as plastic cups, trays, containers, and many consumer electronic packages. The process uses a vacuum forming machine and is an inexpensive way to make thin-walled plastic parts.

Rotational Molding Rotational molding uses a mold that consists of a mold attached to a motor that rotates very quickly. Molten material is placed in the mold and centrifugal force due to the rapid spinning causes the molten material to be distributed uniformly on the inside of the mold, taking its shape. Hollow glass, metal, and plastic objects are sometimes cast using this process.

Structure-Change Processes

Structure-change processes affect the atomic structure of a material. The change can be throughout its mass or just on the surface. Three processes that affect the material's structure are hardening, tempering, and annealing.

Hardening In the process known as hardening, a material's surface or internal structure is made physically harder. Hardening steel requires heating it to a high temperature and then cooling it rapidly. Heat treating a material changes its structure and tensile and fatigue properties. Essentially, when a material is heated, the heat energy causes the atoms of the substance to move around randomly. If the material is rapidly cooled, the atoms are "frozen" in a particular arrangement. Controlling the heating and cooling temperatures and rates determines the microstructure of the material.

When steel (with greater than 0.03% carbon) is heated to about 1,375° F (746° C), its structure resembles a cube with atoms at the corners and centers of each face. This face-centered cubic has open spaces that contain carbon atoms (see Figure 4.63). At this point, the steel is considered a solid solution known as austenite. The amount of carbon in steel determines the steel's hardness. High-carbon steels (about 0.6% carbon) have higher yield strengths and can be made harder than low-carbon steels (about 0.03% carbon).

When suddenly quenched, the carbon atoms in the steel rearrange themselves but do not have time to completely move out of their crystalline structure. The atomic structure changes to a state called *martensite* (see Figure 4.64). In that state, the steel becomes much harder than steel in the austenitic state. (It should be noted that when nonferrous metals, like copper, are heated to red heat and quenched quickly, they actually soften).

Steel and other metals, particularly nonferrous metals, can also be *work-hardened* by deforming the material plastically when working on it. The material can be rolled, pressed, hammered, or bent. The deformation causes irregularities to occur in the crystalline structure of the material. These cause stress to build up in the material, hardening it and reducing its ductility.

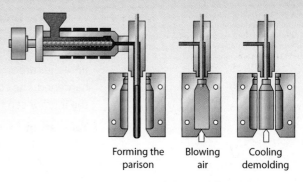

| Forming the parison | Blowing air | Cooling demolding |

Figure 4.61 | Blow molding.

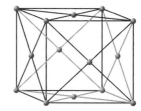

Figure 4.62 | Pills are packaged using a vacuum forming process.

Figure 4.63 | Face-centered cubic.

Figure 4.64 | Martensite structure.

Surface hardening (or *case hardening*) is a process where steel or iron is heated while its surface is in contact with a carbon-based material. The steel is put into a case and carbon-based materials are packed around it. It is then heated in a furnace until the carbon diffuses into the surface of the steel. Only the surface of the steel is hardened. The core is chemically unaffected. The surface is able to resist wear, while the interior can resist impact.

Tempering Martensite-phase steel (steel that is heated and suddenly quenched) is hard, but is extremely brittle and has low ductility. Little or no plastic deformation can take place before it fractures. Tempering can preserve the material's toughness, increase its ductility, and reduce the brittleness. Tempering involves heating hardened steel to a temperature of between 420° to 650° F (216° to 343° C) and soaking it at that temperature for a short time. At the tempering temperature, the carbon particles turn into small carbide particles. The material is then cooled quickly, preserving that new microstructure, with the result that the material is much more able to withstand impact.

Annealing Annealing is a process undertaken to remove stresses in a material. Steel is annealed by heating it above the austenitic point (above 1,375° F or 746° C) and then cooling it slowly. The grain structure becomes more coarse (it turns into a steel called pearlite) and the material then becomes soft and ductile. In annealing, the annealing temperature and the rate of cooling are determined by the material being heat treated.

Annealed materials can be worked much more easily than hardened materials. In fact, steel that will be made into tools is annealed, then shaped, then hardened, then tempered.

To anneal nonferrous materials, such as copper, that have been work hardened, the material would be heated to a red color and then cooled quickly by quenching it in water or oil.

In addition to metals, glass is also heat treated. After glass is melted and formed, it must be heated again to an intermediate temperature (below its melting point) and annealed. It is cooled very slowly to relieve internal stresses. Without this annealing process, the glass could explode into tiny pieces.

Solid-state Phase Change

▷ Normally, the term **phase change** refers to a change in phase from solid to liquid or from liquid to gas. In a **solid-state phase change**, a material remains solid, but its atomic structure rearranges. A unique example and application of solid state phase change is with metals that are called **shape memory alloys**. A shape memory alloy can have one shape at a lower temperature (when it has a martensite cubic structure) and another shape at a higher temperature (when it has an austenitic structure). The shape memory alloy "remembers" the two different shapes that occurred at the two temperatures. By changing the temperature (very often by only a few degrees) the structure of the material, and thus the shape, changes. Using shape memory alloys, eyeglass frames can be totally misshapen and warmed to make them return to their original shape. Shape memory alloys can also be used as actuators. For example, when water reaches a scalding temperature in a bath or shower, a device using a shape memory alloy responds and shuts off the water. They are also used in medical applications as stents (which are small tubes made of wire mesh). These are used to expand arteries to increase the flow of blood. They are inserted when bent, and revert back to their original shape once they reach the body temperature.

Deformation Processes

Deformation processes change the shape of a material *without changing its mass*. Deformation processes include forging, rolling, machine pressing, and drawing.

Forging Forging is a method of changing a metal's shape using hydraulic or pneumatic hammers or large presses (see Figure 4.65). Blacksmiths use forging to make products or parts of products out of iron. Forging makes use of the property of plasticity. Forging can be done while metal is hot or cold. Hot forging is desirable because when metals (primarily ferrous metals) are heated to a cherry red color, they become softer and are easier to shape. Cold forging involves feeding material (often thin bars or wires) into a die to shape the part, for example to form the head of a bolt or a rivet.

Figure 4.65 | Hand forging using a hammer and an anvil.

Rolling Rolling is a process in which a material is passed through a set of rollers to either make it thinner or to shape it (see Figure 4.66). The rolling process can take industrial material such as ingots of metal and shape it into plate, sheet, or foil. Material can be hot rolled or cold rolled. Rolling is also classified according to the temperature of the metal rolled. In hot rolling, the metal can be deformed to a greater degree than with cold rolling, and hot rolled materials retain their microstructures. Cold rolling is done at lower temperatures. As a result, the metal is work-hardened and people must anneal it to form it into products. Not only metals, but also plastic products are produced by rolling. Cellophane tape is one example.

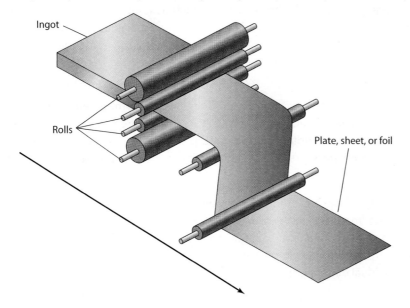

Ingot

Rolls

Plate, sheet, or foil

Figure 4.66 | Rolling.

Figure 4.67 | A machine press.

Machine Pressing *Machine pressing* uses hydraulic presses to stretch a material (often sheet metal) into a desired shape. Often, the press uses top and bottom dies that mate to give the material its final shape (see Figure 4.67).

Extrusion Extrusion is a process whereby a continuous stream of a product is made by squeezing softened material (just below its melting point) through a small opening called a die. The material can be either metal or plastic and the shape of the product takes the shape of the die's inner portion. Extrusion occurs when you squeeze toothpaste out of a tube. Extruded products can be formed into "T" shapes or "L" shapes, and they can be made into a variety of geometric shapes (round, rectangular, octagonal) as cross sections. As the extrusion exits the die, it is carried by a conveyor to a cooling area. Tubing, channels, rods, and pipes are all made by extrusion (see Figure 4.68).

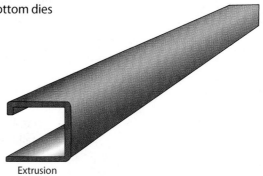

Extrusion

Figure 4.68 | An extruded shape.

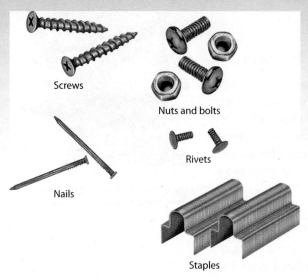

Screws

Nuts and bolts

Rivets

Nails

Staples

Figure 4.69 | Examples of common fasteners.

Drawing Drawing involves pulling a material through a die. The material is always ductile, so that its structure allows plastic deformation. Metal wire and rod is produced by drawing metal through a die. Heated plastic (below the melting temperature) is pulled through a die to make products like drumsticks.

Consolidation Processes

Consolidation processes are used to *combine materials*. These processes can include fastening and joining techniques, making composite materials, and coating materials.

Fastening Techniques Fastening techniques involve using fasteners like nails, screws, nuts and bolts, rivets, and staples (see Figure 4.69).

Joining Techniques Joining techniques include using adhesives, soldering, brazing, and welding. Early adhesives were resins that came from plants. Animal hides and hooves were boiled and their connective tissues were used as glues. Modern adhesives are extremely strong and versatile. Some work by chemically bonding with the surfaces to be joined. Epoxy resins are an example. Other adhesives, such as contact cements, are pressure sensitive and work on the basis of forces causing intermolecular attraction. Some adhesives used in dental applications set up very quickly under ultraviolet light. Hot glues are thermoplastics that are melted and harden between two surfaces when they cool.

Sintering Components can be produced from compacted metal powder, composites, and cements. Sintering is a process where powdered materials are combined with a binding material and pressed into shapes in a mold under heat and high pressure. Most sintered parts are small and weigh less than 5 pounds (2.27 kg).

Composite Materials Another example of a consolidation process is forming composite materials (as explained in Section 1 of this chapter). Composites are formed by combining a matrix material with a reinforcement material. The composite material has properties that are different from and more desirable than the individual materials that comprise it.

Coating Coating materials involves using paints, finishes, and chemical processes. Paints use a pigment (to provide color); a binder (a resin that binds the pigment and imparts the glossy or matte finish to the paint); and the vehicle (which is the solvent that carries the pigment and binder). The vehicle can be water, oil, or alcohol. Finishes (paint can also be considered a finish) include wood finishes, metal finishes, and ceramic coatings. Wood finishes include polyurethane resin, varnish, shellac, and stain.

Metal finishes include *anodizing* (an electrochemical process that thickens the oxide on the surface of nonferrous metals); *galvanizing* (coating ferrous metals with a thin layer of zinc to protect them from corrosion); and *electroplating* (an electrochemical process that deposits metal like silver, gold, and copper on another metallic surface). Ceramic finishing methods include *glazing* and *enameling*, in which powdered glassy materials are dissolved in a solution and painted onto a ceramic object. When heated to their flow point, glazes and enamels adhere to the object's surface.

Summarizing Materials Processing Methods

METHOD	PURPOSE	EXAMPLES
Mass Change	Removes or adds mass	Cladding, Electroplating, Drilling, Sawing, Grinding, Turning
Phase Change	Makes solid parts from liquid or gaseous materials	Chemical Vapor Deposition, Casting, Pressing, Injection Molding, Blow Molding, Vacuum Forming, Rotational Molding
Structure-Change	Affects the atomic structure of a material	Hardening, Tempering, Annealing, Surface Hardening
Deformation	Changes the shape of a material without changing its mass	Forging, Rolling, Machine Pressing, Extruding, Drawing
Consolidation Processes	Used to combine materials	Fastening, Joining, Sintering, Forming Composites, Coating (Painting, Finishing, Anodizing, Galvanizing, Electroplating, Glazing, Enameling)

SECTION THREE FEEDBACK >

1. Give an example of each of the following processes: mass change, phase change, structure change; deformation; and consolidation.
2. Compare grinding and sanding.
3. Explain the process of casting.
4. What is the difference between extruding and drawing material?

SECTION 4: Factors in Selecting Materials

KEY IDEAS >

- Materials are chosen based on their properties, but other factors, such as cost, safety, availability, disposability, and environmental impact, are also considered.
- The choice of material often reflects trading off competing benefits.
- True costs of materials extend beyond the cost of the raw material itself.

Materials are chosen based on their properties, but other factors, such as cost, safety, availability, disposability, and environmental impact, are also considered. The choice of material often reflects balancing competing benefits (see Figure 4.70).

Properties of materials determine their suitability for specific applications. Depending upon the product to be manufactured, certain properties become most important. Table 4.7 compares the properties of aramids, carbon, and glass. Aramids (like Kevlar) have high strength-to-weight properties, are reasonably stiff, are resistant to chemicals, and do not conduct electricity. Carbon-fiber products have high tensile and compressive strength, high stiffness, resist corrosion, but have low impact strength. Fiber glass has good compressive strength and stiffness, good tensile strength and relatively low cost. In choosing among these materials, engineers make trade-offs. Sometimes designers trade off one property at the expense of another. For example, in producing high-strength steel, engineers choose high strength at the expense of ductility. Sometimes people trade off benefits for costs; sometimes companies might trade off safety for cost.

171

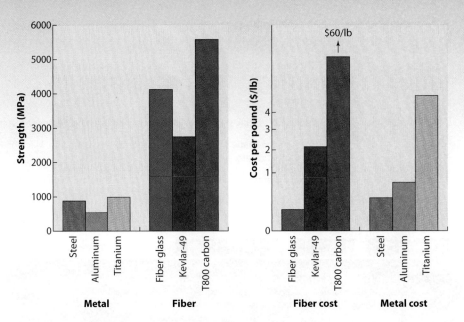

Figure 4.70 | Comparison of cost versus strength for structural materials.

Table 4.7 | Properties of Aramids, Carbon, and Glass

PROPERTY	ARAMIDS	CARBON FIBER	GLASS
Tensile Strength	GOOD	EXC	GOOD
Stiffness	GOOD	EXC	POOR
Compressive Strength	POOR	EXC	GOOD
Flexural Strength	POOR	EXC	GOOD
Impact Strength	EXC	POOR	GOOD
Shear Strength	GOOD	EXC	EXC
Fatigue Resistance	GOOD	EXC	POOR
Fire Resistance	EXC	POOR	EXC
Thermal Insulation	EXC	POOR	GOOD
Electrical Insulation	GOOD	POOR	EXC
Low Thermal Expansion	EXC	EXC	EXC
Low Cost	POOR	POOR	EXC

Figure 4.71 | Ford focus.

Figure 4.72 | Ferrari enzo.

Availability of Materials

Availability of materials is a primary factor in the selection process. When there are only a limited number of producers of a certain material, or when that material is produced in a limited supply, engineers look for alternatives. During World War II, metal was in very short supply and other materials, primarily wood, were substituted.

An adequate supply of materials also depends upon governmental trade policies and international relations. A country may not, for political reasons, wish to engage in trade with another nation although that nation might be a good resource for a particular material.

Because of globalization, however, sources for many materials have increased. Nevertheless, if materials are available locally at comparable prices, they are preferred. Local materials can be transported less expensively than foreign-made materials, and local materials also make use of a resident workforce so that more jobs are available to local workers.

Figure 4.73 | When metal was in short supply during World War II, manufacturers explored making bicycle frames out of wood. This photo shows a prototype made out of laminated wood. These bicycles never were produced in large quantities.

Safety

Safety is an important concern when choosing materials. Prior to the 1980s, asbestos used to be a favored material in the production of automobile brake shoes and construction materials. Asbestos fibers are strong, durable, and resistant to heat and wear. Asbestos use is no longer widespread in the United States because the material has been shown to be a carcinogen (a cancer-causing agent). Other materials such as aramid and carbon fiber, fiberglass, vinyl siding, and wood siding, have been substituted. Brake pads, for example, now are made of Kevlar, or are semi-metallic using copper, brass, and steel-wool flecks held together by a resin.

Disposability and Environmental Impact

Disposability and *environmental impact* drive decisions about materials. People in many nations are becoming more environmentally conscious. Some materials can biodegrade; some, like plastics, virtually never biodegrade. New plastics are being developed, however, that come from renewable raw materials. These *bioplastics* are envisioned as being produced by converting plant sugars into plastic, growing plastic inside microorganisms, and in crops like corn. Table 4.8 shows the length of time it takes for materials to biodegrade.

When choosing materials and processes, engineers also take into account the energy it takes to process materials. These costs can vary widely depending on the current demand, and the political climate in the region in which a particular form of energy is produced. Table 4.9 shows the costs of a variety of energy types. Since these costs are listed in real (1984) dollars (before current inflation), they are most useful as a comparative ratio of the cost of one material to another. As you can see, the costs of using certain materials to provide energy are markedly cheaper than others.

Table 4.8 | Time Required for Material to Biodegrade

Cotton rags	1–5 months
Paper	2–5 months
Rope	3–14 months
Orange peels	6 months
Wool socks	1–5 years
Cigarette butts	1–12 years
Plastic-coated paper milk cartons	5 years
Leather shoes	25–40 years
Nylon fabric	30–40 years
Tin cans	50–100 years
Aluminum cans	80–100 years
Plastic 6-pack holder rings	450 years
Glass bottles	1 million years
Plastic bottles	Forever

Table 4.9 | Table of Energy Costs

GASOLINE COST/GAL	GAS, $/M BTU	HEATING OIL COST/GAL	HEATING OIL $/M BTU	NATURAL GAS COST/1,000 Cu Ft.	NATURAL GAS $/M BTU	ELECTRICITY COST/KWH	ELECTRICITY $ M BTU
140.735	11.328	120.029	8.654	642.477	6.238	4.152	14.101

True Cost of Materials

The true cost of a material includes more than just the cost of the raw material. It also includes the amount of energy and water (a scarce resource in some places) needed to produce it, the amount of pollution generated by the process, and the cost of human time and effort. For example, the energy costs for growing, harvesting, and processing the plants needed to make bioplastic could conceivably consume more fossil fuel than processing plastic in traditional ways.

SECTION FOUR FEEDBACK >

1. Why might a designer or engineer not choose the strongest material to make into an automobile body?
2. Give an example in which safety was traded for cost.
3. How does globalization affect the choice of materials?
4. Although bioplastics might be environmentally desirable, why might they not become commonplace?

CAREERS IN TECHNOLOGY

Matching Your Interests and Abilities with Career Opportunities: Materials Engineering

Materials engineers are involved in the development, processing, and testing of materials used to create a range of products. They work with metals, ceramics, plastics, semiconductors, and composites to create new materials that meet certain mechanical, electrical, and chemical requirements. Materials engineers create and study materials at an atomic level. Most specialize in a particular material.

Significant Points

* Overall job opportunities in engineering are good, but vary by specialty.
* A bachelor's degree is required for most entry-level jobs.
* Starting salaries are significantly higher than those of other college graduates.
* Continuing education is critical for engineers wishing to enhance their value to employers as technology evolves.

Nature of the Industry

Engineers precisely specify functional requirements; design and test materials and components; and evaluate the design's overall effectiveness, cost, reliability, and safety. Most engineers specialize. Ceramic, metallurgical, and polymer engineering are subdivisions of materials engineering.

Working Conditions

Most engineers work in office buildings, laboratories, or industrial plants. Others may spend time outdoors at production sites, where they monitor or direct operations or solve on-site problems. Some engineers travel extensively to plants or work sites. Many engineers work a standard 40-hour week. At times, deadlines or design standards may bring extra pressure to a job, requiring engineers to work longer hours.

Training and Advancement

A bachelor's degree is required for most entry-level jobs. Continuing education is critical for engineers wishing to enhance their value to employers as technology evolves. There are many two- and four-year colleges and universities in the United States that offer programs in materials engineering, materials science, and materials technology.

Outlook

The employment for materials engineers is expected grow about as fast as the average for all occupations through 2014. Although many of the manufacturing industries in which materials engineers are concentrated are expected to experience declining employment, materials engineers still will be needed to develop new materials for electronics, biotechnology, and plastics products. Growth should be particularly strong for materials engineers working on nanomaterials and biomaterials.

Earnings

As a group, engineers earn some of the highest average starting salaries among those holding bachelor's degrees. Materials engineers' starting salaries are among the highest in engineering.

[Bureau of Labor Statistics, U.S. Department of Labor, Occupational Outlook Handbook, 2008–09 Edition, visited January 2008, http://www.bls.gov/oco/]

Summary >

Materials have different qualities and may be classified as natural, synthetic, or mixed. The most common materials from which products are made are wood, metal, plastics, ceramics, and composites. Recyclable materials allow what otherwise would be considered waste to be reprocessed into useable products.

Materials are chosen on the basis of properties that make them appropriate for particular applications. Properties of solid materials can be grouped into the following categories: physical, mechanical, electrical, magnetic, thermal, optical, and acoustic. Mechanical properties include a material's strength, as well as other properties such as its hardness, ductility, elasticity, plasticity, brittleness, toughness, and malleability.

Materials are processed using methods that change basic (raw) materials into industrial materials, and then into finished products. Typical methods of processing materials include mass change, phase change, or structure change;

deformation processes; and consolidation processes. New developments in materials processing will be driven by highly technical processes and the use of new materials.

Materials are chosen on the basis of their properties, but also on the basis of other factors including cost, safety, availability, disposability, and environmental impact. Determining which material to choose often reflects trading off competing benefits. The true costs of materials extend beyond the cost of the raw material itself.

FEEDBACK

1. Give an example of a natural, synthetic, and mixed material.

2. What are the most common materials from which products are made?

3. What is the purpose and value of recycling materials?

4. What is the difference between physical and mechanical properties of material? Give an example of a physical property and an example of a mechanical property.

5. On what basis can we judge the thermal properties of a material?

6. What is the modulus of elasticity?

7. Explain the relationship between elasticity and plasticity of a material.

8. What are the five typical categories of material processing methods?

9. Research two contemporary developments in material process and explain them briefly.

10. On what basis are materials chosen for a particular application?

11. Give an example of a trade-off you might make in selecting a material to build a product.

12. Beside the cost of the raw material itself, what other factors influence the true cost of choosing a material for a particular application?

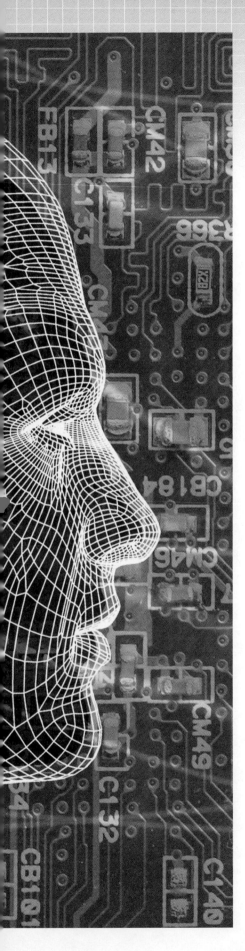

DESIGN CHALLENGE 1:
Glass Technology

• Problem Situation

Materials are processed using methods that change basic (raw) materials into industrial materials and then into finished products. Sand can be processed into glass (primary fabrication), which is then processed into useful products (secondary fabrication). In this activity, you will have the opportunity to engage in both primary and secondary fabrication. You will make glass from raw materials, and then process it into a useful and attractive object.

• Your Challenge

Given the necessary raw materials, you will mix (batch) materials, produce molds, and make a small glass object.

• Safety Considerations

1. Use a great deal of care, as the materials are at extremely high temperatures.
2. Wear eye protection at all times.
3. Wear insulating gloves.
4. Avoid taking crucibles in and out of the kiln. Rapid changes in temperature can cause cracking.
5. If a crucible develops cracks, discontinue using it immediately, and call the instructor to assist you.
6. Hot glass looks just like cold glass. Never leave newly formed pieces out in the open where an unsuspecting individual could touch them.
7. Always use tongs to handle hot crucibles. Never reach into the kiln with your hands to insert or remove a crucible, even while wearing insulated gloves.
8. Finished objects must be properly annealed, or they are liable to explode. Test for stress using a polariscope.

• Materials Needed

1. Enameling kiln
2. Porcelain or graphite crucibles
3. Aluminum or graphite molds
4. Glassmaking chemicals
5. Scales
6. Mortar and pestle
7. Polariscope
8. Drilling and boring tools
9. Cement block or carbon block
10. Tongs
11. Heat-resistant gloves

• Clarify the Design Specifications and Constraints

Your design must be made from a low-melting point borosilicate glass, a glass that can be melted in a copper enameling kiln at a temperature of about 815° C or about 1,500° F. A formula for a suitable glass can be obtained from your instructor.

Molds can be made from graphite, which is easily machined. Dump molds can be formed in the top of a graphite block. A rotational mold made from graphite can be turned on a lathe to a cylinder measuring about 5 cm. (2 inches) outside diameter, and 10 cm. (4 inches) in length. You can use a boring tool to machine the inside to a shape, like a small bell or a salt shaker where the mold has a conical inside shape.

A polariscope can be made from two polarizing filters, or sheets of polarizing film, at 90 degrees to each other. The glass is placed between these two filters or sheets of film. The viewer inspects the glass by looking through the polarized film toward an incandescent light source. A box should be constructed to hold the light bulb and two sheets of film or filters, with enough space between the film sheets to inset and remove the formed glass. If bands of color or streaks of black are seen, the glass should be annealed again.

● Research and Investigate

To complete the design challenge, you first need to gather information to help you build a knowledge base.

1. In your guide, complete the Knowledge and Skill Builder I: Safety considerations.
2. In your guide, complete the Knowledge and Skill Builder II: Making a mold.
3. In your guide, complete the Knowledge and Skill Builder III: Making a polariscope.
4. In your guide, complete the Knowledge and Skill Builder IV: Batching glass.
5. In your guide, complete the Knowledge and Skill Builder V: Melting the glass batch.
6. In your guide, complete the Knowledge and Skill Builder VI: Forming a glass object.
7. In your guide, complete the Knowledge and Skill Builder VII: Annealing the glass object.

● Generate Alternative Designs

Describe two of your possible alternative approaches to making a glass object. Discuss the decisions you made in a) batching the glass, b) designing the mold, and c) building the polariscope. Attach drawings if helpful and use additional sheets of paper if necessary.

● Choose and Justify the Optimal Solution

What decisions did you reach about the design of the glass batch, the mold, and the polariscope?

● Display your Prototypes

Produce your glass batch, mold, polariscope, and final glass object. Include either descriptions, formulas, photographs, or drawings of these in your guide.

● Test and Evaluate

Explain whether your designs met the specifications and constraints. What tests did you conduct to verify this?

● Redesign the Solution

What problems did you face that would cause you to redesign the a) glass batch, b) the mold, and/or c) the polariscope? What changes would you recommend in your new designs? What additional trade-offs would you have to make?

● Communicate Your Achievements

Describe the plan you will use to present your solution to your class. (Include a media-based presentation.)

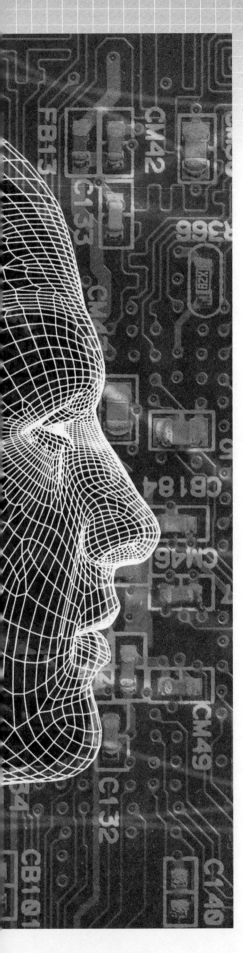

DESIGN CHALLENGE 2:
Testing Materials

• Problem Situation

A wide variety of materials are available to choose from when designing products or systems. Some of these materials have properties that make them more desirable than others. For example, when NASA designs space suits, the materials from which they are to be made must be tested for their ability to withstand very specific stresses. Before they are made into products, materials are tested to determine how well they will survive operating conditions.

• Your Challenge

You will design, construct, and use a piece of materials-testing equipment to test a particular material property (a physical, mechanical, electrical, magnetic, thermal, optical, or acoustic property) of a material you choose to investigate.

• Materials Needed

Various pieces of hardware, such as:
1. wood for frames
2. center punch
3. pulleys
4. motors
5. balances
6. assorted hand tools
7. machines

• Safety Considerations

In this activity, you will be using a variety of materials, hand tools, and machines. Make sure you fully understand all the safety precautions. Discuss these with your team members and your instructor. If you have any questions about procedural safety, make absolutely sure you discuss your concerns with your instructor.
1. Only use tools and machines after you have had proper instruction.
2. Wear eye protection when using tools, materials, machines, paints, and finishes.
3. Make sure that all machines are used with all guards in place.

• Clarify the Design Specifications and Constraints

The apparatus you design must be safe to use. It should include a method of calibrating results. Your results should be consistent with known data about the material and the property you are investigating.

• Research and Investigate

To complete the design challenge, you need to first gather information to help you build a knowledge base.
1. In your guide, complete the Knowledge and Skill Builder I: Properties of materials.
2. In your guide, complete the Knowledge and Skill Builder II: Nondestructive and destructive testing.
3. In your guide, complete the Knowledge and Skill Builder III: Finding known data about materials.

4. In your guide, complete the Knowledge and Skill Builder IV: Testing procedures.

5. In your guide, complete the Knowledge and Skill Builder V: Carrying out the test (Compare the data obtained from the materials tested to anticipated results.)

● Generate Alternative Designs

Determine what you want to test. With your group, brainstorm methods to perform a test to determine properties of the material. Describe two of your possible alternative approaches to making the piece of testing apparatus. Discuss the decisions you made in a) choosing the material to test, b) designing the testing apparatus, and c) collecting data about the material from your test. Attach drawings if helpful and use additional sheets of paper if necessary.

● Choose and Justify the Optimal Solution

What decisions did you reach about the choosing the material to test, b) designing the testing apparatus, and c) collecting data about the material from your test?

● Display Your Prototypes

Construct a functional model of your testing apparatus. Include drawings and sketches that helped you during the construction of the model. Explain how the model works and how you use it to conduct your test.

● Test and Evaluate

Explain whether your testing machine design met the specifications and constraints. How did you verify that it did?

● Redesign the Solution

What problems did you face that would cause you to redesign your apparatus? What changes would you recommend in your new designs? What additional trade-offs would you need to make?

● Communicate Your Achievements

Describe the plan you will use to present your solution to your class. (Include a media-based presentation). Demonstrate the use of your materials testing equipment and explain what you found out about the material under test.

I**F YOU LOOK** around, what do you see? A lot of things—most of them made by humans. We can split our world into the natural (things created by nature) and the human-made. In fact, we humans have created most of what we encounter. We can further divide made things into those that are constructed and those that are manufactured. But, as we will see, the distinction between construction and manufacturing is beginning

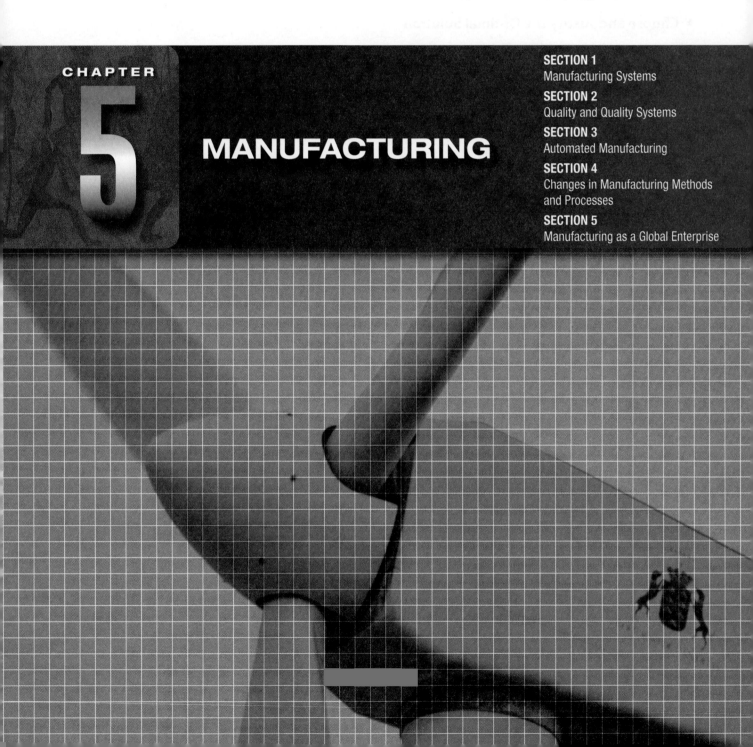

CHAPTER

5

MANUFACTURING

SECTION 1
Manufacturing Systems

SECTION 2
Quality and Quality Systems

SECTION 3
Automated Manufacturing

SECTION 4
Changes in Manufacturing Methods and Processes

SECTION 5
Manufacturing as a Global Enterprise

to blur. In this chapter, we discuss manufacturing, the use of tools, materials, people, and other resources for making things.

Why do we make things? There are a number of reasons, including practical or utilitarian use and consumption or sale. Consider this: How quickly can you find five products around you that are manufactured? How about ten? It's not too difficult, is it? In fact, if you take one of these products—say your cell phone or iPod—and break it down into its component parts, you will find that each of these parts is also manufactured (see Figure 5.1).

Figure 5.1 | Each of the parts that make up an cell phone has to be manufactured.

In the beginning, manufacturing was the work of craftsmen or highly skilled artisans. Pursuit of this vocation required working as an assistant to one of these artisans—serving as an apprentice (see Figure 5.2). The apprenticeship model—individual artisans making things with the help of apprentices—was more of an art than a science. The practice continued until the industrial revolution. With the industrial revolution came the modern manufacturing plant and the use of scientific principles in the creation of goods. When we discuss manufacturing, we include all the processes required for the production of a product and its components. No longer the art of a single craftsperson with very specialized knowledge, manufacturing now involves teamwork and working in collaboration with other disciplines, including materials science, engineering, and design.

As you saw from our earlier "thought experiment," manufacturing impacts nearly every facet of our daily lives and has been and remains a critical piece of the national and global economy, making significant contributions to the national infrastructure and national

Figure 5.2 | In the past, when manufacturing was the work of craftsmen or artisans, neophytes learned their trade by serving as apprentices.

defense. Unfortunately, the rising cost of manufacturing (labor, materials, and physical plant) has forced many companies and even entire industries to close their operations or move them to other nations. This phenomenon—referred to as offshoring—continues to challenge employers and workers alike. New and emerging technologies and approaches to manufacturing have afforded some pockets of growth in advanced manufacturing and employment opportunities, but even in the best of times, manufacturing is not without potentially significant social and environmental consequences.

Introduction

These consequences include clean-up costs and remediation of hazardous waste and health risks to workers. In the United States, we regulate manufacturing industries through labor and environmental laws (Figure 5.3). Working to comply with these laws impacts the cost of manufacturing a product, but it also serves to strike a balance between making a profit, pleasing customers, and protecting the environment and workforce.

Figure 5.3 | In the United States, industries are regulated through labor and environmental laws designed to protect the laborer.

SECTION 1: Manufacturing Systems

KEY IDEAS >

- Durable goods are designed to operate for a long period of time, while nondurable goods are designed to operate for a short period of time.

- Manufacturing systems may be classified into types, such as customized production, batch production, and continuous production.

- Common manufacturing systems include job shop, flow shop, project shop, continuous process, and the newer linked-cell system.

- The pressure to keep costs down and simultaneously meet customer demands is always balanced by the need for quality and reliability.

- The overarching goal in manufacturing is to design and implement manufacturing systems that provide low cost, superior quality, and on-time delivery.

- Chemical technologies provide a means for humans to alter or modify materials and to produce chemical products.

Figure 5.4 | A large, modern power plant like this one can generate and distribute power far more efficiently than older devices, such as waterwheels and steam engines.

Basic machine tools were invented and developed during the first industrial revolution. These led to initial efforts at mechanization and automation and the introduction of factories. Factories allowed us to aggregate resources at a single site with readily available power. The layout of early plants was functional, based on types of machines or operations. Often, factories were placed along rivers, where a waterwheel was used to power overhead shafts that then powered individual machines. Eventually, these methods of power generation were replaced by steam engines and later electric motors. As the modern electric power grid and its supporting infrastructure were developed and built, localized power gave way to power generation that could take advantage of so-called "economies of scale." In other words, a large, centralized power plant (Figure 5.4) could generate and distribute power much more cheaply and efficiently than waterwheels, steam engines, or electric motors at a factory.

What is a manufacturing system? To define a system, we have to clearly identify the system's boundaries or constraints, as well as how well the system responds to external forces. We should have some means of predicting this behavior based on input parameters. Often, we attempt to express the way the system works through a model—usually a mathematical model that can be programmed into a computer and used to control the process. A system has a number of objectives (cost, quality, etc.), but we must optimize the whole, rather than the subsystems. To operate the system, we must gather information and communicate with decision-making processes. Subsystems include people, product design, and materials. Other issues arise from social, political, and business environments that are unique to a particular company.

Figure 5.5 generalizes a manufacturing system. Inputs can include materials, energy, and information. The system itself includes people, machines, and materials-handling equipment. Outputs are another matter. These depend on what we are manufacturing. The most common output is some sort of consumer good, but often the output of one system is the input for another system. Consider the components that make up your electronic devices.

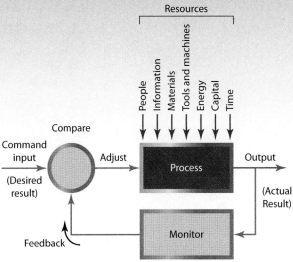

Figure 5.5 | A system takes an input and transforms it into the desired output.

Two of the biggest challenges in any manufacturing system are managing material availability and predicting and reacting to demand. If there's a huge demand for your product, as there has been for the Nintendo Wii, you will want to make as many items as possible to meet that demand. Your ability to make products, however, depends on availability of components. Our outputs or products can be classified into durable and nondurable goods. Durable goods are designed to operate for a long period of time, while nondurable goods are designed to operate for a short period of time. A durable or hard good has a long useful life. Automobiles, appliances, business and electronic equipment, and home furnishings and fixtures are all examples of durable goods (Figure 5.6). Conversely, nondurable or soft goods have a lifespan of less than three years and are often consumed when used once. Examples of nondurable goods include food and beverages, cleaning products, office supplies and paper products, and cosmetics and other personal products.

Manufacturing Systems

Common manufacturing systems include job shop, flow shop, project shop, continuous process, and the newer linked-cell system. Table 5.1 shows a list of manufacturing and service industries. Each of these industries can be classified by one of four classical manufacturing systems, or a combination of the four. Generally, manufacturing systems may be classified into types, such as customized production, batch production, and continuous production.

Figure 5.6 | A Nintendo Wii is an example of a durable good because it is intended to have a long and useful life.

These classifications describe the nature of the process and, in particular, the output or product. The systems include the job shop, flow shop, project shop, and the continuous process. We can build on these classical systems with the newer linked-cell system. The classical assembly line, pioneered by Henry Ford, is an example of a flow shop system.

The job shop is a flexible system because it can support a wide variety of products being manufactured and therefore is useful for small lots or jobs or custom products (Figure 5.7). To support such a wide variety of products, the production equipment must be general purpose and workers must be highly skilled and able to perform a range of different functions. An example of a job shop would be a small machine shop that makes parts for the aerospace industry. In a job shop plant, equipment

185

Table 5.1 | Production and Service Industries and Their Corresponding Manufacturing System

TYPE OF MANUFACTURING SYSTEM	EXAMPLES	
	SERVICE	PRODUCT
Job Shop	Auto Repair Hospital Restaurant University	Machine shop Metal fabrication Custom jewelry Flexible manufacturing systems (FMS)
Flow shop or flow line	X-ray Cafeteria College registration Car wash	TV factory Auto assembly line
Cell or Cellular	Fast-food restaurant Food court at mall 10-minute oil change	Family of turned parts Composite part families Design families Manned cells Robotic cells
Project shop	Producing a movie Broadway play TV show	Locomotive assembly Bridge construction House construction
Continuous process	Telephone company Phone company	Oil refinery Chemical plant

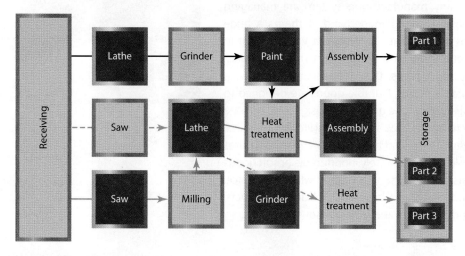

Figure 5.7 | A job shop is a flexible system that can be used to manufacture a wide variety of products. It most often used for small jobs or custom products.

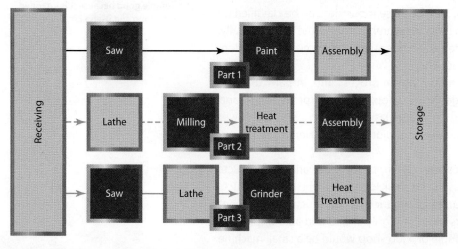

Figure 5.8 | A flow shop, or flow line, system is set up to produce a specific product.

is typically arranged according to function or manufacturing process. While this system works well for small jobs, it does not scale well—meaning that we cannot easily increase the volume or quantity of production.

For higher production runs, we can use the flow shop or flow line system (Figure 5.8). Unlike the more generic job shop approach, the flow shop approach is tailored to produce a particular product. The layout of the manufacturing plant is designed to produce a specific product or family of products. The name even describes how we would like the product to move—flow—through the plant. Often, the factory may have been designed and built for that very product. This type of system requires a large capital investment—the initial cost of the factory, equipment, and tooling. Because of this large initial investment, it is difficult to justify this type of system for small or custom production runs. Workers in this environment are typically less skilled than those in a job shop environment, since most of the production skill is built into the machinery. This system describes the classic assembly line that has enabled mass production and mass consumption. Most factories combine features from both the job shop model and the flow line.

The project shop takes a contrasting approach to the flow shop. Rather than the product flowing from station to station, in the project shop the product remains stationary and people and machinery come to it. This process—though cumbersome—is often necessary because of the size and weight of the product. Industries requiring this approach include the aerospace, shipbuilding, and locomotive industries (Figure 5.9). You will see this technique referred to as fixed-position layout or fabrication. Realize that each of these large manufacturing jobs—ships, aircraft, and locomotives—is made up of smaller subassemblies and components. These smaller pieces are often manufactured in job shop or flow shop environments and brought together for assembly.

In the last of our four classical systems—the continuous process—we take the concept of flow literally. In this system, rather than having a product move or "flow" from station to station, our product actually flows. Examples include food processing, oil refineries (Figure 5.10), and chemical processing plants. The continuous process is an option whenever our product can physically flow—liquids, gasses, and even powders. Chemical technologies provide a means for humans to alter or modify materials and to produce chemical products. These systems are usually the simplest, but also the least flexible for manufacturing. Once a continuous process system is up and running there is very little intervention or "work" needed, except to ensure we have a continuous supply of our raw material.

Figure 5.9 | You can think of a project shop as the opposite of a flow shop. The product remains stationery, with people and machines coming to it.

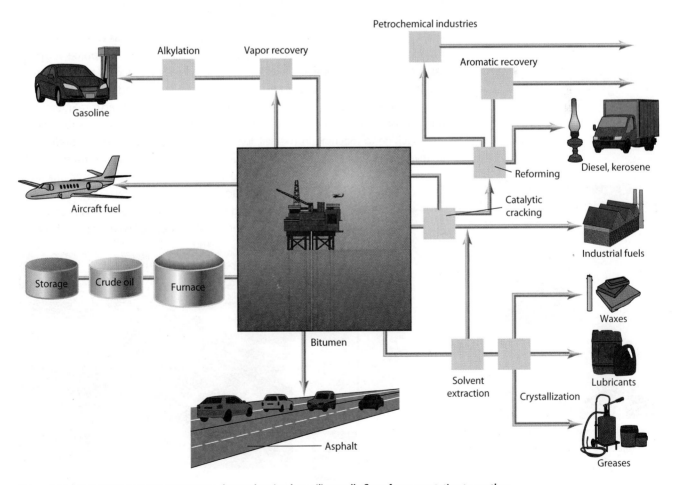

Figure 5.10 | In a continuous process system, the product (such as oil) actually flows from one station to another.

Figure 5.11 | A key-cutting machine uses a jig to create an exact replica of another key.

Jigs and Fixtures Long before the industrial age, in fact, since humans first began to make things, jigs and fixtures have existed to aid skilled craftsmen and the modern factory worker in doing their jobs. A jig is a tool—often viewed as a template or guide—which is used in the trades and in manufacturing to control the location and motion of a tool. There are as many jigs as there are craftsmen and factory workers. It is not uncommon for individuals to create jigs to perform a specific job or function. In addition to increasing productivity, jigs aid a worker in performing repetitious, sometimes mundane tasks and, most importantly, increase the precision and repeatability of an activity. This precision and repeatability enable us to create parts that are exact duplicates of one another. Let's consider what happens when we move into a new house or apartment. We are usually given a key to the front door but not a key for each member of the family. We remedy that situation by a trip down to our local hardware store. An employee at the hardware store takes our key, clamps it into a "key cutting" machine, along with a key blank that matches our key. She starts the machine and a cutting tool follows the profile of our original key to cut into the blank and create an exact duplicate (Figure 5.11). This process takes no more than a couple of minutes and provides us with the additional keys we need. In this example, the key is actually the jig and the "key cutting" machine and the mechanism to clamp down key and blank is a fixture—which we discuss below.

In manufacturing, a fixture is used to hold objects in place so that the object can be machined or assembled. Unlike jigs, which are used to guide a cutting tool along a prescribed path, a fixture holds the part in one place—fixes it—while a tool or cutter moves relative to the part. The primary benefits of using jigs and fixtures include reduced production costs and time, consistent and repeatable quality, and improved safety. With the growth in popularity and use of numerical control (NC) and computer numerical control (CNC), the use of jigs and fixtures is not nearly as widespread.

Linked-cell or cellular manufacturing is a more recent trend. In this approach, a plant floor is composed of a series of linked cells. A cell is usually a U-shaped arrangement of machinery allowing processes to be grouped according to the sequence of operations required to make a product. A cell can be either manned or unmanned (see Figures 5.12 and 5.13). An unmanned cell is a robotic cell. In either case, a worker or robot moves from machine to machine in the cell, loading and unloading parts and performing whatever tasks are required. Often the machines in the cell can perform their tasks without user intervention. The user or robot unloads a previous part, loads a new part, starts the machine and moves to the next machine in the cell. Every complete cycle around the U-shaped cell is a completed part.

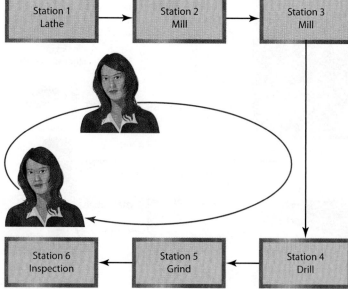

Figure 5.12 | A manned cell is operated by humans.

Figure 5.13 | An unmanned cell, or robotic cell, is operated by robots.

Table 5.2 | Advantages and Disadvantages of the Five Manufacturing Systems

	JOB SHOP	FLOW SHOP	PROJECT SHOP	CONTINUOUS PROCESS	CELLULAR
MACHINES	Flexible, general purpose	Single purpose	General purpose, mobile	Specialized	Simple
LAYOUT	Functional	Flow-based	Fixed-position	Product-based	U-shaped
SETUP (TIME, FREQUENCY)	Long, frequent	Long	Variable,	Custom	Short, frequent
WORKERS	Highly skilled, single function	Lower-skilled, single function	Highly skilled, specialized	Varies	Multifunctional
INVENTORY	Large (variety)	Large (buffer)	Usually large	Very large	Small
LOT SIZE	Small-medium	Large	Small	Very large	Small
LEAD TIME	Long	Short	Long	Very short	Short

Table 5.2 summarizes the advantages and disadvantages of each of these five manufacturing systems. For any of the processes, interchangeability of parts increases the effectiveness of manufacturing processes.

For any of these processes, there is a constant overarching goal to reduce manufacturing costs and make improvements in productivity. For example, on a million-piece run (flow shop or continuous process), saving a couple of seconds per part can provide a tremendous profit. If we consider a million-piece production run and assume a savings of two seconds per part, the total savings would amount to the equivalent of nearly seventy-five eight-hour production shifts. For smaller runs, reductions in time have little impact on the cost, requiring us to find other means of cutting costs. Manufacturing costs are also affected by other factors, requiring that we make efficient use of workers, materials, and machines. Looking at the various systems and the inputs and outputs of a manufacturing system, it's not surprising that we spend a lot of time scheduling and planning. In fact, the complexity of a typical manufacturing system requires considerable attention to scheduling and planning. We must factor into these schedules delays due to design changes, material shortages, emergencies, machine downtime, quality issues, etc. One approach is to create a buffer at each machine or station, which unfortunately results in a large inventory.

Trends in Manufacturing There are a number of new trends in the design of manufacturing systems. There has been an explosion in the number and variety of products available. More choices mean smaller lot sizes and smaller manufacturing runs— not the cheapest way to produce a product. Gone are the days of Henry Ford providing a single Model T version of his product (Figure 5.14). We have to be able to customize, or provide models A through Z and even models we have not yet conceived of. Another factor is the increasing use of manufacturing cells and the implementation of a process called Just-In-Time, or JIT, manufacturing.

Figure 5.14 | Henry Ford started by producing a single version of his product—the Model T. Now the Ford Motor Company offers a wide array of cars and trucks.

Think of how customization and JIT impact your daily life. When you go to your local fast-food restaurant, do you want the standard burger listed on the menu? Perhaps not. Often you may ask for "no onions" or "extra cheese." In doing so, you've forced the restaurant to customize your product. If they don't, they run the risk of losing your business. What about JIT? Consider a late lunch—say three o'clock. Do you want the burger and fries that have been sitting under the heat lamp since the lunch rush? Or do you want your food made fresh? If you're like most people, you prefer a fresh burger and fries, or Just-In-Time burgers and fries. Imagine what these customer constraints do to a business—they're trying to sell their product at the cheapest price possible to maximize their profits.

Other factors that impact how we design manufacturing systems include continued customer demand for better quality, more precision, reduced cost, and on-time delivery, as well as the availability of new materials and manufacturing processes. What about the costs of energy, materials, materials handling, and labor? These are all factors that impact how we set up our manufacturing systems. What about reliability and its implications for potential liability suits? If the pressure to get our burger to our customer cheaply, quickly (JIT), and the demand for customization cause us to undercook the meat, we have opened the door to contamination by bacteria and E-coli and to potential lawsuits. Other businesses are no different—the pressure to keep costs down and simultaneously meet customer demands are always balanced by the need for quality and reliability. The overarching goal in manufacturing is to design and implement manufacturing systems that provide low cost, superior quality, and on-time delivery. There are a number of techniques for achieving this goal.

SECTION ONE FEEDBACK >

1. Identify ten products around you that are manufactured. How many of them do you use daily?

2. Make a visit to your local hardware store to examine the "key cutting" machine. Ask a few questions of the operator, including: How long does it take to duplicate a typical key? What's the failure rate—that is, how often does a customer return with a key that didn't work? How many key are made on a typical day? How much training is required to operate the machine?

3. Of the ten manufactured products identified earlier, how many would you consider durable goods and how many nondurable? Explain your rationale for each.

4. List the four classical manufacturing systems and identify three products that could be produced in each.

SECTION 2: Quality and Quality Systems

KEY IDEAS >

- The pursuit of quality begins at product conception, and follows through the design, analysis, manufacture, and delivery of the product to the customer.

- Materials have different qualities and may be classified as natural, synthetic, or mixed.

- A fundamental challenge in designing and manufacturing a new product is balancing the cost of measuring, assessing, and controlling quality versus the benefits.

- Companies must produce products that are competitive in price and better in performance than others on the market.

- Total Quality Management extends beyond manufacturing a product into design, analysis, marketing, sales, distribution, and billing.

- Servicing machinery keeps it in good operating condition.

- Safety and quality are issues of great concern to all engineers, who must ensure that when products fail, they do not become safety hazards.

Quality is a difficult concept to define, because it means different things to different people. The manufacturing process is where we measure and maintain quality, but it is not where the pursuit of quality begins (Figure 5.15). The design process—starting at conception—has a great impact on quality. Errors or omissions at the design stage can lead to quality-control issues that no manufacturing system can fix. We must build quality into the design. To do so, we have to first understand what quality is. Remember, there is no one definition of quality—it is many things to many people. Each customer or consumer has different expectations of how a product should perform.

Figure 5.15 | Many companies emphasize the quality of their products in their advertisements and promotional Web sites.

If you are like many cell-phone users who have grown up with the technology, entering text into your cell is probably second nature for you—even with a standard 10-key keypad. What other expectations do you have of your cell phone besides texting? You would no doubt like to be able to make calls, but your expectations don't stop there. What about listening to music, watching videos, or customizing your ringtone? These are probably all things you expect from a cell phone. A phone that did not do all these things and do them well would probably be unacceptable to you. Contrast this with your parents' view of a cell phone. What do they typically do with a cell? Not much, you're probably thinking, except make phone calls. This is their primary requirement for a cell phone—no audio, video, or custom ringtones— just calls. On top of that, they expect their cell phone to be easy to use. Easy is also a relative term—easy for you may not be easy for your parents. We need to consider these factors when we explore quality and quality control in manufacturing.

Where does manufacturing fit into the quality equation? Obviously, the manufacturing floor is central to quality. It is where we measure quality, and—if we have done our jobs right—it is where we confirm the quality of our product. As we have learned, quality doesn't start on the plant floor. In manufacturing, we can't fix the design itself; we can only ensure that our product is made to the design specifications. Where we can make an impact is through early involvement—even as early as the concept phase. Every phase of product dev-elopment—conception, design, analysis, manufacture, and testing—is a large, complex discipline in its own right. No one person—even the quintessential lone genius Thomas Edison—can

Figure 5.16 | Even Thomas Edison, the personification of the lone genius, needed help from others to complete the development of his inventions.

know every phase (Figure 5.16). Everyone needs to work as part of a team. Modern manufacturing requires a team approach. Early involvement in the process allows us to make determinations regarding the feasibility of manufacturing the product, required facilities, and equipment. We can also identify potential issues with accuracy, precision, and repeatability. Working with designers, we can help identify appropriate materials and manufacturing processes for these materials. Materials have different qualities and may be classified as natural, synthetic, or mixed. Which materials we select will impact our decision regarding manufacturing. Everyone involved in the manufacturing process has valuable information that can help to build a better quality product.

Engineering always involves a series of trade-offs. For instance, it would be great if you could use your cell phone to make video calls with your friends. But would you opt for this feature if it depleted your battery in four hours? That sort of trade-off—a new feature (video calls) versus reduced battery life—is the type of decision engineers face every day. Quality is no different. A fundamental challenge in designing and manufacturing a new product is balancing the cost of measuring, assessing, and controlling quality versus the benefits. Companies must produce products that are competitive in price and better in performance than others on the market.

Total Quality Management—TQM

Manufacturing high-quality products requires more than instituting quality-control checkpoints. Total Quality Management extends beyond manufacturing a product, into design, analysis, marketing, sales, distribution, and billing. The pursuit of quality begins at product conception, and it follows through the design, analysis, manufacture, and delivery of the product to the customer. Companies have also found that quality improvement and cost cutting often go hand in hand. Having products perform according to specification reduces scrap (out-of-spec parts), repair work on returns, and warranty work in the field. A large manufacturer reported these costs to be 20 percent of revenues before implementing a company-wide Total Quality Management (TQM) program.

There are two basic approaches to quality improvement: benchmarking and employee involvement. In benchmarking, we search for the best product or service in the industry and use that as a basis or benchmark for evaluating the performance of our own product. Once goals for improving service and products have been determined, employees from throughout the company need to be actively involved in achieving these objectives.

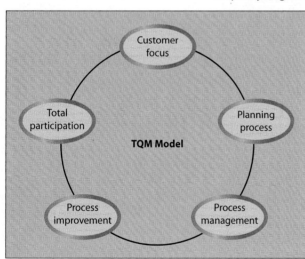

Figure 5.17 | A typical TQM program follows a four-step process.

A typical TQM program follows a four-step process (Figure 5.17). First, we measure and record the performance of various components of the manufacturing process. These components could include cells or individual machines. Tracking performance in this way allows us to isolate small deviations or differences from the norm. Using this data, we can better predict when a machine's operating parameters should be adjusted—before we produce parts that don't pass inspection. As a machine ages, its performance will deteriorate. Although this deterioration is inevitable, our performance tracking will allow us to better anticipate when a machine will need to be serviced or even completely replaced. Servicing machinery keeps it in good operating condition. It also ensures that we can continue to produce high-quality products, while keeping our costs under control (reduced downtime, scrap, substandard quality parts, etc.). The second step of TQM is the design, or

Genichi Taguchi (1924—) is a Japanese engineer and statistician responsible for developing a widely implemented technique for using statistical analysis to improve the quality of manufactured goods. Influenced by the work of W. Edwards Deming of the United States, Taguchi spent twelve years—starting in 1950—at the Electrical Communications Laboratory (ECL) of the Nippon Telegraph and Telephone Corporation developing methods for improving quality and reliability.

redesign, of products making them simpler to assemble. It is easy to observe that a more simple assembly process should, more often than not, result in fewer assembly errors than a more complex assembly process. The third step of TQM is fostering a closer relationship with suppliers. One easy way to do this is to reduce the number of suppliers. As you become a larger and more frequent customer to each supplier, it is in their best interest to maintain the quality of components and materials they provide you. Normally, the quality of what you get from a supplier is out of your control, but if we can build closer relationships and leverage these relationships to improve the quality of our inputs, we can improve our outputs—our products. Lastly, the fourth step of TQM consists of the statistical methods, based on the work of Genichi Taguchi (Figure 5.18), used in the control of manufacturing operations and incorporated to minimize downtime or disruptions to our production operations.

Figure 5.18 | Genichi Taguchi is the Japanese engineer and statistician who developed a widely implemented technique for using statistical analysis to improve the quality of manufactured goods.

Let's look at how we might use TQM in manufacturing. Remember, we're trying to take a holistic or company-wide approach to quality and quality assurance. Our primary tools will be statistical measurements and methods. We will also need a metric—what we are hoping to measure—in this case, quality. Even though we've seen how difficult quality is to define and how subjective it is, we have to make some determination of what our end users expect. We also have to decide how we define failure or deviation from our standard of quality. Ideally, we would like to keep our definition of "failure" very narrow. Why would we want to do this? Imagine we're manufacturing aircraft wings (Figure 5.19). For an aircraft wing, a surface defect or volume defect— even those we can't see with the human eye—could, eventually, lead to small or microscopic cracks and potentially to structural failure of the wing. If we monitor our manufacturing process for these types of defects, rather than obviously visible cracks or imperfections, we can refine the manufacturing process to virtually eliminate these types of defects and improve the quality and reliability of aircraft wings—definitely something customers want.

We usually start by taking a random sample of our product and testing against our metrics. In our earlier example—the aircraft wing—we could begin with visual inspection (looking at it), both on the micro (under a microscope) and macro (by eye) scale. We also have the ability to do what is referred to as destructive and nondestructive testing. In nondestructive testing we might use X-rays or sound waves to examine the part's interior for flaws, but we would do no damage to the part. Destructive testing—as you would expect—involves "stressing" the part until it breaks—what engineers call structural failure. This test is also called a "life" test because we are essentially testing what it takes to bring the product to the end of its useful life.

In a well-designed TQM process, we typically perform all of these tests. It's important to realize that structural failure is not the only thing customers worry about—hopefully we're doing our jobs well enough that users will never encounter structural failure. Other tests might look at how our product performs

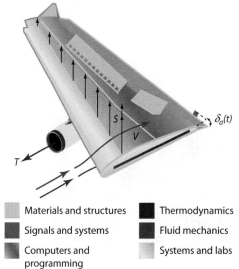

Materials and structures	Thermodynamics
Signals and systems	Fluid mechanics
Computers and programming	Systems and labs

Figure 5.19 | Applying TQM principles can prevent the kind of tiny surface defect that could lead to a structural failure in an airplane wing.

193

in extreme environments—heat, water, humidity, vibration, etc. Let's consider our always-present cell phone again. What sort of things would we look for in our TQM implementation for a cell phone? The next time you have your phone handy, examine it closely and give some thought to what sort of defects would make the product unacceptable.

Here are a few to get you started:

- Defective pixels on the screen
- Poor quality buttons
- Difficult-to-navigate user interface
- Poor call quality
- Obvious defects in casing and construction

You can probably come up with many more. Although some of these problems could be traced to other sources (design, network, supplier, etc.), many, if not most, can be traced back to problems in manufacturing. If we can identify these problems, and, more importantly, fix them, we can reduce the number of defects, increase customer satisfaction, and—most important to our stockholders—increase profits. This process of isolating "failures," identifying their causes, and fixing or improving the process is also called continuous improvement. Because the TQM process is company-wide, it is not uncommon for changes or fixes to propagate all the way back to the design process. When the process is working well, a "TQMed" product is often cheaper to produce because of gains or improvements in efficiency and performance. We can add to these cost savings due to reduced repairs, returns, and replacements.

Safety and Ergonomics

Safety and quality are issues of great concern to all engineers, who ensure that when products fail, they do not become safety hazards. Everything fails at some point—living organisms die, physical devices cease to function. The goal of engineers is to ensure that the failure is safe. What does this mean? When they fail, products should not become safety hazards—an exploding laptop or cell phone battery, an electric motor that bursts into flames. Whether the device is of high quality, often associated with a long lifetime, or low quality, or costly versus inexpensive, its failure should be safe. Quality and the resulting safety are corporate decisions based, in part, on market conditions and cost. A company must be able to design products with a low enough cost to be competitive, or it may soon find itself out of business. For instance, seat belts were not always required in automobiles; they were an extra-cost option. Manufacturers did not believe they could add seat belts as a standard feature and remain competitive. The government stepped in and passed a law requiring them to do so, and the challenge became creating an economical seat belt that meets or exceeds the regulations (Figure 5.20). Making a product safe—protection from electric shock, sharp edges, and moving parts—is not the only consideration.

The product should be comfortable to use, so ergonomics or human factors engineering becomes important. Ergonomics has many implications for us. In addition to designing and manufacturing products that are comfortable to use and conform to our customer's physical constraints, we must also consider workplace ergonomics. In the workplace, we must design tasks and work areas to reduce operator fatigue and discomfort, thereby maximizing the efficiency and quality of an employees' work (see Figure 5.21). Ergonomic analysis requires an understanding of human physiology to harmonize products to our physical abilities so that we are not in conflict with them.

Figure 5.20 | Seatbelts were not always required, but laws were eventually passed that made them mandatory. Manufacturers then had to figure out how to make a safe and useful seatbelt.

Figure 5.21 | This woman is using an ergonomic keyboard and mouse, which are designed to help prevent injury.

1. Explain why it is important to begin our pursuit of quality at the conception stage.

2. Create a table showing the following four typical cell phone users—someone your grandparents' age, your parents' age, your age, and half your age. Fill the table with features that each group would want/expect in a cell phone.

3. Why haven't we seen a modern version of Thomas Edison?

4. Briefly describe the Total Quality Management (TQM) process and its purpose.

5. Research careers in human factors engineering.

SECTION 3: Automated Manufacturing

KEY IDEAS >

- Mass production is the production of large amounts of standardized products on production lines.

- The interchangeability of parts increased the effectiveness of manufacturing processes.

- The goal of automation is for everything—services, manufacture, materials and information handling, assembly and inspection—to all be performed automatically.

- Automation raises several important social issues—in particular, its impact on the workforce.

Mass production is the production of large amounts of standardized products on production lines. This technique was first successfully employed by Henry Ford during the early twentieth century in the production of automobiles (Figure 5.22). An important factor in the success of mass production is the interchangeability of parts. The interchangeability of parts increased the effectiveness of manufacturing processes. Typical of early mass production, Ford's assembly line system used moving tracks or conveyor belts to move unfinished products from station to station, where workers perform simple repetitive tasks. First with the Model T, and later with subsequent models, Ford was able to achieve very high rates of production per worker, high-volume output, and, most importantly, inexpensive automobiles. Modern mass production systems are usually organized into assembly and subassembly lines. Because mass production requires such a large capital investment—machinery and modern robotics—it is best-suited to large, homogenous production runs. The primary benefits of mass production are reduced labor costs, increased production, and a lower cost final product. Limitations include the difficulty in modifying a design or production process and the lack of variety or customization that a mass production line affords.

Figure 5.22 | Henry Ford's assembly line was the first example of mass production.

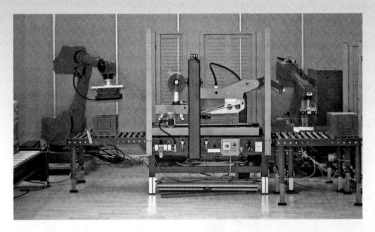

Figure 5.23 | Automation makes a packing line like this one highly efficient.

The term automation was probably first used in the 1950s to describe the automatic handling of materials. It has since become a more generic term describing the use of control systems, usually computers, to control industrial machinery and processes. The goal of automation is for everything—services, manufacture, materials and information handling, assembly, and inspection—to be performed completely automatically (see Figure 5.23). An easy way to consider the move to automatic operation is any human function or attribute that has been replaced by a machine. Mechanization is not quite automation. In mechanization, we augment or supplement human functions with semiautomatic machines, reducing the physical demands on human operators. Automation goes further, reducing not only the physical requirements on humans, but also the mental and sensory demands. The cells we discussed previously are examples of mechanization, not automation. Typically, we classify two forms of automation—hard or fixed-position and soft, flexible, or programmable. Hard automation consists of cams, stops, slides, and hard-wired circuits using relay logic. These machines are programmed with a controller or a handheld control box, which sends a set of instructions telling a machine what, how, and when to do something.

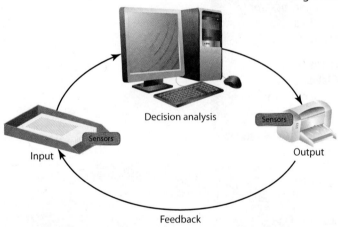

Figure 5.24 | Self-adjusting machines rely on a feedback control loop that provides information about the performance and operation of the machine. The machine then uses this information to control and adjust its operation.

An important class of automatic machines is self-adjusting machinery. These machines implement a feedback control loop—which provides information about the performance and operation of the machine that is used to control and adjust its operation (Figure 5.24). Sensors distributed throughout the machine take readings that are fed back to the controller. If the controller detects that these reading are outside of the norm, adjustments are made to reduce the error. Another important type of automatic operation is adaptive control. Machines that operate this way adapt the process to optimize it in some way. Optimization is a significant improvement over simple error correction. Adaptive control requires a computer in the system, which is programmed with information regarding the process or system behavior. The information is in the form of a mathematical model that describes system behavior, system constraints, and optimization goals. Not all processes are understood well enough to enable this level of automation. The key elements of an adaptive control loop include sensors that measure inputs and outputs of the process (identification), a computer that optimizes the process (decision analysis), and the ability to signal the control to alter system inputs (modification).

Finally, engineers in research and development labs are working on expert systems or artificial intelligence (AI) models for automated manufacturing. These models offer the promise of systems that can "think" and, more importantly, "learn." With expert systems, we attempt to capture the collective knowledge and experience of human experts, providing the AI system with a knowledge base from which to draw experience.

Flexible Manufacturing Systems

Flexible Manufacturing Systems or FMS is an approach to manufacturing a product that focuses on integrating technology to enable more rapid production. With FMS, we use machines to perform repetitive tasks formerly done by humans. Machines can carry out these tasks quickly and efficiently, twenty-four hours a day, seven days a week, with little or no error. FMS makes extensive use of computer numerical controlled (CNC) machines and automated guided vehicles (AGVs). A number of CNC machines are aggregated together to form a cell. Each cell performs a specific function in the manufacturing of a product. AGVs move from cell to cell transporting material, as well as intermediate and finished products. This speed and efficiency does not come without a cost. FMS is not cheap. The machines that make up a flexible manufacturing system cost millions of dollars. A more cost-effective alternative to a full-blown FMS implementation is to introduce a flexible manufacturing cell that mimics FMS, but on a smaller scale. In a flexible manufacturing cell, we produce a portion of the product by machine and the remainder through more traditional techniques.

Computer Numerical Control One critical innovation that has enabled modern automated manufacturing is CNC or computer numerical control (Figure 5.25). The precursor to CNC was numerical control (NC), developed in the 1940s and 1950s by MIT and John T. Parsons. In NC, machines are hard-wired and their operating parameters cannot be changed. A CNC is a computer controller that reads G-code instructions and controls a machine tool to fabricate components by the selective removal of material. Unlike NC, CNC operating parameters can be modified through software.

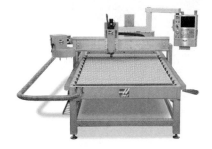

Figure 5.25 | A CNC machine reads G-code instructions and controls a machine tool to fabricate components by the selective removal of material.

ENGINEERING QUICK TAKE

Suppose you just started working for a small manufacturing company that has decided to purchase and implement a CNC machine. This is the first foray into CNC for your company. Your company has budgeted $140,000 for this new venture. As the company's newly hired production manager, you realize that although moving to a CNC system will provide great improvements in productivity, these improvements could be even further enhanced by introducing additional components, including a touch trigger probe (TTP) tool setting and in-cycle gauging equipment. Unfortunately, this additional equipment will require an additional investment of $10,000. Not surprisingly, the board of directors is hesitant to commit to any additional funding. You have made your recommendation to the project team based on your knowledge of production and CNC machines, and you have been asked to make your case to the board of directors. To present your recommendation to the board, you have prepared a brief presentation, shown in Figure 5.26. An engineer's ability to present ideas and recommendations clearly and effectively is a critical job skill.

In your discussion with the board, one of the first things you are asked is, "Why not install the CNC machine and use a portion of the anticipated cost savings to purchase and install the additional components at a later date?" This is the retrofitting scenario. Your response to this would focus first on the required training for all of this new equipment. Employees will be required to attend training session to learn how to operate the new machinery; if you later upgrade the CNC machine, you will have to send these employees for further training. This additional training will incur additional downtime and a corresponding loss of productivity. Another factor is technical support. The support contract will likely only cover the initial installation and not any subsequent modifications. In fact, the machine tool company that is selling you the original CNC setup provides no guarantee that the modifications that would be required would be compatible with the original installation.

Having addressed the issues involved with retrofitting, you now provide the benefits derived from the additional investment that you propose. In discussing the

Presentation to the Board

Recommendations for CNC production

CNC Purchase

We have budgeted $140,000 for CNC

CNC system will provide great improvements in productivity

As production manager, I fully support this purchase

However—these improvements could be even further enhanced by introducing additional components:
•a touch trigger probe (TTP) tool setting, and
•in-cycle gauging equipment
This additional equipment will require an additional investment of $10,000.

Benefits Derived

Benefits derived from the additional investment:
•majority of CNC machine tools are initially set up manually using feeler gauges, gauge blocks, and dial test indicators
•manual setup is prone to operator error and time consuming
•default setup that your company has identified will require these manual operations

Benefits Derived

Benefits derived from the additional investment:
•the $10,000 investment will automate setup using TTP
•a time savings of up to 90% for basic operations
•even greater saving for more complex operations
•typical machine operator makes $24/hour

Retrofitting Scenario

Why not install the CNC machine and use a portion of the anticipated cost savings to purchase and install the additional components at a later date?
We can do this, but ... already sending employees for training; if we retrofit, we would have to retrain all our staff to operate this new equipment.
•additional downtime, and
•a corresponding loss of productivity

Retrofitting Scenario

Technical support:
•The support contract will only cover the initial installation and not any subsequent modifications.
•Machine tool company—selling the original CNC setup provides no guarantee that the modifications that would be required would be compatible with the original installation

Benefits Derived

Typical savings:

	Day	Week	Month	Year
	8 hours	5 days	4 weeks	12 months
Assuming 50% of time is used for setup	3.6 hours x $24/hr = $86.40	$432	$1728	$20,736
Assuming 25% of time is used for setup	1.8 hours x $24/hr = $43.20	$216	$864	$10,368

Benefits Derived

From industry publications
... the time required to set just ten tools manually on a CNC lathe is typically 20 minutes, but by using a touch trigger setting probe this can be carried out automatically in approximately 2 minutes, without operator error ...

More Specifics

additional benefits:
•Reduction/elimination of scrap
•No trial cuts required
•Ability to perform automatic part identification
•Automatic tool wear compensation
•Reduced setup times
•Dramatically improved productivity

Summary

$10,000 investment
More costly if retrofit
Operating—for any length of time—without these critical improvements will significantly impact productivity and company's bottom line.

Figure 5.26 | In business, engineers often have to prepare presentations like this PowerPoint presentation, which argues in favor of automating CNC operations.

benefits of a tool-setting probe, such as a touch trigger probe (TTP), you point out to the board that the majority of CNC machine tools are initially set up manually using feeler gauges, gauge blocks, and dial test indicators. This manual setup is prone to operator error, and it is time consuming as well. The default setup that your company has identified will require these manual operations. The $10,000 investment that you are advocating would automate this process using TTP, providing a time savings of up to 90 percent for basic operations and even greater saving for more complex operations. You then highlight the hourly wage of a typical machine operator and extrapolate the cost and time savings over a day, a week, a month, and a year. You even present published research that validates your point with the following example:

> ... the time required to set just ten tools manually on a CNC lathe is typically twenty minutes, but by using a touch trigger setting probe this can be carried out automatically in approximately two minutes, without operator error ...

You include in your presentation an Excel chart that visually expresses the anticipated savings. When asked for more specifics from the board, you are prepared with a slide that summarizes these and other additional benefits:

- Reduction/elimination of scrap
- No trial cuts required
- Ability to perform automatic part identification
- Automatic tool wear compensation
- Reduced setup times
- Dramatically improved productivity

You complete your presentation by emphasizing again that the equipment and required training you propose will be more costly as a retrofit and that operating the CNC machinery—for any length of time, without these critical improvements, will significantly impact productivity and the company's bottom line.

CNC machines have dramatically changed the manufacturing industry, making complex 3-D surfaces and curves relatively easy to produce and minimizing the number of machining operations that require human interaction. The result is improvements in consistency and quality, reduced operator errors, reduction in time to perform tasks, greater flexibility in the manufacturing process, and a greater variety of products that can be produced. Most modern CAD/CAM (Computer-Aided Design/ Computer-Aided Manufacturing) software can create files (e.g., STL or stereolithography) that can be used to control CNC machines. Using these tools, we can eliminate the need for blueprints and other printed documentation, transforming our design (CAD) directly into a manufactured component (Figure 5.27). The ability of CNC machines to run unattended overnight and on weekends enables us to produce thousands of parts; even allowing us to use a series of lasers and sensors to perform automated quality control.

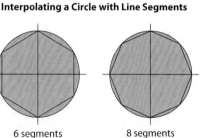

Figure 5.27 | CAD/CAM software eliminates the need for blueprints and other printed documentation.

G-Code Instructions

▷ Moving a machine tool along a straight-line (linear) path is the most basic **G-code instruction** we can create. While some machines are limited to planar (XY) motion with separate height (Z) instructions, others can perform full 3-D movement, that is, simultaneous XYZ motion. CNC machines take advantage of the fact that virtually any motion or path can be constructed from a series of small linear motions—usually a large number of these small motions.

In fact, here's some sample G-code to create a circle (see also Figure 5.28):

```
%
(g02 circle program)
g01 g90 x0 y0 z0 f2
g02 x0 y0 i-.5 j0
g01 g90 x0 y0 z0 f2
%
```

More advanced CNC interpreters combine G-code with logical commands, a technique known as **parametric programming**. Using syntax similar to the BASIC

Interpolating a Circle with Line Segments

| 4 segments | 6 segments | 8 segments | 11 segments |

Figure 5.28 | The most basic G-code can simply move a machine tool along a straight line. Circles like this are more complicated.

programming language, parametric programs incorporate *if/then/else* statements, *loops*, *subprograms, arithmetic operations,* and *variables.* These CNC machines can use a single parametric program to produce a full range of product sizes, by varying a single parameter. For example, to create a threaded bolt, we typically write our G-code to include the size of the bolt (length, diameter, bolt head size) and the thread characteristics (see Figure 5.29). With parametric programming, we create a program using a series of parameters—length, diameter, bolt head size, and thread characteristics. Now we can produce an entire family of bolts using this single program.

12-point screw   Carriage bolt   Elevator bolt

Compressor shear
pin   Ground shoulder
bolt   Special hex flange
screw

Figure 5.29 | You could write G-code that would allow a CNC machine to create any one of these bolts.

Automated Guided Vehicles (AGVs) Another important innovation in modern manufacturing environments is the Automated Guided Vehicle or AGV. An AGV is a mobile robot the helps us to increase efficiency and reduce costs by moving materials around a manufacturing plant or warehouse (Figure 5.30). Early AGVs were nothing more than tow trucks that followed a wire in the floor. Modern automated vehicles instead rely on lasers for guidance—LGVs—and are able to communicate with robots and other LGVs to ensure materials and products are moved smoothly through the warehouse. AGVs and LGVs play a critical role in modern factories and warehouses, safely moving goods to their destinations.

There are a variety of AGV types, including early, simple *Towing* vehicles with 4,000 to 30,000 kilograms (kgs) capacities, to smaller, light-duty *Light Load* AGVs with capacities as small as 200 kgs. More advanced *Unit Load*

Figure 5.30 | An automated guided vehicle, or AGV, is a mobile robot. An AGV can hugely increase the efficiency and reduce costs in a plant or warehouse.

AGVs are equipped with horizontal surfaces or decks, which enable us to load, transport, and transfer materials. In a warehouse environment, *Pallet Truck* AGVs are commonly used to transport palletized goods to and from the manufacturing floor. More flexible varieties include the *Fork Truck,* which can access floor-level, raised, and stacked loads, and the *Assembly Line* AGP, useful for serial assembly applications.

Automation raises several important social issues; in particular, its impact on the workforce. For many, automation harkens back to the Luddites. As discussed in the previous chapter, in the early 1800s, these English textile workers who feared losing their jobs to automated weaving looms engineered a social revolt. The term Luddite has since become a term generically applied to anyone who appears to be against the advance of technology. Another factor in the loss of manufacturing jobs is a phenomenon called outsourcing. Outsourcing—the shift of manufacturing enterprises and jobs to other countries—is an attempt by U.S. manufacturers to lower labor and capital costs. Like automation, outsourcing is often blamed for the loss of U.S. manufacturing jobs, but it is more likely that each has contributed to the demise of U.S. manufacturers.

The counterargument is that automation in fact leads to higher employment. According to this line of reasoning, automation frees up labor from low-skilled jobs and allows workers to pursue higher-skilled, higher-paying jobs. The very machinery responsible for automation requires a skilled workforce to maintain these machines. Although automation has continued to expand and has impacted the global workforce, some would argue that one important outcome of this trend has been an increase in the standard of living for people worldwide. Although it is difficult to quantify—or truly measure—the impact of automation on people's standard of living, it is possible to measure its impact on the manufacturing process. Automation, first introduced to increase productivity and reduce costs, has subsequently become important in quality control and flexible manufacturing. In addition to substantial increases in quality, automation has been important in taking on the role humans formerly played in carrying out hazardous operations. Automation also allows manufacturers the flexibility to easily convert from manufacturing one product to another without completely retooling and rebuilding their production lines.

TECHNOLOGY IN THE REAL WORLD:
Contour Crafting

The machine shown in Figure 5.31, a working prototype of a Contour Crafting machine, is fascinating. Try to figure out what this thing does. The name doesn't provide much help. Contour Crafting, developed by Dr. Behrokh Khoshnevis of the University of Southern California, is a layered fabrication technology that could automate the construction of whole structures and sub-components. Imagine in the near future building a single house or a colony of houses in just a fraction (1/200th) of the time that it takes using conventional techniques. Consider the implications for disaster recovery (think Hurricane Katrina) and low-income housing, as well as the social impacts and benefits of this technology; opportunities for community service, service learning, and addressing poverty; and expanding the boundaries of creativity and innovation in architecture and construction.

Figure 5.31 | Contour Crafting Machine

Under a grant from the National Science Foundation, Khoshnevis and his team are developing nozzle assemblies for full-scale construction applications that have a full six degrees of freedom. This means that these nozzles can move or rotate in three dimensions. Initial research will involve fabrication and testing of full-scale components including walls with built-in conduits and self-supporting roofs. Current Contour Crafting machines use ceramic materials, but are also capable of extruding clay and concrete. Tests will be conducted to evaluate a number of potential contour crafting materials including thermoplastics, thermosets, and ceramics. The new nozzle design has the ability to simultaneously extrude both inner filler material and the outer shell of a structure. Additionally, the nozzle design enables us to embed steel reinforcement as we lay our material. We can easily build a wide variety of complex curved structures.

In manufacturing, automation is well understood, and it has become commonplace. Unfortunately, traditional manufacturing automation methods and technologies, developed and refined for nearly a century, do not easily translate to a typical construction environment. It is difficult to automate the construction of large complex structures with internal elements. Contour Crafting could jumpstart the field of automated construction, enabling the fabrication of not only components, but also entire structures, leading to the automated construction of single houses and even entire neighborhoods (Figure 5.32). This approach is unique because it allows for automatic construction in a single run. It also enables us to embed conduits or channels that can house wiring, piping, and ducting for electrical, plumbing, and air-conditioning systems. In the future, this novel technology could help us establish a presence on the Moon and Mars, permitting us to quickly and inexpensively build living structures for the first inhabitants of these and other planets.

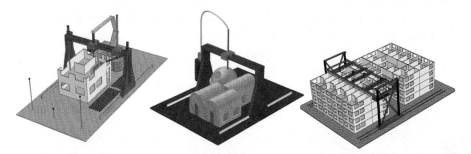

Figure 5.32 | Contour crafting is used to build a traditional, single family home, an adobe home, and a large apartment building.

SECTION THREE FEEDBACK >

1. With Henry Ford and the Model-T as your prototypical example, research and describe another early example of automation.

2. Explain the difference between automation and mechanization.

3. Research and briefly describe an additional instance of technology's displacement of workers, similar to the plight of the British textile worker. Be sure to include the long-term outcome.

4. Research jobs in manufacturing (automobile, aerospace, etc.) and detail the typical entry-level skills required for these jobs.

KEY IDEAS >

- In lean manufacturing, we develop our production schedule based on customer demand and actual needs, keeping inventory, workforce, and other resources lean.

- Rapid-prototyping or rapid manufacturing are becoming increasingly cheaper and more accessible, allowing us to swiftly create a physical model or representation of our product.

- Central to concurrent engineering is a multidisciplinary team (engineering, sales, marketing, manufacturing, consumers, etc.) working together from the beginning to the end of the product life cycle.

- Agile manufacturing provides companies with an organizational strategy and the processes, tools, and training to facilitate quicker decisions and produce a greater variety of new products faster, cheaper, and with greater quality.

- Marketing involves establishing a product's identity, conducting research on its potential, advertising it, distributing it, and selling it.

When we think of traditional manufacturing, we think of the assembly line turning out large numbers of the same part or product. Unfortunately, this approach has its limitations. In particular, we don't always need such a large quantity of product. In addition to the large volume of outputs that the traditional approach generates, we need a large number of resources and inputs (materials, people, energy, etc.) to sustain this process (see Figure 5.33). Let's consider fast food again. How do we know how many burgers or orders of fries to make? Any manufacturing enterprise—burgers, computer chips, and potato chips—develops a production schedule that in part tries to predict or forecast consumer needs. We purchase raw materials, hire people, and buy energy based on these predictions. We then "push" our product through the production line. Unfortunately, when our forecasts don't match reality, we end up with excesses in raw materials, people, energy, and capacity. In our fast food restaurant, this results in stale food sitting under a heat lamp, a freezer full of uncooked food, and employees standing around with nothing to do. In traditional manufacturing the inventory levels may be high, so we need warehouse space to hold products waiting for customer orders—our warehouses are the equivalent to burger warmers. Additionally, we need space for raw materials and goods still in production waiting for the next stage of processing.

In lean manufacturing, we develop our production schedule based on customer demand and actual needs. We keep our inventory, workforce, and other resources lean. What do we mean by "lean"? Lean, in this case, refers to the lack of excess. Lean means we plan short production runs in response to

Manufacturing Systems Diagram

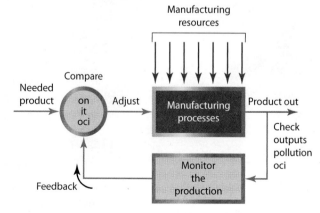

Figure 5.33 | A manufacturing system is made up of inputs, processes, and outputs.

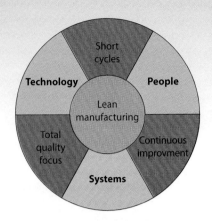

Figure 5.34 | Lean manufacturing requires efficient use of technology, people, and systems.

immediate demands. Rather than "pushing" inventory through our plant—the traditional model—we "pull" inventory through, as we need it. We hire only enough people, purchase enough materials, and build enough capacity to do the job. The challenge is to have an adaptable schedule that can handle customer orders quickly and economically (see Figure 5.34). Our "lean" burger restaurant works this way. Customers enter, order their food, and it is cooked to order. We have the ability to "ramp up" or increase our production during the lunch rush, but we spend the remaining time producing burgers as dictated by customer demand. Lean manufacturing means smaller spaces for raw materials, in-production products, and final products. Since the production run is in response to a customer order, the product moves from stage to stage with little accumulation between stages—there is no excess number of parts. Rather than gathering dust in a warehouse, the final product is shipped immediately to the customer. Of course, this requires very good relationships with suppliers of raw materials, so they have the material needed on hand in the appropriate quality and quantity. There are many benefits to lean manufacturing—improved labor utilization, decreased inventories, and reduced manufacturing-cycle times.

As you would expect, our approach to quality is different for each of these systems. In the traditional manufacturing, the product is randomly sampled and tested, with little worker input into the process. In lean manufacturing, the worker is part of the testing process. The goal is to have the entire product tested as it being manufactured, so workers are empowered to check the product as it is produced. To assist our workers in quality assurance, we often use machines that have tolerance limits built in. If the machine begins to operate or produce products that are out of tolerance, the worker will be alerted or the machine may even shut down. With small production runs, this employee involvement is critical to ensuring product quality, reducing costs, and shortening time-to-market.

Rapid manufacturing

One important trend in manufacturing is rapid-prototyping or rapid manufacturing. Rapid-prototyping and rapid manufacturing are becoming increasingly cheaper and more accessible, allowing us to very quickly create a physical model or representation of our product (Figure 5.35). Imagine that you are tasked with creating a new type of cell phone for "texters" like yourself. Not a traditional cell phone, but something radically different from anything that has come before it. How would you show people your vision—what "your" cell phone would look like? Obviously, you could describe it; you could sketch it on paper or even create a 3-D computer model. Imagine, instead, that you could create a physical model that your friends can hold and manipulate as you describe your vision of next-generation cell phones. Rapid-prototyping allows us to do this—quickly and cheaply. Using these techniques, we can consider and "market-test" a variety of alternative designs.

Certain rapid-prototyping process can even be termed rapid manufacturing, because the quality of the part is acceptable for production, not merely for prototyping. Rapid manufacturing can be applied to the creation of custom or replacement parts, and it can be used in small-batch manufacturing. Methods for rapid prototyping and manufacture include stereolithography (SL), discussed in Chapter 4, as well as selective laser sintering (SLS) and fused deposition modeling (FDM), just to name a few. SLS is similar to SL in that we use a laser to build our part layer-by-layer. The major difference between the two techniques is that with SLS we use a fine powder—metal, ceramic, or plastic—instead of a liquid polymer. With FDM, we use a spool of plastic filament,

Figure 5.35 | Using a rapid prototyping machine, you could create a physical model of a new product. Then you could market test the prototype before you invest in large scale production.

which we heat up to its melting temperature and deposit or lay out layer by layer to build our solid. The filament is heated and extruded, or pushed out, through a nozzle.

Concurrent Engineering

Concurrent engineering is a systematic approach to integrated product design and development (Figure 5.36). Central to concurrent engineering is a multidisciplinary team (engineering, sales, marketing, manufacturing, consumers, etc.), all working together from the beginning to the end of the product life cycle. Marketing involves establishing a product's identity, conducting research on its potential, advertising it, distributing it, and selling it. Often, collaboration software and a set of common or shared design parameters are used to share exchange information, ensuring that any changes impacting other disciplines are identified and addressed.

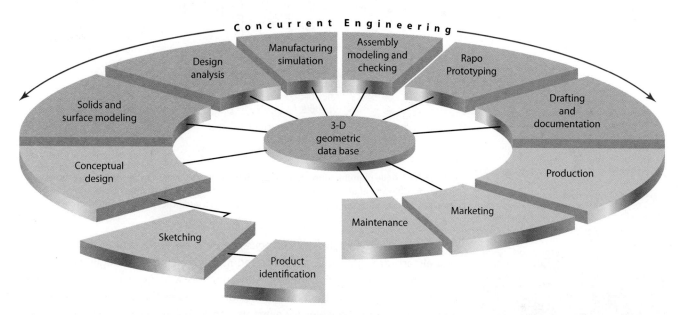

Figure 5.36 | In concurrent engineering, all aspects of the project (conception, materials selection, design, testing, manufacturing, production, market, etc.) are considered as a whole, rather than isolated from one another.

Because everyone is involved from the beginning to the end, concurrent engineering fits very well with TQM programs, allowing high-quality products to be produced more quickly and flexibly than by traditional engineering. For instance, a major aerospace company has found out that changes that cost $1 when made early in the development cycle could rise dramatically to $10,000 during the production phase. Another advantage of concurrent engineering is that the duration of the manufacturing cycle is reduced. A major manufacturer of exercise equipment was able to cut its time to manufacture a new product in half by implementing concurrent engineering and total quality assurance programs.

Agile Manufacturing

Agile manufacturing and its predecessors, i.e., computer-integrated manufacturing, flexible manufacturing, and concurrent engineering, are all ways to increase competitiveness in a global market. Increasingly, companies are challenged to quickly respond to changing customer needs and demands. Agile manufacturing provides companies with an organizational strategy and the processes, tools, and training to facilitate quicker decisions and produce a greater variety of new products faster, cheaper, and with greater quality (see Figure 5.37). When you think of agile manufacturing, focus on terms like "teamwork" (see Figure 5.38), "fast," "customer-centered," and "flexible."

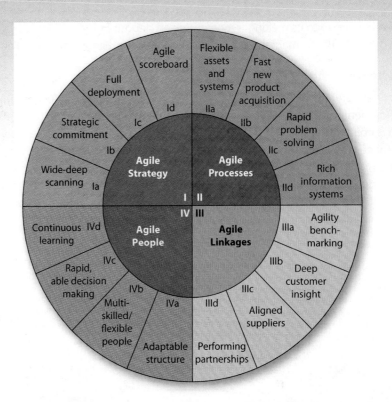

Figure 5.37 | Agile manufacturing brings together a highly skilled, flexible workforce, processes that are flexible and enable rapid action; a strong, company-wide corporate commitment; and a strong alignment and connection to customers, suppliers, and other external and internal partners.

Figure 5.38 | Like the members of crew, workers in an agile manufacturing company have to work together as a team.

To transform themselves into an agile company, enterprises need to build on top of an already strong foundation. As a prerequisite for operating with agility, companies should foster strong relationships with suppliers and even competitors. These partnerships should be built around flexible strategies that can adapt as markets and needs change. Consider, for example, the global virtual company that Boeing has established to produce the 787 Dreamliner (Figure 5.39). Agile manufacturing requires that a company has already mastered advanced manufacturing techniques, such as lean production and concurrent or simultaneous engineering. Facilities should be flexible as well, populated with flexible tooling and automation, and demonstrating strong integration of technology, management, and workforce into a coordinated, independent system.

One key characteristic of agile manufacturing enterprises is the role of the customer. From the beginning of the product-delivery cycle, the enterprise is trying to learn who their customers are, what they say, what they really want, what they will pay for, how to make them happy with the product, and how to improve the product and customize it. They want to gain a deep understanding of the customer, usually through real-time market data collection and analysis. Agile manufacturing environments are dynamic, requiring a multi-skilled, flexible labor force where blue- and white-collar workers have similar educational levels.

Throughout our discussion of quality improvement and manufacturing, we refer over and over to teamwork. Teamwork is a hallmark of modern engineering and manufacturing and is critical for agile manufacturing. Organizations should be flat, rather than hierarchical, with continuous interaction, support, and communications among various disciplines. Within these companies, successful teams are multidisciplinary and include members from different professions (marketing, finance, engineering), and skilled trades and operators. This flat organizational structure enables practical

Figure 5.39 | The Boeing 787 Dreamliner is the most advanced commercial aircraft that exists. It is an excellent example of a product developed through global partnerships and agile manufacturing.

knowledge—details of tooling, fabrication, and programming—to diffuse throughout the organization. This diffusion of knowledge and company-wide sharing of expertise is critical to operating as an agile organization (see Figure 5.37).

Numerical and computational tools for doing analysis have grown at a rapid rate. In an agile manufacturing firm, there is no time to make a number of rapid prototypes, so modeling and simulation of both product designs and processes to manufacture and assemble the products will be key to the on-time delivery of a customized product. Of course, rapid prototyping technology can be used to make some limited-volume products, but it is more likely that product diversity and customization will be achieved by modularity, or by built-in programmable features implemented in designs that will still allow us to deliver products on time.

Nanotechnology Manufacturing

Manufacturing is a common activity. It is through manufacturing processes that raw materials become usable products. Key steps in a typical manufacturing sequence include obtaining the basic materials for processing, modifying the materials through machining processes and assembling various component parts into the final product.

The manufacturing industry familiar to most people is carried out on a large scale. At this level, materials are modified, adjusted, and manipulated by large machinery to create individual component parts. Drilling, cutting, and milling are common manufacturing processes used to alter materials. People and machines assemble or attach the components created through the various manufacturing steps to other components, resulting in products for sale such as automobiles, refrigerators, and televisions. There are also manufacturing processes that produce small components and products, such as watches, DVDs, and books. Still other manufacturing processes, carried out at extremely small scales—the micro and nano levels—require highly specialized tools and techniques. Just as in large scale manufacturing, raw materials are collected and modified so that useful micro and nano scale products are produced. Electronic circuitry formed on integrated circuits is an example of products manufactured at both the micro and nano scales.

The general manufacturing processes you think about are probably examples of top-down manufacturing (TDM). In top-down manufacturing environments, suitable materials are obtained by the manufacturer and then modified by humans and machines until a working product results. Modification includes removing portions of the material until a final desired shape is obtained. If you think about how a block of metal is altered to become an automobile engine, then you are thinking about a TDM process. The engine reaches its proper dimensions after much drilling and machining converts the metal block into the desired shape with the desired tolerances, features, and specifications.

Even at the micro and nanoscale levels, top-down manufacturing occurs. Integrated circuits (ICs) are manufactured through many process steps involving the removal and machining of materials. When it comes to the integrated circuit industry, the word "fabrication" is used, instead of "manufacturing," to describe how circuits are built. The term "fabrication" is commonly used to describe both the manufacturing environment and steps involved in producing an integrated circuit. This is why people involved in integrated circuit manufacturing call the factory where integrated circuits or chips are produced a "chip fab." The chip fab is more than a factory. It is a highly sophisticated complex of buildings, clean rooms, and offices costing several billion dollars to build.

When you study the processes required to build tiny objects, it becomes apparent how difficult it is to build things at the nanoscale (see Figure 5.40 for a better idea of the scale we are talking about). As manufacturing processes are adapted for fabrication of objects at the nanoscale, the laws of physics begin to limit how small we can make things and the reliability of the fabrication processes. Experts predict that the current manufacturing processes used in the integrated circuits industry will reach their limits in roughly ten years; it will then not be possible to make things smaller with current techniques. Consequently, the integrated circuit and semiconductor industry is working to develop other unconventional manufacturing processes.

Bottom-up manufacturing (BUM) is an alternative to top-down manufacturing. In bottom-up nanoscale manufacturing, sophisticated tools are used to coax atoms into place to produce a product. Individual atoms are actually moved one by one into place to create new and different objects. This is truly building things from the bottom up. There are also other kinds of bottom-up processes where atoms, given the right conditions, may assemble themselves into the product desired without using any specialized tools. Bottom-up methods are new and need refinement to become commercially successful, but they are the future in nanoscale manufacturing.

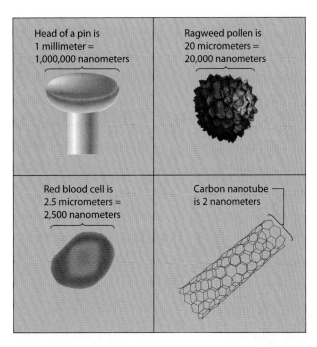

Head of a pin is
1 millimeter =
1,000,000 nanometers

Ragweed pollen is
20 micrometers =
20,000 nanometers

Red blood cell is
2.5 micrometers =
2,500 nanometers

Carbon nanotube is 2 nanometers

Figure 5.40 | These visual cues should give you a sense of the difference between the measurement scales you are used to and nanoscale.

Potential Benefits of Nanotechnology

The application of novel properties and processes associated with the nanoscale promises many benefits for our everyday lives. Highly efficient nanoscale manufacturing processes are envisioned for the efficient use of energy and materials, as well as minimization of environmental impact and waste from industrial productions. Examples that demonstrate the promise of nanotechnology, with projected total worldwide market size of over $1 trillion annually in ten to fifteen years, include the following:

- Materials and manufacturing: The nanometer scale is expected to become a highly efficient length scale for manufacturing once nanoscience provides the understanding and nanoengineering develops the tools. Materials with high performance, unique properties and functions will be produced that traditional chemistry could not create. Nanostructured materials and processes are estimated to increase their market impact to about $340 billion per year in the next ten years.

- Electronics and computing: Nanotechnology is projected to yield annual production of about $300 billion for the semiconductor industry and about the same amount more for global integrated circuits sales within ten to fifteen years.
- Health care and medicine: Nanotechnology will help prolong life, improve its quality, and extend human physical capabilities.
- Pharmaceuticals: About half of all production will be dependent on nanotechnology, affecting over $180 billion per year in ten to fifteen years.
- Chemical plants: Nanostructured catalysts have applications in the petroleum and chemical processing industries, with an estimated annual impact of $100 billion in ten to fifteen years.
- Transportation: Nanomaterials and nanoelectronics will yield lighter, faster, and safer vehicles and more durable, reliable, and cost-effective roads, bridges, runways, pipelines, and rail systems. Nanotechnology-enabled aerospace products alone are projected to have an annual market value of about $70 billion in ten years.

[Courtesy the NaMCATE Project, SUNY at Buffalo]

SECTION FOUR FEEDBACK >

1. What is lean manufacturing?
2. What are the benefits of rapid prototyping?
3. Compare and contrast agile manufacturing and concurrent engineering.
4. Define bottom-up manufacturing.

SECTION 5: Manufacturing as a Global Enterprise

KEY IDEAS >

- Global challenges are forcing many organizations to reshape the way they operate to maintain their global competitiveness.

- In today's global economy, successful manufacturing firms are not those that discover new technology, but instead are those able to identify the needs of customers and bring products that meet those needs to market in a timely manner and at low cost.

It is fairly obvious that you live in a world that is significantly different than that of your grandparents, or even your parents. We develop and use technology that is unlike anything earlier generations could even imagine. The impact of technology

Figure 5.41 | When you think of the modern economy, think of the globe. New technologies affect the entire world.

is no longer limited to our small sphere of experience. Instead, technology impacts the entire globe, and we are, in turn, influenced by technology from all over the world (Figure 5.41). North America, once the undisputed economic and industrial world super power, is being challenged by growing countries such as Japan, Korea, China, and India. These countries are not only growing into economic and industrial powers, but many are growing in population. While the population of North America—a small portion of the world's population—declines with dropping birth rates, many Third World countries are experiencing high growth rates and population explosions. This imbalance compounds the ongoing shift in wealth, industrial power, and technical knowledge from east to west.

Global challenges are forcing many organizations to reshape the way they operate to maintain their global competitiveness. Companies are changing work and management styles to build flatter organizations and encourage more teamwork. They are increasing their use of computers and automation, including computer-aided design (CAD), computer-integrated manufacturing (CIM), Just-In-Time (JIT) inventory control, and computer-control of other operations (purchasing, warehousing, etc.). Companies are working harder than ever to produce world-class products; they are focusing on the global customer, global supply chains, and an increasingly global workforce. To remain competitive nationally and globally, companies have abandoned traditional approaches to manufacturing and implemented flexible, automated manufacturing systems. An enterprise's manufacturing capability must be flexible enough to produce a wide range of quality products, to incorporate new technology into these products quickly and cost effectively, and to introduce them into the market ahead of the competition.

In today's global economy, successful manufacturing firms are not those that discover new technology, but they are those able to identify the needs of customers and bring products that meet those needs to market in a timely manner and at low cost. Technological superiority alone is no longer sufficient for economic prosperity. Leadership in creating new technologies pays off only when that technology is rapidly implemented and embedded in high-quality, cost-competitive, innovative products that meet or exceed customer expectations.

Product life cycles have compressed dramatically. Competitors introduce next-generation products and technology before the former products and technologies are obsolete. Early market entry provides a critical competitive edge, so the motivation to innovate is intense. New technology is developed, embedded in products, and brought to the marketplace quickly. Today's diverse and discriminating customers—who demand products that meet specialized needs and feature the latest technology—heighten this need to innovate. If this were not challenging enough, the fragmenting and rapidly changing global markets have made flexibility and rapid response to change essential manufacturing concepts.

Emblematic of the global nature of manufacturing and its changing methods and processes is the effort made by Boeing Company in its development of the 787 Dreamliner. Boeing is leading a group of over forty global partners distributed through sixteen states and ten countries. Each company is responsible for various subsystems of the Dreamliner 787 aircraft. This global effort is unique in a number of ways, including the fact that partners are not only sharing the design responsibility for their components, but also the risks, as investors in the project. They are part of a global virtual-development team where every aspect of the aircraft and its manufacturing process is designed, created, and tested digitally before anything physically moves into production.

Being able to "virtually" simulate not just the parts, but also the aircraft's processes has been a great benefit for Boeing, giving them more flexibility to make adjustments during the design phase. During a virtual flight test of the 787, the chief pilot noticed some issues related to fin control (see Figure 5.42). Designers were able to evaluate fifty new possible fin configurations, test them, and make the appropriate changes to the rest of the design in only about four weeks. In the past, only three or four new fin configurations may have been evaluated, taking three to four months.

The approach taken with the 787 (Figure 5.43) is in contrast to earlier efforts, where Boeing designed 70 percent of its aircraft and produced 30 percent. In the past, Boeing would bring suppliers and partners to Bellevue, Washington, during the design phase. But that approach is too expensive and impractical for a project of this global scale. Boeing's first foray into virtual development was the 777 aircraft. Using three-dimensional computer-aided design software, Boeing was able to create a digital model for each of the aircraft's nearly 10,000 parts, as opposed to building 10,000 physical prototypes. With a full 3-D product definition, Boeing was able to improve quality and reduce the cost of assembling the aircraft.

Figure 5.42 | CAD software makes it possible to run virtual flight tests of an airplane that hasn't even been built yet.

Airplane Parts Definitions and Functions

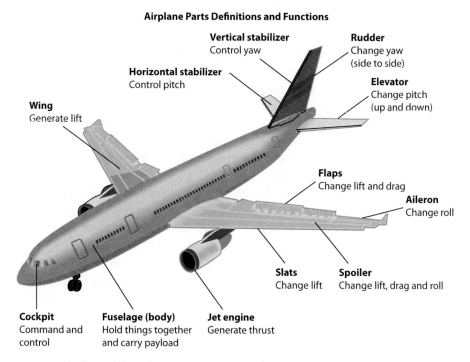

Vertical stabilizer
Control yaw

Rudder
Change yaw
(side to side)

Horizontal stabilizer
Control pitch

Elevator
Change pitch
(up and down)

Wing
Generate lift

Flaps
Change lift and drag

Aileron
Change roll

Slats
Change lift

Spoiler
Change lift, drag and roll

Cockpit
Command and control

Fuselage (body)
Hold things together and carry payload

Jet engine
Generate thrust

Figure 5.43 | It's incredible to think of what was involved in the development of the Boeing 787, when you consider the complexity of a typical airplane and the critical nature of nearly every component.

Based on the success of the 777, Boeing decided to expand the idea even further with the 787 Dreamliner program. With the 787, Boeing leveraged a common digital environment to help a dispersed global design team more effectively collaborate and leverage a single 3-D product definition throughout all phases of the 787's twenty to thirty year life cycle. The approach allows Boeing to more efficiently design and manufacture a product with the highest quality. The approach reduces the amount of data translations and manual processes required, as well.

With the 787, much of the design responsibility fell to the partners, with Boeing focused on system integration and configuration. This large paradigm shift addresses a key problem from past efforts—manufacturers were disconnected from the design, limiting Boeing's ability to make improvements to the product. By making the designer and manufacturer one and the same, Boeing was able to bridge this disconnect, resulting in a lower-cost product that is easier to maintain.

As you can see, manufacturing impacts nearly every facet of our daily lives. It has been and remains a critical piece of the national and global economy and makes significant contributions to the national infrastructure and national defense. Unfortunately, the rising cost of manufacturing (labor, materials, and physical plant) has forced many companies and even entire industries to close their operations or move them to other nations. This phenomenon—referred to as off-shoring—continues to challenge employers and workers alike.

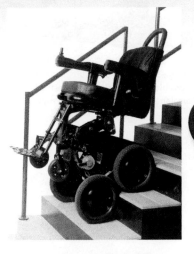

Figure 5.44 | Among other things, Dean Kamen is the inventor of the Segway scooter, a mobile dialysis system, an insulin pump, and an all-terrain electric wheelchair.

TECHNOLOGY AND PEOPLE:
Inventor Dean Kamen

Dean Kamen is an inventor and founder of DEKA Research and Development Corporation. (DEKA is the contracted version of "Dean Kamen.") He is probably best known as the inventor of the Segway human transporter. Kamen, born in 1951 in Rockville Center, New York, holds more than 440 U.S. and foreign patents and was inducted into the National Inventors Hall of Fame in 2005. While still in college, Kamen invented what would become AutoSyringe®, a portable/wearable infusion pump for diabetics, capable of delivering insulin. Another important invention, the Ibot Transporter, is a six-wheeled robotic wheelchair that can climb stairs, negotiate difficult terrain, and even balance on two wheels (Figure 5.44). Kamen was inspired to invent the Ibot after watching a man in a wheelchair unable to get over a curb (see Figure 5.45).

Kamen does not invent for invention's sake, but instead works with a profound belief that technology and creativity are critical for improving life for humankind. At its best, technology, he believes, can provide solutions to the problems presented by pollution, contaminated water, and limited access to electricity. One of the key projects in this area is the development of a Stirling engine, which would generate power while simultaneously purifying water. As a child, Kamen developed an incredible knowledge of the physical sciences through study of primary texts such as Isaac Newton's *Principia Mathematica*. The basement of his house is an incredible workspace, with a foundry, a fully-equipped machine shop, and a computer room.

Figure 5.45 | Dean Kamen demonstrates the iBOT to President Bill Clinton.

Recently, Kamen has taken on the challenge of creating the next-generation prosthetic arm for Iraq War veterans missing either one or both arms. The Luke Arm, as Kamen has dubbed it, would replace technology—which essentially consists of a stick for an arm and a hook for a hand—that hasn't changed much since the American Civil War.

Offshoring

In the developed world, the trend of moving jobs out of the country can be traced back to the 1970s. Offshoring, as this phenomenon is called, involves the transfer of business processes from one country to another. The affected business unit could include production, manufacturing, or services, although the majority of offshoring has impacted manufacturing enterprises through the relocation of factories from the developed to the developing world. The primary reason companies choose to offshore a portion of their operations is to reduce costs, taking advantage of favorable labor, capital, and economic environments. The explosive growth of the Internet

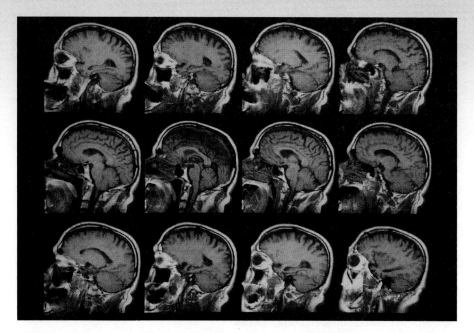

Figure 5.46 | Magnetic resonance imaging (MRI) is a technology that is used to create images of the body, similar to X-rays. MRIs created in one country are often read and interpreted by doctors in another country, with the results sent back to the patient's doctor via the Internet. This practice of transferring businesses processes from one country to another is called offshoring.

has accelerated this trend, affecting many categories of employment, including call centers, computer programming, reading and interpreting medical data (X-rays and MRIs, see Figure 5.46), medical transcription, income tax preparation, and title searching.

Offshoring is often confused with the similar practices of outsourcing or offshore outsourcing. Outsourcing is the movement of an internal business function to an external company, such as hiring a contract to manage and maintain your company's entire information technology (IT) infrastructure. The subcontractor taking on these formerly internal functions may be located in the same country as your company, which would be outsourcing but not offshoring. Moving an internal business unit from one country to another would be offshoring, but not outsourcing. Finally, subcontracting a business function to a different company located in another country would be considered both outsourcing and offshoring.

SECTION FIVE FEEDBACK >

1. Research the origins of five manufactured products you use daily (conception, design, analysis, prototype, manufacture). More specifically, include where these specific functions were performed.

2. Research and identify the "flatteners" Thomas Friedman discusses in his book, *The World is Flat: A Brief History of the Twenty-First Century*. What impact do these have on manufacturing?

3. Interview your grandparents, or someone from that same generation, to determine the scope and scale of the manufacturing industry when they were growing up and what has happened to it since that time.

4. Briefly describe the differences between offshoring and outsourcing.

5. Research NAFTA; describe what it is and detail the arguments for and against NAFTA.

Matching Your Interests and Abilities with Career Opportunities: Aerospace Product and Parts Manufacturing

The way in which commercial and military aircraft are designed, developed, and produced continues to undergo significant change in response to the need to cut costs and deliver products faster. Firms producing commercial aircraft have reduced development time drastically through computer-aided design and drafting (CADD), which allows firms to design and test an entire aircraft, including each individual part, by computer; the drawings of these parts can be sent electronically to subcontractors who use them to produce the parts. Increasingly, firms bring together teams composed of customers, engineers, and production workers to pool ideas and make decisions concerning the aircraft at every phase of product development. Additionally, the military has changed its design philosophy, using commercially available, off-the-shelf technology when appropriate, rather than developing new customized components.

Commercial airlines and private businesses typically identify their needs for a particular model of new aircraft based on a number of factors, including the routes they fly. After specifying requirements such as range, size, cargo capacity, type of engine, and seating arrangements, the airlines invite manufacturers of civil aircraft and aircraft engines to submit bids. Selection is ultimately based on a manufacturer's ability to deliver reliable aircraft that best fit the purchaser's stated market needs at the lowest cost and at favorable financing terms.

Significant Points

* Skilled production, professional, and managerial jobs account for the largest share of employment.
* Employment is projected to grow more slowly in this industry than in industries generally.
* During slowdowns in aerospace manufacturing, production workers are vulnerable to layoffs, while professional workers enjoy more job stability.
* Earnings are substantially higher, on average, than in most other manufacturing industries.

Nature of the Industry

Companies in the aerospace industry produce aircraft, guided missiles, space vehicles, aircraft engines, propulsion units, and related parts. The largest segment of the industry focuses on civilian or nonmilitary aircraft for airlines and cargo transportation companies. Typically, aircraft-engine manufacturers produce the engines used in both civil and military aircraft. The remainder of the industry builds aircraft, helicopters, guided missiles, missile propulsion, space vehicles, and rockets for military and government organizations. A few large firms dominate the aerospace industry, accounting for nearly 63 percent of aerospace manufacturing jobs. These large firms subcontract with smaller firms—fewer than 100 workers, which make up the bulk of the about 2,800 companies in the aerospace industry.

Working Conditions

The average aerospace production employee worked 42.6 hours a week in 2004, compared with 40.8 for all manufacturing workers and 33.7 for workers in all industries. Engineers, scientists, and technicians frequently work in office settings or laboratories, although production engineers may spend much of their time on the factory floor. Production workers, such as welders and assemblers, may have to cope with high noise levels, oil, grease, grime, and exposure to volatile compounds. Heavy lifting is required for many production jobs.

Nature of the Industry

The design and manufacture of technologically sophisticated aerospace products requires a variety of workers. Skilled production, professional and related, and managerial jobs make up the bulk of employment. Of all aerospace workers, 37 percent are employed in production; installation, maintenance, and repair; and transportation and material-moving occupations. The aerospace industry has a larger proportion of workers with education beyond high school than the average for all industries. Professionals, such as engineers and scientists, require a bachelor's degree in a specialized field. For some jobs, particularly in research and development, a master's or doctorate degree may be preferred. Entry-level positions for technicians usually require a degree from a technical school or junior college. Production workers must have minimal skills, mechanical aptitude, and good hand-eye coordination. A high school diploma or equivalent is required, and some vocational training in electronics or mechanics also is favored.

Training and Advancement

Due to the rapid technological advancements in aerospace manufacturing, the industry provides education and training for its workers, including on site, job-related training to upgrade the skills of technicians, production workers, and engineers. Some firms reimburse employees for educational expenses at colleges and universities, emphasizing four-year degrees and postgraduate studies. Unskilled production workers typically start by being shown how to perform a simple assembly task. Through experience, on-the-job instruction provided by other workers, and brief formal training sessions, they expand their skills. Their pay increases as they advance into more highly skilled or responsible jobs. With training, production workers may be able to advance to supervisory or technician jobs. Workers in more highly skilled production occupations must go through a formal apprenticeship. Machinists and electricians complete apprenticeships that can last as long as four years and usually include classroom instruction and shop training.

Outlook

Employment in aerospace manufacturing is expected to grow by 8 percent from 2004 to 2014, slower than the 14-percent growth projected for all industries combined. The outlook for the military aircraft and missiles portion of the industry is better. Employment in this sector is expected to rise. Even though the number of large firms performing final assembly of aircraft has been reduced, hundreds of smaller manufacturers and subcontractors will remain in this industry. In addition to some growth in employment opportunities for professional workers in the industry, there should be job openings to replace retiring workers, especially for aerospace engineers.

Earnings

Production workers in the aerospace industry earn higher pay than the average for all industries. Weekly earnings for production workers averaged $1,019 in

2004, compared with $659 in all manufacturing and $529 in all private industry. Above-average earnings reflect, in part, the high levels of skill required by the industry and the need to motivate workers to concentrate on maintaining high quality standards in their work. Engineering managers, engineers, and computer specialists command higher pay because of their advanced education and training.

[Bureau of Labor Statistics, U.S. Department of Labor, Occupational Outlook Handbook, 2008–09 Edition, visited March 2008, http://www.bls.gov/oco/]

Summary >

Durable goods are designed to operate for a long period of time, while nondurable goods are designed to operate for a short period of time. The manufacturing systems that create these goods may be classified into types, such as customized production, batch production, and continuous production. Common manufacturing systems include the job shop, flow shop, project shop, continuous process, and the newer linked-cell system. Continuous processes usually involve chemical technologies, which provide a means for humans to alter or modify materials and to produce chemical products. For any system, the pressure to keep costs down and simultaneously meet customer demands is always balanced by the need for quality and reliability. The overarching goal in manufacturing is to design and implement manufacturing systems that provide low cost, superior quality, and on-time delivery.

The pursuit of quality begins at product conception, and follows through the design, analysis, manufacture, and delivery of the product to the customer. As materials have different qualities and may be classified as natural, synthetic, or mixed, we must consider materials early in the process. A fundamental challenge in designing and manufacturing a new product is balancing the cost of measuring, assessing, and controlling quality versus the benefits. Companies must produce products that are competitive in price and better in performance than others on the market. Total Quality Management, which extends beyond manufacturing a product, into design, analysis, marketing, sales, distribution, and billing, is one method for balancing quality and benefits. Servicing our equipment is an important consideration that keeps products in good operating condition and helps us maintain quality. Safety and quality are issues of great concern to all engineers, who must ensure that when products fail they do not become safety hazards.

Mass production is the production of large amounts of standardized products on production lines. The interchangeability of parts increases the effectiveness of manufacturing processes. The goal of automation is for everything—services, manufacture, materials and information handling, assembly, and inspection—to be performed totally automatically. Automation raises several important social issues—in particular, its impact on the workforce.

There are a number of important trends and techniques to consider in manufacturing. In lean manufacturing, we develop our production schedule based on customer demand and actual needs, keeping inventory, workforce, and other resources lean. Rapid-prototyping or rapid manufacturing are becoming increasingly cheaper and more accessible, allowing us to very quickly create a physical model or representation of our product. Concurrent engineering is a multidisciplinary team (engineering, sales, marketing, manufacturing, consumers, etc.) working together from the beginning to the end of the product lifecycle. Agile manufacturing provides companies with an organizational strategy and the processes, tools, and training to facilitate quicker decisions and produce a greater variety of new products faster, cheaper, and with greater quality. Marketing involves establishing a product's identity, conducting research on its potential, advertising it, distributing it, and selling it.

Global challenges are forcing many organizations to reshape the way they operate to maintain their global competitiveness. In today's global economy, successful manufacturing firms are not those that discover new technology, but instead are those able to identify the needs of customers and bring products that meet those needs to market in a timely manner and at low cost.

FEEDBACK

1. Identify a career that still requires serving an apprenticeship.
2. Explain why early manufacturing plants were located near rivers.
3. What is the difference between a durable good and a nondurable good?
4. Briefly describe the goal of Just-In-Time manufacturing.
5. If you were to create your "perfect" cell phone, what would it look like? What specific functions would it have? What would this cell phone be able to do that your existing cell phone cannot do?
6. List the four steps in the Total Quality Management (TQM) process.
7. Why is interoperability of parts so important in automated manufacturing?
8. Describe the impact of Computer Numerical Control on the manufacturing process.
9. List four types of AGVs and describe how they are used in automated manufacturing.
10. What are rapid manufacturing and prototyping and what are some applications?
11. What is the goal of concurrent engineering?
12. Describe agile manufacturing and give an example.
13. Why are teamwork and multidisciplinary teams so important in modern manufacturing systems?
14. What is the difference between offshoring and outsourcing?
15. Explain how the Boeing 787 project was different from any other and why it is significant. As needed, use the Internet to research articles on the Boeing 787.

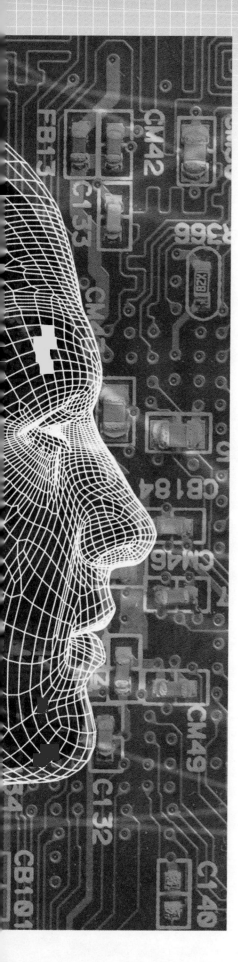

DESIGN CHALLENGE 1:
Manufacturing an Antacid-Powered Rocket

• Problem Situation

We have learned the United States is competing in an increasingly global economy. As the number of scientists, engineers, and technologists in other countries such as China and India is growing, fewer students pursue these careers in this country. Getting young kids interested in careers in science, technology, engineering, and mathematics is a national imperative that many organizations have begun to address. For example, NASA (the National Aeronautics and Space Administration) has developed a number of exemplary materials for this very purpose. One such resource is *3-2-1 POP!*, an activity requiring students to build a small rocket with construction paper or heavy-weight stock paper.

More students might be able to try out the *3-2-1 POP* activity if they had easy access to the material required for the activity. In this Design Challenge, you will develop a means of mass-producing and marketing rocket kits that contain all the material required to build a rocket. You will use the *3-2-1 POP!* activity as a starting point.

• Your Challenge

Given the necessary materials, you will research rockets and propulsion systems, develop a prototype rocket, consider alternative materials and propulsion systems, and finally create the jigs, fixtures, and processes for the small-scale mass production of your chosen prototype. You will then design a packaging and marketing strategy that will excite younger students about careers in engineering and technology and encourage more students to pursue these careers.

• Safety Considerations

1. Use a great deal of care when working with electricity and power tools.
2. Wear eye protection at all times.

• Materials Needed

1. NASA's *3-2-1-POP!* activity
2. Construction paper
3. A variety of heavy-weight (60-110#) paper
4. Flexible plastic sheets
5. A variety of stock pieces of wood and aluminum (block, cylinders, cones, etc.)
6. A variety of fastener materials, including tape, glue, or additional means your instructor may provide
7. Access to a machine shop or hand tools
8. Scissors and an X-acto knife
9. Plastic 35-mm film canisters and other similar small containers with lids
10. Effervescing antacid tablets
11. Plastic packaging bags.
12. Self-adhesive labels
13. Markers
14. Stopwatch

• Clarify the Design Specifications and Constraints

Working as a team, you will create and demonstrate a small-scale mass production unit that produces rocket kits promoting careers in engineering and technology. The rocket should be safe to operate and use no hazardous materials. It should be easy for young people to construct and assemble. The production process should be time and cost effective and provide end-products that are easily reproducible, safe, and of high and consistent quality. Your solution should also include a packaging, labeling, and marketing strategy that is attractive and appealing to young people.

• Research and Investigate

To complete the design challenge, you need to first gather information to help you build a knowledge base.

1. In your guide, complete the Knowledge and Skill Builder I: Web safety.
2. In your guide, complete the Knowledge and Skill Builder II: Shop and tool safety.
3. In your guide, complete the Knowledge and Skill Builder III: Selecting materials.
4. In your guide, complete the Knowledge and Skill Builder IV: Refining an existing design.
5. In your guide, complete the Knowledge and Skill Builder V: Prototyping and testing a design.
6. In your guide, complete the Knowledge and Skill Builder VI: Jigs and fixtures.
7. In your guide, complete the Knowledge and Skill Builder VII: Mass producing a small product.
8. In your guide, complete the Knowledge and Skill Builder VIII: Quality
9. In your guide, complete the Knowledge and Skill Builder IX: Product marketing.

• Generate Alternative Designs

Describe two possible alternatives to your production process. Discuss the decisions you made in (a) materials selection, (b) propulsion system, (c) production process, and (d) marketing strategy. Attach printouts, photographs, and drawings if helpful and use additional sheets of paper if necessary.

• Choose and Justify the Optimal Solution

What decisions did you reach about the design of the rocket, the production process, and the marketing strategy?

• Display Your Prototypes

Produce your completed prototype, packaged, labeled, and ready for distribution. Provide details of the production process, alternatives considered, and average time required to produce a single product.

• Test and Evaluate

Explain if your designs met the specifications and constraints. What tests did you conduct to verify this?

• Redesign the Solution

What problems did you face that would cause you to redesign the (a) rocket, and/or (b) the production process? What changes would you recommend in your new designs? What additional tradeoffs would you have to make?

• Communicate Your Achievements

Describe the plan you will use to present your solution to your class. (Include a media-based presentation.)

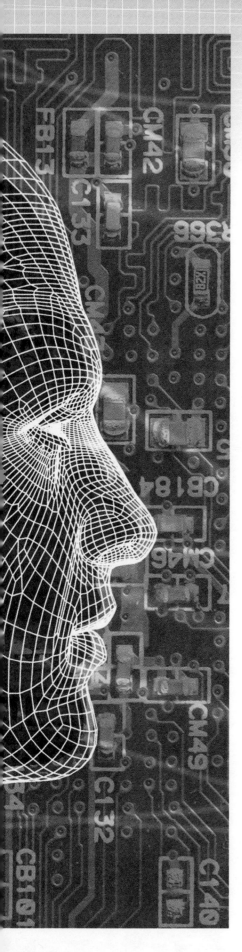

DESIGN CHALLENGE 2:
Prototyping a Manufacturing Cell for Product Assembly

● Problem Situation

One of the most cost-effective ways to bring down the per-unit cost of a product is to automate the assembly process. Often, the cost of producing the individual components is outside of our control, depending instead on a particular vendor and the materials, processes, and procedures they employ. With this constraint, we will attempt to create a method of assembling small products that is cost-effective, rapid, and produces high-quality products. The more we succeed in keeping the costs down for our assembly process and the quality up, the greater the likelihood that we will end up with a profitable product.

● Your Challenge

Given the necessary materials, you will create, test, and refine a manufacturing cell to automate the assembly of multipart ballpoint pens. In the cell, you must be able to quickly sort a bulk container of materials, separating the parts into individual categories. You must then provide a series of operations in the cell that will lead to completed ballpoint pens. Your process should include a mechanism for quality assessment and control.

● Safety Considerations

1. Use a great deal of care when working with electricity and power tools.
2. Wear eye protection at all times.
3. Use insulated tools when working with electric components.

● Materials Needed

1. A variety of ballpoint pens that can be disassembled and reassembled
2. A variety of stock pieces of wood and aluminum (block, cylinders, etc.)
3. Access to a machine shop and/or hand tools
4. Stopwatch

● Clarify the Design Specifications and Constraints

You must first be able to accurately sort a bin of ballpoint pen parts with minimal human intervention. Once the components to the ballpoint pen are sorted, you must be able to quickly and efficiently assemble them. Key metrics for your assembly process include total time for assembly, number of assembly steps required, number of "rejects," and the quality of the products. A quality check must be in place to ensure that the pens function properly—your team will decide what that means. Your prototype assembly should be easily scalable to 1,000, 10,000, and 100,000 without incurring great losses in efficiency or quality.

● Research and Investigate

To complete the design challenge, you need to first gather information to help you build a knowledge base.

1. In your guide, complete the Knowledge and Skill Builder I: Shop and tool safety
2. In your guide, complete the Knowledge and Skill Builder II: Multistep assembly

3. In your guide, complete the Knowledge and Skill Builder III: Automatic sorting
4. In your guide, complete the Knowledge and Skill Builder IV: Jigs and fixtures
5. In your guide, complete the Knowledge and Skill Builder V: Manufacturing cells
6. In your guide, complete the Knowledge and Skill Builder VI: Quality
7. In your guide, complete the Knowledge and Skill Builder VII: Automated assembly

• Generate Alternative Designs

Describe two of the possible alternative approaches to assembling the ballpoint pen. Discuss the decisions you made in (a) sorting pieces, (b) assembling, and (c) quality assurance. Attach printouts, photographs, and drawings if helpful and use additional sheets of paper if necessary.

• Choose and Justify the Optimal Solution

What decisions did you reach about the design of a small-scale assembly process?

• Display Your Prototypes

Produce your manufacturing cell, any jigs and fixtures used, and completed products. As needed, include descriptions, photographs, or drawings of these in your guide.

• Test and Evaluate

Explain whether your designs met the specifications and constraints. What tests did you conduct to verify this?

• Redesign the Solution

What problems did you face that would cause you to redesign the (a) sorting process, (b) the assembly, and/or (c) the quality check? What changes would you recommend in your new designs? What additional trade-offs would you have to make?

• Communicate Your Achievements

Describe the plan you will use to present your solution to your class. (Include a media-based presentation.)

C **ONSTRUCTION HAS BEEN** a human endeavor since prehistoric times, when people needed to build structures to shelter themselves from the environment. From the simple, moveable shelters that nomadic tribes could carry with them to the sophisticated and intelligent buildings of today that use sensors and controllers to monitor and adjust internal conditions, this basic human activity has grown and changed enormously.

CHAPTER

6

CONSTRUCTION TECHNOLOGY

SECTION 1
Examining the Construction Industry and Its Economic Impact

SECTION 2
Learning About Types of Structures

SECTION 3
Maintaining and Renovating Structures

Construction is building or assembling a structure on a site. Today, construction is one of the largest industries in the United States, with about 9 million salaried and self-employed workers. The construction industry comprises establishments that are primarily engaged in the construction of buildings or other engineering projects, such as highways and utility systems. The construction industry includes building new structures (see Figure 6.1) and modifying, maintaining, repairing, and improving existing structures.

Construction offers more opportunities than most other industries for people who want to own and run their own business. Self-employed workers perform work directly for property owners or act as contractors on small jobs, such as additions, remodeling, and maintenance projects.

Figure 6.1 | The artist's conception shows the new Freedom Tower, to be built at the site of the World Trade Center in New York City.

Figure 6.2 | This example shows construction technology in a non-industrialized society.

SECTION 1: Examining the Construction Industry and Its Economic Impact

KEY IDEAS >

- The construction industry is divided into three major segments: general contracting, heavy construction and civil engineering, and specialty trades.

- The infrastructure is the underlying base or basic framework of a construction system.

- Construction projects create social, economic, lifestyle, and environmental impacts.

- Some impacts are positive; some are negative. These impacts must be considered and properly managed.

Figure 6.3 | Constructing a skyscraper is an enormous undertaking.

Construction industry workers are divided into three major segments. General contractors build residential, industrial, commercial, and other buildings, as shown in Figure 6.3. Heavy and civil engineering construction contractors build sewers, roads, highways, bridges, tunnels, and other projects. Specialty trade contractors perform specialized activities related to construction, such as carpentry, painting, plumbing, and electrical work.

Construction is usually performed or coordinated by general contractors who specialize in one type of construction, such as residential or commercial building. They take full responsibility for the complete job, except for specified portions of the work that may be omitted from the general contract. Although general contractors may do a portion of the work with their own crews, they often subcontract most of the work to heavy construction or specialty trade contractors.

Specialty trade contractors usually do the work of only one trade, such as painting, carpentry, or electrical work, or of two or more closely related trades, such as plumbing and heating. Beyond fitting their work to that of the other trades, specialty trade contractors have no responsibility for the structure as a whole. They obtain orders for their work from general contractors, architects, or property owners. Repair work is almost always done on direct order from owners, occupants, architects, or rental agents.

In 1900, 75 percent of urban Americans lived in rented apartments or flats. In contrast, homeownership in the United States now stands at a record high of 68 percent. The demand for residential construction is expected to continue to grow over the coming decade. The demand for larger homes with more amenities, as well as for second homes, will continue to rise. Employment in nonresidential construction is expected to grow a little faster than in the rest of the industry. Industrial construction activity is expected to be stronger, as replacement of many industrial plants has been delayed for years and many structures will have to be replaced or remodeled.

Employment in heavy construction is projected to increase about as fast as the industry average. Growth is expected in highway, bridge, and street construction, as well as in repairs to prevent further deterioration of the nation's highways and bridges.

According to the U.S. Department of Labor Statistics, demand for specialty trade subcontractors in building and heavy construction is rising; at the same time, more workers will be needed to repair and remodel existing homes.

The Construction Infrastructure

The construction infrastructure is the underlying base or basic framework of structures that are essential for people to go about their daily lives and comfortably conduct their affairs. The infrastructure includes airports, bridges, canals, dams, hazardous waste management, pipelines, roadways, tunnels, wastewater and solid waste treatment facilities, and water supply systems. Some of these systems are so essential to our way of life that, in 1996, President Clinton issued Executive Order 13010 concerning the U.S. infrastructure. In the directive, Mr. Clinton defined the critical infrastructure as the part of the infrastructure that is "so vital to the United States that the incapacity or destruction of such systems and assets would have a debilitating impact on security, national economic security, national public health or safety."

Construction projects add infrastructure to the built environment. A great deal of the infrastructure is owned by the local, state, or federal government. Projects to build the infrastructure are not normally built by private, for-profit enterprises, but are undertaken to serve the public.

Figure 6.4 | Public infrastructure projects include roads like this one under construction in El Alto, Bolivia.

Impacts of Construction

Construction affects people, society, and the environment in many ways; some impacts are positive; some are negative. Construction projects add to a region's infrastructure and thereby increase the wealth and assets of a community. These projects can reduce poverty, but they can also create local air, noise, and visual pollution. They can unbalance natural habitats and biodiversity as well. Impacts must be considered, and construction projects must be properly managed to minimize the occurrence of negative effects. One component of proper management is to consider how the construction site (including vegetation) will be restored once construction is completed.

Sometimes, construction projects may interfere with the normal flow of people's lives. People may have to relocate because a new project requires the use of land where they live or work. Responsible project managers try to minimize these dislocations. They try to establish long-lasting relationships with local communities and to respect the environment where they work.

Major construction projects often bring a pool of new workers to a community, and these workers become neighbors to existing residents. The new workers may be immigrants who speak a native language other than English. Schools must accommodate new students. Cultural differences may lead to misunderstandings that have to be resolved by all concerned.

Economic Impacts

Large construction projects generally bring an influx of jobs and money into an area. This can benefit local residents, but it also can cause residential homes to increase in cost while depleting the supply of homes on the market, making it difficult for new residents to afford housing. Because new workers have spending power, local living costs can be inflated.

Sometimes, low-skilled workers arrive on the scene, creating a need for community agencies to provide services. However, new construction projects also rely on a highly skilled labor force from the surrounding areas. Thus, communities with good educational systems tend to benefit because the skilled and educated workforce they produce is seen as an asset by prospective new businesses.

Lifestyle and Environmental Impacts

Although construction projects can enhance our quality of life in many ways, they can also result in some undesirable impacts. Soil erosion, along with air and water pollution caused by carbon-dioxide emissions and industrial waste disposal, can affect agricultural production, wildlife, and people's health. Road construction and traffic-flow projects can cause serious environmental damage. The projects can damage a region's ecology, use up farm land, create a change from a rural to a suburban environment, and cause a need for population resettlement.

One possible negative impact of a building construction project relates to traffic safety. A new shopping mall, for example, will draw increased traffic, which means road accidents could increase. You have undoubtedly heard noise from construction equipment and machinery, and the landscape is often disturbed during the early stages of a project.

Before a large construction project is undertaken, the U.S. government requires that an environmental impact assessment be completed. The assessment details how the project would affect the natural environment. The International Association for Impact Assessment defines environmental impact assessment as "the process of identifying, predicting, evaluating and mitigating the biophysical, social, and other

Figure 6.5 | Construction projects situated near green spaces and bike trails are often more valuable.

relevant effects of development proposals prior to major decisions being taken and commitments made."

Construction projects sometimes consume open spaces and convert them into buildings or roadways. However, real estate developers frequently recognize that when construction projects are sited near green or open spaces, the value of the project tends to rise (Figure 6.5). People find the views and recreational opportunities near green and open spaces to be very desirable.

According to the U.S. Government National Park Service, natural open space and trails are prime attractions for potential home buyers. Almost 80 percent of all home buyers and shoppers rate green and natural open space as either "essential" or "very important" in planned communities. Walking and bicycling paths also ranked high.

Green Construction and Sustainable Development People are becoming more conscious of living in ways that cause minimal harm to the environment. Increasingly, "green construction" is becoming popular. Green construction, often referred to as sustainable development, attempts to substantially limit environmental damage, reduce energy use, and use recycled and recyclable construction materials.

A fascinating sustainable development project is being considered by the nation of Libya in North Africa. The project, called Green Mountain, covers a 5,500-square-kilometer area on the Mediterranean coast. It is an area with a rich Greek and Roman historical heritage and a large and diverse animal and plant population (see Figure 6.6). According to the developer, the goal of the project is "to create and protect in perpetuity the world's first regional-scale, world-class conservation and development area." The project will develop a sustainable infrastructure. It will derive energy from wind and solar power; use organic methods of farming, aquaculture, and food production; develop waste management and recycling facilities that will convert trash into biofuel; use closed-loop water systems; and use energy-efficient transport and housing. An ecological parkland will be included. Green Mountain project managers are committed to the development and use of sustainable and renewable sources of energy and raw materials.

Figure 6.6 | The site of the Green Mountain Project is ecologically diverse.

Building Energy Consumption According to the Oak Ridge National Laboratory (ORNL), a U.S. Department of Energy laboratory, buildings in the United States command 40 percent of the nation's overall energy use. This use ranks above industry, at 32 percent, and transportation, at 28 percent. Buildings demand 71 percent of domestic electric power in the United States and 55 percent of the nation's natural gas—and they produce 43 percent of U.S. carbon emissions.

Because the nearly 5 million commercial buildings and 112 million households use a collective 38.8 quadrillion BTUs of energy each year, curtailing consumption is a tall order. However, conserving energy has enormous potential. Space heating, lighting, cooling, and ventilation demand most of that power, followed closely by water heating. Refrigeration, electronics, computers, and other items add up to their own significant and growing slice of the energy pie.

Buildings' appetites for energy have been on the rise as a result of natural population growth and related development of homes, apartment complexes, shopping malls, schools, office buildings, and health-care facilities. The amount of energy required for each person who occupies these buildings is increasing, as well. Residential floor space per capita in the United States is growing, driven by construction of larger homes as well as a decline in the average number of occupants. The number of power-hungry devices found in today's households—from computers to video games to plasma televisions—is on the rise. As a result, residential energy consumption is expected to grow 1 percent per year until 2025, according to the Oak Ridge National Laboratory.

Zero-Energy Housing At the Oak Ridge National Laboratory, researchers are working on an initiative to develop affordable, near–zero-energy housing by 2020 and zero-energy commercial buildings by 2025.

The research needed to build a zero-energy house is a wasted investment if the house requires a million dollars worth of gadgets to lower energy costs. New energy solutions must be designed for the real world, they must be practical to manufacture, and they must be affordable to most consumers. ORNL and the Tennessee Valley Authority, in partnership with Habitat for Humanity, have constructed a "zero-energy" home for about $100,000. The home, in Lenoir City, Tennessee, uses a combination of insulation, solar panels, energy-efficient appliances, and innovative heating and air conditioning systems that requires a daily energy cost of only 40 cents.

To optimize energy saving, these high-performance buildings must be outfitted with renewable sources of energy, minimizing the demand for fossil fuels. They use alternative energy sources such as heat pump systems that tap geothermal energy in the ground around the building, and solar panels that can support their own energy needs in a way that is affordable, sustainable, and energy efficient. The goal is to develop a building that can actually produce more energy than is needed during certain parts of the day, and then sell the excess back to utility companies.

Oak Ridge Laboratory researchers have worked with industry to develop and demonstrate the energy savings benefits of infrared-blocking pigments used to make dark-colored metal, concrete tile, and asphalt shingle roofing that are highly solar reflective, as well as energy-efficient foam insulation materials that reduce the need for air conditioning.

Virtually airtight houses such as the Lenoir City near-zero-energy home create challenges for equipment that must maintain comfortable temperatures and humidity levels and provide hot water when desired. Airtight homes are ventilated to bring in fresh air and provide indoor moisture in humid climates. To be comfortable year-round, homes like these use either a stand-alone dehumidifier or a heat pump that can provide enhanced dehumidification on demand. In traditional construction, multiple pieces of equipment are used to condition a

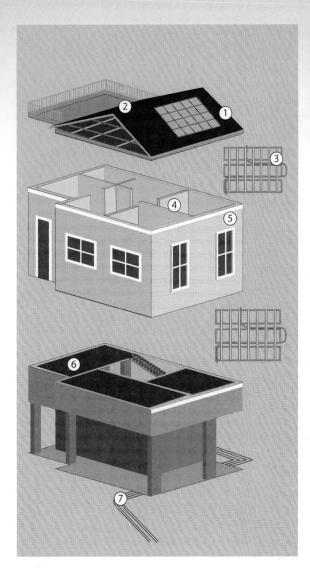

Figure 6.7 | The components of a near-zero-energy home show that alternative energy sources can be used produce heat and electricity.

home's air and water and are not as energy efficient as they could be. For example, cooling and dehumidification units discard heat outdoors while electricity is used to heat water at the same time. The near-zero-energy home uses an integrated heat pump that performs all the needed heating and cooling functions.

To build a near-zero-energy house, the space and water heating and electricity generation must come from alternative energy sources, as shown in Figure 6.7.

The following points provide further explanation for components in Figure 6.7.

(1) In the Lenoir City home, solar panels are used to heat water via the solar hot water heater and generate electricity. Use of these panels helps to offset power use by reducing water heating costs and actually delivers electricity to the grid. The Tennessee Valley Authority pays 15 cents per kilowatt-hour for the solar power.

(2) The Lenoir City homes feature standing seam metal roofs with "cool-colored coatings" that reflect infrared rays, reducing the need for air conditioning. The roofing systems can last up to 30 years, boosting energy efficiency by 150 percent. Ongoing research and development focuses on affordable approaches to roof-integrated natural ventilation, radiant barriers, the addition of thermal mass through concrete tiles, and the use of phase-change materials.

Phase-change materials (as discussed in Chapter 4) change from a solid to a liquid or vice versa. When the phase change occurs, these materials can store a substantial amount of energy. The material used in the walls of the zero-energy home is a paraffin wax that melts at about 75°F and is encapsulated in plastic "balloons." These balloons prevent interaction with other building envelope components and are made from polyethylene that is about 0.005 thick. After these balloons are placed in the wall, the wax will store energy when the outside temperature causes it to melt. The wax releases that energy by solidifying when the temperature drops below its melting point. This can have the effect of delaying heat from entering the building during the hottest part of the day, while allowing heat to be passed along when the outdoor temperatures have dropped during the evening.

(3) The utility wall consolidates most of the home's hot-water plumbing—with the primary bathroom, laundry, and kitchen back-to-back—and enables a home energy savings of 15 percent over a traditionally plumbed house.

(4) The interior design and ventilation design limits ductwork and plumbing lines for most efficient energy use. Using metal studs rather than wood for building interior walls is more cost effective and sustainable.

(5) In the construction of the exterior walls, the home makes use of energy-efficient exterior insulation-finish systems, wall cladding color, the wall's thermal mass, and air tightness. To achieve affordable energy efficiency and durability, passive self-drying wall designs remove moisture that has permeated the wall. Structural insulated panels make up the walls, floors, and roofs of the Lenoir City home and serve both the structural and insulation requirements, reducing the need for space heating and cooling and enabling contractors to easily build airtight structures.

(6) Basement walls are made of exterior-insulated, termite-resistant, reinforced concrete blocks, with fiberglass drainage insulation board on the exterior and an above-grade waterproof covering. The walls are painted with an infrared-reflective

coating. The interior basement wall provides thermal mass for the building, which retains heat energy and provides more uniform and comfortable temperatures.

(7) Geothermal energy is used as an energy source because the ground is warmer than outdoor air in winter and cooler in summer.

(The material in Figure 6.7 is taken from the Oak Ridge National Laboratory Review, Vol. 40, #2, 2007.)

TECHNOLOGY AND PEOPLE:
Jeff Christian

Jeff Christian is an energy efficiency guru and a champion of zero-energy housing. He is also the architect of Oak Ridge National Laboratory's (ORNL) unique living lab of energy-efficient and renewable generating technologies. (See Figures 6.8 and 6.9.)

In 2002, when the U.S. Department of Energy introduced the daunting concept of a house that would produce as much energy as it used, Christian seized the challenge and has not let go. Guided by the belief that energy-efficient homes should be affordable to working families, he partnered with Habitat for Humanity an international organization that strives to eliminate poverty and homelessness, building five homes as a test of the latest renewable energy-producing and energy-efficient technologies. The homes include features such as solar panels, geothermal heat pumps, heat-pump water heaters, airtight, super-insulated walls and roofing panels, and advanced ventilation systems. They demonstrate that such technologies, when properly integrated, work in real-life environments with regular people who can then benefit from the resulting cost savings.

That accomplished, Jeff is now helping transition the concept—and the technologies—to the traditional construction industry, with plans under way to begin building two near-zero-energy homes aimed at traditional home buyers. Christian has brought together a diverse group that includes the nonprofit Habitat organization, utilities, suppliers, contractors, architects, and researchers to demonstrate what he believes is becoming a model for America's homes of tomorrow.

Figure 6.8 | Jeff Christian of Oak Ridge National Laboratory is a great believer in zero-energy housing.

In essence, the houses are airtight and mechanically ventilated. From this clear starting point, the heating and air conditioning ducts are installed inside the conditioned space. From there, the home goes to high-performance windows. As a result of having a well-insulated, airtight home, the heating and cooling distribution system and the mechanical equipment can be downsized.

A real benefit is that many people think a near-zero-energy house needs to be occupied by an engineer, tweaking things at every moment to make the house "go." However, that is simply not true. The ORNL's research offers strong testimony that these homes can be easily occupied by average, nontechnical residents.

(The material for this feature was taken from the Oak Ridge National Laboratory Review, Vol. 40, #2, 2007.)

Figure 6.9 | A near zero-energy home requires a daily energy cost of only 40 cents.

1. Into which three major segments do workers in the construction industry fall?
2. What is meant by the construction infrastructure?
3. List three positive and three negative impacts of construction on people, society, or the environment.
4. Explain what is meant by "sustainable development."
5. What techniques can be used to reduce energy loss in home construction?

SECTION 2: Learning About Types of Structures

KEY IDEAS >

- Products of the construction industry include buildings, industrial plants, roadways, bridges, tunnels, sewers, pipelines, and airports.

- Structures can include prefabricated materials.

- Structures are constructed using a variety of materials, processes, and procedures.

- The design of structures involves a number of requirements that relate to appearance, efficiency, function, safety, strength, longevity, maintenance, and available utilities.

- Design and construction is regulated by laws, codes, and professional standards.

Among the structures produced by the construction industry are buildings, industrial plants, roadways, bridges, tunnels, sewers, pipelines, and airports. These structures are built using a wide variety of building materials and construction processes and procedures. Building materials include concrete, steel, glass, brick, and wood. Construction processes include framing of structures, enclosing and lining structures, and preparing structural foundations. Procedures are often a function of the cost, worker skills, and level of quality desired.

The design of structures involves a number of requirements. These are specified according to desired appearance, efficiency, function, safety, strength, longevity, utilities, and maintenance needs.

Buildings

There are four kinds of building construction: residential, commercial, institutional, and industrial, as shown in Figure 6.10. Residential buildings include houses for one or several families, as well as apartment houses with units that are either rented to tenants or sold as condominiums or cooperative (co-op) units. A condominium (condo) is a community in which the units are individually owned but ownership of common areas like the roof and lobby areas is shared. In a co-op, owners buy shares of stock in the corporation that owns the building.

Houses designed for one or several families are often made from wooden frames. Walls are typically made from 7/16-inch plywood sheathing and are finished with vinyl or aluminum siding or brick veneer. There is a space of 3½ to 5½ inches between the outside wall and the inside wall. The inside wall is usually made from gypsum drywall panels (also known as "plasterboard"). Insulation usually consists of fiberglass batts, loose fill, or blown-in cellulose.

Commercial buildings include offices, retail and wholesale stores, shopping malls, houses of worship, hotels, libraries, and stadiums. Institutional buildings are schools, colleges, universities, hospitals, and correctional facilities (prisons).

The Council on Tall Buildings and Urban Habitat, the acknowledged source of information about tall buildings at the Illinois Institute of Technology, measures the height of a building from the sidewalk level of the main entrance to the structural top. The height includes spires but not television antennas, radio antennas, or flagpoles.

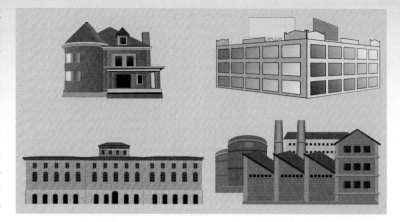

Figure 6.10 | There are four kinds of building construction: residential, commercial, institutional, and industrial.

Figure 6.11 | The Eiffel Tower in Paris is 324 meters (1,063 feet) high at the tip of its antenna. Because of the iron lattice framework construction, the top of the tower only sways about 6.5 cm (about 2.6 inches) from center.

The Effect of Wind on Buildings

▷ Winds can cause shear forces and overturning moments on very tall buildings. (A moment is the turning effect produced by a force at a distance from the axis of rotation: $M = Fs$.) In the United States, tall buildings are allowed to sway 1/500th of their height. For example, the Sears Tower in Chicago is 442 meters (1,451 feet) high. The building sways six inches from either side of its center. When a building sways, it can cause the people inside to feel motion sickness.

Because of its construction as a framework without a solid enclosure, the sway on the Eiffel Tower, shown in Figure 6.11, is only about 6.5 cm (about 2.6 inches) from center.

The tallest building in the world will be the Burj Dubai in Dubai, in the United Arab Emirates on the Persian Gulf. When construction is completed, Burj Dubai will be taller than the KVLY/KTHI television mast in Blanchard, North Dakota, which, at 628.8 meters (2,063 feet) was the world's tallest mast and technically the world's tallest structure, even though it is stabilized by guy-wires.

Although it is still under construction, in September 2007, Burj Dubai became the world's tallest free-standing structure at 555.3 meters (1,822 feet), reaching higher than the CN telecommunications tower in Toronto, Canada. The finished height of the building has not yet been made public, but estimates are that the building and its tower will be 818 meters (or 2,684 feet) high. People standing on the top floor of the building will experience a sway of 1.6 meters (more than five feet).

Burj Dubai (see Figure 6.12) will be at the center of downtown Burj Dubai, a $20 billion, 500-acre development billed as the most prestigious square kilometer on earth. The tower will feature residential, commercial, and retail components, including the world's first Armani Hotel and Armani Residences, exclusive corporate suites, a business center, four luxurious pools and spas, an observation platform on Level 124, and 150,000 square feet of fitness facilities.

Figure 6.12 | When complete, the Burj Dubai will be the tallest building in the world.

Section 2 △ Learning About Types of Structures

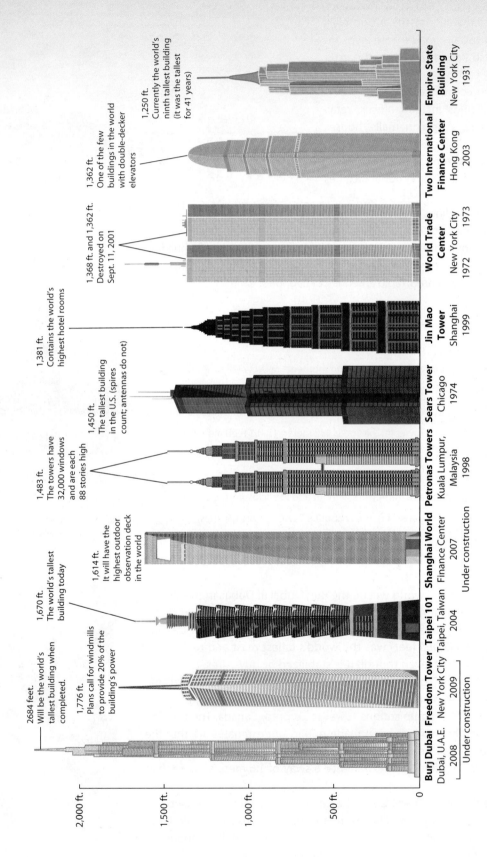

2684 feet.
Will be the world's tallest building when completed.

1,776 ft.
Plans call for windmills to provide 20% of the building's power

1,670 ft.
The world's tallest building today

1,614 ft.
It will have the highest outdoor observation deck in the world

1,483 ft.
The towers have 32,000 windows and are each 88 stories high

1,450 ft.
The tallest building in the U.S. (spires count; antennas do not)

1,381 ft.
Contains the world's highest hotel rooms

1,368 ft. and 1,362 ft.
Destroyed on Sept. 11, 2001

1,362 ft.
One of the few buildings in the world with double-decker elevators

1,250 ft.
Currently the world's ninth tallest building (it was the tallest for 41 years)

Burj Dubai	**Freedom Tower**	**Taipei 101**	**Shanghai World Finance Center**	**Petronas Towers**	**Sears Tower**	**Jin Mao Tower**	**World Trade Center**		**Two International Finance Center**	**Empire State Building**
Dubai, U.A.E.	New York City	Taipei, Taiwan		Kuala Lumpur, Malaysia	Chicago	Shanghai	New York City		Hong Kong	New York City
2008	2009	2004	2007	1998	1974	1999	1972	1973	2003	1931
Under construction			Under construction							

2,000 ft.

1,500 ft.

1,000 ft.

500 ft.

0

Figure 6.13 | The world's tallest buildings include the Sears Tower in Chicago and the Empire State Building in New York City.

Building Energy Consumption

According to the U.S. Government Energy Information Administration's Commercial Buildings Energy Consumption Survey (CBECS), the United States had nearly 4.9 million commercial buildings and more than 71.6 billion square feet of commercial floor space in 2003. Energy consumption can be expressed as the amount consumed within a building (site energy). In its most recent report, CBECS indicates that the total amount of energy consumed by commercial buildings in the United States was 6.5 quadrillion British thermal units (BTUs). The greatest consumption for any energy source was 3.6 quadrillion BTUs for electricity, followed by 2.1 quadrillion BTUs for natural gas. One BTU is the amount of heat needed to raise one pound of liquid water one degree Fahrenheit. The yearly cost of the energy consumed by these buildings is about $107.9 billion.

Stores and service buildings use the most total energy of all the commercial and institutional building types, as shown in Figure 6.14. Offices also use a large share of energy. Education buildings (such as schools and colleges) use 13 percent of all total energy, which is more than all hospitals and other medical buildings combined. Lodging buildings, such as hotels or dormitories, use 8 percent of this energy, while warehouses and food service buildings (like restaurants) each use 7 percent. Public assembly buildings, which can be anything from libraries to sports arenas, use 6 percent; grocery stores and convenience stores use 4 percent.

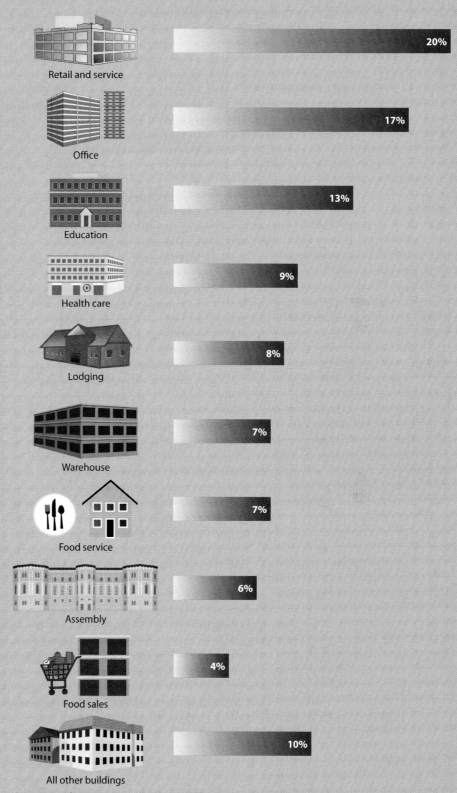

Due to rounding percentages may not add to exactly 100 percent.

Figure 6.14 | Stores, offices, and schools consume the most energy of all commercial and institutional buildings.

All other types of buildings, such as places of worship, fire stations, police stations, and laboratories, account for the remaining 10 percent of commercial building energy.

In buildings, most of the energy is used for space heating and cooling (39%), followed by lighting (23%) and water heating (15%). See Figure 6.15.

Figure 6.15 | A building uses the highest percentage of its total energy expenditure on heating, lighting, and cooling.

Industrial Plants

Industrial plants are factories that manufacture products, power plants that generate energy, petroleum refineries, and wastewater treatment plants. Because these plants often bring jobs and income to a community, local governments compete by providing tax incentives and subsidies to companies to build within their communities. About 400,000 factories exist in the United States. Increasingly, modern factories are clean, energy efficient, and highly productive enterprises.

Laws, Codes, and Professional Standards

▷ Design and construction of structures are regulated by laws, codes, and professional standards. Building codes and standards are set by municipalities, states, or federal governments. They are set to assure a minimum level of health and safety for occupants of buildings and other structures. Codes for construction relate to building practices and procedures, electrical safety and fire protection, and plumbing. Codes also regulate how mechanical equipment is installed, how buildings should be rehabilitated, how hazards should be resolved, and how playgrounds should be designed.

Before new construction or significant remodeling is authorized, a permit from the municipal agency is required. Once the construction is completed, it must be inspected. If the structure passes the inspection construction, officials issue a Certificate of Occupancy documenting that the construction has been completed according to the permit specifications.

The following example of a handrail standard, set by the U.S. Department of Housing and Urban Development, relates to housing for the elderly:

Handrails for exterior steps not attached to dwellings shall be provided in accordance with Uniform Federal Accessibility Standards on both sides of a tenant stairway with a flight rise exceeding 24" and width exceeding 4 ft. and on one side when the width is 4 ft. or less.

Another example of a standard relates to the water supply:

Where adverse impact on local water supplies is possible as a result of the project, the following information shall be provided: A discussion of the location of wells or surface water supply intakes, treatment plants, reservoirs, or other structures of public health significance

Figure 6.16 | Wastewater Treatment Facilities can routinely process million of gallons of wastewater daily.

Roadways

A road is an open way for the passage of vehicles. The oldest road in the world, according to some claims, was a timber track way in Somerset, England built around 3800 B.C.E. It consisted of crossed poles of ash, oak, and lime that were driven into the soil to support a walkway made from oak planks laid end to end. Brick paved streets were used in India as early as 3000 B.C.E.

Paving means covering a roadway with material to make the surface strong and level. Starting about 300 B.C.E., the Romans built more than 50,000 miles of roads paved with stone during their empire's expansion throughout Europe. Some of the roads were paved with cobblestones, as shown in Figure 6.17.

In the United States, roads connect all 50 states, all cities and towns, and virtually all rural communities. A comprehensive highway network allows for business, commercial, and leisure travel. According to the U.S. Government atlas (available at *www.nationalatlas.gov*), the United States has about 4 million roadway miles, and about 2.6 million miles are paved.

The national highway system includes approximately 160,000 miles (256,000 kilometers) of roadway that is important to the nation's economy, defense,

Figure 6.17 | This Paris street is paved with cobblestones.

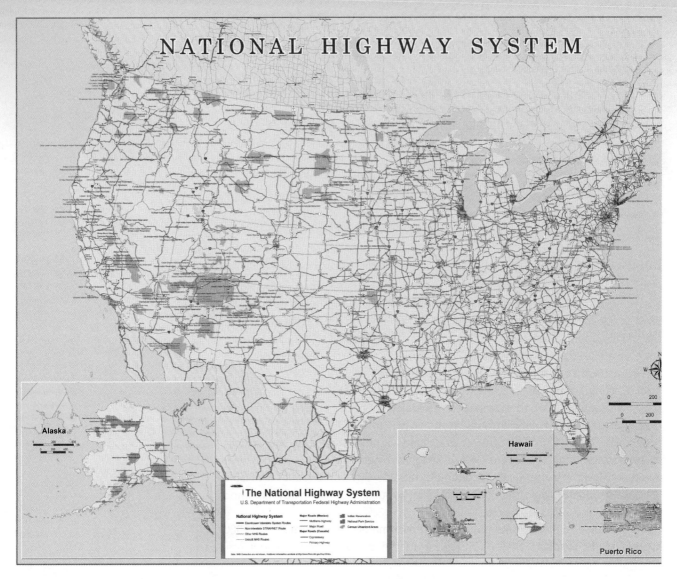

Figure 6.18 | The U.S. National Highway System consists of approximately 160,000 miles of roads.

and mobility (see Figure 6.18). Highways, roadways, and highway bridges are a foundation of the nation's technological infrastructure. The entire highway network has about 600,000 bridges.

In the United States, the Federal Highway Administration (FHWA) of the U.S. Department of Transportation is responsible for overseeing highways, which includes promoting safety, providing technical expertise, developing regulations, and providing development and maintenance of federal roads.

Paving Roads

John Loudon McAdam was born in Scotland in 1757. He developed a design for roadways that used three layers of stones laid over soil, with a crown at the center and ditches at the sides for drainage. Water and stone dust was added and then a heavy roller was used to compact the materials together to make a hard surface. The surface of these roads is called macadam. Later, tar was sprayed on the surface to keep dust from being created. The combination of a tar coating over the macadam is called "tarmac" or coated macadam.

Today, roads are more often paved with asphalt (see Figure 6.19). Asphalt consists of a mixture of aggregate, which can include sand, gravel, stone, waste slag from iron and steel manufacturing, glass, or recycled concrete. Asphalt also includes a binder called bitumen, a viscous black, sticky material obtained from distillation of crude oil. Bitumen is composed primarily of aromatic hydrocarbons (as discussed in Chapter 4, aromatic substances have a very stable atomic structure that often contains six carbon atoms). The difference between coated macadam (tarmac) and asphalt is that the aggregate, filler, and binder give asphalt its strength. Macadam, on the other hand, derives its strength from the interlocking of the stone particles.

Roads are also paved with concrete, a mixture that uses gravel, sand, cement, and water. Workers place concrete into molds called forms. Special equipment is used for paving concrete; these pavers are manufactured in different sizes for different widths of road. Concrete pavers do not compress the material, whereas asphalt pavers finish the material by compacting it to the proper density. To finish concrete, the surface is brought to the required elevation by scraping off (striking) the excess, a process known as screeding.

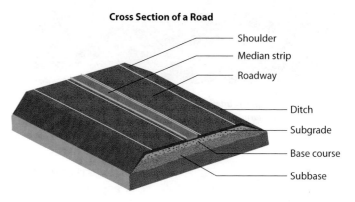

Figure 6.19 | Roads can be paved with asphalt, which is a mixture of different sizes of aggregate and oil-based additives.

Phases of Road Construction

The phases of road construction include planning, road design, earthwork, paving, and finally, opening the road to traffic. Planning includes acquiring the land, identifying environmental concerns, finding funding, and identifying the contractors and engineers. Road design starts with a survey of the area. The properties and drainage capacity of the soil are identified, traffic volume is specified, and effects on the environment and nearby residents are considered. Earthwork establishes the roadway's base layers and foundation, as shown in Figure 6.20. Bulldozers and grading machines level the earth. The earth is then watered down and compacted by heavy machinery. Sewers and drains are installed, and then gravel finally is placed on the road, wetted down, and compacted again.

After the earthwork is completed, the roadway is paved with either asphalt or concrete (see Figure 6.21 for a cross-section view). If asphalt is used, it is applied hot, spread out, and compacted. If concrete is used, it is placed into steel forms to form slabs. Grooves are made between each slab to allow expansion and contraction with temperature changes. Wire and steel dowels are used to connect the slabs.

Before the road can be opened to traffic, it must be tested for quality. Special equipment is used to measure the smoothness of the pavement; otherwise, rough pavement would cause vibration during traffic. Tests are conducted to ensure proper draining. Finally, the road is landscaped and markings are applied as necessary to the pavement.

Figure 6.20 | Earthwork establishes the roadway's base layers and foundation.

Cross Section of a Road

- Shoulder
- Median strip
- Roadway
- Ditch
- Subgrade
- Base course
- Subbase

Figure 6.21 | A cross section shows the layers and construction of a main road.

Section 2 △ Learning About Types of Structures

Costs of Materials

Over the past few years, prices of materials used for road construction have been driven up by demand from rapidly growing nations, such as China. Costs for steel and cement, especially, have increased substantially. According to the Ohio Department of Transportation, "the most important cost drivers of construction cost inflation for the next five years will be energy, steel, and cement. Unlike many other construction materials, these items are impacted by international influences which are difficult to predict."

TECHNOLOGY IN THE REAL WORLD:
Roads Around the World

The following statistics illustrate the importance and enormity of the world's roadway infrastructure.

- Road density refers to the number of linear feet of road per unit area. Road densities vary from a low of around 0.01 km per sq. km to a high of 4.90 km per sq. km (excluding exceptions), with a typical value of 0.20 km per sq. km.
- About 45 percent of the roads in a typical country are paved. The proportion of paved roads varies from 2.5 percent in some developing countries to a high of 100 percent.
- Worldwide, the stock of motor vehicles is growing at nearly 3 percent per year. The number of vehicle kilometers traveled tends to grow somewhat faster than the stock of motor vehicles.
- Industrialized countries typically spend just over 1.0 percent of their Gross Domestic Product (GDP) on roads. Countries with set-aside financing for roadways typically spend over 1.5 percent of GDP.
- Worldwide, between 750,000 and 880,000 people are killed and 23 to 34 million are injured in road crashes each year, costing the global economy about $500 billion per year.
- About 85 percent of these accidents take place in developing and transition countries, with almost half the deaths occurring in the Asia-Pacific region.
- In industrialized countries, only about 15 to 20 percent of fatalities involve pedestrians, nonmotorized vehicles, and motorcycles. In developing and transition countries, the figure is closer to 50 percent; it is as high as 70 percent in Asia.
- In developing and transition countries, road accident rates tend to be 10 to 20 times higher than in industrialized countries and cost 1.0 to 1.5 percent of GNP.

(Material for this feature was taken from the U.S. Federal Highway Administration Web site, www.fhwa.dot.gov, and from The World Bank's Web site, www.worldbank.org.)

Figure 6.22 | A highway interchange system is a good example of the U.S. roadway infrastructure's sophistication.

Bridges

A bridge is a structure that spans a valley, body of water, roadway, railroad tracks, or any other obstruction to continuous travel. Bridges are designed to withstand dead loads and live loads. A dead load is a load that does not change, such as the weight of the structure itself. It includes the deck, sidewalks, parapets, railings, and

Figure 6.23 | This simple bridge is just a plank that spans a creek.

the steel or concrete load-carrying members. These loads account for a significant and considerable percentage (often the majority) of the stress in the load-carrying member. Live loads change; the changes can be due to increased or decreased traffic and pedestrian loading. Other important loads are a function of wind, snow, impact, and natural disasters such as earthquakes. When designing a bridge, strength and safety of the structure are the primary considerations, but cost is very important as well. Bridges are built with safety factors that provide a margin over the theoretical capacity to allow for uncertainty in loading. A bridge normally has a safety factor of about 2.0, meaning that it is built to withstand twice the greatest expected load.

Bridges are designed according to the length of the span and the geologic conditions of the site. A simple beam was the first type of bridge. Logs also were used to span short distances (Figure 6.23). Modern beam bridges are used to span short distances and are constructed using reinforced concrete and steel.

An arch bridge transmits the load from the deck of the bridge outward along the arch to supports (abutments) on both sides. The arch produces horizontal thrusts at the abutments. Arch bridges can form wider spans than an unsupported beam. In an arch, the load at the top is passed to each of the adjacent stones, and then to the abutment, which is anchored to the ground. Arch bridges are among the oldest types of bridges.

Older arch structures used stone or bricks as the construction material. When these masonry arches are properly shaped, large compression forces (which exist in all arches) squeeze the stone or bricks together, making the arch strong.

The Ponte Vecchio in Florence, Italy, spans the Arno River (see Figure 6.24). It was originally built from wood, but after a flood, it was rebuilt in 1345 c.e. entirely out of stone. It is a modern

Figure 6.24 | The Ponte Vecchio in Florence, Italy, is an arch bridge made entirely of stone. It dates back to the fourteenth century.

Figure 6.25 | The first bridge made of iron is located in Ironbridge, England. Because bridges were previously made from wood, the construction used joints that were similar to woodworking joints.

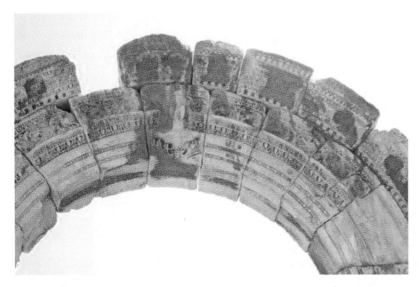

Figure 6.26 | This arch from the Temple of Hadrian (built between 117 and 138 c.e.) is an excellent example of an ancient arch with the keystone clearly evident. The keystone features a carved image of the goddess Cybele.

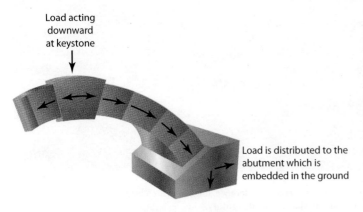

Load acting
downward
at keystone

Load is distributed to the
abutment which is
embedded in the ground

Figure 6.27 | Downward forces in an arch are transmitted to the abutment.

tourist attraction because of its beauty and for the famous art and jewelry shops along its length. Each of the three arches is about 90 feet long.

The world's first bridge made of iron is in the aptly named Ironbridge, England (see Figure 6.25). The Lupu Bridge in Shanghai, China, is the world's longest arch bridge; the arch is 1,815 feet long. Modern arches used for bridges are commonly constructed using either steel or steel-reinforced concrete.

Suspension bridges are constructed to span long distances. The longest suspension bridge in the world is the Akashi-Kaikyō Bridge in Japan. The length of its center span is 1,991 meters (6,532 feet). In a suspension bridge, the bridge deck is suspended from a main suspension cable made from many strands of steel wire. These cables are strung between bridge towers that are embedded into the ground. The cables are in tension that increases with the load on the bridge, and they must be firmly anchored at each end of the span. The towers are in compression. In a suspension bridge, the weight of the deck and its load is transferred to the main suspension cable through suspender rods or cables (see Figures 6.28 and 6.29), then to the tower and its anchor blocks.

Cable-stayed bridges are similar in appearance to suspension bridges (see Figure 6.30). However, suspension bridge cables transmit the load to the

Figure 6.28 | One of the world's most beautiful suspension bridges is the Golden Gate Bridge, which spans San Francisco Bay in California. When built in 1937, it was the longest suspension bridge in the world. The center span is 1,280 meters (4,200 feet) long.

towers and the anchor blocks, while in cable-stayed bridges all the cables are directly attached to the towers, which bear the main load. A cable-stayed bridge design is most often used for span lengths that are shorter than those of suspension bridges. These bridges are generally less expensive to construct than a suspension bridge of equal length, as they use less cable for a given length than a suspension bridge.

Cantilever bridges are constructed using two beams that face each other and are anchored only at the originating end (see Figure 6.31). The other ends of the beams are connected by a third portion, and the connecting span is supported by a column (or pier). The connecting span is usually built to be as lightweight as possible so that it requires the least number of supporting piers.

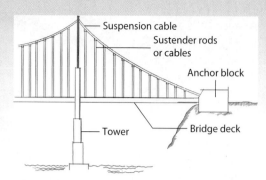

Figure 6.29 | As shown in this schematic diagram of a suspension bridge, the bridge deck is suspended from a main suspension cable made from many strands of steel wire.

Figure 6.30 | The cable-stayed Jätkänkynttilä bridge in Rovaniemi, Finland, spans the Kemijoki river.

Figure 6.31 | Cantilever bridges extend outward and are secured at the ends.

Truss bridges use a framework made of struts that are connected together in a triangular grid. Using triangular struts helps ensure that the resulting structure is rigid. A truss bridge resists bending when subjected to static and dynamic loads (see Figure 6.32 for an example).

Tunnels

Tunnels are built under water and under land. They facilitate transport of people, vehicles, and water, as well as the extraction of minerals. The world's longest underground tunnel connects reservoirs that supply New York City with its water. The tunnel, the New York City—West Delaware water-supply tunnel, is 169 kilometers (105 miles) long. The longest vehicle tunnel is the Seikan rail tunnel that connects the islands of Honshu and Hokkaido in Japan. The tunnel is about 54 km (33.5 miles) long, and about 24 miles of it is under water. The English Channel tunnel (called the Chunnel) is 31 miles long. The Chunnel

Figure 6.32 | The arch of the Sydney Harbor Bridge in Sydney, Australia, is constructed of steel trusses. Under the heaviest allowable load, the deflection at the center of the bridge is 4½ inches, and the maximum thrust at the hinges (the ends of the arch) is 435,000,000 pounds per hinge. The roadway is 150 feet wide, and the total length including the approaches is 3,770 feet.

consists of three separate tunnels—two are for trains and one is a service tunnel. Worldwide, the longest tunnel for road traffic is the Laerdal Tunnel in Norway. It is 24.5 km (15.2 miles) long.

Tunnel Construction There are three steps in constructing tunnels: excavation, supporting the structure, and lining the tunnel (see Figure 6.33). During the first step, engineers dig through the earth with tools and machines. The second step is support; engineers must support any unstable ground around them while they dig. The final step is lining: Engineers add the final touches, such as the roadway and lights, when the tunnel is structurally sound.

Figure 6.33 | Building a tunnel like this one in Caracas, Venezuela, involves three steps: excavation, supporting the structure, and lining the tunnel.

In modern tunnel construction, a tunnel boring machine with a circular diameter and cutting tools on its face is used to excavate the ground. Tunnel boring machines can be used as an alternative to drilling and blasting. The earthen material that is displaced is sent back to the surface through a shaft. If the material being bored is hard rock, the need for supporting the structure is not as great as when the material is soft earth. When tunnels are bored into soft earth, a tunneling shield is installed to keep the earth in place during construction. Tunnel boring machines can bore tunnels up to 15 meters in diameter (about 50 feet).

As the tunnel is bored, pre-cast concrete sections are placed within it to act as a lining. These sections are reinforced with steel and will become permanent supports.

Underwater Tunnels: The Fort McHenry Tunnel

▷   Structures such as tunnels can include prefabricated materials. For example, underwater tunneling uses huge steel tubes that are prefabricated in a factory and transported to their destination above the floor of the waterway. The tubes are then transported by barge and sunk into the proper position.

The Fort McHenry tunnel in Baltimore Harbor is a historic landmark. During the war of 1812, the British bombarded the fort. U.S. soldiers fought the British force of 50 ships. From an American ship in the harbor, a young attorney named Francis Scott Key could see that the American flag still waved over the fort. The sight inspired him to write the lyrics to what became the U.S. national anthem, "The Star Spangled Banner."

In 1985, the 7,200 foot-long Fort McHenry Tunnel was constructed near the fort to link parts of the interstate highway system (see Figure 6.42). Because it was important to preserve the landscape, the tunnel was hidden from the sight and earshot of the fort. The tunnel has eight lanes, carries 70,000 vehicles a day, and is the widest underwater tunnel in the world. The sunken section is 5,400 feet long. It was constructed as two long tubes, each made from 16 individual tubular sections. Each section is 82 feet wide and 42 feet high. Interior and exterior concrete was added to the sections, bringing the total weight of the structure to 31,882 tons.

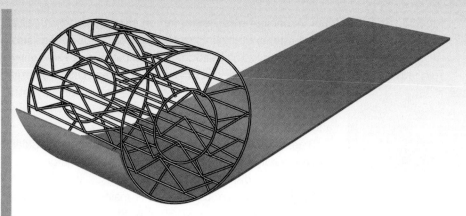

Figure 6.34 | Shaping the module of the Fort McHenry Tunnel involved wrapping a steel plate around a specially designed reel. Thirty-two modules were built, and then two tunnel tubes were constructed by joining sixteen tubes end to end to make each tunnel.

Figure 6.35 | Mont Blanc is the highest mountain in Western Europe. Its summit is 4,807 meters (or about 15,800 feet) above sea level. The tunnel under Mont Blanc was drilled and blasted to connect France and Italy. The tunnel is 11.7 km (7.3 m) long. Construction began in 1957 and ended in 1965.

Sewers and Pipelines

Sewers are pipelines that convey sewage or surface-water runoff from more than one property location. The difference between a sewer and a drain is that a drain only conveys sewage or surface-water runoff from a single property location. The drain ultimately connects to the sewer system (see Figure 6.36).

Sewage is mostly liquid, with the addition of some solid waste material that is the by-product of human activity. It includes household waste products from toilets, washing machines, and dishwashers, along with waste products from industry. To protect the environment and ensure health and safety, the sewage is carried by drain pipes to larger pipelines called sewer systems that lead to sewage treatment plants. In these plants, the sewage is chemically treated, recycled, or disposed of.

Figure 6.36 | This egg-shaped sewer, laid alongside the subway construction under Seventh Avenue in New York City, was made of reinforced monolithic concrete. The massive character of the structure and the elaborate provisions for a water-tight joint when the sewer was extended are clearly shown, along with the nature of the foundation constructed for it.

In municipal areas, rainwater and melting snow flow into catch basins and then into sewers. Normally, this sewage also is carried to sewage treatment plants, but during very heavy rain or snow, sewers can become overloaded. In this case, many coastal cities then discharge some sewage directly into waterways or deep ocean outfalls that are far from swimming and fishing areas.

There are significant environmental issues to consider when discharging sewage. When sewage is discharged into waterways, it is dispersed and diluted, but it is still a pollutant.

Other pipelines are used for water mains and for gas and oil transport (see Figure 6.37). The world's longest pipeline is the Langeled pipeline, which carries natural gas from Norway to the United Kingdom. It is about 1,200 km long (about 750 miles) and carries 70 million cubic meters of natural gas daily.

Figure 6.37 | This pipeline was built to carry water in the Negev desert in southern Israel.

Airports

In some ways, airport construction projects are similar to other large-scale construction projects. However, airports are unique in their scale, as well as in how multiple technological systems must be integrated to assure efficient and safe functioning. Airport infrastructure includes not only structures such as terminals and aircraft hangars, but roadways and runways, parking facilities, transportation systems (monorails, people movers, baggage handling conveyors), communication systems (air traffic control towers and other electronic systems), emergency medical facilities, hotels, and food services. The United States has about 14,000 airports.

Sometimes, because of the size and complexity of airport projects, the demands facing developers resemble the demands of planning a small city. In fact, the proposed Dubai international airport, to be completed in 2018, will cost billions of dollars, cover 140 square kilometers (54 square miles), and house 750,000 people.

Large airports are sprawling; both the aircraft and the structures to house them can be huge. Terminal buildings themselves can cost more than a billion dollars to build, while individual aircraft can cost up to $150 million. The immensity of a large airport project is awesome (see Figures 6.38 and 6.39).

Airports must be built to conform to national and international codes and standards. In the United States, the Federal Aviation Administration specifies standards for airport construction.

Standards for Building Runways

▷ Construction standards for runways are very specific. The following example comes from the Federal Aviation Administration:

212-1.1 This item shall consist of a base course composed of shell and binder constructed on a prepared underlying course in accordance with these specifications and shall conform to the dimensions and typical cross section shown on the plans. The shell shall consist of durable particles of "dead" oyster or clam shell. The base material shall consist of oyster shell, together with an approved binding or filler material, blended or processed to produce a uniform mixture complying with the specifications for gradation, soil constants, and compaction capability. Clam shell may be used only in combination with oyster shell in the proportion up to and including 50 percent. The shell shall be reasonably clean and free from excess amounts of clay or organic matter such as leaves, grass, roots, and other objectionable and foreign material.

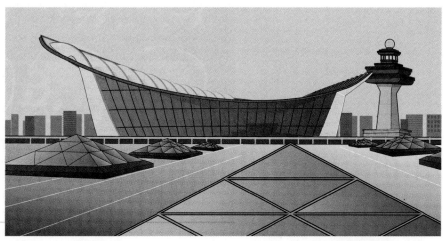

Figure 6.38 | The terminal building at Dulles Airport in Washington, DC, is well recognized for its architecture. It was designed by the well-known Finnish architect, Eero Saarinen, who described the structure as a huge, continuous hammock suspended between concrete trees. Within the concrete are steel cables that support the weight of the concrete roof.

Figure 6.39 | Beijing's new international airport terminal, built for the 2008 Olympics, is the world's largest and most advanced airport building. Its soaring aerodynamic roof and dragon-like form will celebrate the thrill and poetry of flight and evoke traditional Chinese colors and symbols. The terminal encloses a floor area of more than a million square meters and is designed to accommodate an estimated 43 million passengers per year. The terminal was built with a single unifying roof canopy, whose linear skylights are both an aid to orientation and sources of daylight. The color will change from red to yellow as passengers progress through the building.

Section 2 △ Learning About Types of Structures

Substructures and Superstructures

A structure has a substructure and a superstructure. The substructure, often referred to as the foundation, is the part that is in the ground; the superstructure is the part that is visible, and is built upon the substructure.

The purpose of the substructure is to distribute the load of the structure over a large area of the ground to prevent it from sinking. It transmits the load to the rocky, hard soil underneath the structure. It also strengthens the structure so that it can withstand earthquakes or high winds. The substructure becomes a level base upon which the structure can be built. Foundations include the footing and the vertical supports. In construction, the substructures for very tall buildings are huge. For example, the new Freedom Tower will replace the twin towers of the World Trade Center in New York, which were destroyed on September 11, 2001. The substructure will be 11 stories, including utilities, machinery, shopping facilities, subways, and access to the World Financial Center, all of which will be underground.

Several types of footings can be used under different circumstances. The most common type that supports walls and columns is a spread footing, which is used on hard ground and normally built around the entire perimeter of a structure (see Figure 6.40). A spot footing is a single square pad that is about two feet wide and one foot thick (see Figure 6.41). It is used to support a pier or a post.

In cold climates, the footings are always dug so that they are deeper than the frost line (the depth at which the ground freezes). Moisture in the ground causes it to expand when it freezes, which would cause the footing to move or crack.

To construct the substructure of very tall and heavy buildings, vertical holes are dug, sometimes as deep as 100 meters (over 320 feet) into the ground. The holes are filled with concrete that is reinforced with steel rods. These long concrete-and-steel piles are called barrettes. Friction between the barrettes and the soil around them keeps them from sinking further. A very thick concrete slab sits on the barrettes and serves as the foundation for the columns and beams that make up the building frame.

Sometimes, people construct buildings without footings. In such cases, the foundation is made of wooden, steel, or concrete columns called "piles" that are driven into the ground as far as the bedrock. Beams are attached to the tops of the piles to provide a surface upon which the superstructure is built (see Figure 6.42). Piles are used in marshes or other areas where the soil is soft or the water table is high.

Superstructures fall into three general categories: mass superstructures, bearing wall superstructures, and framed superstructures. Mass superstructures are made from large masses of materials and have little or no space inside. Dams and monuments are mass superstructures. They are built from brick, concrete, earth, or stone.

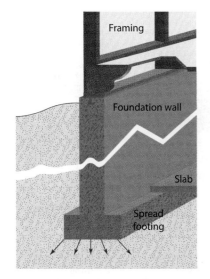

Figure 6.40 | A spread footing is normally built around the perimeter of a structure. The footing is made from poured concrete and is 16 to 24 inches wide and 6 to 16 inches thick.

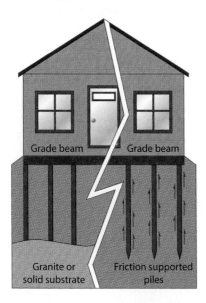

Figure 6.42 | Piles are used when the ground is soft. Piles either rest on bedrock or are driven into the soil far enough that frictional forces prevent them from sinking further into the ground.

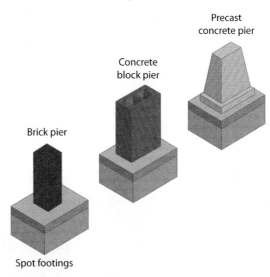

Figure 6.41 | Spot footings support piers or posts.

Bearing wall superstructures enclose a space using walls. They are built from brick, concrete, or stone. For example, the castles built in the Middle Ages are bearing wall superstructures. The walls of some of these castles were 20 feet thick or more at the base. The walls surrounding the Old City of Jerusalem are themselves bearing wall superstructures (see Figure 6.43). The Cathedral of Notre Dame in Paris, France, is also a bearing wall superstructure, as is the Coliseum in Rome (see Figures 6.44 thru 6.46).

Figure 6.43 | The walls surrounding the Old City of Jerusalem are a bearing wall superstructure. The walls of the Old City were destroyed and rebuilt several times. The present walls were constructed during the reign of the Ottoman sultan Suleiman the Magnificent (1520–1566).

Figure 6.44 | The Cathedral of Notre Dame in Paris was completed in the fourteenth century.

Figure 6.45 | The Colosseum of Rome is a bearing-wall superstructure completed in 80 C.E. It could seat as many as 50,000 onlookers at gladiator competitions.

Figure 6.46 | The Blue Mosque in Istanbul, Turkey, is a bearing wall superstructure that was completed in 1616 C.E. The dome rises to 140 feet and is 77 feet in diameter. The mosque can hold about 10,000 people.

Framed superstructures use a framework to support the building. Today, most buildings are framed superstructures (see Figure 6.47). Lumber is used for framing in most houses. In office and apartment buildings, reinforced concrete and steel are used for framing.

Figure 6.47 | The metal-and-glass pyramid that serves as the entrance to the Louvre museum in Paris was designed by architect I. M. Pei. It is an example of a framed superstructure.

Enclosing Framed Superstructures

The building enclosure separates the inside of the building from the outside environment. The building enclosure consists of five components, as shown in Figure 6.48:

- the roof
- the above-grade walls
- windows and doors
- the below-grade foundation wall
- the base floor

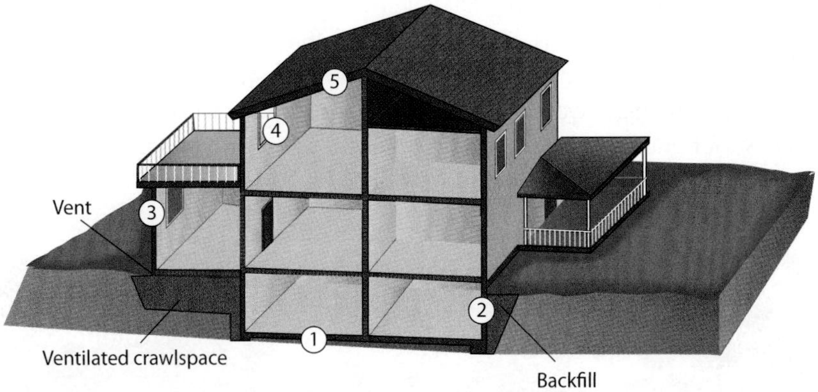

Building Enclosure Components

Vent

3

5

4

1

2

Ventilated crawlspace

Backfill

1. Base floor system(s)
2. Foundation wall system(s)
3. Above Grade wall system(s)
4. Roof system(s)
5. Windows and doors

━━━ Building enclosure
▦▦▦ Interior spatial separators

Figure 6.48 | The building enclosure consists of five components, as shown in the diagram.

Figure 6.49 | The homes and shelters in this Bedouin village are framed with wooden 2×4s and covered with available materials.

Figure 6.50 | PPG Place is a symbol of downtown Pittsburgh, Pennsylvania, and a striking advertisement for the Pittsburgh Plate Glass Corporation. Almost one million square feet of glass was used in the construction enclosure, incorporating 19,750 glass panels.

Framed superstructures must be enclosed. The enclosure serves several purposes, which are broadly categorized as support, control, finish, and distribution. Support functions relate to how the enclosure helps to add structural support to the building. Control functions act to control the separation of the air, moisture, heat, and sound from the outside to the inside. For example, large sections of glass on skyscrapers can be responsible for temperature changes and unwanted glare. The construction of an envelope determines the degree to which the flow of heat, air, noise, driving rain, and water vapor are controlled. Finish functions provide the aesthetic and visual effects. Distribution functions allow utilities and services to be distributed throughout the building.

Enclosures are built from a wide variety of materials, including solid and composite wood, metal, cement, gypsum, plastic, and glass. The materials are chosen for their appearance, cost, and availability, as well as for their physical, mechanical, electrical, magnetic, thermal, optical, and acoustic properties (see Chapter 4, Section 2, "Properties of Materials").

The building enclosure is a topic of discussion among professionals in the construction industry. For example, degradation of enclosure materials due to decay, insect damage, or water leakage around walls and windows can result in high repair costs for building owners and serious disturbances for building occupants.

SECTION TWO FEEDBACK >

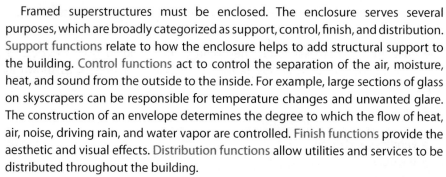

1. What are the main products of the construction industry?
2. Explain the differences among residential, commercial, institutional, and industrial buildings.
3. What causes tall buildings to sway?
4. What are the components of a structural foundation?
5. What would engineers have to consider in deciding between a beam bridge, an arch bridge, or a suspension bridge for use in a specific project?

SECTION 3: Maintaining and Renovating Structures

KEY IDEAS >

- Structures require periodic maintenance, alteration, or renovation to improve them or alter their intended use.

- Because maintenance and repair of structures disrupts people's day-to-day lives, it is important to complete these tasks quickly.

There are limits to the life expectancy of all materials, as shown in Figure 6.51. Some degrade more quickly than others. All structures require periodic maintenance, alteration, or renovation to improve them or to alter their intended use. A design criterion in planning construction projects should relate to how durable a building will be. There is an old adage that still rings true: *Failing to plan means planning to fail.*

Maintenance of Structures

Structures require maintenance because they will degrade even under typical environmental conditions. Bridges collapse; moisture can cause wood to rot; oxidation causes iron to rust; acid rain and other forms of pollution degrade surfaces. According to the 2005 American Housing Survey conducted by the U.S. Department of Housing and the U.S. Census Bureau, about 20 million homes in the United States had water leakage, about 10 million homes had damaged roofs, 5 million had broken windows, and 3.2 million had damaged foundations.

In August 2007, the eight-lane Interstate 35W bridge collapsed during rush hour and fell into the Mississippi River in Minneapolis. (See Figure 6.52.) The bridge carried more than 100,000 vehicles a day and was undergoing repairs when it collapsed. Dozens of vehicles fell; 13 people died and 79 were injured. Parts of the bridge were subsequently judged to be structurally deficient because of corroded bearings and cracks in the metal. Metal components are weakened (called metal fatigue) when heavy vehicles dynamically load a structure. However, in a March 18, 2008 article in the New York Times, the cause of the collapse was attributed to the stress caused by 99 tons of sand placed on the roadway by construction workers directly over two of the bridge's weakest points. According to the U.S. Department of Transportation, the United States is home to almost 80,000 functionally deficient bridges. Not all of these bridges are imminent disasters, but the large number underscores the need for a national strategy that includes consistent maintenance and inspection.

Structures also require maintenance, renovation, or repair due to natural disasters such as earthquakes and hurricanes. These events can damage structural components or cause their complete destruction.

In 1994, for example, sixty people were killed and 7,000 injured in the Northridge earthquake near Los Angeles, California. In addition to the human casualties, there was almost $40 billion of property damage. Engineers learn from these disasters, and the earthquake led to an analysis of building design. Analysts found that for mid-rise buildings, very high strength can actually cause

ESTIMATED LIFE EXPECTANCY AND HOME OWNER MAINTENANCE CHART		
Building Component	**Estimated Life* (years)**	**Homeowner Action**
Concrete/ block foundation	100+	Check for cracks or surface deterioration. Consult a professional if you have any leaking or severe cracking. Check for termite tubes on foundation.
Exposed concrete slabs	25	Inspect for cracking. Seal to prevent water penetration.
Siding (life-span depends on type)	10–100	Clean all types of siding. Paint or seal wood siding (See exterior paints/stains).
Drywall	30–70	Inspect, clean, and paint for aesthetic purposes.
Roofing	15–30	Inspect for missing or deteriorated shingles. Clean to remove mold build up.**
Gutters and downspouts	30	Remove debris.
Insulation	100+	Inspect blown insulation in attic and check floor insulation (crawlspace) to assure that it is in place.
Windows	20–50	Inspect and repair weather stripping. Inspect for broken seals in insulated windows. Clean exterior window frames.**
Exterior doors	20–50	Clean and refinish when necessary (see exterior paints/stains).
Garage doors	20–50	Clean garage door. Lubricate moving parts. Paint or seal as necessary.**
Exterior paints/stains	7–10	Clean and inspect. Repaint and caulk as needed.
Wood floors	100+	Clean and wax.
Carpeting	11	Clean annually.

* All numbers estimated and condensed from NAHB life expectancy survey from [housing, facts, figures and trends] (1997)
** Use care if power washing. The high pressure water can cause more harm than help if not used cautiously

Figure 6.51 | Building materials have different life expectancies. Selecting the best material for a particular construction application reduces overall construction and maintenance costs.

more susceptibility to earthquake damage because the buildings are too stiff. Engineers concluded that if less reinforcing steel were used, the survival rate in these buildings would increase.

People have also had a tendency to overextend the infrastructure by building homes in areas that are too close to the sea because they like the view or because views increase property values. Unfortunately, they risk flooding and storm damage in the process. People tend to move to areas where the climate is generally desirable, but where high winds, hurricanes, and earthquakes may also occur.

Figure 6.52 | The Interstate 35W bridge collapse in Minneapolis killed 13 people.

Other maintenance is simply a function of the natural life cycle of equipment and material. Power, water supply, plumbing, and sewage systems need to be maintained or they will break down. Table 6.1 shows the most common housing breakdowns.

Table 6.1 | Occurrence of Selected Housing Breakdowns in the Last 12 Months, by Year Structure Was Built

SELECTED HOUSING BREAKDOWNS	BEFORE 1920	1920s	1930s	1940s	1950s	1960s	1970s	1980s	1990s
Sewer or septic breakdown	2%	2%	3%	3%	2%	2%	2%	1%	1%
All toilets not working	3%	3%	3%	4%	3%	3%	3%	2%	2%
Water supply breakdown	4%	3%	4%	4%	4%	4%	5%	5%	5%
Uncomfortably cold last winter	11%	9%	9%	8%	8%	7%	6%	5%	6%
Water leak from inside	10%	12%	11%	10%	9%	10%	10%	10%	7%
Fuses blown/breakers tripped	14%	14%	14%	13%	12%	10%	10%	10%	9%
Water leak from outside	20%	18%	17%	15%	13%	11%	10%	10%	9%

Source: U.S. Census Bureau. American Housing Survey: 1999.

Aging Bridges

▷ The present state of U.S. bridges mainly tracks back to age: The combination of weather and vehicle traffic leads to deterioration, including corrosion, fatigue, absorption of water, and loss of pre-stress. In addition, bridges can be damaged by impact, overload, scour, fractures, seismic activity, foundation settlement, cracking, and bearing failure. Bridges, far more than buildings and other structures, are subject to live loads that come and go. These loads include cars, trucks, and people, as well as wind, accumulated snow, and even earthquakes.

Section 3 △ Maintaining and Renovating Structures

Heavy traffic, especially, causes much cyclic loading and deterioration. Fast-moving traffic stresses a bridge horizontally, and the "vehicle bounce" across the bridge increases the vertical loading. The heavier the load, the more damage is caused.

Studies by the U.S. Federal Highway Administration suggest that bridges deteriorate slowly during the first few decades of their 50-year design lives, followed by rapid decline in the last decade. Compounding this issue is the dramatic increase in both the weight and number of heavy commercial vehicles, which impose an exponential increase in damage to the infrastructure.

Figure 6.53 | Inspecting bridge foundations is sometimes dangerous work.

Structural Renovation

Renovation of an existing structure occurs when the structure needs repair or when its style needs updating. A well-known renovation was the work done on the Statue of Liberty to repair damage to its copper surface caused by many years of exposure to salt air.

Renovation or reconstruction is also sometimes desirable when population increases require changes to the supporting infrastructure. After an influx of people into a region, for example, roads must be widened. In that case, existing structures such as bridges and drainage systems may have to be demolished and rebuilt. Airports and seaports are expanded to accommodate an increase in passenger use.

Road Maintenance Under normal use, roadways degrade because of environmental conditions and the constant flow of traffic. To ensure safe and reliable road conditions, people must maintain roadways on a regular schedule. Road maintenance activities include routine maintenance, periodic maintenance, special maintenance, and continued development. Routine maintenance is work that is typically done every year. Examples include patching potholes, cleaning the drains that parallel or cross under the road, and maintaining the landscaping. Periodic maintenance is done every few years. Examples include resealing and resurfacing pavement. Special maintenance is work that cannot be planned in advance. It might include road repairs needed as a result of damage caused by major accidents or severe storms. Development work, such as paving and constructing rest areas, serves to upgrade the roadway.

Accelerated Renovation and Construction

A set of techniques to reduce the time of renovation and construction is called accelerated construction. Accelerated construction uses new technologies and procedures to speed up the construction of major projects, such as roadways and bridges. The idea is to decrease construction time while maintaining quality and safety.

One important approach to accelerated construction is to prefabricate structural elements. These elements can be manufactured off site, transported to the construction site, and installed by construction crews. A case in point is a 4.5-km (2.8-mile) portion of Interstate 15 in the city of Devore, in southern California. The freeway had badly damaged concrete lanes. Using accelerated construction techniques, the roadway was rebuilt in nineteen days instead of the ten months that were originally estimated. The engineers, contractors, and laborers worked around the clock. They used various procedures to facilitate their work, including:

- A moveable barrier system, which provided dynamic lane reconfiguration to minimize traffic disruption

- A rapid-strength concrete mix that made it possible to open the road to traffic 12 hours after its placement
- Incentive/disincentive provisions to encourage the contractor to complete the work on time
- A multifaceted outreach program to gain public support
- Automated information systems to update travelers with real-time work zone travel information

Accelerated Bridge Construction Accelerated Bridge Construction (ABC) is a rapid way to upgrade or repair bridges, which need renovation when they age. ABC is used when construction crews must minimize the impact to the public and the environment while the renovation is in progress. If you have traveled through construction zones in which lanes are closed after midnight to minimize traffic disruptions, you have experienced an approach akin to ABC. ABC takes a systems approach to construction. It involves a great deal of upfront planning to minimize disruptions, including choosing the right contractor and arranging for the necessary permits. During the construction process, components can be prefabricated and procedures can occur concurrently.

In New Haven, Connecticut, for example, a bridge needed to be built that would partially extend over railroad tracks. The Connecticut Department of Transportation specified that, in order to limit disruption to train service, the bridge installation had to be completed in one night. The accelerated construction involved building the bridge section near the final site and then lifting the bridge section into place using a huge crane. The crane itself required more than four weeks of assembly, and its parts were delivered on more than 200 tractor-trailer loads. To see images from the construction project, go to http://www.fhwa.dot.gov/bridge/crane23.cfm.

SECTION THREE FEEDBACK >

1. Give a construction example that validates the following adage: *Failing to plan means planning to fail.*
2. Why do even well-built structures require maintenance?
3. What would cause a building to need renovation?
4. Give two examples of routine road maintenance.
5. What characterizes the idea of accelerated construction?

CAREERS IN TECHNOLOGY

Matching Your Interests and Abilities with Career Opportunities: Careers in Construction Management

Construction managers plan, direct, and coordinate a wide variety of construction projects, including the building of all types of residential, commercial, and industrial structures, roads, bridges, wastewater treatment plants, schools, and hospitals. Construction managers may oversee an entire project or just part of a project. Although they usually play no direct role in the actual construction of a structure, they typically schedule and coordinate all design and construction processes, including the selection, hiring, and oversight of specialty trade contractors.

Significant Points

* Construction managers must be available—often 24 hours a day—to deal with delays, bad weather, or emergencies at the job site.
* Employers prefer people who combine construction-industry work experience with a bachelor's degree in construction science, construction management, or civil engineering.
* Excellent employment opportunities are expected in the future, as the increasing complexity of many construction projects requires more managers to oversee them.

Nature of the Industry

Construction managers are salaried or self-employed managers who oversee construction supervisors and workers. They often go by the job titles of program manager, constructor, construction superintendent, project engineer, project manager, construction supervisor, general contractor, or similar designations. They coordinate and supervise the construction process from the conceptual development stage through final construction, making sure that the project gets done on time and within budget. They often work with owners, engineers, architects, and others who are involved in the construction process. Given the designs for buildings, roads, bridges, or other projects, construction managers oversee the planning, scheduling, and implementation of the project to execute those designs.

Construction managers oversee the selection of general contractors and trade contractors to complete specific pieces of the project, which could include everything from structural metalworking and plumbing to painting and carpet installation. Construction managers determine the labor requirements and sometimes supervise or monitor the hiring and dismissal of workers. They oversee the performance of all trade contractors and are responsible for ensuring that all work is completed on schedule.

Construction managers direct and monitor the progress of construction activities, sometimes through construction supervisors or other construction managers. They oversee the delivery and use of materials, tools, and equipment, and the quality of construction, worker productivity, and safety. They are responsible for obtaining all necessary permits and licenses. Depending on the contractual arrangements, they might also direct or monitor compliance with building and safety codes and other regulations.

Working Conditions

Construction managers usually work out of a main office from which the overall construction project is monitored. Alternatively, they may work out of a field office at the construction site. Managers may travel extensively when the construction site is not close to their main office or when they are responsible for activities at two or more sites. Management of overseas construction projects usually entails temporary residence in another country.

Construction managers may be on call, often 24 hours a day, to deal with delays, the effects of bad weather, or emergencies at the site. Most managers work more than a standard 40-hour week because construction may proceed around the clock. They may have to work long hours for days and even weeks to meet special project deadlines, especially if there are delays. Although the work usually is not considered inherently dangerous, construction managers must be careful while performing onsite services.

Training and Advancement

For construction manager jobs, employers increasingly prefer to hire people with a bachelor's degree in construction science, construction management, or civil engineering, as well as industry work experience. Practical industry experience is very important, whether it is acquired through an internship, a cooperative education program, or work experience in a trade or another job in the industry. As construction processes become increasingly complex, employers are placing a growing importance on postsecondary education.

Many colleges and universities offer four-year degree programs in construction management, construction science, and construction engineering. These programs include courses in project control and development, site planning, design, construction methods, construction materials, value analysis, cost estimating, scheduling, contract administration, accounting, business and financial management, safety, building codes and standards, inspection procedures, engineering and architectural sciences, mathematics, statistics, and information technology.

Outlook

Excellent employment opportunities for construction managers are expected through 2014 because the number of job openings will exceed the number of qualified people seeking to enter the occupation. This situation is expected to continue even as college construction management programs expand to meet the current high demand for graduates. The construction industry does not always attract sufficient numbers of qualified job seekers because it is often seen as having poor working conditions. Opportunities will increase for construction managers to start their own firms. However, employment of construction managers can be affected by the short-term nature of many projects and to cyclical fluctuations in construction activity.

The increasing complexity of construction projects is boosting the demand for management-level personnel within the construction industry. Advances in building materials and construction methods, the need to replace portions of the nation's infrastructure, and the growing number of multipurpose buildings and energy-efficient structures will further add to the demand for more construction managers.

Earnings

Earnings of salaried construction managers and self-employed independent construction contractors vary depending on the size and nature of the construction project, its geographic location, and economic conditions. In addition to typical benefits, many salaried construction managers receive benefits, such as bonuses and use of company vehicles.

According to a 2004 survey, median annual earnings of construction managers were $69,870. The middle 50 percent earned between $53,430 and $92,350. The lowest-paid 10 percent earned less than $42,120, and the highest-paid 10 percent earned more than $126,330.

[Bureau of Labor Statistics, U.S. Department of Labor, Occupational Outlook Handbook, 2007–08 Edition, visited December, 2007, http://www.bls.gov/oco/]

Section 3 △ Maintaining and Renovating Structures

Summary >

The construction industry is divided into three major segments. General contractors build residential, industrial, commercial, and other buildings. Heavy and civil engineering construction contractors build sewers, roads, highways, bridges, tunnels, and other projects. Specialty trade contractors perform specialized activities related to construction, such as carpentry, painting, plumbing, and electrical work.

The construction infrastructure is the underlying base or basic framework of structures that are essential for people to go about their daily lives and comfortably conduct their affairs. The infrastructure includes airports, bridges, canals, dams, hazardous waste management, pipelines, roadways, tunnels, wastewater and solid waste treatment facilities, and water supply systems.

Construction projects create social, economic, lifestyle, and environmental impacts. Some impacts are positive and some are negative. These impacts must be considered and properly managed. People are becoming more conscious of living in ways that do minimal harm to the environment. Increasingly, "green construction" is becoming popular. Green construction, often referred to as "sustainable development," attempts to substantially limit environmental damage, reduce energy use, and take advantage of recycled and recyclable construction materials.

Products of the construction industry include buildings, industrial plants, roadways, bridges, tunnels, sewers, pipelines, and airports. These products and structures can include prefabricated materials and are constructed using a variety of materials, processes, and procedures.

The design of structures involves a number of requirements that relate to appearance, efficiency, function, safety, strength, longevity, maintenance, and available utilities. The design and construction of structures are regulated by laws, codes, and professional standards.

Structures require periodic maintenance, alteration, or renovation to improve them or alter their intended use. Because structure maintenance and repair disrupts people's day-to-day lives, it is important to complete these tasks quickly. Accelerated construction uses new technologies and procedures to speed up the construction of major projects like roadways and bridges. The idea is to decrease construction time while maintaining quality and safety.

FEEDBACK

1. Explain the difference in job responsibilities of general contractors, heavy and civil engineering contractors, and specialty trade contractors.

2. Explain the purpose of President Clinton's 1998 directive regarding U.S. infrastructure, which identified certain structural elements as part of the critical infrastructure of the United States. In what way is a nation's infrastructure *critical?*

3. Choose a current construction project, either local or outside your community. Discuss the project's impact on the economy and the environment. Indicate how these environmental impacts could be managed.

4. Research an example of sustainable development anywhere in the world. Describe how the project was designed to limit environmental damage and reduce energy consumption.

5. Identify the major types of structural products that result from construction work.

6. Give an example of a construction project in which prefabricated materials were used.

7. Explain the difference in construction among mass superstructures, bearing wall superstructures, and framed superstructures. Give a specific example of a structure that fits each category.

8. Draw a diagram illustrating the substructure and the superstructure of a wood-framed single-family house.

9. How would a building contractor or developer determine if a community's building codes were applicable to the construction of a shopping mall?

10. List some examples of routine maintenance for a roadway.

11. Research an example of accelerated construction and explain what procedures were used to speed up that construction.

12. Explain how the choice of enclosure materials for a building can affect not only construction costs, but ongoing maintenance and repair costs as well.

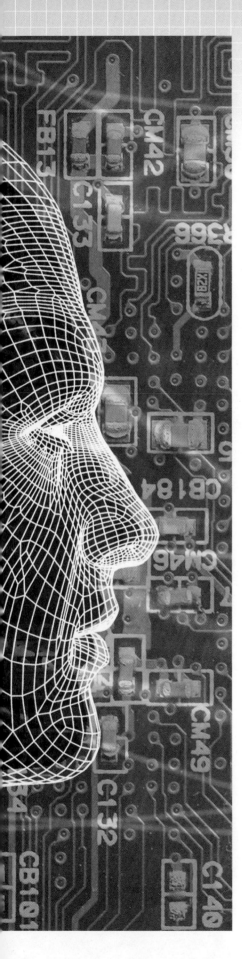

DESIGN CHALLENGE 1:
Bridge Engineering

• Problem Situation

A bridge is a structure that spans a valley, body of water, roadway, railroad tracks, or any other obstruction to continuous travel. Bridges are designed to withstand dead loads and live loads. A dead load is a load that does not change, such as the weight of the structure itself. It includes the deck, sidewalks, parapets, railings, and the steel or concrete load-carrying members. These loads account for a significant and considerable percentage (often the majority) of the stress in the load-carrying member. Live loads change; the changes can be due to increased or decreased traffic and pedestrian loading. Other important loads are a function of wind, snow, impact, earthquake, and temperature variation. When designing a bridge, strength and safety of the structure are the primary considerations, but cost is very important as well.

• Your Challenge

You are asked to design a bridge model that spans a distance of two feet. You will investigate how to make the model strong enough to support the test load. Before you determine which type of bridge to build, you will investigate four types of bridges: a beam bridge, an arch bridge, a suspension bridge, and a truss bridge.

No matter which type of bridge you decide to model, the bridge's deck height will be set at eight inches above the ground. Your bridge must support a load of at least 20 pounds, concentrated at mid-span. The bridge that exceeds the minimum load by the greatest amount for the least cost will be the best design.

• Safety Considerations

1. Only use tools and machines after you have had proper instruction.
2. Wear eye protection when using tools, materials, machines, paints, and finishes.
3. Use caution when loading the bridge deck to the failure point, as components may split and break away.

• Materials Needed

1. Brass fasteners
2. Building bricks to use as embankments
3. Craft sticks (basswood strips, 1/8 inch × 2 inches × 24 inches)
4. Centimeter sticks (1 cm × 1 cm × 40 cm)
5. Large weights (1/2 pound each)
6. Metal nuts
7. Modeling clay
8. Paper clips
9. Popsicle sticks
10. Small weights (100 g to 2 kg)
11. String
12. Tape
13. 30 × 30-inch oak tag sheet
14. White glue
15. Wooden dowels

• Clarify Design Specifications and Constraints

The bridge will be rated according to a figure of merit (F_m), which is defined as F_m = weight/cost. Therefore, the best bridge will support the most weight at the least cost. The weight will be applied at mid-span and will be 20 pounds at minimum. Your bridge must be constructed only with materials specified by your instructor. The bridge must be at least 4½ inches wide so that a four-inch brick can be placed on the bridge deck as the test load.

• Research and Investigate

To complete the design challenge, you need to first gather information to help you build a knowledge base.

1. In your guide, complete the Knowledge and Skill Builder 1: Cost of materials.
2. In your guide, complete the Knowledge and Skill Builder 2: Beam bridges.
3. In your guide, complete the Knowledge and Skill Builder 3: Arch bridges.
4. In your guide, complete the Knowledge and Skill Builder 4: Suspension bridges.
5. In your guide, complete the Knowledge and Skill Builder 5: Truss bridges.

• Generate Alternative Designs

Your group is to choose the type of bridge you want to design and model. Once you determine the type of bridge (beam, arch, suspension, or truss), describe two possible approaches to building that bridge. The approaches may use different materials, different methods of strengthening the bridge, and different systems to support the structure.

• Choose and Justify the Optimal Solution

What decisions did you reach about choosing the model? Why did your group settle on this approach? What trade-offs did you make in coming to this conclusion?

• Display Your Prototypes

Construct a functional model of your bridge. Include drawings and sketches that helped you during the construction of the model.

• Test and Evaluate

Explain how you will test the bridge under load. Explain how you calculated the figure of merit.

• Redesign the Solution

What did you learn through the design and testing of your bridge that would inform a redesign of the model? What additional trade-offs would you have to make?

• Communicate Your Achievements

Describe the plan you will use to present your solution to your class. (Include a media-based presentation.) Demonstrate how you tested your model and what you learned about how the bridge design distributed the load.

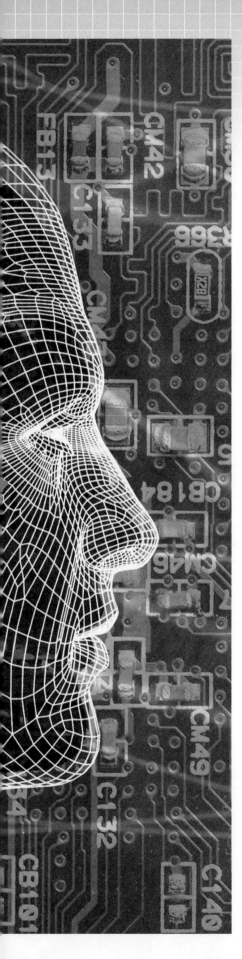

DESIGN CHALLENGE 2:
Emergency Shelter Construction

• Problem Situation

An earthquake has left more than 100 people homeless in central Alaska. As winter is approaching, people need to be housed temporarily for one month while their homes are rebuilt. Although tents that are on hand might be a short-term solution, these lack the insulation needed to help people survive the cold temperatures and snowy conditions. The earthquake location is 200 miles from the nearest city, and snow has made travel impossible. Materials that are available locally must be used because time is of the essence.

• Your Challenge

Your challenge, as one of a team of earthquake victims, is to design and build a rapidly erectable structure that will provide insulation from the cold, withstand snow load, provide air exchange, and be built from materials that are readily available locally.

• Safety Considerations

1. Only use tools and machines after you have had proper instruction.
2. Make sure that if machines are used, all guards are in place.
3. Wear eye protection at all times.
4. Observe safety precautions when using fire-lighting devices.

• Materials Needed

1. Small metal boxes (no larger than 12 × 6 × 8 inches)
2. Small camping stove
3. Stove pipe (4-inch diameter)
4. Tree branches and boughs
5. Tent and tent poles
6. Plastic sheeting
7. Corrugated cardboard
8. Various hand tools
9. Staple gun
10. Canvas tarps
11. Rope and string
12. Fire-lighting device

• Clarify the Design Specifications and Constraints

The shelter must be constructed in seven hours and 30 minutes (equivalent to ten 45-minute class periods); otherwise, survivors will be at risk of hypothermia. The shelter must be large enough for five people to sleep side by side within it.

The shelter must be heated to an inside temperature of 65°F when the outside temperature is −20°F. The design must provide for air exchange. The team may only use materials provided by the instructor.

• Research and Investigate

To complete the design challenge, you need to first gather information to help you build a knowledge base.

1. In your guide, complete the Knowledge and Skill Builder 1: China earthquake, a case study.

2. In your guide, complete the Knowledge and Skill Builder 2: Determining necessary floor space and height.

3. In your guide, complete the Knowledge and Skill Builder 3: Determining the optimal shape of the shelter.

4. In your guide, complete the Knowledge and Skill Builder 4: Insulation materials.

5. In your guide, complete the Knowledge and Skill Builder 5: Heat flow and energy transfer.

6. In your guide, complete the Knowledge and Skill Builder 6: Volume and surface area calculations for various geometric shapes.

7. In your guide, complete the Knowledge and Skill Builder 7: Air exchange, moisture, and condensation.

● Generate Alternative Designs

There are many ways to approach the design of an emergency shelter. Your group will choose its own type of shelter to design and model. Describe two possible versions of a shelter that might satisfy the design criteria and constraints. Sketch each of the design versions.

● Choose and Justify the Optimal Solution

What decisions did you reach that guided your choice of shelter design? Why did your group settle on this approach? What trade-offs did you make in coming to this decision?

● Display Your Prototypes

Construct a functional model of your shelter. Include drawings and sketches that helped you during the construction of the model.

● Test and Evaluate

Show the heat flow calculations you made to indicate the degree to which the shelter will maintain an interior temperature of 65°F when the outside temperature is −20°F. If you live in a cold climate, test the shelter outdoors.

● Redesign the Solution

What did you learn through the design and/or testing of your shelter that would inform a redesign of the model? What additional trade-offs would you have to make?

● Communicate Your Achievements

Describe the plan you will use to present your solution to your class. (Include a media-based presentation.) Demonstrate how you tested the shelter. Explain what you learned about the design of a portable shelter, heat flow, and air exchange.

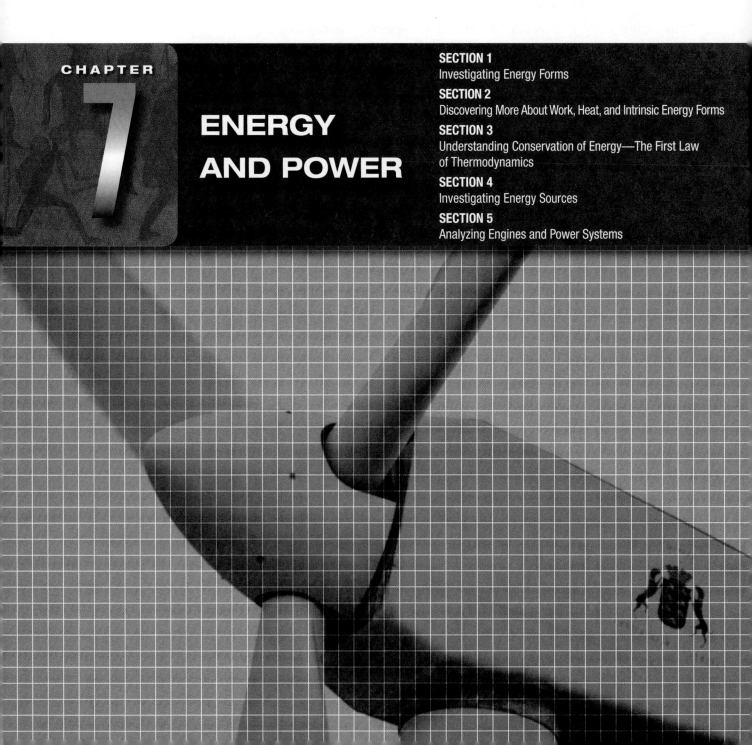

INTRODUCTION

ENERGY SUSTAINS LIFE, so it is not surprising that we use energy all day long. We depend on energy and energy transformations to live. For most of human history, all available energy came from the Sun or from the efforts of humans and animals. In the 1700s, coal and oil provided new sources of energy. We also discovered that our bodies are energy converters, changing

Figure 7.1 | Energy sources include the Sun, wind, fossil fuel, water, and wood.

food energy into chemical energy to power our minds and muscles. Many other systems are also energy converters. When we turn on a light, electrical energy is converted into light energy. The electrical energy itself was created from another energy form—usually oil, gas, or coal, but perhaps wind, solar, or nuclear energy (see Figure 7.1). In cars and buses, the engine converts a hydrocarbon fuel such as gasoline, diesel, or natural gas into mechanical energy.

In this chapter, you will learn about the various sources and forms of energy and how they can be converted into other forms. You will be introduced to the laws of thermodynamics, which explain different types of energy and energy transformations. You will also learn about various types of external and internal combustion power plants.

> # KEY IDEAS >
>
> - The various forms of energy include radiation, thermal energy, electrical energy, mechanical energy, chemical energy, and nuclear energy.
>
> - Energy may be transformed from one type to another type.
>
> - Not all energy transformations can be reversed.

The various forms of energy include radiation, thermal energy, electrical energy, mechanical energy, chemical energy, and nuclear energy. Energy may be transformed from one type to another. For example, atomic energy can be transformed into electrical energy in photovoltaic direct energy conversion. However, not all energy transformations can be reversed; electrical energy cannot be transformed into atomic energy, and it may not be possible to transform one energy form directly to another. Several transformation steps may be necessary.

Radiation

Radiation is energy that is emitted from an object. Examples of electromagnetic radiation are light from the Sun or a light bulb, heat from a heat lamp, and radio waves from an antenna. Electromagnetic radiation can best be described as waves of electric and magnetic energy moving through space. Microwaves, infrared and ultraviolet light, X-rays, and gamma rays are other examples of electromagnetic radiation.

Waves in the electromagnetic spectrum (see Figure 7.2) vary from very long radio waves that can be the length of a building to very short gamma rays that are smaller than the nucleus of an atom.

The amount of radiation emitted by an object depends on (and is proportional to) its temperature. A very hot object emits more energy that a cool object. All objects above the temperature of absolute zero ($-273°$ Celsius) radiate energy.

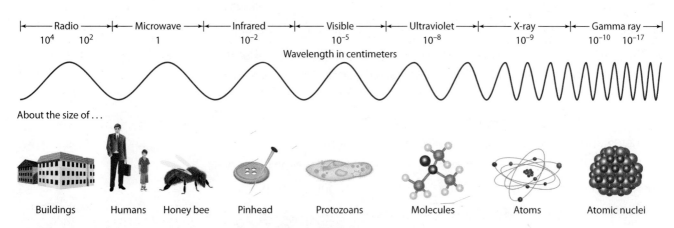

Radio	Microwave	Infrared	Visible	Ultraviolet	X-ray	Gamma ray
10^4 10^2	1	10^{-2}	10^{-5}	10^{-8}	10^{-9}	10^{-10} 10^{-17}

Wavelength in centimeters

About the size of . . .

| Buildings | Humans | Honey bee | Pinhead | Protozoans | Molecules | Atoms | Atomic nuclei |

Figure 7.2 | The electromagnetic spectrum ranges from very long radio waves to very short gamma rays.

Thermal Energy

Thermal energy (also called internal energy) is created by the motion of atoms and molecules that occurs within all matter. Thermal energy refers to the total energy of all the atoms and molecules in an object. For example, when you heat a pot of water on a stove, the heat from the burner adds energy to the water, causing the water molecules to move around more rapidly, increasing the water's thermal energy (see Figure 7.3). Interestingly, even in extremely cold conditions, the atoms and molecules in matter are still moving around a bit. In other words, the matter still contains a very small amount of thermal energy.

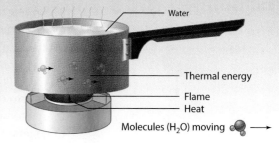

Figure 7.3 | Heat from a flame is transferred to the water, increasing the water's thermal energy.

Electrical Energy

Electrical energy, or electrical work, is caused by the flow of electrical charge through an electrical field. Electrical charge is an accumulation of a huge number of electrons, which are tiny particles with a negative electrical polarity. The basic unit of electrical charge, the coulomb, is equal to 6.28×10^{18} electrons.

When charge flows through an electrical circuit, we have an electrical current. Current, symbolized by i for intensity, is measured in amperes. One ampere of electric current is the flow of one coulomb of charge past a given point in a circuit in one second. An ampere, then, equals one coulomb/second.

Electrons do not move through a circuit for no reason; there must be a force that causes them to move. This force is the voltage, or electrical potential. When this potential exists, an electron from one atom moves to the next atom, displacing and pushing another electron ahead. This electron in turn displaces an electron in the next atom, creating the flow of current. We can make an analogy to a tube of marbles, as illustrated in Figure 7.4.

Figure 7.4 | Marbles entering and leaving a tube provide a model of electron flow through a conductor.

As one marble enters the tube on the left, another leaves the tube on the right, creating the illusion of electron flow, even though the individual electron only moved a short distance. The flow of electrons multiplied by the voltage is electrical power.

Mechanical Energy

Mechanical energy, or mechanical work, is energy that one person or object expends in moving another object. The force used to push a box times the distance the box moved is mechanical work. As another example, a piston that moves in an automobile engine is doing mechanical work.

Chemical Energy

Chemical energy, a type of internal energy, is associated with the molecular and atomic structure of matter. Substances have different abilities to undergo chemical reactions, depending on their chemical energy. For instance, because of its atomic structure, oxygen is a very active element (see Figure 7.5). We know this because many elements and compounds readily oxidize (lose electrons to oxygen). When iron oxidizes, for example, the result is rust. Nitrogen, on the other hand, does not readily interact with other elements or compounds. Its chemical potential, a measure of its chemical energy, is less than that of oxygen.

Nuclear Energy

Nuclear energy is associated with a substance's atomic structure. It, too, is a type of internal energy. Nuclear energy can be produced in two ways: by fusion and by fission. During nuclear fusion, atoms are joined together. Our sun is a giant nuclear

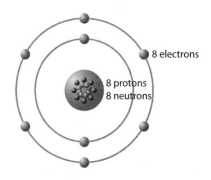

Figure 7.5 | An oxygen atom has eight neutrons, eight protons, and eight electrons.

267

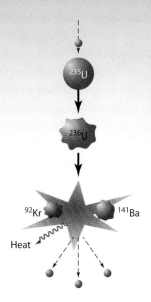

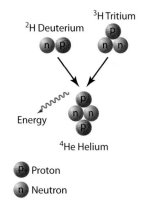

Figure 7.6 | In this example of nuclear fission, a neutron collides with uranium 235, creating an unstable element, uranium 236. Uranium 236 splits into two radioactive elements, krypton and barium, and three neutrons.

Figure 7.7 | In nuclear fusion, smaller nuclei are joined to create a larger nucleus.

fusion reactor running on hydrogen. Each second, it converts 564 million tons of hydrogen to 560 tons of helium through the fusion process.

An atom's nucleus contains protons and neutrons. The nuclei of some atoms, such as Uranium 235, become unstable when hit by an external neutron and split into two different atoms (see Figure 7.6). In the process, they emit more neutrons and energy in the form of light and heat. The new neutrons hit more nuclei and these atoms split, creating a chain reaction. A controlled chain reaction is used to generate energy in nuclear power plants. Without the controls, a devastating explosion could occur. The process of splitting an atom is fission.

A tremendous advantage of nuclear power is that a small amount of matter can create a large amount of energy. However, there is a tremendous disadvantage as well. The atoms created by the splitting are highly radioactive, and they emit dangerous levels of radiation that can kill people. The storage of radioactive waste is a great technological challenge, as the radioactivity persists for tens of thousands of years. Thus, burial and other traditional waste disposal methods cannot be used safely.

The Sun is a giant fusion reaction. We have been trying for decades to design a functioning fusion reactor that could produce virtually unlimited energy for us, but finding and harnessing such a reaction with very little radioactivity has been very difficult to achieve. The reaction combines two isotopes of hydrogen to form helium. These isotopes are deuterium (nucleus of one proton and one neutron) and tritium (nucleus of one proton and two neutrons), as shown in Figure 7.7. The difficulty lies in the process required to fuse the atoms together. The gasses must be heated to hundreds of millions of degrees, so containing the high-temperature gas plasma has been a major problem. In addition, the fusion process requires a great deal of power to start the reaction; the reaction itself, if successful, will produce more power than it consumes, but not significantly more. So, a malfunction can create great power costs for very little gain. The energy potential, however, is almost unlimited because of the virtually limitless fuel sources.

SECTION ONE FEEDBACK >

1. Can you explain the difference between electrical energy and chemical energy?
2. Do you have energy? What kind of energy do you display when you run?
3. Describe three types of electromagnetic radiation. Which has the longest wavelength?

SECTION 2: Discovering More About Work, Heat, and Intrinsic Energy Forms

KEY IDEAS >

- Matter has intrinsic energy associated with its internal structure, its velocity, and its height above the ground.
- Heat and work result from changes in intrinsic energy or cause changes in intrinsic energy.
- Power is the time rate at which work is done.

Heat and work are two energy forms that are sometimes created during a transformation from one type of energy to another (for example, from chemical to electrical). They result from changes in matter's intrinsic energy, or they can cause changes in matter's energy.

Work

When mechanical work is done on an object, energy is transferred from a person or machine to the object. Mechanical work is done whenever a force pushes or pulls an object, causing it to move. The amount of mechanical work done is equal to the distance the object moves times the force applied during the movement:

Work (W) = Force applied to the object (F) × Distance the object moves (D)

$$W = F \times D$$

Imagine that you are pushing a cart in the supermarket. It is easy to push an empty cart, but much harder to push a full one (see Figure 7.8). The force that you exert is greater with the full cart, and so the work you are doing (force × distance) is greater than when the cart is empty. Similarly, when you lift heavy weights, you are doing more work than when you lift light weights or none at all. Again, the force (weight) times the distance you move it is the mechanical work. There are other forms of work, such as electrical work, but they are more difficult to visualize.

Figure 7.8 | The person pushing the empty shopping cart does less work, and therefore exerts less force, than the person pushing the full cart.

Power

Power is the rate at which work is done. For instance, if you push the supermarket cart with a force of 200 newtons, and you move it 30 m, the work is W = (200 N) (30 m) = 6,000 N m, or 6,000 J (1 newton-meter = 1 joule). Power is expressed in terms of work divided by time, or joules/second, which is one watt of power. The faster you do the work, the more power you are exerting. If you pushed the cart for 60 seconds, the power, W, would be W = 6,000 Joules/60 seconds, or 100 watts. If you pushed it for 6 seconds, the power would be 1,000 watts. The work required to move the cart is the same: force times distance. How quickly the cart was moved is an indication of the power supplied. Professional cyclists often train by figuring out how many watts of power they want to expend.

Heat

Heat (symbolized by the letter Q) is thermal energy that is transferred between two objects because of a temperature difference between them. When you toast a piece of bread, as shown in Figure 7.9, the high-temperature wires in the toaster transfer heat to the cooler surface of the bread, raising its temperature. A transfer of energy occurs; in this case, it is heat energy.

Heat energy takes three different forms, depending on how the heat transfer occurs—conduction, radiation, or convection. Thus, heat does not exist within a substance, but occurs because of its temperature and a temperature difference with another substance. If you place your hand over a flame, your hand quickly becomes hot. Why? Heat is transferred from the flame (it has a high temperature of perhaps 1,500 K) by radiation and convection of heat to your hand. The temperature increases on the surface of your hand, you yell "ouch," and then move your hand away.

Conduction is heat transfer within a material caused by a temperature difference on the material's inside and outside surfaces. When you wear a jacket

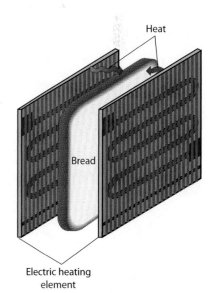

Figure 7.9 | Heat flows from the electric heater elements to the bread via radiation and convection. When the bread's surface temperature reaches about 310 F, the sugars and starches begin to carmelize and the surface turns brown.

Table 7.1 | Thermal Conductivities of Different Materials

MATERIAL	λ W/m-K
Copper	399.0
Steel	43.0
Fiberglass	0.035
Wood	0.069
Water	0.558
Air	0.024
Wool	0.04
Goose down	0.025

in the winter, the temperature of the outside jacket surface is essentially the same as the outside air temperature, but the inside temperature next to your skin is much warmer. Heat transfer occurs because of the temperature difference, but the rate of heat transfer is reduced because of the jacket material. The thicker the jacket is, the less heat is transferred and the warmer you feel. However, some jackets of the same thickness make you feel warmer (have less heat transfer) than others. Why is this?

Look at Table 7.1, which lists the thermal conductivity of various materials. Air has the least thermal conductivity; it does not conduct heat well, but copper conducts heat very well. The insulating effect of the material depends on how much air is trapped. The greater the amount of air the jacket traps, the less heat is conducted, and the warmer the jacket makes you feel. Goose down has a thermal conductivity nearly equal to that of air, so jackets and blankets filled with goose down do not conduct heat well. They enable you to retain heat and stay warm. If the heat loss is less than the energy you generate from your metabolism, you will be warm. If the heat loss is greater than the energy generated, you will feel cold. The thermal conductivity is expressed in terms of watts (W) that are conducted through a one-meter thickness of a substance when there is a one-degree Kelvin temperature difference. The experiments are done on thicknesses of much less than a meter, but the information is converted to this standard format.

The walls of homes and buildings are insulated to reduce the heat flow from the building in the winter and the heat gain in the summer. The R factor is a composite measure used to describe the effect of building insulation, as shown in Figure 7.10. The greater the R factor is, the greater the resistance, and the less the heat flow. The R factor is proportional to $1/\lambda$, so the lower the value of the thermal conductivity, the greater the R factor. The R factor also depends on thickness; the thicker the material is, the greater the R factor.

Radiation heat transfer is the flow of thermal energy between two bodies separated by a distance, such as the bread and electrical wire in a toaster, or the Sun and the Earth. We experience radiation heat transfer when solar radiation is absorbed by our skin and we feel warmer. The energy form is electromagnetic waves that are in the thermal spectrum; these waves are not at the wavelengths of radio or television. An important concept is that the thermal radiation emitted by a surface is determined by its temperature;

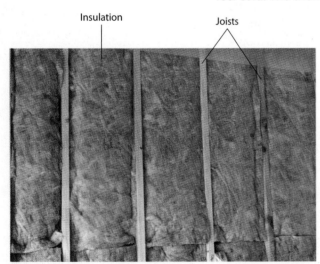

Insulation

Joists

Figure 7.10 | This insulation has been installed into a wall, but it arrived at the building site in rolls. The thicker the insulation is, the greater the R factor. The R factor for a 3½-inch thickness is 13; for a 6-inch thickness, R = 20, and for 12 inches, R = 43.

in fact, the emitted radiation is proportional to the temperature raised to the fourth power. A small change in temperature, therefore, will result in a very large change in emitted radiation.

Convection heat transfer is the transfer between a surface and a fluid; the fluid can be a liquid or gas. Thus, if hot water flows through a pipe, heat is transferred by convection from the fluid to the pipe. You may have figured out that the heat that flows through the pipe walls travels by conduction. Convection occurs again on the outside surface between the pipe and the surrounding air. This is how hot-water heating systems work, as shown in Figure 7.11.

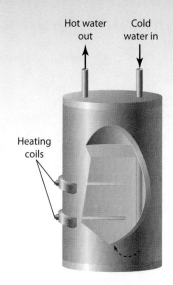

Figure 7.11 | An electric hot water heater works via convection heat transfer between a surface and a fluid.

Intrinsic Energy Forms

Matter has intrinsic energy associated with its internal structure, its velocity, and its height above the ground. A piece of matter's intrinsic energy is the work it can do based on its actual condition, without any supply of energy from outside. The intrinsic energy depends on how fast the matter is moving (kinetic energy), how high above the ground it is (gravitational potential energy), and its molecular structure (internal energy). Heat and work result from changes in intrinsic energy or cause changes in intrinsic energy.

Kinetic energy is the energy an object possesses because of its motion. Imagine that you are walking and have a certain amount of energy because of your motion. Now imagine that you are running. Your energy is greater, of course. Your speed, or velocity, is also greater. So you know intuitively that the expression for kinetic energy must include velocity. Imagine again that you are walking, then running, but this time your younger brother or sister joins you. Are your kinetic energies the same, even though you weigh different amounts? No, you intuitively know that weight, or mass, must be part of the expression for kinetic energy. The formula for kinetic energy is:

$$\text{Kinetic energy} = \tfrac{1}{2}\,\text{mass} \times \text{velocity}^2 \text{ or}$$
$$KE = \tfrac{1}{2}\,m\,v^2$$

Kinetic energy depends on mass and velocity. More importantly, it depends on velocity squared, so as you run faster or as a plane flies faster, the kinetic energy increases dramatically. The mass is in kilograms and the velocity is in meters/second.

If you are standing on the top of a tall building, do you have more energy than when you are standing on the ground? The earth's gravitational acceleration is acting on you at all times. Gravitational potential energy is the energy form associated with a body's relative position to the ground. The higher above the ground it is, the greater the gravitational potential energy. We call the type of energy associated with a position relative to the ground "gravitational potential energy." It is expressed as:

$$\text{Potential energy} = \text{Mass} \times \text{Gravitational acceleration} \times \text{Height}$$
$$PE = m \times g \times h$$

where m is the mass in kilograms, g is earth's gravity ($9.8\,m/s^2$), and h is the height above the ground in meters.

Internal energy, a less tangible form of energy, is associated with a substance's molecular structure. Although we cannot measure internal energy, we can measure changes of internal energy (ΔU):

$$\Delta U = m\,c\,(\Delta T)$$

Figure 7.12 | A runner has kinetic energy.

Table 7.2 | Specific Heats for Various Substances

SUBSTANCE	c kJ/kg-K
Air	1.004
Aluminum	0.963
Brick	0.92
Concrete	0.653
Glass	0.833
Ice	1.988
Steel	0.419
Water (liquid)	4.186
Water (vapor)	1.403
Wood	2.51

where m is the mass in kilograms, c is the specific heat in kJ/kg-K, and ΔT is the change of substance temperature in degrees Kelvin. Table 7.2 indicates some typical values of c for various substances. Think about the terms in the expression for specific heat. The symbols indicate the amount of energy measured in kilojoules (kJ) necessary to increase one kilogram of mass (kg) one degree in temperature (measured in degrees Kelvin). The term "specific" indicates that it applies to one kilogram (a unit mass) of matter.

The table indicates that it takes more than 4 kJ to raise the temperature of one kilogram of water by one degree Kelvin, while it takes about 1 kJ to raise the temperature of a kilogram of air by one degree Kelvin. This amount would be the mass of air in about 850 one-liter bottles at normal room temperature and pressure. The dramatic difference in specific heat values is an indication of how energy is stored in a material's molecular structure. One mode of storage relates directly to temperature increase, but other modes of storage do not. Water stores energy in more modes than air does. Most of the kinetic energy added to air is stored in a mode related to temperature. Other modes of energy storage relate to atoms and molecules rotating and vibrating.

TECHNOLOGY IN THE REAL WORLD: Determining the Energy Usage of an Appliance

Most of us are familiar with a 100-watt incandescent light bulb. Wattage refers to the power the bulb uses in producing light. Imagine that a home has ten 100-watt bulbs:

10 bulbs × 100 watts/bulb = 1,000 watts or 1 kilowatt (kW)

If all 10 bulbs were turned on at the same time, the instantaneous electrical power consumed by the bulbs would be 1 kilowatt. If the bulbs were left on for one hour, the electrical energy used would be 1 kW × 1 h = 1 kW-h (kilowatt-hour).

Utilities charge various rates across the country for kW-h consumption, but a charge of $0.10 to $0.15 per kW-h is typical. In the preceding example, the cost of having the lights on would be 0.10 ($/kW-h) × (1 kW-h) = $0.10, or ten cents. Electrical bills are expensive because we use electricity in so many ways; when we add up all the usage and consumption, the cost can be high.

It is challenging to determine the total kilowatt-hours used by an appliance. For example, a microwave oven may be rated at 1.5 kW, but that rating refers to the maximum power it uses at the highest power setting. This setting may not be typical. As another example, consider a desktop computer; its highest power level may be 250 watts, but only when the computer is operating at its maximum power. The computer often consumes much less power, such as when you are operating the keyboard instead of playing games.

Manufacturers attempt to compute the annual kW-h used by their products. The following table lists typical annual power demands of various types of electrical appliances. You can determine the actual power requirements by looking at the nameplate on the equipment or checking the specifications online.

Extension

Using the information in Table 7.3, calculate the monthly energy consumption in your kitchen. If electricity costs 0.12 $/kw-h, what is the monthly cost of electricity for appliances and lighting in the kitchen?

Table 7.3 | Power Consumption and Usage for Selected Appliances

PRODUCT	POWER CONSUMPTION, IN WATTS	HOURS/MONTH
Air conditioner, central, 2.5 tons	3,500	240–860
Ceiling fan	60	15–330
Clothes dryer (electric)	5,000	6–28
Clothes washer	500	7–40
Coffee maker	900	4–30
Compact fluorescent bulb (60 W equivalent)	18	17–200
Computer and monitor	200	25–160
Dishwasher	1,300	8–40
Fluorescent bulb (2 tubes, 4 ft.)	100	10–200
Food freezer	335	180–420
Incandescent light (60 W)	60	17–200
Microwave oven	1,300	5–30
Range (electric)	12,500	10–50
Refrigerator, frost-free, 17 cubic feet	500	150–300
Television	180	60–440
Toaster	1,150	1–4
Water heater (4 people)	3,800	100–140

SECTION TWO FEEDBACK >

1. If you are "hot," do you "have a lot of heat"?
2. You are riding in a car that has a mass of 1,000 kg and is traveling at 20 m/s. What is its kinetic energy?
3. You have heard of power bars and power tools. Both cars and bulls are powerful. But what *is* power?

KEY IDEAS >

- The First Law of Thermodynamics, the conservation of energy, indicates that energy cannot be created or destroyed. However, it can be converted from one form to another.

- The Second Law of Thermodynamics indicates that not all heat energy can be converted into work and that no energy system can be 100 percent efficient.

Thermodynamics applies two simple yet powerful laws to a wide range of energy systems with applications in transportation, agriculture, and power generation. These systems have major importance to our society.

Conservation of Energy

The First Law of Thermodynamics, the conservation of energy, indicates that energy cannot be created or destroyed. However, it can be converted from one form to another. Energy can change form, but the total amount of energy is constant. This is a powerful statement, and one that we can apply to many situations. For example, imagine that you are sitting in the shade under an apple tree on a warm summer day. An apple falls and hits you on the head, as illustrated in Figure 7.13. Because you are studying engineering and technology, you begin to think about what happened to the apple when it fell. When the apple was hanging on the branch, it contained internal energy. It also had potential energy because it was above the ground, but it did not have kinetic energy because it was not moving. As the apple falls, its kinetic energy is increasing, the potential energy is decreasing, and the internal energy remains the same. After the apple bops you on the head and falls to the ground, it does not have any kinetic energy or potential energy; it only has internal energy, which has increased slightly from the impact with your head and then with the ground. The total energy change that a substance undergoes can be expressed in terms of heat transfer, work, and changes in internal, kinetic, and potential energies.

The First Law describes this energy change. Thus, for the falling apple, the First Law is written as:

$$Q = \Delta U + \Delta PE + \Delta KE + W$$

There is no heat transfer because there is no temperature difference between the apple and the air. There is no work, even though the apple is falling. While it is falling, the change of internal energy is zero, and the apple's structure does not experience change, so the change of potential energy is equal to the change of kinetic energy. When the apple hits the ground, its kinetic energy is zero (it is no longer moving), but its internal energy has been increased by ΔPE.

Figure 7.13 | When an apple hits a man on the head, the apple's kinetic energy is converted to internal energy.

Calculating Horsepower

▷ Did you ever wonder why cars have engines with different horsepower? The more powerful the engine is, the faster the car can go (see Figure 7.14), but even a car with a low-horsepower engine can go 70 mph. Let's investigate what is required to make a 1,200-kg car travel at a speed of 70 mph. The car has work done to it—it is moving through a distance. Start with the First Law:

$$Q = \Delta U + \Delta PE + \Delta KE + W$$

There is no heat transfer in moving the car, so Q is zero; the car does not change its elevation, so ΔPE is zero, and its internal energy does not change ($\Delta U = 0$). The engine is doing work to change the car's velocity (Figure 7.15), thus:

$$W = -\Delta KE$$

The minus sign indicates that the work is going into the car. Positive work occurs when work leaves a device. The engine is doing work (positive), and the car is receiving work (negative).

$$W = -\tfrac{1}{2} m \, (v^2_{final} - v^2_{initial})$$

The velocity needs to be in meters per second (m/s), not miles per hour. One mile per hour equals 0.447 m/s, so 70 mph is $70 \times 0.447 =$ 31.29 m/s. The initial velocity is zero, so the work required to move at 70 mph is:

$$W = -\tfrac{1}{2}\,(1200 \text{ kg})(31.29 \text{ m/s})^2 = -587438 \text{ joules or } -587.4 \text{ kJ}$$

We can eliminate the minus sign because we know that work is going into the car to make it move. Next, how quickly do you want the car to get to 70 mph? In 8 seconds? This is where power (W) becomes important. Power is the rate at which work is done—in other words, work divided by time:

$$\mathbf{W} = \frac{W \text{ (kJ) kilowatts (kW)}}{t \text{ (s)}}$$

For this car, the power required is:

$$\mathbf{W} = \frac{587.4 \text{ kJ}}{8 \text{ s}} = 73.4 \text{ kW}$$

We can convert kW to horsepower (1 kW = 1.341 hp), so $\mathbf{W} = (1.341 \times 73.4) = 98.4$ hp. This value is much less than the engine horsepower found in cars, for two major reasons: One deals with energy transformations and energy losses in the transmission and gearing, and the other deals with drag forces acting on the car's body. When energy is transmitted through gearing, some energy is lost as friction between the gears. (We will examine engines in greater detail in the next section, when explaining a car's energy transformations.) The drag forces are proportional to the car's velocity cubed. When the car travels at higher speeds, the power requirements increase as the cube of the velocity. The drag forces also depend on the car's geometry—in other words, how its aerodynamic shape enables it to minimize drag. Thus, returning to our previous example, the actual horsepower required is much greater than 98.4.

Extension Using an Excel spreadsheet, determine the power (in kW) required to accelerate a car to 10, 20, 30, 40, 50, and 60 miles per hour in 8 seconds. Plot the results and explain them in terms of the equation for work and power.

Figure 7.14 | The more powerful the engine, the faster a car can go; this 2007 Z06 Corvette is fast.

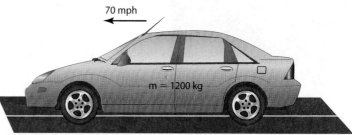

70 mph

m = 1200 kg

Figure 7.15 | The engine is doing work to change the car's velocity.

Figure 7.16 | Traditionally, windmills have been used to generate power for farms.

Windmills are a common way to generate power from the wind. Old pictures of farms frequently show windmills being used to pump water. If you have ever blown on a pinwheel, you have used a type of windmill. Let's think about how the windmill works. Obviously, air needs to be moving to turn the windmill, converting the air's kinetic energy into a force that moves the blades (see Figure 7.16). Of course, the amount of power depends on how large the windmill blades are. When wind flows across the blades of the windmill, generating power, the First Law applies. When there is mass flow, as in the case of windmills, the First Law is written differently to include the mass flow effects. Mass flow describes moving liquids and gasses, like water running through a hose, or air moving through windmills. (The equation is different than the one when there is no mass flow.) The First Law in the case of mass flowing through or across a device such as a windmill is:

$$\text{Energy In} = \text{Energy Out}$$
$$Q + m\,(u + ke + pe + p/\rho)_{in} = W + m\,(u + ke + pe + p/\rho)_{out}$$

where m is the mass flow rate, kg/s; Q is the heat transfer in kilowatts (kW); W is the power in kilowatts (kW); p is fluid pressure in kilopascals (kPa); and ρ is the fluid density in kg/m^3. The p/ρ term occurs because of the flowing air across the windmill's blades, as illustrated in Figure 7.17.

All the energy that is in the wind before it hits the blades—except for the amount that is converted to power—remains in the wind after leaving the blades. The windmill has no heat transfer ($Q = 0$), and there is no change of internal energy of the air ($\Delta u = 0$). The air does not change elevation, so the change of potential energy is zero ($\Delta pe = 0$). Also, the pressure and density of the air do not change as the air flows across the windmill's blades ($\Delta p/\rho = 0$), so the power (W) is equal to the change of air's kinetic energy.

$$W = m\,(ke_{in} - ke_{out})$$

The mass flow rate, m, depends on how fast the air is flowing, its density, and the area that it is flowing through. Thus:

$$m = \rho\,A\,v$$

Notice that the kinetic energy varies with velocity squared, and the mass flow depends on velocity, so windmill power varies as the velocity cubed.

$$W = \tfrac{1}{2}\,\rho\,A\,(v^3_{in} - v^3_{out})$$

The area is the swept circular area of the windmill's blades, which is why windmills have long blades. The value ρ is the air's density (about 1.17 kg/m^3), and the air's velocity is in meters/second. This means we still need to know the exit velocity of the air from the blades, which is very difficult to determine. More advanced engineering analysis indicates that the maximum theoretical power is 59.3% of the energy entering the windmill:

$$W_{max} = \frac{(0.593)(\rho\,A\,(v^3_{in})}{2}$$

In cars, we do not want drag, but in windmills we do. If there were no drag, the air would slip right over the blades. The same factors that affect power in windmills also affect drag forces (hence power dissipation) on cars, which is why drag forces vary as the velocity is cubed.

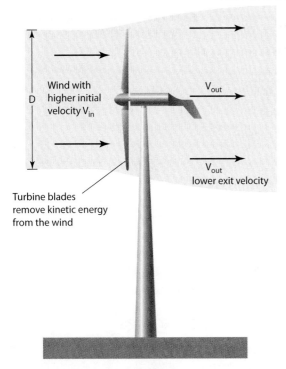

Wind with higher initial velocity V_{in}

D

V_{out}

V_{out}
lower exit velocity

Turbine blades remove kinetic energy from the wind

Figure 7.17 | All the energy in the wind before it hits the blades remains in the wind after leaving the blades, except for the amount that is converted to power.

Second Law

Energy is always conserved, but it does not remain equally valuable to us as it goes from one energy form to another. The Second Law of Thermodynamics discusses energy value: whenever an energy transformation occurs from one form to another, there is a net decrease in energy value. Imagine placing a pot of water on a stove, as illustrated in Figure 7.18. When you ignite the gas flame, the conversion of the gas's chemical energy into the flame's thermal (internal) energy is manifested by a high temperature. The flame heats the pot by convection, the pot transmits the heat by conduction through the metal, and the heat is transferred by convection to the water, increasing the water's internal energy. When the flame is turned off, the pot cools to room temperature. Did the energy disappear? No. The internal energy of the water is transferred as heat to the air in the kitchen, and eventually to the air outside. However, the value of the energy has become very low, compared with the value it had as the chemical energy of the gas.

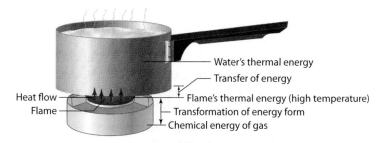

Water's thermal energy
Transfer of energy
Heat flow
Flame's thermal energy (high temperature)
Flame
Transformation of energy form
Chemical energy of gas

Figure 7.18 | **Water heating on a gas stove illustrates the transformation of the gas flame's chemical energy to thermal energy and the transfer of thermal energy from the flame to the water.**

Another important aspect of the Second Law is that not all heat can be converted into work; only a maximum amount can be converted. So, no energy system can be 100 percent efficient. The Carnot Cycle, which we will describe shortly, is a theoretical model of an engine that converts the maximum amount of heat into work. Engines and other power-producing devices, such as gas turbines or power plants, are often compared to the Carnot Cycle.

Multiple Energy Transformations

The internal combustion engine, as illustrated in Figure 7.19, is used to power automobiles, trucks, and many other machines. This engine is an example of a system that contains multiple energy transformations. The engine is a small power plant where fuel is burned.

In the space between the top of the piston and the top of the cylinder, the energy from the burning fuel converts the fuel's chemical energy into thermal energy. The high temperature and pressure are the products of combustion contained in the cylinder. The gas pushes the piston downward, transferring the thermal energy to reciprocating mechanical energy. The crankshaft converts the reciprocating mechanical energy of the piston rod to rotating mechanical energy. The rotating mechanical energy goes through the transmission and differential gears that turn the axles and wheels, thus moving the automobile.

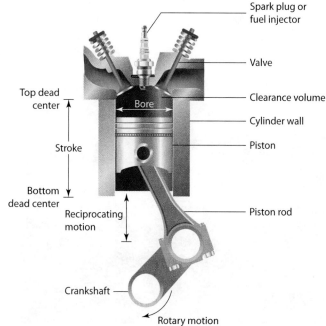

Spark plug or fuel injector
Top dead center
Bore
Valve
Clearance volume
Cylinder wall
Piston
Stroke
Bottom dead center
Reciprocating motion
Piston rod
Crankshaft
Rotary motion

Figure 7.19 | **An automotive engine is a small power plant where fuel is burned.**

Section 3 △ Understanding Conservation of Energy

Fuel is supplied to the engine (chemical energy)

↓

Combustion raises gas temperature and pressure (thermal energy)

↓

Piston moves because of high pressure (mechanical energy—reciprocating motion)

↓

Crankshaft rotates after translating vertical piston motion (mechanical energy—vertical motion converted to rotary motion)

↓

Transmission and differential gears reduce the rpm to the wheels

↓

Wheels convert crankshaft rotary motion (mechanical energy moves vehicle)

Figure 7.20 | Various energy-conversion processes occur in an automobile engine.

Figure 7.20 shows the various energy forms and the transformations of energy that occur. Thermodynamics helps to explain energy and energy transformations.

TECHNOLOGY AND PEOPLE: Ivo Coelho

Ivo Coelho's passion is drag racing. Coelho was born and raised in Portugal, but he never considered the sport until after he immigrated to the United States 15 years ago. He works as a construction mechanic, repairing and maintaining construction equipment, following in his father's footsteps. He can repair most engines, motors, and hydraulic equipment; in addition, his welding skills have been very important in outfitting his car. He must pay for everything himself, so self-sufficiency is important.

Figure 7.21 shows Coelho in his 1978 Pontiac Trans Am, which he bought used in 1991. Little about the car is original at this point: He has replaced the engine, the interior, and parts of the exterior, doing all the work himself. He designed an automatic pneumatic-shift system for the car, as well as a tubular steel body that protects the driver and a sheet metal frame that serves as a dashboard. As you can see in Figure 7.22, the interior of the car is spartan; it is not a car you can drive on the street. Coelho must take the car to drag races in a trailer.

He enters handicap racing events as a Super Pro, in which he not only competes against other drivers, but also against himself. Racers drive in two "time trials" to determine their time on a particular quarter-mile track on a given day. Weather conditions can affect how well a car does when performance is measured by the hundredths of a second. One day, for example, Coelho had trial times of 8.79 seconds and 8.77 seconds. For the competitive race, in which he had to predetermine the time that he would complete the quarter-mile, he chose 8.78 seconds. His competitor also had to project a time for completing the race. Because his competitor had a slower car, he selected (or "dialed in") a time of 9.28 seconds, which was half a second slower than Coelho's projected time. Coelho's competitor was given a half-second head start, at which point the light turned green, clearing Coelho to accelerate his car down the track. If he had been too

Figure 7.21 | Ivo Coelho's dragster has its rear tires screwed to the rim. The tire rippling indicates the large amount of torque being applied.

Figure 7.22 | The interior of the dragster features protective tubing and a pneumatic shifter.

anxious and left too soon, he would have been disqualified (red-lighted); if he won, he would continue to the next round against another car.

His car's 8-cylinder engine has a 14.5 to 1 compression ratio that produces about 900 horsepower. Because of the high compression ratio, he must use very high-octane racing fuel with a 116 octane rating. By contrast, the high-test fuel people buy in gas stations is 93 octane. The higher the octane rating is, the slower the fuel burns; a low-octane fuel tends to explode, but a controlled, rapid burning is preferable.

As with all drag racers, Coelho wants a faster car and a bigger engine. He plans to increase the engine's stroke by increasing the number of cubic inches from 540 to 580. Ultimately, he anticipates being able to produce over 1,000 horsepower.

SECTION THREE FEEDBACK >

1. Explain the relationship between work and power.
2. You leave a hot cup of tea on a table. Where does the energy go?
3. A woman who weighs 50 kilograms is jogging at 5 miles per hour. What is her kinetic energy in kJ? (Remember to convert mph to m/s.)

SECTION 4: Investigating Energy Sources

KEY IDEAS >

- Energy sources are either renewable or nonrenewable.
- Renewable energy resources are often difficult to harness, are not always reliable, and have a low energy value compared with nonrenewable sources.
- Nonrenewable energy sources are being used up at a rapid rate.

Energy sources are either renewable or nonrenewable. Most of the energy we produce is created from nonrenewable sources, such as oil, gas, and coal. Nonrenewable energy sources are being used up at a rapid rate. Renewable sources, on the other hand, replenish themselves. A forest, for instance, is a renewable resource, as shown in Figure 7.23.

As another example, solar energy is renewable and unlimited. (Although the Sun may burn out in several billion years, we will not consider that issue here). As we examine renewable resources, we will find that they are often difficult to harness, not always reliable, and that their energy value is low compared with nonrenewable sources.

Nonrenewable Energy Sources

Oil, natural gas, and coal are called fossil fuels and are nonrenewable energy sources. A fossil fuel is an energy-rich substance formed from the remains of plants and animals that lived millions of years ago.

Figure 7.23 | Forests provide a renewable source of energy. Trees are energy converters; they convert the Sun's energy to cellulose fibers through photosynthesis.

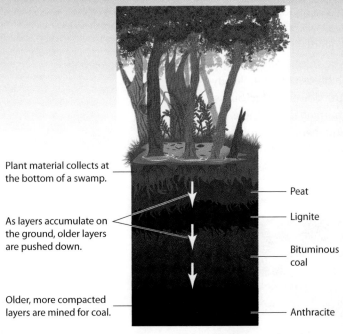

Plant material collects at the bottom of a swamp.

As layers accumulate on the ground, older layers are pushed down.

Older, more compacted layers are mined for coal.

Peat

Lignite

Bituminous coal

Anthracite

Figure 7.24 | **Vegetation is compressed over time to form coal.**

Figure 7.25 | **A methane molecule, written as CH₄, is made of one carbon atom and four hydrogen atoms.**

Fossils were covered by soil but did not rot. Instead, over a very long time, the Earth exerted continuing pressure on them and caused them to change into a fossil fuel. These fuels are made primarily from carbon and hydrogen, so they are called hydrocarbon fuels. During combustion, they break down into simpler molecules. For combustion to occur, the compounds must be able to combine, as indicated by their chemical potentials. We know, however, that gasoline, propane, or wood does not spontaneously react with oxygen.

Three conditions must be met: first, a substance must be able to burn; second, a source of oxygen must sustain combustion; and finally, the temperature must be high enough to start combustion (the ignition temperature). A burning match has a high temperature and can start combustion of a larger fuel source, at first locally near the match. The spreading fire then raises the temperature of more fuel. Most fire extinguishers work by lowering the fuel's temperature or blocking the fuel's access to oxygen.

In a combustion reaction with a hydrocarbon fuel, the carbon is completely burned to form carbon dioxide, and hydrogen is completely burned to form water. Methane, shown in Figure 7.25, is one type of natural gas, a hydrocarbon fuel. Each methane molecule is made of one carbon atom and four hydrogen atoms, written as CH_4. When methane burns with oxygen (air is about 21% oxygen and 79% nitrogen), the combustion equation is:

$$CH_4 + 2O_2 \rightarrow CO_2 + 2H_2O + \text{heat (Q)}$$

In this situation, one methane molecule combines with two oxygen molecules in order to form carbon dioxide and water, as well as to release energy (heat). Other fossil fuels have similar reaction equations. For instance, the octane in gasoline has the chemical formula C_8H_{18}, so eight carbon dioxide molecules and nine water molecules would be formed by burning one molecule of octane. This reaction will have important meaning to us when we examine global warming and air pollution in other chapters. Table 7.4 indicates the heating value (h_{RP}) of various fuels by listing the heat released by completely burning one kilogram of fuel. The units indicate the amount of heat energy, measured in kilojoules (kJ), that is released in the combustion of one kilogram of the fuel.

Notice that gasoline and oil each have about twice as much energy per unit mass as wood or corn cobs. Wood and corn cobs are examples of renewable energy

Table 7.4 | Average Heating Values for Various Fuels

FUEL	kJ/kg (h_{RP})
Coal	27,900
Corn cobs (dry)	21,600
Ethanol	26,750
Gasoline	44,800
Natural gas	57,450
Residual oil	43,000
Wood (dry)	20,350

sources, and oil is a nonrenewable energy source. Thus, the fuel heating values of renewable fuels are much less than those from nonrenewable sources. Also notice that ethanol, produced from corn, does not have a heating value as high as that of gasoline or oil.

Calculating the Heat of Fire

▷   How hot is fire? Can its temperature be predicted? The combustion temperature varies with the type of fuel and the amount of air that is used, as indicated in Table 7.5. Using the minimum amount of air will yield the highest temperature, and using more air will reduce the temperature, since the fuel's energy goes to heating the excess air. If sufficient air is used, temperatures of 1,500 to 2,000 K assure complete combustion for hydrocarbon fuels.

Table 7.5 | The Adiabatic Flame Temperature, the Maximum Possible Temperature That Can Be Achieved by Burning a Fuel, Depends on the Amount of Air That Is Used. In This Table, the Values Are Based on the Minimum Amount of Air That Is Theoretically Needed

FUEL	AIR (DEGREES KELVIN)
Hydrogen	2,385
Methane	2,225
Propane	2,270
Octane	2,275

Coal was one of the first fossil fuels because it is most easily found near the surface of the ground. Coal can be mined from the surface by strip mining, as shown in Figure 7.26, when large swaths of land are uncovered and the coal is removed from slightly below the surface. The land is then covered by the soil and replanted.

Figure 7.26 | Trucks haul coal from a strip mine to a processing plant.

Figure 7.27 | Engineers often work underground in coal mines, taking charge of a host of tasks, including managing the coal handling facilities, maintaining the water systems inside the mine, monitoring ventilation fans and overall inspections. Outside the mine, engineers are charged with other tasks, such as analyzing output and performing feasibility studies to determine whether new mining operations are feasible in a particular site.

Coal mines can also exist deep below the surface of the Earth. In these cases, tunnels are made to dig out the coal and bring it back to the surface (see Figure 7.27).

Coal often contains other elements besides carbon and hydrogen; one of the most common is sulfur. When sulfur burns, it becomes sulfur dioxide (SO_2), which is a gas. However, this gas readily dissolves in water or rain to become sulfuric acid, a primary constituent of acid rain.

Oil is a liquid fossil fuel. In its natural state, it is a very thick and viscous liquid called crude oil (see Figure 7.28). Depending on where it comes from, the oil may contain other constituents such as sulfur. Refineries alter the crude oil through chemical engineering processes to create a variety of hydrocarbon products such as gasoline and diesel fuel.

Uranium is another important nonrenewable energy resource. It is not a fossil fuel, but a nuclear, or atomic, fuel. Matter can be changed into energy through nuclear fission, the splitting of an atom of one type of matter into different forms of matter, generating heat in the process. A small amount of fuel can produce a very large amount of heat energy that, in turn, can generate power. These nuclear reactions are controlled by sophisticated systems to prevent explosions.

Renewable Energy Sources

Some energy sources are readily replaced daily or within a few months or years. These sources are considered renewable, even though it takes time to replace most of them.

Humans and Animals The first source of energy that people used was their own muscle power. Individually, people do not have great strength, but by working together they can accomplish a great deal. For instance, the pyramids of Egypt were constructed by people pulling large blocks of stone up inclines. Many animals are stronger than people, so people learned to domesticate certain animals and use them for plowing fields (see Figure 7.29), and even for pumping water. In some parts of the world, human power and animal power are still important sources of energy.

Figure 7.28 | An oil rig is used to recover oil from beneath the ocean's floor.

Figure 7.29 | Oxen are used to plow fields in developing countries.

Solar Energy The energy from the Sun supports life on Earth. Vegetation depends on the conversion of the Sun's energy into matter through photosynthesis. We also use the Sun's energy, or solar energy, to produce heat and electricity. Solar collectors are used for this purpose, and different types of solar collectors exist for various uses. For example, some solar collectors are placed on rooftops to heat water for showers, washing, and cooking (see Figure 7.30).

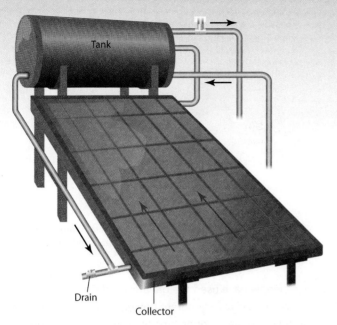

Tank

Drain

Collector

Figure 7.30 | A thermosiphon solar water heater receives cold inlet water. The cold water enters the bottom of the solar collector, passes through tubes, is heated, and is collected and stored in the tank for household use.

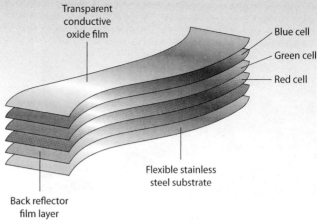

Transparent conductive oxide film

Blue cell

Green cell

Red cell

Flexible stainless steel substrate

Back reflector film layer

Figure 7.31 | In this cutaway view of a Uni-Solar photovoltaic shingle, sunlight is converted directly to electricity in a photovoltaic device. The cells refer to p-n junctions where the energy conversion occurs.

Passive solar designs of homes use the Sun for supplementary heating. Photovoltaic cells convert sunlight into electricity (see Figure 7.31). These solar cells are often used in remote areas that are not served by power lines. They can charge batteries that run radio relay stations and other electrical devices. Photovoltaic cells are also used on most satellites and space vehicles. About 10 percent of incident sunlight is converted into electricity in a photovoltaic cell.

How much energy does sunlight provide? The intensity of sunlight varies, but a typical value for the northeast United States is 1.36 kW/m^2 on a clear, sunny day. This means that a square meter of surface area will receive 1.36 kW of energy. Remember 1 kW is 1 kJ/s. The Sun's intensity varies throughout the day, of course, so at noon the intensity is probably greater than 1.36, and in the morning and evening it is less.

Wind Energy Wind has been used as an energy source for hundreds of years. Windmills have been used to pump water, grind grain, and move ships across the water. Today, wind energy is used to generate electricity. In some locations, farmers can earn income by leasing their land to power companies for wind farms, while still growing crops or grazing animals on the land. However, just as solar energy is impossible when sunlight is obscured by clouds, there is no wind energy when the wind doesn't blow or doesn't blow hard enough. Thus, wind farms are primarily located in places where the wind blows briskly much of the time (see Figure 7.32). Some theoretical studies have indicated that extensive use of wind farms might change local climates because of the wind's energy decrease.

While many people support the idea of windmills, not everyone wants to have them where they can be seen. Because windmills can be 200 to 300 feet tall, they are visible for many miles. For instance, the wind often blows consistently and strongly along the coasts, but many residents and visitors along the seashore do not want to see windmills from the beach. The acronym NIMBY, which stands for "not in my backyard," describes the political and social reaction of people who like a concept in general but do not want it implemented near them.

Figure 7.32 | Wind farms, like this one in Rio Vista, California, are primarily located in places where the wind blows briskly much of the time.

Figure 7.33 | For centuries, people have used the gravitational potential energy of water to generate power. Here, Glen Canyon dam is shown with Lake Powell in the background and the Colorado river in the foreground.

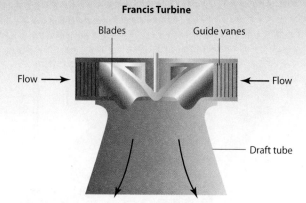

Francis Turbine

Blades   Guide vanes

Flow →   ← Flow

Draft tube

Figure 7.34 | The water from a reservoir passes through a Francis hydraulic turbine, causing it to rotate and turn an electrical generator.

Water Energy For centuries, people have used the gravitational potential energy of water as a means of using water energy to generate power. Water wheels, which once were used to grind grain into flour, are an open type of turbine, just as windmills are. People created dams to provide a height difference for water wheels.

As the dams became larger and the heights greater (see Figure 7.33), enclosed hydraulic turbines were created to better use the water's potential energy. Figure 7.34 shows a hydraulic turbine used in the generation of electricity. The water from a reservoir passes through the turbine, causing it to rotate and turn an electrical generator.

The pull of gravity from the Sun and Moon against the Earth causes tides, which are large movements of water in oceans and tidal rivers. In the Bay of Fundy, which is between the Canadian provinces of Nova Scotia and New Brunswick, tides cause water levels to rise nearly 60 feet in some of the bay's tributaries. When the water flows back to sea, people direct it through turbines to generate power.

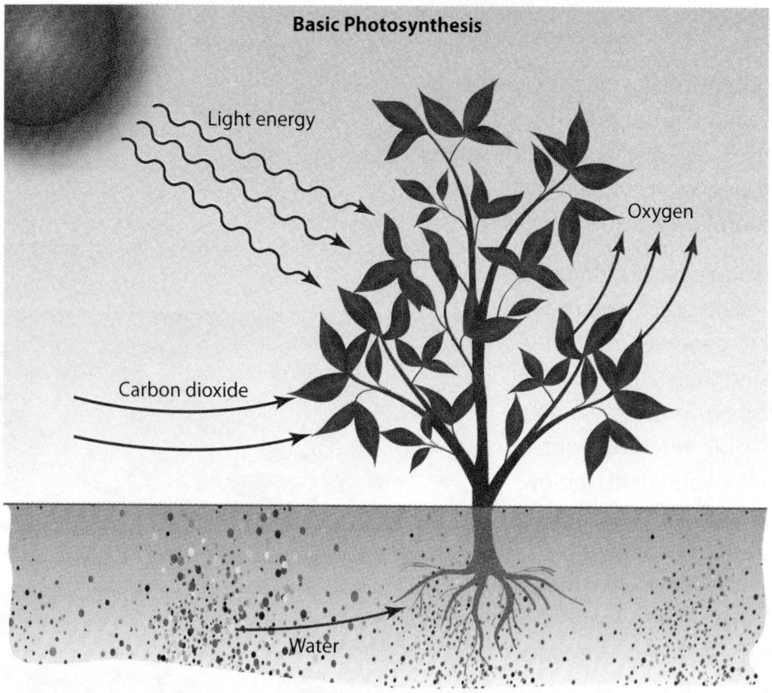

Basic Photosynthesis

Light energy

Oxygen

Carbon dioxide

Water

Figure 7.35 | Photosynthesis is a process of converting the Sun's energy into mass, using carbon dioxide from the air, water, and minerals from the ground. The plants (trees) provide oxygen to the environment as part of the process.

Biomass Biomass is accumulated vegetable and animal matter that can be a major source of renewable energy. For instance, it is possible to burn biomass, such as wood. Wood is renewable, but forests must be managed carefully, because once trees are cut in an area, they need to be replanted. It takes many years before new trees are mature enough to be harvested again for fuel. Vegetation, such as wood, is the result of photosynthesis converting the Sun's energy into matter (see Figure 7.35).

Biomass can be converted to other energy forms. It is a hydrocarbon, so it can be changed from a solid to a gas in a gasification process. This process models rotting in nature, a process by which methane gas is produced. Biomass can be converted to a liquid fuel through fermentation, creating alcohol. For instance, gasohol is a mixture of gasoline and alcohol. Gasoline is a nonrenewable resource, but we can deplete it less quickly if it is mixed with grain alcohol, a renewable resource.

One benefit of renewable energy sources is that they seemingly require no costs. However, the energy value of renewable sources is low compared with nonrenewable sources, so we need much greater amounts of land to gather solar energy for biomass and solar collectors, much greater amounts of space to capture wind energy, and much greater amounts of dammed lakes for hydropower. The renewable sources are also variable; in other words, the Sun needs to shine to provide solar energy, and the wind needs to blow to produce power from windmills.

SECTION FOUR FEEDBACK >

1. Discuss some of the trade-offs in using renewable and nonrenewable energy sources to provide heating for your home.

2. If photovoltaic cells convert 10% of the incident sunlight into electricity, how much electricity could a 6×10-meter roof generate in an 8-hour day?

3. Propane is a hydrocarbon gas, C_3H_8. Write the equation for the complete combustion of one propane molecule.

SECTION 5: Analyzing Engines and Power Systems

KEY IDEAS >

- Power systems have a source of energy, a process, and loads.
- The Carnot power-producing cycle has the highest theoretical efficiency.

Engines produce power from burning fuels. Some engines burn the fuels internally, while others burn the fuels outside the mechanism of the engine itself.

Internal Combustion Engines

We all know that one of the most important power-producing devices is the internal combustion engine, which is used in automobiles and many other machines. It is called an internal combustion engine because combustion occurs within it. The first attempts to make an internal combustion engine were in the mid-1800s, using gunpowder to push the piston upward. The first successful internal combustion engine was invented in Germany by Nicolas Otto in 1876. It was smaller and more efficient than the steam engines of the time. Another early internal combustion engine was developed by Rudolf Diesel, who wanted to operate an engine on powdered coal. Unfortunately, his model exploded. Later designs, in which the engine operated on liquid fuel, were successful.

Figure 7.36 is a schematic diagram of a piston-cylinder internal combustion engine. The engine cylinder has a

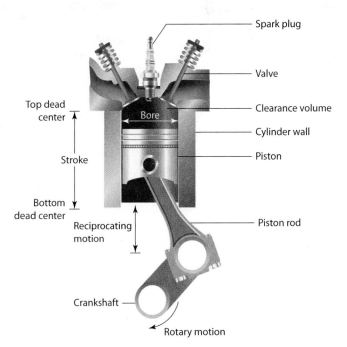

Figure 7.36 | In general, a higher compression ratio in an internal combustion engine will increase the engine's efficiency.

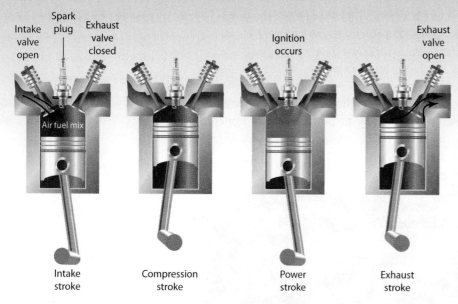

| Intake valve open | Spark plug | Exhaust valve closed | | Ignition occurs | | Exhaust valve open |

Air fuel mix

| Intake stroke | Compression stroke | Power stroke | Exhaust stroke |

Figure 7.37 | **Many spark-ignition and compression-ignition engines operate on a four-stroke cycle.**

bore (its diameter) and a stroke (the maximum length the piston moves in one direction). When the piston is at the topmost extreme, it is said to be at top dead center.

When the piston is at its bottom extreme, it is said to be at bottom dead center. The volume swept by the piston in moving between top dead center and bottom dead center is the displacement volume. The sum of the cylinder displacement volumes is the engine displacement. In automobiles, an engine might be said to have a 2.5-liter or 153-cubic inch displacement. An engine's compression ratio, r, is the cylinder volume at bottom dead center divided by the cylinder volume at top dead center. In general, the higher the compression ratio is, the greater the engine's efficiency. There are limits on the compression ratio for spark-ignition engines, because if the temperature of the air/fuel mixture is too great at the end of the compression stroke, it may prematurely ignite. For both compression-ignition (diesel) engines and spark-ignition engines, there are material strength limitations that curtail the maximum temperature and pressure inside the cylinder. This affects how high the compression ratio may be in diesel engines.

Many spark-ignition and compression-ignition (diesel) engines operate on a four-stroke cycle, as shown in Figure 7.37.

Near the top of the compression stroke, the spark plug fires, and combustion of the air/fuel mixture begins. Fuel combustion continues as the piston moves downward on the power stroke. The pressure would ideally remain constant, but actually it increases, and then decreases once the combustion ceases. The four strokes in the cycle are:

1. Intake stroke: The intake valve is open; the exhaust valve is initially open, but then it closes and the piston moves downward, bringing fresh air/fuel mixture into the cylinder.
2. Compression stroke: Both the intake and exhaust valves are closed, and the air/fuel mixture is compressed by the upward piston movement.
3. Power stroke: Both the intake and exhaust valves are closed; spark ignition and combustion occur, with the resultant pressure increase forcing the piston downward.
4. Exhaust stoke: The exhaust valve is open, the intake valve is closed, and the upward movement of the piston forces the products of combustion (exhaust) from the engine.

A compression-ignition, or diesel, engine has no spark plug. During the compression stroke, air is compressed to high temperature and pressure. Then, a fuel injector injects the fuel at high pressure into the cylinder, where it is ignited in the high-temperature environment. The temperature and pressure at the end of the compression stroke are higher than in the spark-ignition engine.

The four-stroke cycle is a mechanical cycle, and there are two revolutions for every power cycle. One revolution involves intake and compression processes, and the other involves power and exhaust processes.

Figure 7.38 illustrates a four-stroke cycle pressure-displacement (volume) diagram. The area marked W_I represents the work done by the engine during the power stroke, while the area W_{II} represents the work consumed by the engine during the exhaust and intake stroke. The area marked W_{II} has been exaggerated to show the effect of consumed work. Most engines are self-aspirated, or naturally aspirated, which means they do not require a mechanical air supply system. The piston provides the necessary air supply on its intake stroke.

Supercharging and turbocharging of an engine allow a greater mass of air into the cylinder; thus, more fuel can be burned in the cylinder, providing a greater energy release and increasing the power potential of the engine (see Figure 7.39). For a turbocharged engine, the area of W_I on the p-V diagram increases. Thus, the net work will increase as well. In an engine of a given size, the power can be increased by supercharging or turbocharging. A smaller supercharged or turbocharged engine can produce the same power as a larger self-aspirated engine. Many trade-offs must be made in designing an engine to maximize the net work, or power, including timing of intake and exhaust valve opening and closing, the timing of combustion, duration of combustion, and the air mass in the cylinder.

A supercharger is an air pump that is directly connected to the engine, typically by a belt drive. As the engine rotates, the pump compresses air and forces it into the intake manifold. Thus, the air-fuel mixture in a spark-ignition engine or the air in a diesel engine has a greater density, so the mass per unit volume is higher. A turbocharger performs a similar function as the supercharger, but it is not directly coupled to the engine. Instead, a turbocharger relies on the energy from the exhaust to power a turbine, which in turn drives a compressor. At low engine speeds, a turbocharger is not effective because there is insufficient exhaust gas flow. A turbocharger can operate at over 100,000 rpm.

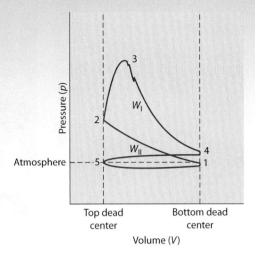

Figure 7.38 | This pressure-volume (p-V) diagram represents a four-stroke spark-ignition engine.

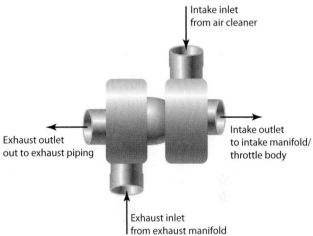

Figure 7.39 | This diagram of a turbocharger indicates the air and exhaust flows.

External Combustion Engines

In external combustion, fuel is burned outside the engine, and its energy is transferred to another liquid or gas within the power system. The liquid or gas converts thermal energy into mechanical energy, producing power. Many power plants use steam as a source of power, as shown in Figure 7.40.

An energy source such as gas, oil, coal, biomass, or nuclear power is used to boil water under pressure in a steam generator, creating high-temperature steam under high pressure. This steam flows through a steam turbine, causing it to rotate an electric generator that produces electricity, which is distributed through power lines. The steam from the turbine is then condensed, and the water is pumped back to the steam generator. This power plant can be easily modeled, as shown in

Section 5 △ Analyzing Engines and Power Systems

Figure 7.40 | A power plant is typically adjacent to a body of water that is a source of cooling water for the plant. The power plant is many stories high, and the smokestacks can extend for hundreds of feet into the air.

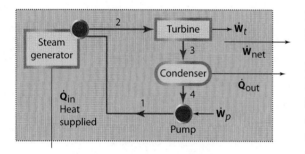

Figure 7.41 | The energy from the heat added to the power plant must be equal to the energy leaving the system.

Figure 7.41. Notice that power systems have a source of energy, processes for converting energy, and energy loads or demands.

The energy from the heat added to the system (from the fuel burned outside the engine), Q_{in}, must be equal to the energy leaving the system. The energy leaving the system includes the net power produced, W_{net}, and the heat output, Q_{out}. Thus, the energy added (Q_{in}) is equal to the power produced and the heat out is equal to the sum of the power produced. This is a simple and far-reaching powerful statement:

$$Q_{in} = W_{net} + Q_{out}$$

This problem also can be solved for the power: $W_{net} = Q_{in} - Q_{out}$. Therefore, the net power is the difference between the heat in and the heat out. It can be determined by knowing the heat flows. The heat in is provided by a variety of fuels—coal, oil, wood, nuclear, or solar.

A power plant converts thermal energy into power, so it is important to know how efficiently this can be done. An ideal model is often used to determine the maximum amount that can be converted; in other words, the maximum efficiency that can be obtained. For all power plants, large or small, the efficiency (eff) is:

$$\text{eff} = \frac{W_{net}}{Q_{in}}$$

The Carnot cycle has the highest theoretical efficiency of any power producing cycle. It can be used to model an ideal power plant, as shown in Figure 7.42.

You can determine the efficiency of a power plant operating on the Carnot cycle in terms of cycle high and low temperatures. The cycle high temperature, or combustion temperature of the gasses, is T_H. The cycle low temperature is the ambient air temperature, T_C. The efficiency is:

$$\text{eff} = 1 - \frac{T_C}{T_H}$$

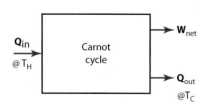

Figure 7.42 | For a power plant operating on the Carnot cycle, you can determine its efficiency in terms of the cycle high and low temperatures.

Heating a Rural School

W_{net} = 1000 kW
8 a.m. to 4 p.m.

W_{net} = 300 kW
4 p.m. to 8 a.m.

Imagine that you live in a rural area. Your local school wants to determine whether it should generate its own electricity by using wood as the fuel supply rather than by purchasing electricity from the public utility. One key question to consider is how much wood is required. You and your engineering/technology team volunteer to determine this amount. The principal says that the school uses 1,000 kW of electricity on school days from 8 a.m. until 4 p.m., and 300 kW from 4 p.m. until 8 a.m. She would like to know the daily and weekly wood requirements.

Figure 7.43 | What is the best source of power for this rural school?

How do you begin? One of the team members remembers that a Carnot cycle produces the maximum efficiency, and suggests using the cycle as a model for the power plant. Another member asks what the high and low temperatures should be. Review the earlier discussion in the chapter about combustion, and ask yourself: What is an indicative value? What is ambient (room) air temperature in degrees Kelvin? The investigations show that T_H = 1,500 K and that room temperature, T_C, is 27 C, or 300 K.

What is electricity? It is a form of power—electrical work divided by time, or W_{net}. Also, you can express efficiency in terms of temperature and in terms of power and heat flow:

$$eff = 1 - \frac{T_C}{T_H} = 1 - 300/1500 = 0.80$$

Because we know the efficiency is 80 percent, we can find the heat required from the other efficiency expression at the two power levels.

$$eff = \frac{W_{net}}{Q_{in}}$$

$$0.8 = \frac{1,000 \text{ kW}}{Q_{in} \text{ kW}}$$

$$0.8 = \frac{300 \text{ kW}}{Q_{in} \text{ kW}}$$

$$Q_{in} = 1,000/0.8 = 1250 \text{ kW}$$

$$Q_{in} = 300/0.8 = 375 \text{ kW}$$

Where does the heat come from? From burning wood. From Table 7.4 we find that the heating value of wood is 20,350 kJ/kg. In other words, one kilogram of wood will produce 20,350 kJ of heat. To find out how much wood is needed, determine the total heat input required at each of the loads and then add the numbers together. In our calculations, the school consumes 1,250 kW for 8 hours and 375 kW for 16 hours. Also, one kW equals one kJ/s, and an hour equals 3,600 seconds.

Total Heat at 1,250 kW

$$Q = (1,250 \text{ kJ/s})(3,600 \text{ s/h})(8 \text{ h})$$

$$Q = 36,000,000 \text{ kJ}$$

Total Heat at 375 kW

$$Q = (375 \text{ kJ/s})(3,600 \text{ s/h})(16 \text{ h})$$

$$Q = 21,600,000 \text{ kJ}$$

The total heat required each day would be the sum of the two numbers, or 57,600,000 kJ. The school needs to burn this amount of wood to heat the school. Let's convert this amount to kilograms and pounds:

$$Q_{in} = m_f h_{RP}$$

$$57,600,000 \text{ kJ} = (m \text{ kg})(20,350 \text{ kJ/kg})$$

$$2,830 \text{ kg}$$

Because 2.2 pounds equals one kilogram, the result is 6,226 pounds, or a little more than 3 tons of wood per day. Typically, the wood is in the form of wood chips, so volume requirements are substantial. Also, the ideal power-plant efficiency we calculated is at least twice that of an actual small power plant, so the actual fuel amount will at least double our result.

Wait, we haven't answered all the principal's questions yet. We have determined the wood requirements for the times when school is in session. On weekends, however, less energy is used, so demands for all 24 hours can be considered as 300 kW.

The total heat for Saturday and Sunday (Q_{SS}) is:

$$Q_{SS} = (375 \text{ kJ/s})(3,600 \text{ s/h})(48 \text{ h}) = 64,800,000 \text{ kJ}$$

The total heat for the five weekdays is:

$$Q_{5D} = (5 \text{ days})(57,600,000) = 288,000,000 \text{ kJ}$$

For the week, the total heat is the sum of the weekend and weekday amounts:

$$Q_{WEEK} = Q_{SS} + Q_{5D} = 64,800,000 + 288,000,000 = 352,800,000 \text{ kJ}$$

$$Q_{in} = m_f h_{RP}$$

$$352,800,000 \text{ kJ} = (m \text{ kg})(20,350 \text{ kJ/kg})$$

$$17,337 \text{ kg}$$

The weekly amount of wood required is 17,337 kg, or 38,171 pounds, or 19 tons.

Extension Using information from the Internet, determine the cords of wood required daily and weekly.

SECTION FIVE FEEDBACK >

1. Describe the four-stroke cycle.
2. If a power plant uses 1,000 kW of heat energy input and delivers 600 kW of heat energy output, what is the power produced? What is the power plant's efficiency?

Matching Your Interests and Abilities with Career Opportunities: Power-Plant Operator

Competition will intensify for jobs as power-plant operators as the power-producing industry consolidates. People with training in computers and automation will be in demand.

Nature of the Industry

Electricity is essential for functioning in today's world. From the moment you awaken to an alarm from a clock radio, to the moment you check out the late news on television or the Internet, you are dependent on electricity. Power-plant operators are at the heart of this electricity-generation process, operating and controlling its machinery. They monitor the equipment—turbines, boilers, generators, and auxiliary equipment—that is used to create electric power, and they monitor the characteristics of the generators' electrical energy (voltage, frequency, current). The operators use computer systems within air-conditioned central control rooms to see and monitor the equipment.

Working Conditions

Power plants operate twenty-four hours a day, seven days a week, and fifty-two weeks a year, so power-plant operators must be on duty at all times. The work is typically divided into three eight-hour shifts or two twelve-hour shifts. Thus, operators might work from 8 a.m. to 4 p.m. for five days, then might work from 4 p.m. to midnight the next week. Operators periodically change shift assignments so that they all share the less desirable times. The work is not physically demanding, but it is mentally challenging; constant attention and vigilance is required in an environment in which the equipment works correctly most of the time.

Training and Advancement

Entry-level laborers can advance to become power-plant operators. Often, employers seek people who have completed college-level technical courses and/ or technological degrees. Entry-level workers begin in the maintenance and repair of equipment. Depending on their aptitude, educational background, and ability, they may advance to become operators. Several years of training are required to become a qualified control-room operator. Utilities often hire from within, so there is limited opportunity to move from one employer to another.

In the case of nuclear power plant operators, the educational background and training is very extensive. Operators must pass Nuclear Regulatory Commission examinations to be licensed. To continue to be licensed, all power plant operators must receive refresher training on simulators that model the plant where the operator works. This training teaches operators how to respond quickly and appropriately to given scenarios, even though they might never occur.

Outlook

The demand for electricity continues, but the construction of new power plants has slowed because of a variety of regulatory and economic challenges. Thus, the workforce for plant operators probably will not expand, but the need for more highly qualified workers will remain. New and upgraded power plants use sophisticated automated equipment, so people with the knowledge and ability to operate in these environments will be in demand.

[Bureau of Labor Statistics, U.S. Department of Labor, Occupational Outlook Handbook, 2008–09 Edition, visited May 2008, http://www.bls.gov/oco/]

Summary >

There are many different forms of energy. These forms include radiation, thermal, electrical, mechanical, chemical, and nuclear. Among these forms of energy, many transformations can take place; for example, atomic energy can be transformed into electrical energy. However, not all energy transformations are reversible; electrical energy cannot be transformed into atomic energy. Furthermore, it might not be possible to transform one energy form directly to another. Several transformational steps may be necessary.

Radiation is energy that is emitted from an object, and thermal energy (also called internal energy) is created by the motion of atoms and molecules that occurs within all matter. Electrical energy, or electrical work, is caused by the flow of electrical charge through an electrical field. Mechanical energy, or mechanical work, is energy that one person or object expends in moving another object. Chemical energy is energy associated with the molecular and atomic structure of matter, and nuclear energy is associated with a substance's atomic structure.

Heat and work are two energy forms that are sometimes created during a transformation from one type of energy to another. Heat is thermal energy that is transferred between two objects because of a temperature difference between them. There are three modes of heat transfer: conduction, radiation, and convection. Conduction is heat transfer within a material caused by a temperature difference between the material's inside and outside surfaces. Radiation is the flow of thermal energy between two bodies separated by a distance, such as bread and electrical wire in the toaster, or the Sun and the Earth. Convection is the transfer of heat between a surface and a fluid or gas. Power is the rate at which work or heat transfer occurs.

Matter possesses three types of intrinsic energy. Kinetic energy is the energy an object possesses because of its motion. Gravitational potential energy is associated with the mass's position relative to the ground. Internal energy, a less tangible form of energy, is the energy often associated with a substance's molecular structure.

Thermodynamics applies two simple yet far-reaching laws to a wide range of energy systems. The First Law of Thermodynamics, the law of conservation of energy, states that in any energy transformation, energy is conserved. Energy can change form, but the total amount of energy is constant. The Second Law of Thermodynamics states that whenever an energy transformation occurs from one form to another, there is a net decrease in energy value.

Energy comes both in nonrenewable and renewable sources. Oil, natural gas, and coal are called fossil fuels and are nonrenewable energy sources. Uranium is another important nonrenewable energy resource. It is not a fossil fuel, but a nuclear, or atomic, fuel. Renewable energy sources, including solar power, wind, water, and biomass, are readily replaced daily or within a few months or years.

One of the most important power-producing devices is the internal combustion engine. In such an engine, combustion occurs from within. In an external combustion cycle, fuel is burned outside the engine, and its energy is transferred to another liquid or gas within the power system. Many spark-ignition and compression-ignition (diesel) engines operate on a four-stroke cycle.

FEEDBACK

1. Describe the relationship between electrical charge and current flow.

2. Research and make a sketch of an iron oxide (rust) molecule.

3. Your mother asks you to carry a gallon water bottle from the car to the house. She carries one as well. You go directly to the house, but your mother gets the mail, checks the flower beds, and then enters to the house. Who did more work in carrying the water? Why?

4. What is heat? Why does heat flow happen?

5. If you double the thickness of insulation, what happens to the heat flow? Does it double, decrease by one-half, or stay the same? Why?

6. Investigate the R factor of building insulation. What values did you find for 3-inch and 6-inch-thick insulation?

7. On a chilly afternoon, you decide to use a portable electric heater to warm the air. Describe the heat transfer processes that occur.

8. A person holds a water balloon that contains 0.5 kg of water. He throws it upward so that it is 5 meters above the ground at its highest point before descending. What is the balloon's potential energy at that point? What is the velocity with which the balloon hits the ground? (In this case, all the potential energy is converted into kinetic energy.)

9. How much heat does it take to increase 2 kilograms of
   (a) air,
   (b) liquid water, and
   (c) aluminum from 27 C to 77 C? (*Hint:* K = 273 + C.)

10. Perform the "Calculating Horsepower" feature earlier in the chapter for a truck that weighs 2,500 kg, travels at a speed of 60 mph, and takes 10 seconds to reach 60 mph.

11. Investigate the size of windmills that are used to generate power. How large are the typical windmills?

12. Assume that a windmill blade has a radius of 50 meters and that the wind is blowing at 15 mph. What would be the maximum theoretical power the windmill could produce?

13. What is the First Law of Thermodynamics?

14. What does the Second Law of Thermodynamics deal with and explain?

15. A student pushes a box with a force of 200 newtons a distance of 10 meters. What is the work done? (*Hint:* A newton-meter is equal to one joule.) If the student accomplished this task in 5 seconds, what was her power?

16. What are the differences between renewable and nonrenewable energy sources?

17. Describe three types of renewable energy sources and how long it takes them to be renewed.

18. What are differences between a spark-ignition (gasoline) engine and a compression-ignition (diesel) engine?

19. Describe and illustrate a four-stroke cycle internal combustion engine.

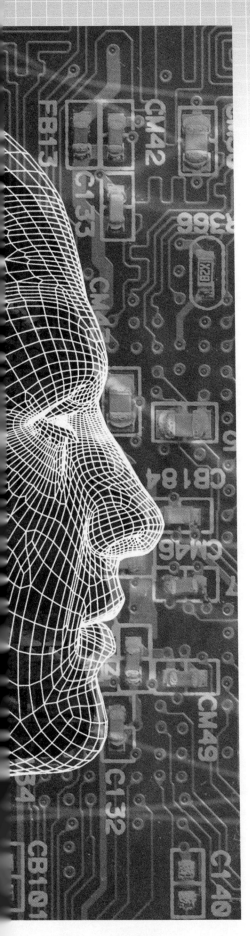

DESIGN CHALLENGE 1:
Windmill Blade Design

• Problem Situation

Windmills have been used for centuries to generate power. For much of this time windmills rotated slowly, generating the power needed to pump water or grind grain. Recent advances in windmill design have allowed windmills with long blades to generate electric power; the blades rotate faster than those of older designs. Windmill-blade design is similar to that for helicopters and airplanes.

• Your Challenge

Design and construct a windmill that produces the maximum power for a given wind velocity. Refer to this activity in the student workbook. Complete the first section of the Informed Design Folio (IDF), stating the design challenge in your own words.

• Safety Considerations

1. Never cut toward your body or point any sharp item at your body or at anyone else.
2. Do not put your fingers near a rotating fan blade.
3. Wear protective glasses at all times.
4. The glue gun can be hot, so use caution.

• Materials Needed

1. Foam board
2. One-inch balsa wood strips
3. Glue gun
4. Sandpaper
5. A hub for connecting blades

• Clarify the Design Specifications and Constraints

To satisfy the problem, your design must meet the following specifications and constraints.

▪ Specifications

1. The windmill will be directly coupled to a DC generator.
2. The wind is generated by a 20-inch box fan placed three feet from the windmill.

▪ Constraints

Refer to the IDF and state the specifications and constraints. Include any others that your team or your teacher included.

• Research and Investigate

1. Refer to the Student Workbook and complete KSB 1, History of windmills.
2. Refer to the Student Workbook and complete KSB2, Constructing a Simple hub.
3. Refer to the Student Workbook and complete KSB 3, Blade design.

• Generate Alternative Designs

Refer to the Student Workbook and describe modifications to the standard design solutions. Use sketches to indicate each solution's strengths and weaknesses. Your sketches should show sufficient detail, such as blade shaping and number. Include justifications for the different solutions.

- ## Choose and Justify the Optimal Solution

Refer to the Student Workbook. Explain why you selected a particular solution and why it was the best.

- ## Display Your Prototypes

Construct the windmill blades and rotor. Include drawings of the pieces you used to construct the blades and windmill. Also include a sketch of the final design or take a photograph of your model and place it in the Student Workbook.

In any technological activity, you will use seven resources: people, information, tools/machines, materials, capital, energy, and time. In the Student Workbook, indicate which resources were most important in this activity and how you made trade-offs among them.

- ## Test and Evaluate

Did your new design meet the initial specifications and constraints? Explain the tests you conducted and the experiments you performed to verify your results.

- ## Redesign the Solution

Respond to the questions in the Student Workbook about how you would redesign your solution based on knowledge and information that you gained during the activity.

- ## Communicate Your Achievements

In the Student Workbook, describe the plan you will use to present your solution to your class and show what handouts or graphs you will use. (You may include PowerPoint slides.)

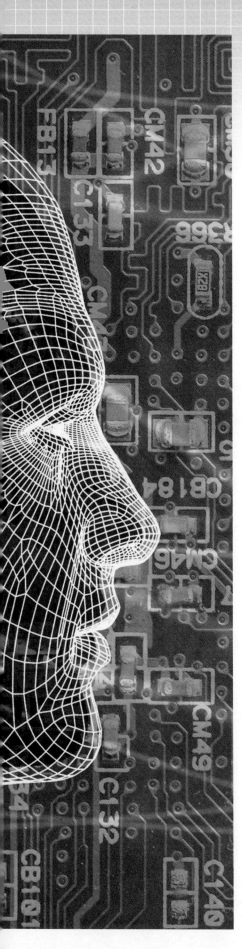

DESIGN CHALLENGE 2:
Cooling a Hot Dog

• Problem Situation

During the summer, temperatures can reach uncomfortably high levels, especially for our four-legged friends who live in doghouses. In this challenge, the doghouse is not located near an electrical outlet. However, photovoltaic shingles generate electricity when the Sun hits them, so you can use this resource to provide Fido with a cooling fan powered by the shingles.

• Your Challenge

Design and construct a model doghouse that uses photovoltaic shingles to provide electricity for powering a fan.

• Safety Considerations

1. Never cut toward your body or point any sharp item at your body or at anyone else.
2. Do not put your fingers near a rotating fan blade.
3. Wear protective glasses at all times.
4. Use hammer and nails with care.

• Materials Needed

1. One photovoltaic (PV) shingle, approximately 85 inches long
2. Wiring for the shingles
3. Nails
4. Wood
5. Small DC motor
6. Fan blade

• Clarify the Design Specifications and Constraints

To satisfy the problem, your design must meet the following specifications and constraints.

■ Specifications

The design should provide circulating air flow in the doghouse.

■ Constraints

Refer to the IDF and state the specifications and constraints. Include any others that your team or your teacher included.

• Research and Investigate

1. Refer to the Student Workbook and complete KSB 1, History of photovoltaic power.
2. Refer to the Student Workbook and complete KSB2, Positioning of the solar cells.
3. Refer to the Student Workbook and complete KSB 3, Wiring fundamentals.

• Generate Alternative Designs

Refer to the Student Workbook and describe modifications to the standard design solutions. Use sketches to indicate each solution's strengths and weaknesses. Your sketches should show sufficient detail, such as the slope of the roof and wiring diagrams. Include justifications for the different solutions.

- ## Choose and Justify the Optimal Solution

Refer to the Student Workbook. Explain why you selected a particular solution and why it was the best.

- ## Display Your Prototypes

Construct the doghouse with the PV shingles. Include a sketch of the final design or take a photograph of your model and put it in the Student Workbook.

In any technological activity, you will use seven resources: people, information, tools/machines, materials, capital, energy, and time. In the Student Workbook, indicate which resources were most important in this activity and how you made trade-offs among them.

- ## Test and Evaluate

Did your new design meet the initial specifications and constraints? Explain the tests you conducted and the experiments you performed to verify your results.

- ## Redesign the Solution

Respond to the questions in the Student Workbook about how you would redesign your solution based on knowledge and information that you gained during the activity.

- ## Communicate Your Achievements

In the Student Workbook, describe the plan you will use to present your solution to your class and show what handouts or graphs you will use. (You may include PowerPoint slides.)

TRANSPORT DEALS WITH the movement of goods or people from place to place. The ability to do this efficiently is so vital to how we live that society has created an entire organizational system devoted to that task: the transportation system. As illustrated in Figure 8.1, the transportation system is made up of the various means of moving goods or people from place to place, combined with a

CHAPTER

8

TRANSPORTATION

SECTION 1
Understanding the Global
Transportation Picture

SECTION 2
Examining Transportation Systems

SECTION 3
Investigating New Technologies

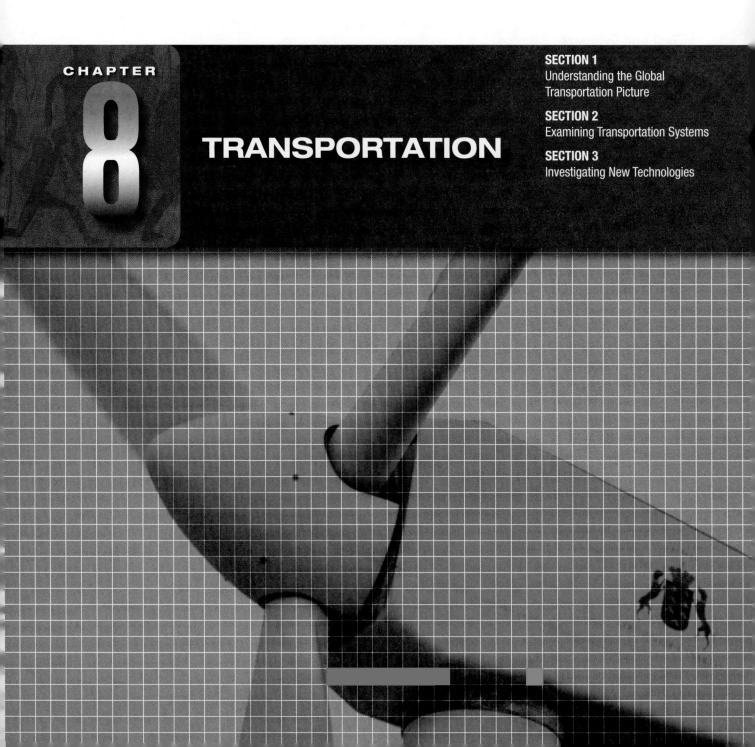

large and highly integrated infrastructure of support components, such as roads, docks, railroads, and air corridors.

Users of the transportation system employ a number of distinct modes of operation that include sea, road, rail, and air transportation systems. The users of such systems often have varied needs. Sometimes the cost of transporting goods or people is a significant factor; sometimes, when time is critical, cost is not as important as speed. The transportation industry does its best to offer flexibility. Providing a workable solution to the varied needs of those using the transportation system is often a compromise between three key factors:

- speed
- cost
- volume

Although it is not always possible to generalize, the following also could be said to apply:

- The higher the speed, the higher the cost
- The larger the volume, the lower the cost

Figure 8.1 | The transportation system is made up of the vehicles and infrastructure that move people and goods around the world.

Given the above, users of the transportation system have to select the most suitable mode of transport for their particular needs. The considerable growth in global trading patterns, as well as greater worker mobility and expanded leisure opportunities, drives continued growth in this industry year-on-year.

Increasing demand for transportation services creates a need for the manufacture of all forms of vehicles (cars, trucks, rail cars, ships, and aircraft). The transportation system also needs energy to function. The increased demand for transportation is having a significant effect on the consumption of the world's energy resources. In 2007, transportation alone consumed nearly half of the 86 million barrels per day of oil consumed worldwide. Since the production of U.S. domestic oil has greatly decreased, largely due to exhaustion of supplies, the United States has, for some time, been a net importer of oil to provide the energy needs of the nation.

Oil consumption has a significant impact on the environment. As fossil fuels, such as gasoline, jet fuel, and diesel oil are burned to power the

Introduction

transportation system, the exhaust fumes include a large volume of carbon dioxide and other pollutants, as well as gasses that are thought to contribute greatly to the increase in global warming and air pollution. These two major factors—the need to reduce consumption of oil and the need to reduce exhaust gas emissions and pollution—will have a significant effect on how people design, make, and use transportation systems in the future.

In this chapter, we will discuss improvements to the transportation system over the centuries. We will also examine the current global transportation system, its impact on the world, and the factors that have influenced its development. We will review the various transportation systems and look at new technologies that will affect future developments in this industry.

SECTION 1: Understanding the Global Transportation Picture

KEY IDEAS >

- Transport of various items and people extends back to the beginning of civilization. The need to move people or carry goods from place to place gave birth to transportation.

- Transportation plays a vital role in the operation of other technologies, such as manufacturing, construction, communication, and agriculture.

- Prior to the beginning of the Industrial Revolution, water was the only realistic transportation method for moving large volumes of products.

- Beginning in the early 1800s, the steam locomotive grew

to dominate long-distance movement of people and goods until the advent of the internal combustion engine era in the early twentieth century.

- The mechanization of travel started a period of growth and development that continues to this day and into the future in terms of improvement of the three key attributes of the transportation industry: speed, volume, and cost.

- Transportation services and methods have led to a population that is regularly on the move.

- Transportation has a significant impact on the environment.

Transport of various items and people extends back to the beginning of civilization. The need to move people or carry goods from place to place gave birth to the earliest transportation systems. In addition, transportation has played a vital role in the operation of other technologies, such as manufacturing, construction, communication, and agriculture. Over time, humans developed ways of making

Figure 8.2 | Oxcarts, such as this one carrying a farmer in India, are still used in developing countries around the world.

the process of moving goods or people easier. They developed the sled to drag objects too heavy to carry; they developed the wheel and, probably not long after, the cart to make the process of moving goods and people over land much easier. Once animals were domesticated, the movement of loads too heavy for humans to handle became possible by harnessing animals to carts. The animals used for this purpose tended to be docile, powerful, and well-suited to hauling a very large load. They were, however, very slow. An oxcart, similar to the one shown in Figure 8.2, has a top speed of 1–2 mph.

When humans moved away from hunting and gathering, they turned to planting crops, raising animals for food, and making articles for trade. In order to trade effectively, people needed ways to move their goods to new markets. For 6,000 years, human- or animal-powered carts and boats traveled via roads, rivers, and canals to distribute goods or to move people. The process was quite successful, but quite slow. The Industrial Revolution changed everything. Beginning in England in the late 1700s, the Industrial Revolution marked the transition from an agrarian economy to an economy built around machines and industry. The volume of goods being produced after the start of the Industrial Revolution was such that products needed to be moved quickly and efficiently to people across the globe who would buy and use them.

Prior to the Industrial Revolution, water was the only realistic transportation method for moving large volumes of products. Improvements such as building a system of locks on rivers in order to move boats between different water levels and developing canals to link towns not connected by rivers expanded the naturally flowing river-based distribution system. Canals, such as the one shown in Figure 8.3, are manmade waterways between two of more points. Although more efficient than the oxcart, the water-based distribution system was limited. It could not easily cross natural barriers, and you cannot build a canal without water. In addition, many parts of the world did not have access to the large volume of water needed to support a canal-based transport system.

Section 1 △ Understanding the Global Transportation Picture

Figure 8.3 | The Bridgewater Canal in northwest England served as a commercial waterway until the 1870's.

Mechanization of Travel

By 1802, Richard Trevithick of Cornwall, England, had employed the steam locomotive as the source of motive power to move a vehicle. A steam engine uses high-pressure steam to power reciprocating pistons that move the wheels of the locomotive. After this milestone event, the power of fossil fuels would be used to generate power for transportation in a way never imagined before. The steam locomotive boasted a pulling force that far exceeded that of oxen, and it could travel at a much higher speed. Figure 8.4 shows an example of a steam locomotive that is still running at the Steamtown National Historic Site in Scranton, Pennsylvania.

Beginning in the early 1800s, the steam locomotive grew to dominate long distance movement of people and goods until the advent of the internal combustion engine era in the early twentieth century. The use of steam power was not limited to railroads. Shipping companies discovered that the steam engine enabled the industry to move from slow and haphazard sail power, to large metal-hulled steam ships that carried greater loads of freight and passengers at much higher speeds and reduced journey times. The mechanization of travel sparked a period of growth and development that continues to this day and into the future in terms of improvement of the three key attributes of the transportation industry: speed, volume, and cost.

Figure 8.4 | This steam locomotive is still running at the Steamtown National Historic Site in Scranton, Pennsylvania.

Global Trade

Very few products are now made in one location, or even in one country. The growth of companies that trade in many countries, as well as the cost savings achieved by moving manufacturing or assembly to countries with different economic climates, has led to a massive growth in international trade. In order to meet the demands of international trade, transportation of manufactured goods has grown by 50 percent worldwide since 1990. This trend shows no sign of slowing, as trade with the most populous countries in the world, China and India, is expanding every year.

Passenger Travel

Not only are we seeing growth in the movement of goods, but more people are traveling than ever before. Transportation services and methods have led to a population that is regularly on the move. Worldwide business and leisure travel is growing at a rate of 15 percent annually. A traveler can now make complex travel arrangements, including booking international flights, hotels, and car rentals, all from a home computer via the Internet (see Figure 8.5). The travel industry was not always so accommodating. Each company had it own booking system, and no common method of exchanging information between companies existed. As the use of computerized reservation systems (CRS) became more widespread in the 1950s, the airline industry realized that by integrating their individual company booking systems, they would be much more powerful as a whole if the booking process could automatically and easily find the shortest journey time by using flights from multiple companies. The interconnected booking capability of the airline industry proved to be a significant factor in its growth and is a prime example of the way technology has influenced the transportation sector.

Figure 8.5 | Using the Internet, travelers can book trips via the computerized reservation systems of airlines, cruise lines, and car rental agencies.

Freight

The movement of goods and materials is the largest sector of the transportation industry. Although some high-value products need to be transported quickly, in general the freight industry is willing to accept longer journey times in exchange for reduction in transportation costs.

Containerization is the process of placing goods in a secure movable container (see Figure 8.6) that does not have to be unloaded and repacked at every transfer point. Prior to containerization, freight was unloaded manually from rail or truck and then manually loaded in small batches into the hold of a cargo ship by a dockside crane. This was a lengthy and expensive process that had to be repeated in reverse at the other end of the voyage. In 1956, when the first U.S. container service was inaugurated, the cost of hand loading freight was $5.86 per ton. The first container ship carrying a total of 58 containers to leave the Port of Newark was loaded at a cost of $0.16 per ton.

The development of the standardized shipping container significantly changed the world of manufactured freight transportation. The ability to place goods in a secure container at the place of manufacture, and, without any additional handling and repacking, to have the goods shipped to the customer using standardized truck, rail, and ship systems, was truly revolutionary.

The widespread use of containers for movement of almost all nonbulk goods has led to the expansion of intermodal shipping, which means that goods are shipped using more than one mode of transport without additional handling. For example, a container can be filled with manufactured items at a factory, securely locked, and sealed. It can then be loaded on a truck, which transports it to a rail-freight depot. At the rail depot, the container can be loaded onto a flat car and taken to a container port for loading onto a container ship for transport to anywhere in the world. All of this movement occurs without any additional handling of the goods inside the container.

Figure 8.6 | Shipping containers like the ones shown here can be loaded with freight at a factory, shipped by track or rail to a port, and then lifted into a ship's hold.

Section 1 △ Understanding the Global Transportation Picture

Figure 8.7 | Malcom McLean's vision created an industry comprising thousands of container ships, carrying hundreds of thousands of tons of freight between ports throughout the world that are equipped to move standard containers easily from ship to railcar or truck.

A high school graduate from Texas, Malcom McLean started a small freight-hauling company with one truck and built it into one of the largest freight haulage companies in the United States.

While running his company, McLean had observed that the loading and unloading of ships was very inefficient. Every item had to be unloaded from a truck onto a hoist, and then transferred into the hold of a ship where it was then unloaded once more. This labor-intensive system, which employed longshoremen to do all the loading and unloading, is known as "break-bulk loading." McLean realized that if the entire truck trailer could be loaded by crane into the hold of the ship, this would greatly speed up the loading and unloading of ships, and, at the same time, reduce breakage and theft. He wasn't allowed to put his ideas into practice at that time because the interstate commerce laws did not allow a trucking company to own a shipping company.

McLean sold his trucking company in 1955 and moved to New York, where he bought a small shipping line. It wasn't too long before he converted a ship to carry pre-packed freight in standard-sized shipping containers and trailers. Although not the first to transport freight pre-packed in some form of shipping unit, he was one of the first to use a sealed container that could be packed at the point of manufacture and not opened until delivered to its destination. His standard-size shipping containers could be carried by truck, railcar, or ship, and the contents did not need handling at any point during the journey.

On April 26, 1956, one of McLean's converted ships, the Ideal-X, was loaded at the Port of Newark, New Jersey, and departed for the Port of Houston, Texas, carrying fifty-eight 35-ft containers. At that time, break-bulk loading of a ship using cranes, slings and longshoremen on the dock and in the hold of the ship cost $5.86 per ton. Using his pre-packed containers, it cost McLean only 16 cents per ton to load the Ideal-X, a saving of 97 percent.

At the same time that McLean was developing his container ships, the shipping industry was trying to increase its efficiency by increasing the size of ships and docks and the size and capacity of cranes that loaded ships, as well as making other mechanical improvements to increase the handling capacity for break-bulk loading of ships. The willingness to use more mechanization in U.S. ports, when combined with McLean's pre-packed containers that could be loaded so efficiently onto purpose built ships was a setting that would lead to a revolution in the shipping industry for the next twenty years.

The containerization revolution started quite slowly, with the first ship configured to carry only containers, the Gateway City, commencing regular service between New York, Florida, and Texas in 1957. In 1964, McLean opened a 100-acre container port at Elizabeth, New Jersey, to handle the growth of U.S. container traffic. Even with the development of large container-handling facilities in the U.S., the growth of the worldwide container market was slow until the late 1960s. Many ports in other countries did not have the cranes to lift containers on and off ships, and change was slow to come to an industry steeped in tradition. Also, the dock workers' unions, both in the United States and abroad, initially resisted the idea of containerization because it appeared to threaten their jobs. As U.S. involvement in the Vietnam War grew in the 1960s, McLean's success in supplying U.S. forces in Vietnam persuaded the rest of the world of the potential of containerized shipping.

In addition, to assist the growth of containerization across the shipping world, McLean chose to make his patents for his standardized containers available without royalty payment to the shipping industry. He knew that for the industry to grow, containers must be manufactured to the same exact standard dimensions in all countries without any legal impediment to adopting his standard. The move toward greater standardization accelerated the intermodal transportation industry. By the end of the 1960s, McLean's company, SeaLand Industries, owned and used 27,000 standard containers, thirty-six container ships, and container handling facilities in over thirty port cities.

Oil is often transported by pipeline to a port, where it is loaded into an oil tanker. The tanker will then transport the oil to a distant location where the oil is transshipped by pipeline or offloaded to a depot to be loaded on to a rail or road tanker. Mission critical or high-value freight is often carried by air. A number of freight companies such as FedEx or UPS operate very tightly integrated company-based systems, owning and operating all aspects of the business. This means that they are able to deliver goods to anywhere in the world with extraordinary speed, although the cost of such services is considerably higher than the cost of moving the same goods by bulk shipping.

Environmental Impact

Transportation has a significant impact on the environment. From the exhaust emissions of automobiles, trucks, trains, ships, and aircraft, to the use of yet more land to support our need for roads, runways, and railroads, to the noise created by all these various transport systems sharing our living, working, and leisure space, transportation has substantial effect on the world. As shown in Figure 8.8, transportation accounts for 27 percent of U.S. greenhouse gas emissions and 80 percent of urban air pollution.

Vehicles powered by the internal combustion engine produce a large number of pollutants. The emissions from all forms of transport are significant, but it is clear that carbon dioxide (CO_2) by far forms the majority of the output. This gas is thought to be one of the major causes of global warming. In 2004, the transportation sector in the U.S. emitted 331 million tons of carbon dioxide (see Figure 8.9). Although the U.S. comprises only 5 percent of the world's population, we produce 21 percent of the world's output of greenhouse gasses.

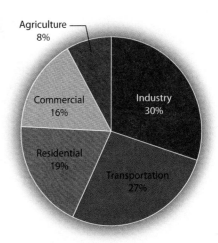

Figure 8.8 | This pie chart, created by the United States Environmental Protection Agency (EPA) shows the various factors contributing to U.S. greenhouse gas emissions in the year 2000. The transportation system accounted for 27 percent of all greenhouse gas emissions in the United States.

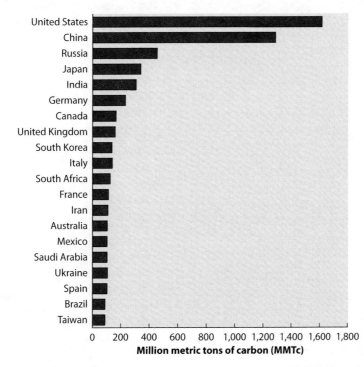

Figure 8.9 | As of 2004, the United States was the world's largest producer of carbon dioxide emissions from automobiles.

Section 1 △ Understanding the Global Transportation Picture

Aircraft emissions contributed 3.5 percent to greenhouse gas emissions in 2005. A recent research paper has suggested that aircraft emissions, since they are released in the atmosphere, amplify the effect of the carbon dioxide, nitrogen oxide, and water vapor emitted. If that is the case, then the total warming effect of aircraft emissions is 2.7 times greater, or 9.45 percent in total. The transportation industry is slowly responding to this challenge by developing cleaner technologies and alternate fuels.

SECTION ONE FEEDBACK >

1. Why is containerization so important to manufacturers who export goods?
2. How do computerized reservation systems help the traveler?
3. When would a manufacturer use air freight to deliver its goods?
4. What is the most significant factor in the growth of the global transportation system?
5. Describe the significance of the Industrial Revolution.
6. Describe three environmental impacts of the global transportation system.
7. What was the percentage of worldwide carbon dioxide emission produced by the United States in 2004?

SECTION 2: Examining Transportation Systems

KEY IDEAS >

- Intermodalism is the use of different modes of transportation, such as highways, railways, and waterways as part of an interconnected system that can move people and goods easily from one mode to another.

- The vast majority of manufactured and semi-manufactured goods exported from one country to another are carried on specialist container vessels in standardized containers.

- The high costs of other means of travel have recently spurred interest in developing high-speed intercity train services in the United States.

- Even though the technology to build better roads existed in the early 1900s, it was the massive growth in automobile ownership from the 1920s onwards that created a nationwide demand for a comprehensive network of roadways.

- The interstate system of limited-access highways offers rapid travel over long distances between major cities across the nation.

- Increasingly, space is being commercialized, and a large number of communication companies can exploit the very wide footprint or viewable area of a single satellite placed in Earth's orbit.

The transportation industry serves the needs of a wide and diverse group of people, many with differing needs. A multinational manufacturer of large systems wants parts moved quickly and cheaply from various countries to the point of assembly. Such a manufacturer then often wants finished goods shipped to the customer in the most secure and economical way. A business traveler might see travel time as time that could have instead been used to transact sales and so prefers the shortest journey time. The leisure traveler relaxing on a cruise ship is less concerned with speed, but rather with comfort and accommodation. The complex and multifaceted needs of users of the transportation system has led to the development of a wide range of modes of transportation. Each mode of transportation offers its own opportunities and challenges. Intermodalism is the use of different modes of transportation, such as highways, railways, and waterways as part of an interconnected system that can move people or goods easily from one mode to another.

Water

Water covers 71 percent of the Earth's surface. If a buoyant article is placed in water, it will require much less force to move that article compared with moving it on a solid surface, such as the ground. These factors made waterways the primary means of transporting people and goods from the Stone Age until the advent of mechanized transport systems developed during the Industrial Revolution. The wooden ships of the eighteenth century were the peak of the technological development at that time. The trading ships weighed some 100–300 tons, and were built to carry goods across seas and oceans. The advent of the Industrial Revolution allowed the use of iron for manufacturing the ships' hulls and steam to propel the ships. From this point on, the size of ships grew considerably and journey times decreased.

Freight Carrying Ships As you can see in Figure 8.10, freight-carrying vessels comprise the majority of the world's shipping fleet. They vary in size from small inland vessels weighing 100 tons or less, used for off loading bigger vessels or for local transport on small lakes and rivers, to very large ships weighing over 300,000 tons. The volume of freight cargo has been growing steadily for many years. As the size of a ship increases, so does its efficiency, since it can carry more goods in a single journey. Since the 1950s, technology and shipbuilding techniques have evolved to make it much more practical to build and operate larger ships; however, there are some practical limitations to ship size.

For a large ship to be able to navigate safely, it must sail in water deep enough to allow the vessel to float without grounding on the ocean or sea floor. There are a number of places in the world where the depth of the shipping lanes is reduced by natural geophysical factors. This has limited the maximum size of the current generation of ships. For example, the Straits of Malacca, the main shipping channel between the Indian and Pacific Oceans, located between Malaysia and Indonesia, carries over one quarter of the world's bulk and manufactured goods annually. This 700-mile long navigation channel is 2.6 miles wide at its narrowest point, and 130 feet deep at its shallowest point. This limits the size of ships using the Straits of Malacca to approximately 300,000 tons.

There are other significant navigational barriers found across the world that limit ship size to much smaller than 300,000 tons. Many of the barriers are manmade

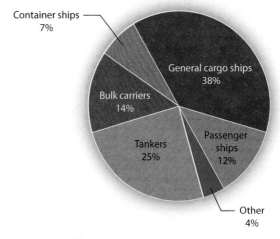

Figure 8.10 | This pie chart shows the percentages of different types of ships in use today around the world. Cargo ships are the most common, followed by tankers.

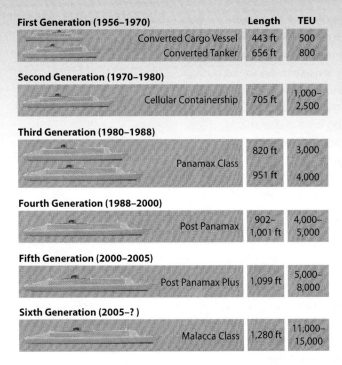

		Length	TEU
First Generation (1956–1970)			
	Converted Cargo Vessel	443 ft	500
	Converted Tanker	656 ft	800
Second Generation (1970–1980)			
	Cellular Containership	705 ft	1,000–2,500
Third Generation (1980–1988)			
	Panamax Class	820 ft	3,000
		951 ft	4,000
Fourth Generation (1988–2000)			
	Post Panamax	902–1,001 ft	4,000–5,000
Fifth Generation (2000–2005)			
	Post Panamax Plus	1,099 ft	5,000–8,000
Sixth Generation (2005–?)			
	Malacca Class	1,280 ft	11,000–15,000

Figure 8.11 | Here you can see how container ships have increased in size since 1956. One TEU (twenty-foot equivalent unit) is roughly equal to the capacity of one twenty-foot long shipping container. First generation ships could carry 500 to 800 TEUs. Modern ships can carry up to 8,000 TEUs.

canals, such as the Panama or Suez canals, that were constructed at a time when the technology did not exist to build very large ships or to construct and operate the complex port and dock facilities that large ships demand.

Container Ships The vast majority of manufactured and semi-manufactured goods exported from one country to another are carried on specialist vessels in standardized containers. Containers range from twenty feet in length up to fifty-three feet. To make it easier to calculate container loading, the industry measures container volume in twenty-foot equivalent units (TEU). Container-based freight has tripled in the last twelve years, and this growth is predicted to continue in the future. Container ships range from feeder ships carrying 500 to 1,000 TEU containers up to those carrying 9,000 TEU or more containers.

Tankers, Liquid Natural Gas, and Bulk Carriers Currently, the tanker and bulk carrier fleets comprise the largest proportion of the freight fleet. The products they carry—mostly oil, pressurized or liquefied gasses, metal ore, or grain products—benefit from movement in bulk. Like all freight carrying ships, there are benefits from scale, so, not surprisingly, such ships have grown in size over time. The largest ship in the world is an oil tanker weighing 500,000 tons, although the vast majority of the tanker fleet ranges from 279,000 to 320,000 tons due to the navigation limitation mentioned earlier.

Cruise Ships and Liners The $13-billion cruise line industry has been the fastest growing segment of the travel and tourism industry. Worldwide, it has grown from 0.5 million passengers a year in 1970 to 9.5 million passengers in 2000. As the industry seeks to increase revenue by carrying more passengers per cruise ship, the size of the average ship has grown from 75,000 tons in the 1970s to over 200,000 tons and more. The desire to provide passengers with sea views has created ships that are tall and wide, thus providing more external surface area for cabin windows. Liners are ships designed to make ocean crossings at speed in all weather, unlike cruise ships that are designed and operated for passenger comfort. The Queen Mary 2, shown in Figure 8.12, is one of a new generation of "cruise liners" that can operate safely and at speed when crossing oceans and also act as a cruise ship when necessary. The long and sleek design of the Queen Mary 2 reflects its need to be able to handle the severe weather that can be encountered when crossing oceans.

Warships Warships can vary from the very small beach-assault craft up to 100,000 ton nuclear-powered aircraft carriers. Warships tend to have very specific roles. For example, the amphibious assault ship, shown in Figure 8.13, is designed to allow troops to land using small amphibious craft carried within the hull of the larger vessel. The aircraft carrier carries a large number of aircraft and their support infrastructure. The aircraft carrier can be nuclear powered, giving it great endurance since the power plant requires very infrequent refueling.

Leisure Craft Leisure craft can range from the low cost small recreational watercraft up to luxury ships owned by the very wealthy. Personal watercraft are small vessels that the rider sits on or stands on, shown in Figure 8.14. They are generally

Figure 8.12 | The Queen Mary 2 is a liner designed to cross the oceans safely at speed, but it is also designed to be able to function as a cruise ship when needed.

Figure 8.13 | This amphibious assault ship plays a vital role in the U.S. Navy.

constructed from plastic moldings and buoyant materials. Off-shore racing boats are designed to go fast in all sea conditions. These boats race from point to point or over an extended circuit course and can exceed 190 mph. Luxury yachts are very large ships owned by a single person. The largest personal vessel is currently the Al Salamah (456 feet long—a football field is 360 feet long) owned by the Prince Sultan bin Abdul Aziz of Saudi Arabia. Equipped with a helipad and an indoor swimming pool, this diesel-engine boat can travel at 21 knots (nautical miles per hour).

Marine Propulsions Systems Marine vessels require both a means for propelling the vessel forward and a power source for that propulsion system. Advancements in both technologies have resulted in more efficient and effective water transportation. With the replacement of sail with steam-powered paddle propulsion systems, ships had the ability to travel independently of the direction of the wind.

Figure 8.14 | Personal watercraft are popular and relatively low cost vehicles. They are designed to carry one to three people.

Paddle wheels quickly gave way to the much more efficient propeller screw propulsion systems, which proved to be much more efficient than side wheel paddles. A screw propeller has angled blades that rotate and push against the water to propel a ship. Similarly, the early reciprocating steam engines gave way to much more powerful steam turbines, which used hundreds of small blades placed in the path of jets of very high-pressure steam to rotate the propeller shaft. Although more efficient, steam turbines were expensive to manufacture and to maintain. Today, most large freight-carrying ships are powered by large and efficient diesel engines, propeller screws for propulsion (see Figure 8.15), and a rudder for direction control.

The cruise industry has fostered the development of the propulsion pods as a replacement for the usual engine-propeller and shaft-propeller screw arrangement. Propulsion pods contain high-efficiency electric motors that turn a propeller. A pod-equipped ship will usually use two or four pods, in combination with maneuvering systems such as bow thrusters. Pods, such as the one shown in Figure 8.16, can be fixed or can rotate through 360 degrees to maneuver the ship. Using rotating pods in combination with powerful bow thrusters, a very large cruise ship can be navigated from the bridge by means of a joystick-controlled integrated maneuvering system.

Figure 8.15 | This ship's screw propeller is easy to see as the ship sits in dry dock.

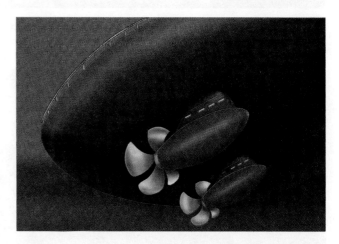

Figure 8.16 | Propulsion pods each have a motor attached to the screw propeller. Pods are usually used on a ship in groups of two or four.

Queen Mary 2

▷ In May 2007, the Queen Mary 2 was docked at her berth in New York by the use of the integrated maneuvering system, without the aid of any tugboats (they were there just in case, but were not needed). The 150,000 ton, 1,315-foot-long Queen Mary 2 (like many other modern cruise ships) does not have a traditional engine room or a propeller shaft taking the power to propellers mounted at the stern of the ship. Instead it uses four propulsion pods. The two forward pods are fixed and pull the ship through the water; the rear pods can rotate through 360 degrees and are used to steer the ship.

The Queen Mary 2 is the first cruise liner to make use of modified, combined diesel and gas turbine engines linked to an electric propulsion system. Since this propulsion system requires electric motors for propulsion, it generates its own electrical power by connecting the four large diesel and two gas turbine engines to generators. For low-speed operation, the diesel engines are used. For high-speed operation, the diesel and gas turbine engines are used.

Warships often need to be able to depart rapidly from port, or to proceed at high speed, so the propulsion system cost and the fuel cost per mile are secondary considerations. Warships often use the most sophisticated propulsion systems: nuclear powered aircraft carriers use nuclear energy to generate steam that drives turbines; nuclear submarines use nuclear power to generate electricity to drive the propeller. Many smaller warships use combined diesel and gas turbine (CODAG) propulsion. In a CODAG propulsion system, the diesel engine is used for normal cruising and the diesel and gas turbine engines are used for higher speeds or for rapid departure from port.

A freight-carrying ship needs to be very economical to operate, so the more expensive propulsion systems found in warships and cruise liners are not likely to be employed to transport freight. The high-efficiency, single large diesel engine driving a single propeller is likely to remain the favored means of propulsion for some time. The freight industry is, however, interested in looking at other augmentive propulsion systems that would draw from other sources of energy, such as wind power, to supplement the diesel-engine-powered propulsion systems. The SkySail system has been developed for use on vessels up to 500 tons, and it has proved capable of reducing fuel cost per mile (see Figure 8.17).

Figure 8.17 | Sail-assisted propulsion systems can be fitted to conventional ships to reduce fuel consumption.

Water Jet Propulsion A water jet propulsion system, illustrated in Figure 8.18, has an engine that drives a water pump, which forces water out of a nozzle. This jet of water propels the vessel forward. The advantage of the water jet propulsion system is that is does not have any external moving parts (such as propellers) and so can be used near people or in very shallow water.

Rail

The early wooden rails that allowed carts carrying coal to be hauled from mines by coal miners, and later, by horses, evolved into steel rails that allowed steel-wheeled wagons to be hauled with much greater

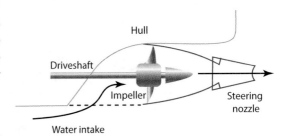

Figure 8.18 | A personal watercraft uses a water-jet propulsion system. The system shown employs a pressurized jet of water to propel and steer the vessel.

Section 2 △ Examining Transportation Systems

efficiency. This made it easier to haul coal over short distances using animal power, but the high cost of the rail infrastructure meant that the railroad was limited to distances of 5 to 10 miles at most. The advent of the steam locomotive changed that considerably. Suddenly, there was a means of harnessing considerable power to move people and goods over great distances. The railroad suddenly was a major competitor to the canals, and it had the advantage of being able to connect cities at a much lower cost than could canals.

Passenger Travel Until the 1940s, rail carried more passengers than any other form of transport in the United States. From that time onwards, the development of a comprehensive interstate network and the resulting convenience of road travel, along with the speed and affordability of air travel, have all greatly reduced the use of the national rail network for medium- and long-distance passenger travel.

The high cost of other means of travel has recently spurred interest in developing high-speed intercity train services in the United States. This interest is also driven by the fact that rail travel is relatively fuel efficient and has a much smaller impact on the environment than expansion of the road or air travel networks. The added benefit of reducing direct greenhouse gas emissions and reducing automobile accident rates are also factors to be considered.

High Speed Trains The highly traveled east-coast corridor between Washington and Boston is an ideal location for high-speed train travel. A high-speed train (HST) is a train that travels at top speed in excess of 90 mph. The Amtrak Acela high-speed train (HST) runs at speeds up to 150 mph from Washington, DC, to Boston. The Acela train uses tilting railcar technology to increase the ability of the train to take curves at high speed on the existing track. The Acela journey time of 6 ½ hours from Washington to Boston is quicker than flying in most cases. The Acela depends on the use of a highly automated, computer-controlled positive control system that uses track sensors and computer systems to control all the trains using the various sections of track to ensure that each train safely runs on its own track. The positive control system allows the train to work with the command, signaling, and control systems operated by six different rail track companies along the route of its journey, as well as being able to switch between different electrical power systems along the journey.

Very High Speed Trains In countries where major cities are closer and the passenger rail system is well developed, the rail system already provides a viable alternative to air travel. The French and the Japanese have built a comprehensive, very-high-speed train system (VHST). The Japanese Shinkansen, or bullet train, is shown in Figure 8.19. VHST systems use specially constructed tracks that are segregated from general and freight traffic and are designed exclusively for high-speed use. A semi-integrated HST/VHST system now serves Belgium, France, Denmark, Germany, Italy, the Netherlands, Spain, Switzerland, and the United Kingdom.

Rapid Transit Systems In urban areas, rail-based rapid transit systems have proved to be an invaluable addition to the public transportation system. Rapid transit systems are high-frequency railways that usually run on tracks that are separated from other forms of transport, sometimes using underground or elevated tracks. Large cities served by such systems have seen ridership grow over time as the inconvenience and cost of driving in cities increases. The Metro in Washington, DC, and the subway in New York City are well known examples of such systems. Developing rapid transit systems is complex and expensive and few cities have commenced construction of such a system in the last twenty years.

Figure 8.19 | The Japanese Shinkansen system extends for some 1,500 miles. These "bullet trains" average speeds of 200 mph.

Light rail Many cities have followed the lead taken by San Diego in 1981 when it developed its light-rail public transportation system. Light-rail systems use electric-powered trams or streetcars and often run in the medians of city streets (see Figure 8.20). Light rail is now found in many U.S. cities, and many more cities are developing such systems.

Freight Railroads in the United States are still principally used for moving freight traffic. Rail companies carry 16 percent of the nation's freight by weight and 28 percent by value—more than any other single form of transport. The rail system is well suited to heavy and bulky commodities and to the shipment of goods over longer distances. Coal, chemicals, intermodal containers, and automobiles comprise the majority of commodities shipped by rail.

Figure 8.20 | This Minneapolis Light Rail train crosses many intersections on its route.

Road Travel

The earliest roads were often muddy and deeply rutted as a result of the traffic traversing them, and when the weather was bad, journey times increased greatly. The Romans realized that, to move their armies from place to place quickly, they needed good roadways that were unaffected by weather conditions. Accordingly, they constructed roads of stone or deep beds of crushed rock that withstood the weight of their carts and drained when wet (see Figure 8.21).

Even though the technology to build better roads existed in the early 1900s, it was the massive growth in automobile ownership from the 1920s onwards that created a nationwide demand for a comprehensive network of roadways. As shown in Figure 8.22, the annual production of vehicles in the major manufacturing countries continues to grow. As a consequence, the number of vehicles on the road multiplies year by year, as do the miles driven by those vehicles.

Figure 8.21 | Cardo Maximus, the main Roman road in Jerash, Jordan, is still paved with its original stones.

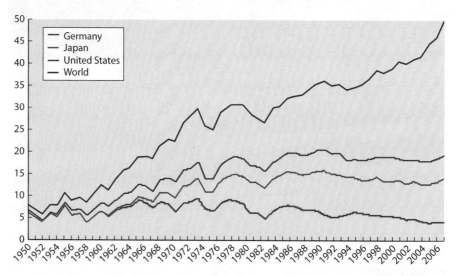

Figure 8.22 | This graph shows automobile production activity in the United States, Japan, and Germany from 1950 through 2006. The y-axis shows the number of automobiles expressed in millions.

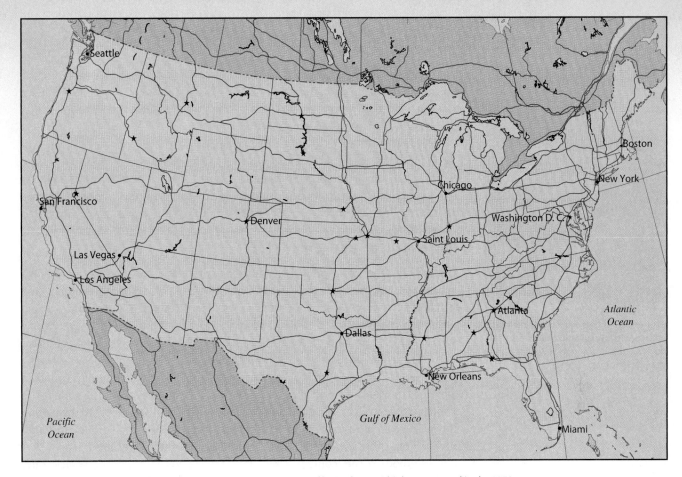

Figure 8.23 | Construction of the United States interstate system of limited-access highways started in the 1950s.

Road Infrastructure The ability to bear heavier loads has been combined with improvements in highway design so that traffic travels faster and can flow smoothly around and over natural obstructions on purpose-built highways, such as the interstate system. The highway system offers rapid travel over long distances between major cities across the nation. Many larger urban areas use limited-access divided highways to speed traffic flow around and across their cities in conjunction with the interstate system.

The greatly increased volume of road traffic is creating its own problems, with traffic overwhelming the main streets and highways of cities and urban areas. In the past, the solution was to build new roads. However, competition for land in and around cities and towns is fierce. Developers want to build homes and businesses, and the community often wants more leisure and recreation space. This conundrum is leading to the situation where urban planning has to be much better coordinated so that the growth in homes and businesses is coordinated with integrated transportation planning. An integrated approach to infrastructure development means that communities can explore integrating transit accessibility as a means of reducing reliance on the automobile, including rail-based rapid-transit systems and dedicated bus lanes.

A number of larger cities have realized that they are now reaching the capacity of their road-based transportation infrastructure and that building more roads is impossible, so they are resorting to new approaches to reduce vehicular traffic. They are charging the user to use the road by setting up road pricing systems.

Figure 8.24 | Cities such as London that use road pricing systems hope to cut down on traffic in congested areas during peak traffic times.

Road pricing operates in a very similar fashion to a toll road, in that users pay a fee to use the roadway. In cities using such schemes, vehicles are charged fees to enter the most congested parts of the city. London, England, has operated a Congestion Charging Zone for a number of years (see Figure 8.24). Cars and trucks entering the city during the working day are charged a fee. The system is automated for frequent users, in the same way as some toll roads in the United States use systems such as New York State's E-ZPass automated charging system.

After four years of road pricing, London has seen a 30 percent decrease in the number of vehicles entering the center of the city (from 260,000 to 190,000 vehicles each day), and journey times for those entering the zone have dropped by 15 percent. Those no longer using their cars seem to have moved to public transportation or car sharing to get to work. A similar scheme has been considered for New York City, which would extend the New York State E-ZPass system to parts of lower Manhattan during the workday.

Automobile The popularity of the modern automobile powered by an internal combustion engine (ICE) was born from the use of efficient assembly and manufacturing processes, which allowed the vehicles to be offered to a working person at a price they could afford. Up to that time, the automobile was a luxury only the rich could afford, and the vehicles were built using techniques that carried over from an earlier time when the wealthy purchased handmade horse-drawn carriages.

The annual production of vehicles in the major manufacturing countries continues to rise. The manufacturers of modern automobiles employ a wide range of technologies and materials to produce innovative cars that appeal to the buying public. The cars produced range from exotic sports cars like the Ford GT, to tough but luxurious off-road vehicles like the Hummer H2, to small city cars.

These modern cars contain a wealth of technology to make the vehicle safer and more efficient. The modern electronic engine management system measures multiple inputs, including barometric pressure, air temperature, engine temperature, exhaust gas temperature, engine air flow, engine speed, and engine load, and from these measurements, the system can calculate the amount of fuel to be injected into the cylinders at any time. The system also monitors the output of oxygen in the exhaust gasses to ensure that the correct mixture of fuel and air is being burned in the cylinders and to reduce the output of unburned hydrocarbons. Increasing fuel economy and reducing greenhouse gas emissions are high on the agenda of the automobile manufacturers. Unfortunately, the modern gasoline-powered, internal-combustion engine is very close to its optimum in terms of both economy and emissions. A number of automobile manufacturers are developing small cars for city driving. The reduced weight, combined with the use of small, fuel-efficient engines, means that these "microcars" are very economical to run (see Figure 8.25).

Figure 8.25 | The Smart Car started as an idea from the founder of the Swiss watch manufacturer, Swatch, who worked with Daimler-Benz to develop the car for production.

Freight Trucks The demand for hauling freight by truck is increasing. And the number of large trucks on the roads has increased to meet this demand. Over 70 percent of freight in the United States is moved by large truck—that is, a truck that can carry more than 4 tons. Freight trucks range in size from the single-unit truck carrying up to 51,000 lbs, to the much more common semi-trailer or short double-trailer truck carrying up to 80,000 lbs of freight (see Figure 8.26).

Buses Buses serve as the backbone of mass transit in most U.S. cities. City buses are vehicles specially designed for the stop-start pattern of urban routes. School

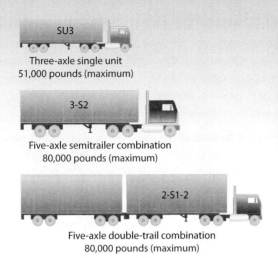

SU3

Three-axle single unit
51,000 pounds (maximum)

3-S2

Five-axle semitrailer combination
80,000 pounds (maximum)

2-S1-2

Five-axle double-trail combination
80,000 pounds (maximum)

Figure 8.26 | This figure shows the standard interstate configurations for freight trucks.

Figure 8.27 | This Los Angeles city bus is powered by compressed natural gas. The bus, featuring sixty seats and three doors, is part of the city's high-capacity vehicle program. Compressed natural gas produces fewer emissions than diesel.

buses serve a similar function for many students throughout the United States, as well. Buses designed for long distance travel serve travelers in the United States and throughout the world. Almost all buses use diesel engines, although some cities are experimenting with the use of compressed natural gas as a fuel for the diesel engine, as shown in Figure 8.27.

Diesel Engine The diesel engine is significantly more efficient than a gasoline engine producing the same power (46 percent of the energy contained in diesel fuel is converted to useful work, compared to 15 percent of the energy available from gasoline fuel). This reduces fuel consumption by a significant margin when comparing gasoline powered engines producing the same power (tests show fuel consumption reductions of up to 40 percent in ideal situations). There are some drawbacks, however. The diesel engine is heavier and more expensive to manufacture than the gasoline engine. In general, for uses in which an engine needs to generate high power outputs for long periods of time, the diesel engine is the preferred choice. However, the diesel engine generates a significant proportion of the nitrogen oxides and particulates found in the atmosphere.

Air

Compared to the water, rail, and road transportation systems, passenger and freight air services are a fairly new development. However, the growth of air travel has been very rapid. The aviation industry grew quickly in the 1920s, to the point where more aircraft were sold than automobiles in 1929. The crash of the markets in the 1930s brought a halt to the growth in general aviation for some time. Air passenger travel and air freight conveyance began growing rapidly again in the 1970s, as shown in Figure 8.28.

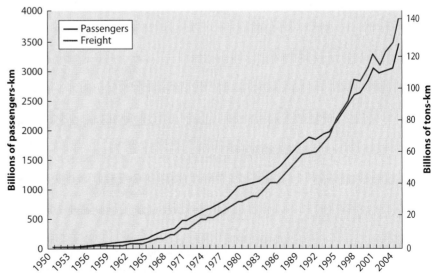

Figure 8.28 | Air transportation of people and goods has been growing rapidly since the 1970s, with slowdowns occurring during economic recessions and geopolitical instability.

Figure 8.29 | The Cirrus range of modern two- and four-seat general aviation aircraft are built using composites and modern avionics.

New technology and the use of modern composite materials (materials that combine chemically and mechanically to enhance the features of their component parts—for example, carbon fiber and epoxy resin) has allowed the development of a new generation of general aviation aircraft that are lighter, faster, and more fuel efficient (see Figure 8.29).

Commercial Aviation The commercial aviation industry was born as a result of public interest in this fast and convenient way to travel, plus the rapid development of reliable, multi-engine aircraft that could travel significant distances quickly and safely. The first airliners were propeller-driven aircraft with a seating capacity of ten to fifteen passengers, a top speed of 150–200 mph and a range of 500–800 miles. In 1935, the Douglas DC-3 revolutionized air travel with its ability to carry thirty-two passengers at a top speed of 240 mph at a range of over 1,000 miles. In 1930s, even the advanced-for-its-time DC3 took over twenty-one hours, with three stops for fuel, to fly thirty passengers from Los Angeles to New York. Today, it takes a Boeing 777 a little over five hours to fly over three hundred people on the same journey.

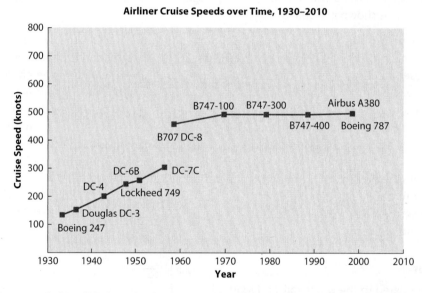

Figure 8.30 | This graph shows the changes in average cruise speeds in the commercial aviation industry from 1930 to 2010.

From those early beginning, the commercial aviation industry has grown considerably and continues to flourish. In 2006, the U.S. fleet of commercial aircraft was comprised of some 6,630 passenger aircraft and 1,000 cargo aircraft. The remarkable leaps in speed and efficiency of aircraft have reduced the cost of flying considerably. As a case in point, the relative cost of flying from New York to London has been reduced by a factor of ten since the early 1950s.

Aircraft Propulsion Aircraft propulsion systems have advanced at a remarkable pace. The early engines were heavy and unreliable. They had low power outputs and high fuel consumption. Gradual growth in engine power allowed the size of the airframe, the plane's outer structure, to increase. Passenger capacity increased as a result. By the early 1950s, the Douglas DC7 could carry 100 passengers at a top speed of 355 mph. Powered by four radial engines, each rated at 3,400 BHP, it had a range of 4,600 miles. This range allowed it to break a significant commercial aviation barrier: the capacity to cross the Atlantic Ocean without refueling.

The development of the turbojet engine gave aircraft designers a much lighter and more powerful engine. The turbojet engine uses a form of kerosene as fuel and mixes the fuel with compressed air inside a combustion chamber (see Figure 8.31). The rapidly expanding gasses drive the compressor and then exit the engine at high pressure through the exhaust nozzle, providing forward thrust for the aircraft. The latest generation of turbojet engines, known as high-bypass turbofans, uses a large bypass fan to force air out of the engine nozzle to provide propulsion (see Figure 8.32). All modern jet-powered commercial aircraft use high bypass turbofan engines, which are much more efficient and somewhat quieter than non-bypass turbojets.

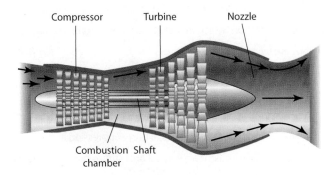

Figure 8.31 | The light but very powerful turbo-jet engine changed the face of commercial aviation.

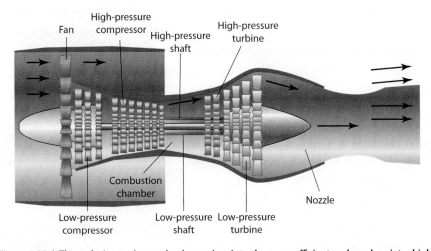

Figure 8.32 | The turbojet engine evolved over time into the more efficient and much quieter high bypass turbofan.

Modern Airliner Design Over time, the need to carry more people faster and more economically, while producing less pollution, has driven the demand for new aircraft. This same demand has created a market for more powerful and more fuel-efficient engines. Since an aircraft has to lift all its airframe and power plant into the air, it makes good sense to reduce the weight of the airframe and engines as well.

Boeing 787

▷ The Boeing 787 is a new twin-engine, wide-body medium-to-long-distance jet aircraft. The most striking feature of its design is the very high proportion of composites used in the construction. By weight, 50 percent of the aircraft is constructed using **composite materials** (80 percent by volume). Composite materials are a combination of two or more components, elements, or constituents, such as carbon fiber that is reinforced with plastic resins such as epoxy or polyurethane.

Composite construction offered many benefits to the designers of the Boeing 787, including:

- Made the aircraft strong and reduced the weight by up to 40,000 lb.
- Allowed the designers to improve the interior arrangements and increase the height of the windows, so passengers can look across the seatbacks and see out.
- Allowed an increase in cabin pressurization, without increasing material fatigue stresses.
- Increased cabin pressurization, which makes, it more comfortable for passengers since they feel like they are flying at a lower altitude.
- Reduced maintenance, particularly that associated with metal corrosion.
- Allowed the development of an aerodynamically efficient carbon-fiber wing.

Besides the extensive use of composites, the 787 uses very advanced flight control systems. They offer the ability to undertake vertical gust suppression and maneuver load reduction. If the aircraft enters turbulent air, vertical gust suppression will automatically dampen as much of the turbulence as possible. Maneuver load reduction uses the same systems to reduce the mechanical load on the wings and other control surfaces, caused when turning quickly or encountering turbulence. The engines are designed to be 8 percent more fuel efficient, and when combined with the reduced weight of the aircraft and its improved aerodynamics, it results in an aircraft that is 20 percent more fuel efficient. As a result, it produces 20 percent less pollution and greenhouse gas emissions.

ENGINEERING QUICK TAKE

Just How Much Freight Can a FedEx Aircraft Carry?

The Cessna Caravan, such as the one shown in Figure 8.33, is a single-engine turbo-propeller high-wing aircraft used for a variety of tasks. The U.S. Army uses it as an ambulance; some companies use it as a short haul corporate aircraft; and many companies use it to haul air cargo because of its rugged construction and its ability to lift heavy loads.

Even though it can carry a lot of freight, it is possible to overload the Caravan (just as it is possible to overload any aircraft in the world—they all have a maximum takeoff weight). The pilots of the caravan have to ensure that they are able to take off safely and have sufficient fuel to make the journey to their next stop. In this exercise, you will see just what the pilots have to do to ensure that the aircraft can make its journey safely.

Figure 8.33 | The Cessna Caravan is a turbine-powered aircraft that is often used to move freight from smaller airports to major freight hubs.

Since the Caravan must lift the weight of the airframe, the propulsion system, and its fuel into the air, as well as the weight of passengers or cargo, the pilot must be certain that the aircraft does not exceed the maximum takeoff weight.

The following lists the aircraft weight and distance information for a standard cargo Caravan:

Minimum empty weight (without fuel, freight, or passengers, but including all other items needed to fly such as engine oil, avionics, radios, etc.) = 4,570 lbs
Maximum takeoff weight = 8,750 lbs
Maximum landing weight = 8,500 lbs
Maximum fuel capacity = 335.6 gallons of Jet A fuel
Maximum usable fuel = 332 gallons
Maximum payload (the maximum weight of cargo that can be loaded, including the weight of the crew) = 3,980 lbs
Weight of Jet A fuel = 6.5 lbs per gallon
FAA required fuel reserve = enough fuel for 45 minutes cruise speed flying time
Fuel Burn (cruise) = 52 gallons/hour
Airspeed (cruise) = 175 knots
Nautical Mile (Knot) = 1.15 miles
Crew Weight (average for single pilot operation) = 200 lb

- Given the above, what is the maximum weight of cargo that can be carried at takeoff with full tanks?
- In the event of an emergency, how long would an aircraft with the maximum fuel load have to fly to reduce the weight of fuel so that the aircraft was at or just below the safe maximum landing weight?
- Assuming the aircraft's tanks were filled, what is the maximum range of the aircraft (including the fuel reserve) in miles (not nautical miles)?
- How far could the aircraft travel, at cruise speed, using the 45-minute reserve fuel load?
- How far could the aircraft fly in miles (not nautical miles) if loaded with the maximum payload?
- How far could the aircraft fly carrying a load of 2,000 lbs? Calculate the distance with and without the 45-minute reserve.

Space

The atmosphere above our heads extends into deep space. At night we can see distant stars and our own moon. Increasingly, we can see other objects that have been placed in orbit in the upper atmosphere that have been launched from Earth. For an object to remain in Earth orbit, it has to be placed in an orbit. Low Earth orbits (LEO) are in the range of 200–1,200 miles. Satellites in LEO will orbit the Earth about every 90 minutes. Most communications, broadcast, and weather satellites need to be placed in geostationary orbits. A geostationary orbit is one in which a satellite moves above the Earth's equator at a speed matching the Earth's rotation. These satellites maintain a constant position relative to a point on the globe while orbiting some 22,400 miles above the Earth's equator (see Figure 8.34).

A number of developed countries have military space programs and place surveillance and other satellites in orbit. Increasingly, space is being commercialized, and a large number of communication companies can exploit the very wide

Figure 8.34 | The GOES-10 (Geostationary Operational Environmental Satellite) is in orbit above the Earth's equator. Its position and movement allow it to detect natural hazards, such as severe storms, floods, drought, landslides, volcanic ash clouds, and wildfires.

footprint or viewable area of a single satellite placed in Earth orbit. From a single geostationary satellite, it is possible to broadcast a large number of audio or video channels to subscribers on Earth.

Humans in Space For a number of years, human space travel was more about testing the technology of near-earth orbits and firing the imagination of those on earth than it was about undertaking serious research or exploration of space. However, it was not long after the first human traveled into space that scientists and astronomers sought access to space to undertake extended research on topics that could not be easily studied on Earth.

The Soviet Union and the United States experimented with space stations in the 1970s and 1980s with the Salyut and Skylab programs. These were single-launch stations that could not be refueled and so had a limited life. The International Space Station (ISS), shown in Figure 8.35, is a research facility built as a collaborative venture between the space agencies of the United States, Russia, Japan, Canada, and the European Space Agency. Unlike earlier space stations, it can be expanded and is easily refueled and re-supplied by robot rockets that automatically dock to the station on a regular basis. The construction of the modular space station started in 1998 and is still progressing.

Commercial Space Travel Until recently, space travel by humans was limited to space agencies. A new development has seen wealthy individuals purchase flights to the ISS; however, access is so limited it cannot be classed as commercial space

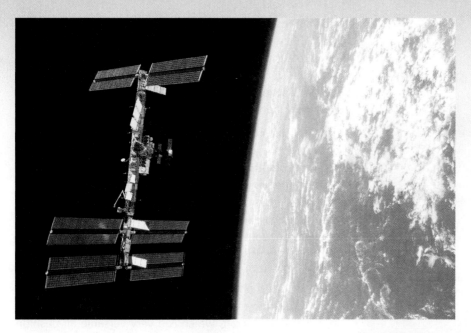

Figure 8.35 | The International Space Station is in a low-Earth orbit at an altitude of approximately 210 miles above the Earth.

travel. To date, five such space tourists have visited the space station. The recent sub-orbital flights by Scaled Composites' SpaceShipOne won the $10 million dollar X-Prize for the first privately developed and manned spacecraft to reach an altitude of over 60 miles twice within fourteen days.

The success of SpaceShipOne has spurred interest in a commercial service for space travelers who would like to make the same journey. The Virgin Galactic space tourism company, part of the conglomeration that also owns Virgin Atlantic Airways, is taking bookings from space tourists for flights in 2009, at a price of $200,000 for each brief flight into low Earth orbit space. The new spacecraft will be based on the technology used in SpaceShipOne (Figure 8.36). The first production model of SpaceShipTwo will be named VSS (Virgin space ship) Enterprise prior to commercial launch. SpaceShipTwo and its new carrier aircraft, White Knight 2, will be approximately three times the size of the original craft. SpaceShipTwo is designed to reach an altitude of over 80 miles. Virgin Galactic is already planning the development and launch of an orbital craft, SpaceShipThree and intends to extend its destinations to include travel to the moon.

Figure 8.36 | SpaceShipOne was carried underneath the fuselage of its launch plane, White Knight, to an altitude of 50,000 ft. and then released. It then fired its own rockets to reach an altitude of over 350,000 ft. (70 miles above the Earth).

Other Forms of Transport

In conjunction with the most common transportation systems, there are a number of other means of moving goods, materials, and people that are not so easily classified. Such systems often use a combination of transportation technologies in new or innovative ways to achieve their ends.

Pipelines A pipeline is the most economic means of transporting a fluid (a gas or a liquid) from one place to another for an extended period of time. Pipelines can be very small, for local distribution, or they can range up to 48 inches in diameter for interstate distribution. The contiguous forty-eight states are crossed by pipelines carrying fuel and gas. Most are local distribution pipelines for water and natural

Figure 8.37 | Pipelines are very expensive to build and require constant maintenance to ensure that they do not deteriorate over time.

Figure 8.38 | A hovercraft is a type of air-cushion vehicle (ACV). It travels on a controlled cushion of air that allows it to travel over water and a wide variety of terrain.

gas, but many are state-wide or inter-state and carry oil-based fuels and natural gas. The 799-mile-long Trans-Alaska Pipeline System pipeline, which is shown in Figure 8.37, is 48 inches in diameter and runs from Prudhoe Bay (which is located above the Arctic Circle) to Valdez. This pipeline can carry up to two million barrels of crude oil a day.

Air Cushion Vehicles Most people have seen how a sheet of paper, when dropped, can move horizontally over the floor and can sometimes seem to float for a great distance. This floating effect is well known to aircraft engineers. It is used to aid aircraft when landing, when the cushion of air under the aircraft wings allows the aircraft to float downwards more slowly than would otherwise occur at a higher altitude. In the 1960s, a British inventor applied the above principle of generating a controlled cushion of air under a vehicle so it could be lifted off the ground, combined with an aircraft-like propeller system to provide directional propulsion to invent the hovercraft, which is a very versatile air-cushion vehicle (ACV), pictured in Figure 8.38.

All current hovercraft designs use the same basic principles. The craft is surrounded by a very heavy-duty skirt of material that traps air so the craft will float on the air from the lift fans. Engine-driven fans generate lift by directing air under the craft that is trapped by the flexible skirt, and additional engines driving aircraft-like propellers provide directional propulsion. Although the hovercraft can move easily over smooth water or relatively flat land at high speed (up to 60 mph), it cannot easily move over objects or terrain that exceeds the depth of the flexible skirt, so rough water or very rocky terrain limit its ability to operate.

Local Movement Systems Elevators and escalators have a long history of moving people from one level to another with a minimum of personal effort. Moving walkways are a more recent innovation that allow large numbers of people to be moved horizontally or on a slight incline. The moving walkway offers travelers a convenient means of moving horizontally and at slight inclines or grades over distances of up to half a mile (see Figure 8.39).

Figure 8.39 | Moving walkways are usually found within large buildings such as airports, malls, or convention centers, where people need to move over significant distances at a safe speed. They use rotating belts running over rollers or glide ways to move people at walking speed (2–3 mph).

Human-Powered Vehicles Bicycles and other human-powered vehicles offer the opportunity to travel at much higher speed or with less effort. The bicycle has evolved considerably from the original metal-framed pedal-powered vehicle that was first seen in the mid-1800s (see Figure 8.40). The current Tour De France racing bicycle has a frame of carbon fiber composite construction. The entire bicycle can weigh as little as 5 lbs. The most popular form of bicycle is the mountain bike, which is rugged and capable of being used both on and off the road. The mountain bike is equipped with a wide range of gear ratios that allow the bike to be used both on level terrain, using a high ratio of wheel rotation to pedal rotations, or on steep inclines, using a low wheel-to-pedal ratio.

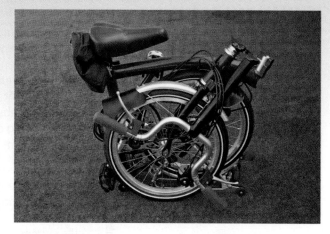

Figure 8.40 | The commuter or traveler who wants the convenience of a bike without the inconvenience of a full-sized one can make use of the folding bicycle.

SECTION TWO FEEDBACK >

1. What is the largest ship that can use the St. Lawrence Seaway?

2. What are the benefits of a propulsion pod to a captain of a ship?

3. Using the Internet, research the journey time for the Acela High Speed Train between Baltimore and Boston, and compare that with scheduled air service between the same cities. Don't forget to make allowances for travel to and from the airports in both cities.

4. Why does high-speed train travel offer environmental benefits?

5. What role does an engine-management system perform in a modern automobile?

6. What innovative material helps make the Boeing 787 so fuel efficient?

7. Research the type of satellite that would be placed in a geostationary orbit and those that are placed in low Earth orbit. Provide at least two examples for each orbit pattern.

SECTION 3: Investigating New Technologies

KEY IDEAS >

- The design of intelligent and non-intelligent transportation systems depends on many processes and innovative techniques.

- Since worldwide demand for oil is continuing to rise and oil reserves are declining, it is vital that the transportation industry increase the fuel efficiency of all of its vehicles or develop alternate fuels or new technologies to propel the vehicles, vessels, and aircraft of the future.

- The energy balance is the ratio of fossil-fuel energy used to make the fuel (energy input), and the energy in the fuel (energy output). If an alternative fuel has an energy balance of less than one, it is not a viable fuel, since more that one unit of fossil-fuel energy has been used to make one unit of alternate energy.

- Biofuels are fuels derived from renewable biological resources, such as plant material and organic waste products.

- The electric motor employs electric energy to generate mechanical energy in order to propel a vehicle. It is much more efficient than the internal combustion engine at converting energy into mechanical motion.

- The combined effects of increasing demand for transportation services and the environmental impact of the current systems are forcing the United States to reconsider how it can sustain its growth in this sector. The United States faces a diminishing domestic oil supply and a growing domestic demand. At the same time, there is a need to examine the environmental impact of this prodigious appetite for oil. The design of intelligent and non-intelligent transportation systems depends on many processes and innovative techniques.

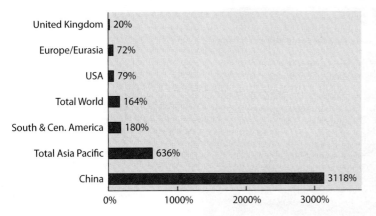

Figure 8.41 | This chart shows the growth in oil consumption for the years 1965 to 2000 in several parts of the world. Note China's relatively large increase in oil consumption.

Oil Dependency

The continued growth of the transportation industry is driving the domestic demand for oil. The transportation sector is one of the largest consumers of oil products in the United States. This usage will continue to rise in the future unless steps are taken to reduce consumption. United States oil production has steadily declined in the last forty years, since oil wells began to run dry. Increasingly, the United States must rely on imported oil to make up the shortfall. Increased oil demand is also seen elsewhere in the world. As can be seen in Figure 8.41, the growth in consumption in China is likely to continue to the point where they will soon be the largest consumer of worldwide oil production.

As the demand for oil is growing, there are a number of studies that show that worldwide oil production is likely to peak in the near future and then begin to decline. Since worldwide demand for oil is continuing to grow and oil reserves are declining, it is vital that the transportation industry increase the fuel efficiency of all of its vehicles or develop alternate fuels or new technologies to propel the vehicles, vessels, and aircraft of the future.

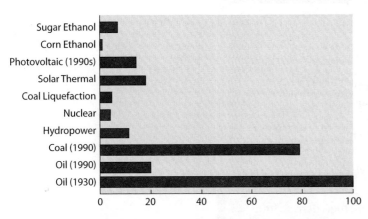

Figure 8.42 | This graph shows the energy balance or EROI (Energy Return on Investment) for various forms of energy. The EROI for an energy source is a ratio consisting of the amount of energy obtained from that source to the amount of energy expended to obtain it.

Alternate Fuels

What can alternate fuels offer us in the future? If we look towards the production of alternate fuels that can be used with current engine technologies, a number of potential sources can be explored. Large-scale production of agricultural crops that are converted into bioethanol and biodiesel is now underway in many countries across the globe. Small-scale production of gasoline and diesel oil from coal has been seen in the United States and Canada, as well. In the alternate fuel world, not all things are equal. For example, to make ethanol from plants requires energy to carry out the processing steps to produce a usable fuel. The Energy Return on Investment (EROI) is the ratio of the quantity of energy produced compared to the quantity of energy used to make the alternate fuel. If an alternative fuel has an EROI of less than one, it is not a viable fuel since more than one unit of fossil fuel energy has been used to make one unit of alternate energy.

Biofuels Biofuels are derived from renewable biological resources, such as plant material and organic waste products. Corn-based ethanol is an alcohol-based fuel produced by fermenting and distilling crops such as corn, barley, and wheat. In 2008, 12 percent of the U.S. corn production was used to meet 2 percent of the U.S. gasoline need. Although widely promoted as a means of reducing dependence on foreign oil, corn-ethanol is not likely to provide more than a small percentage (less than 10 percent at best) of the gasoline needed in the United States in the future.

Cellulosic ethanol is produced from agricultural plant wastes, sawdust, paper pulp, switchgrass, other "biomass" materials, and even municipal waste. The fact that cellulosic ethanol can be made from such abundant sources of raw materials makes this a very attractive alternative to corn-ethanol. Small-scale plants are in operation to help develop suitable technologies for large scale production of this alternative source of fuel.

Although steps are being taken to increase the efficiency of production, in 2007, corn ethanol was rated with an energy balance of between 1.0–1.5. Sugarcane has a much higher EROI of 6.0–8.0. Why the difference? Sugarcane can be fermented into ethanol directly, while the starch in corn has to be converted into a sugar before it can be fermented, a process that requires energy provided from other sources, including those that use fossil fuels. The high crop density that can be achieved with sugarcane and the very high use of surplus fiber leftovers, such as stalks, are a source of fuel for heating. There are some significant effects from the rise of interest in biofuel production, including the possibility of impacting world hunger and the destruction of the tropical forest (see Figure 8.43). The diversion of large quantities of plant products, particularly corn, from the world's food market to make biofuel can

Figure 8.43 | The deforestation in Brazil—and other countries that seek to create agricultural land for growing crops to be used in the production of biofuels—is leading to the loss of swathes of natural vegetation that had previously served to remove large quantities of carbon dioxide from the atmosphere.

also have a significant impact in terms of food prices elsewhere. If the production of corn ethanol grows at the current rate, it is very likely to lead to food shortages across the world, particularly in underdeveloped and poor countries.

Biodiesel is manufactured by processing oil-rich crops like canola and sunflower seeds, or by refining vegetable oils, animal fats, or recycled restaurant cooking oil combined with diesel fuel in various proportions. Although the energy balance for biodiesel manufactured from canola oil is in the 2.5 range, the quantity of biodiesel that can be produced by growing crops is physically limited by the land available to grow such crops. Research is being undertaken in the production of biodiesel fermented from algae that is sustained with waste energy from power plants. If the process can be developed to commercial production levels, it is hoped that yields of 5,000 gallons of biodiesel per acre could be achieved, which is considerably higher than any other plant-based yield.

The use of alternate fuels in the aviation industry, with its need for a powerful fuel for jet engines, is limited. However, the research into the use of algae to produce biodiesel has spawned some promising developments that might lead to a bio-source for jet fuel.

Compressed Natural Gas Compressed natural gas is gas that is pressurized to about 3,600 pounds per square inch and stored in large tanks. If this fuel technology is further developed, a local distribution pipeline would be needed to deliver the compressed natural gas (CNG) to vehicle filling stations. Currently, filling a vehicle with CNG calls for special precautions and training.

Liquid Petroleum Gas (LPG) Liquid petroleum gas is gas produced by refining oil or processing natural gas and then liquefying it for storage in tanks. It can be distributed by pipelines or by specialized trucks. In the United States, LPG is often known as propane. In certain countries (Australia in particular), vehicle manufacturers have produced dual-use vehicles that can use LPG and gasoline.

Hydrogen There is an increased interest in the use of hydrogen as a fuel: it is abundant, it emits water vapor, and it has the potential to replace oil as the primary fuel for the future. However, although abundant, hydrogen gas is not easily produced. Research is being undertaken to develop lower-cost methods of producing hydrogen, including the use of biomass gasification and the use of a solar-thermal process, where the energy of the Sun is concentrated on water vapor in a solar reactor vessel.

Hydrogen is difficult to store. For this reason, research is being undertaken to improve the quantity that can be carried in a hydrogen-powered vehicle. Hydrogen also needs special fueling equipment, and it takes longer to refuel with hydrogen compared to gasoline. Also, hydrogen lacks the extensive distribution infrastructure that is commonplace for gasoline and diesel fuels. However, if the extensive research being undertaken on developing a viable hydrogen fuel strategy is fruitful, then it is possible that new solutions to the issues of production, storage, and refueling can be developed.

Electric Vehicles

Electricity is a concentrated and powerful form of energy, and it can deliver more power in a smaller volume than similar mechanical systems, as shown in Figure 8.44. This very efficient energy form is being used in automobiles to power a wider range of accessories and systems than ever before. The electric motor utilizes electric energy to generate mechanical energy to propel a vehicle, and it is much more efficient than the internal combustion engine at converting energy into mechanical motion.

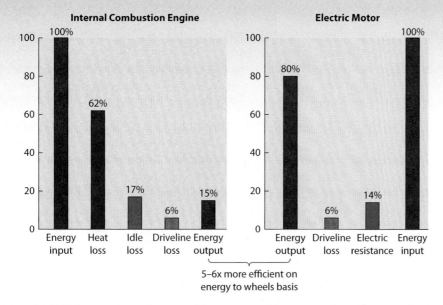

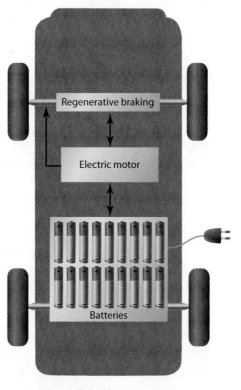

Internal Combustion Engine

Electric Motor

5–6x more efficient on energy to wheels basis

Figure 8.44 | This graph shows that most of the energy used by an internal combustion engine is lost as heat and other inefficiencies, leaving only 15 percent of the energy left to power a vehicle. By contrast, 80 percent of the energy produced by an electric motor is available to power a vehicle.

Battery-Powered Electric Vehicle (BEV) Battery-powered electric vehicles (see Figure 8.45) need to be plugged in to an external power source to recharge. BEVs were very popular in the early 1900s, as they proved to be easier to drive and maintain than the internal-combustion engine cars of the day. Improvements in the reliability of gas powered cars, as well as the development of electric starters, led to the eventual decline of these early electric vehicles. Today's plug-in battery-powered electric vehicles use energy stored in rechargeable battery packs to power highly efficient electric traction motors. They also use regenerative braking. In this type of braking system, the kinetic or motion energy of the vehicle is used to drive the electric motor to recharge the batteries and at the same time to slow the vehicle. The cost of operating an electric vehicle is about 3¢ per mile, compared to 12¢ per mile for a gasoline-powered car. The distance a BEV can travel on one charge is dependent upon the type of battery used. Some current models can achieve up to 200 miles on one charge.

Hybrid Technology Hybrid electric motive power is not new; it uses the same technology used in modern cruise ships and diesel locomotives. In a hybrid electric vehicle (see Figure 8.46), an internal combustion engine drives a generator, and the

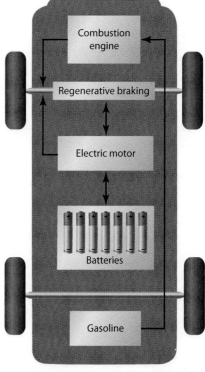

Figure 8.45 | A modern, plug-in battery-powered electric vehicle uses energy stored in rechargeable battery packs. Regenerative braking captures the vehicle's kinetic energy to recharge the batteries and slow the vehicle at the same time.

Figure 8.46 | An electric vehicle includes an internal combustion engine that drives a generator; the generator, in turn, is used to drive traction motors.

power from the generator is used to drive traction motors. Hybrid electric vehicles offer the benefits of more efficient electric traction without the range limitations of battery-powered electric vehicles. Hybrids often combine electric motors with a means of generating electrical power using a conventional gasoline or diesel engine.

There is a wide range of ways that such technologies can be combined, so hybrid vehicles fall into a number of sub-classifications.

- **Series Hybrid Technology**—The series hybrid electric vehicle exclusively uses traction motors for driving its wheels. The series hybrid stores electric energy in batteries and uses it to power its traction motors. The traction motor can receive power either from the batteries or from a combination of batteries and an onboard generator powered by an internal-combustion engine. The batteries are recharged by the engine-driven generator or from regenerative braking.
- **Parallel Hybrid Technology**—In the parallel hybrid, both the electric-traction motor and the internal-combustion engine can drive the wheels.
- **Hybrid Vehicles**—A full hybrid can be set in motion from a complete stop using stored battery power driving the traction motor. This ability to undertake an electric launch without having to use the gasoline or diesel engine, even if only for a short while, distinguishes a full hybrid from a mild hybrid.
- **Mild Hybrid**—In a mild hybrid, the vehicle cannot operate using batteries alone. The internal combustion engine needs to operate at times the vehicle is moving. The system does use regenerative braking, and adds features such as automatic engine stop-start when at a stop.
- **Power Hybrid**—A power hybrid has all of the features of a mild hybrid, with the exception that it is often equipped with a high horsepower internal-combustion engine.
- **Series-Parallel Hybrids**—In series-parallel hybrids, both the engine and traction motor can provide power to the wheels, but, in addition, the engine can be operated independently of the drive train to drive the generator.
- **Plug-in Hybrid Vehicle (PHEV)** The plug-in hybrid electric vehicle (see Figure 8.47), with its ability to store charge in more efficient batteries, combined with the ability to charge the batteries at home from a 110v outlet, offers the driver the ability to travel to and from work without using the engine at all. For longer journeys, the engine would then operate to charge the batteries or to provide supplementary traction when needed.
- **Fuel Cell Electric Vehicle (FCEV)** Fuel cell electric vehicles (FCEV) are powered by electricity and use electric traction (see Figure 8.48). Unlike hybrid vehicles, they create their electrical power by the use of fuel cells. Fuel cells create electricity by a chemical process that uses hydrogen and oxygen.

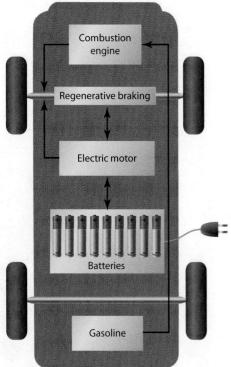

Figure 8.47 | A plug-in hybrid vehicle combines the benefits of a plug-in battery-powered vehicle with the ability to use the engine to recharge batteries on the road.

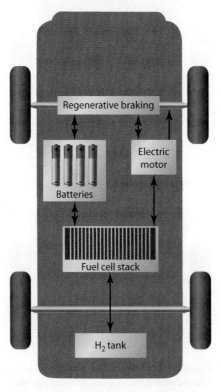

Figure 8.48 | Fuel cell electric vehicles are powered by fuel cells, which create electricity via a chemical process.

The hydrogen gas is stored onboard the vehicle in high-pressure tanks and the oxygen needed is obtained from the air. This technology offers a significant opportunity to reduce pollution and emissions.

TECHNOLOGY IN THE REAL WORLD:
Hydrogen and the Car of Tomorrow

Honda recently released hydrogen-powered FCX Clarity in California for drivers who want to lease the car for a three-year period. The FCX Clarity uses fuel-cell technology where the hydrogen is converted into electricity to power the traction motors.

The FCX Clarity uses a number of other energy saving techniques such as regenerative braking. The car is said to be twice as efficient as the current generation of full hybrid vehicles, and four times as efficient as a comparable gasoline-powered car.

The performance of the FCX Clarity is comparable to a car with 2.4 liter engine, and develops 134 BHP and 189 ft/lbs of torque. This compares well to the Honda Accord, with its 2.4 liter inline four engine developing 177 BHP and 161 ft/lbs of torque. The fuel consumption is listed as equivalent to 68 miles per gallon. The high-pressure hydrogen tank will hold enough hydrogen to travel 270 miles on a single tank.

Currently, leasing the FCX Clarity is restricted to a select few who live near the hydrogen refueling stations in California (there are three so far, in Torrance, Santa Monica, and Irvine). It is anticipated that as more hydrogen-powered cars take to the roads, the number of hydrogen refueling stations will grow.

In the meantime, in an attempt to limit the impact of the current limited number of hydrogen refueling stations and to develop an alternative approach to producing hydrogen, Honda has partnered with a U.S. technology company, PlugPower Inc., to produce a home energy station that uses a fuel cell system that converts natural gas into hydrogen, which, in turn, can be used to fuel a hydrogen-powered car. The unit also serves as a source of heat and electricity for residential use. This innovative system would ensure that the owner of the FCX Clarity could refuel the car overnight from the home-heating system.

SECTION THREE FEEDBACK >

1. Why is the United States so dependent on foreign oil? Investigate the various factors that have led the country to the point that it is dependent on foreign oil.

2. As a country, the United States is a very large consumer of energy. Are there any actions that we as individuals can take to reduce the consumption of energy in the United States?

3. Describe the concept of energy balance as it relates to alternate energy.

4. Using the Internet, examine the effect on our world food supply of using a large percentage of the U.S. corn crop for the production of ethanol.

5. What are the benefits of cellulosic-ethanol and algae-based biodiesel? Using the Internet, research these technologies and produce a paper predicting when they will have an impact on the need to find economic alternate-energy sources.

6. Why should we look to electric motors for powering our road vehicles in the future?

Matching Your Interests and Abilities with Career Opportunities: Airline Pilot

Pilots are highly trained professionals who fly either fixed-wing airplanes or helicopters to carry out a wide variety of tasks. Most are airline pilots, copilots, and flight engineers who transport passengers and cargo. One out of five pilots, however, is a commercial pilot involved in tasks such as crop dusting, spreading seed for reforestation, testing aircraft, flying passengers and cargo to areas not served by regular airlines, directing firefighting efforts, tracking criminals, monitoring traffic, or rescuing injured persons.

Significant Points

* Pilots usually start with smaller commuter and regional airlines to acquire the experience needed to qualify for higher-paying jobs with national or major airlines.
* Many pilots have learned to fly in the military, but growing numbers have college degrees with flight training from civilian flying schools that are certified by the Federal Aviation Administration (FAA).
* Earnings of airline pilots are among the highest in the nation.

Nature of the Industry

Except on small aircraft, two pilots usually make up the cockpit crew. Generally, the most experienced pilot, the captain, is in command and supervises all other crew members. The pilot and the copilot (also known as the first officer) share flying and other duties, such as communicating with air traffic controllers and monitoring the instruments.

Before departure, pilots must plan their flights carefully. They thoroughly check their aircraft to make sure that the engines, controls, instruments, and other systems are functioning properly. They also make sure that baggage or cargo has been loaded correctly. They discuss their flight plans with flight dispatchers and aviation weather forecasters to find out about weather conditions en route and at their destination. Based on this information, they choose a route, altitude, and speed that will provide the safest, most economical, and smoothest flight. When flying under instrument flight rules—procedures governing the operation of the aircraft when there is poor visibility—the pilot normally files an instrument flight plan with air traffic control so that the flight can be coordinated with other air traffic.

Takeoff and landing are the most difficult parts of the flight and require close coordination between the pilot and first officer. For example, as the plane accelerates for takeoff, the pilot concentrates on the runway, while the first officer scans the instrument panel. To calculate the speed they must attain to become airborne, pilots consider airport altitude, outside temperature, airplane weight, and wind speed and direction.

Unless the weather is bad, the flight itself is usually relatively routine. Pilots, with the assistance of the autopilot and the flight management computer, steer the plane along their planned route, monitored by the air traffic control stations they pass along the way. Pilots regularly scan the instrument panel to check fuel supply; engine condition; and the air-conditioning, hydraulic, and other systems. In contrast, because helicopters are used for short trips at relatively low altitude, helicopter pilots must be constantly on the lookout for trees, bridges, power lines, transmission towers, and other dangerous obstacles. Regardless of the type of aircraft, all pilots must monitor warning devices designed to help detect sudden shifts in wind conditions that can cause crashes.

Pilots must rely completely on their instruments when visibility is poor. On the basis of altimeter readings, they know how high above ground they are and whether they can fly safely over mountains and other obstacles. Special navigation radios give pilots precise information that, with the help of special maps, tells them their exact position. Other sophisticated equipment provides directions to a point just above the end of a runway and enables pilots to land entirely independent of an outside visual reference. Once on the ground, pilots must complete records about flights and aircraft maintenance status for their company and for the FAA.

Education and Training

Although some small airlines hire high school graduates, most airlines require at least two years of college and prefer to hire college graduates. In fact, most entrants to this occupation have a college degree. Because the number of college-educated applicants continues to increase, many employers are making a college degree an educational requirement. For example, test pilots often are required to have an engineering degree.

Pilots also need flight experience to qualify for a license. Initial training for airline pilots typically includes a week of company indoctrination; three to six weeks of ground school and simulator training; and twenty-five hours of initial operating experience, including a check-ride with an FAA aviation safety inspector. Once trained, pilots are required to attend recurrent training and simulator checks once or twice a year throughout their careers.

All pilots who are paid to transport passengers or cargo must have a commercial pilot's license with an instrument rating issued by the FAA. Helicopter pilots also must hold a commercial pilot's license with a helicopter rating. Airline pilots must fulfill additional requirements. Captains must have an airline transport pilot's license. Applicants for this license must be at least twenty-three years old and have a minimum of 1,500 hours of flying experience, including night and instrument flying, and they must pass FAA written and flight examinations. Usually, they also have one or more advanced ratings, depending upon the requirements of their particular job. Because pilots must be able to make quick decisions and accurate judgments under pressure, many airline companies reject applicants who do not pass required psychological and aptitude tests. All licenses are valid so long as a pilot can pass the periodic physical and eye examinations and tests of flying skills required by the FAA and company regulations.

Working Conditions

Because of FAA regulations, airline pilots flying large aircraft cannot fly more than 100 hours a month or more than 1,000 hours a year. Most airline pilots fly an average of 75 hours a month and work an additional 75 hours a month performing nonflying duties. Most pilots have a variable work schedule, working several days on, then several days off. Most spend a considerable amount of time away from home because the majority of flights involve overnight layovers. When pilots are away from home, the airlines provide hotel accommodations, transportation between the hotel and airport, and an allowance for meals and other expenses.

Commercial pilots also may have irregular schedules, flying thirty hours one month, ninety hours the next. Because these pilots frequently have many nonflying responsibilities, they have much less free time than do airline pilots. Except for corporate flight department pilots, most commercial pilots do not remain away from home overnight; however, they may work odd hours.

Commercial pilots face other types of job hazards. The work of test pilots, who check the flight performance of new and experimental planes, can be dangerous. Pilots who are crop-dusters may be exposed to toxic chemicals and seldom have the benefit of a regular landing strip. Helicopter pilots involved in rescue and police work often have to undertake hazardous flights in dangerous conditions. Although flying does not involve much physical effort, the mental stress of being responsible for a safe flight, regardless of the weather, can be tiring. Pilots must be alert and quick to react if something goes wrong, particularly during takeoff and landing.

Earnings

Earnings of aircraft pilots and flight engineers vary greatly depending on whether they work as airline or commercial pilots. Earnings of airline pilots are among the highest in the nation and depend on factors such as the type, size, and maximum speed of the plane and the number of hours and miles flown. For example, pilots who fly jet aircraft usually earn higher salaries than do pilots who fly turboprops. Airline pilots and flight engineers may earn extra pay for night and international flights. In May 2004, median annual earnings of airline pilots, copilots, and flight engineers were $129,250. Median annual earnings of commercial pilots were $53,870 in May 2004.

Airline pilots usually are eligible for life and health insurance plans. They also earn retirement benefits, and, if they fail the FAA physical examination at some point in their careers, they will receive disability payments. In addition, pilots receive a daily expense allowance, or "per diem," for every hour they are away from home. As an additional benefit, pilots and their immediate families usually are entitled to free or reduced-fare transportation on their own and other airlines.

Job Outlook

Job opportunities are expected to continue to improve with regard to regional airlines and low-cost carriers, which are growing faster than job opportunities with major airlines. Opportunities with air cargo carriers also should rise because of increasing security requirements for shipping freight on passenger airlines, growth in electronic commerce, and increased demand for global freight. Business, corporate, and on-demand air-taxi travel should also provide some new jobs for pilots.

Pilots attempting to join major airlines will face strong competition, as those firms tend to attract many more applicants than the number of job openings. Applicants also will have to compete with laid-off pilots for any available jobs. Pilots who have logged the greatest number of flying hours using sophisticated equipment typically have the best prospects. For this reason, military pilots often have an advantage over other applicants.

In the long run, demand for air travel is expected to grow along with the population and the economy. In the short run, however, employment opportunities of pilots generally are sensitive to cyclical swings in the economy. During recessions, when a decline in the demand for air travel forces airlines to curtail the number of flights, airlines may temporarily furlough some pilots.

[Bureau of Labor Statistics, U.S. Department of Labor, Occupational Outlook Handbook, 2008–09 Edition, visited June 25, 2008, http://www.bls.gov/oco/]

Summary >

The transportation sector is always evolving, often in response to the impact of new technologies or new demands from customers. Increasingly, it responds to major external factors, such as the increasing scarcity of oil—combined with its increasing cost—and the significant environmental impact of the industry. The modern transportation industry faces significant challenges, and it is likely that scientists and technologists will be developing a wide range of new and innovative systems to meet the needs of shippers and travelers for some time to come.

Although science and industry can do much to meet future transportation needs, it is likely that users of transportation services will also have to make significant adjustments. People must be willing to change how they live, produce goods, work, and travel to ensure that transportation becomes fully cost efficient and environmentally friendly.

FEEDBACK

1. Identify the major modes of the current transportation system. In addition, list the major benefits and limitations of each system in terms of its role in the transportation system.

2. Within the United States it is suggested that water-based transportation might have a role in regional distribution of movement of freight products, such as oil. Research the Atlantic Intercoastal Waterway, and explain how such a resource might distribute freight more economically than road transport. Also, explain the limitations of the current waterway.

3. Research the largest cruise liner currently in use. List the key features of the vessel, including number of cabins, maximum number of passengers, number of crew, tonnage, maximum speed, and type of main engines and propulsion system.

4. Over a four-week period, graph the fuel consumption of your primary vehicle (if you do not have a vehicle that you can use, chart the primary vehicle used by your family or a friend). The chart must show fuel purchased, cost per gallon, and the odometer reading at the point of each fill. Calculate the miles per gallon and the cost per mile for each fill. Try to start your chart with the tank as close to empty as possible.

5. Research driving methods and vehicle operation techniques that could be used to improve your fuel consumption. List the benefits and drawbacks for each method.

6. Research the variations in the price of a gallon of regular unleaded gasoline for the last ten years. Show your findings in a graph, and explain your findings fully.

7. Ultra-low sulfur diesel fuel is increasingly being used throughout the world. After consulting the Internet, explain why it is important that such fuels are used in terms of reducing pollution and improving cardiovascular health.

8. The Acela high-speed train connects the busy northeast corridor of cities: Research where other high-speed train systems might be used in the United States. In your research, list the cities, the population of each city, the relative distances between the cities, and, if you are able to do so, list the number of air passengers traveling between your selected cities each year.

9. There is a renewed interest in traveling to outer space, in general, and to the Moon, in particular. Research Internet sites, including NASA, to determine how the lunar landing program is developing. Include in your research detailed information on the launch vehicles, the lunar-based facilities, and the long-term goal of such an exploration program.

335

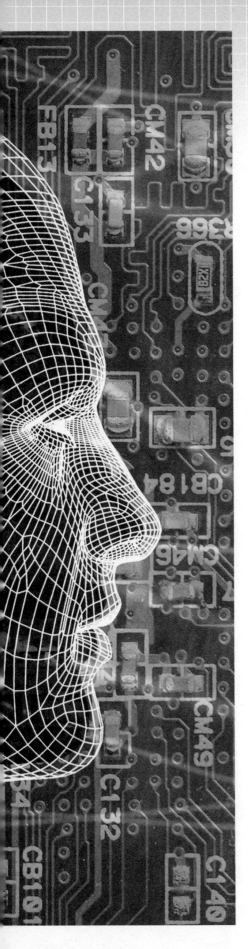

DESIGN CHALLENGE 1:
Building a Battery-Powered Electric Vehicle

• Problem Situation

Transportation of people and goods is essential for business, commerce, and leisure sectors to continue to be viable. Fossil fuel is a finite resource. The use of renewable or alternate energy sources is vital to our economy's future. The Sun is a vast source of energy that powers life on earth. It is now possible to harness that energy in ways that allow it be used to power vehicles. Solar cells can be used to collect energy from the Sun. That energy can then be turned into electrical power that can be used to charge batteries or power traction motors. Since solar power alone is not a viable energy source for operational road vehicles (just imagine what would happen after sundown), we must look to the development of a *Plug-In Electric Vehicle* model, using a stationary charging station coupled with a battery-powered electric vehicle.

• Your Challenge

As a team, you are to design and build a rechargeable-battery–powered electric vehicle that can carry a load of 1 lb. for a distance of thirty feet and can climb an incline of 1:50 that measures six feet in length as part of the thirty-foot distance. The vehicle must use rechargeable cells that are charged using a solar recharging station designed by the team. Using your *Student Activity Guide*, state the design challenge in your own words.

• Safety Considerations

1. Only use tools and machines after you have had proper instruction.
2. Wear eye protection at all times. Be particularly careful about using cutting tools, spraying paint, and applying finishes.
3. Make sure that all machines are used with all guards in place.
4. Be careful with electrical and electronic devices. Make connections first, and then turn on power. Take care to ensure there are no short circuits.
5. If using a power supply, be sure that the combination of battery chargers, batteries, and other devices that are connected to a power supply do not exceed the current capacity of the supply.

• Materials Needed

1. One 1.5v solar cell panel (250 ma minimum)
2. One small high-efficiency low-voltage electric motor (4.5–6.0v)
3. Two 4-cell battery holders (one for the vehicle and one for the charging station)
4. Four rechargeable AA-type cells
5. One pack of small assorted pulleys
6. One set of axles and wheels to suit vehicle
7. Insulated wire

You can also use other construction materials as needed to construct the chassis, steering system, and component retaining items.

• Clarify the Design Specifications and Constraints

To solve the problem, your team must design a rechargeable, battery-pack–powered vehicle. You can use up to four 1.5v rechargeable battery cells on the vehicle. The solar panel must be used to charge the rechargeable batteries. The duration of the charging period must be sufficient to fully charge the batteries. The chassis can be designed using any suitable materials and must have the ability to be steered. The steering can be fixed in place when undertaking the challenge. The transmission system can be designed using any suitable materials, including the assorted pulley pack.

• Research and Investigate

To complete this design challenge, you need to gather information on solar cells and how they can be used to charge rechargeable batteries. You will need to look at how cells can be connected to provide the voltage for a small electric motor. To ensure that the motor is able to power the vehicle, you will have to investigate a pulley-based transmission system since the motor produces most of its power and torque (turning force) at high RPM (revolutions per minute). Your vehicle will also need some form of steering, which must be adjustable before each run.

You will find a series of Knowledge and Skill Builder exercises in the *Student Activity Guide* that will help you in this endeavor.

• Generate Alternate Designs

In the *Student Activity Guide*, describe two or more possible solutions that might be used to meet the solar-car challenge. Discuss the decisions that you made.

• Choose and Justify the Optimal Solution

In the *Student Activity Guide*, explain why you selected the particular rechargeable car design solution, and why it is the best choice among the alternate designs. In any technological activity, you will use seven resources: people, capital, time, information, energy, materials, and tools and machines. In your guide, indicate which resources were most important in this activity, and how you made tradeoffs between them.

• Display Your prototype

To test your ideas and designs, you can experiment with the solar-panel recharging station design, as well as with the electric motor, power system and transmission, prior to mounting the various systems in a vehicle. From your tests, you can modify your chosen designs. Include descriptions and either photographs or drawings in the *Student Activity Guide*.

• Test and Evaluate

How will you test and evaluate your final design? In the *Student Activity Guide*, describe the testing procedures you will use. Indicate how the results will show how your rechargeable car design solves the problem and meets the specifications and constraints.

• Redesign the Solution

Respond to the questions in the *Student Activity Guide* about how you would redesign your rechargeable car solution, based upon the knowledge gained from the testing and evaluation process.

• Communicate Your Achievements

In the *Student Activity Guide*, describe the plan you will use to present your rechargeable car solution to the class. Show what handouts and/or PowerPoint slides you will use.

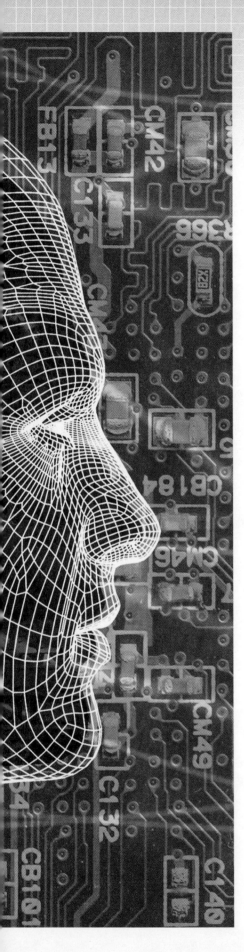

DESIGN CHALLENGE 2:
Designing a Hovercraft

• Problem Situation

Sometimes it is essential to be able to traverse a variety of terrains. When a search-and-rescue team needs to reach swampland or when communities are partially flooded, having a means of transport that can cross land, water, swamp, and paved roads can be lifesaving. The air cushion vehicle, often known as a hovercraft, is a very adaptable means of transport, and it is unique in its ability to travel over a wide variety of terrains. However, the hovercraft does present some real challenges to its designers: It looks like a land vehicle, but it must be able to cross water. The propulsion and control system look like they have come from an aircraft.

• Your Challenge

As a team, you are to design and build a hovercraft that can transport a driver and a supply load across a smooth surface. The total load must equal 150 lbs. The vehicle must be able to hover, plus move in any direction as controlled by the driver. The vehicle must be able to maneuver around the course, using parking lot cones as per the diagram in the *Student Activity Guide*. There is no time limit, but if this is a team activity, then time to complete the course could become a factor in determining success. Using your *Student Activity Guide*, state the design challenge in your own words.

• Safety Considerations

1. Only use tools and machines after you have had proper instruction.
2. Wear eye protection at all times. Be particularly careful about using cutting tools, spraying paint, and applying finishes.
3. Make sure that all machines are used with all guards in place.
4. If you are handling large sections of plywood, ensure that you wear gloves. Get assistance with maneuvering large boards.
5. Be careful with all electrical connections. Make connections first, and then turn on power.
6. If you are using your hovercraft outside of the classroom or workshop area, ensure that every precaution is taken to avoid water. The electrical cables must not contact any wet surfaces, puddles, etc.
7. All electrical connections must be made using a Ground Fault Circuit Interrupt (GFCI) protected power strip.

• Materials Needed

1. One sheet of 8 foot by 4 foot by ½" plywood sheet
2. Five lengths of 8 foot by 6 inch by 1 inch pine board
3. Two electric leaf blower unit
4. One large sheet of 20 mil (or thicker) plastic tarpaulin or plastic sheet.
5. One 4-outlet Ground Fault Circuit Interrupt Protection (GFCI) power strip with ON OFF control
6. One hairdryer with cold air setting or small workshop vacuum with a blower outlet for prototype development

You can also use other construction materials as needed.

• Clarify the Design Specifications and Constraints

To solve the problem, your hovercraft must use some form of air cushion, but the design of the cushion and how the cushion is secured is left open for the design team. The wood-panel chassis size can be varied a little, but the overall shape should remain roughly circular.

The hovercraft must be steerable. The propulsion is likely to be provided by one of the leaf blowers. The team must design a steering system that will direct the energy from the leaf blower propulsion motor. The steering system can use moveable rudders controlled using a joy stick, or a pivoting system that rotates the leaf blower mounting assembly.

The team might look at experimenting with control systems that allow for the hovercraft to be slowed as it moves forward. Don't forget, hovercrafts do not have brakes, so be sure to ask yourselves, "How do they stop?"

The lift is likely to be provided by a leaf-blower unit. This can be mounted where convenient to the driver. The driver must be able to turn off the lift motor quickly in the event of an emergency: Removal of lift will bring the hovercraft to a swift and controllable halt very quickly.

In the event that a load is needed to make up weight, it can comprise canned goods or similar items. The load can be stored anywhere on the hovercraft in such a way that it does not interfere with the steering or operation of the craft and does not upset the balance of the craft.

All electrical connectors MUST be fed through a ground-fault, circuit-interrupt–protected power strip. The power strip MUST be mounted within easy reach of the driver so the entire hovercraft can be quickly shut down in the event of driver difficulty. A minimum of three people should be monitoring the power cords being used to provide power from the wall outlet to the moving hovercraft. The person observing the wall outlet should be prepared to disconnect the power in the event of difficulties.

• Research and Investigate

To better complete the design challenge, you need to gather information on how a hovercraft works, as well as on how the skirt materials can be best fixed to the body, in order to design a solution. You will find a series of Knowledge and Skill Builder exercises in the *Student Activity Guide* that will help you in this endeavor.

• Generate Alternative Designs

In the *Student Activity Guide*, describe two or more possible solutions to the design of the hovercraft.

• Choose and Justify the Optimal Solution

Using the *Student Activity Guide*, explain why you selected the solution you did, and why it is the best choice. In any technological activity, you will use seven resources: people, capital, time, information, energy, materials, and tools and machines. In your guide, indicate which resources were most important in this activity, and how you made tradeoffs between them.

• Display Your Prototype

The design of the hovercraft skirt is critical to the development of a successful hovercraft. Using a 10"–12"-diameter wooden disk with a hole cut in the base, it will be possible to design and test a variety of skirt designs. A hairdryer set to blow cold air can be used as the source of the air blowing into the hovercraft base. From your tests, you can modify your chosen design in the *Student Activity Guide*.

• Test and Evaluate

How will you test and evaluate your final design? In the *Student Activity Guide*, describe the testing procedures you will use. Indicate how the results will show that the design solves the problem and meets the specifications and constraints.

• Redesign the Solution

Respond to the questions in the *Student Activity Guide* about how you would redesign your hovercraft and what further improvements could be made, based upon the knowledge gained from the testing and evaluation process.

• Communicate Your Achievements

In the *Student Activity Guide*, describe the plan you will use to present your completed hovercraft design solution to the class, as well as the improvements that you would make in the future. Show what handouts and/or PowerPoint slides you will use.

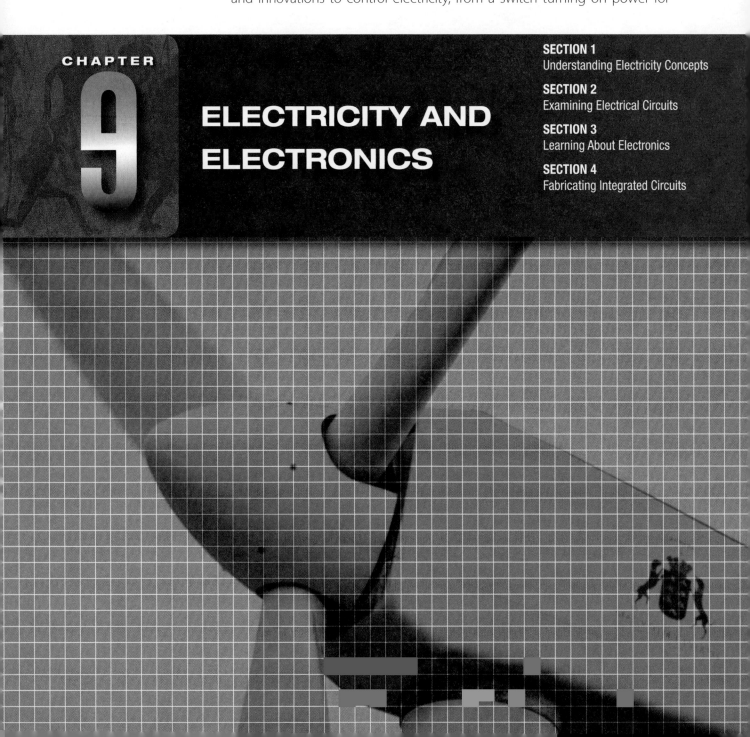

INTRODUCTION

PREVIOUSLY YOU LEARNED that one useful form of energy is electrical energy. Electricity's powerful force is caused by electrons moving from one point to another. Learning the characteristics of electricity makes it possible for us to direct how and when electrons flow, which enables us to use electricity beneficially. Engineers, technicians, and scientists have developed many devices and innovations to control electricity, from a switch turning on power for

CHAPTER

9

ELECTRICITY AND ELECTRONICS

SECTION 1
Understanding Electricity Concepts

SECTION 2
Examining Electrical Circuits

SECTION 3
Learning About Electronics

SECTION 4
Fabricating Integrated Circuits

lights to a computer commanding roving robots on Mars. There are countless other examples of innovations in the field of electricity and electronics that make our lives interesting, safe, and productive.

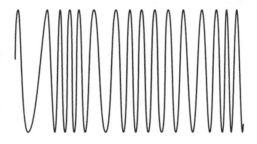

Figure 9.1 | Signals generated by computer circuits (shown on the right) are different from electrical signals generated by an FM radio station (shown on the left).

You will learn in this chapter that electricity is generated in many forms and that each form has unique characteristics. As an example, the signals generated by computer circuits are different from the electrical signals generated by an FM radio station, as shown in Figure 9.1. Engineers and technicians learn about the signals, circuits, devices, formulas, and equipment that are specific to the field of electricity and electronics. In the next few chapters you will learn some of the same basics.

Electronics deals with devices and specialized circuitry to control electricity. Semiconductor devices such as the transistor have revolutionized the industry and form the foundation for electronic products. Integrated circuits, which are complex devices built using millions of transistors, are the backbones of such modern products as the computer, cell phone, and Internet router (see Figure 9.2 for an example of a transistor and an integrated circuit).

Transistor   Integrated circuit

Figure 9.2 | Transistors and integrated circuits are vital to modern electronics devices.

The subjects of electricity and electronics are fascinating to learn. The next several sections and chapters introduce you to electricity concepts and interesting applications in the computer and communications fields.

SECTION 1: Understanding Electricity Concepts

KEY IDEAS >

- We use the energy from electrons to do work.
- Atomic structure determines an element's electrical properties.

Have you ever been "zapped" by static electricity? Shocking, isn't it? When it happened, you felt the force of electricity as electrons moved from your body to the object you touched. This force is what enables electricity to do work for us. This section examines atomic structure, the starting point for understanding the nature of electrical force.

What Is Electricity?

A fan turns and blows cooling air, as shown in Figure 9.3. A spinning motor causes fan blades to move air, but electricity causes the motor to spin. The motion derived from a reaction between electricity and magnetism provides the energy that causes the motor to spin. The motor is simply a device created to convert electron energy into rotary motion. The value of electricity is that we use the energy from electrons to do work.

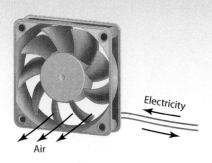

Figure 9.3 | Electricity provides energy to move air.

Motors, which are found in fans, refrigerators, robots, conveyors, and other products, are just one type of electrical device invented to productively use electricity. Electricity also powers our computers, televisions, DVD players, and MP3 players. It runs traffic lights, street lights, house lights, and flashlights. Electricity is a powerful and useful force used in many of the things we take for granted every day. Understanding electricity requires that we investigate elements of nature we cannot even see. We will learn that atomic structure determines an element's electrical properties.

The primary component of electricity, the electron, is a negatively charged particle found in all atoms. Atoms are the basic component of the approximately 115 elements that make up all matter. One way to understand how an atom's structure relates to electricity is to use a model that shows how the atom is put together. The model we use is a drawing of an atom (see Figure 9.4), which consists of a central area called the nucleus. The nucleus contains two of the three main atomic particles: the proton and the neutron. From an electricity standpoint, the charge on these two particles is important. The proton has a positive charge and the neutron is neutral, which means it has no charge. The notion of charge is important because electricity flows from one charge level to another. Clearly, the nucleus has only one particle with charge, so we need to find another particle with a different kind of charge.

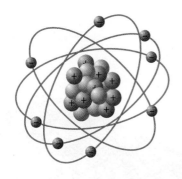

Figure 9.4 | Atoms are the basic components of all matter.

Now focus on the third atomic particle, the electron. Each atom may have one or several electrons. They are not trapped within the nucleus with the protons and neutrons. Electrons instead orbit the nucleus, much as a satellite orbits the earth. In terms of size, electrons are the smallest of the three particles and are relatively far away from the nucleus. They exist at specific distances from the nucleus, and each particular atom has a certain number of electrons that are a certain distance from the nucleus. For instance, the atom carbon has six electrons and the silicon atom has fourteen electrons. Every atom has a matching number of electrons and protons. Scientists use this number to classify the atom.

The structure of one iron atom, for instance, is the same as the structure of every other iron atom, but it is different from the structure of a silicon or hydrogen atom. Just as the number of electrons varies from one element to another, so do the placements or orbits of the electrons. Atoms can have many orbits for their electrons, and the greater the number of electrons, the greater the number of orbits. Each orbit can hold several electrons, and the exact number can be determined mathematically. If you looked at models of each atom in the periodic table of elements, you would notice that electrons populate the orbits closest to the nucleus. Each orbit can hold a certain number of electrons; only when a given orbit holds as many as nature allows will we see a more distant orbit used. An important point to appreciate is that the distance between the nucleus and an electron in orbit is related to the electron's energy. The further an electron is from its nucleus, the greater its energy.

While an atom may have many electrons distributed across several orbits, of particular interest are the electrons in the outermost orbit of atoms (called the valence orbit). Electrons in the valence orbit are the furthest from the nucleus, and they possess the greatest energy, as shown in Figure 9.5. These electrons also react with electrons from other atoms and thereby help determine how one element reacts with another. Valence electrons can be moved from their orbit by externally applied energy from heat or a battery. Once outside their orbits, these electrons are free to move away from their parent atom. The movement of these free electrons is electricity.

Every material is made up of many atoms. Atomic forces bond atoms to one another and create the solids, liquids, and gasses we use to manufacture the products we want and need. For materials to be useful in electrical devices, certain properties are necessary. For example, some materials easily permit electricity to flow through them, but other materials inhibit the flow of electricity. We classify materials as conductors, insulators, or semiconductors, depending on how they permit electricity to flow.

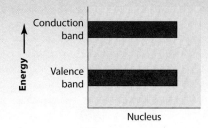

Figure 9.5 | Electrons in the valence orbit are the furthest from the nucleus and possess the greatest energy.

Conductors, Insulators, and Semiconductors

A material is said to conduct electricity if it easily allows electricity to flow. Such materials are conductors. Metals such as copper, iron, and gold are very good electrical conductors and are frequently used in electrical circuits. We can see why metals are good electrical conductors by examining the atomic structure of metal atoms. Specifically, the number of electrons in an atom's valence orbit determines its conductivity (see Figure 9.6). Remember that valence electrons have high energy because of their great distance from the nucleus. This explains why it is easy for valence electrons to become the moving electrons in electricity. A valence electron requires very little additional energy to move from its orbit to a higher energy level outside the atom, which is called the "conduction band." This means that a valence electron in a conductor may free itself from its attraction to the nucleus of its parent atom. At room temperature, thermal energy alone provides many valence electrons with the boost they need to jump to the conduction band. Here, the freed electrons move easily throughout the metal's structure. For example, in a length of copper wire made from billions of atoms, millions of "free electrons" are available for conduction.

Electrons in the conduction band are available to support the flow of electricity. When a conductor is connected to the electrical force provided by a battery, for instance, conduction band electrons begin moving and electrical flow is achieved. Electrical conductors have a natural supply of electrons ready to go.

You must realize that sometimes a flow of electricity is undesirable. For example, imagine a table lamp plugged into an electrical outlet with two bare wires. Wires that are not covered by insulating plastic create a dangerous situation by exposing people to a serious shock hazard. Therefore, we sometimes need materials that are the opposite of conductors; in other words, materials that resist the movement of electrons and inhibit the flow of electricity. Such materials are insulators because they do not readily conduct electricity. Materials with insulating properties protect

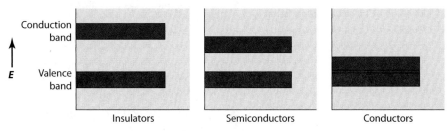

Figure 9.6 | Different conduction and valence energy bands exist for insulators, semiconductors, and conductors.

people and equipment from electric currents. Certain elements are inherently insulators, but most practical insulating products are combinations of several insulating materials.

Because we investigated how conductors function at the atomic level, we will do the same with insulators to see why some atoms do not provide free electrons to the conduction band. The reason for this insulating behavior again resides in the makeup of the valence orbit.

Argon, for example, is an insulating element. Argon's valence orbit has eight electrons, far more than in a metal. In nature, eight electrons happen to be the precise number needed to fill all the available energy levels in the valence orbit. In an atom with eight electrons, a single electron requires a significant amount of energy to become free and move to the conduction band. At room temperatures, very few electrons attain enough energy to escape to the conduction band. Therefore, insulators have few free electrons; without free electrons, there is no flow of electricity even when electrical force is applied.

The third class of electrical materials is semiconductors. As the name implies, semiconductors are not good conductors of electricity, nor are they good insulators. Nevertheless, they are extremely useful electrical materials.

Silicon atoms, as an example, are semiconductors. Silicon atoms at room temperature have a small number of free electrons, an insufficient number to conduct electricity well. Semiconductor materials have four valence-orbit electrons, which you will notice is exactly half the number between the full orbit of insulators and the nearly empty orbit of conductors. When semiconductor materials combine with other elements, manufacturers can create devices such as transistors and integrated circuits. These devices are discussed later in this chapter.

SECTION ONE FEEDBACK >

1. Can you explain the difference between conductors and insulators?
2. Why does it make sense that electricity is the movement of electrons rather than the movement of protons?
3. Why is copper used to connect electrical devices together?
4. What part of an atom determines the electrical property of a material?

SECTION 2: Examining Electrical Circuits

KEY IDEAS >

- Schematics are circuit diagrams that show how electrical components are wired together.

- Ohm's Law is a simple equation, which is algebraically arranged to solve for any of three properties: voltage, current, or resistance of a circuit

- Electronic gear is built using two basic circuit forms called the series circuit and the parallel circuit. All complex circuits are combinations of the two.

- Electricity exists in two basic forms: direct current (DC) and alternating current (AC).

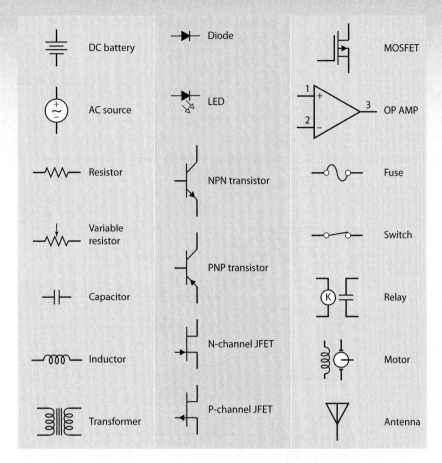

DC battery	Diode	MOSFET
AC source	LED	OP AMP
Resistor	NPN transistor	Fuse
Variable resistor	PNP transistor	Switch
Capacitor	N-channel JFET	Relay
Inductor	P-channel JFET	Motor
Transformer		Antenna

Figure 9.7 | **Schematics make it easy for people to analyze, assemble, test, and troubleshoot circuits.**

Our investigation of atomic structure gave us the background to understand electricity. Engineers and technicians use this knowledge to analyze how electrical circuits function. Using specialized equipment and components, they are able to design, test, and troubleshoot real-life products and systems.

Electrical and Electronic Symbols

The many parts used in the electronics industry are represented by symbols that can be used in diagrams showing how an electrical circuit is wired. Such diagrams are called schematics, which are blueprints for electrical and electronic circuits. Schematics are circuit diagrams that show how electrical components are wired together. On the schematic, symbols represent parts, and lines represent wires. These schematics make it easy for people to analyze, assemble, test, and troubleshoot circuits. When schematics are created with a computer, engineers use specialized software programs to simulate how the circuit will operate. This is handy in the design process because errors are spotted long before actual hardware is assembled.

Figure 9.7 shows many of the electrical symbols used in schematics. Each symbol's shape depicts a unique electrical component. Numbers and other identifying information, placed next to the symbol on the schematic, show important electrical characteristics of the component. For instance, consider the schematic for the common circuit shown in Figure 9.8.

This circuit uses several symbols to show how two typical 1.5-volt batteries are connected to a switch and a 3-volt light bulb. Later we will see exactly how this circuit operates, but for now, you see that the schematic shows how to wire or

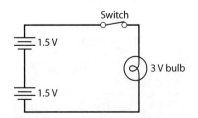

Figure 9.8 | **This simple electrical schematic shows how two typical batteries are connected to a switch and a light bulb.**

interconnect the components. Can you change the circuit schematic to show how two 1.5-volt bulbs could be used instead of the 3-volt bulb?

Block Diagrams

▷ A type of schematic called a block diagram shows the connections of a product's major subsystems. A subsystem is the circuitry for an important portion of a product. Block diagrams do not show each electrical component in the design, but they do give an overall picture of how the product subsystems fit together. The block diagram for a video camera, shown in Figure 9.9, is an example showing the important subsystem circuits needed to create a useful product.

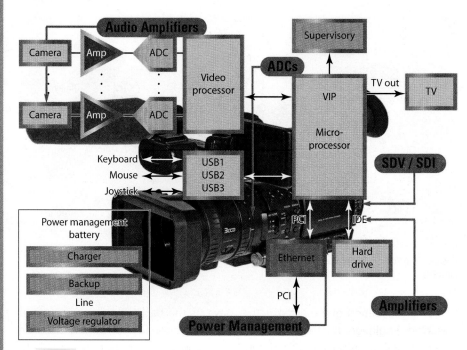

Figure 9.9 | A block diagram shows how subsystem circuits connect to provide the important functions for a complex electronic product, such as a video camera.

Voltage, Current, and Resistance

Engineers and technicians figure out how electricity flows and reacts by acquiring knowledge about electrical circuits. This section begins by defining a few important terms.

We previously learned that conduction band electrons move easily through conductive materials by the application of external energy. In an electrical circuit, a voltage source provides the energy to move electrons. Voltage is the electrical pressure that pushes electrons through an electrical circuit. The common battery is a voltage source. You probably know that batteries come in different sizes, shapes, and voltages; some common battery voltages are 1.5 V and 9 V. Notice that the letter V is used to represent voltage. Some books also use the letter E to represent a voltage source.

Batteries (see Figure 9.10) produce energy through a chemical reaction. The battery's terminals are the two points at which wires are connected. The electrons move rapidly from the negative terminal toward the positive terminal when a complete circuit or electrical path is in place. We use the term circuit to describe device connections that form paths along which electricity flows. In a later section of this chapter, we will take a closer look at the characteristics and operation of electrical circuits.

Figure 9.10 | Batteries are a common source of electrical voltage.

Consider a 9 V battery installed in a clock. It produces an electron flow in the clock's circuitry. With this image in mind, we introduce the conventional term for the flow of electrons, which is current. The greater the number of electrons flowing in a circuit, the greater is the current.

The letter I (for intensity) is used in formulas and on schematics to represent current. We measure the amount of current flowing in amperes, or amps for short. Typical circuits have current levels that range from several amps to as little as one-millionth of an amp (stated as one microamp, or 1μA, see Figure 9.11 for other common prefixes). A 100-watt light bulb, for instance, draws approximately 0.8 amps.

Prefix	Prefix Value	Prefix Symbol	Prefix Name	Example
Femto	10^{-15}	f	Quadrillionth	25 femtosecond = 25×10^{-15} s
Pico	10^{-12}	p	Trillionth	33 picojoules = 33×10^{-12} j
Nano	10^{-9}	n	Billionth	65 nanometers = 65×10^{-9} m
Micro	10^{-6}	μ	Millionth	45 microamps = 45×10^{-6} A
Milli	10^{-3}	m	Thousandth	12 milliseconds = 12×10^{-3} s
Kilo	10^{3}	k	Thousand	26 kilovolts = 26×10^{3} V
Mega	10^{6}	M	Million	115 megawatts = 115×10^{6} W
Giga	10^{9}	G	Billion	2.3 gigahertz = 2.3×10^{9} Hz
Tera	10^{12}	T	Trillion	80 terabits = 80×10^{12} b

Figure 9.11 | The electronics industry uses many common metric prefixes.

When a circuit is formed, electrons flow from the negative voltage source terminal, through the circuit components, to the positive voltage source terminal. When electrons flow through the components within a circuit, useful things happen. For example, if the electronic components in a circuit provide amplification, the flow of electrons might enable us to hear the music recorded on a CD.

You may be wondering what limits the amount of current flowing through a circuit. This is a good question to ponder because too much electric current can be a safety concern. For example, fire can break out if too many electrical devices are plugged into the same extension cord (see Figure 9.12). In this example, each electrical device operates by drawing current, and all current flows through the

Figure 9.12 | An overloaded circuit can be dangerous.

same extension cord. As electrons flow, energy is released as heat, and too much heat causes materials to ignite.

Resistance controls current levels. Increasing resistance will decrease current, so when little current flows, we say resistance is high. When current is high, resistance is low. Because all materials have resistance, so do electrical components. Resistance is measured with a unit called the ohm, and is represented by the letter R. The Greek letter omega (Ω) represents the amount of resistance in ohms. For example, we might say that a certain electrical component has a resistance of 2,500 Ω, or that R = 2,500 Ω.

It is useful to visualize electric current as billions of particles moving through a pipe. Something pushes the particles to get them moving, and something else is needed to control how many particles will move. In electrical circuits, voltage sources do the pushing, and resistance controls the number of moving particles or current.

A second useful visualization is comparing electricity to water, as in Figure 9.13. Water flow through a pipe is similar to current flow through a wire. Water moves through a pipe because of pressure exerted on the water by gravity or a water pump. This pressure is like a battery's voltage potential pushing electrons. The diameter of a pipe in a water system creates resistance, which determines the amount of water flowing. In a circuit, both the resistance created by the wire, and the components used, control the amount of current flowing. When you consider this water analogy, it is easy to recognize that a wider pipe lets more water flow than a narrow-diameter pipe. Keep this analogy in mind when analyzing electrical circuits.

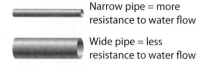

Narrow pipe = more resistance to water flow

Wide pipe = less resistance to water flow

Figure 9.13 | Water flow through a pipe is similar to electrical-current flow through a wire.

Ohm's Law

In our previous discussion, we determined that voltage, current, and resistance are all related. George Ohm mathematically related these three properties through experiments he conducted in the 1820s. The results, now called Ohm's Law, form the basis of most circuit design.

Ohm's Law is a simple equation that is algebraically arranged to solve for any of three properties: voltage, current, or resistance of a circuit (see Figure 9.14). The three forms of Ohm's Law are:

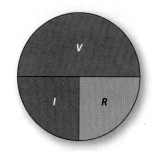

Figure 9.14 | Ohm's Law can be presented by a circle.

$$V = I \times R \qquad \text{(voltage = current multiplied by resistance)}$$
$$I = V/R \qquad \text{(current = voltage divided by resistance)}$$
$$R = V/I \qquad \text{(resistance = voltage divided by current)}$$

An example will show how to use Ohm's Law. Assume that a 9 V battery is powering an MP3 player. When the player is on, a current of 0.01A flows from the battery. What resistance does the MP3 player present to the battery so that the current is limited to 0.01A? Using Ohm's Law, we can calculate that the MP3 player resistance is:

$$R = V/I$$
$$R = 9 \, V/0.01A$$
$$R = 900 \, \Omega$$

Electrical Circuits

Electronic gear is built using two basic circuit forms called the series circuit and the parallel circuit. All complex circuits are combinations of the two.

A series circuit has one path for electrons to follow. A flashlight is an example. Electrons begin their journey from a voltage source and follow a single path. Batteries supply electrons with energy to flow through wires interconnecting an on-off switch and a light bulb. By tracing the connections in Figure 9.15, you will see that electrons

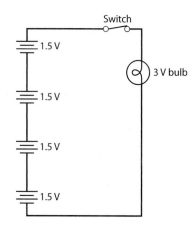

Figure 9.15 | This circuit schematic of a flashlight shows that electrons only flow through circuit components one at a time.

ENGINEERING QUICK TAKE

How Much Current Is Safe?

The electrical circuitry in a house is protected by fuses or circuit breakers. These protective devices open or interrupt the current flow in a circuit when current exceeds a preset level. Careful planning is needed to ensure that the wiring circuits in homes and businesses are safely designed and installed. If you were designing the wiring for a new home, how would you determine how much current is too much?

It is common to use the standard incandescent light bulb as a measure. For example, a circuit designer will use Ohm's Law to calculate how many lights can be plugged into a typical house circuit before its current requirements are exceeded.

First, a bit of information is necessary. A 100-watt incandescent bulb operates on 120 V. The resistance of this bulb is 144 Ω. A typical circuit breaker interrupts current flow at 20 A.

Using this information, calculate how much current each bulb requires.

$$I = V/R$$
$$I = 120V/144\,\Omega$$
$$I = 0.83A$$

Each bulb requires 0.83 A of current. All current for the lights on this circuit will flow through the 20A circuit breaker. To see how many bulbs can be on without tripping the breaker, perform the following calculation:

Number of bulbs = 20 A/0.83 A
Number of bulbs = 24.096
Number of bulbs = 24

It appears that twenty-four bulbs will use a bit less than the 20A of current needed to open the circuit breaker.

only flow through circuit components one at a time. The circuit operates by closing the switch to turn on the light. The light glows as electrons flowing through the bulb's filament resistance give up their energy in the form of heat and light.

A parallel circuit is an arrangement in which two or more paths exist for electrons to follow. In Figure 9.16, for example, two bulbs are wired in parallel. Current from the voltage source flows through each bulb so that both are on simultaneously. In this circuit, the flow of electrons leaving the battery divide at the wire junction, so some current flows through one bulb and the remainder flows through the other bulb. We say that each bulb is a different branch of the parallel circuit. Past each bulb, the two current paths recombine and electrons return to the other voltage source terminal. If the bulbs were identical, each would receive half the available electrons. In parallel circuits in which each branch is not electrically identical, meaning each branch has different resistance values, electrons flow through the branches in differing amounts. A mathematical relationship called Kirchhoff's Current Law describes the fact that the total current flowing into parallel branches is equal to the sum of the current in each parallel branch. Ohm's Law is also used to predict exactly how much current will flow through each branch.

In all circuits, the voltage potential of the source is distributed across each component in the circuit. We say voltages develop across components as current

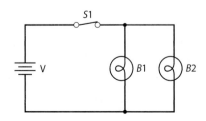

Figure 9.16 | A parallel circuit provides two or more paths for electrons to follow.

Section 2 △ Examining Electrical Circuits

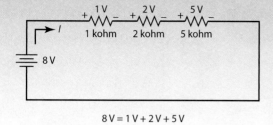

$$8V = 1V + 2V + 5V$$

Figure 9.17 | Voltage dropped across each component in a series circuit adds up to the source voltage.

flows through the components. The voltage developed across a component is called a voltage drop.

In series circuits, voltage drops add up to the source voltage. Observe, in Figure 9.17, that the individual voltage drops across each component mathematically add to equal the voltage of the source. This mathematical relationship is Kirchhoff's Voltage Law (KVL). The size or magnitude of each voltage drop depends on the resistance of any individual component relative to all the resistances in the circuit. This means that the larger the resistance is, the larger the voltage drop.

Power

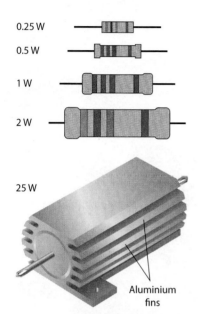

Figure 9.18 | A physically large resistor dissipates more power than a smaller resistor, as indicated in this illustration of resistor power ratings.

As current flows, the devices' resistance releases energy from electrons. Some of the energy provides the intended work of the circuit. Additionally, a significant amount of the energy is released as heat. A simple light bulb illustrates this concept. When a light is on, some of the electrical energy produces visible light, but a good amount of energy also radiates as heat. Components might release enough heat to cause a nasty burn. Electrical components have a power rating that designates the rate at which energy is dissipated by a particular component. The power rating is just one consideration when selecting electrical components for use in circuitry. Resistors, as an example, come in different physical sizes. A physically large resistor will dissipate more power than a smaller resistor. Even though resistors may have the same resistance value, the power rating may determine if it is usable in a circuit. It is important to select components with proper power ratings because an undersized component will not release heat fast enough to keep working properly.

The power in electrical circuits is easy to compute. The amount of power is expressed in watts (W). Like Ohm's Law, three variations of the power formula exist.

$$P = I \times V$$
$$P = I^2 \times R$$
$$P = V^2/R$$

Any of the formulas is used to calculate power in watts. To understand this concept, calculate the electrical power of a 120 V light bulb drawing 0.833A of current.

$$P = I \times V$$
$$P = 0.833 \times 120$$
$$P = 99.96 \sim 100\,W$$

Alternating Current (AC)

Electricity exists in two basic forms: direct current (DC) and alternating current (AC), as shown in Figure 9.19. In our discussion of electricity, we stated that current flows from a battery's negative terminal, through circuit components, and back to the battery's positive terminal. This scenario is true for direct current (DC) electricity. DC is characterized by a constant level of voltage and current, with current flowing in one direction only.

Alternating current (AC) is characterized by changes in the level and polarity of electrical signals over a period of time. The term signal represents any changing electrical quantity such as voltage, current, or electric field. Changing signals are extremely useful in conveying information. Simply turning a signal on and off, for example, is sufficient to represent information.

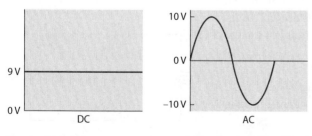

Figure 9.19 | As you can see in these images of a DC signal and an AC signal, AC voltage is more complicated than DC voltage.

Many signals are AC, including radio, TV, cellular communications signals, and computer signals, as well as signals generated by a microphone. The most common AC signal is the voltage present at an electrical outlet. Your power company provides the AC voltage at an outlet. The voltage level is standardized throughout the United States, which is why you can plug in an appliance anywhere in the country and it will operate properly. Additionally, the outlet voltage changes polarity 60 times a second, a fact we will explore in more detail shortly. In the meantime, you can see from this brief introduction that AC voltage is certainly more complicated than DC voltage.

A thorough understanding of the sine wave is the key to unlocking the nature of AC signals. The sine wave is the most fundamental AC signal in nature. Complex AC signals are made from individual sine waves, much as light is divided into colors with a prism. By studying the sine waveform, engineers and technicians can understand the characteristics of complex signals.

Looking at the sine wave in Figure 9.20, you can see why the word "wave" is used when referring to this signal—it looks like a wave in water. Notice how the sine wave is drawn on a simple graph. The vertical axis shows the amount of signal available. In electrical terms, the y-axis typically represents voltage or current, with the largest value of y indicating the peak value of the signal. We refer to the peak value as the amplitude of the signal. Using Figure 9.20, we say the amplitude of the signal is a 10 V peak.

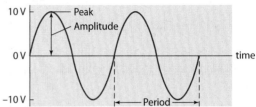

Figure 9.20 | A sine wave is the most fundamental AC signal in nature.

The horizontal or x-axis represents time and provides information on how many times per second the AC signal changes polarity. Notice that by a change in polarity, we mean the signal is positive for a certain amount of time and then negative for another increment of time. This means that, in an AC circuit, current will flow first in one direction and then in the opposite direction.

The number of times an AC signal changes per second is important. For example, the AC signal at an electrical outlet changes 60 times per second, or once every 0.0166 seconds, whereas an AC signal from an FM radio station can change 100 million times per second, or once every 0.00000001 seconds. We refer to the rate of signal change as frequency, and we use time information from the horizontal axis to determine the signal's frequency.

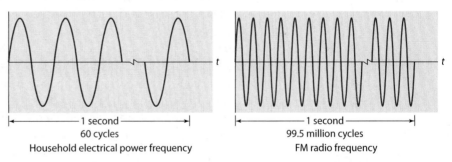

Figure 9.21 | In this comparison of low- and high-frequency waves, the AC signal at an electrical outlet changes once every 0.0166 seconds, while an AC signal from an FM radio station can change once every 0.00000001 seconds.

AC Signal Measurements

There are several ways to state the amplitude of an AC signal. To examine the conventional ways, use the electrical outlet waveform shown in Figure 9.22.

Notice that the positive peak reaches 169.7 V, and the negative peak value reaches –169.7 V, which are standard voltages in the United States. Because the waveform is continually changing voltage over time, any of several possible values could specify the waveform amplitude, but because the peak in the waveform is easily recognized, the positive peak voltage is a standard way to represent the entire signal voltage.

Over time, the waveform voltage rises from zero volts to the peak voltage of +169.7 V and then back to zero volts. The wave next changes polarity and becomes negative, heading to a negative peak voltage of –169.7 V.

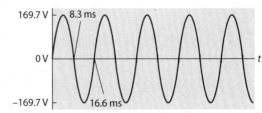

Figure 9.22 | The amplitude of an AC signal is represented by this 60-Hz waveform of a home electrical outlet.

Finally, the wave returns to zero volts and becomes repetitive from this point. That is, the sine wave does exactly the same thing repeatedly. We will get back to the repetition shortly, but for now we will continue examining the voltage levels.

It is perfectly acceptable to say that the amplitude of the AC waveform is 169.7 V, which is called the peak voltage. Engineers and technicians understand that we do not need to refer to the negative peak voltage because it is the same as the positive peak. Another way to indicate the size or magnitude of the voltage is by stating the peak-to-peak voltage. Can you figure out why the peak-to-peak voltage of the waveform in Figure 9.22 is 339.4 Vp-p?

AC signal magnitudes are often represented in RMS values. RMS stands for Root-Mean-Square, a mathematical way to relate the power content of AC and DC signals. RMS is a very important way to represent AC signals in the electrical industry. In standard practice, pure AC signals easily convert to RMS values when you multiply the signal's peak amplitude by 0.707. The constant 0.707 is the RMS mathematical conversion for sine waves, so even though the words Root-Mean-Square may sound intimidating, it is a fairly simple conversion to carry out. For example, the RMS value of the electrical outlet voltage is:

$$169.3 \text{ Vp} \times 0.707 = 119.7 \text{ Vrms}$$

This amount is rounded to 120 Vrms. Therefore, when someone says your outlet voltage is 120 V, you will know the reference is to the RMS voltage value. This is why "120 V" is stamped on light bulbs and most of the electrical gear you plug into outlets. RMS is also important because almost all AC electrical test meters display voltages in RMS values.

Figure 9.22 also shows a sine wave repeating its shape over time. You should expect your outlet to provide a continuous stream of sine waves (unless you pop a breaker or forget to pay the power company). Each repetition of a sine wave is called a cycle, and each cycle lasts the same amount of time. In this example, one cycle lasts for 0.0166 seconds. This may seem like an odd number, but, at this rate, 60 cycles occur in one second. We say the frequency of the wave is 60 cycles per second. In the electronics industry, the term hertz (Hz) means cycles per second, so you will see the waveform frequency written as 60 Hz. Because we have stated that a certain number of cycles occur per second, it is apparent that frequency and time are mathematically related. Frequency (f) is easily determined from time by dividing 1 by the time (period) of one full cycle:

$$f = 1/T$$

Furthermore, time or period (T) of one cycle is also computed from the frequency:

$$T = 1/f$$

Using the sine wave example given:

$$f = 1/T$$
$$f = 1/0.0166 \text{ sec}$$
$$f = 60 \text{ Hz}$$

Frequency is a very important characteristic of AC signals. In other chapters, you will see how frequency relates to bandwidth. Frequencies are conveniently grouped into ranges, as shown in Figure 9.23, and different frequency ranges have useful characteristics.

For example, frequencies in the range of 100 million Hz are in the VHF band, and these are used in communication systems. Some TV signals, for instance,

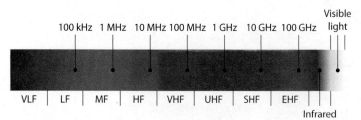

- **AM radio**—535 kilohertz to 1.7 megahertz
- **Short wave radio**—bands from 5.9 megahertz to 26.1 megahertz
- **Citizens band (CB) radio**—26.96 megahertz to 27.41 megahertz
- **Television stations**—54 to 88 megahertz for channels 2 through 6
- **FM radio**—88 megahertz to 108 megahertz
- **Television stations**—174 to 220 megahertz for channels 7 through 13

Figure 9.23 | Electromagnetic frequencies are conveniently grouped into ranges, or bands, each of which has useful characteristics.

are in this range. Compare the high frequencies of the VHF band to the low frequency of 60 Hz, which is in the SLF band. Many electrical applications exist because designers take advantage of the characteristics that different frequencies provide.

It is worthwhile to mention that high frequencies are often used to carry information. When you listen to an FM radio, for example, the song you hear was created from many complex AC signals that represent the music. The musical sounds are created by musicians and then converted to AC signals by microphones, guitar pickups, and other electronic music devices. The AC signals produced by these devices vary, such that the signal amplitude increases as music volume increases and frequency increases as musical pitch increases. These musical signals are now information in electrical form. At the FM radio transmitter, the musical signals, which are relatively low frequencies, are combined with a very high-frequency AC signal (see Figure 9.24). This combination process is known as frequency modulation (FM). With frequency modulation, the musical information moves wirelessly from an FM station to many radio receivers simultaneously. This process is one of the many fascinating things about electronics.

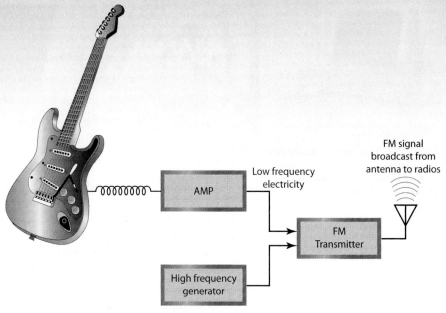

Figure 9.24 | Two electrical signals are converted into an FM signal.

Test Equipment

▷ Engineers and technicians need equipment to test circuit performance and troubleshoot broken products. A typical electronics laboratory will have many kinds of **test equipment** for these purposes. Some equipment is used to test almost every circuit, while other pieces are specialized and used in specific applications.

The power supply, the digital multimeter (DMM) and the oscilloscope are among the most common pieces of test equipment (see Figures 9.25–9.27).

A power supply converts the AC voltage at a wall outlet into DC voltage. The DC voltage then powers a circuit. A power supply used for laboratory work is adjustable. With the twist of a knob, the DC output voltage can be changed through a range, such as from 1 VDC to 30 VDC. Power supplies also provide controls so that output currents may be limited. Many bench-top power supplies have several outputs so that multiple voltages can be used at one time.

The digital multimeter is the work horse of the electronics industry. The meter has a control knob which sets the meter to read AC or DC voltages and currents. The control knob also permits the meter to read resistance. Depending on the meter's complexity, it may also be used to check continuity, test diodes, measure transistor betas, and determine capacitor values. The DMM is small enough to be held in the hand. Two probes connect the meter to the circuit under test.

Figure 9.25 | A bench-top power supply like this one typically has several outputs so that multiple voltages can be used at one time.

Figure 9.26 | A digital multimeter (DMM) can read AC or DC voltages, currents, and resistance. Some can also check continuity, test diodes, measure transistor betas, and determine capacitor values.

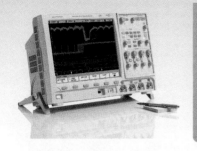

Like a multimeter, an oscilloscope measures electronic signals but has a display screen showing the electrical signal. The picture of the signal on the oscilloscope screen permits the operator to measure amplitude and frequency. Oscilloscopes have several input channels so that several signals may be viewed at once. This makes it convenient to compare a circuit's input signal and the resulting output signal. Oscilloscopes are complex devices—people require some training before they can use them effectively.

Figure 9.27 | An oscilloscope is a complex device that measures electronic signals—like a multimeter—but it features a display screen showing the signal.

Electrical Components

Too many electrical and electronic components are available for circuit design to discuss in this book, but we will discuss some of the common ones. Engineers and technicians must know what is available, and they use many resources to find the right components for a circuit design. It is important to know where to find information about parts. People who work in the electronics industry constantly read about new developments and products. They attend seminars and take courses to keep current. Learning is a necessary part of any technology or engineering career.

Finding parts for products takes work, so engineers and technicians spend time developing contacts with vendors and researching what is available. The Internet, as you might expect, is a great source for finding parts and good prices. Price is particularly important because designers often buy ten thousand units of the same part. The major component companies provide catalogs to the electronics industry. In addition, many component manufacturers provide free samples of their parts to engineers and technicians for testing and evaluation.

Now we will look at the characteristics of several important parts.

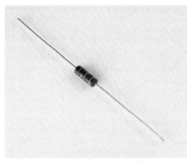

Resistor Resistors are parts designed to have specific values of electrical resistance. They come in an astounding array of resistance values, so a circuit designer can buy almost any resistance value he needs. Resistors (see Figure 9.28) are used when a specific value of electrical resistance needs to be inserted into an electrical path. They come in many sizes and shapes, and the physical size and shape directly relate to the resistor's power-handling capability. A 100-ohm resistor, for example, is available with power ratings that range from a few milliwatts to hundreds of watts. Resistors are built from a combination of conductive and insulating elements that are mixed together to create specific resistance values.

Some types of resistors have adjustable resistance values. These variable resistors are called potentiometers. The volume controls on radios, stereos, and TVs are often potentiometers. A potentiometer's resistance will change in response to a person who is listening to a radio adjusting the volume knob. This change in resistance changes the radio signal's voltage (according to Ohm's Law), which in turn raises or lowers the sound level.

Figure 9.28 | Fixed and variable resistors possess an astounding array of resistance values.

Capacitors Capacitors are components constructed with alternating layers of insulating and conducting materials. Capacitors store and release electrical energy, and their relatively simple construction provides amazing results (see Figure 9.29). Energy is stored when electrons accumulate within the capacitor structure.

Electrons flowing in a capacitor circuit collect on the conducting surfaces of the capacitor. These surfaces are called plates. A layer of insulating material called "dielectric" separates plates from one another. The dielectric material prevents electrons from flowing from plate to plate and determines how much charge can be stored. Electron charge will deposit onto the capacitor plates at a specific rate. The total charge stored within the capacitor increases as more electrons accumulate on the plates. You may be familiar with the large capacitors used in car

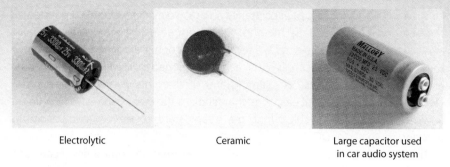

Electrolytic   Ceramic   Large capacitor used
in car audio system

Figure 9.29 | Capacitors store and release electrical energy.

audio systems. The charge that the capacitors store provides a momentary boost in energy to the audio amplifier. In this way, the audio system can obtain the spikes in power needed by the speaker system to reproduce low-frequency bass notes.

Capacitors store energy, a process called charging, and release energy, a process called discharging. Both processes are predictable. The charging and discharging periods span a length of time that is determined by the circuit capacitance and resistance values. These periods can be accurately calculated and used to our advantage in timing circuits. A typical example is a motion-sensitive outside lighting system, in which the motion of a person who crosses in front of the light is the trigger to turn on the light. The light remains on for an interval of time before automatically turning off. In this system, a capacitor charge or discharge time determines how long the light stays on.

The following formula is used to calculate the capacitor voltage, as determined by its charge time. The formula shows that capacitor charging follows an exponential curve:

$$Vc = 1 - e^{-t/RC} \text{ (Vc = 1 minus natural log e raised to the power } -t \text{ divided by R times C)}$$

Similar formulas calculate the capacitor discharge voltage. Other variations of these formulas determine voltage for situations in which the capacitor may already be partially charged. You can see in Figure 9.30 that the capacitor's charging and discharging voltage changes from moment to moment.

Capacitors have another useful function: the ability to sort out AC and DC signals. Capacitors respond differently to DC than to AC. When both signals are present in a circuit, a capacitor will block DC current flow but permit the AC current flow. We can understand how this happens by referring to the previous discussion of charging and discharging. The actual charging and discharging of the capacitor occurs relatively quickly, and the time it takes to do so is called the transient state. As you have learned, the duration of the transient state is

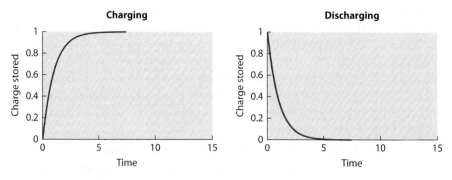

Figure 9.30 | These graphs show how capacitors charge and discharge exponentially.

controlled by the amount of capacitance and resistance in a circuit. Regardless of the duration, charging or discharging will be complete at some specific time.

After the transient state ends, we say the capacitor has entered a steady state. In the steady state, the capacitor is either at a fully charged or fully discharged voltage level. Because the process of charging and discharging involves movement of electrons, no movement of electrons characterizes the steady state. No DC current flows in the capacitor circuit during the steady state, so we say the capacitor is blocking the path for DC.

The situation is different with AC currents. An alternating signal is never constant, so a capacitor subjected to AC is always charging or discharging at a rate dependent on the frequency of the AC signal. With low frequencies (DC, by the way, is the lowest frequency at 0 Hz), capacitors dramatically impede current flow. As a signal frequency increases, the capacitor more easily permits current flow.

In the electronics field, the term impedance describes the opposition to current flow in an AC circuit. Impedance is similar to resistance and is measured in ohms. However, impedance is far more complicated. A capacitor's impedance is called capacitive reactance (Figure 9.31) and is represented by the symbol Xc. Capacitive reactance is calculated using the following formula:

$$X_C = 1/2\pi fC \text{ (Xc = 1 divided by 2 times 3.14 times frequency times}$$
capacitance, where f = frequency and C = the value of the capacitor)

Because of the inverse relationship between Xc and frequency or capacitance, Xc decreases as frequency or capacitance increases.

Engineers and technicians study the transient and steady state responses of capacitors to DC, as well as the frequency sensitivity of capacitors to AC signals. These are important responses for engineers and technicians to understand.

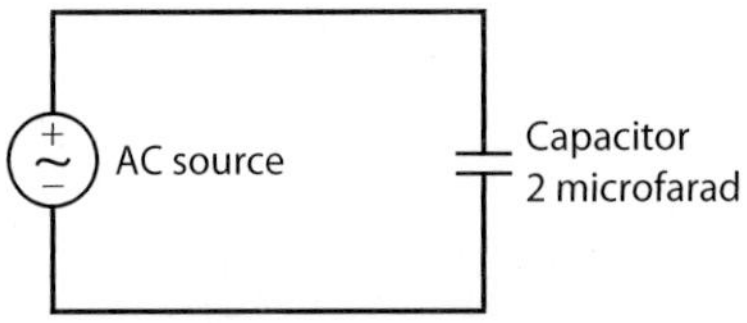

Inductors Inductors have many electrical similarities to capacitors. They are made by coiling wire, and they operate by generating a magnetic field as current flows through the coils (see Figure 9.32). This property of electromagnetism means that a magnetic field is generated by electricity. The reverse property is also true; current is generated when wire is moved through a magnetic field. These two effects, electricity generated from magnetism and magnetism generated from electricity, are fundamental to the creation of many helpful devices, including the generators supplying electricity to our homes and businesses. The response from inductors to a particular electrical signal stems from this interaction as well.

Like capacitors, inductors store and release energy. Because of current flow, energy is stored in an inductor as a magnetic field. Energy returns to the inductor circuit when current flow changes or stops, which causes the magnetic field to increase or decrease in strength. The changing magnetic field induces an opposing current flow into the inductor. This happens because a collapsing or expanding magnetic field changing around a wire has the same effect as if wire were moved through the field.

The study of an inductor's response to DC and AC is beyond this chapter's scope. However, some of the inductor's similarities to the capacitor can be identified. With DC, the inductor has a transient state and a steady state. Creating a magnetic field in an inductor takes time. This time is the inductor's transient state; the resistance and inductance in the circuit control the transient state time. In the steady state, inductors freely pass DC currents but inhibit, or choke off, AC currents. Inductors present an opposition to AC signals called inductive reactance, which is represented by the symbol X_L. The formula is:

$$X_L = 2\pi fL \text{ (X}_L = 2 \text{ times pi times frequency times inductance)}$$

Here, f = frequency in hertz and L = inductance in henrys. Inductive reactance increases with frequency and with the amount of inductance. This response to frequency is opposite to that of the capacitor, so high frequencies are blocked.

Frequency	$X_C = 1/2\,\pi fC$
200 Hz	398 ohms
2,000 Hz	39.8 ohms
20,000 Hz	3.98 ohms

Figure 9.31 | Capacitive reactance decreases as frequency increases.

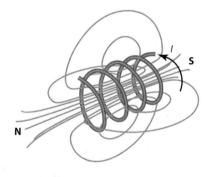

Figure 9.32 | Electricity flowing in a coil of wire produces magnetism.

Batteryless Flashlights

▷ Maybe you have seen advertisements on TV for flashlights that work without batteries (see Figure 9.33). All you have to do is shake the flashlight to generate the electricity needed for the LED (light emitting diode) bulb to light. This device uses the electromagnetic properties of inductors to generate electricity. A magnet in the flashlight is free to slide back and forth within the

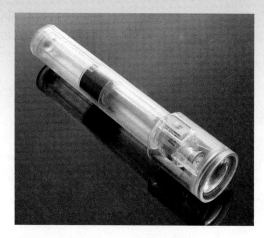

Figure 9.33 | **A batteryless flashlight uses the electromagnetic properties of inductors to generate electricity.**

inductor. The magnet's magnetic field cuts through the wires of the coil as the magnet is shaken. This action induces a voltage across the inductor. Current flows due to the developed voltage and is stored in a capacitor. The LED uses the stored charge in the capacitor as its power source.

Transformers Transformers are devices that use electricity and magnetism. They function like two inductors formed together (see Figure 9.34). The magnetic field created in one of the inductors overlaps the coils of the second inductor. Under proper conditions, this induces a voltage into the second inductor. Two inductors positioned together make up a transformer. The two sets of coil windings are identified as the primary winding and the secondary winding.

Transformers are used to step up or step down voltage or current levels. That is, the voltage or current on the primary side changes to a different level on the secondary side. To do this, one need only control the number of turns of wire in each coil of the transformer. More turns in the secondary side of a transformer than in the primary side will, for example, step up or increase the voltage at the secondary terminals. This means voltage levels are increased or decreased with a transformer. This is the most common purpose for a transformer.

You may have noticed large, round transformers mounted on power poles. These transformers usually reduce the higher voltages needed to transfer electricity across the country to the lower voltage levels used in homes. You also may have plugged a power pack into a wall outlet to run a laptop or video game system. The power pack's transformer lowers the outlet voltage to an appropriate level for the electronic circuitry in your system.

Transformer symbol

Figure 9.34 | **Various transformers are represented by the transformer symbol.**

Switches Switches are electromechanical devices used to turn the current flow to a device or circuit on or off. They come in many sizes and shapes for just about any application imaginable (see Figure 9.35). A typical wall switch used to turn on lights is one application; the keys on a computer keyboard are another application. Each key is

DIP switch   Push button   Keypad

Figure 9.35 | **Switches are used to turn current flow on or off.**

Section 2 △ Examining Electrical Circuits

a separate switch, and typing on a key turns the switch on. A computer chip in the keyboard senses this switch action and then interprets the movement as a letter or number.

Buying Electrical Parts

▷ Now you are familiar with some electrical components. As mentioned earlier, many companies sell these parts, and you might be amazed at the selection available. For example, Figure 9.36 shows part of a page from a typical catalog. The information shown is just a small sampling of the selection available for one type of capacitor; the catalog has ten pages for capacitors alone.

SAL-A 123 Axial Solid AL Electrolytic Capacitors

FEATURES:
- Polarized aluminum electrolytic capacitors, solid electrolyte MnO2
- Axial leads, aluminum case, ceramic seal, blue insulation sleeve.
- SAL-A: standard version
- Extremely long useful life: 20,000 hours at 125°C
- Extended usable temperature range up to 200°C
- Excellent low temperature impedance and ESR behavior
- Charge and discharge proof, application with 0 Ω resistance allowed
- Reverse DC voltage up to 0.3 x UR allowed
- AC voltage up to 0.8 x UR allowed
- Advanced technology to achieve high reliability and high stability

APPLICATIONS:
- EDP, telecommunications, general industrial, automotive, military and space
- Smoothing, filtering, buffering, timing
- For power supplies, DC/DC converters

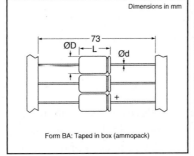

Dimensions in mm

Form BA: Taped in box (ammopack)

DESCRIPTION	VALUE
Case sizes (ØD$_{nom}$ x L$_{nom}$ in mm)	6.7 x 15.3 to 12.9 x 32.0
Rated capacitance range (E6 Series), C$_R$	1.5 to 2200 µF
Tolerance on C$_R$	±20%
Rated voltage range, U$_R$	6.3 to 40V
Category temperature range	-55°C ~ 125°C
Usable temperature range	-80°C ~ 200°C
Endurance test at 155 and 125°C	5000 and 8000 hours
Useful life at 40°C, I$_R$ applied	450,000 hours
Shelf life at 0 V, 125°C	500 hours
Based on sectional specification	IEC 384-4/CECC 30300
Detail specification	IEC384-4-2/CECC 30302
Climatic category IEC 68 (DIN 40040; NF C20-600)	55/125/56 (FKD; 434)
Approvals	CNET LNZ 44-04 COS-C (PTT) Gam-t-1 (MIL) CECC 30302-003

WV (VDC)	Cap. (µF)	Leakage Current (µA)	tan δ	Ripple Curr. (mA)	Dimensions (mm) L	D	d	Digi-Key Part No.	Cut Tape Pricing 1	10	Digi-Key Part No.	Tape and Box Pricing 100	500	1,000	5,000	Vishay Part No.
6.3	330	104	0.180	330	23.3	9.3	0.6	4267PHCT-ND	10.26	94.02	—	—	—	—	—	2222 123 13331
10	33	17	0.180	63	15.3	6.7	0.6	4271PHCT-ND	8.33	76.37	—	—	—	—	—	2222 123 14339
16	33	55	0.140	89	20.4	7.6	0.6	4284PHCT-ND	8.67	79.45	—	—	—	—	—	2222 123 15339
	68	110	0.140	180	20.4	7.6	0.6	4286PHCT-ND	9.21	84.43	—	—	—	—	—	2222 123 15689
	330	500	0.160	510	32	10.3	0.8	4290PHCT-ND	13.40	122.82	—	—	—	—	—	2222 123 15331
25	100	250	0.160	250	23.3	9.3	0.6	4299PHCT-ND	11.26	103.24	—	—	—	—	—	2222 123 16101
35	15	30	0.120	53	20.4	7.6	0.6	4310PHCT-ND	9.08	83.27	—	—	—	—	—	2222 123 10159
	100	200	0.160	220	32	12.9	0.8	4315PHCT-ND	15.53	142.40	—	—	—	—	—	2222 123 10101
40	3.3	13	0.120	16	15.3	6.7	0.6	4318PHCT-ND	8.17	74.86	4318PHTB-ND	639.70	2615.00	4063.00	3834.00/M	2222 123 17338
	68	270	0.160	170	32	10.3	0.8	4326PHCT-ND	12.48	114.37	4326PHTB-ND	977.40	3995.50	6207.00	5858.00/M	2222 123 17689

ASM 021

Axial Standard Miniature Electrolytic Capacitors

FEATURES:
- Polarized aluminum electrolytic capacitors, non-solid • Axial leads, cylindrical aluminum case, insulated with a blue sleeve • Charge and discharge proof • Case Ø10mm x 30mm to 21mm x 40mm with pressure relief • Taped versions up to case Ø15mm x 30mm available for automatic insertion • Miniaturized, high CV-product per unit volume

APPLICATIONS:
- General purpose, industrial, automotive, audio-video • Coupling, decoupling, smoothing, filtering, buffering and timing • Portable and mobile equipment (small size, low mass) • Low mounting height boards, vibration and shock resistant.

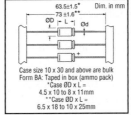

Dim. in mm

Case size 10 x 30 and above are bulk
Form BA: Taped in box (ammo pack)
*Case ØD x L = 4.5 x 10 to 8 x 11mm
**Case ØD x L = 6.5 x 18 to 10 x 25mm

DESCRIPTION	VALUE	
Case sizes (ØD$_{nom}$ x L$_{nom}$ in mm)	4.5 x 10 to 10 x 25	10 x 30 to 21 x 40
Rated capacitance range, C$_R$	0.47 ~ 15,000µF	
Tolerance on C$_R$	±20%	
Rated voltage range, U$_R$	6.3 ~ 100V	
Category temperature range	-40°C ~ 85°C	
Endurance test at 85°C: U$_R$ = 6.3 ~ 25V	1,000 hours	5,000 hours
U$_R$ = 40 ~ 100V	2,000 hours	5,000 hours
Useful life at 85°C	2,500 hours	8,000 hours
Useful life at 40°C, 1.4 x I$_R$ applied	70,000 hours	200,000 hours
Shelf life at 0 V, 85°C	500 hours	500 hours
Based on sectional specification: U$_R$ = 6.3 ~ 25V	IEC 384-4/CECC 30300 GP grade	IEC384-4/CECC 30300 LL grade
U$_R$ = 40 ~ 100V	LL grade	LL grade
Climatic category IEC 68 (DIN 40040)	40/085/56 (GPF)	

WV (VDC)	Cap. (µF)	Leakage Curr. (µA)	tan δ	Ripple Curr. (mA)	Dimensions (mm) L	D	d	Digi-Key Part No.	Cut Tape Pricing 1	10	100	Digi-Key Part No.	Tape and Box Pricing 500	1,000	5,000	10,000	Vishay Part No.
6.3	470	22	0.250	260	11	8.0	0.6	4000PHCT-ND♦	.79	6.78	53.11	4000PHTB-ND♦	217.00	337.00	318.00/M	294.00/M	2222 021 33471
	1000	42	0.250	440	18	8.0	0.8	4001PHCT-ND♦	.91	7.77	60.87	4001PHTB-ND♦	249.00	387.00	365.00/M	336.00/M	2222 021 33102
	2200	87	0.290	710	25	10	0.8	4002PHCT-ND♦	1.32	11.31	88.60	4002PHTB-ND♦	362.00	563.00	531.00/M	490.00/M	2222 021 90589
10	100	10	0.200	100	10	4.5	0.6	4003PHCT-ND♦	.56	4.83	37.84	4003PHTB-ND♦	—	240.00	227.00/M	209.00/M	2222 021 34101
	220	17	0.200	160	10	6.0	0.6	4004PHCT-ND♦	.62	5.31	41.60	4004PHTB-ND♦	—	264.00	249.00/M	230.00/M	2222 021 34221
	330	24	0.200	230	11	8.0	0.6	4005PHCT-ND♦	.79	6.78	53.11	4005PHTB-ND♦	217.00	337.00	318.00/M	294.00/M	2222 021 34331
	470	32	0.200	310	18	6.5	0.8	4006PHCT-ND♦	.79	6.78	53.11	4006PHTB-ND♦	—	337.00	318.00/M	294.00/M	2222 021 34471
	1000	64	0.200	550	18	10	0.8	4008PHCT-ND♦	1.04	8.94	70.03	4008PHTB-ND♦	286.50	445.00	420.00/M	387.00/M	2222 021 34102
	1500	94	0.230	690	25	10	0.8	4009PHCT-ND	1.32	11.31	88.60	4009PHTB-ND	362.00	563.00	531.00/M	490.00/M	2222 021 90525
	2200	136	0.250	800	30	12.5	0.8	4011PHBK-ND♦†	1.84	15.81	123.85	—	506.25	786.56	742.26/M	684.42/M	2222 021 14222
	3300	202	0.270	1000	30	12.5	0.8	4012PHBK-ND♦†	2.02	17.28	135.36	—	553.33	859.70	811.27/M	748.06/M	2222 021 14332
	4700	286	0.290	1180	30	15	0.8	4013PHBK-ND♦†	2.35	20.10	157.45	—	643.63	1000.00	943.67/M	870.13/M	2222 021 14472

♦ RoHS Compliant   † Bulk

(Continued)

Figure 9.36 | A wide variety of capacitors are available.

1. Draw a schematic for the switches, outlets, and lights in your classroom.
2. If a flashlight is powered by two 1.5 V batteries and has 0.02 A of current flowing through the bulb, what is the resistance of the bulb?
3. How many cycles per second is the frequency of your local AM radio station?
4. Name an important characteristic of capacitors.
5. What component is made by wrapping fifty turns of wire around a wooded dowel?

SECTION 3: Learning About Electronics

KEY IDEAS >

- When a semiconductor element is combined with small amounts of other elements, beneficial electrical properties result in the newly formed material.

- Semiconductor materials are used to create semiconductor devices such as diodes and transistors.

- Two basic kinds of transistors exist: the bipolar junction transistor (BJT) and field effect transistor (FET).

- Transistors are often used as electronic switches.

Electronic components are made from many different materials. Silicon is primarily used to make electronic devices. This section discusses silicon's use as an electrical material and how it is used to make transistors.

Semiconductors

Semiconductor devices are created from elements such as silicon and germanium, which are neither good conductors nor insulators of electricity. Based on this fact, it appears that these elements are not very useful electrically, but nothing could be further from the truth. When a semiconductor element is combined with small amounts of other elements, beneficial electrical properties result in the newly formed material. Semiconductor materials are used to create semiconductor devices such as diodes and transistors.

In order to manufacture semiconductor devices, two types of silicon based semiconductor materials are created. These materials are simply called N-type and P-type.

N-type material is formed by combining silicon atoms with small amounts of another element that possesses a specific atomic structure (see Figure 9.37). In the semiconductor manufacturing industry, adding materials to silicon is called doping. Doping radically alters the electrical properties of the resultant material.

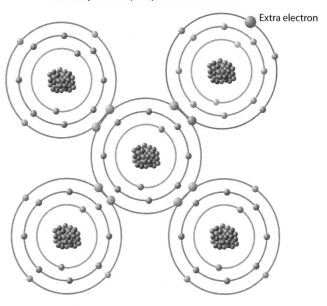

N-type Silicon:
Silicon crystal with phosphorous introduced

Extra electron

Figure 9.37 | Because electrons carry a negative charge, the resulting material is called N-type material.

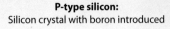
P-type silicon:
Silicon crystal with boron introduced

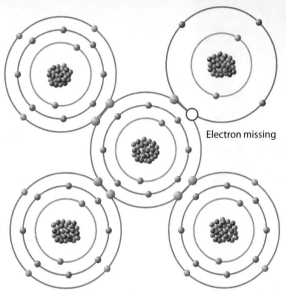

Electron missing

Figure 9.38 | The doping process creates an excess of energy levels, which have a positive charge associated with them in P-type material.

To understand N-type material, recall that an atom consists of a nucleus surrounded by electrons and that the number of electrons in the outer orbit of an atom determines its electrical properties. Conductors typically have few electrons in the valence orbit, while insulators have many electrons in the valence orbit. A silicon atom possesses four outer electrons, making pure silicon neither a good conductor nor a good insulator.

When silicon atoms are combined with other atoms such as phosphorus or arsenic, which are elements of five valence electrons each, a unique electrical material is created. In nature, atoms can bond or link to each other by sharing electrons, which is why solid materials exist: Atoms are able to attach to each other. In a combination of silicon and arsenic atoms, four separate silicon atoms are able to bond to one arsenic atom. The bonding occurs when valence electrons are shared among the five atoms. These strong bonds are called covalent bonds. The material that forms from this mixture is made up of a regular and orderly crystalline structure. However, within this structure, each arsenic atom has one electron unable to bond with an electron from nearby silicon atom. The arrangement of atoms in the crystalline structure makes such bonding impossible. The nonbonded arsenic electron is loosely attached to the arsenic atom and is readily available to become a free electron in the conduction band. Keep in mind that although our description pertains to one atom, it actually happens to millions of atoms at the same time. The result is that many free electrons are now available. Because electrons carry a negative charge, the resulting material is called N-type. It is the availability of many free electrons that makes N-type semiconductor material very conductive—far more conductive than silicon alone.

P-type material is produced similarly to N-type. The difference is that the doping element used has three electrons in the valence orbit. Boron is often used as the dopant for P-type material (see Figure 9.38). When small amounts of boron combine with silicon, a reaction occurs at the valence-orbit level. The three boron electrons bond with electrons in three adjoining silicon atoms. Due to the crystalline arrangement of the atoms, a fourth nearby silicon atom is also available to form a bond. Because boron only donates three electrons to this process, one remaining silicon electron is unable to form a strong covalent bond. However, a vacant energy level is available for an electron to fill and thereby form a bond. An electron can make its way to the vacant energy level and fill it. Keep in mind that this electron movement happens throughout the valence energy levels.

Any vacant energy level is a hole and is a positive charge because holes will attract electrons. Electrons from nearby atoms can simply move from one atom to the vacancy in another. This is equivalent to the hole moving because when an electron leaves one atom to fill a hole in another atom, the resultant movement creates a hole, which another electron can then occupy. This hole-current flow occurs because the doping process creates an excess of energy levels, which have a positive charge associated with them in P-type material.

Transistors: The BJT and FET

The term *semiconductors* most commonly refers to the devices made from N-type and P-type materials. The most common semiconductor device is the transistor; it is technically possible to simultaneously build millions of these components on integrated circuits, resulting in the development of sophisticated and helpful products. Computers, cell phones, PDAs, health equipment, and Internet routers are just some of the products built around transistors. Clearly, semiconductor devices have changed the electronics industry and our world.

Two basic types of transistors exist: the bipolar junction transistor (BJT) and the field effect (FET). Figure 9.39 shows the common electrical symbols used for BJTs and FETs. With additional circuitry, both transistor types can operate as amplifiers, which are circuits increasing the amplitude of voltage signals. Amplifiers, for instance, permit a very small signal to be transmitted from a satellite and received on earth, resulting in a great-looking picture on a high-definition TV.

Transistors are often used as electronic switches. This may not seem very important to you at first, but consider the significant difference between a mechanical light switch and a transistor switch. The light switch is turned on or off by a person, whereas a transistor switch is turned on or off by a low-voltage signal. The low-voltage signal can come from computers, sensors, or many other sources. Controlling switching with an electronic signal makes the digital age a reality, for the essence of a digital signal is simply turning something on or off. Transistor switches are the heart of computers.

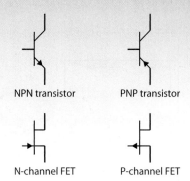

NPN transistor   PNP transistor

N-channel FET   P-channel FET

Figure 9.39 | This figure shows schematic symbols for bipolar and FET transistors.

Field Effect Transistors (FET)

Field effect transistors are physically smaller than BJTs, and millions of them are packed onto the small area of an integrated circuit. There are several types of FET designs; we will concentrate our transistor study on the metal oxide semiconductor FET, or MOSFET. This transistor is used as a switch, and it is relatively simple to fabricate.

Examine how an FET is constructed, and how it operates as a switch. Transistors are made from N-type and P-type semiconductor material. Figure 9.40 shows how N-type and P-type materials combine to build an N-channel MOSFET. This transistor has three important sections: the source, drain, and gate. Notice that the source and drain are both small N-type areas embedded into a larger P-type block. The P-type material therefore separates the source and drain from each other. Constructed over this area is the gate. The gate is fabricated from semiconductor material, but is deliberately isolated from the P-type region between the source and drain. A small layer of silicon dioxide, a type of glass, electrically insulates the gate from the rest of the structure. This means no electricity can flow between the gate and the rest of the MOSFET. Keep in mind that, in the modern world of nanotechnology, the gate structure may only be 45 billionths of a meter wide. On this extremely small scale, it is possible to pack millions of transistors on a very small integrated circuit (see Figure 9.41).

In order to operate as an on-off switch, the transistor must control when current will flow between source and drain. When current flows, the transistor is on; when current does not

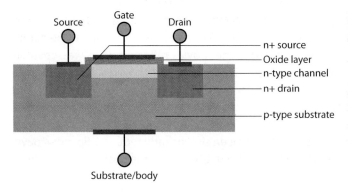

Figure 9.40 | An N-channel MOSFET has three important sections: source, drain, and gate.

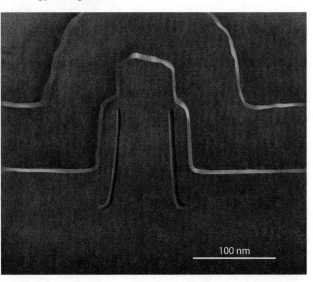

100 nm

Figure 9.41 | This is what a MOSFET structure looks like under an electron-microscope. It is possible to pack millions of transistors on a very small integrated circuit.

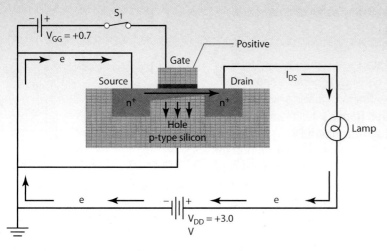

Figure 9.42 | When current flows, the FET is on. If you remove the gate voltage, the free electrons dissipate back into the P-type region, and current flow ceases.

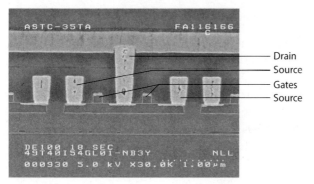

Figure 9.43 | The CMOS circuit is the foundation of all circuitry used in computer chips.

flow, the transistor is off. Preventing current flow is easy, as this is the device's natural state. By simply keeping the gate voltage at zero, the MOSFET enters the off state. Creating current flow requires applying a positive potential to the gate. When this happens, free electrons within the P-type material drift toward the glass layer. (There are always some free electrons in P-type.) After a sufficient number of electrons migrate to this area and line up along the insulating glass layer, a conducting N-channel forms. The positive gate voltage has forced free electrons to line up as a bridge between the source and the drain terminals. The channel can now carry current. When current flows, the FET is on. If we remove the gate voltage, the free electrons dissipate back into the P-type region, and current flow ceases (Figure 9.42). Now the transistor is off.

The MOSFET is very small and simply designed. Manufacturers fabricate P-channel devices just like N-channel devices. The difference is that the channel is formed using holes rather than electrons. Operation is very similar; a negative gate voltage turns the channel on, and the lack of any gate voltage turns the channel off.

Connecting N-channel devices to P-channel devices forms a complementary metal oxide semiconductor (CMOS) circuit. This circuit (see Figure 9.43) is the foundation of all circuitry used in computer chips. We will discuss logic circuits created by this technology in the next chapter.

TECHNOLOGY IN THE REAL WORLD:
Electronics Equals Small

Since the invention of the transistor in 1948 and the development of the integrated circuit in 1959, electronics products have become a necessary part of our lives. Products we use every day and take for granted did not exist when the transistor and integrated circuit were invented. No one had heard of or used a Bluetooth earpiece in the 1980s, for example.

The tiny size of modern products, along with their light weight and small power requirements, has made it possible for large, bulky items to be reduced to portable, pocket-sized objects. Have you ever seen a movie in which the actors communicated using a car telephone? The movies reflected what was possible in early communications equipment. The telephone was an actual table-top phone connected to a very large transmitter, which took up lots of space in the automobile. Additionally, the power requirements were also large and required connections to an automobile battery. With the development of transistors and integrated circuits, size and power has been significantly reduced. So, nowadays, we just flip open a cell phone to talk, text, and use the Internet just about anywhere we like.

A more recent improvement in technology, made possible by electronics, is the computer mouse. The early mice were electromechanical devices (see Figure 9.44).

The movement of the mouse depended on the motion of a ball on a mouse pad. The ball's motion was converted to an electrical signal through variable resistors, which adjusted the resistance as the user moved the mouse. These early mice were heavy, and they worked poorly because dust and dirt entered and clogged the mechanical mechanism.

The modern mouse is completely electronic (see Figure 9.45). This newer mouse uses an LED to shine light onto a surface. A CMOS sensor detects the image provided by the LED light. The signal from the sensor is sent to a special computer chip called a DSP, or digital signal processor. The DSP determines how far the mouse has moved based on the image provided by the sensor. Everything in the mouse is electronic and is therefore lightweight and reliable.

Figure 9.44 | The first computer mice were electromechanical devices.

SECTION THREE FEEDBACK >

1. Explain how N-type semiconductor material is created.
2. Name the three parts of a field effect transistor.
3. Describe how an FET operates as a digital device.
4. What are the devices needed to make a CMOS circuit?

Figure 9.45 | The modern computer mouse is completely electronic.

SECTION 4: Fabricating Integrated Circuits

KEY IDEAS >

- ICs begin their formation as wafers of doped silicon.

- ICs are comprised of thousands, and sometimes millions, of transistors built on a single very small piece of silicon.

- On each wafer, numerous ICs are built.

- The chip fab is a special manufacturing environment in which wafers become functional integrated circuits.

The electronic products we use daily are designed with complex circuitry. Because of the integrated circuit, it is possible for products to be powerful, yet small and energy efficient. This section examines the manufacturing of integrated circuits and how millions of microscopic devices are built simultaneously.

Integrated Circuits

Transistors, resistors, and other parts are called discrete parts. Discrete parts are soldered together on circuit boards to form an electronic system. Systems built with discrete parts tend to be physically large. The trend in the electronics industry is to make products small, light, and power efficient.

A revolution in the electronics industry came about with the invention of the integrated circuit (IC). ICs are comprised of thousands, and sometimes millions, of transistors built on a single very small piece of silicon. The FET described earlier is one of the most commonly built IC devices. FETs can function as transistors or as resistors, making IC fabrication easier, because the same basic steps build both devices. Fabrication is the process of making an integrated circuit.

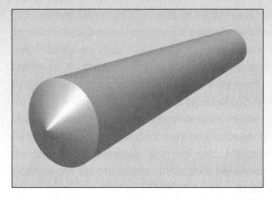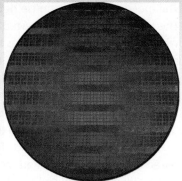

Figure 9.46 | The silicon used to manufacture integrated circuits is grown to form a cylinder; individual wafers are cut from the cylinder.

Wafers

ICs begin their formation as wafers of doped silicon. First, silicon is melted and purified. Dopants are added as needed to convert the refined silicon into N-type or P-type material. Over a lengthy period, the molten silicon is slowly withdrawn from its heated container. This causes the melted silicon mixture to cool. As it solidifies, a long silicon cylinder forms. This cylinder is sawed into thin disks or wafers (see Figure 9.46). Many ICs will be built upon the wafers.

Wafers are processed in a chip fabrication plant (see Figure 9.47) or chip fab. The chip fab is a special manufacturing environment in which wafers become functional integrated circuits. The chip fab contains many clean rooms, which are specially constructed rooms that contain highly filtered air and vibration-dampening flooring. Dust and other impurities will render an IC useless if permitted to contaminate the silicon, and because a speck of common dust is dramatically larger than a transistor, a special environment is necessary.

As in any manufacturing process, the individual elements of each transistor are built one at a time. What makes integrated circuit manufacturing unique is

Figure 9.47 | The AMD Chip Fab facility in Dresden, Germany, produces integrated circuits.

that millions of the same transistor parts are built simultaneously. For example, multiple wafers are processed at the same time. On each wafer, numerous ICs are built. On each IC, a million FETs are under construction. In addition, all this fabrication takes place at the same time. Talk about mass production! This is one reason that the price of electronic gear decreases over time. It is very cost-effective to build products from ICs.

IC fabrication is divided into two general processing stages. The front end of the line (FEOL) is where the transistors are created. The back end of the line (BEOL) is where transistors are interconnected to form working circuitry.

Front End of the Line

The transistor construction process begins with the wafer acting as the substrate, or foundation, for the entire IC. The wafer is placed in a special oven, called a "diffusion furnace." There, under specially controlled temperature and pressure, silicon and doping materials are added. Given some time, atoms fall upon the wafer, resulting in a new layer just a few hundred atoms deep. This new layer of semiconductor material is called the epitaxial (epi) layer, and it becomes the channel area for many FETs.

Thousands of individual transistor features are created on the wafer. Masking is the technique that defines exactly where the features go (see Figure 9.48). A mask is a special glass covered with opaque lines in certain areas and clear glass in other areas. The mask shows where features of the FET, such as the source and drain, are placed on the IC.

The wafer, covered with its new epi layer, is transported to a process machine where a light-sensitive liquid chemical called photoresist is applied. A mask is placed over the coated wafer, and then the wafer is exposed to ultraviolet light, which will cause the photoresist to harden. The photoresist hardens only in the areas where the mask permits light to pass. In the areas where the mask prevents light from passing, the photoresist remains unexposed and liquid. This process is similar to that used to develop a conventional photograph.

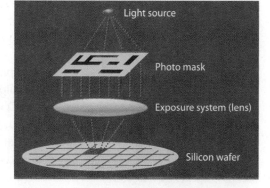

Figure 9.48 | The masking process defines where features of the FET are placed on the integrated circuit.

Chemical solvents dissolve the unexposed photoresist. Only the exposed and hardened photoresist remains, protecting portions of the silicon epi layer. Now, the next step begins. The opened areas on the wafer are placed in an ion implantation machine, which blasts ions of dopant materials into the wafer's uncovered areas, creating the sources and drains of each FET. An ion implant converts a section of P-type material into N-type material or vice versa.

Additional masking steps take place to define the placement of insulating glass and the gate structure itself. The same steps are repeated many times as each feature is created. It is remarkable that the process can create gate feature sizes as small as 45 billionths of a meter, and that IC manufacturers are working to achieve feature sizes as small as 10 billionths of a meter.

Back End of the Line

After FETs are fully formed on the IC, they must be interconnected to create functional circuits. For example, the FETs on a computer processor chip and a cell phone controller chip are formed in the same way. The manner in which the FETs interconnect is what distinguishes one chip from another.

To create a circuit, each terminal of the FET—gate, source, and drain—must have a conducting connection made to other devices. ICs are built

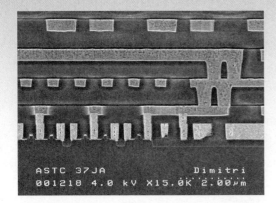

ASTC 37JA   Dimitri
001218 4.0 kV X15.0K 2.00μm

Figure 9.49 | This figure shows a cross section of an integrated circuit.

in layers, including metallization layers, which are designed for interconnections and formed from aluminum or copper. Interconnection layers, created with the masking process, resemble a sandwich of metal and insulation (see Figure 9.49). Finally, ICs are tested one by one while still on the wafer to ensure that they work. Next, the finished wafers are sawed to separate the individual ICs into chips. These chips are then packaged. This means that the chip is sealed, and external pins are added to connect the chip to other parts on a circuit board.

The preceding description briefly explains the key steps in manufacturing integrated circuits. Actual manufacturing takes two to three weeks to complete and involves more than 5,000 individual steps.

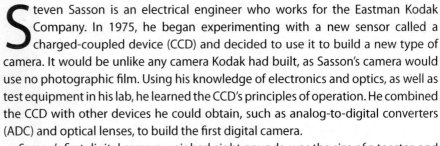

TECHNOLOGY AND PEOPLE:
Engineering Behind the Digital Camera

Steven Sasson is an electrical engineer who works for the Eastman Kodak Company. In 1975, he began experimenting with a new sensor called a charged-coupled device (CCD) and decided to use it to build a new type of camera. It would be unlike any camera Kodak had built, as Sasson's camera would use no photographic film. Using his knowledge of electronics and optics, as well as test equipment in his lab, he learned the CCD's principles of operation. He combined the CCD with other devices he could obtain, such as analog-to-digital converters (ADC) and optical lenses, to build the first digital camera.

Sasson's first digital camera weighed eight pounds, was the size of a toaster, and required a TV to view the image (see Figure 9.50). It took 23 seconds for this camera to take a picture and another 23 seconds to see it on the TV. The picture resolution was a mere 0.01 megapixels—not much compared to the multi-megapixel cameras we can buy today.

Steven Sasson still works for Kodak. In a 2006 interview, he said, "I have had a great career at Kodak. I had the opportunity to see the birth of digital photography and see it mature. From the time when we had several arguments on whether digital photography can ever be a reality to actually seeing digital photography completely changing the world of photography has been a special privilege to me."

Figure 9.50 | Steven Sasson holds the first digital camera.

SECTION FOUR FEEDBACK >

1. If the overall size of an integrated circuit FET is 95 nanometers, how many transistors can be placed in a row on a silicon wafer that is 300 mm wide?
2. At what point in the IC manufacturing process does FEOL end?
3. What makes an iPod chip and an HDTV chip different when they are both made from FETs?
4. How is a wafer first formed?
5. What process defines where specific features are placed on an integrated circuit?

Matching Your Interests and Abilities with Career Opportunities: Engineering Technicians

Electrical and electronic engineering technicians make up 34 percent of all engineering technicians. Employment of engineering technicians often is influenced by the same local and national economic conditions that affect engineers; as a result, job outlook varies with industry and specialization. Opportunities will be best for people who have an associate degree or extensive job training in engineering technology.

Nature of the Industry

Electrical and electronics engineering technicians help design, develop, test, and manufacture electrical and electronic equipment. This includes communication equipment, navigational equipment, and computers, as well as radar, industrial, and medical monitoring or control devices. These technicians may work in product evaluation and testing, using measuring and diagnostic devices to adjust, test, and repair equipment.

Working Conditions

Most engineering technicians work at least 40 hours a week in laboratories, offices, manufacturing or industrial plants, or construction sites. Some may be exposed to hazards from equipment, chemicals, or toxic materials.

Training and Advancement

Although it may be possible to qualify for certain engineering technician jobs without formal training, most employers prefer to hire someone with at least a two-year associate degree in engineering technology.

After completing the two-year program, some graduates get jobs as engineering technicians, whereas others continue their education at four-year colleges. Many four-year colleges offer bachelor's degrees in engineering technology, but graduates of these programs often are hired to work as technologists or applied engineers, not as technicians.

Outlook

Opportunities will be best for people with an associate degree or extensive job training in engineering technology. As technology becomes more sophisticated, employers will continue to look for technicians who are skilled in new technology and require a minimum of additional job training. An increase in the number of jobs related to public health and safety should create job opportunities for engineering technicians who have the appropriate training and certification. Overall employment of engineering technicians is expected to increase about as fast as the average for all occupations through 2014.

Earnings

Median annual earnings of electrical and electronics engineering technicians were $46,310 in May 2004. The middle 50 percent earned between $36,290 and $55,750. The lowest 10 percent earned less than $29,000, and the highest 10 percent earned more than $67,900. Table 9.1 shows median annual earnings in the industries that

Table 9.1 | Earnings of Engineering Technicians

Federal government	$64,160
Wired telecommunications carriers	$51,250
Architectural, engineering, and related services	$44,800
Navigational, measuring, electromedical, and control instruments manufacturing	$42,780
Semiconductor and other electronic component manufacturing	$41,300

employed the largest numbers of electrical and electronics engineering technicians as of May 2004.

Bureau of Labor Statistics, U.S. Department of Labor, Occupational Outlook Handbook, 2008–09 Edition, 2/2008, http://www.bls.gov/oco/

Summary >

All materials are composed of atoms. One atomic component accounts for electricity

Every atom is made of protons, neutrons, and electrons. Some of the electrons in atoms move away from their parent atom by the application of external energy. This flow of electrons is electricity.

Materials are classified by their ability to permit electricity to flow. Conductors are materials through which electricity easily flows. Insulators are materials that present a great opposition to the flow of electricity. Semiconductors are somewhat conductive.

Electron flow in an electrical circuit is called current. Voltage provides the energy to make current flow possible. The electrical resistance of the devices in the circuit regulates the current in a circuit. All materials have some amount of resistance.

The equation $I = V/R$ is one form of Ohm's Law. This equation tells us that, with a fixed voltage, an increase in resistance will decrease the current flow. Conversely, decreasing resistance increases the level of current.

Useful electronic products are made from complex circuits. The circuits are the interconnection of components such as resistors, capacitors, inductors, and transistors in series and parallel arrangements. Good engineers and technicians understand electrical and electronic theory, and they use this knowledge to design and test electronic products.

Silicon is the main material used in semiconductors. When small amounts of other materials are added to silicon, N-type and P-type semiconductor materials are formed. These materials then are used to make transistors and ICs.

Many types of transistors are manufactured. Bipolar transistors include the NPN and PNP. Another type of transistor is the field effect transistor (FET). Transistors are used as amplifiers or as switches.

Many ICs (integrated circuits, or chips) are formed on a wafer. Within each IC, millions of FET transistors are fabricated. Each transistor functions as an on-off switch. When interconnected to each other at the IC's metallization layer, complicated circuits are formed. Computer chips are common examples of the parts produced by the integrated circuit fabrication industry.

FEEDBACK

1. What is electricity? What is necessary for electricity to flow through a circuit?

2. Research how circuit breakers work.

3. An MP3 player draws 300 mA of current from a 12 V battery source. What is the resistance of the MP3 player?

4. Draw a series circuit schematic consisting of a 25 V battery and five 100-ohm resistors.

5. Using the series circuit in Question 4, determine the current provided by the battery to the resistors.

6. Using the series circuit in Question 4, confirm that the voltage across each 100-ohm resistor will be 5 V.

7. For a series circuit with a fixed voltage source, what happens to the current when the resistance is decreased?

8. It is a hot day. You buy four fans and plug them into outlets. The outlets are powered by a circuit protected by a 20 A circuit breaker. If each fan is rated at 250 W and the outlet voltage is 120 V, will the circuit breaker interrupt current flow?

9. Do research to find out why the frequency of the AC voltage in a typical home outlet is 60 Hz.

10. Investigate what the outlet frequency is in France, Australia, Japan, and Russia.

11. Students can sometimes talk to the astronauts on the International Space Station using amateur radio. A certain group of students use a radio transmitter operating at 145.8 MHz for this communication. How much time does it take for one cycle of this signal to complete?

12. High-voltage transmission lines carry electricity from power-generating stations to cities. If a line is specified as a 25,000-Vrms line, what is the line rating in voltage peak and voltage peak-to-peak?

13. Investigate how inductors and capacitors store electrical energy.

14. A transformer is used in a video-game system that requires 12 V to operate. The transformer is plugged into the house outlet voltage on one side and connected to the game system on the other side. Is this transformer stepping up or stepping down the voltage?

15. A potentiometer has 15 volts across it. Between the center terminal and one end, a technician measures 3.4 V. What voltage will the technician measure between the center terminal and the other terminal of the potentiometer?

16. Use online research to determine how Moore's Law applies to the integrated circuits industry.

17. Do online research to find out how many transistors are in the following microprocessor chips:
    a. 8085
    b. 8086
    c. Pentium III

18. Restaurants depend on electricity to power cooking equipment, cash registers, refrigerators, freezers, security systems, lights, and the HVAC (heating, ventilation, air conditioning) system. Visit a local fast-food restaurant and list all the electrical devices you notice. Using online resources, determine the restaurant's power requirements for the devices in your list.

DESIGN CHALLENGE 1:
Electrical Wiring Design

• Problem Situation

Your kitchen is outdated—it needs a face-lift. An important part of the renovation is to rewire the electrical circuits for the updated kitchen appliances. Unfortunately, your home breaker box is nearing its electrical capacity, and your budget does not allow for an upgrade in the electrical service. You will have to make do with what is available.

The existing breaker box has two 15 A circuit breakers and two 20 A circuit breakers that are available to use for the kitchen's new electrical requirements. You must design the wiring plan to use the available breaker capacity, minimize the amount of wiring needed for the new circuitry, and above all, create a safe, non-overloaded electrical system.

• Your Challenge

You and your teammates first must lay out the updated kitchen physical wiring plan on graph paper. Next, you will prototype the system using standard electrical devices.

The plan will accommodate five overhead lights controlled by a switch. Eight new electrical outlets will also be included in the plan. Small appliances for the kitchen will include a food processor, a can opener, a small flat-screen TV, a toaster, and a radio. Additionally, large appliances such as a refrigerator, microwave oven, dishwasher, and garbage disposal are required. Any other appliances and devices found in a standard kitchen are currently powered by the existing kitchen wiring, so you do not need to include these in your wiring plan.

• Safety Considerations

1. Always exercise caution when working with electricity.
2. Assume that circuits are energized.
3. Follow your instructor's safety guidelines for your lab.

• Materials Needed

1. Graph paper
2. Hook-up wire
3. 12 V bulbs
4. Switches
5. 120 V standard electrical outlets
6. 120 V standard switches
7. 12 V batteries or power supply

• Clarify the Design Specifications and Constraints

Your design will first be detailed on graph paper. Select a teammate's kitchen floor plan to use for your design. On the graph paper, draw the entire electrical system using standard schematic symbols. The design must ensure that no circuit will be overloaded.

Build a prototype of your design using 12 V light bulbs to represent the kitchen lights and appropriately sized resistors to represent the appliances. Devise tests to determine whether your system design meets the needs and safety requirements for the kitchen renovation.

Build a second prototype using standard electrical outlets and switches. If a circuit panel is available, tie each branch circuit into the panel breakers. Your instructor will determine if this prototype will be powered with 120 V or 12 V.

● Research and Investigate

To complete the design challenge, you need to first gather information to help you build a knowledge base.

1. In your guide, complete the Knowledge and Skill Builder I: Household appliance electrical requirements.
2. In your guide, complete the Knowledge and Skill Builder II: Circuit breakers and standard electrical wiring devices.
3. In your guide, complete the Knowledge and Skill Builder III: Voltage, current, resistance, and power.
4. In your guide, complete the Knowledge and Skill Builder IV: Series and parallel circuits.

● Generate Alternative Designs

Describe two of your possible alternatives to the wiring layout. Discuss the decisions you made in (a) distribution of the wiring layout and (b) assignment of devices to circuit breakers.

● Choose and Justify the Optimal Solution

Explain why you selected a particular solution and why it was the best.

● Display Your Prototypes

Construct your chosen design, and include either photographs or drawings to explain its operation.

● Test and Evaluate

Explain whether your design met the specifications and constraints. What tests did you conduct to verify this?

● Redesign the Solution

Explain how you would redesign your kitchen electrical system based on the knowledge and information you gained during this design challenge.

● Communicate Your Achievements

Describe the plan you will use to present your solution to the class. Include a media-based presentation.

DESIGN CHALLENGE 2:
Transistor Control of an LED

• Problem Situation

You tell your friends that you read that transistors are the heart of all electronic equipment and that a typical integrated circuit may contain more than a million transistors. Your friends ask you to show them how a single transistor works. They don't think you can do this. You decide to prove them wrong by showing how a simple transistor can control an LED.

• Your Challenge

LEDs require certain amounts of current flow for light to be visible. Your challenge is to show your friends that an LED will not glow when connected to a low-level signal, but it will when the signal is amplified by a transistor. In this way, you will demonstrate to your friends how the transistor operates. You must design the proper circuitry needed for the transistor to function reliably in this application, and you must create a source for the low-level signal.

• Safety Considerations

1. Always exercise caution when working with electricity.
2. Assume that circuits are energized.
3. Follow your instructor's safety guidelines for your lab.

• Materials Needed

1. Various colored LEDs
2. Resistor assortment
3. NPN transistor, such as a 2N3904
4. FET, such as a 2N7000
5. DC voltage or battery
6. Square-wave voltage source
7. Proto-board for circuit construction

• Clarify the Design Specifications and Constraints

Show how an LED will glow when an appropriate voltage is applied to the device. Devise a test to confirm exactly what voltages and currents are needed for light to appear. Show how this varies depending on the LED color. From this testing, select a voltage or current level that is insufficient to light up the LED. You will design a transistor circuit that can use this low level to light the LED.

Design the transistor circuit so that the low-level signal is the input to the transistor. You will design the remaining circuitry so the transistor can sufficiently amplify the signal to light the LED. Because a variety of transistor devices are available, justify which transistor component you ultimately select.

Because you really want to impress your friends, consider modifying the input signal so the LED does more than simply light up.

• Research and Investigate

To complete the design challenge, you need to first gather information to help you build a knowledge base.
1. In your guide, complete the Knowledge and Skill Builder I: LEDs.
2. In your guide, complete the Knowledge and Skill Builder II: Bipolar transistors.

3. In your guide, complete the Knowledge and Skill Builder III: Field effect transistors.

● Generate Alternative Designs

Describe two of your possible alternatives to the transistor circuit. Discuss the decisions needed to use a bipolar and a field effect transistor.

● Choose and Justify the Optimal Solution

Explain why you selected a particular solution and why it was the best.

● Display Your Prototypes

Construct your chosen design and include either photographs or drawings to show how it operates. Use data from test equipment to prove amplification is taking place.

● Test and Evaluate

Explain whether your design met the specifications and constraints. What tests did you conduct to verify this?

● Redesign the Solution

Explain how you would redesign your circuit based on the knowledge and information you gained during this design challenge. What additional features would you have added to the design if you had more time?

● Communicate Your Achievements

Describe the plan you will use to present your solution to the class. Include a media-based presentation.

INTRODUCTION

MOST PEOPLE ARE familiar with computers. They know that computers can quickly evaluate mathematical equations. They understand that computers control complex rocketry systems, are used in movie special effects, and that computers, in fact, control almost every electrical device we use. Cars, washing machines, and cell phones all rely on small computers

CHAPTER 10

COMPUTERS AND COMPUTER ARCHITECTURE

to function. You are probably very familiar with the common desktop and notebook computers used each day in schools, homes, and businesses.

All computers are designed with similar common features. When designing new computer systems, engineers and technicians combine special digital logic circuits, called computer hardware, which create the unique functions necessary for computing. Computer hardware, shown in the representative sample in Figure 10.1, is needed in order for computer instructions or software

Figure 10.1 | Most computer hardware is formed on microchips like this one.

Machine level code (8051 microcontroller)

```
LOC      OBJ          LINE    SOURCE
  0000                  1     ORG 0000H
                        2
                        3
                        4
0000     758902         5           MOV TMOD, #02H
0003     758C00         6           MOV TH0, #00H
0006     D28C           7     REPEAT:    SETB TR0
0008     308DFD         8           JNB TF0, $
000B     C28C           9           CLR TR0
000D     C28D          10           CLR TF0
000F     80F5          11           SJMP REPEAT
                       12
                       13     END
```

High level language code (C++)

```
#include <iostream>
#include <iomanip> // Required for formatting
using namespace std;
// This program converts Celsius temperatures to Fahrenheit
int main()
{
  const int MAX_CELSIUS = 50;
  const int START_VAL = 5;
const int STEP_SIZE = 5;

int celsius;
float fahren;

cout << "DEGREES    DEGREES\n"
     << "CELSIUS    FAHRENHEIT\n"
     << "                      \n";

celsius = START_VAL;
cout << setiosflags(ios::fixed)
     << setiosflags(ios::showpoint)
     << setprecision(2);

while (celsius <= MAX_CELSIUS)
{
  fahren = (9.0/5.0) * celsius + 32.0;
  cout << setw(4)   << celsius
       << setw(13) << fahren << endl;
  celsius = celsius + STEP_SIZE;
}
```

Figure 10.2 | Computer instructions tell the computer circuits how to operate: (a) Machine-level instructions operate directly on the hardware. (b) High-level languages must be translated or compiled into many machine-level instructions.

Introduction

to be processed or executed within the machine (see Figure 10.2). Sequences of individual computer instructions make up a computer program, which is the set of commands defining what a computer does at any given time.

In this chapter, we will examine computer architecture—that is, we will learn how computer hardware is organized. We will also see how computers do many fascinating tasks simply by manipulating ones and zeros.

SECTION 1: Digital Logic

KEY IDEAS >

- Basic digital logic functions are the building blocks of computer hardware design.
- Digital system design is based on the binary number system.
- Computer hardware is composed of key architectural elements including registers, counters, and ALUs.

Digital logic devices are used to create digital systems, including computers, MP3 players, high-definition TVs, and many other products. Digital logic devices are specialized electrical circuits built from transistors, diodes, and resistors. By arranging these electrical components in various ways, logic functions are created. Logic functions determine how digital signals are processed. When logic functions are combined, they produce digital systems. Basic digital logic functions are the building blocks of computer hardware design.

The computer is a common digital system and is the primary subject of this chapter. In order to understand how computers are designed, we need to learn about various common logic devices and about the binary number system.

Logic Families

Digital logic devices are useful in digital design because they exist in logic families. A logic family is a group of digital circuits with similar electrical properties. Logic families may contain thousands of specialized logic devices. Digital system designers like working with logic families since they can concentrate their design efforts on the logical design of a system rather than on the electrical design. By eliminating the need to consider the voltages and currents in a circuit, for example, the designer is free to interconnect devices at the logic level to build the digital system desired. Manufacturers consider electrical requirements when they create logic families, giving the digital designer the freedom to treat digital devices as interconnecting building blocks. Figure 10.3 shows how various electronic parts are connected to create a logic device.

CMOS (complementary metal oxide semiconductor) is one circuit configuration used in constructing logic devices. Another widely used circuit form is TTL (transistor-transistor logic). CMOS and TTL logic families

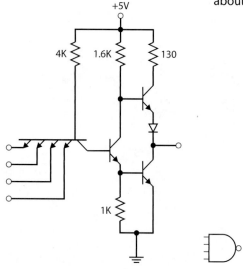

Figure 10.3 | Digital devices, such as this 4-input NAND gate, are constructed from electronic components.

have been available for many years. Each has unique electrical characteristics, and both create similar logic devices or functions. When designers interconnect CMOS devices, for example, each device understands the electrical signals sent to it from other CMOS devices. Each device understands the signals from others in an electrical sense; thus, logic design becomes relatively easy.

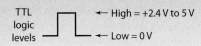

Figure 10.4 | Logic levels in the TTL system consist of two distinct ranges of voltages.

Voltage signals in digital systems are far different from voltage signals in a radio or a table lamp. Many analog systems use a range of voltages to function, but digital systems use only two distinct voltages. Typical voltages in a TTL-based digital system are, for example, 0 V and 5 V. This means each TTL logic component produces an output of either 0 V or 5 V and responds only to inputs of either 0 V or 5 V. This is fairly simple and straightforward and leads us to consider that a digital device is either off (0 V) or on (5 V). Any two voltages can represent on and off conditions, but certain voltage levels (such as 0 V and 5 V) are standardized by industry. Another common set of standardized voltages used in digital systems is 3.3 V and 0 V.

The logic devices are actually more flexible and variable than mentioned. Using TTL again as an example, the output of a TTL part is considered on if its output voltage is anywhere between 2.4 V and 5 V. We do not care if the output voltage is 2.6 V or 3.7 V, or any other number of volts, for each value between 2.4 V and 5 V is considered to be in the on condition. The off condition exists if the output is anywhere between 0 V and 0.4 V. The voltage ranges make it easier to interconnect parts, since a unique, specific voltage level is not necessary (see Figure 10.4).

The key point is that digital logic devices produce and respond to only two different conditions, on and off. Moreover, because it is easy to build circuits that respond only within two different voltage ranges, the digital parts are inexpensive.

Binary Numbers

Digital system design is based on the binary number system (also called base two), which uses only two numbers, zero and one. This should seem reasonable as digital devices use two voltages. When using the binary number system with digital devices, we say devices that are on are in the on state or high state or one state. The binary number 1 (one) usually signifies the on state. Any device that is off is considered to be in the off state or low state or zero state, and this state is usually designated by the binary number 0 (zero).

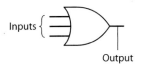

Figure 10.5 | A logic circuit, such as this 3-input OR gate, has signal input wires and a signal output wire.

A single one or zero is called a bit; a bit is a single binary digit. Using binary numbers and digital devices, designers develop digital systems based on simple on-off responses. In actual practice, the binary numbers used will consist of more than one bit. Eight bits grouped together are common in digital systems. Eight bits make up a byte.

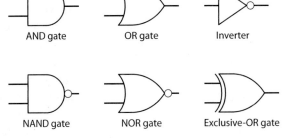

Logic circuit functions are defined by inputs (the wires leading to a circuit) and outputs (the wires coming from a circuit). The signal on an input or output line is always either at the one or zero level. No other condition is possible. Most digital circuits also have multiple inputs (see Figure 10.5). Depending on the one-zero values of these inputs, the circuit output responds with the appropriate one or zero level. The output level is determined by the circuit's logical function. Figure 10.6 shows the electronic symbols used to represent many of the digital devices you will study in this chapter.

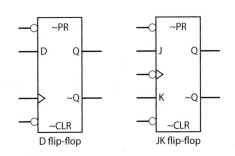

Figure 10.6 | Shown here are some of the digital logic device symbols used in the computer and electronics industry.

Binary numbers can represent many different things in digital systems. A binary number may express a specific numerical value,

Section 1 ▵ Digital Logic

or it may convey other information, such as a letter, a color, or a sound level. For example, binary numbers conveniently represent letters and numbers on a computer keyboard, which typically has 104 keys. Each individual key is a simple on-off switch, and all are identical, but the keyboard's function is to produce unique signals for each key. There must be a means to differentiate one key symbol from another.

Every key switch could be wired to the computer, but it is senseless and unnecessary to install one wire per key, or 104 wires total. (If you take a moment to check your computer keyboard connector, you will see that it does not contain 104 pins.) The binary number system makes it easier to represent all the possible key combinations with fewer wires. For example, the letter "A" is represented as eight binary bits (00011110); "B" is represented as a different grouping of eight binary bits (00110000). Each of the 104 keys has its own unique 8-bit binary code (called a scan code), meaning that only eight bits are needed to carry the keyboard information.

Let us examine why a few bits can carry so much information. The secret is to understand the relationship between the number of bits and the combinations of binary numbers the bits produce. We already know that one bit may represent a one or a zero. Based on this fact, two bits will represent exactly four combinations of ones and zeros (00, 01, 10, 11). Taking this concept a little further, you will see that there is always a mathematical relationship between the number of bits and the number of combinations. This relationship is determined with the following formula:

$$\text{Number of Combinations} = 2^N$$

where N = number of bits.

As an example, using four bits we determine the number of combinations:

$$\text{Number of Combinations} = 2^4$$

$$\text{Number of Combinations} = 2 \times 2 \times 2 \times 2$$

$$\text{Number of Combinations} = 16$$

By running a few numbers, we see that each additional bit doubles the number of combinations. Can you figure out how many combinations are possible with 16 bits?

When designing digital circuits, binary numbers are listed in a counting sequence. This makes it easy for the designer to track individual input combinations and the meaning for each in a particular design. We increase a binary count the same way we increase a decimal count, one value at a time. In addition, just as the number of digits determines the largest decimal number one may represent, the number of bits determines the largest binary number one may represent. For example, using three decimal digits, you can count from zero to 999, which represents one thousand unique decimal values. The binary counting sequence using four binary bits is shown in Figure 10.7. Four bits permits us to count from 0000 to 1111, a range which represents sixteen unique binary values.

Each bit in a binary number has a numerical value or weight based on its position in the number. This, again, is similar to the decimal system. For example, the decimal number 234 reads four times one, plus three times ten, plus two times one hundred. The weight of each decimal position increases by ten for every decimal position to the left. In the binary system, the weight *doubles* for every binary position to the left.

For binary numbers, the value of the rightmost bit position is always one. The value of the next position to the left is two. Moving further left, the weights of the

Binary Numbers	Decimal Numbers
0000	0
0001	1
0010	2
0011	3
0100	4
0101	5
0110	6
0111	7
1000	8
1001	9
1010	10
1011	11
1100	12
1101	13
1110	14
1111	15

Figure 10.7 | Binary numbers are counted in sequence, just as decimal numbers are.

positions are 4, 8, 16, 32, 64, 128 and so on. Using this knowledge, we can easily count in a binary sequence and have a method for relating the numerical values of binary and decimal numbers. For example, the decimal value of the four-bit binary number 0110 is found by adding the binary weights of each one-level bit position (we ignore the zero bits). Thus, 1×2, plus 1×4, equals 6. A binary 0110 is equal to a decimal six.

Important Logic Functions

Just a few basic logic functions form the foundation for most digital designs. These logic functions are defined by the output signal produced for certain combinations of input signals. We discuss these functions next.

AND Gate The AND gate is the digital device producing the AND function. The AND gate produces a one-level output signal when all the gates' inputs are at one levels. When any AND gate input is zero, the output is zero. For example, an AND gate with four input wires may receive any of sixteen possible binary input combinations. Only the combination consisting of all ones produces an output level of one. Each of the fifteen remaining input combinations has at least one of its input wires equal to zero, so each combination produces a zero output level. Typical devices have between two and sixty-four inputs.

Incidentally, the term "gate," when used with logic devices, has a different meaning than the term "gate" as it was used in the last chapter in the discussion of FETs. An FET's gate is a physical part of the FET. In logic devices, the term "gate" refers to the control a logic device has on input signals.

How is something as fundamental as an AND gate used? An example is the familiar CTRL-ALT-Delete key sequence you might type on your computer. An AND function is used to tell the computer that all three keys have been pressed simultaneously. The computer recognizes the sequence only when CTRL, ALT, and Delete are pressed at the same time. All three keys must provide a high-level signal at the same moment for the sequence to be recognized. Figure 10.8 shows a logic schematic illustrating this AND operation.

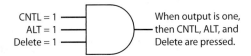

Figure 10.8 | This simple circuit shows how an AND Gate is used to detect the Ctrl-Alt-Delete key sequence.

OR and Exclusive-OR Gates The OR gate produces another primary logic function. The OR gate output is one if even one of its inputs is one; the output is zero only when all the inputs are zero. Figure 10.9 shows that the output is one when input A or B or C is one. You can also see that the OR gate produces an entirely different result than the AND gate.

An important variation of the OR function is the Exclusive-OR (EX-OR) gate, which produces a one output when either of the two inputs is one. When both inputs are ones or both are zeros, the EX-OR produces a zero output. The EX-OR function is used extensively in mathematical, error checking, and encryption circuitry.

Inverter, NAND, NOR Changing a signal's level from one to zero or vice-versa is an important logical operation accomplished by a circuit called an inverter or NOT gate. The inverter has a single input and a single output, and its sole function is to flip the logic level. When inverters are combined with ANDs and ORs, two additional logic functions result, the NAND and NOR.

2-Input OR

A	B	Y
0	0	0
0	1	1
1	0	1
1	1	1

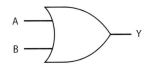

(a)

2-Input Exclusive-OR

A	B	Y
0	0	0
0	1	1
1	0	1
1	1	0

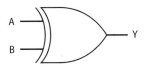

(b)

Figure 10.9 | This illustration shows the symbol and truth table for a 2-input OR gate (a), and the symbol and truth table for an Exclusive-OR gate (b).

3-Input NAND Gate

A	B	C	Y
0	0	0	1
0	0	1	1
0	1	0	1
0	1	1	1
1	0	0	1
1	0	1	1
1	1	0	1
1	1	1	0

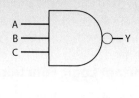

(a)

3-Input NOR Gate

A	B	C	Y
0	0	0	1
0	0	1	0
0	1	0	0
0	1	1	0
1	0	0	0
1	0	1	0
1	1	0	0
1	1	1	0

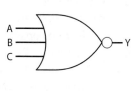

(b)

Figure 10.10 | This illustration shows how the symbols and truth tables are represented for a 3-Input NAND gate (a) and a 3-Input NOR gate (b).

NAND stands for 'Not AND' and NOR stands for 'Not OR'. These devices produce the inverted responses of the AND and OR gates as shown in Figure 10.10. NANDs and NORs are powerful logic functions. It is possible to design and build an entire digital system just from NAND or NOR devices (lots of them, of course). You may even own devices built with these circuits, such as MP3 players and USB thumb drives, which use flash memory chips based on NAND technology.

Flip-Flops The ability to store ones and zeros is also important in a digital system design. Memory circuits do this. Digital systems need memory to store any quantity of information, from a single bit to many gigabits.

Flip-flops are digital circuits that store a single bit (see Figure 10.11 for an example of a flip-flop). They respond to special timing or triggering signals that indicate exactly when a bit is ready to be stored (for instance, when numbers change on a digital clock). The flip-flop dutifully retains the stored value until another trigger signal causes a change to occur.

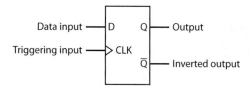

Figure 10.11 | A D-Flip-Flop has inputs labeled D and clock (CLK), as well as two complementary outputs labeled Q and not Q.

Small numbers of flip-flops combined together are called registers. Registers are common storage areas in computers and typically store between 8 and 64 bits. Memory devices, on the other hand, are built to store many millions of bytes of information at a time. Memory chips can be thought of as large numbers of flip-flops organized to work together to efficiently store massive amounts of binary information. Memory devices are discussed in detail in another section of this chapter.

Combinatorial and Sequential Circuits

Digital circuits are classified by their operating characteristics. There are two general classes of digital circuitry—combinatorial and sequential circuits.

In combinatorial circuits, the output is directly dependent on circuit input signals. Sequential circuitry contains storage elements so that the output level is dependent on existing input signals as well as previously stored data. Sequential circuits store data in flip-flops or memory.

In Figure 10.12, a truth table shows every binary combination possible for the number of inputs in a combinatorial circuit. Truth tables are created to detail how the circuit should react for every input combination. In this example, A, B, C, and D are the names assigned to the circuit inputs. The output is identified with the name Y. Since there are four inputs, this circuit has $2^4 = 16$ input combinations. Notice how the input combinations are listed in an ascending binary count. Each binary number represents a possible set of signals for the circuit. Here, the truth table describes how the circuit reacts for each set of input values.

This truth table shows that output Y will produce a one level only when combinations ABCD = 0110 or ABCD = 1100 are present at the circuit's inputs. All other input combinations are designated to produce a zero output response. AND, OR, and Invert gates are used, as shown in Figure 10.12, to create this circuit. After the truth table is created to describe the circuit response, AND gates and inverters are used to decode the combinations that will produce the one-level output. An OR gate further combines the two decoded signals into one final output. The resulting circuit produces a one output only for two specific input combinations. So, by default, the same circuit produces a zero level output for all remaining input combinations.

The example in Figure 10.12 shows the most common type of combinatorial circuit structure. The circuit is comprised of AND gates connected to an OR gate. Circuits with this structure are called AND-OR networks.

Engineers use many techniques to design combinatorial circuits. It is possible to design two circuits that produce the same output, but one will require fewer parts, so digital designers employ specialized minimization techniques to reduce circuitry size without changing its function. For this purpose, computer aided design is a helpful tool.

One common technique in designing and simplifying combinatorial circuitry is Karnaugh Mapping (K-map). K-maps graphically rearrange truth table combinations to make circuit minimization obvious to the trained eye. This is important because circuits with fewer parts are smaller, more energy efficient, and more reliable.

Sequential circuit design is much more involved than combinatorial design and requires specialized design knowledge. The counting circuit described in the next section is a simple example of a sequential circuit. The microprocessor chip, discussed later in this chapter, is an example of a highly complex sequential system.

As you have seen, basic digital functions combine to create complex circuits. Complex circuits, in turn, combine to produce computers and other sophisticated digital systems. Computer hardware is composed of key architectural elements, including registers, counters, and ALUs. Within these systems, certain circuits tend to recur. We will examine these important commonly used circuits next.

A	B	C	D	Y
0	0	0	0	0
0	0	0	1	0
0	0	1	0	0
0	0	1	1	0
0	1	0	0	0
0	1	0	1	0
0	1	1	0	1
0	1	1	1	0
1	0	0	0	0
1	0	0	1	0
1	0	1	0	0
1	0	1	1	0
1	1	0	0	1
1	1	0	1	0
1	1	1	0	0
1	1	1	1	0

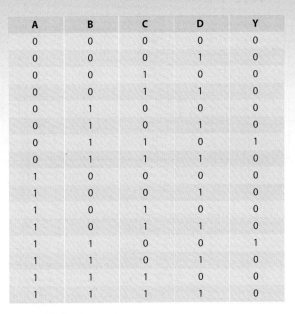

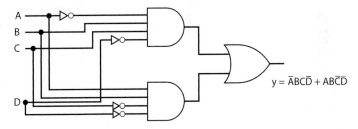

$$y = \bar{A}BC\bar{D} + AB\bar{C}\bar{D}$$

Figure 10.12 | This combinatorial circuit is designed to produce the outputs levels shown in the truth table for each corresponding input combination.

Counters

Counters are digital circuits that react to input signals and consequently produce a binary counting sequence. Some counters are designed as up counters, which produce a binary counting sequence from zero to some final maximum value.

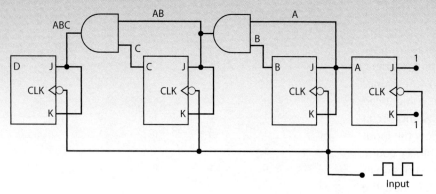

Figure 10.13 | A, B, C, and D are the four outputs of this 4-bit Binary Up Counter.

Down counters provide a count sequence, which decreases from a maximum value to zero. A common microwave oven uses a down counter. For example, after cooking time is set on the microwave, the oven's digital display, controlled by a down counter, counts down to zero. The counter's zero value visually informs the user that cooking time has ended. The counter may also generate other signals that turn off the heat, ring a bell, or flash a message.

Counters use flip-flops in their designs. Figure 10.13 illustrates the circuitry for a 4-bit up counter. 4-bit means that four flip-flops are used to build the counter and each provides an output signal. The number of flip-flops used determines the counting range. The counter in Figure 10.13 counts from 0000 to 1111, which comprises sixteen different counts. We refer to individual counts as states. In digital systems, each counter state may drive other circuitry in the system. Counter circuits are particularly useful for controlling the sequencing of events. In the microwave oven example, the zero state of the counter signals additional circuitry to shut down the oven.

So far, we have not defined what the counter is actually counting. We often refer to the counter's input signal as a clock pulse. The actual clocking signal may be a periodic electronic timing signal or a signal generated by a sensor. When a clock pulse arrives at the counter's input, the counter advances its count by one state. For example, if the counter is currently in its fifth state, a clock pulse will advance the count to the sixth state.

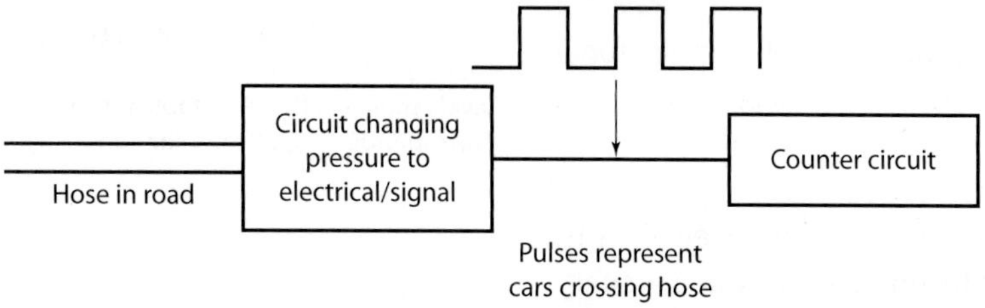

Here is a simple example of actual event counting. Have you ever driven over a hose placed across a road by the highway department? The hose is an input device used to help count the number of cars using a highway. A lack of pressure in the hose might produce a digital zero while an increase in pressure could produce a digital one level. The weight of your car compresses the hose as you drive over it, changing the air pressure within the hose. A pressure sensor detects the air pressure change and converts the change into an electrical signal. These changing signals become clocking signals for a counter. In this arrangement, the counter counts cars as the changes in pressure are detected (see Figure 10.14).

Figure 10.14 | A clocking signal derived from pressure changes in a hose can trigger a counter and thus be used to count automobile traffic.

Arithmetic and Logic Unit

A critical task for any computer is performing mathematical operations. The digital circuitry responsible for mathematics is the Arithmetic and Logic Unit, or ALU. The ALU circuitry usually performs mathematical operations such as adding, subtracting, multiplying, and dividing. Complex ALUs perform high-level mathematical operations, as well. Logical operations such as AND, OR, and Invert are also implemented within the ALU. Mathematical circuitry uses many logic gates. In most computer processing chips, a significant amount of the circuitry is devoted to mathematical operations.

We can demonstrate the use of basic logic gates in mathematical circuitry by examining the simplest addition circuit, the half adder. Figure 10.15 shows this circuit, along with its truth table, which shows all four possible additions. The half

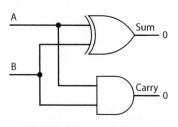

A	B	Carry	Sum
0	0	0	0
0	1	0	1
1	0	0	1
1	1	1	0

Figure 10.15 | This half adder circuit requires only two gates to add two one-bit numbers together.

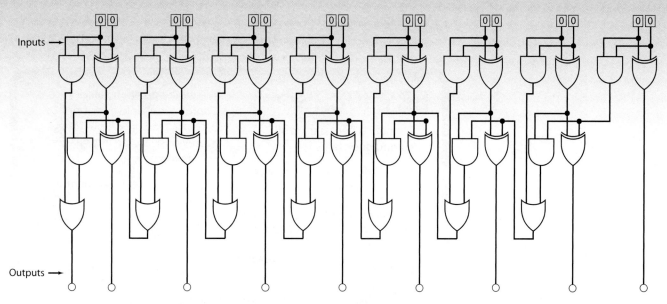

Inputs →

Outputs →

Figure 10.16 | Large scale adders, such as this 8-bit adder, are formed by cascading many half and full adders.

adder adds two individual bits together. A two-bit answer is produced on the half adder outputs.

The half adder circuit must have two outputs for results because the largest result in binary requires two places (1 + 1 = 10). These outputs are named sum, which is the least significant bit of the result, and carry, which is the most significant bit of the result. The sum is generated using an Exclusive-OR gate, while the carry is generated using an AND gate.

Figure 10.16 shows the circuitry for a larger but relatively simple 8-bit adder (for example, 10000111 + 10000100 = 100001011). An 8-bit adder is capable of adding any two 8-bit numbers together. Notice how the number of logic gates needed for this level of addition has increased dramatically. Because logic gates work at electronic speeds, this adder easily performs over 1.5 million additions per second. Nonetheless, in modern computing systems, this is slow. So, for faster computation, even more circuitry is required, adding substantially to the amount of the overall ALU circuitry. You can imagine how many gates would be necessary to add two 64-bit numbers together at high speeds, which is commonplace in desktop microprocessor chips. Also, keep in mind that we are examining only adder circuits. A full ALU will have considerable additional circuitry to support all the other mathematical and logical functions required.

SECTION ONE FEEDBACK >

1. Calculate the decimal value of the binary number 10101.
2. Sketch the truth table for a 5-input OR gate.
3. Calculate the number of input combinations a 64-input AND gate will have. How many of these combinations produce a one-level output?
4. How could a counter circuit be used to count paint cans moving along a conveyor line in a factory? How could the counter help group eight cans together for packaging?
5. Prove that the 8-bit adder circuitry shown in Figure 10.16 adds 10010111 + 11001010 and produces an answer of 101100001.

KEY IDEAS >

- Memory is necessary in all computer systems.

- Many different memory technologies exist.

- Memory chips contain electrical circuitry capable of storing a one or a zero.

- A write operation places information into a memory chip; a read operation retrieves previously stored information from memory.

- Volatility describes a memory chip's ability to retain information once it is stored.

Computer systems cannot function without memory devices. Many different memory technologies exist, each possessing characteristics desirable for specific applications. Typical computer systems use several of these memory technologies.

Data in Memory Chips

Memory chips contain electrical circuitry capable of storing a one or a zero. Stored information is called data. Electrical signals activated memory's storage circuitry to accept digital values as data and then store the data at specific locations inside the memory. This storage capability of a memory device is similar to that of a flip-flop. The difference between the flip-flop and memory is that a flip-flop stores one bit of information, whereas a memory chip stores millions of bits of data (see Figure 10.17). This is why the computer industry expresses memory sizes in Megabytes (MB) (approximately a million bytes) and Gigabytes (GB) (approximately a billion bytes). Integrated circuit fabrication makes it possible for a single memory chip to hold vast amounts of data. Memory is necessary in all computer systems.

The enormous storage capacity of a memory chip requires mechanisms to track and retrieve stored information, similar to locating papers in a file cabinet. To understand how memory works, we will discuss the key memory concepts of organization, reading, writing, and addressing.

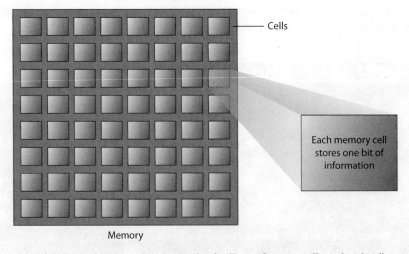

Figure 10.17 | Individual memory chips are made of millions of storage cells, and each cell may store a one or a zero.

Memory Cell

Memory chips store one bit of information within an electrical circuit called a cell. A memory chip has millions of cells, but a typical memory operation (such as storing or retrieving) requires using only a few cells at a time. It is common, for instance, to store or retrieve eight bits at a time. The cell construction varies from technology to technology. Some cell structures are flip-flops, including that of an important type of memory called static RAM (SRAM). The cells in another memory technology, called dynamic RAM (DRAM), are constructed from one transistor and one capacitor. There are many other technologies available. Each has interesting operating characteristics, but the key work of any cell is to store and retain a single binary bit.

As mentioned, a useful memory chip has millions of cells, and keeping track of the cells is important. Memory chips are manufactured with cells organized in rows and columns. This organizational pattern also describes the functional specification of the memory chip. For example, a chip that stores or retrieves eight bits at one time and has 16 million groups of eight cells is called a 16M $\times$ 8 memory chip. The 16M represents the chip's 16 million locations or rows, and the 8 represents eight columns, which is the number of bits of information available at one time. In this example, the eight bits of data are the word size, or amount of data used at any one time. Once you understand memory organization, the naming convention of chips is easy to interpret. A memory chip must have the circuitry to select any one of its row locations at any time. Addressing circuitry, discussed in more detail shortly, accomplishes this task. Addresses are the numerical way to locate information in a memory chip.

What Is a Million?

▷ In the computer memory business, one million is an approximate amount.

With the binary number system, the closest value to one million is 1,048,576

because 2^{20} equals this number. It is more convenient to say "one million" than "1,048,576."

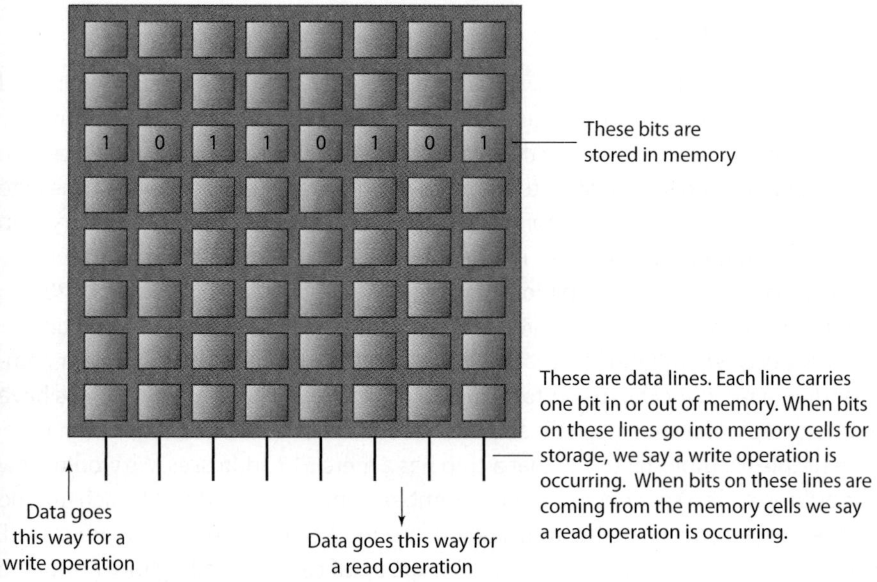

These bits are stored in memory

These are data lines. Each line carries one bit in or out of memory. When bits on these lines go into memory cells for storage, we say a write operation is occurring. When bits on these lines are coming from the memory cells we say a read operation is occurring.

Data goes this way for a write operation

Data goes this way for a read operation

Figure 10.18 | Data flows into a memory chip during a write operation and flows out of the memory chip during a read operation.

Writing and Reading

In memory systems, Writing is the act of storing information, and reading is the act of retrieving it (see Figure 10.18). There are pins on a memory chip exclusively for data. For instance, an 8-bit memory chip has eight data pins, which connect external data to the memory chip. Electrical signals are sent to the memory chip control pins, commanding the chip to store information in selected or addressed cells. The timing of signals ensures that the write operation occurs at the correct moment. The designer is responsible for choreographing the relationship between all signal and data lines.

Eventually, a read operation retrieves previously stored information from memory. Electrical signals command the chip to copy information from addressed memory cells and place it onto data pins for another device, such as the CPU, to use. The read operation is the inverse of the write operation (see Figure 10.19).

These are address lines. The number of lines is related to the number of rows in the memory chip. The binary pattern on these lines determines which row is selected for a read or write operation.

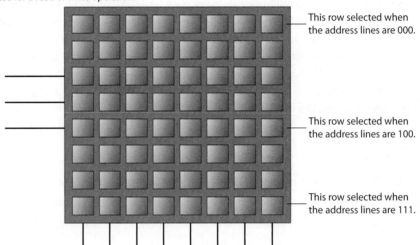

This row selected when the address lines are 000.

This row selected when the address lines are 100.

This row selected when the address lines are 111.

Figure 10.19 | Each memory operation involves only a few cells, which are selected by the addressing circuitry of the chip.

Memory Addressing

Let us examine in more detail how addressing circuitry controls the cells used in read and write operations. A certain number of cells is used for any one operation, and the number of cells used is based on the way the chip is organized. For example, a certain chip manufactured to read or write eight bits at a time reads or writes data in one byte quantities, and each one byte storage location has a unique address. A memory chip capable of storing a million bytes is organized as a 1M × 8 chip. This means there are approximately one million 8-bit locations or addresses on the chip. Figure 10.20 lists common memory chip sizes according to their address and word size organization.

Previously, we mentioned that electrical control signals are necessary in commanding a memory chip to perform a read or write operation. Memory chips have numerous pins for these signals, and many carry the chip's addressing information. As a simple example, let us say that a chip has a mere 512 addresses. We would view the addresses as a sequence of unique numbers ranging from 0 to 511. Each unique address represents or points to a specific group of cells in which data are stored. Therefore, in order to activate any specific group of cells for reading or writing, the specific cell address must be supplied to the chip. If the chip stores eight bits per address, then each address points to eight specific cells.

Addresses are binary numbers, and there is a relationship between the number of addresses on a memory chip and the number of pins needed to represent those addresses. Recall that in the binary system, $2^9 = 512$. This tells us that the memory chip requires nine pins just for the addressing information. Since we keep track of addresses in a binary counting order, the first address in this example is 000000000, while the last is 111111111. The 9-bit combinations from 000000000 to 111111111 represent all 512 addresses.

It is helpful to visualize memory addresses as house addresses. Within the memory chip, an electrical version of mail delivery takes place. In the same way your mail carrier reads and decodes the address on an envelope and delivers it correctly, addressing circuits decode each address supplied on the memory chip's address lines and deliver a signal, activating the appropriate cells.

Memory Characteristics

Modern computing systems use several memory technologies, since no single technology meets all the requirements of a computer system. The key technology considerations in selecting a memory chip for design are speed, density, and volatility.

Speed The speed of a memory chip determines how quickly a read or write operation occurs. Computer chips carry out read and write operations in nanoseconds. We define the overall speed of memory chips by their access time, the time it takes for data to appear during a read operation or the time it takes to store data during a write operation. (A write operation places information into a memory chip.) Ideally, memory chips should work as fast as any of the other components of a computer system. Unfortunately, this is currently not technically or economically possible. Therefore, designers use various memory technologies in computer system designs.

Density Density refers to the number of cells per chip. The cell's electrical structure determines its physical size, which, in turn, determines the number of cells possible on a chip: The more cells, the greater the density of the chip. Building computer systems with fewer chips means using fewer parts, thus keeping costs down.

Volatility Volatility describes a memory chip's ability to retain information once it is stored. Your home computer, for example, has both volatile and nonvolatile memory chips. The nonvolatile chips retain their data whether the machine is on or off. The volatile chips lose information as soon as the power is cut.

Take a closer look now at how these three characteristics affect a home computer system.

Unfortunately, no single memory technology provides high speed, great density, and nonvolatility all in one chip. Computer designers use the memory technologies best suited for specific computer functions.

Consider the process of turning on your computer. When power is first applied, we say the system boots. This means that the computer reads simple instructions from a memory chip in order to prepare the computer for use. Initialization programs, which test the machine's components and start some of the computer's hardware, are stored in nonvolatile memory. Therefore, every time the system starts, the same instructions run first, readying the machine for your use.

Say you start a video game. The game program is loaded from a hard drive into dense memory chips so that the millions of complex instructions defining the game are ready for the computer to execute. You surely want the game to run fast, so some of the game instructions are stored in high-speed memory chips in order to keep up with the computer's speedy processor chip. The three memory characteristics of speed, density, and volatility have all affected the operation of a computer system.

Total Capacity	Organization
256k	32k × 8
1M	128k × 8
2M	128k × 16
2M	256k × 8
4M	256k × 16
4M	512k × 8
8M	512k × 16
16M	512k × 36
4M	1M × 4
8M	1M × 8
16M	1M × 16
36M	1M × 36
16M	2M × 8
32M	2M × 16
4M	4M × 1
32M	4M × 8
64M	4M × 16
128M	4M × 36
64M	8M × 8
128M	8M × 16
128M	16M × 8

Figure 10.20 | Memory chip sizes are organized by address range and word size.

RAM and ROM

Semiconductor memory technologies fall into two main classifications, Random Access Memory (RAM) and Read Only Memory (ROM) (see Figure 10.21 for a more in depth classification of memory technologies). RAM devices, which make up the main memory in a computer system, tend to be volatile. RAM chips directly connect to microprocessor chips because RAM can keep up with the microprocessor's

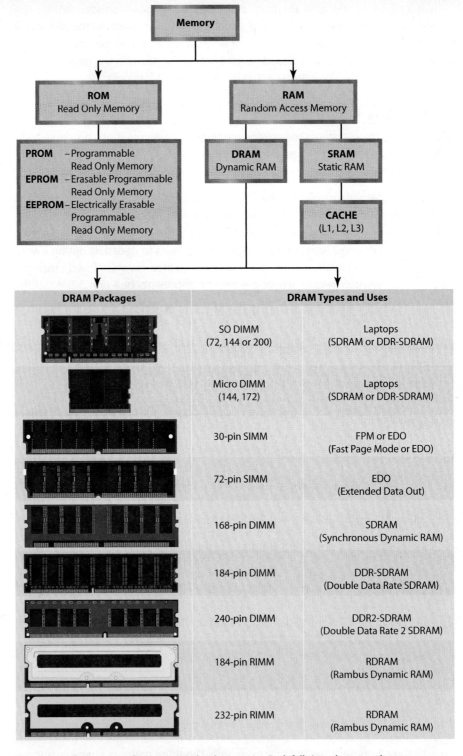

DRAM Packages	DRAM Types and Uses
SO DIMM (72, 144 or 200)	Laptops (SDRAM or DDR-SDRAM)
Micro DIMM (144, 172)	Laptops (SDRAM or DDR-SDRAM)
30-pin SIMM	FPM or EDO (Fast Page Mode or EDO)
72-pin SIMM	EDO (Extended Data Out)
168-pin DIMM	SDRAM (Synchronous Dynamic RAM)
184-pin DIMM	DDR-SDRAM (Double Data Rate SDRAM)
240-pin DIMM	DDR2-SDRAM (Double Data Rate 2 SDRAM)
184-pin RIMM	RDRAM (Rambus Dynamic RAM)
232-pin RIMM	RDRAM (Rambus Dynamic RAM)

Figure 10.21 | Many specific memory technologies exist. Each falls into the general category of RAM or ROM memory.

operating speed; RAM's best asset is that read and write operations occur equally fast. ROM chips have unequal read and write access times, so ROM technologies are not practical as computer main memory. The significant advantage of ROM technologies over RAM is nonvolatility. Once data are stored on a ROM chip, it stays there with or without electrical power. ROMs do need power to read data, but do not lose data without power. This allows you to turn off your computer knowing it will work when you turn it on again.

SRAM

Static RAM (SRAM) memory chips are among the fastest RAM memory technologies available (Figure 10.22), so SRAM is the memory technology most often connected to the microprocessor. In fact, the powerful microprocessor chips used in home computers and workstations frequently have SRAM memory fabricated right on the microprocessor chip. Microprocessors execute instructions very quickly with this architecture, since the instructions are stored in the SRAM right next to the processor. In the field of computer engineering, cache memory is the name given to high speed SRAM used this way.

Figure 10.22 | Static RAM (SRAM) chips are noted for their fast access speeds.

The main disadvantage of SRAM is its relatively low density compared to other technologies. Each SRAM storage cell requires a considerable amount of circuitry to function, and each cell takes up a fair amount of space. Consequently, a chip will hold fewer cells. This lack of storage capacity makes SRAM expensive for large memory systems.

DRAM

Dynamic Random Access Memory (DRAM) chips are used when large amounts of memory are needed (Figure 10.23). DRAM chips are dense because each storage cell consists of only one transistor and one capacitor. This is a small-sized cell, making it possible to pack millions of cells on a single chip. If you buy a home computer, you will find it important to consider the amount of DRAM memory you buy.

Figure 10.23 | Dynamic RAM (DRAM) chips are noted for their great storage capacity.

DRAM chips have the peculiar characteristic of losing information after a short time, even with power applied to the chip. If a one value is stored in a DRAM cell, for instance, the information lasts for two to four thousandths of a second and then is gone. The short storage time occurs because of the way DRAM stores data. Ones are stored as an electric charge in microscopic capacitors. Although it is easy to charge the capacitor to a level representing the one value, the small capacitor size prevents much charge from being stored. Furthermore, the capacitors' poor retention characteristic means that charge will "leak off" after a short amount of time. In spite of its short storage time, the DRAM chip is still useful.

DRAM refresh circuits control DRAM memory systems to counteract the leakage problems. Using refresh circuits, data are periodically read and rewritten so information is retained as long as each cell is periodically used. The refresh circuitry is complex and makes DRAM memory designs complex. Even with this limitation, the low cost of DRAM makes it the best choice for large memory system applications.

DRAMs are typically connected together on small circuit cards called DIMMs (Dual In-line Memory Modules), as shown in Figure 10.24. A DIMM plugs into a connector slot in a computer system so that the user can control the amount of memory.

Figure 10.24 | Here, eight DRAM chips are connected to operate together on a module.

EEPROM

Electrically Erasable Programmable Read Only Memory (EEPROM) is a ROM technology. An EEPROM stores information in a special cell structure known as

Figure 10.25 | The Electrically Erasable Programmable ROM (EEPROM) chip can hold data even without power applied.

a floating gate. The floating gate is made from semiconductor material and may hold electrons, much like a capacitor. It is either charged or uncharged, yielding the two distinct binary levels. Unlike the capacitor used in a DRAM memory chip, the floating gate will not lose any charge and so is nonvolatile.

EEPROMs require elevated voltages for write operations in order to force charge into the floating gate, and they also require specialized equipment called device programmers to store data. Once programmed with data, the EEPROM chip is installed in a computer system (see Figure 10.25). If the information on the EEPROM needs updating, the chip must be removed from the computer system and returned to a device programmer for modification. This is called erasing the chip. Obviously, this process is very time consuming, so designers often build the device-programming circuitry into the computer. However, the write operation, whether done in-system or out-of-system, is slow. Additionally, the device programming circuitry is costly.

The EEPROM is electrically erasable. This key feature differentiates the EEPROM from the older EPROM (Erasable Programmable Read Only Memory), which can be erased only when out-of-circuit and by exposure to ultraviolet light. Only after this slow erasing process is the old EPROM chip reprogrammable.

FLASH Memory

Flash memory technology is similar to that of EEPROM and currently dominates the nonvolatile memory market. Flash stores data in many digital products, some of which are shown in Figure 10.26. Digital cameras use flash chips to store images;

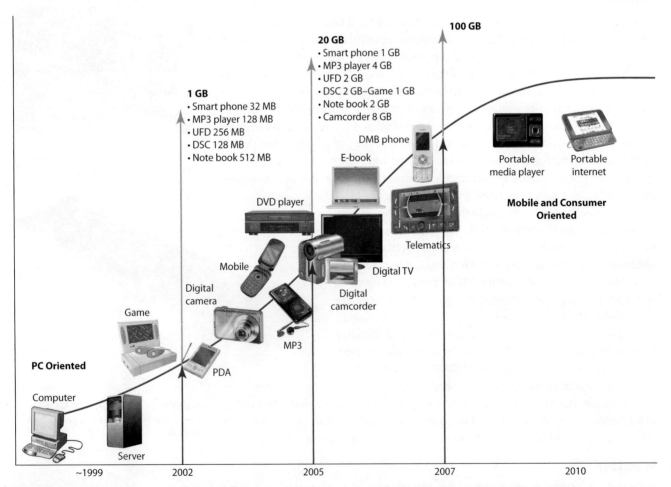

Figure 10.26 | Flash memory has operational characteristics similar to that of EEPROM and is the memory technology currently dominating the nonvolatile memory market.

MP3 players use flash to store music and video files; your USB thumb drive uses flash to store your term paper. In complex equipment, flash stores programming and operating system instructions. As in the EEPROM, flash write time is slower than read time. Unlike EEPROM, in which one can erase a single byte at a time, flash can be erased only in large blocks of information.

Flash is dense and therefore useful for applications in which large amounts of storage are necessary and slower write speeds tolerable. In an MP3 player, for example, a delay is acceptable when downloading a music file. In listening to a song, however, we expect no delays and appreciate flash's fast read operation.

MRAM

Magnetoresistive Random Access Memory (MRAM) is a promising nonvolatile memory technology. MRAM stores information magnetically, whereas the other ROM technologies store it electrically. MRAM read and write cycles are fast, which overcomes one of the serious limitations of other nonvolatile technologies. Also, MRAM densities are very high, giving this technology the potential to be the perfect memory chip. MRAM is still in development.

ENGINEERING QUICK TAKE

Increasing Memory System Capacity

Computer systems require large amounts of memory. Usually, one memory chip is insufficient to meet all requirements, so designers interconnect individual memory chips to create a larger memory system.

Memory capacity may be increased in two ways. If the designer's goal is to read or write more information at one time, she has to increase the data or word size of the memory system. If the goal is to increase the number of storage locations, she must increase the address range. Of course, she can also increase both dimensions if needed.

Let us assume we are designing a small computer needing a 512k $\times$ 8 memory system. The memory system design will use as many 256k $\times$ 1 memory chips as necessary because we have determined that these are readily available and inexpensive. We also know that a memory chip with 256k addresses will have 18 address lines (2^{18} = 256k). Since we are expanding the memory to 512k, we will need more addressing lines. But, how many will we need? Based on the fact that each additional bit doubles the number of combinations, we use one more bit, changing 18 to 19, and calculate that 2^{19} = 512k. We have now determined that the specified 512k addresses require a total of 19 address lines in order to access every storage location in the system. We also notice that the specifications call for eight data lines.

Our next calculation will show the number of 256k $\times$ 1 chips needed for this system. When we multiply 512k $\times$ 8, we get 4,096k. This informs us that the memory system has 4,096k total cells.

We determine the number of cells in each chip by multiplying 256k $\times$ 1 and getting 256k cells per chip.

We next divide the total number of cells required by the number of cells per chip (4,096k / 256k). This tells us that the system needs sixteen chips (4,096k / 256k = 16).

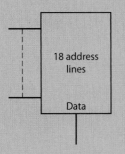

Figure 10.27 | It requires 18 address lines to fully address a 256k memory chip.

Next, we determine how the chips should be interconnected. Figure 10.27 is a representation of the address and data lines on a single 256k × 1 chip. Since the chip has 256k addresses, 18 address lines are present per chip. We know that the overall memory system requires 19 total address lines. It looks as if we have far more address lines than we can possibly use since sixteen chips with 18 address lines each equals a total of 288 address lines. We will need to solve this dilemma.

There is also a single data line on each of the sixteen chips, and we need to provide eight total data lines for the system. Once again, we seem to have more lines than we need.

Figure 10.28 shows how we will build the memory system. Of the sixteen chips, eight of them will connect their data lines to the other eight. As two lines connect, they fuse into one data line. Eight individual data lines result.

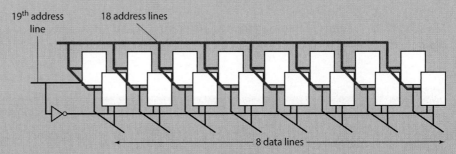

Figure 10.28 | Sixteen 256k × 1 chips are connected this way to form a 512k × 8 memory system.

We connect one address line from the first chip to the corresponding address lines on all the other chips. As we do this for each line on every chip, we combine lines to create eighteen total address lines. You can see that all the individual chips' lines are simply connected in parallel.

However, the system requires one additional address line to bring the final system total to nineteen. Each memory chip has a chip-select input line, which we can use as an address line. This nineteenth address line is the most significant one because it determines which half of the sixteen chips operates together at any one time. If all sixteen chips were active simultaneously, data from one-half of the system would electrically conflict with data from the other half, so it is critically important in this design that only eight chips are active at once. When chip-select is active, the chip places data on its data lines; otherwise, the lines are electrically deactivated.

SECTION TWO FEEDBACK >

1. What memory technology is most useful for booting a computer system?
2. How many storage cells are in a 4M × 8 flash chip?
3. What is the word size for a 4M × 8 flash chip? How many address lines does this chip have?

KEY IDEAS >

- Computer systems use microprocessor chips as the main computing device.

- The heart of any microprocessor chip is the central processing unit, or CPU.

- The main function of the CPU is to process instructions.

- Processing each instruction occurs in two phases called fetch and execute.

- Instructions are binary commands directing the CPU to carry out relatively simple actions.

What do you think of when you hear the word "computer"? Do you visualize the desktop machines in your computer lab at school? Maybe the wireless notebook your friend uses to check e-mail comes to mind. Possibly, you are thinking about the device controlling the airbag in your car. It is possible that you know your cell phone has a computer chip controlling how you download, organize, and listen to music.

Figure 10.29 | Embedded processors are computer chips placed inside a product in order to control the product's functions.

Microcontrollers and Microprocessors

Familiar computers, such as desktops and notebooks, are obvious to most people. Other computers are not so obvious. Many common products are computer controlled, but do not appear computer-like to the user. Computer chips hidden inside a product (joystick, coffeemaker, windshield wipers) are embedded processors (see Figure 10.29), sometimes known as microcontrollers. Sales of embedded processors exceed those of any other computing chips. Usually, an embedded processor is programmed to do very specific things, and the user of the product is unaware of its existence. For instance, when sending a text message, are you thinking, "I'm using a computer"? Probably not, because you are concentrating on the keys you press to compose the message. An embedded processor is working behind the scenes to handle the many technical processes required to compose and send the message. Of course, desktop and notebook computers also need computing chips to function. In these general-purpose computers, the digital hardware doing the main computing is the microprocessor. Computer systems use microprocessor chips as the main computing device.

The modern microprocessor is an integrated circuit containing millions of transistors that form the digital logic architecture of the chip. The prefix "micro" distinguishes these chips from older, larger computers. When integrated circuit technology advanced to the point where millions of transistors per chip became feasible, the microprocessor era began, and the computer on a chip became reality. The Intel Pentium and the AMD Opteron lines of chips are examples of microprocessors.

The next section defines more specifically the key hardware found inside a modern computer microprocessor chip.

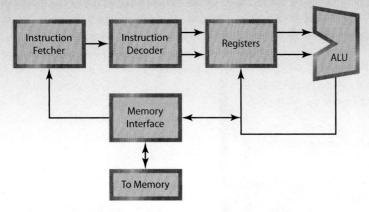

Figure 10.30 | The central processor unit of a processor chip performs several key functions, as detailed in this block diagram.

The CPU

The heart of any processor chip is the central processing unit, or CPU (see Figure 10.30). The CPU circuitry performs computing functions at high speeds. These computing functions may include a simple process, such as moving a byte of information from one part of the CPU to another, or a more complicated computing function, such as multiplying two 64-bit numbers.

The structure of logic circuitry in a CPU defines its architecture. Computer architecture for a Pentium chip in a desktop computer is different from the architecture of an embedded processor used to control a microwave oven, just as the architecture of a castle is different from that of a log home.

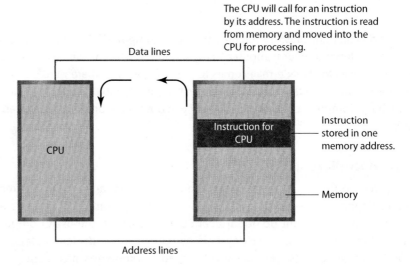

The CPU will call for an instruction by its address. The instruction is read from memory and moved into the CPU for processing.

Figure 10.31 | CPUs begin to execute instructions after first reading the instruction code from memory.

Instruction Code	Instruction Binary Pattern
MOV A, R0	11101000
ADD A, R6	00101110
RR A	00000011

Figure 10.32 | These examples of basic microprocessor instructions show how individual patterns of ones and zeros represent specific CPU operations.

Computer Instructions

The main function of the CPU is to process instructions. Instructions are binary commands directing the CPU to carry out relatively simple actions. Each specific binary pattern of the microprocessor defines a small task. When many instructions occur in sequence, we say the CPU is executing a program. A program running on a computer defines what the computer can do at that moment. The CPU architecture determines whether the instructions are simple or complicated.

Instructions are very specific patterns of ones and zeros, and each pattern triggers the CPU to perform a specific operation. Every microprocessor has an instruction set, which is the total of all its commands. A typical instruction set may have several hundred instructions.

When a computer runs a program, the program's instructions are stored in memory and called by the CPU one after another for execution (see Figure 10.31). A computer designer, programmer, or computer engineer creates the sequence of instructions executed (see Figure 10.32). The person or team responsible for making the computer work at this level, called the hardware level, must have an in-depth understanding of CPU architecture. Binary instructions, or machine-level instructions, are part of the hardware. They are different from instructions used in

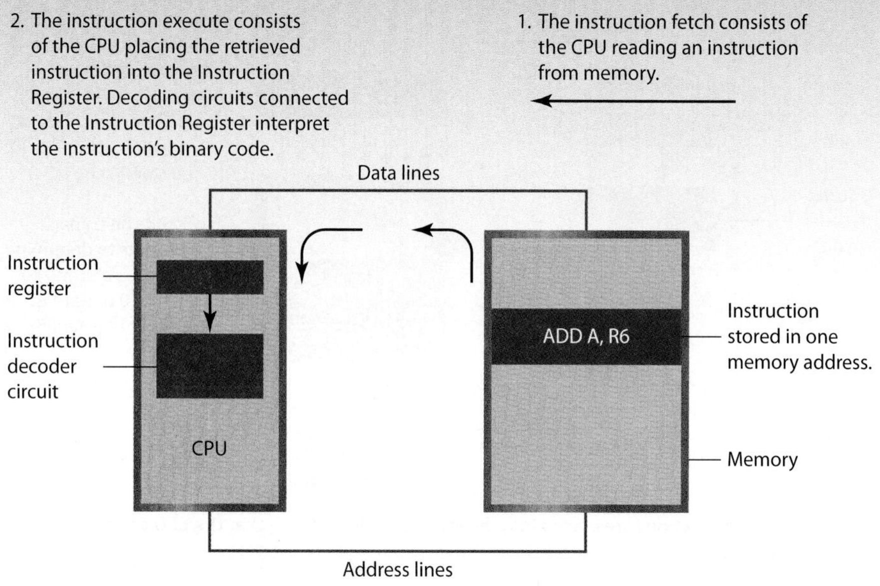

2. The instruction execute consists of the CPU placing the retrieved instruction into the Instruction Register. Decoding circuits connected to the Instruction Register interpret the instruction's binary code.

1. The instruction fetch consists of the CPU reading an instruction from memory.

Data lines

Instruction register

Instruction decoder circuit

CPU

ADD A, R6

Instruction stored in one memory address.

Memory

Address lines

Figure 10.33 | **The overall pattern of CPU operation is to fetch and then execute an instruction.**

high-level programming languages such as Visual Basic, C++, or Java. Programs written using a high-level programming language are converted by compilers into appropriate sequences of machine-level instruction. It is machine-level instructions that actually execute within a microprocessor (see Figure 10.33).

Digital Signal Processor (DSP)

▷ Typical microprocessor chips all have basic operational characteristics, such as the ability to move information from register to register. Sometimes, microprocessor designers customize parts of a chip's architecture to optimize the chip's performance. A **Digital Signal Processor (DSP)** is such a microprocessor. Instructions for DSPs are designed around a logic architecture specifically created to enhance the mathematical capabilities of the chip.

DSPs are used in products that require continuous computer processing. For example, audio filtering in hearing aids eliminates extraneous sound and noise information that the user of the hearing aid would find objectionable. A DSP can be programmed to do this filtering. In this application, the DSP constantly receives audio information as an input and uses complex mathematical operations to examine the audio and sort out the useful and non-useful sound information. A DSP excels at such a task compared to a standard microprocessor because of the detailed high-level mathematical functions built into the chip.

Instruction Execution

Let us examine a microprocessor whose instruction set consists of instructions made from just eight binary bits. This machine could have up to 256 individual instructions. We will examine what happens as the CPU processes these instructions.

In general, processing each instruction occurs in two phases called fetch and execute. The CPU fetches, or obtains, the first instruction from memory by reading

397

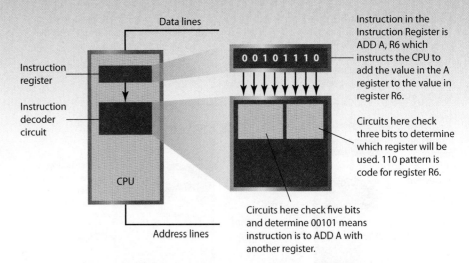

Data lines

Instruction in the Instruction Register is ADD A, R6 which instructs the CPU to add the value in the A register to the value in register R6.

Instruction register

0 0 1 0 1 1 1 0

Instruction decoder circuit

Circuits here check three bits to determine which register will be used. 110 pattern is code for register R6.

CPU

Circuits here check five bits and determine 00101 means instruction is to ADD A with another register.

Address lines

Figure 10.34 | Instructions are placed into the Instruction Register so that the CPU circuitry can determine what to do.

the memory chip holding the instruction. The CPU does this by first generating the address where the instruction is located and then placing the address on the memory chip's address lines. The CPU then activates signals necessary to carry out a read operation. During this fetch phase, the CPU is merely locating and acquiring the instruction.

The fetched instruction from memory enters the CPU on eight wires called the data bus. Bus is a word used to identify a common group of wires. The CPU captures the instruction's bit pattern and stores each bit in a special set of flip-flops called the Instruction Register (see Figure 10.34). Once the instruction is transferred from memory to the CPU, the instruction fetch is complete.

Remember that, in our example, the CPU could have fetched any of 256 unique instructions. A good part of the CPU's logic circuitry is devoted to decoding the various instruction bit patterns in order to determine which instruction it has received. The circuitry designed to decode the instruction bit patterns is called the Instruction Decoder. Its function is to examine the levels of the instruction pattern and then activate the specific parts of the CPU that carry out the command.

Assume that we want to move a data byte from one part of the CPU to another. The command is fetched and then held in the Instruction Register. Some of the instruction bits signify that data is to be moved. Other bits tell where the data is located, and still others identify where the data will go. Just eight instruction bits encode all this information. Once the instruction decoder examines these bits, additional control logic circuits are activated to move the data (see Figure 10.35 for an example of this process). After instruction execution is complete, the CPU will automatically

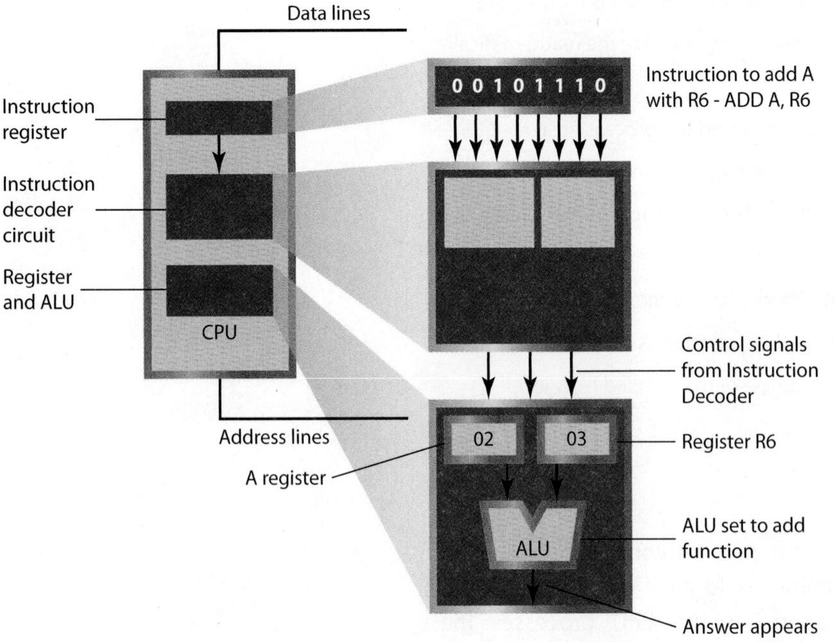

Data lines

Instruction register

0 0 1 0 1 1 1 0

Instruction to add A with R6 - ADD A, R6

Instruction decoder circuit

Register and ALU

CPU

Control signals from Instruction Decoder

Address lines

A register

02   03

Register R6

ALU set to add function

ALU

Answer appears here, 02 + 03 = 05

Figure 10.35 | This picture shows how the various parts of the CPU work to execute an addition instruction.

begin fetching the next instruction stored in memory. The fetch-execute cycle occurs repeatedly in a computer and governs the basic timing of all computer operations.

Every instruction in a processor chip exists by deliberate design. Computer engineers determine that an instruction, such as moving data, is a process they want the CPU to carry out, and they purposely design the CPU circuitry to do this task. The specific design of a computer's architecture evolves from the instructions the computer engineer wants the chip to execute.

Registers

CPUs are equipped with places that temporarily store data. The most common locations are the general purpose registers, which are groups of flip-flops built for this purpose. A byte of data, for instance, can easily be placed in a general purpose register to be used later by the CPU. Many microprocessors also use specific addresses in SRAM as general-purpose registers.

A general purpose register common to most machines is the accumulator, or A-register. In a simple CPU, the A-register is eight bits wide. If we were to specify moving data to the A-register from the B-register (another common register name), we would need a specific binary pattern of instruction to do so. After decoding the instructions, the CPU would carry them out. As a result, the 8-bit data contained in the B-register would be copied to the A-Register. This seemingly simple task is the kind of process carried out by machine-level instructions.

ALU

The CPU also contains ALU circuitry, which typically supports add, subtract, multiply, and divide operations. The ALU also makes possible instructions that compare the magnitudes of two numbers, and it supports logical operations, such as ANDing and ORing. The computations supported by arithmetic circuits vary greatly from microprocessor model to model, and because arithmetic circuitry is complex, mathematically powerful chips cost more.

There is also microprocessor circuitry designed for electronic communication with external devices (such as connecting your computer to a printer). Serial ports, found on many computers, are common examples of communication circuitry. Additional hardware and instructions are usually necessary to use the communications features.

Every microprocessor includes circuits that control the timing requirements of all CPU processes, such as fetch and execute. In a microprocessor, nothing happens randomly. Every operation is synchronized to a master clock.

Interrupts

Another important feature of microprocessors is the interrupt logic, which permits the processor to shift its attention from one program to another. Interrupts make it possible for many external devices to work with one microprocessor. The effect is that many devices seem to work simultaneously, but only one device has the processor's attention at any moment. An external device (one not on the microprocessor chip) needing attention interrupts the CPU only when action is necessary. For example, each key struck on a keyboard generates an interrupt, and the CPU responds to each by executing an interrupt program that accepts the character code of the key. Once the interrupt program completes, the CPU returns to whatever program was previously running. Since CPUs are faster than typing, the interrupt system permits the slow keyboard device to communicate with the CPU only when necessary. This is an efficient way to use computer-processing resources.

TECHNOLOGY AND PEOPLE:
Master of the Electromagnetic Wave

Joseph Iannotti (Figure 10.36) is a senior engineer in the RF (radio frequency) and photonics lab at General Electric Research and Development Center. He designs systems that make use of electromagnetic waves, from medical equipment to airport security devices. They are not usually products you would buy at a store, but high-value systems that companies are willing to pay a lot of money for. "Some of the stuff I work on costs billions of dollars," Iannotti says.

The first step for a new project is talking to the customers to see what they need. "They want to know how they can get their product to be better than anybody else's," Iannotti says. "I try to figure out what technology we can use to differentiate their system."

Iannotti got interested in engineering when he helped his dad build their house. "I thought it was pretty cool the way home wiring worked," he says. He was also drawn to the many gadgets that appeared in the early 1980s, like computers and videocassette recorders.

Iannotti earned an associate degree in Electrical Technology from Fulton-Montgomery Community College. He continued his studies at the Rochester Institute of Technology, graduating with a degree in electrical engineering technology. He got his first job as a design engineer.

"I screwed up on the first project they gave me, but I finally did it right," he says. "Failures happen, but the main thing is to learn from them."

Since then, Iannotti has participated in many different projects. He helped modernize aging military systems and worked on ways for satellites to communicate with one another.

"I never get bored because I'm doing different things all the time," he says.

According to Iannotti, the two things that make a great engineer are a love of math and a love of solving problems. "The problems can be as abstract or as real as you like," he says. "There are a lot of people at my company who never leave the lab. There are other people who never leave their office—they just think a lot."

Education	A.A.S.—Electrical Technology, Fulton-Montgomery Community College, Johnstown, NY B.S.—Electrical Engineering Technology, Rochester Institute of Technology, Rochester, NY
Job Description	Design Engineer at a research and development lab developing and inventing products, which use radio frequency and photonics to communicate and sense.
Advice to Students	"Understand the basics, enjoy what you do, and try to work with the best people you can!"

Figure 10.36 | Mr. Joseph Iannotti is an electrical engineer working in a research and development lab.

SECTION THREE FEEDBACK >

1. Sketch a diagram showing how memory and a microprocessor communicate.
2. What is an instruction set?
3. What is the purpose of the instruction decoder logic in a CPU?
4. Where are the instructions for a computer program stored prior to execution?

KEY IDEAS >

- Complete computer systems consist of many elements including a microprocessor, memory, and I/O.

- In a home computer system, the main PCB is the motherboard, which holds many key components, including memory, the microprocessor, and a chip set.

- The purpose of the chip set is twofold. It improves performance and reduces the number of parts needed.

- Computer owners can add to their systems by plugging small, printed circuit cards into the expansion slots.

- Interrupts are key structural features in computer architecture that ensure efficient operation.

In most cases, a microprocessor chip is not a complete computer system. A complete, functional computer needs additional parts and assemblies. If you were to remove the cover from a home computer, you would recognize this immediately. You would see many integrated circuits, several circuit cards, and a few metal boxes connected by wire cables and printed circuit wiring. In order for the entire system to work, all these devices must communicate with each other. Complete computer systems consist of many elements, including a microprocessor, memory, and I/O.

Computer Subsystems

Figure 10.37 illustrates several of the critical subsystems found in a typical home computer. The CPU, part of the microprocessor chip, coordinates most computer activity. As previously discussed, the CPU processes instructions received from memory. Figure 10.37 shows many forms of memory in a typical system. The memory technology used depends on the memory application, the memory speed required, and the volatility of the memory.

As a computer operates, the microprocessor interacts with external devices such as keyboards, monitors, disk drives, DVD writers, USB drives, and network interface cards. These external devices are known as peripheral devices, or, more specifically, I/O devices. I/O stands for input/output and describes components sending information to the CPU (input) or receiving information from the CPU (output).

You see from Figure 10.38 that both I/O devices and memory share a common set of wires with the CPU. The number of wires varies depending on the complexity of the system, but it typically ranges from eight to sixty-four wires. Notice a number of the wires labeled with the word "bus." A bus is several wires used for a similar purpose. For example, the data bus shown is eight bits wide. This means eight wires carry the data information between the CPU and memory or between the CPU and I/O. Each wire carries one of the eight data bits. Instead of referring to each individual wire, it is easier to refer to the data bus in general. The size or width of the data bus is an important factor in determining how fast a computer will operate. The more wires there are in the data bus, the more information

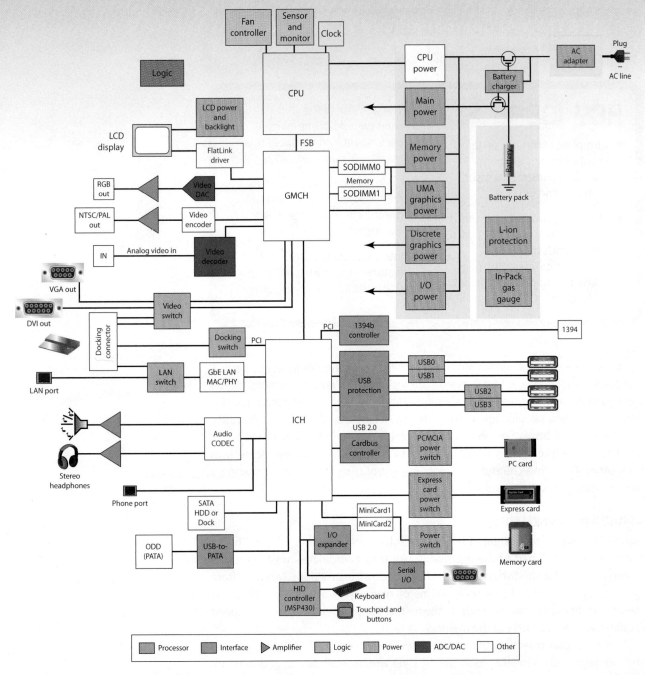

Figure 10.37 | This block diagram shows the many important subsystems necessary in a notebook computer.

each computer operation can process. For systems using bus structures, as most do, it is important to ensure that only one device actually uses the bus at any one time. Otherwise, severe electrical and timing problems will result. For this reason, computer engineers carefully design their systems and use components specifically created for bus structures.

If you could view the signals on a data bus, you would find either binary instructions or binary data passing back and forth between the components attached to the bus. Since systems operate at electronic speeds, any particular piece of information exists on the bus for only a few nanoseconds. You can appreciate just how important timing is when devices operate so quickly.

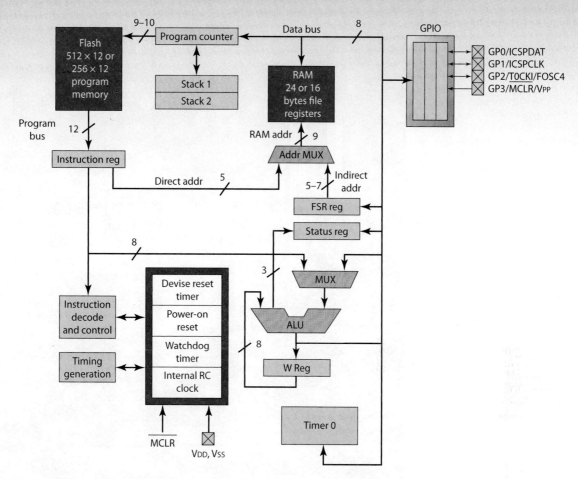

Figure 10.38 | Common groups of wires, called busses, are used in computer systems on computer chips so that data can be shared by many devices.

The other important bus shown in Figure 10.38 is the address bus. This bus consists of many wires, commonly between sixteen and thirty-six. On these wires, the CPU provides the addressing information necessary to access any specific piece of information.

Computer systems also have many dedicated signals controlling the timing and sequencing of operations in the system. For example, a signal is necessary to inform memory and I/O devices whether they are needed at any particular moment. Since both devices connect to the same bus, but only one at a time may use the bus, selection control is essential.

Motherboard

When you look closely at pictures of integrated circuits, you will see many small pins. In order for a computer system to work, the pins of many ICs and electrical components must be interconnected to form the necessary circuits. These wiring interconnections are made on a printed circuit board (PCB) that connects components efficiently and provides beneficial electrical properties. For instance, PCBs keep wiring as short as possible and help reduce signal interference.

The PCB is constructed of nonconductive materials coated with copper metal. Some PCBs are several layers thick. Copper is removed in selected areas of the board by an etching process. After etching, thin copper lines or wires called traces remain on the board. You can easily see the traces on a PCB running from one component to another.

Figure 10.39 | The majority of circuitry in a computer system is interconnected on the system motherboard.

Holes drilled through the copper traces permit insertion of integrated circuits and other components, interconnecting the small pins of individual parts and holding them in place with solder. On some PCBs, very small surface-mount components are soldered directly to the traces without the need to drill holes.

In a home computer system, the main PCB is the motherboard, which holds many key components (see Figure 10.39), including memory, the microprocessor, and a chip set. The chip set is often two ICs, one called Northbridge, and the other called Southbridge. The Northbridge chip handles the microprocessor's connections to system memory and graphics memory; the Southbridge chip controls the I/O devices supporting serial communications, USB, and other I/O operations. Figure 10.40 illustrates this with a simplified block diagram.

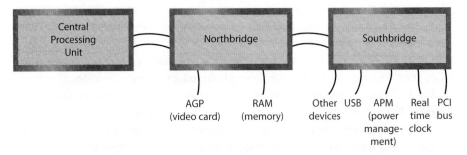

Figure 10.40 | A computer chip set simplifies the construction and assembly of computer systems.

Figure 10.41 | Motherboard expansion slots make it possible for additional equipment to be added to a computer system.

The purpose of the chip set is twofold. It improves performance and reduces the number of parts needed, which means increased reliability. Computer speeds also increase when circuitry connections are on as few chips as possible.

Of course, computer technology is always changing as designers strive to make systems faster and more useful. An example is a new series of chip sets from Intel called the Intel Hub Architecture, which will make systems even faster.

The motherboard also holds special electrical and mechanical connectors called expansion slots. Computer owners can add to their systems by plugging small printed circuit cards into the expansion slots (see Figure 10.41). For example, a gaming system requires superior video capabilities, and, in order to keep up with game action, a gamer might want to add a high performance video card to his system.

The peripheral components interface (PCI) bus is an example of a common expansion slot that is standard in home systems. This 47-pin bus interface connects devices such as sound cards and disk drives to the motherboard hardware.

TECHNOLOGY IN THE REAL WORLD:
USB Connections

Do you download pictures from digital cameras to personal computers? You could be using a USB connection to do this. Figure 10.42 shows the symbol for USB. You might see this logo near some of your computer connections. The thumb drive you use for file storage (also known as a memory stick, flash drive, or jump drive) uses USB to store files.

USB stands for Universal Serial Bus. The word serial means to send or receive many bits, one bit at a time. The word "bus" in USB tells how many devices share a wire. The idea behind a bus in a digital system is to have many devices attached together via a common wire (or groups of common wires). This saves a lot of wiring and works well as long as only one device uses the bus at a time. The USB does require specific wiring and connectors, defined voltages on specific pins, and distinct signaling specifications that are agreed upon by the sender and receiver.

Figure 10.42 | The trident symbol indicates a USB connection.

Section 4 △ Components of a Computer System

Figure 10.43 | A USB Type A connector is usually plugged into a computer.

Figure 10.44 | A USB Type B connector is usually plugged into devices.

Therefore, USB is actually the bus system permitting devices to connect with a computer. Almost any device connected to a computer can be made to work using a USB connection.

You don't need to be a computer expert to make such connections. USB is known for its easy plug-and-play characteristics. Plug-and-play means that devices can connect or disconnect from a computer while the machine is running. The computer is designed to support USB and has the necessary USB software on it (Windows, for example), so the instant a USB-capable device is plugged into a socket, it is recognized by the computer and able to work. USB is also fast and transmits information up to 480 Mbits (million bits) per second.

USB has several kinds of connectors. Figure 10.43 shows a type A connector, and Figure 10.44 shows a type B connector. The computer end of a USB connection, often called the host controller, uses the type A connector. The devices that attach to the computer use the type B connector. As USB has become more popular, designers have created mini-versions of these connectors so that smaller devices, like digital cameras, can use USB.

A USB connector holds only four wires. Two wires are for power: one is a +5 volt wire and the other a ground wire. The other two wires, labeled D− and D+, work together to carry a single bit of information. The latter two wires implement a kind of signaling called differential signaling. This method helps reduce electrical noise and makes the connection more reliable.

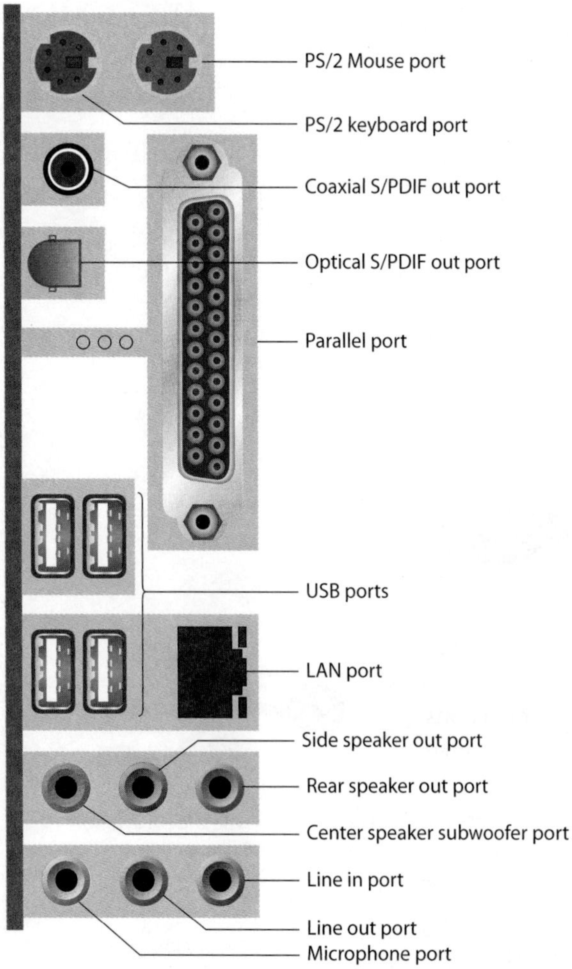

When a USB device is connected to a computer, a process known as device enumeration begins, which permits the computer and USB device to communicate. USB is available in several speed versions—the computer determines the speed needed. The computer also reads device information to determine the kinds of drivers required for the device. A GPS unit needs different support than does a keyboard, for example. USB allows that multiple devices be connected, and since the computer assigns a 7-bit address to the device, up to 128 devices may be connected at one time (2^7 equals 128). Hubs are used to interconnect the extra devices.

I/O

The CPU is constantly obtaining instructions from memory to determine which processes to carry out. Many of these instructions command the CPU to communicate with external devices through the Input/Output (I/O) circuitry. Like memory, I/O devices are addressable and may receive or send data to the CPU. Because there are many I/O devices in a computer system, they need addresses as identifiers. When information is sent to a computer monitor screen for display, the CPU sends the display data to the monitor electronics via the monitor's I/O circuits. The monitor has a specific I/O address, and only when the CPU addresses the monitor does the exchange of data begin.

A keyboard illustrates another basic I/O action. A binary code is transferred to the computer system each time a key is pressed. The I/O keyboard address selects the keyboard for service, enabling the CPU to read the keyboard data. Think of any piece of equipment you might attach to a computer system, and you will most likely have identified gear using the computer's I/O system (see Figure 10.45).

— PS/2 Mouse port

— PS/2 keyboard port

— Coaxial S/PDIF out port

— Optical S/PDIF out port

— Parallel port

— USB ports

— LAN port

— Side speaker out port

— Rear speaker out port

— Center speaker subwoofer port

— Line in port

— Line out port
— Microphone port

Figure 10.45 | I/O connectors found on home computers make it easy for the user to plug in equipment.

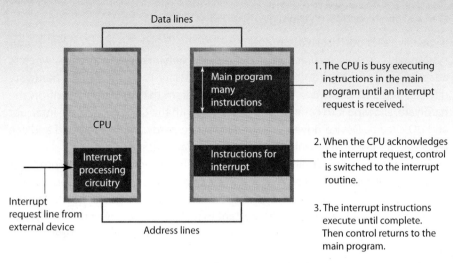

Data lines

CPU

Interrupt processing circuitry

Interrupt request line from external device

Address lines

Main program many instructions

Instructions for interrupt

1. The CPU is busy executing instructions in the main program until an interrupt request is received.

2. When the CPU acknowledges the interrupt request, control is switched to the interrupt routine.

3. The interrupt instructions execute until complete. Then control returns to the main program.

Figure 10.46 | Interrupts help the microprocessor handle many different tasks.

Interrupts

The microprocessor is a busy chip. The CPU is always executing instructions for the main system programs, but also has to keep track of externally attached devices. Because CPU time is a precious resource, methods have been developed to free the CPU from dealing with external devices until a specific need arises.

Assume you are typing an e-mail message on your computer keyboard. Every time you press a key, binary information representing that key's letter transfers from the keyboard to computer memory, so that your e-mail program can use it. The CPU controls this data transfer, but it is a waste of CPU time to constantly check the keyboard for each typed character. It is more efficient for the CPU to respond to the keyboard only when you actually press a key. Computer designers engineer this response by designing the keyboard so that every keystroke sends a special signal to the CPU. This signal starts a process known as an interrupt.

Interrupts are key structural features in computer architecture that ensure efficient operation. Even if you type quickly, say one character every 0.2 seconds, a reasonably fast CPU can execute over 2 million instructions between keystrokes. It does not make sense to waste this computing power looking for keyboard activity, so instead the keyboard circuitry generates an interrupt request signal each time a key is pressed. The CPU notices the request, suspends its current work, acknowledges the inter-rupt, and, through the I/O system, receives keyboard data. Once finished with that key, the CPU picks up exactly where it left off prior to the interrupt. Many other devices also use the interrupt system.

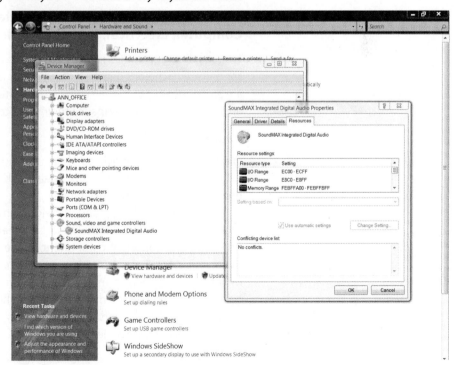

Figure 10.47 | Computer users can configure and troubleshoot system hardware using features in their computer's operating system.

Section 4 △ Components of a Computer System

Other Devices in a Computer System

If you have ever purchased a computer, you are aware that you may customize your system with devices such as mice, sound cards, network interface cards, wireless adapters, and speakers. Software is usually necessary to operate these hardware devices, and a device driver is the software program that specifically controls the hardware, allowing it to communicate properly with the CPU through the interrupt and I/O circuits. Device drivers are supplied when hardware is purchased and can be updated through a download from the manufacturer's Web site.

Every system device requires electrical power. The system power supply produces the necessary voltages and currents for the system. Computer power supplies plug into household outlets and convert AC into the regulated DC power needed by the machine's electronic devices. It is important to make sure the power rating, given in watts, is sufficiently large to supply all the electrical needs of the computer.

SECTION FOUR FEEDBACK >

1. What are the key differences between the microprocessor chip used in a home computer system and the processor used in an embedded system?
2. What memory technology will you find in the main memory of a computer?
3. How would you attach a writing tablet to a computer? Besides the tablet, what else would you need to make the tablet operate?

CAREERS IN TECHNOLOGY

Matching Your Interests and Abilities with Career Opportunities: Computer, Electrical, and Electronic Engineers

Overall job opportunities in engineering are expected to be good, but will vary by specialty. A bachelor's degree is required for most entry-level jobs. Starting salaries are among the highest of all college graduates. Continuing education is critical for engineers who want to enhance their value to employers as technology evolves.

Nature of the Work

Engineers apply the principles of science and mathematics to develop economical solutions to technical problems. Their work is the link between perceived social needs and commercial applications.

In addition to design and development, many engineers work in testing, production, or maintenance. These engineers supervise production in factories, determine the causes of component failure, and test manufactured products to maintain quality. They also estimate the time and cost to complete projects. Some move into engineering management or into sales. In sales, their engineering backgrounds enable them to discuss technical aspects and assist in product planning, installation, and use. Supervisory engineers are responsible for major components or entire projects.

* Computer hardware engineers research, design, develop, test, and oversee the installation of computer hardware and supervise its manufacture and installation.

* Electrical engineers design, develop, test, and supervise the manufacture of electrical equipment. Electrical engineers specialize in areas such as power systems engineering or electrical equipment manufacturing.
* Electronics engineers are responsible for a wide range of technologies, from portable music players to the global positioning system (GPS). Electronics engineers design, develop, test, and supervise the manufacture of electronic equipment, such as broadcast and communications systems.

Working Conditions

Most engineers work in office buildings, laboratories, or industrial plants. Some engineers travel extensively to plants or worksites. Many engineers work a standard 40-hour week. At times, deadlines or design standards may bring extra pressure to a job, requiring engineers to work longer hours.

Training and Advancement

A bachelor's degree in engineering is required for almost all entry-level engineering jobs.

Most engineering programs involve a concentration of study in an engineering specialty, along with courses in both mathematics and the physical and life sciences.

In addition to the standard engineering degree, many colleges offer 2- or 4-year degree programs in engineering technology. These programs, which usually include various hands-on laboratory classes that focus on current issues in the application of engineering principles, prepare students for practical design and production work, rather than for jobs that require more theoretical and scientific knowledge.

Engineers should be creative, inquisitive, analytical, and detail oriented. They should be able to work as part of a team and to communicate well, both orally and in writing. Communication abilities are important because engineers often interact with specialists in a wide range of fields outside engineering.

Beginning engineering graduates usually work under the supervision of experienced engineers and, in large companies, also may receive formal classroom or seminar-type training. As new engineers gain knowledge and experience, they are assigned more difficult projects with greater independence to develop designs, solve problems, and make decisions. Engineers may advance to become technical specialists or to supervise a staff or team of engineers and technicians. Some may eventually become engineering managers or enter other managerial or sales jobs.

Outlook

Overall engineering employment is expected to grow by 11 percent over the 2006–16 decade, about as fast as the average for all occupations. Engineers have traditionally been concentrated in slow-growing manufacturing industries, in which they will continue to be needed to design, build, test, and improve manufactured products. Overall job opportunities in engineering are expected to be favorable because the number of engineering graduates should be in rough balance with the number of job openings over this period.

Earnings

Earnings for engineers vary significantly by specialty, industry, and education. Even so, as a group, engineers earn some of the highest average starting salaries among those holding bachelor's degrees.

The following tabulation shows average starting-salary offers for engineers, according to a 2007 survey by the National Association of Colleges and Employers.

CURRICULUM	BACHELOR'S	MASTER'S	PH.D.
Aerospace/aeronautical/astronautical	$53,408	$62,459	$73,814
Agricultural	49,764	—	—
Architectural	48,664		
Bioengineering and biomedical	51,356	59,240	—
Chemical	59,361	68,561	73,667
Civil	48,509	48,280	62,275
Computer	56,201	60,000	92,500
Electrical/electronics and communications	55,292	66,309	75,982
Environmental/environmental health	47,960		
Industrial/manufacturing	55,067	64,759	77,364
Materials	56,233		
Mechanical	54,128	62,798	72,763
Mining and mineral	54,381	—	—
Nuclear	56,587	59,167	—
Petroleum	60,718	57,000	—

[Bureau of Labor Statistics, U.S. Department of Labor, Occupational Outlook Handbook, 2008–09 Edition, visited May 12, 2008, http://www.bls.gov/oco/]

Summary >

Computers are manufactured using electronic components, such as resistors and transistors. Computer design is accomplished at the logic level, using logic devices grouped in families with similar electrical characteristics.

The binary number system provides a mathematical foundation for logic circuits that use binary levels as inputs and outputs. Binary values called "bits" simplify computer design, as only one of two possible values can exist on any logic device input or output. The two values representing on and off are referred to as high and low levels.

Logic devices include AND, OR, NAND, NOR, invert, and flip-flop. These fundamental logic devices combine to create complex digital circuitry. Computer engineers interconnect basic logic devices into combinatorial and sequential circuits; this circuitry is the heart of digital computers. Common digital circuits include the adder, ALU, and counter.

Modern computing systems require memory chips to store data and programming instructions. No memory technology is the perfect combination of speed, density, and nonvolatility, so computer systems use several memory technologies.

Microprocessor chips contain the computation circuitry for computers. The term microprocessor includes embedded processors used in simple products, microcontrollers used in complex systems, and advanced microprocessors used in desktop computers. All microprocessor-type chips contain common features such as the CPU, I/O, and interrupts. Many microprocessor chips have on-chip memory as well.

The microprocessor internal architecture determines the level of performance. The faster instructions reach the CPU, the better the overall performance will be. Instructions received by the CPU go into the Instruction Register for decoding. Since machine level instructions are just unique binary patterns, they are easily decoded. Once decoded, the CPU carries out the task signified by the instruction. CPUs undergo a constant fetch-execute process in the course of running a computer program.

Microprocessor chips are only one component in a typical desktop or laptop computer system. These systems require support chips to interconnect memory and I/O. The chips in a computer system communicate on a motherboard. Computer systems also have many peripheral devices attached, such as mice, keyboards, and video cards. The I/O and interrupt circuitry coordinate how peripheral devices share the CPU resources.

FEEDBACK

1. An AND gate logic device has six input lines. How many combinations of ones and zeros produce a one level output? How many combinations will produce a zero output from the device?

2. Find Web sites pertaining to building your own computer. Identify three important factors to keep in mind when buying a motherboard.

3. AGP, DIMM and PCI are connector slots found on motherboards. Find information on these slots, including the technical specifications of each.

4. Look at the memory modules in your own computer. How many memory chips are on the module? Can you determine how the chips interconnect on the module in order to expand the overall memory size?

5. Research online to find out what differentiates PC100 memory modules from PC133 memory modules.

6. Design a burglar alarm system using logic devices, so that if any one of three doors opens, a binary signal will trigger an alarm. Assume the doors are equipped with magnetic switches, such that a closed door switch provides a zero signal to your circuit, and an open door switch provides a one signal to your circuit. Also, assume the alarm activates with a one-level signal.

7. Look up the term K-map online to see if you can find an explanation for designing logic circuits with the K-map. Also, search for a K-map program to download. You should find free programs readily available. Use the program to design the circuit discussed in Figure 10.12.

8. Do you have a digital camera? What kind of memory device does your camera use to store pictures? What is the size of the memory device?

9. Using the Internet, find a data sheet for a DRAM chip. Can you figure out the relationship between the number of address lines and the number of addresses?

10. Look up information on the ASCII code. If you were to design a memory system to store ASCII codes, what word size would you want the memory system to have?

11. Look on the Internet for the instruction set of the 8051 microcontroller. What kinds of instructions are possible with this popular 8-bit machine?

12. Go to a computer manufacturer's Web site, and design your own computer system. Explain the reasons for your selections of memory, power supply, microprocessor, and peripheral devices.

13. Look up information on video-graphics cards. Determine what kind of memory technology these cards use. Are there any other significant features of these cards related to microprocessors?

411

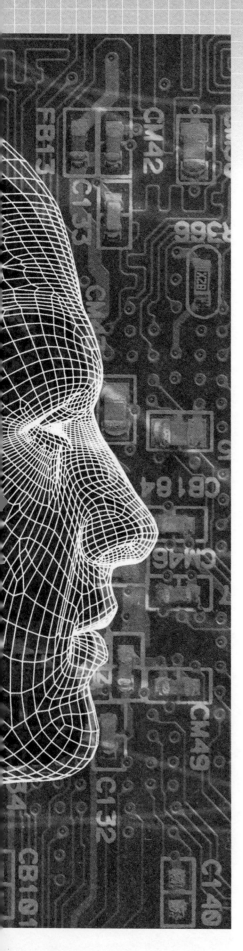

DESIGN CHALLENGE 1:
Digital Adder Design

- ### Problem Situation

You and your classmates decide to start a digital design business. You must be able to prove to potential clients that your team can design and build digital circuitry. The group decides that demonstrating the operation of a 12-bit adder will show off the design skills of your company.

- ### Your Challenge

Your challenge is to design a working adder capable of adding two 12-bit numbers and to display the result of the addition on LEDs. You will need to simulate the design, as well as determine how to drive the LEDs with digital signals. You will also have to generate the two 12-bit binary numbers for addition.

- ### Safety Considerations

    1. Always exercise caution when working with electricity.
    2. Assume circuits are energized.
    3. Follow your instructor's safety guidelines for your lab.

- ### Materials Needed

    1. LEDs
    2. Resistor assortment
    3. TTL or CMOS ICs
    4. Circuit Breadboard
    5. DC Voltage source
    6. Switches
    7. Simulation software (such as DigitalWorks)

- ### Clarify the Design Specifications and Constraints

Draw the complete adder schematic on paper. Identify the individual digital IC part numbers needed for the circuit as well as the total number of ICs needed. Once you are convinced of the design's correctness, create the circuit using the schematic capture feature of the circuit simulation software you are using. Simulate the adder to confirm that your design adds properly, and then gather the parts and build the system. It is often advantageous to "bring up" the circuit in small sections. Follow good wiring practices. Short wires are best. Caution: LEDs are analog devices. Digital signals will not necessarily light an LED. Switches are also analog devices and need to be configured to produce digital levels.

- ### Research and Investigate

To complete the design challenge, first gather information to help you build a knowledge base.
    1. In your guide, complete the Knowledge and Skill Builder I: Digital gates.
    2. In your guide, complete the Knowledge and Skill Builder II: LEDs and switches in digital applications.
    3. In your guide, complete the Knowledge and Skill Builder III: Half and full adders.

- ### Generate Alternative Designs

Describe two of your possible alternatives to the adder design. Discuss the factors needed to use switches and LEDs as digital devices.

- ## Choose and Justify the Optimal Solution

Explain why the solution you selected is better.

- ## Display Your Prototypes

Construct your chosen design. Use the working model and simulation results to show how the adder operates.

- ## Test and Evaluate

Explain how your design met the specifications and constraints. What simulated and actual tests did you conduct to verify this?

- ## Redesign the Solution

Explain how you would redesign your circuit based on the knowledge and information that you gained during this design challenge. What additional features would you add to the design if you had more time?

- ## Communicate Your Achievements

Describe the plan you will use to present your solution to your class. Include a media-based presentation.

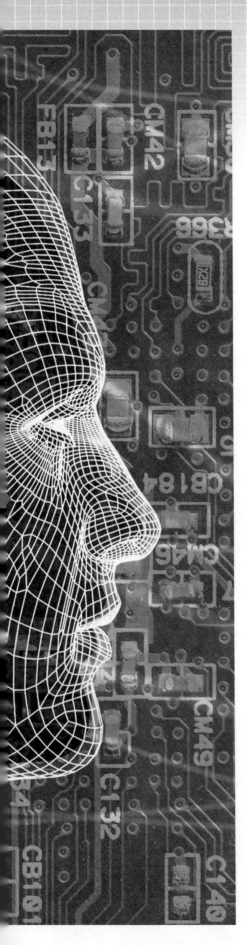

DESIGN CHALLENGE 2:
Storage for a Digital Adder

● Problem Situation

Your digital design company is off to a rousing start. Several clients are impressed with your adder. One client wants to view on LEDs the results of three consecutive additions. You tell the client, "No problem." Now your company's designers need to perform more circuit design work. The pressure is on!

● Your Challenge

Your challenge is to add storage capacity to the 12-bit adder previously designed. You will need to devise ways to add the storage as well as a method to display the results of three consecutive additions. A key concern is how to keep track of each addition result.

● Safety Considerations

1. Always exercise caution when working with electricity.
2. Assume circuits are energized.
3. Follow your instructor's safety guidelines for your lab.

● Materials Needed

1. 12-bit adder from previous design challenge
2. LEDs
3. Resistor assortment
4. TTL or CMOS ICs
5. Circuit breadboard
6. DC Voltage source
7. Simulation software (such as DigitalWorks)

● Clarifying the Design Specifications and Constraints

Modify the adder schematic on paper. Consider which devices are capable of storing digital information. Be sure that the storage circuitry for the first addition result will not interfere with the other two sets of storage circuitry. Identify the individual digital IC part numbers needed for the storage as well as the total number needed. Once you are convinced of the modified design's correctness, use the schematic capture feature of the circuit simulation software you are using to simulate the circuit. Gather the additional parts and build the system.

● Research and Investigate

To complete the design challenge, first gather information to help you build a knowledge base.

1. In your guide, complete the Knowledge and Skill Builder I: Latches and flip-flops.
2. In your guide, complete the Knowledge and Skill Builder II: Digital registers.
3. In your guide, complete the Knowledge and Skill Builder III: Decoders.

● Generate Alternative Designs

Describe two of your possible alternatives to the adder storage design. Discuss the decisions needed for the control of the storage circuitry.

• Choose and Justify the Optimal Solution

Explain why the solution you selected was better.

• Display Your Prototypes

Construct your chosen design. Use the working model and simulation results to show how the adder stores addition answers.

• Test and Evaluate

Explain how your design met the specifications and constraints. What simulated and actual tests did you conduct to verify this?

• Redesign the Solution

Explain how you would redesign your circuit based on the knowledge and information that you gained during this design challenge. What additional features would you add to the design if you had more time?

• Communicate Your Achievements

Describe the plan you will use to present your solution to your class. Include a media-based presentation.

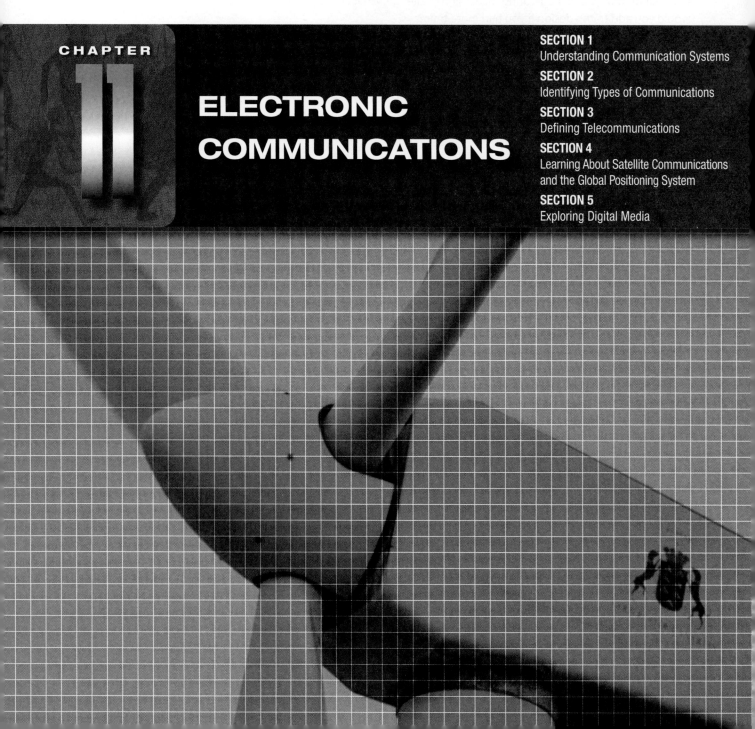

INTRODUCTION

S THIS SCENE familiar to you? You are sitting at the computer with multiple instant messaging windows open. You are simultaneously texting on your cell phone and listening to your iPod. This human imperative to communicate is not new. It can probably be traced back to the time of cave paintings, at least. Your generation, however, is unique in its capacity to communicate. Whether it is communicating with your friends, your parents, or your teachers, you engage in some sort of communication on an almost constant basis.

CHAPTER

11

ELECTRONIC COMMUNICATIONS

Communications is as old as humankind and is a critical aspect of what makes us who we are. Suppose we go through a typical day and determine how many of your interactions can be classified as communications. We can begin with the morning ritual of hearing the alarm clock, hitting the snooze button, and repeating the cycle until we get out of bed. This type of communication is one-way, machine-to-human, yet it is communication nonetheless. You may have had a parent stick his or her head in your room and try to wake you up, and you may have tried to do the same with a younger brother or sister—all oral communications. As you're getting ready, you might listen to the radio or watch television, which are examples of one-way broadcast communications—intended either to entertain us or inform us. As an alternative, you may be listening to your favorite morning playlist on your iPod. During this time, you may have checked your cell phone for text messages from your friends, texted them, used a laptop or desktop computer to send or check e-mail, or had a quick instant messaging (IM) session with a friend (see Figure 11.1). E-mail, text messaging, and instant messaging are all two-way electronic communications that have quickly become primary ways we communicate.

Figure 11.1 | **Two girls share an instant message.**

Notice that in our scenario, it is still morning. We haven't had breakfast, gone to school, or done any of the other activities that make up a typical school day. That's a lot of early morning communication—even for your generation. What about "older" forms of communication? (see Figure 11.2 for an example of really old forms of communication!) On a typical day, what's the likelihood that you would listen to the radio or watch television? What about reading a newspaper? Your household probably subscribes to a newspaper, but how often do you use it to communicate? We'll end our illustration here, before you go to school, but you can see that if we followed you for a whole day, your communication would probably be nonstop. You might be surprised by how much your life is centered on communication—which we'll leave for an end-of-chapter exercise.

Figure 11.2 | **Cave painting is an ancient form of communication.**

KEY IDEAS >

- Information and communication technologies include the inputs, processes, and outputs associated with sending and receiving information.

- Information and communication systems allow information to be transferred from human to human, human to machine, machine to human, and machine to machine.

- Communication systems are made up of a source, encoder, transmitter, receiver, decoder, storage, retrieval, and destination.

- At the source, a modem modulates an analog carrier signal to encode digital information. At the destination or receiver, a modem demodulates the signal to decode the transmitted information and receive the message.

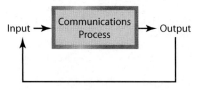

Figure 11.3 | A communications block diagram can represent communications in terms of input, output, the communications process, feedback.

We can represent any communication system with some basic building blocks, as shown in Figure 11.3. For instance, consider an instant messaging (IM) application. You sit at your computer and type or input your message into an IM client. Your computer is connected to a network, so your message can be transmitted to your friend across town or even across the world. This portion of the communication system is the process; when your friend receives the message, it is displayed or output on his or her computer. If your friends do not understand the message, they could respond with a question or a request to repeat the message. Here your friends are providing feedback. Any of the information and communication technologies we use include the inputs, processes, and outputs associated with sending and receiving information.

Although our block diagram is simplistic, it is useful: It can be applied to almost any type of communication. Let's consider an example. If you have musical talent, you've probably taken piano lessons. Consider what happens when you play the piano (see Figure 11.4). You may be reading the sheet music for the song you're trying to learn. This sheet music is your input. The process involves reading the notes, mentally mapping those notes to the correct keys on the piano, and then striking the proper keys. As you become a better musician, the process becomes less about memory and hand-eye coordination and becomes more automatic. The output from the process is the sound of the piano. The feedback here is your response—or your teacher's response—to the notes (the output). If your output is correct—that is, if it matches the input (the sheet music)—the feedback is likely positive. If your output has errors, the feedback communicates that to you and requires you to alter your process—hopefully to improve your output. The scenario we've described is a human-to-human exchange of information; however, information and communication

Figure 11.4 | A child playing the piano is actually a form of communication.

systems also allow information to be transferred from human to machine, from machine to human, and from machine to machine.

Let's look at the middle portion—the communication process—in a little more detail. We can break this process—or any process, for that matter—into five elements: a source, an encoder, a channel, a decoder, and a receiver. You, your computer, and your IM client are the source of a message. Your computer is connected to some sort of modem—short for modulator/demodulator. The modem is responsible for converting the message from your computer into a form that can easily be transmitted—in this case, converting the computer's binary message (ones and zeroes) into an analog signal ready for transmission. Here the modem is acting as an encoder. The message travels over phone lines, copper cables, and fiber-optic cabling, bouncing from router to router until it reaches its destination network. This element of the communication process—the communication channel—could be simple or very complex, depending on the path the message must take and the variety of networks that have to be traversed. A modem at the receiving end—at least the demodulator part of the modem—decodes the message into a form the computer can understand. In this case, the message is converted from an analog signal to a binary signal and your friends' modem is acting as a decoder. Finally, your friend—the receiver—reads the message on his or her computer using an IM client. Similar processes can be identified for other forms of communication, including e-mail and written and spoken communications. If we combine the elements of our communications process with the receiver and transmitter and the ability to store and retrieve information, we can further generalize communication systems as consisting of a source, encoder, transmitter, receiver, decoder, storage, retrieval, and destination (see Figure 11.5).

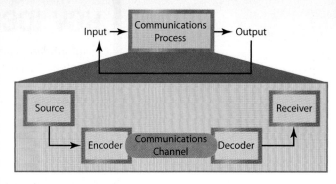

Figure 11.5 | Communication process consists of a source, encoder, transmitter, receiver, decoder, storage, retrieval, and destination.

Figure 11.6 | On the receiving end, analog waves enter a modem and digital signals are output.

At the source, a modem modulates an analog carrier signal to encode digital information; at the destination or receiver, this process is reversed. There the modem demodulates the carrier signal to decode the transmitted information and receive the message (see Figure 11.6).

SECTION ONE FEEDBACK >

1. Sketch a block diagram of the communication system that makes up an instant message with a friend. Sketch a similar diagram for the preceding example of the piano.
2. Create and label a block diagram for the communication process—the middle portion of our system—for the preceding piano example.
3. What is a channel? Give three examples.
4. Explain the purpose of an encoder and decoder in a communications system.
5. What does a modem do?

Section 1 ⊿ Understanding Communication Systems

KEY IDEAS >

- Information and communication systems can be used to inform, persuade, entertain, control, manage, and educate.

- There are many ways to communicate information, such as graphic and electronic means.

- Technological knowledge and processes are communicated using symbols, measurements, conventions, icons, graphic images, and languages that incorporate a variety of visual, auditory, and tactile stimuli.

- Graphical analysis and presentation can be divided into two general types: information that is qualitative in nature, and information that is quantitative.

- Oral presentations to bosses, teachers, customers, or colleagues at meetings or conferences can be used to share information, or sell an idea, a product, or even yourself (for example, a job interview).

We communicate for a number of reasons. Communication, regardless of what form it takes, can be used to inform, persuade, educate, and entertain. Some communications are generic and can be used for any of these purposes; others have been developed to achieve one or more of these objectives. We can classify communications into two distinct categories: graphic and electronic. Take a look at each.

In graphic communications, we use words and pictures to convey our message. Graphic communications that use images or icons are a particularly powerful means of engaging an audience and sharing information. For example, consider the iconic examples of the Apple and Windows logos. Apple and Microsoft have spent a lot of money on these logos to make sure that we recognize them. Graphic communication is particularly useful for presenting technological knowledge and processes, which are communicated using symbols, measurements, conventions, icons, graphic images, and languages that incorporate a variety of visual, auditory, and tactile stimuli.

In electronic communications, we use electrical signals, pulses of light, or radio waves to carry our messages. Computers use any of these electronic techniques (electricity, light, radio waves) to send and receive messages. As devices become increasingly more powerful, we find that cell phones, smart phones, and personal digital assistants (PDAs) can send and receive electronic messages and display them as text and graphics. As networks get faster and provide greater capacity, these devices access even greater amounts of information. Many of these handheld devices can be used for receiving and watching video, and a select few can send and receive live video. Consider the tremendous impact that devices such as the Apple iPhone, Amazon Kindle e-book reader, and Nokia's N95 cellular phone have on how, when, and where we access data (see Figure 11.7).

How we communicate our message depends on the communication medium. In electronic communications, the medium is copper wire, glass fiber, or radio waves through air. The medium can also be considered the container of the message.

Figure 11.7 | Handheld devices such as the iPhone, the Kindle, and the N95 provide instant access to information and communication.

In graphic communications, these containers could include books, magazines, newspapers, and other written communication. Media for electronic communication include Web pages, radio, television, e-mail, instant messaging, and texting, just to name a few. Of these communications outlets, the ones with large audiences are often called mass media, and communications with smaller audiences are called micromedia.

Graphical Communication

You probably don't give much thought to the old saying, "A picture is worth a thousand words." However, that old adage really comes to life when you consider the explosion in popularity of digital cameras, camcorders, and cell phones that can take pictures and even record video, as well as video and image sharing sites such as YouTube and Flickr. These devices and sites are a testament to the power of graphical communication. In fact, we borrowed language from the world of infectious diseases to describe an incredibly popular video as a "viral" video. Try searching YouTube for the terms "Mentos" and "Diet Coke" and you'll see examples of "viral" videos. Graphic communication is not limited to pure entertainment—there's even a YouTube-like site for education called TeacherTube, and Apple's popular iTunes music store is also host to iTunes University, which provides video access to classroom content from colleges and universities.

When we use graphic communication in engineering and technology, we often refer to it as technical communication—implying that we are sharing technical information through graphs, graphics, and other visual tools. In technical communication, we use graphs to display a trend or a relationship between variables that is not apparent from looking at numerical data. In many cases, it is more important to show this general trend than to focus on the actual data. Graphic communication, when appropriate, is supplemented with written and oral communications. The result of combining graphic, written, and oral communications is typically more effective than any of the three alone. Graphics need not be artistic masterpieces or high-end, computer-generated images and animations. Although these have their time and place, a quick hand-drawn sketch is often sufficient to communicate your point. Graphics and imagery are not only important for communicating ideas to others, they also assist you in formulating and refining your own ideas and complement your verbal description. One of the most common drawings you will encounter in engineering and technology is the blueprint (see Figure 11.8). Although they are not actually "blue" anymore, these technical drawings, created by skilled draftspeople using Computer-Aided Design (CAD) systems, are an important way of communicating and sharing technical information. You may never actually draw and create a set of blueprints, but you may be responsible for reading and approving these drawings, so you must be able to read and understand them.

Figure 11.8 | A blueprint is one of the most common drawings you will encounter in engineering and technology.

In graphical communication, you must consider not only the type of information that you are presenting, but also your intended audience. A very technical audience may expect to see detailed numerical or quantitative information, whereas a nontechnical audience may not be as interested in the actual data, but instead in data trends and their implications. With this in mind, we can divide graphical analysis and presentation into two general types: one in which the information is qualitative in nature, using drawings, bar graphs, and pie charts; and one in which the information is quantitative, using tables and line graphs. For some presentations, it may be appropriate to provide your audience with a general understanding of the magnitude of the terms, without presenting specific numerical data.

Have you ever sat through a presentation, or even given one, and wondered how many more tables, pie charts, or bar charts you would have to look at? Well, it turns out there's a whole universe of visualization methods that most of us didn't even know existed. At *www.visual-literacy.org*, you'll find a periodic table of visual-ization methods to help identify and select alternative ways to present your data (see Figure 11.9).

Notice that the figure is color-coded and uses a number of symbols to help differentiate each of the methods. The various colors allow you to quickly determine which method to use, depending on what you are trying to visualize (data, information, concept, strategy, and so on), and the text color identifies methods that are appropriate for either process or structure visualization. The chart is interactive: If you mouse over any of the elements of the table, a small window will appear with a sample of the visualization method. This chart is not only a great resource, it's also a great example of graphical communication. For example, the periodic table entry for the treemap is shown in Figure 11.10.

According to the periodic table, the treemap is used for information visualization, can be used to provide both an overview and detail, and represents convergent thinking. If you do a little research, you can find examples of treemaps—something you've probably never heard of before. For example, *www.marumushi.com/apps/*

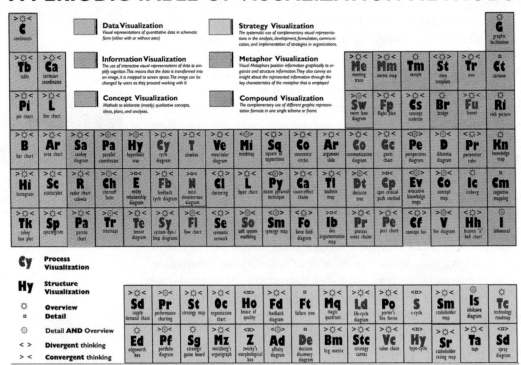

Figure 11.9 | The periodic table of visualization methods gives you alternatives for presenting your message.

newsmap/newsmap.cfm is a site that presents news aggregated together as a treemap (see Figure 11.11).

Take some time to explore the Newsmap site. The different colors represent different types of news (world, national, sports, entertainment, and so on). Different shades of color indicate the age of the news (brighter shades are less than 10 minutes old, while increasingly darker shades represent older news). Finally, the size of each rectangle is related to the number of news stories found for each topic.

Figure 11.10 | A treemap can be used to visualize information.

Figure 11.11 | You can use a newsmap to peruse the headlines in a novel way.

ENGINEERING QUICK TAKE

Creating a Treemap

Now look at how your data might become treemaps. Imagine that you are asked to discuss the table of data shown in Figure 11.12.

The table details a number of companies, their market sectors, and market caps—a measure of a company's size calculated by multiplying the current stock price by the number of outstanding shares. This is a fairly simple example, but it is not hard to get lost in a table, even one this small. Using a spreadsheet program such as Microsoft Excel, we "normalize" the data by dividing by the largest value and use these normalized values to generate rectangles with corresponding areas. We can then create a treemap for the two sectors using different-sized rectangles to indicate the relative market cap of each company. The result might be something similar to the treemap shown in Figure 11.13.

STOCKS			
Company	**Sub Sector**	**Market Cap**	**Performance ID**
Amer.Italian Pasta	Food Products	834 USD	−1.75%
Campbell Soup	Food Products	11359 USD	−0.99%
Dean Foods	Food Products	1621 USD	−1.40%
General Mills	Food Products	12282 USD	0.12%
Heinz	Food Products	14646 USD	−0.28%
Hershey Foods	Food Products	8686 USD	−1.14%
Kellogg	Food Products	12114 USD	−0.06%
Ralcorp Holdings	Food Products	61402 USD	0.37%
Ralston Purina	Food Products	582 USD	−0.66%
Sara Lee	Food Products	9448 USD	0%
Suiza Foods	Food Products	1678 USD	0.75%
Coca-Cola	Soft Drinks	120708 USD	−0.22%
Coca-Cola Ent.	Soft Drinks	7616 USD	−0.44%
Pepsi Bottling Group	Soft Drinks	7040 USD	−1.86%
Pepsi Co	Soft Drinks	86699 USD	0.42%

Figure 11.12 | This table of stock data might look more appealing as a treemap.

Figure 11.13 | In this illustration, the stock data from Figure 11.12 has been converted to a treemap.

I think you would agree that this view of the data is a lot more visually appealing and much easier to understand. There's nothing wrong with tables, but take a look at your data and at some of these techniques. Could you and your communications benefit from a fresh new presentation? Try one.

Oral Communication

You will frequently need to make oral presentations to bosses, teachers, customers, or colleagues at meetings or conferences. In the presentation, you could be sharing information, or selling an idea, a product, or even yourself. (A job interview is nothing more than a presentation in which you're selling yourself.) For any oral presentation, it is normal to be a little anxious at first, but you can take steps to improve your oral communication skills. A great first step is to take a speech class or to join a local chapter of an organization such as Toastmasters International. These steps, along with taking every opportunity to speak to a group, will provide you with experience, and build poise and confidence. After a while, the nervousness will disappear after the first minute or two, especially if you are well prepared and you know the topic you are presenting.

Preparation Very few people can easily give an impromptu presentation—standing up and talking about a subject without first preparing what they want to say. Just as in written communication, you must have a clear idea of what you want your message to be. It may be helpful initially to jot down the facts, create an outline, and identify a theme that connects the facts. With this preparation, you can begin to build your oral or written presentation. As mentioned before, it's important to know your audience and prepare the presentation to suit their background. You should not give too elementary a talk, or one that is too sophisticated; in either case, you will lose contact with the audience. Remember that communication is a two-way street—transmission and reception. In most cases you want to split your presentation into 15- to 20-minute chunks—any longer and you may lose the audience. It should be apparent from your presentation that you are secure and in control of the subject matter. You will be asked questions, which is often the most informative and important part of a presentation. You must be prepared to expand on given topics.

How should you give your talk? You should go over what you want to say many times before actually making the presentation. Once you believe you have all the information needed to make the presentation, prepare note cards from the outline and give a trial talk to yourself. Time it. Remember to speak clearly and not too fast. Once you are satisfied that all the information you want to convey is in your presentation material, try the presentation out on a friend or coworker.

Visual Aids In giving an oral presentation, visual aids can be a tremendous help in making your point or providing complementary details. The most common— and probably unpopular—method of making a presentation is to use Microsoft's PowerPoint program or Apple's KeyNote software. Harvard Professor Edward Tufte refers to all of these applications as "slideware," and cautions against their indiscriminate use. In particular, Tufte worries about a phenomenon many have taken to calling "Death by PowerPoint." There is nothing wrong with using slideware, but it is important to consider how you are using it. Rather than dumping all of your content into PowerPoint, consider creating a very visual slide deck with a limited amount of text (see Figure 11.14). Remember, you are giving the oral presentation, not the software. You should know your topic well, and the slides should provide visual cues to help guide your presentation. Too often, presenters put the content in the slides, do not prepare adequately, and do not thoroughly understand what they'd like to say, so they end up reading the slides. Your audience can read the slides—they don't need you to read to them. If you'd like them to read, prepare a more in-depth and detailed paper or manuscript as a companion piece for your presentation.

Here are some key points to remember when using PowerPoint:

1. Keep it visual—use large text, high-quality images, and pictures that tell a story.
2. Just because the templates use bullets doesn't mean you have to.
3. If you do use text, make it big. Use 24-point type at a minimum. (72 points is about an inch.)
4. Avoid unnecessary animations and transitions—they don't really add anything to what you have to say.
5. Pick a background that is uncluttered and provides a good contrast to your text color.
6. Preview your slides and familiarize yourself with the location at which you are presenting.
7. Take your time—don't rush.
8. Make eye contact; stay connected with your audience.
9. Don't read the slides.
10. Know your topic.
11. Buy yourself a "clicker" and get out from behind the podium.
12. Practice, practice, practice.

(a) Slides from Microsoft's Bill Gates

(b) Slides from Apple's Steve Jobs

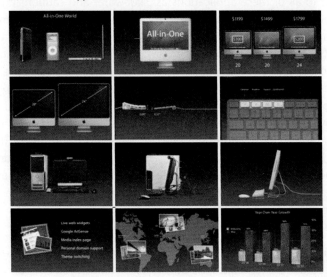

Figure 11.14 | The Presentation Zen web site provides examples of (a) poorly designed slides, and (b) well-designed slides.

Also consider props as visual aids. If you have a 3D model, you can hold it up and describe it or pass it around the room. It will help you to connect with the audience and make the content more tangible to them.

Presentation Techniques Just as the appearance of your written report affects how people view its contents, your own appearance affects audience reaction to what you have to say. Not only should you look presentable, neat, and well dressed, you should behave properly. Speak clearly and loudly enough so the audience can hear you; people do not want to strain to understand what you are saying. Speak slowly enough so they can follow. You know what you want to say and what the connections are, but the audience does not, and you must allow time for these connections to occur.

Talking quickly and softly are signs of nervousness that you can be aware of and correct. Avoid saying "uh," "like," and "you know." Often you will have a podium to hold your notes and to grasp initially, but don't give it a death grip. Maintaining eye contact with the audience throughout your talk is important. Look at people directly, and shift your attention from one person to another. This keeps you involved with the audience, makes you aware that your message is getting through, and lets the audience know that you are engaging individual members constantly.

TECHNOLOGY AND PEOPLE:
Hans Rosling

Hans Rosling is a professor of global health at Sweden's Karolinska Institute. After becoming a licensed physician in 1976, Rosling served as District Medical Officer in northern Mozambique from 1979 to 1981. In 1981, he discovered an outbreak of a formerly unknown paralytic disease, and the investigations that followed earned him a Ph.D. at Uppsala University in 1986. He spent two decades studying outbreaks of the disease in remote areas across Africa and supervised more than ten Ph.D. students. His research group named the new disease "konzo," the local designation by the first affected population. As an academic, Rosling has focused on creating and teaching new courses on global health, and he has coauthored a textbook on global health.

Rosling is a dynamic and enthusiastic presenter (see Figure 11.15). You can tell he loves what he teaches; his enthusiasm is infectious. The focus of his work has been global health and poverty trends, and common misconceptions regarding differences between developing countries and the so-called western world. For Rosling, it is not enough to know his topic or to be passionate about it—he is dedicated to changing how data is presented. Rosling makes his case using statistics drawn from United Nations data and illustrated by visualization software

Figure 11.15 | Hans Rosling helped to develop visualization software through his nonprofit organization, Gapminder.

Section 2 △ Identifying Types of Communications

he developed through his nonprofit organization, Gapminder. Anyone who has seen Rosling present his work marvels at how the software takes flat data and brings the data to life. It is a great example of how technology, at its best, can help to elucidate rather than obfuscate.

Trendalyzer, the software from Gapminder, is free and can be loaded with any data (see Figure 11.16). To take it for a spin and watch Rosling make a presentation, visit *www.gapminder.org/world*. You can vary the data for both the x and y axes and animate the results over time. So, whether you like to watch great presenters in action, are interested in global issues, or just want to see new technologies for teaching and learning, give this video a look. Not surprisingly, Gapminder was purchased by Google in March 2007. Here's an excerpt from the announcement:

> *Gapminder's Trendalyzer software unveils the beauty of statistics by converting boring numbers into enjoyable interactive animations. . . .*

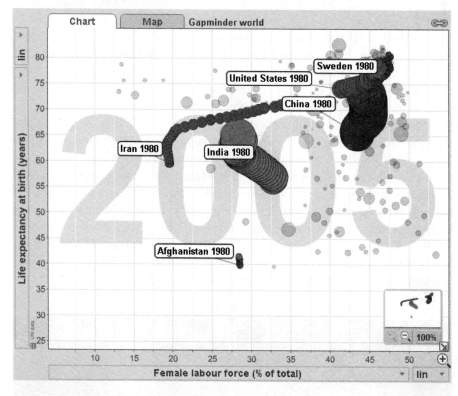

Figure 11.16 | Gapminder's Trendalyzer software can unveil the beauty of statistics, such as this graph of life expectancy of the female labor force.

SECTION TWO FEEDBACK >

1. What types of communication are there?
2. What is the difference between communicating qualitative information and quantitative information?
3. Using the periodic table of visualization methods at *www.visual-literacy .org*, research and briefly describe two alternative ways to present quantitative information.
4. Use the Internet to find a video of a well-known presenter, and share a video with your classmates that is either an example of a well-done or poor presentation.

KEY IDEAS >

- Telecommunications is a very broad term that implies transmission of messages at a distance. It could include transmission of signals via smoke (smoke signals), sound (drums), flags (semaphore), and even reflected sunlight (heliograph).

- Modern telecommunications involves some combination of an electronic transmitter and receiver.

- Communication technology and the ability to communicate electronically provide a competitive advantage.

- Digitizing voice data allows telecommunications carriers to use a single common digital infrastructure for voice, video, and data.

- A basic communications system consists of three things—a transmitter to send the message, media over which to send the message, and a receiver of the information.

When you hear the term **telecommunications**, you and most others probably focus on the *tele* prefix and associate the term with telephones and telephony. In fact, telecommunications is a very broad term that implies transmission of messages at a distance. In the broadest of definitions, telecommunications could include transmission of signals via smoke (smoke signals), sound (drums), flags (semaphore), and even reflected sunlight (heliograph). In our discussion of modern telecommunications, we will restrict ourselves to communications that involve some combination of an electronic transmitter and receiver. Examples include two-way communication devices such as telephones and computers and one-way broadcast communications via radio and television.

Society and Telecommunications

Telecommunication, the ability to connect to others with voice, video, and data, has become so commonplace that we perceive it as we do electricity: We expect it to always be there. Telecommunications is not just a critical part of society, it is also an important part of our global and national economy. Some estimates put the global telecommunications market at 3 percent of the gross world product. Just as communications technologies have revolutionized our daily lives, these technologies have dramatically altered the way we work and how we do business. In particular, telecommunications technologies have empowered businesses—small, medium, and large—to operate in a global environment. For that matter, even individuals can leverage technology to become part of the global economy. Amazon.com and Wal-Mart are two retailers that have integrated telecommunications into virtually every aspect of their business. Figures 11.17, 11.18, and 11.19 demonstrate the shift that is occurring away from older media companies and older ways of communicating.

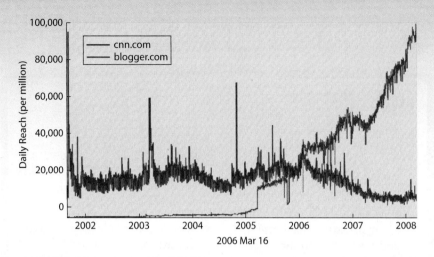

Figure 11.17 | Modern communication is shifting from old media (such as CNN.com) to new media (such as blogs).

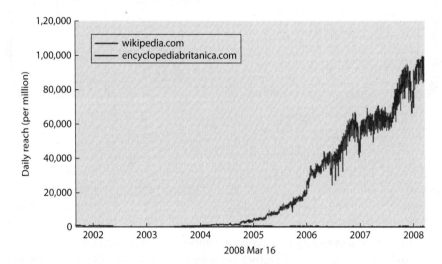

Figure 11.18 | The shift away from traditional encyclopedias to a user-created, online encyclopedia has been dramatic.

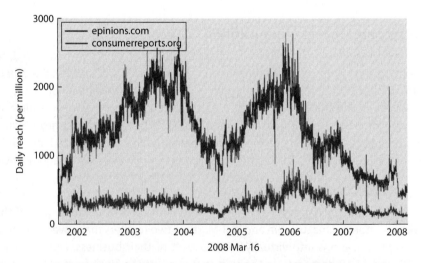

Figure 11.19 | Another dramatic shift is from product reviews and testing written by experts to online user reviews.

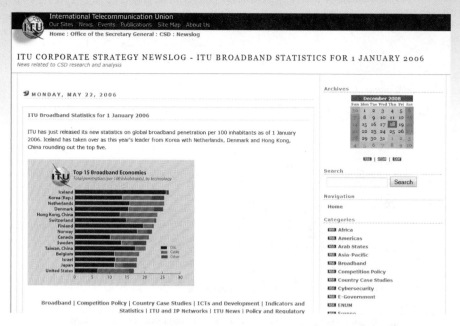

Figure 11.20 | Surprisingly, the United States is not one of the leading countries in broadband Internet adoption.

Communication technology definitely provides a competitive advantage. Consider working on a homework assignment. If you have access to a computer and a high-speed Internet connection, you will probably use those resources in researching and completing your homework. Imagine the competitive disadvantage faced by classmates who don't have access to similar resources. This technology gap between the "haves" and the "have nots" is often referred to the digital divide. This disparity is evident in Figure 11.20, which was compiled using data from the International Telecommunication Union (ITU). A striking point in this data is that the United States is not at the top in high-speed or broadband Internet adoption—instead it is in the middle of the pack.

Even more alarming is the gap between the industrial world and developing countries. ITU has compiled a Digital Access Index (*www.itu.int/ITU-D/ict/dai/*). This data, illustrated in Figures 11.21 and 11.22, shows the developing world lagging behind the rest of the globe in fixed telephony and Internet adoption.

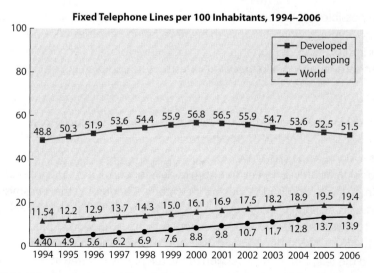

Figure 11.21 | The developing world lags behind the rest of the world in fixed telephone use.

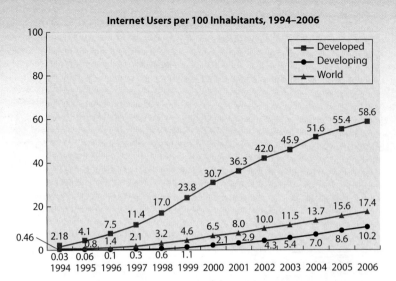

Figure 11.22 | Internet access is also less frequent in the developing world than in other countries.

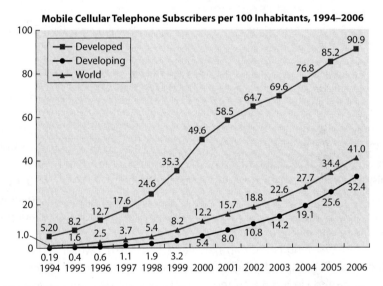

Figure 11.23 | Worldwide cellular phone adoption could help address communication disparities among countries.

One possible solution in the developing world is the introduction of mobile or cellular communications. Because many developing countries have no communications infrastructure, it is often more cost effective to build a wireless communications infrastructure and skip the wired infrastructure completely (see Figure 11.23).

Telephones

Early telecommunications took the form of telegraphy. Samuel Morse is credited with developing a version of the electrical telegraph and the series of dots and dashes—Morse code—that we associate with the telegraph. Morse's telegraph was not significantly different from those developed by his contemporaries. His most important contribution was his new code. Telegraphy, the transmission of messages as a series of dots and dashes, was eventually replaced by telephony, or what we've come to know as the telephone. In 1876, working independently of each other, Alexander Graham Bell and Elisha Gray almost simultaneously invented the telephone. The invention of the telephone ushered in the era of voice communications, which remains central to our lives.

In the first analog or circuit-switched telephone networks, callers' lines were physically connected with a wire. As you might remember from pictures of old-fashioned switchboard operators (see Figure 11.24), they would move cabling on a switchboard to establish a physical circuit between callers. As technology advanced, these manual switching systems gave way to mechanical systems and finally to electronic switching. The switches establish a circuit or electrical connection between the two callers. The telephone number dialed determines the circuit's path. A small microphone in the handset is used to convert voice into an electrical signal suitable for transmission over the circuit. At the receiving end, this electrical signal is converted back to voice by a speaker in the handset. To enable a two-way conversation, we actually create two circuits.

Most residential wired telephones are still analog devices, although the increasing adoption of Voice over Internet Protocol (VoIP) has led to greater use of digital handsets by business and residential customers. The fact that most of our phones are still analog has not stopped telecommunications companies from converting their aging analog equipment to a mostly digital infrastructure. It is unlikely nowadays that a call will travel from source to destination as an analog-only signal. Digitizing our voice data allows telecommunications carriers to use a single common digital infrastructure for voice, video, and data. Mobile or cellular phones have had a significant impact on the telecommunications market. Many consumers have abandoned their traditional landline in favor of a mobile phone. As we saw from the ITU data earlier, much of the developing world is leading the way in mobile phone

Figure 11.24 | Manual switchboard operators moved cabling to establish a physical circuit between callers.

adoption. The dominant protocol for communication across the Public Switched Telephone Network (PSTN) is Asynchronous Transfer Mode, or ATM. ATM is well suited for voice communications and enables us to simultaneously send voice, video, and data over the same network. Telecommunications companies are already beginning to transition their networks from ATM to newer technologies such as Multiprotocol Label Switching (MPLS).

Basic Elements of Communications

When we describe communications, it is convenient to use a "systems" approach in which we consider the entire system that enables us to communicate. As we delve into these systems, we begin to break them down into their basic building blocks or components. To understand communications, it is important to be able to work at both the systems level and the component level. To create a basic communications system, we need three things—a transmitter to send the message, media over which to send the message, and a receiver of the information. All communication systems, regardless of their purpose or complexity, can be modeled with these core components. When you're traveling in a car and listening to the radio, you are receiving a broadcast signal—a signal intended for many recipients. In this example, the radio station has a very powerful transmitter and antenna or broadcast tower at its facility. The transmission medium in this case is the air around us, and the receiver is the radio in your car or home. When you tune your radio to your favorite station, you are tuning the receiver to the specific frequency of that radio station.

Radio is an example of one-way broadcast communications. Many other forms of communication are two-way, and require both a receiver and transmitter on each end of the conversation. A device that can act as both a *trans*mitter and re*ceiver* is often called a transceiver. If you think of the devices you typically use over the course of the day, you realize that there are many transceivers around us. In fact, if you group together a number of transceivers and connect them with some sort of communications media, you have built a network.

Analog Versus Digital Signals

Electronic communication can take the form of either an analog signal or a digital signal. To better understand the difference between the two, consider a typical garden hose you might use to wash a car or water the lawn. If you move the hose up and down, you create a wave of water that looks a lot like an analog signal. You'll notice that you get a continuous stream of water that forms a smooth curve (Figures 11.25 and 11.26). Looking at the wave or waveform, we can identify a few important characteristics. The first is the height of the curve. If we draw a horizontal line through its middle, the distance from the line to the top of the curve is called the amplitude of the waveform—we could use the symbol *A*. Notice also that the wave repeats itself; the time that it takes to repeat is known as the period of the signal and is usually represented with the symbol *T* (see Figure 11.27). Finally, how tightly the curve's peaks are grouped together is related to the frequency (*f*) of the signal. Using our hose analogy, if we move the hose slowly, the peaks will be far apart—a low frequency (see Figure 11.25). If we move the hose faster, the waves are spaced more closely together, corresponding to a higher frequency (see Figure 11.26). To measure frequency, we count the number of peaks during a given interval of time (see Figure 11.28). We usually measure frequency using the unit of hertz (*Hz*). If we count ten peaks per second, our frequency is 10 Hz. The frequency (*f*) can also be calculated from the period T, using the inverse or reciprocal relationship shown in the following equation:

$$f = \frac{1}{T}$$

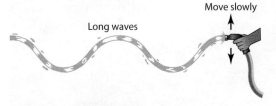

Figure 11.25 | **If you move a garden hose slowly up and down, the water stream shows low-frequency peaks.**

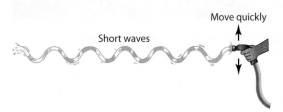

Figure 11.26 | **By moving the hose faster, you create waves of higher frequency.**

Now consider a digital signal. Simply put, a digital signal is a series of ones and zeroes. If we consider our garden hose again, imagine the stream of water you create if you alternately cover and uncover the end of the hose (see Figure 11.29). Unlike the continuous stream we get with an analog signal, here we get short bursts of water—each burst corresponding to a binary one and each period with no water a binary zero. When working with

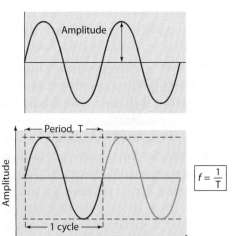

Figure 11.27 | **The flow from the hose can be depicted as a waveform that shows amplitude (A) and period (T).**

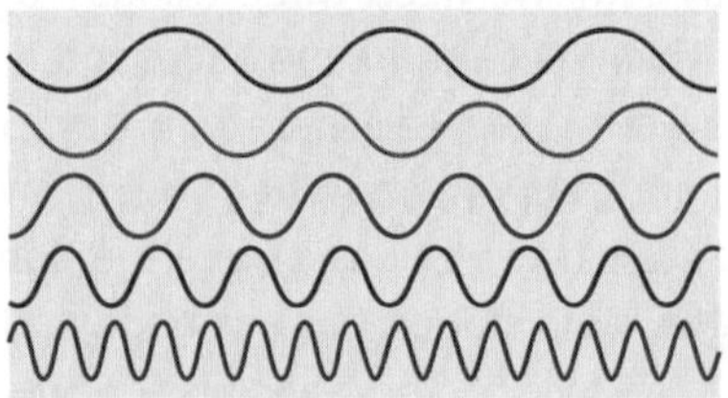

Figure 11.28 | **Analog signals have varying frequency.**

Figure 11.29 | You can use the stream from the hose to simulate binary 1s and 0s.

Figure 11.30 | A light switch is a simple example of a toggle.

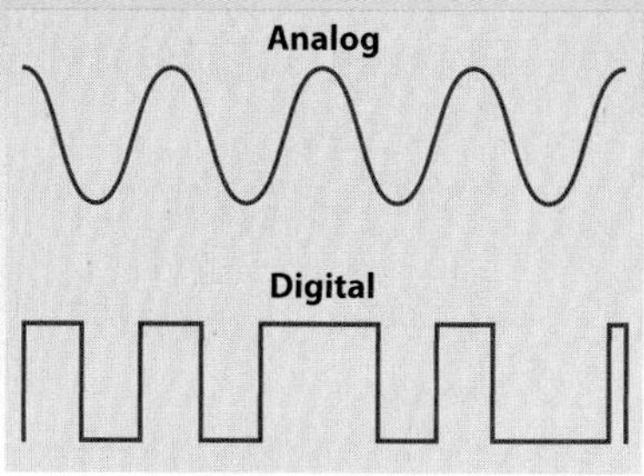

Figure 11.31 | The difference between an analog signal and a digital signal is dramatic.

binary numbers, it is useful to consider the concept of a simple toggle switch. A toggle switch varies between two states or values, like the on and off states of a typical light switch in your house (see Figure 11.30).

While listening to your car radio, you are acutely aware of a key disadvantage of an analog signal: its susceptibility to interference from other signals and to noise (see Figure 11.31).

Channels

If you watch television or listen to the radio, you are already familiar with the concept of a communications channel. In these industries, each network or station broadcasts its own signal in a very specific communications channel—usually associated with a frequency. You "tune" your radio or TV—each of which is a receiver—to watch or listen to the particular channel you're interested in. The use of channels allows us to have multiple stations, traveling either through the air or through the coaxial cable of your cable-TV service. This technique is known as multiplexing (see Figure 11.32). We slice up the available frequencies, with each "conversation" assigned its own frequency. We bundle or multiplex all the channels together, then send them over a medium and unbundle or demultiplex them at the receiving

Multiplexing-Demultiplexing
at fixed aggregate capacity (circuit–switching style)

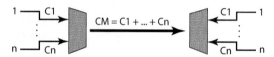

Figure 11.32 | Multiplexing enables the simultaneous transmission of several messages along a single communications channel.

Digital Telephony

Telephony quality voice $\lessapprox 3.5$ KHz $\to \approx 8000 \frac{\text{samples}}{\text{second}}$

8000 samples/s $\times$ 8 bits/sample = 64,000 bits/s... one digital telephony circuit

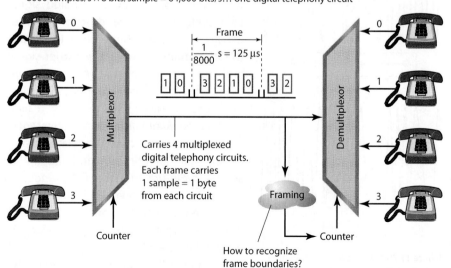

Figure 11.33 | In frequency division multiplexing (FDM), frequency channels are bundled together, sent over a medium, and unbundled or demultiplexed at the receiving end.

435

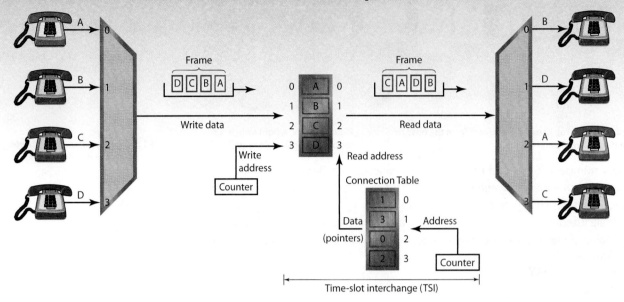

Figure 11.34 | In time division multiplexing (TDM), time is "chopped" into slices or slots and each channel receives its own time slot.

end. More specifically, this is referred to as frequency division multiplexing (see Figure 11.33). Another technique, typically used when only a limited number of frequency channels are available, is time division multiplexing (see Figure 11.34), in which we "chop" time into slices or slots and assign each conversation its own time slot.

Carrier Waves

We have discussed the concept of communication using either analog or digital signals, but how do we send a message using these techniques? To better understand how a message is sent, we need to introduce two new concepts: a carrier wave and modulation. The analog signal or waveform that we described earlier usually acts as our carrier. It is important to differentiate between our carrier signal and our actual message. As you would expect, the carrier or carrier wave "carries" our message. How? Via a technique called modulation. Radio broadcasts, for instance, are usually modified or modulated using either AM (amplitude modulation, as shown in Fig-ure 11.35) or FM (frequency modulation, as shown in Figure 11.36). If you remember the analog waveform from the garden-hose discussion, for AM we modify the height or amplitude A of our carrier wave. For FM we modify the frequency f of our carrier wave.

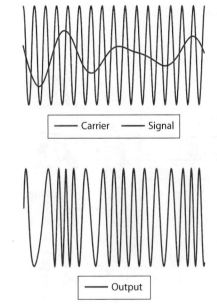

Figure 11.36 | In frequency modulation (FM), the frequency of the carrier wave is modified.

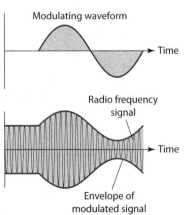

Figure 11.35 | In amplitude modulation (AM), the height or amplitude of the carrier wave is modified.

1. How is telecommunications similar to electricity?
2. Give three examples of how your typical day would be different if you did not have access to telecommunications services.
3. What is the digital divide?
4. Explain the difference between a digital signal and an analog signal.
5. What is a transceiver?

SECTION 4: Learning About Satellite Communications and the Global Positioning System

KEY IDEAS >

- In communications, a satellite is a manmade object positioned in the Earth's orbit to facilitate communication on the Earth. A satellite usually travels in either a geostationary, elliptical, or low Earth orbit (LEO).

- A satellite constellation is a group of satellites working together.

- The Global Positioning System (GPS) uses a constellation of at least twenty-four medium Earth orbit satellites to transmit microwave signals to a GPS receiver.

- Civilian GPS is only accurate within 15 meters because of a combination of factors: errors due to atmospheric conditions, multipath effects, clock drift in the satellite's onboard clock, selective availability (intentionally introduced errors), and relativistic errors.

In astronomy, a satellite is a celestial body orbiting the Earth or some other planet. In the context of communications, a satellite is a manmade object positioned in the Earth's orbit to facilitate communication on the Earth. Satellites are usually categorized by their orbits, some of which include geostationary, elliptical, or low Earth orbits (LEO). Satellites have a variety of applications, particularly when traditional wireline communications are impossible or impractical.

On October 4, 1957, the former Soviet Union launched the first satellite, Sputnik I, into space. This dramatic event caught the United States by surprise and ignited the space race between the United States and the Soviet Union. Although few people realized it at the time, launch of Sputnik I also heralded a new era of communications. Sputnik I (see Figure 11.37) was equipped with an onboard, battery-powered radio transmitting at 20.005 and 40.002 MHz, and it traveled at 29,000 kilometers

Figure 11.37 | Sputnik I ignited the space race between the old Soviet Union and the United States.

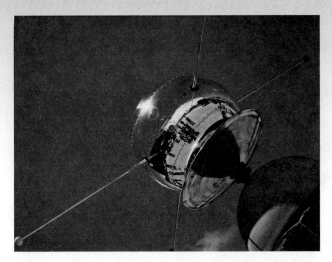

Figure 11.38 | One U.S. response to Sputnik was Vanguard I, which Soviet Premier Nikita Khrushchev referred to as "the grapefruit satellite."

(18,000 miles) per hour. Amateur radio operators from around the world monitored the historic signals for twenty-two days, until the batteries finally ran out. After six months in orbit and nearly 37 million miles traveled, Sputnik I fell out of orbit and burned up while reentering the Earth's atmosphere.

In response to Sputnik I, the U.S. Army's Ballistic Missile Agency (ABMA) and Jet Propulsion Laboratory (JPL) built Explorer I in eighty-four days and launched it into orbit on January 31, 1958. Explorer I was followed by Vanguard I on March 17, 1958. Vanguard I, though small—six inches in diameter—was significant because it was the first solar-powered satellite. Its diminutive size reportedly led Soviet Premier Nikita Khrushchev to refer to Vanguard I as "the grapefruit satellite" (see Figure 11.38). Vanguard's revenge is that it remains the oldest artificial satellite still in space. Probably the most significant and long-lasting outcomes of the response to Sputnik were the creation, in 1958, of ARPA (the Advanced Research Projects Agency) and NASA (the National Aeronautics and Space Administration), and an increase in spending on scientific research and education.

It did not take long for communications satellites to appear. Courier 1B was the first active repeater satellite, launched in October 1960. Built by the Army's Fort Monmouth Laboratories in New Jersey, Courier 1B was powered by 19,000 solar cells and nickel cadmium (NiCad) storage batteries (see Figure 11.39). Launched in 1962, AT&T's Telstar, designed to transmit telephone and high-speed data communications, was the first active communications satellite (see Figure 11.40). Telstar received microwave signals from a ground station, then amplified and rebroadcast the signals.

Figure 11.39 | Courier 1B was the first active repeater satellite.

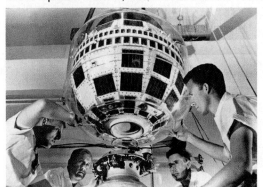

Figure 11.40 | Telstar was the first active communications satellite.

Satellite Orbits

As we've said, satellites are often characterized by their orbits (see Figure 11.41). Consider a geosynchronous orbit. The key part of this term to remember is *synchronous*. You can understand how a satellite behaves in geosynchronous orbit by imagining that each day you can go to the same location at the same time and observe the satellite in the same spot in the sky. With a geostationary orbit, the critical thing to remember is *stationary*. To an Earth-based observer, geostationary satellites appear to be fixed or stationary relative to the Earth. The geostationary orbit is a special case of a geosynchronous orbit, in which the orbit is circular and the satellite passes directly over the equator. Geostationary orbits are particularly useful for communications satellites. Whereas a satellite in geosynchronous orbit requires roughly one day to complete an orbit, a semisynchronous orbit requires only twelve hours to orbit the Earth. Satellites that are part of the Global Positioning System all follow semisynchronous orbits.

A circular orbit at about 250 miles above the Earth's surface is called a low Earth orbit (LEO). Because of their low altitude, these satellites can complete an orbit in

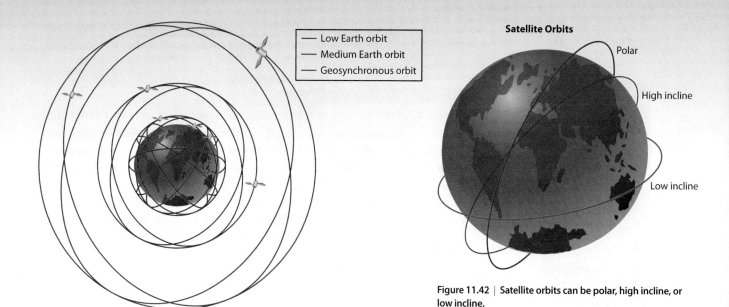

— Low Earth orbit	
— Medium Earth orbit	
— Geosynchronous orbit	

Satellite Orbits

Polar

High incline

Low incline

Figure 11.42 | Satellite orbits can be polar, high incline, or low incline.

Figure 11.41 | Various types of orbits include geosynchronous, medium Earth, and low Earth orbits.

ninety minutes and are only visible from within a radius of roughly 600 miles. The low altitude means that the satellite's position is changing quickly and each satellite has a relatively small footprint, requiring a large number of satellites to provide appropriate coverage. These satellites do have advantages, however; for example, they are cheaper to launch and require less transmission power. In addition to their orbits, satellites can be characterized by their inclinations, the path relative to the Earth's equator (see Figure 11.42).

A satellite constellation is a group of satellites working together (see Figure 11.43). One such constellation is the sixty-six-satellite constellation that supports the Iridium satellite phone (see Figure 11.44).

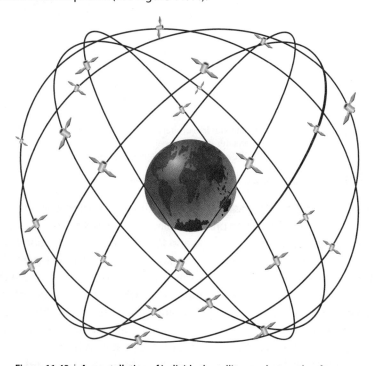

Figure 11.43 | A constellation of individual satellites works together for a common purpose.

Applications of satellite communications include international telephony, both fixed and mobile, satellite television and radio, amateur radio, satellite broadband access, and the Global Positioning System.

The Global Positioning System

The Global Positioning System (GPS) uses a constellation of at least twenty-four medium Earth orbit satellites to transmit microwave signals to a GPS receiver (see Figure 11.45). GPS enables a user to accurately determine location, speed, direction, and time. The GPS system was developed by the U.S. Department of Defense and is managed by the U.S. Air Force. Prior to 1983, access to the GPS system was restricted primarily to military applications. That year, President Ronald Reagan issued a directive opening the GPS system to free civilian use. The directive was a response to the downing of Korean Air Lines Flight 007 after it strayed into Soviet airspace. Opening GPS to free civilian use led to an incredible explosion in the industry and a wide variety of GPS receivers.

Figure 11.44 | A satellite constellation supports satellite phones, such as the one pictured.

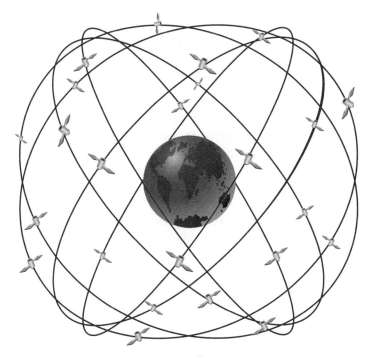

GPS nominal constellation
24 satellites in 6 orbital planes
4 satellites in each plane
20,200 km altitudes, 55 degree inclination

Figure 11.45 | A constellation of satellites helps to operate the Global Positioning System.

Simplified Method of Operation Although the actual process is much more complex and involved, this section provides a simplified look at how GPS works. A GPS receiver identifies satellites within range and then calculates its position relative to three or more of these satellites. Locating a point in three-dimensional space requires at least three reference points—in this case, three satellites. Each satellite in the GPS constellation has a highly accurate atomic clock. Each satellite provides continuous transmissions of the exact time, its location, and general system health. The receiver measures the reception time of each message, which allows the receiver to calculate the distance to each satellite. Knowing the distance to at least three satellites, and its own (we are referencing the GPS receiver) positions, the receiver computes its position using trilateration. Although the original system called for twenty-four satellites, there are thirty-one actively broadcasting satellites in the GPS constellation, each making two complete orbits (semisynchronous) per day. This arrangement provides us with some redundancy and ensures that at any time, at least six satellites are always within the line of sight of any point on the Earth's surface.

You can purchase dedicated GPS receivers or devices with an integrated GPS receiver. In fact, many cars, phones, and even watches have built-in GPS functionality. At a minimum, a GPS receiver system includes an antenna, a receiver, a processor, and a clock. Stand-alone GPS units often add an LCD screen to view navigation information and a speaker to provide audible directions. Modern GPS receivers typically have twelve to twenty channels, enabling the receiver to monitor multiple satellites simultaneously.

Accuracy and Error Sources GPS receivers are commonplace—often they are installed in your car at the factory. To calculate its position and keep us from getting lost, a GPS receiver needs three pieces of information: the current time, the position of the satellite, and the delay of the signal. The accuracy of the receiver is most dependent on the satellite position and the signal delay. Commercial (nonmilitary) GPS can measure the signal delay within 10 nanoseconds or 10×10^{-9} seconds—a small number. For signals traveling at the speed of light, this results in an error of about 3 meters— the smallest possible error for commercial or consumer GPS. This is also the minimum error possible using only the GPS C/A (Coarse/Acquisition) signal. When combined with other potential sources of error, civilian GPS is accurate to within 15 meters.

Other sources of error include atmospheric conditions (see Figure 11.46), which can alter the speed of GPS signals passing through the various layers of the Earth's atmosphere. Atmospheric errors are largest when satellites are near the horizon. Multipath effects, always an issue with wireless signals, also degrade the accuracy of GPS calculations (see Figure 11.47). Multipath is a phenomenon in which wireless signals bounce off obstructions between the source and

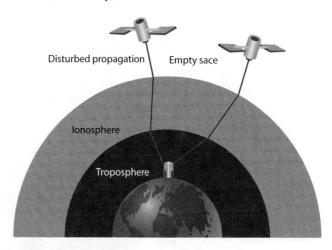

Figure 11.46 | Atmospheric effects and errors have the greatest effect on GPS accuracy when satellites are near the horizon.

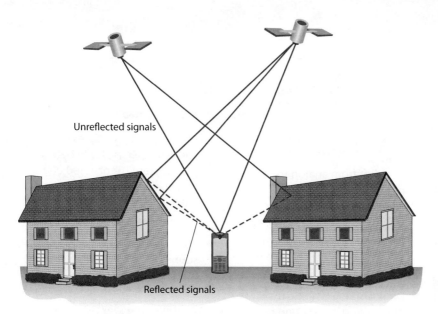

Figure 11.47 | Multipath effects result in the receiver getting multiple versions of the same message.

destination. This results in the receiver getting multiple versions of the same message, which can lead to inaccuracies. The messages are received at different times because they take different paths. Interestingly, these effects are somewhat mitigated by a moving car. It seems that the duplicate or false messages are less likely to reach the destination when it is moving.

Additional sources of error include drift in the satellite's onboard clock, selective availability (an error intentionally introduced to reduce the accuracy of civilian GPS systems), and relativistic effects. Selective availability (SA) originated to ensure that military GPS could maintain its performance advantage over civilian products. SA is no longer enabled on civilian GPS systems. Relativistic errors are due to our assumption that we observe satellites from a stationary reference frame, when in fact the Earth is moving. The result is that satellite-based clocks will advance slightly faster than Earth-based clocks. The difference is less than 40 µs, or 40×10^{-6} seconds per day. The GPS system is also subject to interference and jamming. Potential sources of interference include solar flares and naturally occurring electromagnetic radiation. Other consumer electronic devices can also be unintentional sources of interference.

Applications GPS was originally developed by the military, but because it has both civilian and military applications, it is considered a dual-use technology. Military applications include navigation, fleet management, target tracking, missile and projectile guidance, search and rescue, reconnaissance, and map creation. The primary civilian applications of GPS are surveying and navigation.

SECTION FOUR FEEDBACK >

1. List three satellite orbits and describe their differences.
2. Use the Internet to research prices and features of GPS receivers. Create a table that documents your research.
3. Research the multipath error and provide a brief explanation.
4. When and why was GPS technology made available for civilian use?
5. Do you encounter GPS technology in a typical day? Explain.

SECTION 5: Exploring Digital Media

KEY IDEAS >

- The shift from analog to digital information has forever altered the way we view sound, images, and video, and has opened up entirely new methods of communication and connectivity.

- Digital media is made up of ones and zeroes, and is measured in bits, bytes (8 bits), kilobytes (10^3 or 1,000 bytes), megabytes (10^6 or 1,000,000 bytes), and gigabytes (10^9 or 1,000,000,000 bytes or 1,000 megabytes).

- With analog media, we record sound by "scratching" an analog signal—created by your voice or a musical instrument—onto a surface and playing it back

- by following the grooves we created.

- The MP3 file format uses a compression algorithm to reduce the size of a song while retaining near-CD sound quality, compressing a 32-MB song to 3 MB.

- Digital cameras and camcorders function by focusing light onto a small semiconductor image sensor that filters the light into the three primary colors, records the colors, and combines them to create a full-color image.

Your generation, probably more so than any other, is immersed in digital media. Most of the devices you use are built around the same basic process: converting analog information (represented by a smooth wave—remember the garden hose) into digital information (represented by ones and zeroes, or bits). This shift from analog information to digital information has forever altered the way we view sound, images, and video. It has opened up entirely new methods of communication and connected us like never before. Consider photography alone. In the past, we used a very expensive single lens reflex (SLR) camera, a moderately priced camera, or even an inexpensive disposable camera. All of these examples used film as their medium (see Figure 11.48). How did we use these? We first had to load the film—usually in rolls of twenty-four or thirty-six exposures—and then we took our photographs. We had no ability to preview what the pictures would look like, so it was fairly common to take a number of shots of the same scene to make sure we got the picture we wanted. For a brief event, we probably did not finish exposing the roll of film, so we put the camera back on a shelf until our next special event gave us an opportunity to take more pictures. Eventually, we finished the roll of film and took it to the local drug store, where we dropped it off for processing. Processing, which could take anywhere from one hour to a day or a few days, involved having our film developed using a chemical process, and then ordering paper prints to be made from those negatives—another chemical process. With the time required to finish the roll and process the film, it was not unusual to wait several days before you could see the pictures of your special event. Compare that process to today's method. Today, when we talk about a "camera," we are usually referring—by default—to a digital camera, or even to a cell phone with a built in camera. Today, we can take as many pictures as our media—such as a memory card—can hold. We can see the pictures immediately, keeping the ones we like and deleting the ones we don't like. We can share pictures with our friends and relatives by e-mailing them from our computer or even directly from our cell phones. Today, we have an unprecedented ability to capture, modify, and share digital information.

Figure 11.48 | Before digital cameras, you needed a camera and roll film to take photo.

We call it digital media because it is made up of ones and zeroes. When we refer to digital media, we refer to bits, bytes (8 bits), kilobytes (10^3 or 1,000 bytes), megabytes (10^6 or 1,000,000 bytes), and gigabytes (10^9 or 1,000,000,000 bytes or 1,000 megabytes). With improving technology and decreasing costs, we have even begun to refer to terabytes—10^{12} = 1,000,000,000,000 bytes. That's a lot of data! The decimal number system to which you are accustomed uses the ten digits (0, 1, 2, 3, 4, 5, 6, 7, 8, 9) and powers of 10. For example, the number 18,432 is calculated as follows:

1	8	4	3	2
10^4	10^3	10^2	10^1	10^0
1 × 10,000	8 × 1,000	4 × 100	3 × 10	2 × 1
10,000 + 8,000 + 400 + 30 + 2 = 18,432				

This may seem pretty simple, but only because you are so familiar with the base 10 system. Let's take a look at base 2 or binary numbers, the system we use for digital media and computers. In base 2, each binary digit has two possible values: 0 and 1. Computers use base 2 because it is so easy to represent the two positions *0* and *1* by *off* and *on* in electronics. A typical binary number is:

10110101

which is 8 binary digits or bits, or 1 byte. Using the same technique that we use for decimal numbers, we can calculate the decimal equivalent of the binary number as follows:

1	0	1	1	0	1	0	1
2^7	2^6	2^5	2^4	2^3	2^2	2^1	2^0
1×128	0×64	1×32	1×16	0×8	1×4	0×2	1×1
$128 + 0 + 32 + 16 + 0 + 4 + 0 + 1 = 181$							

These ones and zeroes—binary data—become important when we start talking about digital media: audio, images, and video. First, consider audio. When your parents refer to music, do they talk about vinyl records or albums? Sorry, they can't help it. That's what they grew up with. Albums are vinyl disks containing analog recordings of music. CDs, introduced in the early 1980s, heralded the era of digital media and signaled the decline of analog media—the cassette tapes and vinyl records to which your parents grew up listening.

From Thomas Edison's first analog recordings on tinfoil cylinders to the vinyl recordings that dominated the 1960s and '70s, analog media had its day. For over 100 years, we recorded sound by "scratching" an analog signal—created by your voice or a musical instrument—onto a surface and playing it back by following the grooves we created. Sound waves are captured by some sort of receiver, converted to vibrations, and "scratched" on our media. Playback reverses the process—the grooves cause a needle to vibrate, the vibrations are converted to sound waves, amplified, and heard through speakers. The problem with analog media is that, unless you invest in a high-quality or high-fidelity audio system, the sound quality can be poor and sometimes noisy. Additionally, because we repeatedly follow the grooves in a record, the album will eventually wear out, further reducing sound quality.

With a digital recording, we attempt to approximate a recording (our analog signal) with a digital signal. If we do it right, we can get a digital recording that is virtually indistinguishable from the original, and because it is made of ones and zeroes, it will never wear out. To accomplish this, we convert the analog wave— your voice or your favorite band—to a series of ones and zeroes and record that sequence. The device we use for this process is called an ADC or an analog-to-digital converter. As you might expect, in order to reverse the process and listen to our recording, we use a DAC or digital-to-analog converter, which gives us back our analog wave. The analog wave produced by the DAC is amplified and fed to the speakers to produce the sound.

How closely our analog wave (produced by the DAC) will approximate the original analog signal depends on a number of factors, particularly the sampling rate of our ADC—the higher the sampling rate is, the better quality our recording will have. We can get a better understanding of sampling rates if we consider our original sound wave. With sampling, we take instantaneous readings of our analog signal. If we take a relatively small number of readings per second, we get a crude representation of our original curve. As we increase our sampling rate, we take more readings per second and get a more accurate representation of our original curve. Another variable that we can control is the sampling precision, which represents how closely we can measure

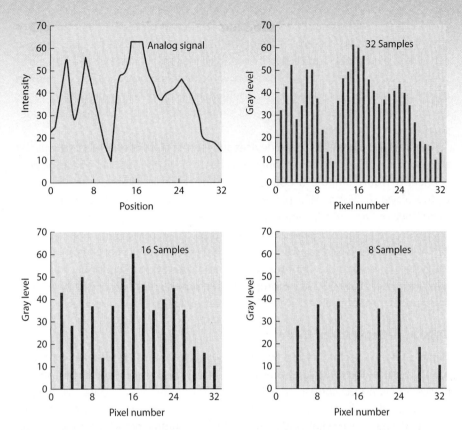

Figure 11.49 | Sampling allows you to convert your analog signal to digital. Notice that as you increase the sampling rate (8, 16, and 32) you get a digital signal that more closely matches your original analog signal.

each sample. For example, if our precision is 10, we may take samples of 3, 4, and 5, but with a precision of 20, our samples may be 6, 7, and 9—giving us a more accurate sample. As the sampling rate and sampling precision increase, the output of our DAC more closely matches the source wave (see Figure 11.49).

What are typical values for sampling rate and precision? If we consider music CDs, the sampling rate used is 44.1 kilohertz, or 44,100 samples per second; the precision used is 65,536—giving sound quality that is virtually indistinguishable from the original. To get this quality, we have to store a lot of data. Let's see just how much data. The numbers generated by analog-to-digital conversion are stored as bytes, and our CD precision requires 2 bytes—16 bits or $2^{16} = 65,536$. Assume that we are recording music in stereo and that our CD can hold up to 74 minutes of music.

44,100 *samples*/(*track* × sec) × 2 *bytes*/*samples* × 2 *tracks* × 74 min × 60 sec/min
= 783,216,000 *bytes*

That is approximately 783 MB of data. Assuming an average 3-minute song:

44,100 *samples*/(*track* × sec) × 2 *bytes*/*samples* × 2 *tracks* × 3 min × 60 sec/min
= 31,752,000 *bytes*

That is about 32 MB per song. At that size, if we were storing music on our computers or on a portable player, we would run out of space quickly. One thousand songs would consume an incredible 32 gigabytes of space. The solution to our storage problem comes in the form of the MP3 file format.

The MP3 file format uses a compression algorithm to reduce the size of a song while retaining near-CD sound quality. With this file format, our 32-MB song compresses to 3 MB and our 1,000-song library goes from 32 GB to 3 GB—a tremendous savings in space. How do we compress a song by a magnitude of

10 (32 MB to 3 MB) without a significant loss of quality? The MP3 file format takes advantage of certain characteristics of the human ear to reduce file size. Because the human ear cannot hear certain frequencies, we can eliminate these frequencies without reducing the quality of the song, while retaining other sounds the human ear hears much better than others. Additionally, we can eliminate other sounds because when two sounds are playing simultaneously, humans hear the louder sound but not the softer one.

The MP3 file format has made devices such as the iPod, iPhone, and other small and portable MP3 players not only possible, but wildly popular. These devices connect to your computer's USB port to transfer music, videos, and other data to the player. A typical MP3 player consists of a microprocessor, storage media (hard drive or flash drive), a liquid crystal display (LCD) screen, and a digital signal processor (DSP). The microprocessor handles all of the device's input and output, controlling the LCD screen, which displays video and song information, and the DSP, which is responsible for processing the audio and video. To play back our MP3 files, we process with DSP to decompress, then run the resulting binary data (ones and zeroes) through a digital-to-analog converter (DAC) and an amplifier.

Figure 11.50 | Digital cameras and video cameras provide yet another means of communicating and sharing information.

Digital Cameras and Camcorders

Digital cameras and camcorders (Figure 11.50) both function by focusing light onto a small semiconductor image sensor. Most often, this sensor is a charge-coupled device (CCD). A CCD consists of a 1-cm panel of hundreds of thousands of light-sensitive diodes called photosites. Each photosite measures and records the amount of light at a particular point, and converts this information into an electrical charge. Photosites are only capable of measuring light intensity, resulting in black and white images. To create a color image, we need not only the total light levels, but also the levels of each of the three colors of light: red, green, and blue. Most sensors filter the light into these three primary colors, record the colors, and combine them to create a full-color image.

High-end cameras and camcorders use three sensors, as well as three filters with a beam splitter to direct light to the different sensors. The advantage of this method is that the camera records each of the three colors at each pixel location. Unfortunately, cameras that use this method tend to be bulky and expensive. A cheaper and more practical method places a three-color filter array over each photosite. With this technique, we get enough information near each sensor to interpolate or estimate the true color at that location. This method requires only one sensor and can record all color information (red, green, and blue) simultaneously, resulting in smaller, cheaper cameras.

SECTION FIVE FEEDBACK >

1. Describe how you typically use digital photography.
2. Use the Internet to research the cost of a typical digital camera memory card, as well as the space, in megabytes, that a print-quality photograph will consume on the card. Using this information and the price of a roll of film, determine the cost savings of digital media.
3. Chart your typical daily interactions with digital media.
4. Interview your parents and grandparents regarding the evolution of home movie recorders and players, including their size, media type, costs, and so on.

Matching Your Interests and Abilities with Career Opportunities: Telecommunications

Telecommunications carriers are expanding their bandwidth by replacing copper wires with fiber-optic cable. Fiber-optic cable, which transmits light signals along glass strands, permits faster, higher-capacity transmissions than traditional copper-wire lines. In some areas, carriers are extending fiber-optic cable to residential customers, enabling carriers to offer cable television, video on demand, high-speed Internet, and conventional telephone communications over a single line. However, the high cost of extending fiber to homes has slowed deployment. In most areas, wired carriers are instead leveraging existing copper lines that connect most residential customers with a central office, to provide digital subscriber line (DSL) Internet service. Technologies in development will further boost the speeds and services available through a DSL connection.

Changes in technology and regulation now allow cable television providers to compete directly with telephone companies. An important change has been the rapid increase in two-way communications capacity. Conventional pay television services provided communications only from the distributor to the customer. These services could not provide effective communications from the customer back to other points in the system due to signal interference and the limited capacity of conventional cable systems. Cable operators are implementing new technologies to reduce signal interference and increase the capacity of their distribution systems by installing fiber-optic cables and improving data compression. This allows some pay television systems to offer two-way telecommunications services, such as video on demand and high-speed Internet access.

Cable companies are increasing their share of the telephone communications market by using high-speed Internet access to provide VoIP (voice over Internet protocol) (see Chapter 12). VoIP is sometimes called Internet telephony because it uses the Internet to transmit phone calls. While conventional phone networks use packet switching to break up a call onto multiple shared lines between central offices, VoIP extends this process to the phone. A VoIP phone breaks a conversation into digital packets and transmits the packets over a high-speed Internet connection. Cable companies use the technology to offer phone services without building a conventional phone network. Wireline providers' high-speed Internet connections also can be used for VoIP, and cellular phones are being developed that use VoIP to make calls using local wireless Internet connections. All major sectors of the telecommunications industry are using VoIP already or will increasingly use it.

Wireless telecommunications carriers are deploying several new technologies to allow faster data transmission and better Internet access that should make them more competitive with wireline carriers. With faster Internet connection speeds, wireless carriers are selling music, videos, and other exclusive content that can be downloaded and played on cellular phones. Wireless equipment companies are developing the next generation of technologies that will allow even faster data transmission. The replacement of landlines with cellular service should become increasingly common because advances in wireless systems will provide ever faster data transmission speeds.

Significant Points

* Telecommunications includes voice, video, and Internet communications services.

* Employment will grow because technological advances will expand the range of services offered.
* With rapid technological changes in telecommunications, people with up-to-date technical skills will have the best job opportunities.
* Average earnings in telecommunications greatly exceed average earnings throughout private industry.

Nature of the Industry

The telecommunications industry delivers voice communications, data, graphics, television, and video at ever-increasing speeds and in an increasing number of ways. Whereas wireline telephone communication was once the primary service of the industry, wireless communication services, Internet service, and cable and satellite program distribution make up an increasing share of the industry. The largest sector of the telecommunications industry continues to be wired telecommunications carriers. Companies in this sector mainly provide telecommunications services via wires and cables that connect customers' residences or businesses to central offices maintained by telecommunications companies.

These new services are made possible through the use of digital technologies that provide much more efficient use of the telecommunications networks. The transmission of voice signals requires relatively small amounts of capacity on telecommunications networks. By contrast, the transmission of data, video, and graphics requires much higher capacity. This transmission capacity is referred to as "bandwidth." As the demand increases for high-capacity transmissions—especially with the rising volume of Internet data—telecommunications companies have been expanding and upgrading their networks to increase the amount of available bandwidth.

Wireless telecommunications carriers, many of which are subsidiaries of the wired carriers, transmit voice, graphics, data, and Internet access through signals over networks of radio towers. The signal is transmitted through an antenna into the wireline network. Increasing numbers of consumers are replacing their home landline phones with wireless phones.

In an increasingly competitive market, traditional telecommunications companies are beginning to offer new services such as cable television, video on demand, and high-speed Internet. These are in addition to their existing telephone communications offerings. To be able to offer these additional services over a single line, these carriers must increase their available bandwidth. To do this, they are replacing their aging copper infrastructures with faster, high-capacity fiber-optic cable, which transmits light signals along glass strands. This new high-speed infrastructure is even finding its way into the homes of residential customers, for example Verizon's FiOS, or Fiber Optic Service.

Working Conditions

The telecommunications industry offers steady, year-round employment. Workers in this industry are sometimes required to work overtime, especially during emergencies, such as floods or hurricanes, when employees may need to report to work with little notice.

Installation, maintenance, and repair occupations account for one in four telecommunications jobs. One of the most common occupations is telecommunications line installers and repairmen, who work in a variety of places, both indoors and outdoors, and in all kinds of weather. Their work involves lifting, climbing, reaching, stooping, crouching, and crawling. They must work in high places such as rooftops and telephone poles, or below ground with buried lines. Their jobs bring them into proximity with electrical wires and circuits, so they must take precautions to avoid shocks. These workers must wear safety equipment when entering manholes and test for the presence of gas before going underground.

Telecommunications equipment installers and repairmen, except line installers, generally work indoors—most often in a telecommunication company's central office or a customer's home or place of business. They may have to stand for long periods, climb ladders, and do some reaching, stooping, and light lifting. Following safety procedures is essential to guard against work injuries such as minor burns and electrical shock.

Most communications equipment operators, such as telephone operators, work at video display terminals in pleasant, well-lighted, air-conditioned surroundings. The rapid pace of the job and close supervision may cause stress. Some workplaces have introduced innovative practices among their operators to reduce job-related stress.

Training and Advancement

Training is a key component in the careers of most workers in the telecommunications industry. Due to the rapid introduction of new technologies and services, the telecommunications industry is among the most rapidly changing in the economy. This means workers must keep their job skills up to date. From managers to communications equipment operators, increased knowledge of both computer hardware and software is of paramount importance. Telecommunications industry employers now look for workers with knowledge and skills in computer programming and software design; voice telephone technology, known as "telephony"; laser and fiber-optic technology; wireless technology; and data compression. Several major companies and the telecommunications unions have created a Web site that provides free training for employees, enabling them to keep their knowledge current and helping them to advance.

The telecommunications industry offers employment in jobs that require a variety of skills and training. Many jobs require at least a high school diploma or an associate degree, in addition to on-the-job training. Other jobs require particular skills that may take several years of experience to learn completely. For some managerial, professional, and maintenance and repair jobs, employers require a college education.

Outlook

Greater demand for an increasing number of telecommunications services will cause overall employment to increase in the telecommunications industry. In addition, many job opportunities will result from the need to replace a large number of workers who are expected to retire in the coming decade. Employment in the telecommunications industry is expected to increase by 5 percent from 2006 to 2016, compared with 11 percent growth for all industries combined. The building of more advanced communications networks, such as fiber-optic lines, faster wireless networks, and advanced switching equipment, will increase employment, particularly in the near term. In the long term, employment gains will be partially offset by the improved reliability of advanced networks, which is expected to reduce maintenance requirements. With a growing number of retirements and the continuing need for skilled workers, good job opportunities will be available for people with up-to-date technical skills. Jobs prospects will be best for those with two- or four-year degrees.

Earnings

Average weekly earnings of nonsupervisory workers in the telecommunications industry were $963 in 2006, significantly higher than average earnings of $579 in private industry. Most full-time workers in the utilities industry receive substantial benefits in addition to their salaries or hourly wages. This is particularly true for workers who are covered by a collective bargaining agreement. Twenty-two percent of employees in the telecommunications industry are union members or are covered by union contracts, compared with about 13 percent for all industries. Most telecommunications employees belong to the Communications Workers of America or the International Brotherhood of Electrical Workers.

[Bureau of Labor Statistics, U.S. Department of Labor, Occupational Outlook Handbook, 2008–09 Edition, 2/2008, http://www.bls.gov/oco/]

Summary >

Information and communication technologies include the inputs, processes, and outputs associated with sending and receiving information. These systems allow information to be transferred from human to human, human to machine, machine to human, and machine to machine. Communication systems are made up of a source, encoder, transmitter, receiver, decoder, storage, retrieval, and destination. At the source, a modem modulates an analog carrier signal to encode digital information. At the destination or receiver, a modem demodulates the signal to decode the transmitted information and receive the message.

Information and communication systems can be used to inform, persuade, entertain, control, manage, and educate. There are many ways to communicate information, such as graphic and electronic means. Technological knowledge and processes are communicated using symbols, measurements, conventions, icons, graphic images, and languages that incorporate a variety of visual, auditory, and tactile stimuli. Graphical analysis and presentation can be divided into two general types: information that is qualitative in nature, and information that is quantitative. Oral presentations to bosses, teachers, customers, or colleagues at meetings or conferences can be used to share information, or sell an idea, a product, or even yourself (a job interview, for example).

Telecommunications is a very broad term that implies transmission of messages at a distance. Telecommunications could include transmission of signals via smoke (smoke signals), sound (drums), flags (semaphore), and even reflected sunlight (heliograph). Modern telecommunications involves some combination of an electronic transmitter and receiver. Communication technology and the ability to communicate electronically provide a competitive advantage. Digitizing voice data allows telecommunications carriers to use a single common digital infrastructure for voice, video, and data. A basic communications system consists of three things—a transmitter to send the message, media over which to send the message, and a receiver of the information.

In communications, a satellite is a manmade object positioned in the Earth's orbit to facilitate communication on the Earth. A satellite usually travels in either a geostationary, elliptical, or low Earth orbit (LEO). A satellite constellation is a group of satellites working together. The Global Positioning System (GPS) uses a constellation of at least twenty-four medium Earth orbit satellites to transmit microwave signals to a GPS receiver. Civilian GPS is only accurate within fifteen meters because of a combination of factors: errors due to atmospheric conditions, multipath effects, clock drift in the satellite's onboard clock, selective availability (intentionally introduced error), and relativistic errors.

The shift from analog to digital information has forever altered the way we view sound, images, and video, and it has opened up entirely new methods of communication and connectivity. Digital media is made up of ones and zeroes. It is measured in bits, bytes (8 bits), kilobytes (10^3 or 1,000 bytes), megabytes (10^6 or 1,000,000 bytes), and gigabytes (10^9 or 1,000,000,000 bytes or 1,000 megabytes). With analog media, we record sound by "scratching" an analog signal—created by a voice or a musical instrument—onto a surface and playing it back by following the grooves we created. The MP3 file format uses a compression algorithm to reduce the size of a song while retaining near-CD sound quality, compressing a 32-MB song to 3 MB. Digital cameras and camcorders function by focusing light onto a small semiconductor image sensor that filters the light into the three primary colors, records the colors, and then combines them to create a full-color image.

FEEDBACK

1. Use the Internet to research cellular telephony and create a block diagram showing how you make a call from your cell phone to a friend's cell phone.

2. What role does a modem play in our communications system? Do we always need a modem or similar device?

3. How are we able to compress digital data and make it so much smaller than its analog equivalent?

4. When would we use qualitative information instead of quantitative information? Give an example.

5. Using the periodic table of visualization methods at *www.visual-literacy.org,* research and briefly describe two alternative ways to present qualitative information.

6. Using Hans Rosling's Trendalyzer software (see *www.gapminder.org/world*), research statistics in global poverty and health and use the software to make a brief presentation to your classmates.

7. Research the digital divide. How many Americans are at a competitive disadvantage? Suggest steps we could take to improve the situation.

8. Create a multimedia presentation for a nontechnical audience that explains the difference between analog and digital signals.

9. Using a notebook or journal, record every activity in a typical day in your life that could be classified as communication of any kind.

10. Create a multimedia presentation that details three types of satellite orbits.

11. Create a multimedia presentation explaining how GPS technology works.

12. Identify three applications of GPS technology, excluding personal use in travel.

13. Interview one of your grandparents or another relative about their experiences growing up with photography and film. Either record the audio of your interview or summarize the interview in a multimedia presentation.

14. Interview your parents to determine what type of audio media they have (CDs, vinyl records, and so on) and how many of each kind.

15. Gather information from five of your classmates regarding the number of songs they have on their iPods or other MP3 players. Using the average of these numbers, determine how many CDs would be required to hold the same number of songs. Using the volume of a CD case, the number of CDs required and the volume of an MP3 player, create a graphic to compare their relative sizes (for example, 1,000 songs on an iPod shuffle versus 1,000 songs on, say, 100 CDs—assuming 10 songs per CD).

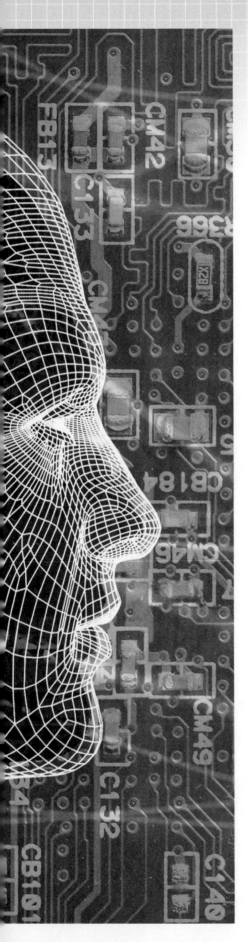

DESIGN CHALLENGE 1:
Securing a Door Electronically

• Problem Situation

Individuals and organizations are becoming increasingly concerned about security, personal safety, and identity theft. To address these growing concerns, many people are implementing electronic access control methods. One such method involves the use of radio frequency identification (RFID) tags that are programmed to provide access to a building. In this activity, you will have the opportunity to install an electronic door strike that is controlled by an RFID reader, as well as set up and program RFID access tags that will unlock the door.

• Your Challenge

Given the necessary materials, you will modify an electronic deadbolt to unlock via RFID, install the modified deadbolt on a door, set up a PC to control the deadbolt, populate a database on the PC to determine access control, and program RFID tags to unlock the door.

• Safety Considerations

1. Use caution when working with electricity and power tools.
2. Wear eye protection at all times.
3. Use insulated tools when working with electric components.

• Materials Needed

1. Two general-purpose circuit boards
2. A 5 V reed relay
3. A 5 V DPST or DPDT relay
4. An electronic door strike
5. Phidgets USB RFID reader from Phidgets USA.com
6. Phidgets RFID kit
7. Plastic project box
8. A Kwikset Powerbolt 1000 electronic keypad deadbolt
9. Soldering iron and solder
10. Desoldering wick or solder vacuum
11. Hot glue gun and glue
12. Drill and drill bits

• Clarify the Design Specifications and Constraints

Your design must be made from off-the-shelf products that you can modify with typical tools found in an electronics shop. The prototype need not be waterproof, but consider how a production model might need to be waterproof.

A PC with a database must be used to determine and control RFID tags—and users—that can or cannot unlock the door. The database should include a means of limiting access during off hours. The database should be clearly defined so that anyone with access to it can easily enable or disable an RFID tag.

• Research and Investigate

To complete the design challenge, you need to first gather information to help you build a knowledge base.

1. In your guide, complete the Knowledge and Skill Builder I: Safety considerations.

2. In your guide, complete the Knowledge and Skill Builder II: Understanding RFID.

3. In your guide, complete the Knowledge and Skill Builder III: Soldering and desoldering.

4. In your guide, complete the Knowledge and Skill Builder IV: Electronic switches and relays.

5. In your guide, complete the Knowledge and Skill Builder V: Using databases.

● Generate Alternative Designs

Describe two of the possible alternative approaches to electronically securing a door. Discuss the decisions you made in (a) modifying the deadbolt, (b) connecting the PC, and (c) building the database. Attach printouts, photographs, and drawings if they are helpful and use additional sheets of paper if necessary.

● Choose and Justify the Optimal Solution

What decisions did you reach about the design of the RFID-enabled deadbolt?

● Display Your Prototypes

Produce your RFID-enabled deadbolt, access tags, and access database. Include descriptions, formulas, photographs, or drawings of these items in your guide.

● Test and Evaluate

Explain whether your designs met the specifications and constraints. What tests did you conduct to verify this?

● Redesign the Solution

What problems did you face that would cause you to redesign the (a) RFID-enabled deadbolt, (b) the PC-to-deadbolt link, and/or (c) the access database? What changes would you recommend in your new designs? What additional trade-offs would you have to make?

● Communicate Your Achievements

Describe the plan you will use to present your solution to your class. (Include a media-based presentation.)

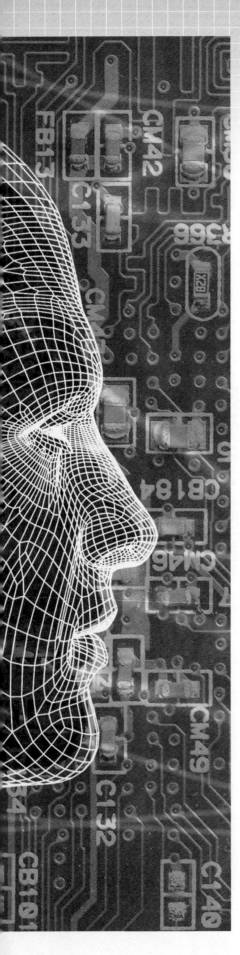

DESIGN CHALLENGE 2:
Viral Marketing Video for YouTube

- ### Problem Situation

Video has become an important means of communication and sharing information. This is due to the proliferation of low-cost, powerful video cameras, easy-to-use video editing software, and Web-based video sharing Web sites such as YouTube. In this activity, you will research, storyboard, and create a video that focuses on a career in engineering and technology.

- ### Your Challenge

Given the necessary materials, you will research careers in engineering and technology, develop a plan for a video that describes such a career, create a storyboard, and use appropriate software and hardware to create a video. The goal of the video is to educate younger students about careers in engineering and technology and to encourage more students to pursue these careers.

- ### Safety Considerations

Take appropriate precautions when sharing information about yourself or your classmates on the Internet.

- ### Materials Needed

1. Posterboard or 8 ½ × 11-inch storyboard paper
2. Markers
3. PC with video and image editing software (such as Movie Maker or iMovie)
4. Digital camera capable of recording small video clips
5. Video camera (optional)
6. YouTube account (optional)

- ### Clarify the Design Specifications and Constraints

A PC with movie editing software must be used to create a 1- to 2-minute video that describes and promotes careers in engineering and technology. The video should include still images, brief video clips, and informational slides. The video should include musical accompaniment and narration, if appropriate.

- ### Research and Investigate

To complete the design challenge, you need to first gather information to help you build a knowledge base.

1. In your guide, complete the Knowledge and Skill Builder I: Web safety.
2. In your guide, complete the Knowledge and Skill Builder II: Fair use and copyright.
3. In your guide, complete the Knowledge and Skill Builder III: Developing a plan.
4. In your guide, complete the Knowledge and Skill Builder IV: Researching video and video hosting.
5. In your guide, complete the Knowledge and Skill Builder V: Researching a career.
6. In your guide, complete the Knowledge and Skill Builder VI: Making a simple storyboard.
7. In your guide, complete the Knowledge and Skill Builder VII: Practice recording.

8. In your guide, complete the Knowledge and Skill Builder VIII: Practice editing.

9. In your guide, complete the Knowledge and Skill Builder IX: Producing a video.

10. In your guide, complete the Knowledge and Skill Builder X: Uploading a video to YouTube (optional).

• Generate Alternative Designs

Describe two possible alternatives to your video. Discuss the decisions you made in (a) storyboarding, (b) career research, and (c) editing. Attach printouts, photographs, and drawings if helpful and use additional sheets of paper if necessary.

• Choose and Justify the Optimal Solution

What decisions did you reach about your designs of the choice of career, the storyboard, and the video?

• Display Your Prototypes

Produce your video with a supporting storyboard and documentation.

• Test and Evaluate

Explain whether your designs met the specifications and constraints. What tests did you conduct to verify this?

• Redesign the Solution

What problems did you face that would cause you to redesign the (a) career plan, (b) the storyboard, and/or (c) the video? What changes would you recommend in your new designs? What additional trade-offs would you have to make?

• Communicate Your Achievements

Describe the plan you will use to present your solution to your class. (Include a media-based presentation.)

YOU USE NETWORKS every day. What do you typically do over the network? You probably send messages back and forth with your friends, usually in the form of text messages and instant messages and, more rarely, as e-mails. Perhaps you reserve e-mail for communicating with your parents or teachers. You also probably spend a lot of time on the Internet doing research for your school work, building your MySpace or Facebook pages, or just playing games. Regardless

CHAPTER

12

DATA NETWORKING AND COMMUNICATION

Figure 12.1 | When you send instant messages, you are sharing resources—in this case an IM server—on a network.

of which of these activities you are doing, you are using the network to share resources—usually in the form of Web, messaging, and e-mail servers (see Figures 12.1 and 12.2).

Modern networks as you know them are a relatively new thing. At the beginning of the computing age, the only computers available were mainframes. A mainframe is a very large computer, often taking up an entire room (see Figure 12.3). Mainframes were not very fast or powerful by today's standards. They typically performed only basic arithmetic calculations that you can do today with a five-dollar calculator. In fact, any one

Figure 12.2 | Text messages travel from cell phone to cell phone via a network.

Figure 12.3 | Mainframe computers like this one had less processing power than a cheap calculator you could buy at the drugstore today.

457

Figure 12.4 | Modern devices like these are much more powerful than the room-sized mainframe computers that were used at the beginning of the computer age.

of the small electronic devices you use every day (cell phones, Sony PSPs, calculators, or MP3 players) is more powerful than an early mainframe (see Figure 12.4). Because of the size and cost of these computers, very few people had access to these machines. If you were lucky enough to work for a large research university, the government, or a large corporation, you may have had access, but even then only limited access.

SECTION 1: Networking Technologies and Applications

KEY IDEAS >

- Information and communication systems allow information to be transferred from human to human, human to machine, machine to human, and machine to machine.

- Information and communication systems can be used to inform, persuade, entertain, control, manage, and educate.

- There are many ways to communicate information, such as graphic and electronic means.

- The Open System Interconnect model is a seven-layer model that helps us understand how networks operate, and the TCP/IP model is a four-layer model that more closely represents how real networks work.

Figure 12.5 | A dumb terminal was essentially a monitor that allowed the computer user to interact with the mainframe. Multiple dumb terminals attached to a single main frame allowed multiple users to share the mainframe's processing power.

The first networks evolved to allow great numbers of people to access mainframes. Usually, a room adjacent to the mainframe would contain a number of devices that could connect to the mainframe and share computing resources. These devices were known as "dumb" terminals. This term was used because these devices had no computing power of their own, only the ability to connect to the mainframe and use its computing capability (see Figure 12.5). These "dumb" terminals allowed multiple people to connect to the mainframe at the same time, a process that has become known as time-sharing.

Networks did not evolve much from that era until 1984, a significant year in the evolution of networks because it ushered in the introduction of the personal computer (PC). International Business Machines (IBM) began selling the first computer that was compact enough and inexpensive enough for smaller organizations and companies—even for home users (Figure 12.6). As personal computers proliferated, people began to connect them to printers and use PCs to work collaboratively. Unfortunately, sharing printers or files was not a simple matter. To share files, we used what has been called a "sneaker net." The term sneaker net means exactly what it implies. If Alice wanted to share a file with her co-worker Bob, she saved the file to a floppy disk, removed the floppy disk from her computer, and walked across the room to Bob, who inserted the disk into his computer and opened the file. It was not exactly a fast and efficient way of sharing information. Printing was an even bigger problem. If there was only one printer, only the computer directly connected to it could print—meaning we had to rely on sneaker net. The only alternative was to purchase a printer for every computer user who needed to print.

Although we view networks as a part of the PC revolution of 1984, the technology that formed the foundation of networks has its roots in the early 1970s. In the late 1960s and early 1970s a government agency, the Defense Advanced Research Projects Agency (DARPA) funded research that still forms the basis for modern networks (see Figure 12.7). If you research DARPA, you will find that over the years the agency name has switched back and forth between DARPA to ARPA.

Figure 12.6 | Although not much to look at by today's standards, the first IBM personal computer launched a computer revolution. It was introduced in 1984.

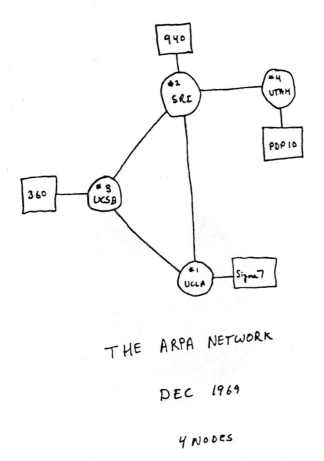

Figure 12.7 | The modern Internet has its roots in the ARPANET, the first modern data network. Conceived in this original 1969 handwritten sketch, ARPANET's design allowed for multiple paths between sites, so that if one site was destroyed or went offline, the remaining sites could continue to communicate with one another.

Regardless of what we call it, this agency was responsible for building ARPANET, the first modern data network and precursor to today's Internet. ARPANET was designed and built during the Cold War between the United States and the former Soviet Union. The goal of ARPANET was to connect strategic defense sites, allowing them to communicate with one another and providing redundant communication in the event of a feared nuclear attack. The network was designed to provide multiple paths between sites, so that if one site was destroyed or went offline, the remaining sites could continue to communicate with one another. This work was refined over the next few years, until a collection of rules for communication, or "protocol suite," was developed. The suite, commonly known as TCP/IP or Transmission Control Protocol/Internet Protocol, was completed in 1978 with the suite's fourth iteration or version. TCP/IP version four—from 1978—is still the standard protocol used on the Internet.

Figure 12.8 | The TiVo device or Digital Video Recorder (DVR) is a small, specialized computer with a large hard drive that retrieves show information from the Internet and lets you record your favorite shows or even pause live TV.

There are many ways to communicate information, such as graphic and electronic means, but networks have become the primary way most of us communicate. Like any information and communication system, networks allow us to inform, persuade, entertain, control, manage, and educate. Although most communication through networks is human to human, networks have also become an important mechanism for transferring information from human to machine, machine to human, and machine to machine. Through these interactions, networks have made their way into almost every aspect of existence, whether it is programming our TiVo to record our favorite show (see Figure 12.8), or receiving location and travel information from a global network of satellites via our in-car global positioning device (see Figure 12.9).

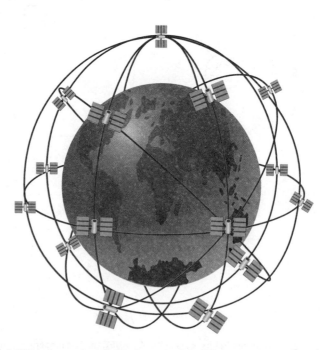

Figure 12.9 | A GPS device, such as the kind commonly built into new cars, retrieves location and travel information from a global network of satellites.

According to the *New Oxford American Dictionary,* a communications protocol is "a set of rules governing the exchange or transmission of data electronically between devices." So a protocol suite, such as TCP/IP, then, is a collection of rules for communicating between two or more devices. Think of a protocol as a language. For two people to communicate in the most effective fashion, they must both speak the same language. Likewise, for two devices to communicate, the devices must use a common protocol.

Modern Ethernet

We have already touched upon how important networks and the Internet have become in your daily life. You depend on these technologies to help keep you connected with friends, family, and the world at large. When you have a school assignment, how often do you find yourself on the Internet looking for information? The majority of the world is no different from you in this regard. Most of us depend on networks and the Internet. We do online banking and shopping; we look up movie times and maps to unfamiliar destinations. Increasingly, we rely on the Internet for our news and our entertainment. Networks have changed the way we work as individuals and have dramatically altered how businesses operate. In fact, it is increasingly difficult to find a business that does not depend on networks. When you consider the role networks play in banks, schools, retail stores, and for architects and engineers, it becomes painfully obvious that the network has become—for businesses—a mission-critical application. Network down time can be linked to lost productivity and, more importantly, to lost revenue. The information transmitted on our networks is not limited to data. Because of the cost savings, it is becoming increasingly common for these networks to transmit voice and video as well. The ongoing convergence of voice, video, and data onto this one network only serves to underscore the crucial roles played by networks and connectivity.

TECHNOLOGY AND PEOPLE

(adapted from http://www.adaptivepath.com/ideas/essays/archives/000385.php)

Jesse James Garrett

Information architect Jesse James Garrett (see Figure 12.10), president and co-founder of Adaptive Path (a consulting firm focused on helping companies improve Web-based interfaces and usability), is credited with coining the term "AJAX" in a February 2005 essay at www.adaptivepath.com. Garrett is the author of *The Elements of User Experience,* a book that has influenced Web designers, software developers, and industrial designers. He has been featured in a number of publications, including *The New York Times, The Wall Street Journal,* and *BusinessWeek.* In 2006, Jesse received *WIRED* magazine's Rave Award for Technology and has been named one of *PC World* magazine's *"50 Most Important People on the Web," eWeek* magazine's *"Top 100 Most Influential People in IT,"* and one of *Software Development Times'* top 100 technology industry leaders.

AJAX, an acronym for Asynchronous JavaScript and XML, describes a suite of technologies that has fundamentally changed the way we interact with the

Figure 12.10 | Information architect Jesse James Garrett is the author of *The Elements of User Experience,* a book that has influenced Web designers, software developers, and industrial designers. He coined the term "AJAX."

Web. The goal of AJAX is to provide the richness and responsiveness of desktop applications in Web-based applications. The technologies that, according to Garrett's definition, comprise AJAX include:

- standards-based presentation using XHTML and CSS
- dynamic display and interaction using the Document Object Model (DOM)
- data interchange and manipulation using XML and XSLT
- asynchronous data retrieval using XMLHttpRequest
- and JavaScript binding everything together

As shown in Figure 12.11, traditional Web browsing was built upon the classic client-server model. In this model, users run Web-browser software on their computer (the client), which accesses information on a Web server. The classic client-server model works like this. User actions (in a browser on the client system) initiate a Hyper Test Transfer Protocol (HTTP) request back to a Web server. Based on this request, the server may perform a number of tasks, including data retrieval, number crunching, etc. The server then returns that information (e-mail, weather forecasts, movie times, etc.) as a Hyper Text Markup Language (HTML) page to the client. (See Figure 12.11.) While this model has worked well for building out an Internet focused on content, it works poorly when we look to the Web as a platform for building, hosting, and running software applications.

By contrast, AJAX applications reduce the "click-and-wait, click-and-wait" interactions of traditional Web pages by introducing an intermediary—an AJAX engine—between user and server. At the start of the session, the browser loads

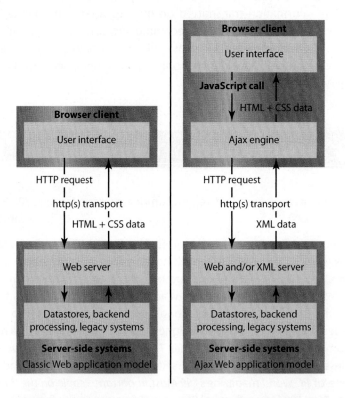

Figure 12.11 | Traditional Web browsing is based on the classic client-server model, in which the Web-browser software on the client computer requests HTML files from a Web server. In the AJAX model, an AJAX engine on the client computer decides whether to send the request to the Web server or to handle it locally on the client computer. For many actions, such as refreshing a Web page, a request to the server is not required, and so AJAX-based pages load much faster and allow much more dynamic user interaction than the tradition client-server model.

the AJAX engine instead of a Web page. The AJAX engine is transparent to the user as JavaScript code "living" in a hidden HTML frame. Acting as an intermediary, this engine is responsible for rendering the user interface and communicating with the server. The AJAX engine speeds up user interaction with the interface and the application by handling all of this asynchronously (on one side) without the constant intervention of the server. The result is a much more pleasant user experience; the user never has to stare at a blank page or watch an hourglass icon while a page reloads (see Figure 12.12).

Every user action that normally would generate an HTTP request instead creates a JavaScript call to the AJAX engine. The AJAX engine decides how to handle this request. For requests that do not require communication with the server, the engine responds directly. For other requests requiring processing or information from the server, the engine requests the information from the server—again asynchronously—without impacting a user's interaction with the application.

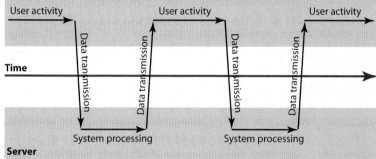

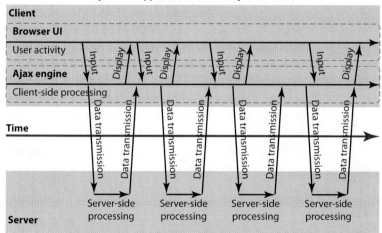

Figure 12.12 | In the classic Web model, users spend a fair amount of time waiting for the Web server to respond. In the AJAX model, users receive the information they want much more quickly.

Although there have been a number of networking standards developed, Ethernet has become and remains the dominant standard. The Ethernet standard defines how computers connect to one another, communicate, and share resources. Before we can dive into a discussion of Ethernet, we first need to learn some basic terminology and develop a basic understanding of networking. The purpose of Ethernet is to link computers that are in the same building. Because, geographically,

Figure 12.13 | While working at Xerox Corporation's Palo Alto Research Center in 1973, Bob Metcalfe figured out a way for devices on a network to communicate over a single shared cable. He is considered the father of Ethernet technology.

this is a small area, we often refer to these Ethernet networks as Local Area Networks or LANs. For instance, the computers you use on your campus are linked through an Ethernet network and therefore belong to a single LAN.

Bob Metcalfe is considered the father of the Ethernet (Figure 12.13). In 1973, while working at Xerox Corporation's Palo Alto Research Center, Metcalfe designed and tested the first Ethernet network. Metcalfe not only developed the physical method of connecting or cabling devices, but also the rules or standards that defined how devices communicate. Although Metcalfe's original Ethernet definition was only intended to connect a computer to a printer, it was comprehensive enough that it was easily expanded to allow all devices on a network to communicate over a single shared cable. This novel approach allowed for network expansion and for the introduction of new devices, all without the need for modifying the existing network or existing devices.

Now, let us move on to some basic nomenclature, or terminology. In Ethernet, media or medium refers to the conduit or pathway over which we transmit our data. The most common types of media are copper wire, optical or glass fiber, and air for the transmission of wireless signals. In Ethernet, we often refer to the medium as a shared medium, which relates to the way Ethernet operates. Ethernet allows multiple devices to share the medium (copper, glass, or air) and contend for the ability to transmit a message over that medium.

Another important concept in Ethernet is that of a segment—alternatively called an Ethernet segment, a broadcast segment, or a network segment. In this context, a segment is a shared piece of network media. Devices on a segment contend for that shared media, but do not share the media with other network segments. The devices that are connected to a segment and contending for the shared media are often referred to as hosts or nodes. The shared nature of Ethernet makes it impractical to send messages in their entirety. Instead, we break a message up into smaller pieces called frames and send our message as a series of frames that are received and reconstructed into our original message at the destination. Using frames ensures that one node does not monopolize the shared medium and increases the likelihood that the transmission will be successful. An unsuccessful transmission is often due to a collision, which results when two or more nodes attempt to transmit simultaneously. When this happens, the data "collides" on the shared medium and both frames are lost. To help avoid these potential "collisions," we implement a protocol called CSMA/CD (see Figure 12.14).

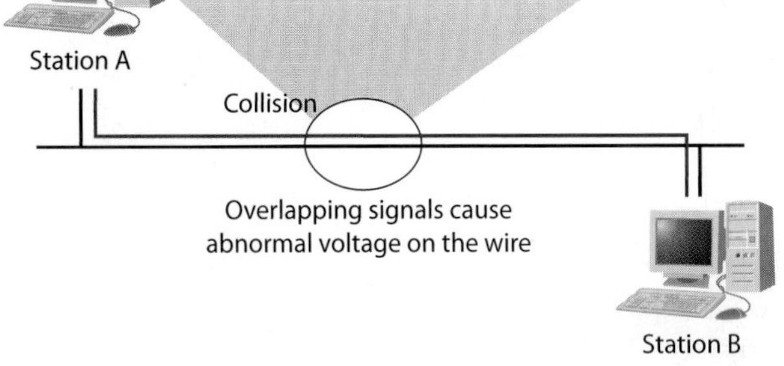

CSMA/CD Imagine that you are attending a friend's party. Everyone at the party is sitting or standing and engaged in some conversation. If it is a crowed party, there are many conversations happening at the same time. Unfortunately, if the noise and the other conversations around you are too loud, they will often drown out your conversation. It is not uncommon for someone in such a situation to say, "Could you repeat what you just said? I couldn't hear you." A crowded party is not much different from the Ethernet (see Figure 12.15). Imagine each Ethernet segment as its own party. To deal with this room full of "partygoers," Ethernet implements a protocol, or set of rules, called **C**arrier **S**ense **M**ultiple **A**ccess with **C**ollision **D**etection (CSMA/CD). Let's work our way through that acronym to see how this protocol works (see Figure 12.16).

Figure 12.14 | Data packets collide when two or more nodes attempt to communicate simultaneously. The CSMA/CD protocol prevents collisions.

Figure 12.15 | Have you ever been to a crowded party where you had to say "Excuse me, I can't hear you?" That's similar to what happens when data packets collide on a network.

Ethernet networks use Carrier Sense to manage a process called contention. Contention implies that each host connected to the Ethernet segment contends for use of that shared medium. Just as you listen at a party for an opening in a conversation—an opportunity to speak—each node on an Ethernet segment "listens" to the shared medium while waiting for an opportunity to "speak." Multiple Access is pretty straightforward because it is a consequence of an Ethernet segment being a shared medium. So every computer connected to that segment shares access to the media. This means that we have to take turns communicating so that everyone gets a chance to talk. In our party example, everyone who is part of the conversation—the segment—can hear what is being said. If what is being said is intended for you, you respond; if not, you continue to listen, monitoring the conversation.

Finally, Ethernet involves something called Collision Detection. If we consider our party again, what happens when people are all talking at the same time? Usually the message gets garbled and lost. When this happens, polite behavior dictates that everyone stops talking and someone takes the opportunity to begin speaking again. At parties, we don't have any formal protocols that we can rely on to resolve this situation. Fortunately, in Ethernet, the Collision Detection portion of CSMA/CD takes care of this. As they transmit, nodes listen to the shared medium for collisions. A collision is apparent when a host "hears" its own message garbled or senses a spike on the line

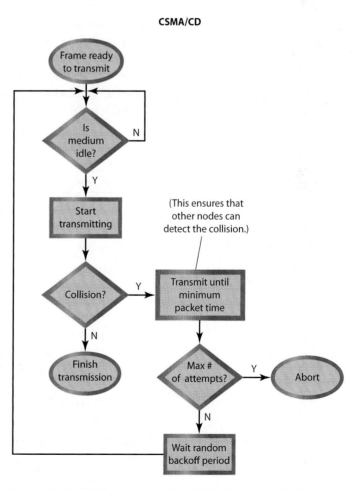

CSMA/CD

Figure 12.16 | This flowchart shows how CSMA/CD prevents data collisions.

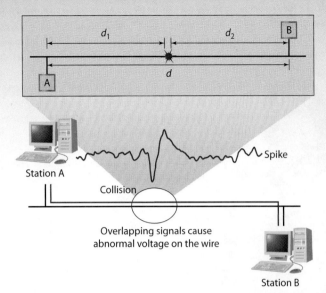

Figure 12.17 | A collision becomes apparent to a host if one of the host's own messages becomes garbled on the network, or if the host senses a voltage spike.

(e.g., a voltage spike for copper), as shown in Figure 12.17. Once a collision is detected, the nodes on the segment all stop transmitting and are each assigned a random wait time. When a node's wait time has expired, it checks to see if the medium is available. If the medium is available and the node has a message to send, it begins transmitting its message.

The OSI Seven-Layer Model

In the early days of networking, each manufacturer had its own closed, or proprietary, standards. Closed standards are intended to protect a company's intellectual property, but in networking the lack of openness meant that if you purchased some portion of your hardware from one vendor and some portion of it from another, there was no guarantee that the products from different vendors would work together. To alleviate this problem, the International Standards Organization (ISO) in 1987 proposed and in 1989 established a standardized networking architecture/structure called the Open System Interconnect or OSI model. This model defines seven specific layers—the OSI layers—that split up the responsibilities for getting data from the host computer to the destination computer (see Figure 12.18). Each layer provides a framework for the design of network drivers, protocols, applications, and hardware. It is not uncommon to develop and use an acronym to remember the layers of the OSI Model. The memory cue I use is the sentence "**P**lease **D**o **N**ot **T**hrow **S**liced **P**izza **A**way," which helps me remember Physical, Data-Link, Network, Transport, Session, Presentation, and Application. Also keep in mind that how you view the OSI model depends on what you do; software developers typically start at the Application Layer (Layer Seven) and work their way down, while hardware people usually start at the Physical Layer (One) and work their way up.

Layer Seven	Application
Layer Six	Presentation
Layer Five	Session
Layer Four	Transport
Layer Three	Network
Layer Two	Data Link
Layer One	Physical

Now look at each layer in greater detail.

1. The Physical Layer's job is to transmit or receive the data—the ones and zeroes—across the medium. This layer is responsible for converting a stream of ones and zeroes, such as 0011011011001100, into either an electrical signal (copper), pulses of light (fiber optic), or radio waves (wireless). This is one of the strengths of the OSI model; layers two through seven do not "care" what the physical transmission medium is, they do their jobs and let the Physical Layer worry about the media.

Now let's start moving our data up the layers.

2. The Data-Link Layer takes the bits—the ones and zeroes from layer one—and builds frames from them. Frames begin with a header. This header contains important information, in particular the source and destination **M**edia **A**ccess

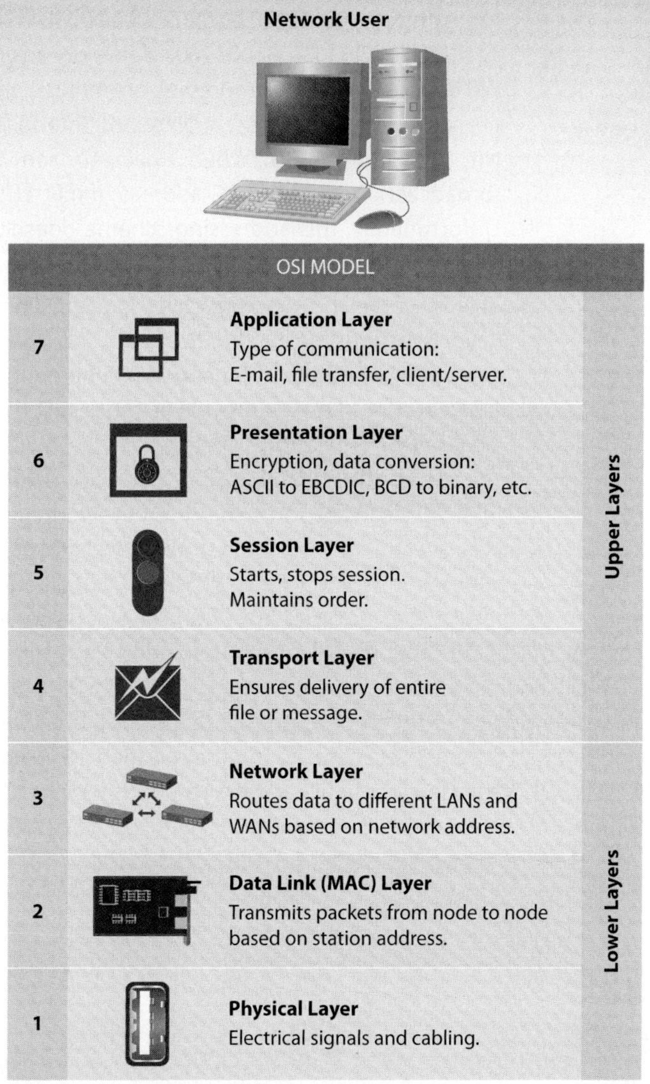

Network User

OSI MODEL

7	**Application Layer** Type of communication: E-mail, file transfer, client/server.
6	**Presentation Layer** Encryption, data conversion: ASCII to EBCDIC, BCD to binary, etc.
5	**Session Layer** Starts, stops session. Maintains order.
4	**Transport Layer** Ensures delivery of entire file or message.
3	**Network Layer** Routes data to different LANs and WANs based on network address.
2	**Data Link (MAC) Layer** Transmits packets from node to node based on station address.
1	**Physical Layer** Electrical signals and cabling.

Upper Layers (layers 7–4)
Lower Layers (layers 3–1)

Figure 12.18 | **The Open System Interconnect (OSI) model helps you understand network communications by establishing seven layers that describe specific functions that occur at each layer. A device or protocol has to only worry about operating at a single layer and being able to communicate its information with adjacent layers.**

Control (MAC) addresses. The frame—also sometimes called a data packet—contains the actual data of your message. The network data frame, or packet, also includes a checksum for error checking, and the data itself.

At the Data-Link Layer, we use MAC Media Access Control addresses. The MAC address is a unique, 48-bit (48 ones and zeroes) address, also called the physical address. An address with 48 ones and zeroes would be hard to work with. Imagine trying to send out invitations to a party, but instead of using:

John Doe
110 Street Road
Any Town, Any State 00000

you had to use all ones and zeroes. You probably wouldn't invite many people. To simplify addressing, we usually show MAC addresses using six colon-separated pairs of hexadecimal digits, each pair representing eight bits. For instance, my MAC address is 00:17:f2:ea:fd:4c, which when converted to binary

Figure 12.19 | Each network card has its own, unique MAC address.

(ones and zeroes) becomes 00000000:00010111:11110010: 11101010:11111101:01001100, which doesn't exactly roll off the tongue. MAC addresses are unique in that they are associated with each Network Interface Card, or NIC (see Figure 12.19). When a packet is sent to all hosts (broadcast), a special MAC address (ff:ff:ff:ff:ff:ff) is used. Unfortunately, this addressing scheme doesn't work well for larger networks.

3. This is where the Network Layer comes in. The Network Layer is responsible for addressing the data so that the data can get from network to network. As before (in the Data-Link Layer) we have to provide an address in the header or beginning of the data packet. In this case, however, we use the source and destination Internet Protocol (IP) addresses. The devices that form the Internet—routers—reside on this layer of the OSI model. We have moved to a more robust addressing scheme (IP versus), but we still need MAC addresses on our networks to get messages from one node to another. The protocol that makes this happen is Address Resolution Protocol (ARP). When we have the IP address of our destination (e.g., 192.168.104.10), but not its MAC address, we send out an ARP request. An ARP request is a broadcast message to all nodes on an Ethernet segment, asking the question *"Who has IP address 192.168.104.10?"* The node with that IP address responds with its MAC address. As this information is received, nodes build a local ARP cache, a temporary table that maps IP addresses to MAC addresses (see Figure 12.20).

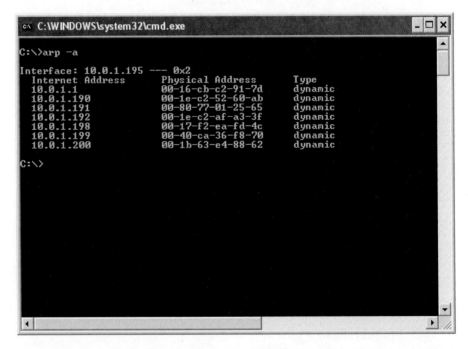

Figure 12.20 | You can use the command "arp—a" to display a table of local MAC addresses learned through the Address Resolution Protocol (ARP).

4. As we move up the OSI model, we next encounter the Transport Layer. If you are the source station, this layer is responsible for taking your messages, chopping them into smaller pieces or segments, numbering the segments, and preparing them for transmission. On the other end of the conversation—the destination—the Transport Layer collects the pieces of the conversation and reassembles them into the original message. Two important protocols reside at this layer—the Transmission Control Protocol (TCP) and the User Datagram Protocol (UDP). There are a number of distinctions between these two protocols. We will highlight some of the important differences. Generally, TCP is referred to as a connection-oriented, reliable protocol, while UDP is considered connectionless and unreliable. Let's first address connection versus connectionless. In TCP, before we begin sending data, we first make sure that the destination for our data is available and we establish a connection. As you would expect, in UDP we do not worry about whether the destination is there, and we don't establish a connection. As for reliability, TCP inserts a calculation code in each piece of data. If the calculation code doesn't match what the destination expects, the destination discards that data and requests that that portion be resent. In TCP, the destination also sends an acknowledgement or ACK message for each packet that it receives. As the sending station, you wait for those ACKs. If you don't get an acknowledgement, then you resend that portion of the message. UDP does no error checking and does not send any acknowledgements if a packet fails to reach the destination For these reasons, you would think that no one would use UDP, but UDP has some very important applications. Among these is streaming audio and video—applications where the latency inherent in TCP (setting up the connection, error checking, resending dropped packets) would render the content unusable.

5. Thus far we have discussed sending data between a source and destination. It is important to realize, however, that this single communications session is just one of many. What is a "session"? Consider just one of the applications

Figure 12.21 │ The Session layer of the OSI model makes it possible for you to conduct multiple Instant Messaging conversations at once.

Section 1 △ Networking Technologies and Applications

you regularly use—instant messaging. Every conversation you engage in through your instant messaging client software is treated as a separate session (see Figure 12.21). It is the Session Layer that is responsible for managing these sessions. When you begin a new session, the Session Layer creates that session, and as you continue to IM, this layer manages and maintains these sessions. Think of the session layer as the traffic cop of the OSI model. Just as a traffic cop stands at an intersection directing traffic flow, the Session Layer is responsible for managing our network traffic—our conversations. When you finish your conversation, this layer is responsible for properly tearing down or terminating the session.

6. One of the things you probably love about the Internet and networking is the ability to view images, watch videos, and share these images and videos with friends. In the OSI model, the Presentation Layer is responsible for taking the information—whatever it is—and making it "usable" or "presentable" to the application layer. Types of information include images, video, audio, and even the secure connection that you use to do online shopping or check e-mail. This layer accomplishes things like data encoding and decoding, compression and decompression, and data encryption and decryption.

7. We all know what applications are. Any executable program that you run on your computer is an application. At the Application Layer, we use a slightly different definition. Here an application is any program that uses the services of the network. For instance, the Web browser and instant messaging applications you use are all network applications. In fact, because you can embed a Web link or e-mail address in a Microsoft Word document, even Word is a network application (see Figure 12.22). This layer is where you and I connect to the network through these applications.

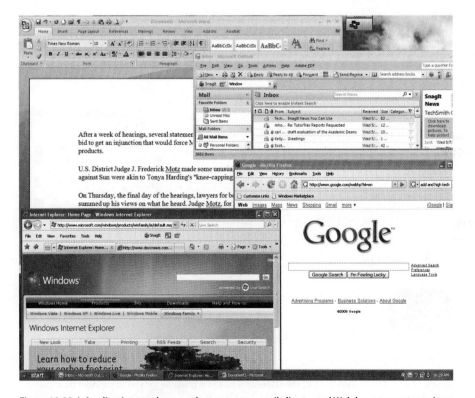

Figure 12.22 | Applications such as word processors, e-mail clients, and Web browsers operate in the Application layer of the OSI model.

TCP/IP

The OSI model is great for describing and learning about networks in an abstract way, but, unfortunately, it doesn't provide us with any real detail on how networks work and how to perform various functions at various layers. For that we go back to the TCP/IP model, developed through the work of DARPA and in support of ARPANET. As we have said, TCP/IP has become the dominant set of protocols or rules on which the Internet is built. Like the OSI model, the TCP/IP suite or stack can be viewed as a layered approach to networking (see Figure 12.23).

As with the OSI model, the TCP/IP Application layer facilitates network communication for applications. In TCP/IP, the Transport layer is responsible for source-to-destination message transfer. The layered approach frees the Transport layer from the functions of the lower layers, such as the networking technology and the transmission media. Key functions at this layer include error control, fragmentation, and flow control. The two most common protocols at this layer are the connection-oriented Transmission Control Protocol (TCP) and the connectionless User Datagram Protocol (UDP). The TCP/IP network layer is responsible for moving data within the network and between networks. This layer performs the "Internetworking," or routing, functions of OSI's Network layer as well as functions of the Data-Link layer. The IP portion of this protocol suite routes packets from source to destination and provides support for a variety of routing protocols. The Physical layer in TCP/IP performs a similar function to the OSI model, transmitting bits over our communications media (copper, fiber, or wireless).

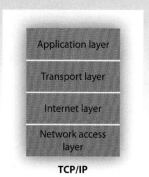

Figure 12.23 | The TCP/IP model shows how the various TCP/IP protocols work together to transmit data across a network.

OSI Versus TCP/IP

Although both are layered approaches, there are a number of differences between the OSI model of networking and the TCP/IP model (see Figure 12.24). First and foremost is the fact that OSI is primarily a teaching tool that helps us to learn and understand how networks operate, while TCP/IP more closely represents real-world networks and architecture, but is more difficult to understand and examine. Another critical difference is that the OSI model has seven layers while the TCP/IP model has only four. To begin to compare the two, we can draw some rough correlations between the layers.

In TCP/IP, the functions of OSI layers 5–7 (Session, Presentation, and Application) are performed by a single layer—the TCP/IP Application Layer. In both models, the role of the Transport Layer is roughly analogous and the TCP/IP Internet Layer performs more or less the same functions as the OSI's Network Layer. Lastly, the functions of the OSI Data-Link and Physical Layers are bundled together in TCP/IP's Network Access Layer. Keep in mind that there is not a perfect one-to-one correspondence. For instance, some of the functions of the Session Layer show up in TCP/IP's Application layer, while others show up in the Transport Layer.

Network User

OSI MODEL			TCP/IP
7	**Application Layer**	Type of communication: e-mail, file transfer, client/server.	FTP, Telnet, HTTP, SNMP, DNS, OSPF, RIP, Ping, Traceroute
6	**Presentation Layer**	Encryption, data conversion: ASCII to EBCDIC, BCD to binary, etc.	
5	**Session Layer**	Starts, stops session. Maintains order.	
4	**Transport Layer**	Ensures delivery of entire file or message.	TCP (delivery ensured) UDP (delivery not ensured)
3	**Network Layer**	Routes data to different LANs and WANs based on network address.	IP (ICMP, IGMP, ARP, RARP)
2	**Data Link (MAC) Layer**	Transmits packets from node to node based on station address.	
1	**Physical Layer**	Electrical signals and cabling.	

Figure 12.24 | To understand how a network functions, it is helpful to compare the layers of the OSI model with the layers of the TCP/IP model.

Standards

Standards have been critical to the continued growth and popularity of networking and its underlying technologies. Standards help to ensure industry acceptance and interoperability, support evolution, and encourage innovations and improvements in functionality. A standards body that has been central to this effort is the Institute of Electrical and Electronics Engineers, or IEEE (pronounced "I-triple-E"). In February 1980, the IEEE created the 802 working group to establish and maintain standards in networking technologies. The IEEE created 802.X subcommittees to address different aspects of networking. For example, Ethernet and CSMA/CD are covered by the IEEE 802.3 standard.

TECHNOLOGY IN THE REAL WORLD: Voice over Internet Protocol

One area where networks and networking have had a particularly significant impact is in telephony. Telephony is a term that comes from our traditional telephone system and the telephone companies. In telephony, we are bombarded by acronyms, including POTS—Plain Old Telephone System, and PSTN—Public Switched Telephone Network. A more recent acronym—VoIP (pronounced "voipe"—like voice with a *p*) is our focus here. VoIP, which stands for Voice over Internet Protocol, is the term we use to describe internet telephony—a method for taking traditional phones calls and moving them across the Internet, rather than through traditional phone lines. This requires us to convert our analog phone call to data—the digital ones and zeroes we transmit across networks. VoIP has already revolutionized how we make phone calls and required phone companies to respond to new competition, and to rethink how their own networks operate.

Traditional phone calls use a method called circuit switching, where a connection is established and maintained between caller and receiver for the duration of the call (see Figure 12.25). While this system has worked well for many years and is very reliable, there are inefficiencies associated with circuit switching. Consider when you call your friend Mary across town and her mother has to put the phone down and find Mary to get her on the phone. Because—in circuit switched environments—we maintain the connection, the time the phone is not being used—no one is talking—is all wasted capacity on our PSTN (Public Switched Telephone Network). Now consider that wasted capacity replicated in hundreds of thousands of other calls, and you'll start to get a sense of the inefficiency of our POTS.

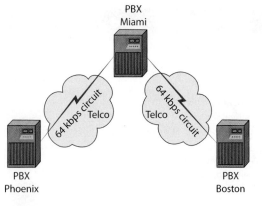

Figure 12.25 | Traditional phones rely on circuit switching, in which the connection between caller and receiver is maintained throughout the entire call. This form of switching has proven to be inefficient in modern communications.

Next opportunity you get, ask your parents or grandparents about how expensive long distance called used to be—back in the "good ole days." There are many reasons why long distance was so expensive, but one factor was the circuit-switched method. Let's look at an example. Let's say that your grandparents lived in New York in the 1950s and wanted to call some relatives in Kansas City, Missouri, 1,200 miles away. (See Figure 12.26.) In the 1950s, a circuit-switched call meant that physical pieces of copper wire were being connected end-to-end to establish and make a call. In our example, this means 1,200 miles of copper wire stretched out between New York and Missouri. Assuming today's cost of $60 per 500 feet of copper wire, we can calculate just how much that wire would cost.

$$1{,}200 \text{ miles} \times 5{,}280 \text{ feet/mile} \times \$60/500 \text{ feet} = \$760{,}320$$

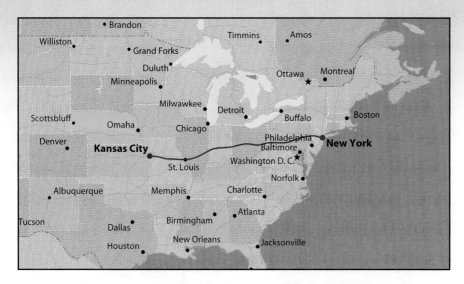

Figure 12.26 | In the 1950s, a circuit-switched telephone call from New York City to Kansas City required 1,200 miles of copper wire, which, in today's dollars, would cost about $800,000. No wonder long distance calls were expensive back then.

So your grandparents were essentially "renting" $800,000 dollars worth of copper for the duration of their call—good reason to keep the calls short. Modern phone systems have improved significantly, but the inherent inefficiencies of the circuit switched method still exist.

In VoIP, we use packet switching (the same method used on data networks) as opposed to circuit switching. Instead of transmitting a call as a continuous analog stream, we send small digital packets (made up of ones and zeroes) of voice. In fact, if no one is talking, nothing is sent. Another major difference is that instead of establishing and maintaining a single circuit for the duration of the call, each of our small voice packets is routed from source to destination by the many routers that make up the Internet. As with any data packet, these voice packets will follow different paths to get from source to destination. If you are talking—transmitting— the information on your end is chopped into small data packets (usually smaller than other data packets), a source address and sequence number are added and sent to a nearby router. Since this is data now, the router will determine the best path and send to the next router. Our little voice packets will "hop" from router to router until they reach their destination, where the sequence numbers will be used to reassemble the data into our voice call. This method of transmitting voice data is very efficient.

Unfortunately, your regular telephone is unable to make calls over the Internet. To do this, we can take one of three approaches. The first is to keep your existing analog phones and use an adapter to convert the analog phone call to digital ones and zeros for transmission across a data network. A second method— common in business and industry—is to replace all of the analog phones with what are called IP Phones. IP Phones are designed to connect to a data network; no adapters are necessary (see Figure 12.27). Another unique feature of these phones comes from the fact that they are "data" phones. They can receive network data just as a computer might. Many IP Phones have a small LCD (liquid crystal display) that can display information received by the phone, including weather, stock quotes, etc. The last method for using Voice over IP is to download software on your computer and make calls using the software, a microphone, and speakers or headphones. This type of VoIP "device" is often referred to as a "soft" phone, because there is no phone hardware—the phoning is all done

473

Figure 12.27 | An IP phone transmits calls over the Internet, instead of over the Public Switched Telephone Network (PSTN).

Figure 12.28 | Using Voice over IP (VoIP) technology, you can make calls on your computer using special software, a microphone, and speakers. Skype is the most popular VoIP software.

through software (see Figure 12.28). The most popular software for doing this is Skype (www.skype.com), which allows you to make free calls to other Skype users and very inexpensive calls to traditional phone numbers.

VoIP's primary advantage over traditional circuit-switched telephony is the savings in call capacity. With packet switching, several telephone calls now occupy the amount of space formerly required for a single circuit-switched call. This capacity savings translates to a cost savings for the end user and for businesses adopting VoIP. VoIP is not without its disadvantages, particularly reliability. Reliability is something that is inherent in PSTN. You pick up the headset; you hear a dial tone; you make a call. Everything works—even when we have power outages. We have grown accustomed to, even dependent upon, this reliability. One particular weakness of VoIP is its dependence on the power you provide. If you lose power, you lose your phone. Another serious issue is the inability to route 911 calls. Lastly, VoIP is subject to the same issues that we face when transmitting any data over the Internet. When we are sending an e-mail or an Instant Message, these issues are not particularly troubling. However, when we are routing voice, network performance issues become increasingly important. As data travels across the network, any delays that are encountered introduce latency, which becomes apparent through gaps in the conversation. Smaller delays and out-of-sequence packets result in jitter—similar to the "choppiness" you may see when watching online video. Finally, any data transmission is going to be subject to garbled data and dropped packets. With e-mail and other traditional data, we can retransmit a dropped packet and maintain the integrity of our message. Unfortunately, with voice we don't have that luxury. A dropped packet or packets, along with jitter and latency, can all contribute to poor call quality.

Even if you do not think you are using VoIP, you probably are. Phone companies have all begun using Voice over IP to introduce the same efficiencies and cost-savings that any user would want. These companies use IP gateways to transition circuit switched calls onto an IP network (packet-switched) and then back again onto a circuit near the destination. This technique has resulted in considerable savings and improved bandwidth. It is inevitable that all circuit-switched networks will eventually be replaced with packet-switching technology. If VoIP hasn't made its way to your home or school yet, don't worry; it's only a matter of time.

SECTION ONE FEEDBACK >

1. Explain the difference between packet-switched and circuit-switched networks.
2. Explain why we would use UDP with streaming audio and video.
3. In a typical day, how many different networks and network devices do you use?
4. What network-enabled applications do you use daily?
5. What is a MAC address? Find the MAC addresses for network devices you use daily.

SECTION 2: LANs, WANs, and Networking Devices

KEY IDEAS >

- A WAN or Wide Area Network connects together LANs or Local Area Networks, which are groups of computers connected together in a small geographic region.

- The most common network topology is the hybrid star-bus, which combines a physical with a logical bus.

- A switch is a multi-port bridge that connects two or more network segments and serves to segment a larger network.

- Routers are devices that use IP addresses to connect networks together and route information from one network to another (also called "Internetworking").

LANs

A Local Area Network (LAN) is a group of computers connected together using any of the media (copper, fiber optic, or wireless) that we discussed earlier. Usually, we connect these computers together to either access a shared resource, such as a printer, a Web, e-mail, or a database server (see Figure 12.29). The computers connected together at your campus form a Local Area Network. Even a wireless network that you set up at home is a LAN (see Figure 12.30), set up so that you, your friends, your parents, and your brothers and sisters can get on the Internet or print reports for school or work. How many hours a day do you spend on some sort of LAN?

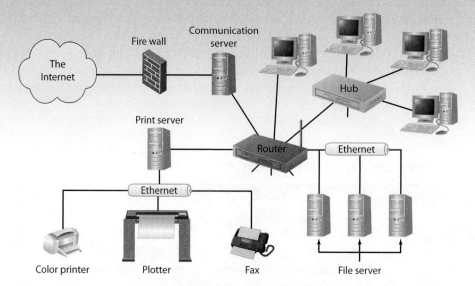

Figure 12.29 | A Local Area Network (LAN) is a group of computers connected via media such as copper wire, fiber-optic cable, or a wireless connection. The purpose of a LAN is to allow users to share resources such as printers or an Internet connection.

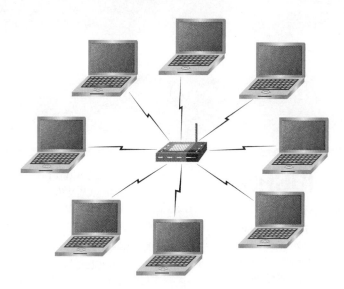

Figure 12.30 | A wireless LAN works just like any other LAN, except that the media connecting the computers is a wireless signal.

Topology

A network's topology is a description, or map, of how computers in a network are connected together and communicate. We can look at physical topologies (how the computers are physically connected to one another), and we can consider logical topologies, which describe how the computers share information or communicate with one another. The most common topologies are the bus, ring, star, and mesh topologies (see Figure 12.31). Each has its advantages and disadvantages in both a physical and logical sense.

In networks, physical and logical topologies were originally identical—meaning that a group of computers connected together in a physical bus topology also used that logical topology to communicate with one another. Historically, bus topology has been the most popular. Unfortunately, this physical arrangement of computers had a significant weakness that limited its use and application. Because computers on the bus topology are all connected to the same wire or bus, the bus became a single point

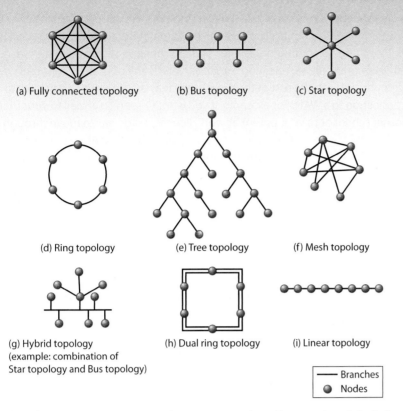

(a) Fully connected topology

(b) Bus topology

(c) Star topology

(d) Ring topology

(e) Tree topology

(f) Mesh topology

(g) Hybrid topology
(example: combination of
Star topology and Bus topology)

(h) Dual ring topology

(i) Linear topology

— Branches
● Nodes

Figure 12.31 | When constructing a network, engineers can choose from a variety of physical layouts, which are known as topologies. Each physical topology has its own advantages and disadvantages.

of failure. In practice, this means that if a single computer or node loses its connection to the bus, or the bus has a break in it, the entire network stops working. To avoid this limitation of the bus topology, we have moved to hybrid network topologies, which combine the best features of a physical topology with a logical topology. The most common hybrid topology is the star-bus (see Figure 12.32). The star-bus hybrid topology combines a physical star topology with a logical bus topology. The star provides the fault tolerance we require, meaning that if one computer fails, that computer falls off the network, but the remaining computers are still able to communicate. The logical bus allows us to communicate via the dominant network standard Ethernet.

WANs

Local Area Networks are great, but how does the information get from the LAN to your friends at another school or your grandparents in another state? This is where Wide Area Networks (WANs) come in. WANs connect LANs together and are responsible for routing information from one LAN to another (see Figure 12.33). This connecting of LANs is sometimes called internetworking, and the devices that make this possible are the routers we talked about earlier. Routers learn information about the network they are connected to and pass or route information from network to network until the message gets to your friend or your grandparent.

LANs Versus WANs

When you think about networks, you usually consider the actual computers you have access to, whether at home, at school, at a

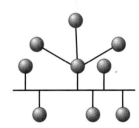

Figure 12.32 | A hybrid topology combines a physical topology with a logical topology. The most popular hybrid topology is the star-bus hybrid topology.

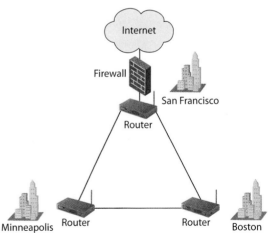

Figure 12.33 | A Wide Area Network (WAN) is a network made up of interconnected LANs. Corporations often have WANs that span multiple cities or states. Network traffic traveling out of the WAN to the Internet has to pass through a firewall, which is a piece of hardware (or a combination of hardware and software) designed to protect a network from security breaches.

477

library, or some other location. These computers are grouped together in what we refer to as LANs. Typically, LANs will be confined to a single building, allowing the computers to communicate with one another and to access shared resources such as printers, file servers, and e-mail servers. If any of these computers want to communicate with the rest of the world—such as the Internet—they have to have a connection to a WAN. For example, if you want to send an e-mail to your friend at another school or to your parent at work, your message has to move through your

LAN until it gets to your school's WAN connection, which will route the message to your friend's or parent's WAN, where it will make its way onto their LAN and eventually to their computer. In fact, there are hundreds of thousands of LANs all over the world, connected together by WANs. This worldwide combination of local and wide area networks gives you the opportunity to access people, places, and information from virtually anywhere in the world. At home, your WAN connection might be provided by your cable television company through a cable modem or through your telephone company by either a DSL modem (see Figure 12.34) or a dial-up modem or even a high-speed fiber-optic connection.

Figure 12.34 | A WAN connection at your home might be provided by through a cable modem (shown here on the left) or a DSL modem (shown on the right).

Building the Network: Devices

In the physical bus topology we discussed earlier, a single shared cable serves as the basis for a complete Ethernet network. In fact, even using the more popular star-bus hybrid topology (physical star; logical bus) we are communicating as if we are all connected to the same single shared cable. This concept of a shared communication medium—copper, glass fiber, or radio waves—is central to how Ethernet works. Unfortunately, this "shared" medium introduces some limitations into our networks. Fortunately, there are a number of devices we can introduce into our networks to address many of these issues.

Repeaters and Amplifiers We are all familiar with the situation of trying to call out to someone but being too far away for him or her to hear us. As the destination gets farther and farther away from the source of the signal, the signal starts to decay or die out (see Figure 12.35). This phenomenon—called attenuation—occurs not only in our example with sound waves, but also with electrical signals, light pulses, and radio waves. To overcome this problem in networks and extend the reach of our network, we can use either a repeater or an amplifier (see Figure 12.36). Placed

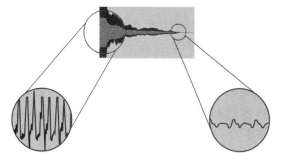

Figure 12.35 | This illustration shows the decay of a signal of the word "Hello." The signal starts out strong and regular. Through the process of attenuation, it decays into a weaker, erratic signal.

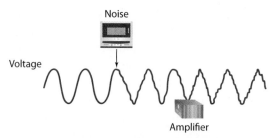

Figure 12.36 | An amplifier increases the strength of a signal that has decayed, so that the signal can cover a longer distance. Unfortunately, any noise or artifacts that have been introduced into the signal will also be amplified.

along the path between source and destination, a repeater takes a signal that is dying or decaying and regenerates it, allowing it to continue toward its destination (see Figure 12.37). An amplifier performs a similar function by increasing the signal strength so that our signal can cover a longer distance. One important distinction between the two is that a repeater regenerates our original signal, while the amplifier takes the existing signal—including any noise or other artifact that may have been introduced—and increases the signal strength. It is important to keep in mind that both amplifiers and repeaters have no built-in intelligence. These devices cannot examine the data—all they do is amplify or regenerate the signal and send it on.

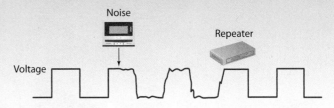

Figure 12.37 | A repeater takes a signal that is dying or decaying and regenerates it, allowing it to continue toward its destination.

Bridges On the surface, a bridge may seem to you to be very similar to a repeater because it connects two network segments or pieces together (see Figure 12.38). In fact, however, a bridge operates very differently from a repeater. While a repeater has no inherent intelligence or ability to examine the data it is receiving, a bridge does have that ability. Bridges examine network traffic—looking at the source and destination MAC or physical addresses. If these addresses are on the same network segment, the bridge does not forward the data; if they are on different segments, the bridge forwards the data to the segment where the destination computer is.

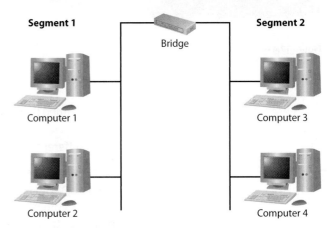

Figure 12.38 | A bridge is a type of hardware that connects two network segments.

Switches In a LAN, switches are the most common and most important devices (see Figure 12.39). In fact, what we've referred to as Modern Ethernet is sometimes referred to as Switched Ethernet. If you understand how a bridge works, then it is fairly easy to understand switching, since a switch is essentially a multi-port bridge. While a bridge has two ports and connects together two network segments, a switch has many ports—8, 16, 24—and connects together as many network segments as there are ports. Switches play a critical role in building out LANs. If we were to examine the physical layout for the computer classrooms in your school, you would probably find that the computers in your classroom are all connected to a switch, which is itself connected to another switch (Figure 12.40).

Figure 12.39 | A switch, the most common and most important device on a LAN, is used to connect multiple network segments into a single network.

Routers As we mentioned earlier, routers play a critical role in internetworking or connecting together LANs. (You can see an example of a router in Figure 12.41.) While bridges and switches look at the physical or MAC address of data, routers look at IP addresses. The MAC or physical address works well for moving data around a LAN, but is very inefficient at moving data from LAN to LAN. A router is nothing more than a very specialized computer that looks at the IP (sometimes called logical address) of data and then determines the best path or route to get the data to its intended destination. Once the router has determined the best path, it sends the data to the next router in this path. Each hand-off from one router to the next is called a "hop." Using path determination and the destination IP address, we are able to hop from router to router until we get to a router attached to the destination network or LAN. Best path determination—the job of the router—depends on a number of criteria or metrics that routers are

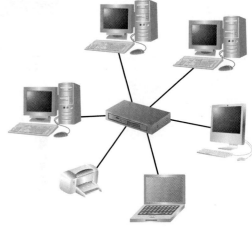

Figure 12.40 | In school computer labs, the individual computers are often connected to a switch. That switch, in turn, is often connected to another switch.

Figure 12.41 | A router is used to connect a LAN to a WAN. The Internet is a network of routers.

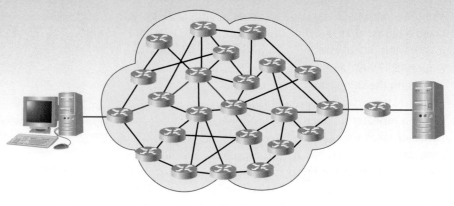

Figure 12.42 | Routers on the Internet are responsible for choosing the path by which information travels between a source computer and a destination computer.

programmed to use. Because network and network traffic are dynamic, constantly changing, it is not surprising that data that is part of the same conversation often takes different paths (see Figure 12.42).

Gateways Computers and routers that sit at the boundary between networks are acting as gateways. Because of their location at the entrance to the network, gateways often will also be running as a firewall to help protect the network from malicious traffic. Within a network, we configure a default gateway address. Any packet that is destined for an external network is sent first to the gateway, which determines how to route the packet to its destination network. Likewise, traffic coming into our network enters through the gateway.

SECTION TWO FEEDBACK >

1. Explain the difference between a signal that has been amplified and one that has been repeated. Sketch what you might expect the two waveforms to look like.

2. Sketch the topology of your school network and identify any networking devices.

3. Determine your gateway address. Is this address associated with a router?

4. How many routers are in your school network and where are they located?

SECTION 3: IP Addressing

KEY IDEAS >

- An IP address is a 32-bit binary number that computers read as a string of 32 ones and zeroes.

- For convenience, we write IP addresses as a set of four decimal numbers or "dotted decimal" notation.

- To communicate from network to network, we use source and destination IP addresses.

- An IP version 6 address has 128 bits or 128 ones and zeroes.

You may not normally give this much thought, but with all of the millions of computers connected to the hundreds of thousands of networks, how do you find your friends? Furthermore, how do you find a Web site that you would like to visit among the millions of Web sites out there? The answer lies in the network address or IP address we discussed earlier. We can consider public IP addresses and private IP addresses. If you have a high-speed Internet connection and a wireless router at home, you're already using public and private addresses. In fact, the network you use at school uses both private and public IP addresses. At home, your cable or digital subscriber line (DSL) modem connects to the Internet and is assigned a public IP address by your ISP or Internet Service Provider (see Figure 12.43).

For example, my public address may be *68.45.39.245*, and my private address may be *10.0.1.198*. Another way to think of these is as a WAN (public) address versus a LAN (private) address. When I communicate over the Internet or browse Web sites, all the Internet sees is my public or WAN address. No matter how many computers I add to my LAN, each is viewed by the Internet as coming from *68.45.39.245*. So how does a message get from the Internet to my laptop rather than someone else's? That's the job of the router and our IP addresses.

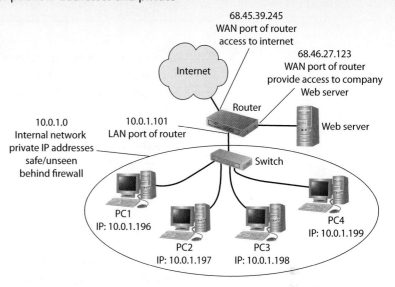

Figure 12.43 | A computer on a small network might have a private IP address that is only used by devices on that LAN. The LAN itself can have a public IP address, which is the IP address all the computers on the Internet use to communicate with computers on the LAN. To be seen from the outside world, a Web server must also have a public IP address.

Now let's explore IP addresses further. The numbers we use—*68.45.39.245*—are not how computers and networks view IP addresses. Remember, computers communicate in binary—a string of ones and zeroes. The way we humans write and communicate IP addresses is referred to as "dotted decimal" notation. Computers see IP addresses as a 32-bit binary number or a string of 32 ones and zeroes (see Figure 12.44). We use the dotted decimal convention because it is much easier for humans to remember an address like *68.45.39.245* than it is to remember its binary equivalent:

01000100.00101101.00100111.11110101

Each group of eight ones and zeroes is referred to as an "octet."

DNS Servers

▷ Although the address *68.45.39.245* is much easier to remember than *01000100.00101101.00100111.11110101*, it is still more than most people are able or willing to remember. To make IP addresses even easier to remember, we rely on a worldwide system of servers, collectively the Domain Name System (DNS), that store a map of domain names, such as www.google.com, to their IP address. In fact, we can use these DNS servers to determine the IP address of a domain name. These are many sites that will provide this service. Try doing a Web search for "DNS lookup" to find one. One such site you might find is http://www.zoneedit.com/lookup.html. This site tells us that www.google.com has an IP address of 72.14.253.103. Every time we attempt to visit a Web site, our default DNS is contacted and starts the process of determining the IP address of the site we wish to visit.

01000100.00101101.00100111.11110101

Figure 12.44 | This is an IP address expressed in binary notation. You probably wouldn't enjoy having to type a number like this into your browser every time you wanted to go to a Web site. Thanks to a worldwide system of servers called the Domain Name Service (DNS), you need not do so.

The table below is useful for examining an individual octet and converting from binary to decimal. Since we are working in binary, each position can take a value of either *1* or *0*, or *on* or *off* using our light switch analogy.

2^7	2^6	2^5	2^4	2^3	2^2	2^1	2^0
128	64	32	16	8	4	2	1

If all of the positions are turned off—binary *0*—our decimal equivalent is 0. With all the positions turned on—binary *1*—we get a decimal value of 255, giving us 256 values from 0 to 255. Use this table to do some quick calculations. First, look at converting a binary octet to its decimal equivalent. Consider the binary number *10011011*. We begin by filling in the positions in each of the eight columns, adding together values from any column that has a binary 1 or is on.

2^7	2^6	2^5	2^4	2^3	2^2	2^1	2^0
128	64	32	16	8	4	2	1
1	*0*	*0*	*1*	*1*	*0*	*1*	*1*

| 128 | + 0 | + 0 | + 16 | + 8 | + 0 | + 2 | + 1 | = 155 |

To convert from decimal to binary (consider the decimal *104*), we can work through the table in the opposite way. Working from left to right, compare the given number 104 to the decimal equivalents (*128, 64, 32, ... , 1*) in the table. If the number is greater than that in the table, fill that position with a *1* and subtract the value from the table from our given number. If the number is less than the value in the table, fill that position with a binary *0* and move to the next column. We repeat this process for each column, working our way from left to right. Since *104* is smaller than 128, we place a zero in that position and move to the next column.

2^7	2^6	2^5	2^4	2^3	2^2	2^1	2^0
128	64	32	16	8	4	2	1
0							

In the next column, 104 is larger than 64, so we place a 1 in this position and subtract 64 from 104 (*104 − 64*), leaving us with 40.

2^7	2^6	2^5	2^4	2^3	2^2	2^1	2^0
128	64	32	16	8	4	2	1
0	*1*						

For the next column, 40 (what we have left from the last step) is larger than 32, so we place a 1 in that column and subtract 32 from 40.

2^7	2^6	2^5	2^4	2^3	2^2	2^1	2^0
128	64	32	16	8	4	2	1
0	*1*	*1*					

Continuing that process from column to column until we have no remainder left results in the following table:

2^7	2^6	2^5	2^4	2^3	2^2	2^1	2^0
128	64	32	16	8	4	2	1
0	*1*	*1*	*0*	*1*	*0*	*0*	*0*

This gives us *01101000* as the binary equivalent of 104.

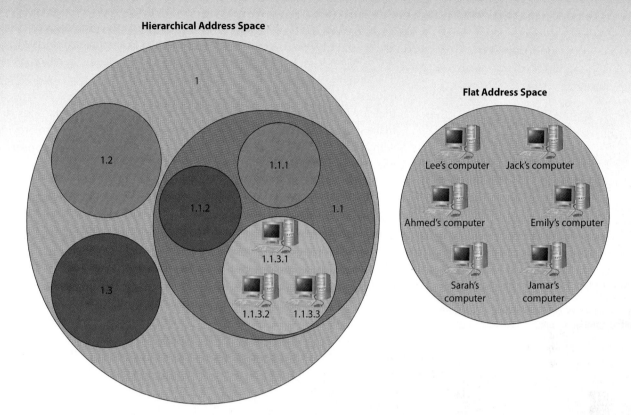

Hierarchical Address Space

1
1.2
1.1.1
1.1.2
1.1
1.1.3.1
1.3
1.1.3.2   1.1.3.3

Flat Address Space

Lee's computer   Jack's computer
Ahmed's computer   Emily's computer
Sarah's computer   Jamar's computer

Figure 12.45 | A flat address space requires each node or computer to have a unique name or address. For this reason, flat address spaces do not scale or grow well and are usually only appropriate for small networks. Instead, IP addresses are hierarchal, so that we have all the 10 addresses at the top level or hierarchy; then all the 10.0; then the 10.0.1 addresses; and finally our node—10.0.1.198. Hierarchal address spaces scale well, allowing you to easily grow your network.

MAC or physical addresses, which we discussed earlier, form a flat address space, which cannot be organized in a hierarchical way. In contrast, IP addresses form a hierarchal address space, which is much better suited to communicating within a large distributed network. (Refer to Figure 12.45 to see the difference.) The most common example of a hierarchal address space is the telephone network. Consider what happens when we dial the number *1-516-463-6600*. When we make a call, we start with the country code (1 for the United States) as the top of the hierarchy. Then we dial *516*, which takes us to Nassau County, Long Island, New York, then *463* for Hempstead, and finally *6600*, which identifies the actual phone line we are calling. Fortunately for us, with IP addresses we only have to work about a two-level hierarchy. The first level defines the network, and the second level defines the host.

When we work with IP addresses, we sometimes refer to them by classes (also called classful IP addressing). Early networks were exclusively classful. The network classes are defined by which portion of the address is reserved for the network address and which is the host address. The network portion of the address identifies the network to which the address belongs, and the host portion identifies the actual computer on the network. For example, the following address would be split as shown below:

$$\underbrace{192.\ 168.\ 10.}_{Network}\underbrace{104}_{Host}$$

For very large networks, we usually use Class A addresses. For Class A addresses, the first octet (eight bits) is reserved for the network address and the last three octets (twenty-four bits) are reserved for host bits. In fact, Class A addresses have a first octet between *1* and *126*, which corresponds to a binary leading bit of 0. This gives only 126 Class A networks available. Given the number of host bits (*H*) available, we can calculate the number of hosts using the following equation:

$$Hosts = 2^H - 2$$

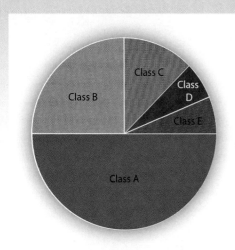

Figure 12.46 | This chart provides you with a sense of how many network addresses are available in each class (A, B, C, D, or E). There can be many Class A networks, but each can have only a very limited number of computers or hosts—254.

For Class A networks, this means that each of these 126 networks can have up to $2^{24} - 2$ or *16,777,214* hosts. Each computer in an organization is a host, such that if you have a Class A network, you can assign nearly 17 million public IP addresses to computers on your network. Do you remember my public IP address? As you would expect, *68.45.39.245* is a Class A address. This means that my Internet Service Provider (ISP) owns a Class A network. Does this make sense? Consider the number of home users that an ISP has to provide with a unique public IP address. In my case, the network address is: *68.0.0.0* and the host portion of my address is *45.39.245* (the last three octets).

Class B addresses have a first octet that ranges from *128* to *191*, which corresponds to binary leading bits of *10*. Another important difference between Class A and Class B networks is that here the first two octets, or sixteen bits, are reserved for network addresses, leaving only sixteen bits for host addresses. As you would expect, this means that we can have a greater number of Class B networks *16,384* or 2^{14}, each with $2^{16} - 2$ or 65,534 hosts. Class B addresses are appropriate for large corporations and universities.

Class C networks have an address that starts with *192* to *223* or leading bits *110*. These addresses are appropriate for small- to medium-sized businesses, because the first three octets, or twenty-four bits, are reserved for the network address and the last octet (eight bits) for the hosts. This means that there are *2,097,152* or 2^{21} Class C networks, each with *254* or $2^8 - 2$ hosts.

Most of the addresses you will encounter will fall into these three classes (A, B, and C). Other classes exist, but they are reserved for special purposes. Class D, for instance, is used for multicasts. In multicast communications, we send a message to a group of addresses or hosts, as opposed to sending it to everyone (broadcast) or just a single host (unicast). For a Class D address, the first four bits of the address are *1110* and the remaining twenty-eight bits identify the group of hosts that is the destination of our multicast message. Finally, Class E is reserved for experimental purposes. You'll recognize a Class E address because it begins with binary 1111 or decimal 240–255. Class E addresses are also referred to as "limited" broadcast. As with Class D, the remaining twenty-eight bits are used to identify the destination addresses. The table below provides an overview of the Class A through E networks (see also Figure 12.46).

CLASS	LEADING BITS	1ST OCTET	NETWORK BITS	HOSTS BITS	# OF NETWORKS	# OF HOSTS PER NETWORK
A	0	1–126	7	24	128	16,777,214
B	10	128–191	14	16	16,384	65,534
C	110	192–223	21	8	2,097,152	254
D	1110	224–239	4	28		
E	1111	240–255	4	28		

Subnet Masks and Subnetting

In all likelihood, the type of network that you will encounter most often is a Class C network. Unfortunately, because there are only 254 hosts or nodes available, a Class C network may not be suitable for most real-world networks, other than a home network or a small office network. It is impractical to have two distinct Class C networks—your ISP will charge you for each. A potential solution is to split our network into subnets (see Figure 12.47). We typically use a shorthand notation that can also be applied to classful routing. Rather than express our IP address and subnet mask as:

IP: 192.168.10.104

SM: 255.255.255.0

we write our address using the notation *192.168.10.104/24*, where 24 corresponds to the number of *1* bits in the subnet mask. For subnetting, 192.168.10.104/27 indicates a Class C address that has been subnetted by borrowing three bits (*27 − 24*).

Subnetting

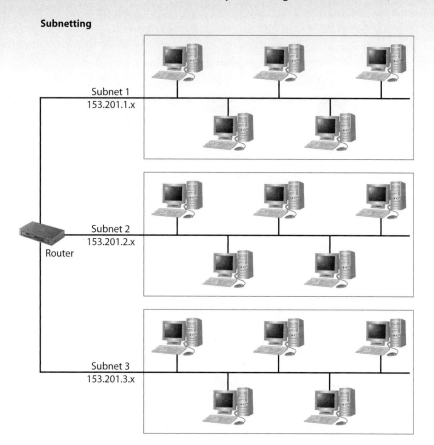

Figure 12.47 | Subnetting makes it possible to divide a network into multiple parts, with each part assigned a separate subnet IP address. This technique allows you to break a large network into a number of smaller subnetworks that are easier to manage and results in a more efficient network overall.

ENGINEERING QUICK TAKE

You have just been hired as a network administrator by ACME Corporation. Your first task on the job is to set up a network for a new office in Alabama. Based on your research of similar ACME corporate offices around the country, you determine that you will provide access for four distinct groups: Sales, Human Resources, Management, and Engineering. Based on similar offices, you expect to have twenty users in Sales, ten in Human Resources, twenty-five in Management and fifteen in Engineering. You contact InterNic and are assigned a network IP address of *192.168.10.0*. Based on your research, you've determined that you need at least four subnets (Sales, Human Resources, Management, and Engineering) and at least twenty-five hosts per subnetwork. With this information in hand, we need to determine:

- The class of your network
- Which subnet masks will meet your requirements
- The subnet addresses, usable host addresses, and broadcast address for each subnet

By inspection, we see that this is a Class C address; the first three octets are the network portion of the IP address and the last octet is the host portion. For subnetting, we have eight bits available—we have to borrow at least two bits, and we must leave at least two host bits remaining. Label the number of host bits as H and the number of borrowed bits as B. What are our options?

Remember that using B and H we can calculate the number of subnets created and the number of hosts per subnet from the following expressions:

$$Subnets = 2^B - 2 \text{ and } Hosts = 2^{(H-B)} - 2$$

B	H	SUBNETS	HOSTS/SUBNET
2	6	$2^2 - 2 = 2$	$2^6 - 2 = 62$
3	5	$2^3 - 2 = 6$	$2^5 - 2 = 30$
4	4	$2^4 - 2 = 14$	$2^4 - 2 = 14$
5	3	$2^5 - 2 = 30$	$2^3 - 2 = 6$
6	2	$2^6 - 2 = 62$	$2^2 - 2 = 2$

Reviewing our table of results, we see that the only alternative that fits our scenario is number two—borrowing three bits. If we borrow three bits, the last octet for our subnet mask becomes *11100000* or 224 in decimal, making our subnet mask 255.255.255.224. This subnet mask will give us six subnets with thirty hosts per subnet. To determine our subnet, broadcast, and host address, we use the technique called the magic number. We can calculate the magic number by subtracting the last non-zero octet of our subnet mask from 256. In this case,

$$Magic\ Number = 256 - 224 = 32$$

Using the magic number, we can write the following subnet addresses:

192.168.10.0	unusable subnet

192.168.10.32	
192.168.10.64	
192.168.10.96	
192.168.10.128	6 useable subnets
192.168.10.160	
192.168.10.192	

192.168.10.224	unusable subnet

Looking at subnets, we can determine our corresponding host and broadcast addresses.

SUBNET NETWORK ADDRESS	HOSTS	BROADCAST
192.168.10.32	*192.168.10.33-62*	*192.168.10.63*
192.168.10.64	*192.168.10.65-94*	*192.168.10.95*
192.168.10.96	*192.168.10.97-126*	*192.168.10.127*
192.168.10.128	*192.168.10.129-158*	*192.168.10.159*
192.168.10.160	*192.168.10.161-190*	*192.168.10.191*
192.168.10.192	*192.168.10.193-222*	*192.168.10.223*

Notice that the magic number 32 shows up repeatedly in these calculations.

Limitations of Classful Addressing

Addressing using Class A, B, and C is also called classful addressing. Unfortunately, IP version 4—the protocol that defines classful IP addressing—was designed with no anticipation of the explosion in networks and the Internet. There are a number of factors that contributed to this problem. Among these is the choice of a 32-bit address, which gives us 2^{32} or 4,294,967,296 available addresses. While this may seem adequate, when you consider the growth of the Internet and the proliferation of network-enabled devices (cell phones, PDAs, laptops, etc.) it becomes apparent that we are running out of IP addresses (see Figure 12.48). Another factor is the implementation of classful A, B, and C addresses with fixed octets and a rigid boundary between the network and host portions of an IP address. Although this implementation is easy to understand and implement, it does not provide us with enough flexibility in creating and assigning network and host addresses. There are a number of technologies that have been created to address these limitations. These include Classless Inter-Domain Routing (CIDR), Network Address Translation (NAT), and Internet Protocol version 6 (IPv6).

Figure 12.48 | It is no wonder we are running out of IP addresses. Companies such as LG Electronics are designing network-connected bathroom mirrors that can show the news, weather, and traffic, as well as network-connected kitchen appliances that can be controlled remotely and e-mail you operational and diagnostic information.

CIDR CIDR is an acronym for Classless Inter-Domain Routing, pronounced "cider." CIDR was created in response to the limitations of classful (A, B, and C) and ignores the fixed network-host boundary of classful addresses. In fact, a CIDR address sets a rather arbitrary boundary between the network and host portion of an IP address. Eliminating this constraint of classful addressing allows us greater flexibility in using and assigning IP addresses. With CIDR, we can combine Class A, B, and C addresses to create a larger address space. With CIDR, we typically use a shorthand notation that can also be applied to classful routing. Rather than expression our IP address and subnet mask as:

IP: 192.168.10.104
SM: 255.255.255.0

we write our address using the notation *192.168.10.104/24*, where the 24 corresponds to the number of *1* bits in the subnet mask. When subnetting, 192.168.10.104/27 indicates a Class C address that has been subnetted by borrowing three bits (27 − 24).

IPv6 Internet Protocol version 6 (IPv6) has been developed to eventually replace IPv4. IPv6 addresses were developed to address the shortage of IPv4 addresses, providing a much larger pool of available addresses. While IPv4 has 32-bit addresses, IPv6 supports 128-bit addresses. IPv6 supports 2^{128} or 3.4×10^{38} unique addresses. This is a much larger address space that allows greater flexibility in assigning addresses and eliminates the need for complex workarounds such as Network Address Translation (NAT).

SECTION THREE FEEDBACK >

1. Determine the IP address of a computer at your school. Is this a public or private address? If private, what is the public address?
2. Rewrite your IP address from (1) in binary.
3. Determine the IP address for your DNS server(s).
4. Is it possible to write an IPv4 address as IPv6? If so, demonstrate with the address you found in problem 1.

HEY IDEAS >

- As greater numbers of people gained access to networks and the personal computer proliferated, security threats to networks, both unintentional and malicious, became increasingly common.

- To properly secure a network, one must develop and enforce an organization-wide security policy that implements internal and external security measures.

- Passwords, permissions, and user access control are important internal mechanisms for securing a network.

- Security vulnerabilities are often the result of improper configurations, or poorly implemented hardware and software security measures.

Figure 12.49 | The best way to limit internal security threats is to limit physical access to a computer.

Network Threats

Remember that in the beginning of the computing age, only a select group of people within an organization had access to computers—then called mainframes. Furthermore, these early computers were so expensive that only large corporations, government agencies, and research universities could afford them. Limited access to these computing resources served to insulate them from any threats. The fewer the number of people that can gain access to a computer, the less is the likelihood of anyone trying to gain unauthorized access to computing resources. In fact, in the early days of computing, security was not even a major concern. As we started to provide increased access to computing resources—first through local "dumb" terminals, then through remotely connected "dumb" terminals, and, finally, starting in 1984, through dial-up and high-speed access via the personal computer, the potential for security issues, both intentional and unintentional, grew. The explosion in popularity of networking and the Internet have also increased the number of potential threats to our networks.

In the context of a network, we define a threat as any type of activity—accidental or malicious—that prevents end users from accessing shared resources and fundamentally impacts the productivity of our organization. Put simply, a threat is anything that limits your ability and my ability to get work done. Threats typically can be isolated as originating from two potential sources—either internal to an organization or originating externally. Internal threats include a wide range of potential activities, including unauthorized user or system access, destruction of property or data, and theft and introduction of malware—malicious software (viruses, worms, spyware, etc.) into the network. The most common reasons for internal security breaches are unintentional user error, disgruntled employees, or corporate espionage. External threats typically exploit vulnerabilities in your network infrastructure—hardware, software, or configuration. Some of these threats originate from casual hackers with no malicious intent, but the threats to be most wary of are threats from crackers—typically malicious users trying to break into a system for fame or financial gain.

Internal Security

Internal security begins with the development and enforcement of organization-wide policies. These policies would include guidelines for password selection, strength, and expiration. It is also important to both set and enforce controls for user permissions and user accounts. Permissions ensure that employees can only access files that they should have access to. For instance, an employee working in Engineering should not have the same permissions as someone in Human Resources. User account controls (UACs) limit when and from where a user can connect to the network. A typical scenario would be to limit access (see Figure 12.49) after working hours and on weekends and to restrict accessing the network remotely (i.e., from home or on a business trip).

External Security

Just as with internal security measures, external security policies should be well defined in a corporate policy document. The most important aspect of external security is to restrict physical access to the premises and the network to authorized employees only. External security should also include proper set up and maintenance of physical and software firewalls. Policies should be in place to limit or preclude employee installs of software and browsing of questionable Web sites. Finally, the organization should have a detailed backup and disaster recovery plans in the event of a security breach or inadvertent loss of data. (See Figure 12.50.)

 ELEMENTS OF EXTERNAL SECURITY

• A corporate policy document
• Proper setup of physical and software firewalls
• Proper maintenance of physical and software firewalls
• Limitation on employees' ability to install software and browse questionable Web sites
• A detailed backup and disaster recover plan

Figure 12.50 | To ensure a network's external security, these elements are essential.

SECTION FOUR FEEDBACK >

1. Explain why modern networks are much more susceptible to attacks than mainframe computers of the past.
2. What factors lead to internal security vulnerabilities?
3. Do a Web search for "strong passwords" and describe some characteristics that make a password strong.
4. Describe two critical pieces of external security.
5. Search the Web for sample organizational security policies.

CAREERS IN TECHNOLOGY

Matching Your Interests and Abilities with Career Opportunities: Computer Support Specialists and Systems Administrators

Networks have become mission-critical applications in virtually every industry. There are many opportunities in this industry. Entry-level network and computer systems administrators are involved in routine maintenance and monitoring of computer systems, typically working behind the scenes in an organization. After gaining experience and expertise, they often are able to advance into senior-level positions, in which they take on more responsibilities. Administrators may become software engineers, actually involved in the design of the system or network and not just its day-to-day administration.

Persons interested in becoming a computer support specialist or systems administrator must have strong problem-solving, analytical, and communication skills, because troubleshooting and helping others are vital parts of the job. The constant interaction with other computer personnel, customers, and employees requires computer support specialists and systems administrators to communicate effectively on paper, via e-mail, or in person. Strong writing skills are useful in preparing manuals for employees and customers.

Significant Points

* Rapid job growth is projected over the 2004–14 period.
* There are many paths of entry to these occupations.

* Job prospects should be best for college graduates who are up to date with the latest skills and technologies.
* Certifications and practical experience are essential for persons without degrees.

Nature of the Industry

Computer support specialists provide technical assistance, support, and advice to customers and other users: answering telephone calls, analyzing problems, and resolving recurring difficulties. Network administrators and computer systems administrators design, install, and support an organization's local-area network (LAN), wide area network (WAN), network segment, Internet, or intranet system. They provide day-to-day onsite administrative support for software users, maintain network hardware and software, analyze problems, and monitor the network to ensure its availability to system users. Systems administrators are the information technology employees responsible for their organization's efficient network use.

Working Conditions

Computer support specialists and systems administrators normally work in well-lighted, comfortable offices or computer laboratories. They usually work about 40 hours a week, but that may include being "on call" via pager or telephone for rotating evening or weekend work if the employer requires computer support over extended hours. Overtime may be necessary when unexpected technical problems arise.

Training and Advancement

Many employers prefer to hire persons with some formal college education. For applicants without a degree, certification and practical experience are essential. Some jobs require a bachelor's degree in computer science or information systems, while a computer-related associate's degree may be sufficient for others. Regardless of their preparation, persons working in these fields must keep their skills current and acquire new ones in response to changing and improving technology.

Outlook

Employment of computer support specialists is expected to increase faster than the average for all occupations through 2014, as organizations continue to adopt increasingly sophisticated technology and integrate it into their systems. Employment of systems administrators is expected to increase much faster than the average for all occupations as firms continue to invest heavily in securing computer networks. The information security field is expected to generate many opportunities over the next decade as firms across all industries place a high priority on safeguarding their data and systems.

Earnings

Median annual earnings of computer support specialists were $40,430 in May 2004 and $58,190 for network and computer systems administrators.

Summary >

Information and communication systems allow information to be transferred from human to human, human to machine, machine to human, and machine to machine. Information and communication systems can be used to inform, persuade, entertain, control, manage, and educate. There are many ways to communicate information, such as graphic and electronic means. The Open System Interconnect model is a seven-layer model that helps us understand how networks operate, and the TCP/IP model is a four-layer model that more closely represents how real networks work.

A Wide Area Network (WAN) links Local Area Networks (LANs), which are groups of computers connected together within a small geographic region. The most common network topology is the hybrid star-bus, which combines a physical with a logical bus. A switch is a multi-port bridge that connects two or more network segments and serves to segment a larger network. Routers are devices that use IP addresses to connect networks together and route information from one network to another—also called internetworking.

An IP or Internet Protocol address is a 32-bit binary number that computers read as a string of 32 ones and zeroes. For convenience, we write IP addresses as a set of four decimal numbers or "dotted decimal" notation. To communicate from network to network, we use source and destination IP addresses. An IP version 6 address has 128 bits or 128 ones and zeroes.

As greater numbers of people gained access to networks and the personal computer proliferated, security threats to networks, both unintentional and malicious, became increasingly common. To properly secure a network, one must develop and enforce an organization-wide security policy that implements internal and external security measures. Passwords, permissions, and user-access control are important internal mechanisms for securing a network. Security vulnerabilities are often the result of improper configurations or poorly implemented hardware and software security measures.

FEEDBACK

1. Interview one of your grandparents or another relative about the older telephone networks and making long-distance calls.

2. We have learned that the IEEE 802.3 standard defines Ethernet and CSMA/CD. Research another of the IEEE 802.x standards and write a one-page written description.

3. Find three Web sites that make extensive use of the new Web-interaction model AJAX.

4. Use your school's public IP address to determine the school's Internet Service Provider (ISP).

5. Research a device called a hub. Can you find any hubs in your school network? How about for purchase at online computer stores?

6. Research the Routing Information Protocol (RIP). Draw and annotate a diagram describing how RIP works.

7. One interesting proposed application for IP version 6—since it provides a virtually inexhaustible supply of IP addresses—is to "weave" a network of IP addresses into a soldier's uniform. This would mean that every part of the uniform of every soldier would have a unique IP address, which would allow us to remotely assess battlefield injuries by examining the "lost" or missing IP address. Think of two additional ways you could use IP version 6 address-space.

8. Describe a scenario in which it would be important for a computer to have a public IP address.

9. Research methods of transitioning from IP version 4 (32 bits) to IP version 6 (128) bits and provide a brief overview. Do the computers and networking devices in your school network support IPv6?

10. Does your school have a security policy? If so, detail three key parts of the policy. If your school does not have a security policy, use a template from the Web to create a draft school security policy.

11. Search the Web for information on a type of password cracking technique called a "dictionary" attack. Describe why this type of attack makes strong passwords critical.

12. Research an infamous network attack and put together a five-minute presentation describing and detailing the attack.

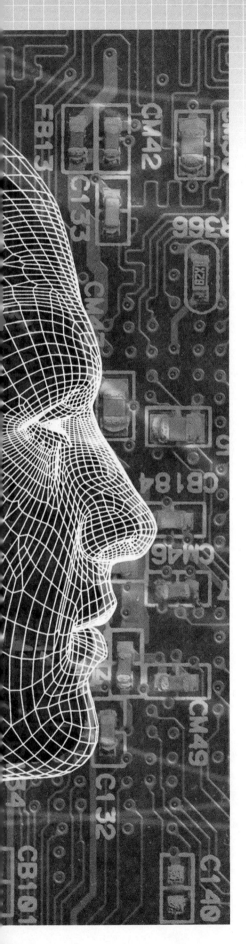

DESIGN CHALLENGE 1:
Using a Network to Transfer Data to New Computers

● Problem Situation

You work in a small law firm employing between twenty and thirty employees and a corresponding number of computers. The computers vary in age from two years to five years and use a variety of operating systems. The company has made plans to upgrade all the computers and build a network. As part of this process, new computers must be set up and many years of user data must be transferred from the old computers to the new. Most of the budget is allocated to purchasing new computers, as well as to designing, building, and securing the network. In this activity, you will research, test, document, and recommend methods for transferring large quantities of data between computers.

● Your Challenge

Given the necessary materials, you will research methods of transferring data from one computer to another. You will implement and test a number of transfer solutions, benchmarking each and properly documenting your work. Based on your research and testing, you will recommend the appropriate method for completing the ultimate transition from old computers to new. You must be able to account for the wide variety of operating systems and computer architectures.

● Safety Considerations

As a portion of this activity may require Web-based research, be sure to take appropriate precautions when sharing information about yourself or your classmates on the Internet.

● Materials Needed

1. 2 or more computers, with network interface cards (NICs)
2. A variety of Ethernet cables, including straight-through or patch, rollover, and crossover
3. A small 8-port hub
4. A small 8-port switch
5. Network performance measurement software—Ixia's free QCheck software is recommended, available at http://www.ixiacom.com/products/display?skey=qcheck. Alternatively, students can used built-in network troubleshooting tools, such as *ping* and *traceroute*, or research other free or trial alternatives via a Web-based search.

● Clarify the Design Specifications and Constraints

Using the provided equipment, you will implement and test a variety of data transfer solutions between two computers. You solution must be valid for a wide variety of operating systems and computer architectures. Other than the equipment you have on hand, you do not have a budget available to purchase any additional equipment or software. You solution will be judged on its flexibility, ease of implementation, speed/throughput, and cost.

• Research and Investigate

To complete the design challenge, you need to first gather information to help you build a knowledge base.

1. In your guide, complete the Knowledge and Skill Builder I: Web safety.
2. In your guide, complete the Knowledge and Skill Builder II: Operating systems.
3. In your guide, complete the Knowledge and Skill Builder III: Network cabling: Unshielded twisted pair (UTP).
4. In your guide, complete the Knowledge and Skill Builder IV: Network devices: NICs, hubs, and switches.
5. In your guide, complete the Knowledge and Skill Builder V: IP addresses.
6. In your guide, complete the Knowledge and Skill Builder VI: Data: UDP versus TCP.
7. In your guide, complete the Knowledge and Skill Builder VI: Network troubleshooting: Testing throughput.

• Generate Alternative Designs

Assuming that you are given a moderate budget, describe possible alternatives to your data transfer solution. Provide information regarding any additional hardware or software that you would purchase to implement an alternative solution. Discuss the decisions you made in (a) testing, (b) research, and (c) implementation. Attach printouts, photographs, and drawings if helpful, and use additional sheets of paper if necessary.

• Choose and Justify the Optimal Solution

What decisions did you reach about the design of the choice of data transfer solution?

• Display Your Prototypes

Produce a brief report documenting alternatives tested, test results, method selected, and justification.

• Test and Evaluate

Explain whether your designs met the specifications and constraints. What tests did you conduct to verify this?

• Redesign the Solution

What problems did you face that would cause you to redesign the (a) hardware/cabling used, (b) the network topology used (if any), and/or (c) the final solution? What changes would you recommend in your new designs? What additional trade-offs would you have to make?

• Communicate Your Achievements

Describe the plan you will use to present your solution to the class. (Include a media-based presentation.)

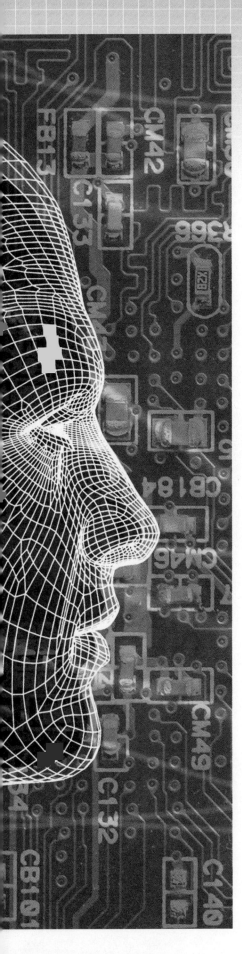

DESIGN CHALLENGE 2:
Getting Wired

● Problem Situation

You work in a small law firm, with between twenty and thirty employees and a corresponding number of computers. To date, each employee has worked on a stand-alone computer, also called a work station. None of these work stations can communicate with one other or the outside world. Because file sharing, printing, and Web research (online legal databases) have become increasingly important, the firm's partners have decided to invest in building a network. The primary considerations for moving in this direction are the gains in efficiency and productivity that the firm expects to see. The company has already upgraded all the computers, and you were personally responsible for moving user data from the old computers to the new. Because this is a relatively small network, the budget allocated to the project is modest. In this activity, you will research, test, document, and recommend methods for building a small network supporting this law firm.

● Your Challenge

Given the necessary materials, you will research methods of building a small network that will allow users to share a network connection, printers, files, and folders. You will implement and test a number of solutions, benchmarking each and properly documenting your work. Based on your research and testing, you will recommend the appropriate method for building the network. You must be able to account for the wide variety of operating systems and computer architectures and provide capacity for future growth. Your team will submit a formal proposal describing a recommended network layout and configuration.

● Safety Considerations

As a portion of this activity may require Web-based research, be sure to take appropriate precautions when sharing information about yourself or your classmates on the Internet.

● Materials Needed

1. 2 or more computers, with network interface cards (NICs)
2. A variety of Ethernet cables including straight-through or patch, rollover, and crossover
3. A small 8-port hub
4. A small 8-port switch
5. A small 4-port router
6. A network printer—*optional*
7. A parallel or USB printer—*optional*
8. One speed network connection
9. Floor plan diagrams of the building in which the new network will be installed
10. Various colored pencils and drawing tools for cut sheets/wiring diagrams
11. Network performance measurement software—Ixia's free QCheck software is recommended, available at http://www.ixiacom.com/products/display?skey=qcheck. Alternatively, students can used built-in network troubleshooting tools, such as *ping* and *traceroute*, or research other free or trial alternative via a Web-based search.

● Clarify the Design Specifications and Constraints

Your proposed system must reflect state-of-the-art technology and performance. Your designed system should provide maximum bandwidth for the proposed price.

Your team must address the needs of the new network regarding bandwidth, security, reliability, and speed, and prices must be competitive. The quality and content of presentations should be appropriate for an audience of professionals.

● Research and Investigate

To complete the design challenge, you need to first gather information to help you build a knowledge base.

1. In your guide, complete the Knowledge and Skill Builder I: Web safety.
2. In your guide, complete the Knowledge and Skill Builder II: Operating systems.
3. In your guide, complete the Knowledge and Skill Builder III: Bandwidth and frequency of digital signals.
4. In your guide, complete the Knowledge and Skill Builder IV: Network cabling: Unshielded twisted pair (UTP).
5. In your guide, complete the Knowledge and Skill Builder V: Network devices: NICs, hubs, and switches.
6. In your guide, complete the Knowledge and Skill Builder VI: Network devices: Routers, gateways, and firewalls.
7. In your guide, complete the Knowledge and Skill Builder VII: IP addresses.
8. In your guide, complete the Knowledge and Skill Builder IIX: IP subnetting.
9. In your guide, complete the Knowledge and Skill Builder IX: Dynamic host configuration protocol (DHCP).
10. In your guide, complete the Knowledge and Skill Builder X: Network troubleshooting: Testing throughput.

● Generate Alternative Designs

Describe possible alternatives to your network design. Provide information regarding any additional hardware or software that you would purchase to implement an alternative solution. Discuss the decisions you made in (a) testing, (b) research, and (c) implementation. Attach printouts, photographs, and drawings if helpful and use additional sheets of paper if necessary.

● Choose and Justify the Optimal Solution

What decisions did you reach about the design of the choice of network?

● Display Your Prototypes

Produce a brief report documenting alternatives tested, test results, method selected, and justification.

● Test and Evaluate

Explain whether your designs met the specifications and constraints. What tests did you conduct to verify this?

● Redesign the Solution

What problems did you face that would cause you to redesign the (a) hardware/cabling used, (b) the network topology used (if any), and/or (c) the final solution? What changes would you recommend in your new designs? What additional trade-offs would you have to make?

● Communicate Your Achievements

Describe the plan you will use to present your solution to your class. (Include a media-based presentation.)

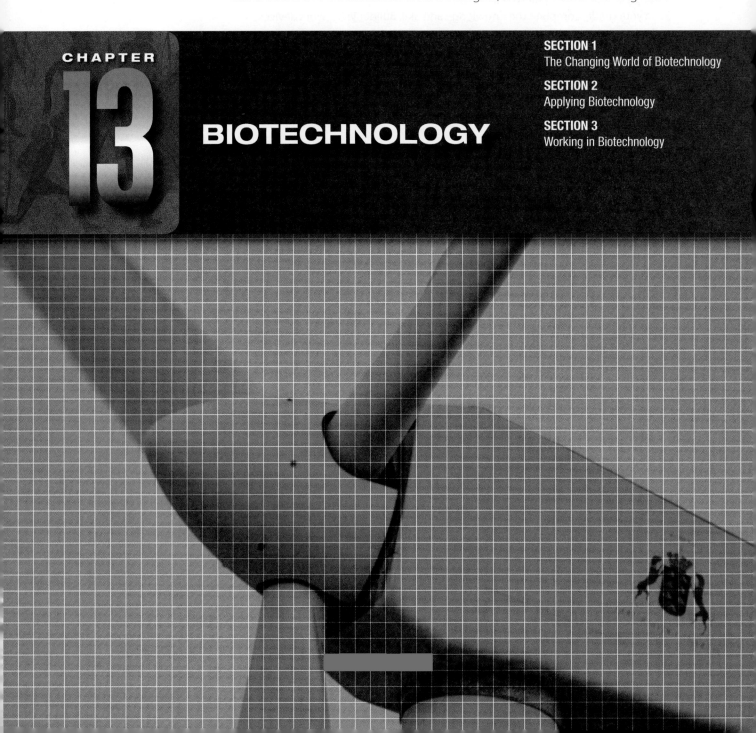

INTRODUCTION

B IOTECHNOLOGY IS BROADLY defined as the use of living organisms or the products of these organisms to benefit humankind. Plant and animal breeding that produces hybrid corn and cows, respectively, are examples of biotechnology. In today's technological world, biotechnology companies not only need scientists who understand engineering, but also engineers who understand science. To be successful biotechnologists, these scientists and engineers

CHAPTER

13

BIOTECHNOLOGY

SECTION 1
The Changing World of Biotechnology

SECTION 2
Applying Biotechnology

SECTION 3
Working in Biotechnology

need to understand how to manage projects as well as the economic value of the products they help to develop.

Even if you decide not to become a biotechnologist, you will still need to understand topics related to biotechnology in order to make informed decisions regarding food selection, diagnostic tests, and treatments for diseases. Understanding biotechnology information will allow you to independently evaluate the risks and benefits of modern biotechnology products.

For example, we know that applying pesticides to crops is expensive; pollutes the surrounding environment, leading to wildlife death; and increases the number of pesticide-resistant insects. For this reason, some biotechnologists have argued in favor of genetically-modified (GMO) pest-resistant crops, in which the pesticide is actually a part of the plant, making the application of pesticides unnecessary. They argue that GMO pest-resistant crops eliminate the negative side-effects associated with pesticide application. Opponents state that GMO crops are a danger, claiming that they will result in wide-spread killing of beneficial pollinating insects such as butterflies (see Figure 13.1) and cause life-threatening allergic responses in individuals who are allergic to the pesticide-containing produce and food products. As a potential consumer of GMO pest-resistant crops, you should ask this question: "What type of testing has been done to prove the safety of the GMO pest-resistant crops?"

Figure 13.1 | GMO corn kills pests, but does it also harm caterpillars that turn into beneficial butterflies?

Another example is the use of drugs, known as statins, to lower cholesterol in a person's bloodstream to prevent the build up of plaque. It is scientifically accepted that the buildup of plaque can lead to heart disease and eventually to heart attacks. However, it is also accepted that statins may cause liver and muscle damage in some people. Presently people taking these drugs are being advised to get tested once a year to determine if their liver is being affected, and inform their doctor if they develop muscle weakness. Patients who do not develop these side-effects generally continue to take cholesterol-lowering drugs indefinitely. As a potential consumer of such drugs, you should ask this question: "Is this testing and the qualitative knowledge about muscle weakness adequate for detecting all possible long-term negative effects?" When new drugs are released to the public, the long-term side effects are often poorly understood. In many cases, scientists and doctors don't discover the long-term side effects until many people have taken the drug for a long time. When you start taking a new drug, you should ask this question: "Do the benefits of taking this new drug outweigh the possible known and unknown long-term side effects of this drug?"

When trying to evaluate the risks and benefits of using new technologies, it is often helpful to list them. Table 13.1 is an example of such a list.

Think about doing this the next time you are faced with evaluating a new technology.

Table 13.1 | New Technologies: Risks and Benefits

CHOICES	BENEFITS	RISKS
GMO pest-resistant crop vs. large-scale use of pesticides	Decreased costs, no large-scale wildlife death, or contamination of water sources	Allergic response in consumers of product, death of beneficial insect pollinators
Cholesterol-lowering medication vs. no medication	Decrease in heart disease and heart attacks, decrease in associated health care costs	Liver disease and other unknown side-effects from long-term use

In this chapter, you will learn about the changing world of biotechnology. You will examine biotechnological applications; explore issues related to bioethics; and become acquainted with the business and regulations surrounding biotechnology.

The final topic is biomimicry, and like biotechnology its origins are from nature. However, it is different from biotechnology in that it does not use the natural world, it is inspired by it. Biomimicry is the utilization of designs and technology found in nature for the purpose of making life easier for people but also more environmentally friendly.

SECTION 1: The Changing World of Biotechnology

KEY IDEAS >

- Traditional biotechnology started with the beginning of civilization
- Large historical events, such as wars, and paradigm shifts in technology rapidly accelerate the development of biotechnology

Traditional biotechnology started with the beginning of civilization. In this section, we examine some early forms of biotechnology. Then we move on to explore the recent rapid advances in biotechnology.

Plant and Animal Domestication

People have tried to make life better for themselves through breeding and selecting animals and plants. This process is known as selective breeding, and probably started during the Mesolithic Period (9000 to 2700 BCE) of the Stone Age with the domestication of the wolf. Scientists have named the initial transitional wolf to dog animal "proto-dog"; and continued selective breeding of this animal has produced the many dog breeds of today.

How does selective breeding work? Originally, people wanted dogs that could help with work, such as hunting and herding. For example, suppose you wanted a dog that could herd goats. You would start by finding a male and female that had shown some ability to do so. Presumably their offspring would have inherited both parents' ability to herd goats, making it a better goat herder than either parent. Then you would breed this dog with another dog that was known to be good at herding goats, and so on, causing the goat-herding trait to intensify in successive generations.

The small, fearless, and intelligent terrier is a good example of a dog breed that owes its distinctive look and behavior to careful breeding (see Figure 13.2). The different types of terriers were bred to hunt foxes and other small animals in their holes and, as a result, had to be fairly aggressive animals. Another breed, the husky, was originally developed by the indigenous people of the far North to pull sleds and herd reindeer. These are also one of the breeds employed for the Iditarod, the sled dog race across Alaska. These days, both terriers and huskies are most often kept as pets by people in many countries and do not necessarily have to do any work.

Figure 13.2 | Selective breeding has produced hundreds of different dog breeds, each one adapted for specific tasks. (a) Terriers were bred to crawl down holes and pull out foxes for sport or to remove them from farms. (b) The Siberian husky was bred for pulling sleds and enduring cold weather. (c) Today, both breeds are kept as pets.

Garbage Triggered the Domestication of the Wolf

▷ How was the wolf domesticated, leading to the first proto-dog? Studies conducted on Russian fox farms by scientist Dmitri Belyaev in the 1950s seem to have provided an answer. Fur farms found it difficult to work with wild silver foxes. Belyaev thought that if he did selective breeding with captive foxes that were more docile, eventually he would select for a tamer, easier-to work with, silver fox. He identified foxes that were more tolerant of humans than other foxes by sticking his hand in their cages and picking the foxes that approached to sniff his hand. Within one generation, the foxes became tamer. However, the tamer foxes also developed physical traits, unlike their parents, such as floppy ears, curly tails, barking like a dog, and strange coat colors. In fact, these foxes had exactly the same kinds of qualities we see in domesticated dogs, but never in wild foxes or wolves. Why should selecting for tamer animals also create individuals with all these unusual

Section 1 △ The Changing World of Biotechnology

physical qualities? The theory is that by selecting for individuals that are friendlier and less suspicious of humans, you also affect some aspects of the developmental process; most likely because there is no selection for maintaining the traits that were necessary for survival. The foxes could remain "puppy-like" and survive. Linked to the genes that control this extended puppyhood are other genes that affect coat color and structural features such as ears and tail.

Based on this behavioral research on foxes and on their own observations of wild dogs at garbage dumps, scientists such as Raymond Coppinger theorize that wolves that were willing to be near people to get food from the village garbage dump would self-domesticate. In fact, if you think about it, any wolves that were willing to frequent the garbage dumps of ancient man or take food from a human would be closer to a reliable food source and therefore more likely survive when food in the wild was scarce. They would remain healthy enough to successfully reproduce, passing on the "tameness traits." These wolves would be the ancestors of the proto-dog. Over time, humans would exploit the superior senses of this proto-dog to track food, and the proto-dog would depend on them for shelter and food when fresh, wild food was scarce. This evolving relationship probably led to the development of the domesticated dog.

Figure 13.3 | Tame silver foxes are dog-like in the lack of "fox" smell.

Figure 13.4 | Common food-fermentation products are soy sauce, wine, beer, bread, sauerkraut, yogurt, and cheeses.

Fermentation

Another form of biotechnology that has been around for thousands of years is the use of microorganisms in food, especially microorganisms such as yeast or bacteria that can ferment. The term "fermentation" refers to yeast or bacteria breaking down sugars, in the absence of oxygen, for the purpose of generating energy for themselves. Such sugars include those found in fruits such as grapes, milk, or bread. Depending on the type of microorganism and the sugar used, examples of fermentation products are ethanol, carbon dioxide, lactic acid, acetone, and butyric acid. Ancient people did not understand that organisms fermented sugar to obtain energy; they only knew that the process could be exploited to prepare food and, in some cases, preserve it; for example, fermentation in bread-making is the reason that bread rises. Mold and bacterial fermentation in cheese produces chemicals that make the sharp taste you taste in cheese, and wine is the result of yeast fermentation of the sugars found in grapes. (See Figure 13.4 for other examples of food-fermentation products.)

Most likely, ancient people discovered the benefits of the fermentation process by accident when food was contaminated by these microorganisms. Some examples of early uses of fermentation in food production are as follows:

- In 4000 BCE, Egyptians used yeast to bake leavened bread (see Figure 13.5) and to make wine.
- As early as 6000 BCE, Indians made yogurt, produced through the acidification of milk. Acidic "cooking" techniques rely on acids, which are produced by microorganisms as they ferment sugars, to kill bacteria. These techniques led to the development of both yogurt and sauerkraut.

Figure 13.5 | The use of yeast to make bread has been around for thousands of years, even as far back as ancient Egypt.

Advances in Modern Biotechnology

Everyday life events, such as food spoilage, found ancient people exploiting biotechnology. As town and city populations increased and countries competed for products, the public, governments, and industry realized that biotechnology could solve larger-scale problems. In fact, large historical events, such as wars, and paradigm shifts in technology, rapidly accelerate the development of biotechnology. (A paradigm shift is a change in the way one thinks.)

One example of the way that the needs of businesses can spur new developments in biotechnology occurred in the mid-1800s. Louis Pasteur, a French scientist, was approached by wine growers to figure out a way to prevent undesirable products from appearing in wine. Pasteur theorized that bacteria were contaminating wine after the yeast had finished fermenting the sugar. He invented a method, known as pasteurization, of heating the wine to kill the undesirable bacteria. Pasteurization is not intended to kill all microorganisms in food. Instead, the process aims to reduce their number so they are unlikely to cause disease or produce undesirable products. Today, pasteurization is done on a large scale and is used to extend the shelf life of many different food products (see Figure 13.6).

In the following sections, you will learn about some other events that accelerated the development of biotechnology.

Figure 13.6 | Pasteurization kills only some microorganisms, not all of them. That is why there are expiration dates on pasteurized products so that people know when they should throw them out because they may be contaminated after that date.

Industrial Fermentation and Penicillin World War I (1914 to 1918) led to the rise of industrial fermentation as a vital process around the world. Industrial fermentation is fermentation on a huge scale, for example, not just one bakery making bread for a small town, but a company making bread for people across the United States. Countries at war were not selling products to each other so they had to develop other resources within their own countries. For example, Max Delbruck, a biologist in Germany who later won the Nobel Prize for his work in genetics, grew yeast on a huge scale during the war. As a result, he was able to meet 60 percent of Germany's animal-feed needs. Because Germany could not get glycerol, a hydraulic fluid, from the countries who were now enemies, they used lactic acid, another fermentation

Figure 13.7 | During World War 1, this anaerobic (without oxygen) fermentation process was used primarily for producing acetone in the making of explosives. Over the years, the process was improved to boost yield and produce different mixtures of solvents.

product, instead. On the Allied Powers' side, the Russian chemist Chaim Weizmann figured out how to use starch to eliminate Britain's shortage of acetone, a key raw material in explosives. He developed an industrial fermentation process where corn, which contains starch, was fermented to produce acetone (see Figure 13.7).

As a result of these successes, scientists realized that biological processes could be exploited and scaled up to produce resources. In 1917, Karl Ereky coined the word "biotechnology" in Hungary to describe a technology based on converting raw materials into a more useful product. For him, biotechnology was not limited to industrial fermentation; he also felt that biotechnology applied to mass production of animals and crops. He built a slaughterhouse for 1,000 pigs and also a fattening farm with space for 50,000 pigs, raising over 100,000 pigs a year. The enterprise was enormous, becoming one of the largest and most profitable meat and fat operations in the world. In order for his endeavor to succeed, he needed the mass production of animal feed that Max Delbruck designed.

World War II (1939–1945) spurred the development and production of the world's first manmade antibiotic, penicillin. Penicillin had been discovered before World War II, and by 1941, it was being produced on a small scale. However, the massive number of casualties caused by World War II pushed governmental funding and enabled scientists and engineers from both the United States and England to figure out how to mass produce the antibiotic. Their eventual success meant that the number of soldiers who died from pneumonia in World War II dropped dramatically compared to the number in World War I (less than 1 percent as compared to 18 percent). Penicillin was considered the wonder drug of its time.

The Foundations of Molecular Biology Once industrial fermentation and other biotechnology processes became a common way to produce needed food and medicine, scientists worked to find ways to make other products, such as proteins in cells. This search gave rise to the next major technological advancement, the making of recombinant DNA. Recombinant DNA technology is the science of extracting DNA from different organisms and combining it at the molecular level to produce proteins, which become the basis for new foods, drugs, and even completely new organisms.

The technology behind recombinant DNA had its beginnings in 1953, when scientists James Watson and Francis Crick published a short paper in the journal *Nature*. Their paper described the structure of DNA, a nucleic acid chemical in a cell's nucleus that carries genetic (hereditary) characteristics. A later paper explained how DNA, because of its double-stranded structure, can be duplicated in the cell. Then, around 1958, Crick proposed that most cells, whether bacteria, plant, or animal cells, use a similar process to make proteins. The following formula sums up this process:

DNA → (RNA) → Protein

At the time the intermediate chemical, known as RNA, had yet to be identified. What scientists knew was that DNA in a cell was like a huge library, and this library was found in every cell that had a nucleus. This DNA library contains all of the information that cells need to make products such as proteins. The information is composed of four different types of nucleic acids commonly represented as the symbols A (adenine), T (thymine), G (guanosine), and C (cytosine). It is the linear arrangement of these nucleic acids that determines the products that a cell makes such as proteins (see Figure 13.8). For example, the linear sequence AATTGGGTTAAG produces a different protein than the linear sequence AATGGGTCCTTA.

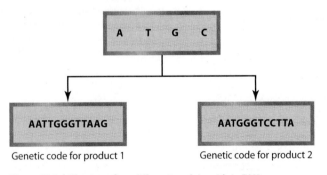

Genetic code for product 1   Genetic code for product 2

Figure 13.8 | There are four different nucleic acids in DNA, symbolized by the letters A, T, G and C. The linear sequence of these nucleic acids determines the products made by a cell and is commonly referred to as the "genetic code" for these products.

What they did not know is how this DNA information is decoded to make products such as proteins. Crick proposed that an intermediary product from DNA carries the information for whatever product a cell needs to make at a given moment. In 1959, Servo Ochoa identified this intermediary product as RNA, for which he won the Nobel Prize in Medicine. RNA is another type of nucleic acid found in cells, and it represents only one of several books in the huge DNA library of a cell. So when a cell needs to make a particular protein, instead of taking the DNA book out of the library (where it could get lost or destroyed), the information in the DNA book is transcribed into an RNA message known as mRNA (see Figure 13.9). The process that makes the message is known as transcription. When mRNA is made, scientists say that a particular DNA code (as represented by the symbols AATGGT, etc., in a linear sequence) is expressed.

At this point, the cell's protein manufacturing machinery, called a ribosome, can now decode the mRNA information and use it to make the protein or proteins that the cell needs at that moment (see Figure 13.10). The process of decoding or expressing the information carried by the mRNA is known as translation. Interestingly, even though every nucleus in an organism contains the same DNA library, cells only use part of that library based on their function in the organism. For example, a particular type of cell in the pancreas makes the protein insulin. In fact, they are the only cells in that organism that make this protein.

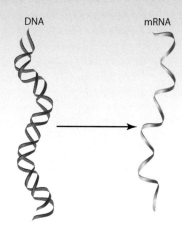

DNA   mRNA

Figure 13.9 | When DNA is expressed, mRNA is made based on the linear sequence found in the DNA. This process is called transcription.

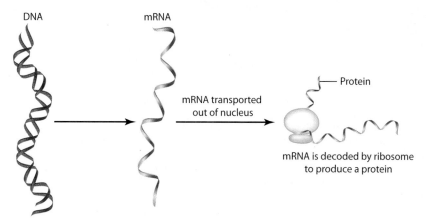

Figure 13.10 | In human cells, the mRNA is transported out of the nucleus and then decoded by protein-making machinery known as ribosomes. This process is called translation.

Just like a huge library, DNA is organized so that the information that is used all the time is found easily and other information not used is packaged away for safe keeping; in fact, this packaging is amazing in that, if you placed all of the human DNA in a cell end to end, it would be twelve feet long (see Figure 13.11).

In nuclei, just like in rows of bookcases, the DNA library is organized into collections called chromosomes. Chromosomes are composed of double-stranded DNA tightly packed and wound around proteins. The information is further organized into discrete packets known as genes, which are similar to books found on the shelves of the bookcases. Each gene is composed of a linear sequence of the four different nucleic acids and represents a book containing a set of instructions and blueprints for a particular protein. Genes are located on either strand of the double-stranded DNA and are found in the exact same location on a chromosome for a particular species.

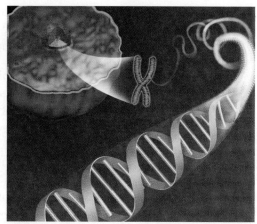

Figure 13.11 | In the nuclei of plant and animals cells, DNA is linearly arranged into genes, which are packaged on chromosomes. The amount of DNA that has to be organized and packaged in a human cell would be twelve feet long if stretched out end to end.

Because there are no nuclei in bacteria, the mRNA is translated as it is being made (transcribed). Because there are nuclei in animal and plant cells, the mRNA is made and transported to the decoding machinery. As a result, it takes longer for mRNA in humans to be translated than in bacteria. This is one of the factors contributing to the rapid growth of bacteria: Some bacteria can divide in two every twenty minutes. Biotechnologists take advantage of this fact and transfer DNA from other organisms, such as humans, into bacteria so that commercial products can be made swiftly and in large amounts. One example of such a product is human insulin. (See Figure 13.12.)

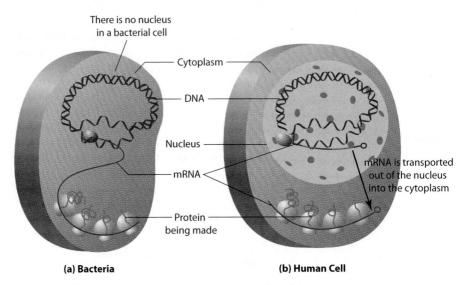

There is no nucleus
in a bacterial cell

Cytoplasm

DNA

Nucleus

mRNA

mRNA is transported
out of the nucleus
into the cytoplasm

Protein
being made

(a) Bacteria

(b) Human Cell

Figure 13.12 | Bacterial cells can make proteins much faster than human cells because transcription and translation can occur at the same time. In a human cell the mRNA must be transported out of the nucleus before it can be translated so transcription and translation cannot occur at the same time. Biotechnologists transfer genes from other organisms into bacteria so they can make a commercial product, such as human insulin, quickly and in large amounts.

Once biotechnologists learned to use cellular processes to make protein products quickly and in large amounts, products such as human proteins became available for scientific study or as commercial products. The next section reviews this ground-breaking technology.

Recombinant DNA: Development of the Technology In 1974, Herb Boyer of University of California, San Francisco (UCSF), and Stanley Cohn of Stanford University, applied for the first patent in recombinant DNA technology. Boyer and Cohn developed an rDNA procedure for the purpose of producing a commercial product. They put the human gene that "codes for" (contains the information for) insulin into bacteria for the purpose of commercializing human insulin. Because bacteria grow very rapidly compared the human cells, and because, in 1974, only bacteria cells could be grown in large fermentors, it was (and still is) very inexpensive to grow bacterial cells compared to other cells. Therefore, by putting the human gene that contains the code for insulin in a bacterium, these two scientists could cheaply make large amounts of this specific protein. Besides understanding the processes of transcription and translation, these two scientists also had to understand other cellular processes and use other proteins found in cells so they could harvest and manipulate the DNA found there. The additional technology included:

1. The discovery and purification of bacterial enzymes that cut DNA at very specific sites. These enzymes are known as the restriction enzymes, which are "molecular scissors" that are used to cut DNA in rDNA technology.

2. The discovery and purification of plasmids, DNA that can be engineered to "shuttle" DNA or genes from other species into bacterial cells.

3. The chemical technology to isolate intact DNA from any type of cell so that genes on that DNA can be cut out for the purpose of placing them in plasmids.

4. The discovery and purification of enzymes known as ligases that join cut DNA together.

Using the technology and molecular tools described above, as well as previously developed industrial fermentation technology, biotechnologists grow large amounts of genetically-modified bacteria, known as the recombinant bacteria, to produce large amounts of proteins (see Figure 13.13), such as human insulin.

This technology signaled a paradigm shift, in that biotechnology moved from traditional, fermentation-based techniques to molecular biology-centered techniques. At this point, scientists realized that they could engineer microorganisms to make commercial products and large amounts of proteins for scientific study. The patents acquired by Herb Boyer and Stanley Cohn eventually had more than 200 licensees—biotechnology and pharmaceutical companies—and earned Stanford University and the University of California at San Francisco more than $100 million in royalties. With the advent of rDNA and its commercial applications, the term "biotechnology" became commonly used by both the public and scientists.

Creating Guidelines for Recombinant DNA Technology The advent of recombinant DNA also sparked concern in scientists about the negative possibilities associated with this ground-breaking technology. Paul Berg, a biochemist at Stanford who was among the first to produce a recombinant DNA molecule in 1972, wrote a letter, along with ten other researchers, to the journal *Science*. In the letter, Berg and the other researchers urged the National Institutes of Health to regulate the use of recombinant DNA technology. They also urged scientists to halt most recombinant DNA experiments until they better understood whether the technique was safe. These concerns eventually led to the 1975 Asilomar Conference, where 100 scientists, newspaper reporters and governmental officials gathered to discuss the safety of manipulating DNA from different species (see Figure 13.14). The meeting resulted in a set of NIH guidelines. NIH has revised the document, "Guidelines for Research Involving Recombinant DNA Molecules," several times since 1976.

PCR Technology The development of Polymerase Chain Reaction (PCR) technology, in the mid-1980s, allowed scientists to both detect small amounts of specific DNA sequences, such as genes, and also duplicate them many times

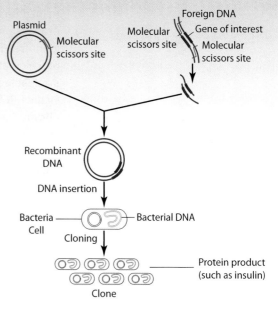

Figure 13.13 | The basic tool kit of the early rDNA biotechnologist contained several items: "molecular scissors" known as restriction enzymes that cut DNA at specific sites, plasmids, bacteria that carry the plasmids, and the DNA of cells from which the scientist isolated foreign pieces or genes. The foreign gene is placed onto plasmids and the plasmids are placed into bacteria. The bacteria divide and making lots bacteria containing the plasmid. The foreign gene on the plasmid is expressed and the foreign protein is made. Examples of gene products are human insulin and enzymes for cleaning clothes. Once harvested and purified, these products could be commercially sold.

Figure 13.14 | Some of the participants of the Asilomar Recombinant DNA Conference (from left to right) were Maxine Singer, Norton Zinder, Sydney Brenner, and Paul Berg. Thirty years ago, participants from different nations were engaged in debates about whether recombinant DNA research was too dangerous to be allowed to continue. Even today, as it continues to permeate people's lives, the public, governments, and researchers worry about the use of this technology.

Section 1 △ The Changing World of Biotechnology

to increase their DNA concentration so they could be genetically engineered into other organisms, such as bacteria. PCR technology means that a person can be identified from the smallest amount of DNA, even the amount resulting from touching an object or licking a postage stamp. For this reason, PCR technology is invaluable in forensics applications, where it is important to identify suspects or victims. PCR can also be used to identify genetic diseases in humans, such as cystic fibrosis; identify the presence of viruses, such as HIV; identify human and animal ancestors; and produce targeted DNA that can be put into plasmids to form rDNA.

You can consult several Web sites in order to watch PCR in action: Go to Bio-Rad Educational Resources, Dolan DNA Learning Center Animations and search for PCR animation, or just search the Internet using the phrase "PCR animation." (See also Figure 13.15.)

DNA Sequencing and the Human Genome DNA sequencing is the act of determining the linear sequence of DNA. Once DNA sequencing techniques were developed for individual cells like bacteria, scientists realized that they would be

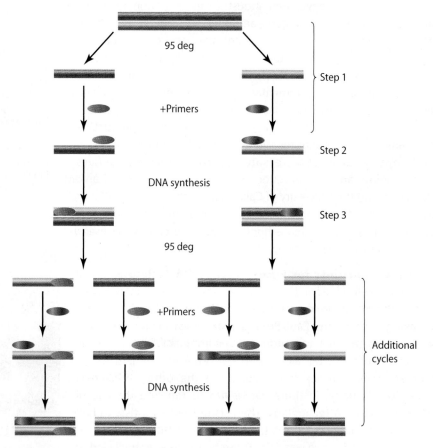

Figure 13.15 | This diagram illustrates the three-step process of DNA amplification by the Polymerase Chain Reaction (PCR).

STEP 1   (denaturation) unzips the double-stranded DNA (represented by red and blue lines) into two separate strands of DNA by heating the reaction mix to 95 degrees Celsius.

STEP 2   (annealing) isolates the DNA target region the researcher wants to multiply, by attaching 2 linear pieces of DNA known as **primers** (purple and peach pieces) that bind to either side of the target location and cooling the reaction.

STEP 3   (extension) involves heating the mix to 72 degrees at which point an enzyme, in the reaction mix, adds new DNA to the primers. In the figure, the new DNA is indicated by the upper red and blue bars attached to the primers. These three steps are repeated 25–40 times to produce millions of exact copies of the target region of DNA and each time it is reproduced is known as one **cycle**.

able to use automated computer technology to determine the DNA sequence of any genome, which is the entire DNA in a cell. This also meant that scientists could determine the sequence of genes that code for proteins that are made incorrectly, causing genetic diseases such as sickle cell anemia and cystic fibrosis. Of course, scientists also thought of sequencing the human genome, but they realized that it would be an enormous undertaking and wondered if it would be scientifically worthwhile. They were unsure whether the benefits of knowing this information would justify the cost to carry out the project.

Why was this considered such an enormous undertaking? Recall from Figure 13.11 that the DNA in every cell is about twelve feet long if it is stretched out end to end. You have also learned that DNA itself is made up of four different nucleic acids that vary because they contain four different nucleic acids adenine (A), thymine (T), guanosine (G), and cytosine (C). These nucleic acids are called "bases," and they occur in pairs because chromosomal DNA is double-stranded. The bases on the opposite strands actually pair with the bases found on the other strand, in a manner similar to the function of a zipper. Thymine is always paired with adenine and guanine is always paired with cytosine. The total number of base pairs in one human genome, in one cell, is approximately 2.0×10^7, or about 20 billion base pairs—an enormous amount of DNA. These 20 billion base pairs are divided up among forty-six chromosomes, or twenty-two pairs plus two sex chromosomes. By contrast, the bacterium *E. coli* that lives in your gut only has one chromosome for a total of 4.5×10^3 base pairs. You can see why sequencing the human genome (22 pairs + 2 sex chromosomes) was considered a huge task (see Figure 13.16).

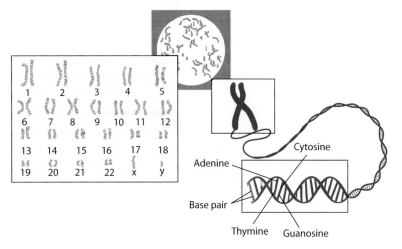

Figure 13.16 | The human genome is made up of 23 pairs of chromosomes; half from the father and half from the mother. One pair is known as the sex chromosomes (x and y); these chromosomes determine whether a person is going to be male, which is xy, or female, which is xx. Each chromosome is made up of double-stranded DNA packaged tightly together.

Scientists proposed the complete sequencing of the human genome, called the Human Genome Project (HGP), in 1986. The project began in October, 1990. It was an international effort that was projected to take fifteen years. The U.S. Human Genome Project (HGP) consisted of the Department of Energy Human Genome Program and the National Institutes of Health's National Human Genome Research Institute.

In order to sequence the staggering amount of DNA in the human genome, the project was divided up among different research groups in several countries and each group was assigned a chromosome or piece of a chromosome to sequence. The Human Genome Project identified approximately 30,000 genes in human DNA.

To learn more about the Human Genome Project, go to *http://www.genome.gov*. There are several educational sites that explain automated sequencing and show how it works. One of them is at the Dolan DNA Learning Center site: Search using the key phrase "cycle sequencing."

The DNA in your cells is not only the blueprint for your life; it also holds the history of your life because your DNA came from your mother and father and their ancestors. Through the Human Genome Project, scientists discovered that your DNA can tell you where your ancestors originated in the world. This has allowed scientists to trace the movement of ancient people worldwide. You can even purchase a kit to have your DNA traced (Figure 13.18).

Figure 13.17 | In law enforcement, forensic scientists use 13 different DNA sites in the human genome to determine identity. The identity of these sites are revealed by the technique DNA sequencing and can literally be "read" by scientists.

Figure 13.18 | Your genome, in every cell of your body, except sperm or egg, is composed of the same DNA and is the blueprint for your life. Your genome contains most of the blueprint of your parents and their parents; in short, your ancestors for thousands of years. People are now paying to have that blueprint investigated to clarify their ancestry.

Bioinformatics Due to the need for rapid storage and manipulation of sequence information, the Human Genome Project also led to the rapid development of sequencing software tools and gave birth to the field of bioinformatics. Bioinformatics is defined as the collection, organization, and analysis of large amounts of biological data, using networks of computers and databases. All of the DNA sequence information generated in the world can be found via several portals, accessible globally. In the United States, scientists typically visit the GenBank Web site at the National Center for Biotechnology Information (NCBI) to view these data (see Figure 13.19). GenBank is the DNA database for the National Institutes of Health (NIH) and is available for anyone to view. It is constantly updated.

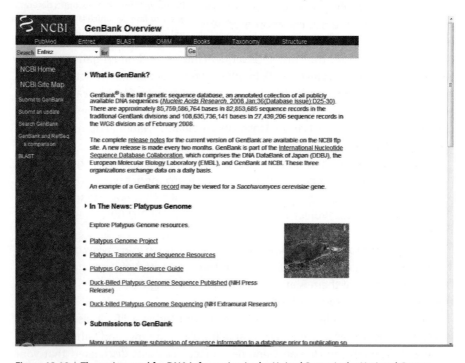

Figure 13.19 | The main portal for DNA information in the United States is the National Center for Biotechnology Information, or NCBI. The other two major portals are the European Molecular Biology Laboratory EMBL, in the United Kingdom and the DNA Database of Japan (DDBJ) Anyone can access this information.

Projects Related to the Human Genome Project Building upon the success of recombinant DNA technology, the Human Genome Project has ushered in the Genomics Age. In this time period, a large number of whole genomes in several species, including humans, are being sequenced, and numerous new government-funded projects, commercial ventures, and fields of study are starting, all as a result of the information generated by the Human Genome Project. If you want to learn what genomes have been recently sequenced, search the Internet for the Web site "Genome News Network," a site independently managed by the J. Craig Venter Institute.

Functional genomics is the study of how genes interact, what proteins they make, when they make them, and in what cells. For example, in a person who has cancer, genes that should not be expressed are expressed, while genes that are normally expressed are turned off, so it is important for scientists to understand what mechanisms and processes control gene expression. The HGP discovered many of these genes and their sequences.

As part of this research, scientists developed a special technique called microarray technology, which allows them to quickly analyze, in one test, which genes are expressed. For example, scientists might use a microarray analysis of gene expression in a person's cancer cells, comparing their cells to another person's normal cells isolated from the same area of the body such as the breast. Information from this test helps the physician determine the best drugs for treating the cancer. The test involves several steps: (1) Genetic material from the patient's cancer cells is isolated and marked with a fluorescent red dye. (2) Genetic material from individuals who do not have cancer is isolated from the same type of tissue and marked with fluorescent green dye. (3) Both groups are mixed and added to a prepared slide containing dots of genetic material, such that each dot represents a gene that is known to change as a result of cancer and any that they want to know if they change as a result of cancer. The results are then interpreted as follows. (1) If the red-labeled genetic material binds to a gene spot on the slide, then that spot will be red, indicating that this particular gene is expressed in the cancer cells. (2) If a green labeled genetic material binds, then that indicates that gene is active in normal cells. (3) If a gene is active in both cancer and normal cells, then the spot binds both fluorescently labeled DNA and appears yellow. See Figure 13.20 for a visual representation of this process.

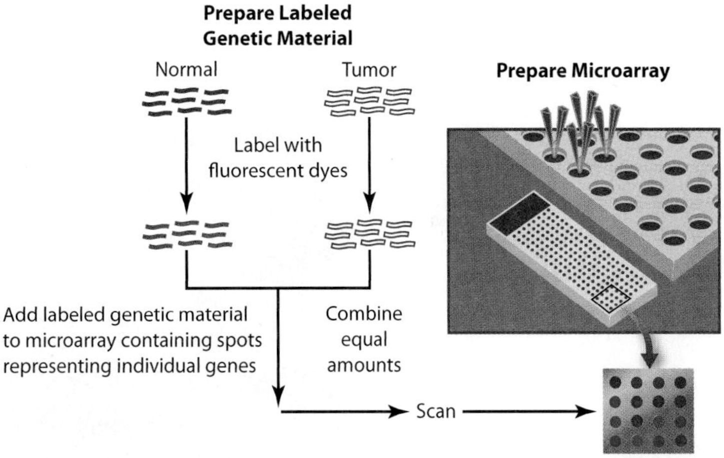

Scientists can use the knowledge they obtain from microarray technology to determine what genes are only active in cancers, and in what type of cancers, to help determine the best treatment. At the present, microarray technology is not suitable for diagnosis and treatment for the general public, but this may change as the technology is improved.

Figure 13.20 | Microarray technology is a new way of studying the expression of large numbers of genes. Genetic material, representing different genes, is placed in spots on a glass slide. Every spot represents a different gene, In this example involving cancer research, each spot represents a gene that physicians think are changed as a result of breast cancer. Genetic material is isolated from the patient's cancer cells and labeled with a red fluorescent dye. Then genetic material is isolated from normal cells in the same tissue and labeled with a green fluorescent dye. Both of these samples are added to the slide containing spots representing different genes. If a spot turns red, then that gene is expressed in that patient's breast cancer. If the spot turns green, then that gene is only active in noncancerous cells. If the spot turns yellow, then the gene is expressed in both cancer and normal cells.

Another new field that has come about as a result of the HGP is known as genetic anthropology. Genetic anthropology is an emerging discipline that combines DNA and physical evidence to reveal the history of ancient human migration. It seeks to answer the question, "Where did we come from, and how did we get here?"

One offshoot of the new interest in genetic anthropology is the Human Genome Diversity Project, which focuses on determining the ancestral lineage of individual people. This information is used to map human migration during ancient times and also collects DNA information that differentiates people. However, as previously stated in the caption for Figure 13.18, the largest difference is going to be how the DNA is expressed between individuals.

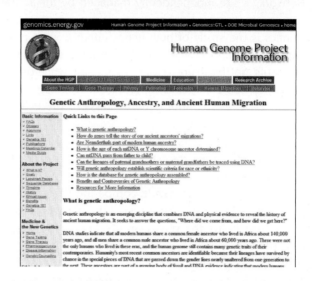

Figure 13.21 | Information for the public sector on the Human Genome Project; its associated projects, such as genetic anthropology; and its applications can be found at the Department of Energy Web site.

SECTION ONE FEEDBACK >

1. Define biotechnology and give examples of traditional biotechnology and molecular biology-based biotechnology.

2. Provide three examples of how historical events and paradigm shifts in technology led to biotechnological advances

3. Using the Internet, list two types of pasteurization and how they are different. Also list four types of pasteurized foods that you eat.

4. Define recombinant DNA (rDNA).

5. List three outcomes of the Human Genome Project and explain how one of these outcomes has revolutionized how scientists think about our genome or another animal's genome.

6. One ethical question that was hotly debated at the Asilomar Conference was, "Should genes or parts of genes be patentable, since a gene is a product of nature?" Look up the requirements for a patent and explain why you think Genentech was allowed to patent recombinant insulin.

7. Visit the Access Excellence Web site at *www.accessexcellence.org/AB*. Find a topic of biotech that interests you or has impacted your life. Write a short paragraph on that topic explaining why it interests you or how it has affected your life.

KEY IDEAS >

- Biotechnology has applications in such areas as genetic engineering, agriculture, pharmaceuticals, food and beverages, medicine, energy, the environment, and forensics.

- The field of genetic engineering focuses on inserting DNA from one organism into another organism.

- Biomimicry is the utilization of designs found in nature to benefit people.

Biotechnology has applications in such areas as genetic engineering, agriculture, pharmaceuticals, food and beverages, medicine, energy, the environment, and forensics. In this section you'll learn about some important uses, or applications, for biotechnology—for example, the genetic engineering of human insulin, the first commercially made recombinant DNA protein, and how rDNA techniques are being used to treat changes in genes that cause diseases such as cystic fibrosis. You will also learn how biotechnology is being used to clean up the environment, to produce energy, and to identify criminals and victims and to set free wrongly convicted individuals. Finally, you will learn about biomimicry, another way of copying nature's design and technology to improve human lives, a goal it shares with biotechnology.

Commercial Applications of Recombinant DNA

Important applications of biotechnology can be found in the field of genetic engineering. As you have learned, the field of genetic engineering focuses on inserting foreign DNA from one organism into another organism in order to create a commercially viable product. Since the resulting protein is made from rDNA, it is now known as a recombinant protein. A company will select and create a recombinant protein based on its commercial value. Medically important proteins can have tremendous value.

One of the first recombinant proteins on the market was recombinant human insulin. Insulin is a hormone that lowers the level of glucose (a type of sugar) in the blood. It is made by a particular type of cell in the pancreas and released into the bloodstream when the glucose level goes up, such as after eating. Insulin helps glucose enter the body's cells, where it can be used for energy or stored for future use. A person who is diabetic either does not produce insulin or does not properly respond to insulin so that glucose is not taken up by the body's cells. Complications of diabetes are kidney disease, nerve disease, blindness, amputation, and heart disease. According to the World Health Organization (WHO), 3 million people die from diabetes-related illnesses every year. Furthermore, the WHO predicts that the number of diabetics in the world will grow to approximately 360 million by the year 2030.

511

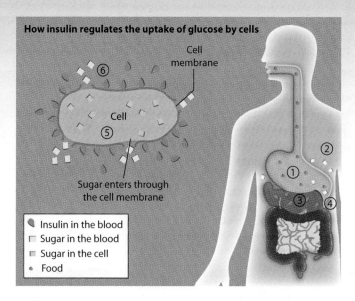

How insulin regulates the uptake of glucose by cells

Cell membrane

⑥

Cell

⑤

Sugar enters through the cell membrane

⬦ Insulin in the blood
▢ Sugar in the blood
▢ Sugar in the cell
• Food

②
①
③ ④

Figure 13.22 | (1) Carbohydrates in the stomach are broken down into sugars, such as glucose. (2) Glucose enters the bloodstream, increasing the level of glucose in your blood. (3) When your body senses an increase in glucose, it sends a signal to your pancreas. (4) The pancreas responds to the signal by making insulin, which is released into the bloodstream. (5) Insulin acts by allowing the glucose to go from the bloodstream into cells, such as muscle cells; this decreases the concentration of glucose in the bloodstream. (6) The level of glucose drops in the bloodstream; if it drops too far, you get sleepy. In diabetic patients it can drop so low they go into a coma.

Recombinant human insulin (rhInsulin) was made first by Genentech and approved by the U.S. government for public use in 1982. Before the human gene for insulin was cloned, it took seven to ten pounds of pancreas from approximately seventy pigs or fourteen cows to purify enough insulin for one year's treatment for a single diabetic. In addition, diabetics had to worry about having adverse reactions to nonhuman insulin. This is no longer the case, thanks to the development of recombinant human insulin. (See Figure 13.22.)

Case Study: rhInsulin as a Commercial Product Because diabetes is such a serious and widespread genetic illness, it was clear that choosing insulin as the first human protein to be genetically-engineered for mass production would have tremendous economic and social benefits. Just as with other commercially viable products, the company had to research the product and the market to make sure that rhInsulin would both make money and be a major benefit to a large segment of the worldwide human population.

But there are many such proteins. Why should this particular protein be selected for genetic engineering? First, it is a very small protein, making it easier to genetically engineer. Secondly, unlike other proteins, it is composed of one unit instead of multiple units, so it was reasonable to expect that it would self-assemble correctly in bacteria.

Once the research and development group in the company determined how to genetically-engineer it they also had to figure out how to mass produce it. They did this on a small scale, a process called small-scale manufacturing. After the company worked out all of the manufacturing problems, they scaled up the process and transferred it to another part of the company known as large-scale manufacturing. (See Figure 13.23.)

Early on in the process, the company applied for a patent and also applied to the appropriate U.S. governmental agencies for pharmaceutical use of this product. To

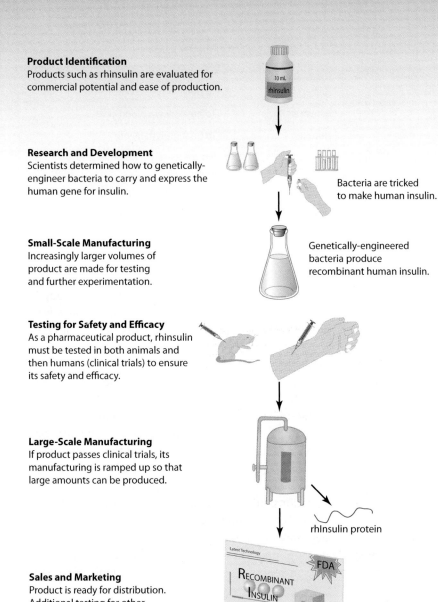

Product Identification
Products such as rhinsulin are evaluated for commercial potential and ease of production.

Research and Development
Scientists determined how to genetically-engineer bacteria to carry and express the human gene for insulin.

Bacteria are tricked to make human insulin.

Small-Scale Manufacturing
Increasingly larger volumes of product are made for testing and further experimentation.

Genetically-engineered bacteria produce recombinant human insulin.

Testing for Safety and Efficacy
As a pharmaceutical product, rhinsulin must be tested in both animals and then humans (clinical trials) to ensure its safety and efficacy.

Large-Scale Manufacturing
If product passes clinical trials, its manufacturing is ramped up so that large amounts can be produced.

rhInsulin protein

Sales and Marketing
Product is ready for distribution. Additional testing for other applications is done.

Figure 13.23 | rhInsulin was first developed in the research and development (R&D) part of a company. R&D also figures out how to make large amounts of the recombinant protein. The manufacturing part of the company produces large amounts of the product, in this case rhInsulin. Quality assurance (QA) tests it to make sure it is produced to specifications. The fill and packaging department ensures that it is correctly labeled, filled, and packed. Since it is a pharmaceutical product, besides Good Manufacturing Practices the company must comply with FDA regulations.

be used as a pharmaceutical product by humans, rhInsulin had to be tested first in animals and eventually on a small scale in humans. This last series of tests is known as clinical trials. After the product passes clinical trial testing, it is released to the market. But even at that point, information on side effects is collected from patients and doctors to ensure product safety.

Other types of enzymes and protein products were also inserted into single-celled organisms so they could be mass produced. Genencor was the first company to develop and mass produce an engineered enzyme to be used in detergents. As indicated in Table 13.2, today companies like Genencor produce a variety of recombinant products.

Table 13.2 | Two Biotech Companies and Examples of Their Products

COMPANY	PRODUCTS	USE
Genentech	Herceptin®	To prevent breast cancer
	Activase®	To break down blood clots
	Nutropin®	To treat growth failure
Genencor	Purafect® Purastar®	Enzymes used in detergents to digest stains
	Puradax®	Prevent fuzz (piling) in fabrics
	Stargen™	To produce fuel ethanol

Applications in Medicine and Pharmaceuticals

You have already read about one medical application for biotechnology—the development of recombinant proteins, such as recombinant insulin. Other advances in biotechnology will soon affect how doctors and patients handle all types of prescription drugs.

If you are like the overwhelming majority of people in the United States, at some point in your life you have taken a prescription drug. Although your doctor may have considered your medical history when he or she selected the drug, it is not likely that the doctor could fully predict how you would react to the medication before you took it. In fact, because of inherited variations in your genes, your ability to metabolize any given drug and the side effects you may experience from that drug differ greatly from those of other people. But thanks to advancements in biotechnology, doctors will increasingly have the ability to prescribe medications, adjust dosage, and select treatments on the basis of the patient's personal genetic information. The Human Genome Project and related projects have provided the DNA-sequence information necessary to make this possible. In turn, microarray technology will make it possible to determine how each person reacts to different drugs, making personalized medicine a reality.

Other medical and pharmaceutical applications for biotechnology include the previously mentioned PCR-based diagnostic technologies and the possibility of actually fixing genetic defects.

TECHNOLOGY IN THE REAL WORLD:
Treating a Genetic Disease with Biotechnology

Cystic fibrosis (CF) is the most common genetic (inherited) disease in the United States today. There are about 30,000 people in the United States who are affected with the disease: about 1,000 babies are born with CF each year. CF occurs mainly in Caucasians, who have a northern European heredity, although it also occurs in African Americans, Asian Americans, and Native Americans. Approximately 1 in 31 people in the United States are carriers of the cystic fibrosis gene. These people are not affected by the disease and usually do not know that they are carriers.

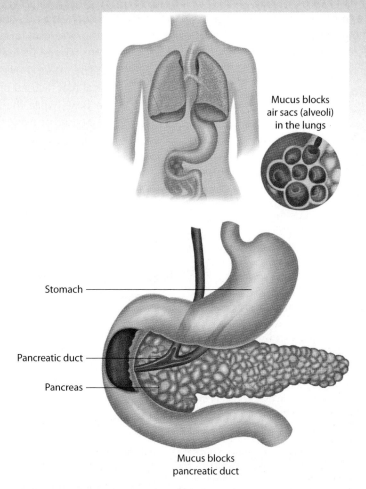

Mucus blocks
air sacs (alveoli)
in the lungs

Stomach

Pancreatic duct

Pancreas

Mucus blocks
pancreatic duct

Figure 13.24 | The genetic disease cystic fibrosis causes the accumulation of a thick mucous layer in the lungs and intestines. As a result, the person has a predisposition to bacterial lung infections and an inability to properly digest foods. The lung infections produce scarring and decreased lung function, eventually resulting in congestive heart failure.

Cause and Symptoms

CF is caused by a change in a gene product. The normal expression of the gene produces a protein that transports chloride in and out of cells. This protein, known as the chloride transporter (CFTR), is found on the surface of the cells that line the lungs and other organs. The defective CF protein is unable to transport chloride properly, and, as a result, the amount of sodium chloride (salt) is increased in bodily secretions, and the secretions are thick and sticky. This is why the skin of babies with this defect tastes salty.

Common symptoms vary, but, in general, CF patients develop major respiratory infections as a result of the thick secretions; have difficulty digesting foods because the thick secretions block the pancreas, which prevents the release of digestive enzymes; and, if male, are usually sterile due to the lack of secretions. Even though supportive care, such as antibiotics, has increased the quality and quantity of life for these patients, the average life span is thirty-five years. (See Figure 13.24.)

Research and Treatment

CF research has accelerated sharply since the discovery of the effects of a defective chloride transporter in 1989. In 1990, scientists successfully genetically engineered a gene that corrected the defective chloride transport mechanism and added it to CF cells in the laboratory. Clinical trials were initiated with an aerosol form of the corrected DNA packaged inside. Test subjects breathed in the aerosol DNA. The idea was that the corrected DNA would merge with test subject's own cells, providing the necessary information to correct the defect. The first clinical trial focused on fixing cells in the nose, while the second clinical trial focused on fixing the cells in the lungs. In both cases, the clinical trials were successful, and symptoms of the disease disappeared from the treated area. (See Figure 13.25.)

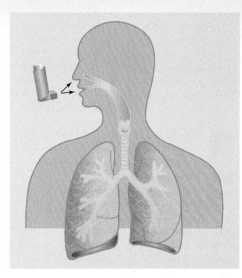

Figure 13.25 | In cystic fibrosis treatment, the defective gene is replaced in the cells lining the inside of the nose and lungs; membrane-bound DNA-containing packets that fuse with cells are delivered using an inhaler. This treatment fixes the cells lining the nose and lungs, but does not cure the disease since the "fixed" cells eventually die and are replaced with defective cells. To fix the error permanently, the gene would have to be replaced in all of the cells of the developing fetus

So far, the treatment seems promising, but before it can be released to the general public, some questions need to be answered: How long will the positive effects of a treatment last? Are there any long-term negative side effects of using the membrane-bound DNA delivery system? Is there any way to use this technology to fix defective cells elsewhere in the body, such as the digestive tract? How much will this treatment cost?

It is likely that this treatment will be released to the public in the next five years. If it can prevent lung failure, the leading cause of death in CF patients, their life spans may increase considerably.

Environmental Applications

Biotechnology is used to clean up the environment, a process known as bioremediation. For example, microorganisms are used to clean up oil spills. Through biological processes, plants and microorganisms can break down oil into carbon dioxide and water, convert dangerous pesticides into something less toxic, remove harmful contaminants such as radioactive elements from the environment, or change harmful contaminants into a form that can be easier to remove (see Figure 13.26).

Figure 13.26 | A process called phytoextraction is one way researchers are beginning to use plants in bioremediation. Some plants, such as sunflowers, can pull lead, uranium, arsenic, and other contaminants out of the soil. The plants can then be harvested and safely destroyed, a new crop planted, and the process repeated until the soil is sufficiently recovered.

One of the worst oil spills in history occurred in 1989, when an oil tanker named the Exxon Valdez ran aground off the coast of Alaska. Both physical and biotechnological methods were used to clean up the oil (see Figure 13.27). Examples of physical methods were oil booms, which were used to contain the oil and keep it from spreading. Birds and other sea life were also cleaned by hand. These methods were useful but only cleaned up a small fraction of the oil. Biotechnological methods were developed to clean up the majority of the spill or enhance the effectiveness of physical methods. The biotechnological method was the use of bacteria that was genetically engineered to degrade (break up) oil. Bacteria that were naturally found in the sand of the Alaskan coast would have eventually cleaned up the oil, but the genetically modified bacteria did it much faster. The only limitation to using bacteria to clean up oil was that the bacteria need oxygen to breakdown the oil and survive. This meant that the bacteria could only clean up oil on the surface of the shoreline, which was open to the air (and thus to oxygen). Oil that had seeped into the ground was not affected by the bacteria.

While many natural organisms already clean up the environment, they work slowly, so researchers and scientists continue to study these processes in order to develop genetically-engineered organisms that clean up faster and more completely.

Energy Applications

With known oil reserves diminishing and petroleum prices rising, biofuels have been proposed as an answer for energy. Biofuels are fuels that are produced from organic (living) compounds. Two types of biofuel are fermented corn or algae. Gasses produced by microorganisms are also a viable biofuel. Methane, which is a major component of natural gas, is produced by certain microorganisms in environments in which oxygen is absent. The largest source of methane on earth is methane hydrates, a combination of methane and water found on the ocean floor, produced by microorganisms at the bottom of the ocean. If we could economically harvest these methane hydrates, we would have a long-term source of energy. (See Figure 13.28.)

Hydrogen is another potential fuel derived from microbes. It is an attractive fuel alternative since its combustion produces only water and energy. Several microbial species, an alga, *Chorella* and a bacterium, *Clostridium*, can produce hydrogen gas for prolonged time periods, and more species are being screened from genomic data bases for the enzymes capable of hydrogen production. You will learn more about the future of biofuels in Chapter 14, when you study chemical technology.

Applications of DNA Identification

As you have probably seen on police shows on television, DNA technology is often used to identify criminals or victims based on a small sample of hair, blood, or other organic substance (see Figure 13.29). People who use fingerprints, hair, footprints, blood, and DNA samples to identify people are known as forensic scientists.

Any type of organism can be identified by examining DNA sequences unique to that species. However, identifying individuals within a species is less precise. As part of the President Initiative to provide DNA technology for the purpose of identifying people, forensic scientists developed DNA identification testing for thirteen human DNA sites that are known to vary from person to person. This information is used to create a DNA profile of that individual. Statistically, there is an extremely small chance that another person has the same DNA profile for all thirteen particular sets of DNA sequences (see Figure 13.30). Information on these thirteen sites can be found at the CODIS Web site, managed by the FBI. The CODIS Unit manages the Combined DNA Index System (CODIS) and the National DNA Index System (NDIS) and provides the CODIS

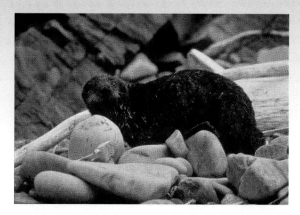

Figure 13.27 | Three days after the tanker *Exxon Valdez* ran aground off the coast of Alaska, oil started to pool in the rocks on the shore and show up on wildlife such as otters. Genetically engineered bacteria were used to help clean up what turned out to be one of the worst environmental disasters in history.

Figure 13.28 | Another alternative energy source is hydrogen gas. Algae can be grown to produce it.

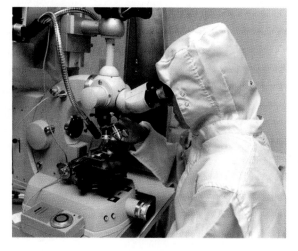

Figure 13.29 | A forensic scientist examines evidence with an optical microscope. By courtesy of U.S. Department of Energy.

Section 2 △ Applying Biotechnology

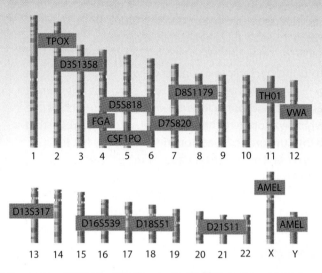

Figure 13.30 | The system of DNA identification used by the FBI requires polymerase chain reaction (PCR) amplification of areas of human DNA in which a short 4-base long sequence is repeated over and over. Such sequences are called "short tandem repeats," or STRs. The FBI has named this system CODIS for Combined DNA Index System.

Program to federal, state, and local crime laboratories in the United States. It also works with international law enforcement crime laboratories to exchange and compare forensic DNA evidence from violent crime investigations.

DNA identification can be used for a variety of purposes in forensics and family law. Among other things, it can be used to:

- Identify potential suspects whose DNA may match evidence left at crime scenes
- Exonerate persons wrongly accused of crimes
- Identify crime and catastrophe victims
- Establish paternity and other family relationships

DNA identification is also used in other fields, including wildlife conservation, medicine, and agriculture. For example, DNA identification can be used to:

- Identify endangered and protected species as an aid to wildlife officials (could be used for prosecuting poachers, see Figure 13.31)
- Detect bacteria and other organisms that may pollute air, water, soil, and food
- Match organ donors with recipients in transplant programs
- Determine pedigree for seed or livestock breeds
- Authenticate consumables, such as caviar and wine

In addition, people who work in wildlife protection use PCR-based technology to identify illegally obtained protected-animal parts for the prosecution of poachers and dealers who would sell these parts.

Figure 13.31 | Humpback whales are a protected species, however, they are still hunted illegally and their meat shows up in markets, especially in Asia. Earthtrust's Whale DNA Project examines DNA samples from whale meat found in markets to determine if it comes from humpback whales.

Applications Inspired by Biomimicry

Biomimicry is the use of designs found in nature to benefit people. It is based on the idea that the process of natural selection has already produced successful and environmentally sound design solutions and that people simply need to recognize and adopt these designs in order to produce environmentally safe and energy efficient buildings and commercial products. Biomimicry is not biotechnology because it is not a manipulation of nature. However, similar to biotechnology, it does take its inspiration from nature to solve human problems.

Figure 13.32 | The Eastgate Centre building in Harare, Zimbabwe, is modeled after tall termite mounds that are naturally cooled by the movement of air throughout the mound. As hot air in the mound rises, it draws cool air in from the bottom.

For example, Eastgate Centre, a shopping center and office block in Harare, Zimbabwe, has a mechanical cooling system made up of vents and flues that remove hot air from the structure. The design of this system is based on the natural cooling mechanism found in giant termite mounds (see Figure 13.32). Scientists studying these mounds found that as the hot air rises out of them, cooler air is drawn in at ground level, keeping the mounds cool. The temperature inside a termite mound hovers around 31° C all day and all night, while the temperature outside the mound can vary between 3° C and 42° C. Thanks to a building design that mimics the cooling system found in a termite mound, Eastgate Centre uses only 10 percent of the energy of a conventional building its size, leading to a projected saving of 3.5 million dollars in air conditioning costs in the first five years. As a result, it can rent space for 20 percent less than a newer building next door.

PureBond®, a type of adhesive, is also a product of biomimicry. Unlike most adhesives, which are made from formaldehyde (a potentially lethal chemical), PureBond® is formaldehyde-free. It was inspired by the ability of mussels to cling to rocks despite the constant pounding of ocean waves. Kaichang Li, an associate professor at Oregon State University's College of Forestry, was crabbing off of the Oregon coast when he noticed how mussels could withstand the impact of waves breaking on the shore (Figure 13.33). Li and his team decided to replicate the adhesive by blending protein-rich soy flour containing the same amino acids used by the shellfish. Recognizing the potential value of such a product, a company named Columbia Forest Products offered to fund Li's research. PureBond® is more expensive than formaldehyde-based, traditional adhesives but it is stronger and more water resistant, cures rapidly, and meets California's strict regulations for formaldehyde emissions in composite wood.

Figure 13.33 | Mussels being pounded by waves provided the idea behind new glue.

Another example of biomimicry has resulted in products that make it possible to keep buildings, and other items that are exposed to weather, clean without the use of harsh cleaning agents or power washing. Scientists noticed that the surface of plants and insect wings are not smooth, but dimpled and covered in a waxy substance. They found that, unlike smooth surfaces, which can cause water and dirt to spread and cling, surfaces covered with these microscopic structures stayed relatively clean. Dirt rests lightly on these microscopic structures and is washed away when it rains (see Figure 13.34).

Several companies have created products based on this self-cleaning phenomenon, including self-cleaning roof tiles by a company named Erlus and a glass coating system by a company named Ferro. Another product based on this form of biomimicry is Lotusan®, an exterior building paint developed by a company named Sto. All three of these products were developed to make maintenance easy: Surfaces can be cleaned without the use of detergents or sandblasting. Plus, according to the Sto® Web site, Lotusan® resists the growth of algae, mold, and mildew. Products like these may lead to the development of other products with dirt-repelling properties, such as pipes that resist clogging. The possibilities seem wide-reaching, and they greatly reduce the use of harsh, environmentally damaging detergents.

Figure 13.34 | The leaves of lotus plants have a self-cleaning property. Their microscopic structure and surface chemistry mean that the leaves never get wet. Rather, water droplets roll off a leaf's surface like mercury, taking mud, tiny insects, and contaminants with them.

SECTION TWO FEEDBACK >

1. Search the Internet to find out how oil spills are cleaned up by U.S. government agencies. Write a short paragraph explaining the steps these agencies take when a large oil spill is discovered off U.S. coasts.

2. Design an advertising brochure for Dog DNA testing, entitled "Who Is Your Dog?" The brochure should target the general public in terms of information and include a basic description of how the test is done. There are companies that already do this for both humans and animals, so searching the Internet will provide information to help you design this brochure.

3. Define biomimicry and describe one example of how it has been used.

4. Search the Internet for the term "biomimicry," and find an example not discussed in this text of how it is being used to benefit people.

5. Go to the Dolan DNA Learning Center DNAi site (*http://www.dnai .org*) and review the information on the Romanov family. Write a short paragraph on how DNA was used to identify them.

KEY IDEAS >

The biotechnology work place is found in companies, colleges, and governmental agencies. Companies focus on products or services that make money; college labs focus on the advancement of knowledge and the training of students; and governmental agencies and governmental agencies usually utilize biotechnology in the performance of a service.

- While an individual drug company identifies and manufactures a new product, bringing a product to market is actually a collaboration of several companies and governmental agencies.

- Biotechnology products are subject to regulation by multiple government agencies.

- Whether in an academic or industry lab setting, people are expected to adhere to certain ethical standards.

The biotechnology work place is found in companies, colleges, and governmental agencies. Companies focus on products or services that make money; college labs focus on the advancement of knowledge and the training of students; and governmental agencies usually use biotechnology in the performance of a service.

Biotechnology Companies

Stages in a company's drug manufacturing process include product identification, research and development, manufacturing and quality assurance, and sales and marketing. First, a company surveys the marketplace and identifies the product it wants to make. Second, the research and development group, known as R&D, determines how to make the product. If it is a pharmaceutical product, R&D must prove that the product is safe to ingest and is effective without serious side effects. R&D also determines how to manufacture large amounts of the product, creating a series of rules, or protocols, that explain exactly how to manufacture the product. This is also known as determining the correct product synthesis procedure; not only should it work, but it should also be cost-effective. Third, the manufacturing department is responsible for following the protocols established by R&D for the large-scale synthesis of the product.

Fourth, the Quality Assurance (QA) department monitors the process to make sure that the product is made according to specifications. In the United States, the means by which QA monitors the manufacturing process is usually determined by federal regulations in addition to internal company guidelines. Other departments in a biotechnology company include marketing (which is responsible for making the product attractive to consumers and disseminating information about the product) and technical services (which takes calls from customers when they have questions or problems with products).

Biotechnology companies fall into one of the several categories listed in Table 13.3.

Table 13.3 | Types of Biotechnology Companies

TYPE OF COMPANY	ACTIVITY
Medical/Pharmaceutical	Produce medicines, either natural or recombinant, from plants, animals, and fungi; vaccine, gene therapy, prosthetics, engineered organs, and tissues
Agricultural	Involved in transgenic or genetically engineered plants and animals, breeding of livestock and plant crops, aquaculture (animals or plants grown in water), and marine (ocean) biotech, horticultural products, production of plant fibers (for example, plant materials used for textiles) and biofuels, pharmaceuticals in genetically engineered livestock and plants
Industrial	Produce fermented foods and beverages, genetically engineered proteins for industry, DNA identification/fingerprinting of endangered species, production of plastics made from live organisms (biopolymers) the use of organisms to detect toxins (i.e., biosensors), and the use of organisms to clean up pollution (bioremediation)
Contract Manufacturing	Produce genetically engineered materials on a large scale; Includes research and development (R&D) to produce the materials
Contract Analytical Service	Conduct quality testing on a range of materials or products (e.g., purified water, air, final products, wastewater, raw materials)
Research Reagents and Kits	Make kits for university, company, and federal labs doing research or testing materials, such as in analytical services
Sequencing and Data Management	Collect, assemble, store and distribute sequence information whether it is DNA, RNA, or protein data.
Diagnostics	DNA sequencing, genetic testing and screening; DNA identification, forensics, tests using antibodies (proteins that animals make to identify and protect against foreign proteins and pathogenic organisms), and other molecular biology techniques (e.g., using PCR)

Not all biotechnology companies make products with a serious, life-saving purpose in mind. Some make life-style drugs (e.g., drugs that do not treat a disease but make a person look better, such as promoting hair growth in bald men) or even pets. How would you like to have your own genetically engineered pet? How about a fish that glows (see Figure 13.35)? Yorktown Technologies, LP has mass produced the GloFish ™. Look up this item on the Web to learn how you can purchase your own glow fish. The gene responsible for the light came from a species of jellyfish. Do some Web research and then see if you can outline the steps that the company used to make the GloFish™.

Figure 13.35 | A gene isolated from jellyfish produces a fluorescent green color. To get other colors such as red, the gene is altered. Originally the gene was used in research to follow the movement of other recombinant genes or to tag abnormal cells, such as cancer cells. Now, this gene and its altered forms are available in fish, producing neon-colored fish.

The Product Pipeline

While an individual drug company identifies and manufactures a new product, bringing a product to market is actually a collaboration of several companies and governmental agencies. The process these organizations use to create a drug is commonly called the product pipeline. Because there are so many organizations and processes involved, the cost of producing a new drug is very high. The Pacific Research Organization reported that in 2006, it cost between $800 million and $1 billion to bring a drug to market. The generic drug development cycle is outlined in Figure 13.36.

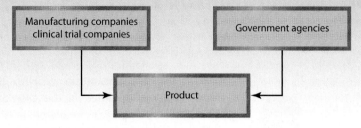

Figure 13.36 | Companies and governmental agencies collaborate to produce a product. This collaboration defines the drug-product pipeline.

Many drugs fail the safety and efficacy tests known as clinical trials. Clinical trials are conducted under the strict guidelines of the federal government's Food and Drug Administration (FDA). The three rounds of clinical trials test the drug, each time with a larger group of people. Even so, the largest group of people being tested will always be the general public when the drug is released to the market. Sometimes serious side effects do not show up until a drug is released. For example, a drug might cause a deadly side effect in one out of a hundred thousand people. Such a rare side effect might not occur during the comparatively small clinical trials, but can occur after the drug is released to a market larger than a hundred thousand people.

Regulation of Biotechnology Products Manufacturing companies establish their own regulations, known as current good manufacturing practices (cGMPs), and this is viewed as the minimum amount of regulation. These regulations are developed by company employees and made available in writing to anyone who wants to know what a manufacturing company does to ensure a quality product. For example, a biotechnology manufacturing company that makes research kits would most likely establish cGMPs for each product made. It can be as simple as how much information about a product is provided on its label, to as complex as how each product is produced, including how and when it is tested during production. If the product could possibly harm the environment, then federal agencies, such as the Environmental Protection Agency (EPA), become involved in its regulation. If the product is a pharmaceutical, then the Food and Drug Administration (FDA) becomes involved in its regulation. The FDA performs regular audits of the company to ensure that guidelines are being followed. In short, the exact federal agency assigned to oversee, or regulate, a particular product depends on the nature of the product and its intended use. In general, additional layers of regulation are added the closer a product comes to being used by humans. Research kits used by scientists in academic institutions are the least regulated and pharmaceuticals are the most regulated. These regulations do not necessarily apply to products sold overseas; as other countries have their own agencies.

As you have learned, if the product is a recombinant DNA plant or animal (for example, a recombinant corn plant that carries a pesticide-coding gene), then the Environmental Protection Agency (EPA) is involved as the product may harm the environment. The EPA is a federal agency that is charged with safeguarding human health and the natural environment. The United States Department of Agriculture (USDA) also becomes involved when a recombinant plant or animal is going to be ingested by people or animals. The USDA is a federal agency that develops quality standards for food and animal products; safeguards human, animal, and plant health; promotes the export of U.S. agriculturally produced products; and ensures the quality and safety of imported agriculture products. If the recombinant plant or animal is producing a pharmaceutical product that is to be used by humans, then the FDA is involved. All companies, including companies outside the biotechnology field, must follow regulations set down by the Occupational Safety and Health Administration (OSHA). These regulations safeguard employees working at companies. In the case of biotechnology companies, they particularly protect employees who work in laboratories. For example, one standard OSHA regulation requires all employees to wear closed-toed shoes and eye protection when working in a laboratory.

As described earlier, biotechnology products can be subject to simultaneous regulation by multiple government agencies. For example, biotech researchers at the Boyce Thompson Institute for Plant Research in Ithaca, New York, are studying ways to fortify raw foods like bananas and potatoes with vaccines to provide painless, inexpensive protection against disease. For example, the hepatitis B vaccine has been placed in bananas (see Figure 13.37), so people just have to eat that fortified fruit to develop protection against or immunity to hepatitis B. Three agencies are involved in regulating this product: the FDA, the EPA, and the USDA. Since vaccines are pharmaceutical products, the FDA is involved; to make sure the plant or fruit cannot harm the environment, the EPA is involved; and to ensure that the growing of the plant and its distribution meets specific standards, the USDA is involved.

Figure 13.37 | Bananas are being genetically engineered to carry a vaccine. All a person has to do is eat such a product and he or she is protected against a certain disease.

ENGINEERING QUICK TAKE

Conversions Within the Metric System and Preparing Dilutions

A) Conversions

Biotechnology laboratories use metric measurement units. Usually volumes are measured in liters, milliliters, and microliters, and distance is measured in centimeters and millimeters. After working in a laboratory for a while, you will be able to easily visualize the units that you use every day, and convert from one unit to another (e.g., microliters to milliliters). Use the Larger-Smaller Rule for conversions within the metric units. According to this rule, if you are going from a smaller to a larger unit, then you move the decimal to the left. If you are going from a larger to a smaller unit, then you move the decimal to the right.

For example, if you are converting from meters (m) to millimeters (mm), then you are going from a large unit to a small unit. Therefore, you need to move the decimal to the right (multiplying). Here is one conversion: 2.35 liters is equal to 2,350 milliliters, or, if you know scientific notation, 2.35×10^3 milliliters.

Now, try going the other way. If you are converting from millimeters (mm) to meters, you are going from a small unit to a large unit. Therefore, you need to move the decimal to the left (dividing). Here is a sample conversion: 534 milliliters is 0.534 liters or, in scientific notation, 5.34×10^{-1} liters.

Another technique for performing conversions is called cancellation of units. Using this technique, you can convert from larger to smaller volumes and vice versa simply by canceling the units, leaving the final number in the desired units as shown below. In this case, you want to convert liters to milliliters, so only milliliters or mL should be the units left in the answer.

$$\frac{2.35 \, \cancel{L} \times 1000 \text{ mL}}{1 \, \cancel{L}} = 2.35 \times 1000 = 2,345 \text{ mL}$$

$$\frac{2,345 \, \cancel{mL} \times 1 \text{ L}}{1000 \, \cancel{mL}} = \frac{2,345 \text{ L}}{1000} = 2.345 \text{ L}$$

Milliliters (mL) cancel out, leaving the unit Liter (L) in the answer.

Other common volume measurements in biotech labs are milliliters (mL or 10^{-3} L), or microliters (uL or 10^{-6} L). Biotechnologists have to be able to quickly and accurately convert between such measurements.

B) Preparing Dilute Solutions from Concentrated Solutions

Besides conversions, you must also be able to make solutions (a mixture of a solvent (the larger volume of liquid) and a solute (the smaller liquid volume or a solid) for your experiments, which means you will have to know some applied chemistry. One type of common solution made in the laboratory is one from a concentrated stock solution. It saves the biotechnologist time and space to have several stock solutions already prepared, so that all that is necessary is to do a quick dilution of one to prepare a solution for a project or experiment. For example, it is common to have on hand 10X stock solutions that are ten times more concentrated than normally used. To prepare a 1X solution of a particular volume, the biotechnologist uses the following equation:

The $C(i) \times V(i) = C(f) \times V(f)$ Equation where i = initial values; f = final values; C = concentration; and V = volume.

How does this equation work? Suppose you have a 10X stock solution, and you want to make up 100 mL final volume of a 1X solution. You need to figure out how much of the stock solution you need and how much water to use when diluting it in order to get the correct final concentration and final volume.

You know: C(i) = 10X, C(f) = 1X and V(f) = 100 mL. What you do not know is V(i), or the amount of the concentrated solution to use. The amount of water will be the rest of the volume, such that V(i) + amount of water for the dilution = 100 mL. Now, plug in your values and solve for the unknown value, as follows:

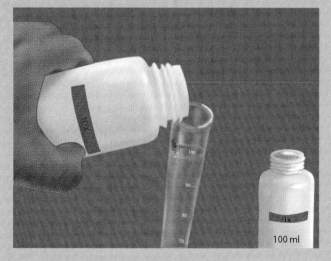

1. $(10X) \times$ (unknown V(i) = (1X) $\times$ (100 mL)
2. To solve for the unknown value, divide both sides by 10X (removes 10X on the left side of the equation).
3. Unknown V(i) = (1X) $\times$ (100mL)/(10X)
4. Cancel Units in the denominator and numerator (cross out X's).
5. Solving for the value, you get 10 mL of 10X solution.
6. 10 mL of 10X stock + ? water = 100 mL so the answer is 90 mL of water. (See Figure 13.38.)

It is very common to have to make a solution and do a unit conversion at the same time. For example, how would you prepare a 1X solution final volume = 1.0 mL starting with a 5X solution of 1 L?

Figure 13.38 | To save time and space in the laboratory, it is common for biotechnologists to make up concentrated solutions, similar to the concentrated solutions of orange juice you can purchase at the grocery store. Then when they want to make a solution they remove the correct amount and dilute it by adding purified water the same way you would dilute a concentrated juice. The solution in this picture, labeled 10X, is 10 times more concentrated than normal.

Bioethics Whether it is an academic or industry lab setting, biotechnology personnel are expected to adhere to certain ethical standards. In the biotechnology industry, it is common to refer to ethical standards as bioethics. You have already read about the Asilomar Conference, where scientists voluntarily came together to discuss the ethical implications of genetic engineering and passed guidelines for research in this area. Likewise, the scientists who organized the Human Genome Project realized that the genetic information and new technologies generated by the HGP would raise critical bioethical questions, including:

- Is it right to use fertilized embryos for stem cells?
- How careful do we have to be about protecting information gleaned through genetic testing?
- Should we test drugs meant for humans on animals first?
- Is there anything wrong with genetically modifying crops and livestock?

In order to address these and other questions, 3 percent of the HGP's annual budget was devoted to the study of ethical, legal, and social issues surrounding the availability of genetic information. People have strong feelings about such issues, and governments have passed laws reflecting these feelings. For example, the distribution of genetically modified seed was banned by a number of countries, and opponents charge that such bans will undermine those countries' efforts to end hunger.

How are bioethical questions answered? Answers to these questions cannot necessarily be tested in an experiment; many of the questions raised in bioethics are subjective. Plus, the answers can vary, depending on a person's age or personal experiences. So it is important for every person who will be affected by a particular biotechnology to know as much as possible about it. That way, we can all make informed decisions ourselves, and not leave that decision to other sources such as TV or newspaper articles.

There are many strategies for examining a bioethical issue, ranging from "the common good approach" (i.e., the needs of the many outweigh the needs of the individual) to the "virtuous approach" (i.e., what kind of person would I be if I do this?). Whatever strategy seems most valuable to you, it is a good idea to be systematic. When examining a bioethical issue, try to follow these steps:

1. Identify the bioethical problem or issue and learn everything you can about it.
2. Make sure that your information sources, whether books, magazines, or Web sites, are reputable. That is, make sure they have been reviewed and endorsed by reputable sources. Examples of reputable sources include textbooks, peer-reviewed research journals such as *Science,* or a government-run Web site such as the Human Genome Project Web site: (http://www.ornl.gov/sci/techresources/Human_Genome/home.shtml or search the Internet using the key term "Homan Genome Project Information" to find this site).
3. List all possible solutions to the issue.
4. Considering legal, financial, medical, personal, social, and environmental aspects, identify the pros and cons of adopting one solution or position as opposed to another.
5. Based on the pros and cons for each solution, rank all solutions from best to worst.
6. After reviewing all of the information, decide if the problem is important enough to take a position. If it is, decide what your position is and be prepared to describe and defend it.

Would you like to be a scientist? If you ask Dr. Larry Britton (Figure 13.39), a researcher at the University of Texas, he would tell you how hard his work is. But he would be quick to talk about how rewarding it is when an experiment works.

How did he get interested in science? Britton was a naturalist even as a child, always dissecting frogs. He was also a natural experimenter, once even making homemade gunpowder. When one of his playmates fell ill one day and was later diagnosed with polio, Britton started thinking about how something as small as the polio virus could cause life-changing events. Polio, which could render a child paralyzed and unable to breath without the help of a machine known as an iron lung, was greatly feared when Britton was a child in the late 1940s early 1950s. The first polio vaccine was issued in 1955, and an improved version was released in 1957.

In his undergraduate college days, Britton thought he would go to medical school, and started taking the necessary classes. But he soon realized he didn't want to be a medical doctor. To discover what he wanted to do, he took a variety of science courses with labs. When he took a microbiology course, he realized immediately that he wanted to become a microbiologist. He received a Bachelor of Science (B.S.), a Master of Science (M.S.), and eventually a Doctor of Philosophy (Ph.D.) in Microbiology.

Britton's first position, in the Microbiology Department at the University of Texas as an assistant professor, involved researching basic science questions. His work focused on how microorganisms dealt with the generation of toxic oxygen compounds as a result of metabolism. Britton then moved into the chemical industry, where he solved applied problems with financial implications, such as making sure that the synthetic chemicals were biodegradable. Looking for a change from applied research in a company setting, he then moved back to a combination of applied and basic research at the University of Texas. In his current research in the Petroleum Institute at the University Texas, Britton evaluates ways to recover the remaining oil from depleted oil fields, which may still contain as much as 50 percent of their original contents. This work requires him to apply his knowledge of microbiology, biotechnology, and chemistry.

So what makes a good scientist? According to Britton, a good scientist needs "a never ending curiosity about how things work." He points out that "good researchers tend to have the same thought processes as good automobile mechanics. They look at problem, whether it's a phenomenon in the biological world, or a broken automobile the same way and come up with a plan to solve the problem or answer the question." Not surprisingly, Britton is very good at fixing things, including automobiles.

Figure 13.39 | Dr. Britton built a photobioreactor, so-called because it makes a large amount of biological organisms that use light (photo) as energy. In this case the organism is an algae and it gets it carbon from either the carbon dioxide in the air or from other sources such as those found in sewage. Algal-growing photobioractors can be used to clean up sewage-contaminated water sources. When the algae is grown to capacity, it can be harvested and processed to make a biofuel.

Has this chapter convinced you that you need to understand biotechnology and its applications? You should now realize that biotechnology permeates a broad range of situations, including your own medical well-being. As you read the next three chapters, you will notice that topics and examples in one chapter could have easily fit in another chapter. For example, genetic engineering now affects the chemical, agricultural, and medical industries. As a result of this interchange between scientific disciplines, careers no longer remain fixed in one industry, let

527

alone scientific discipline. For example, a microbiologist trained to study microbes that make products who is working in the chemical industry could transfer that knowledge to environmental microbiology and work in industrial waste cleanup. He could then find himself in the petroleum industry working on developing techniques to enhance oil recovery. This is an actual example of a career path. As a student, you will start in a certain field will but will most likely follow a career path that spans several disciplines. As Louis Pasteur said, "Chance favors the prepared mind," and in this technological world, a person with a prepared mind understands how technology impacts life and uses this understanding to chart any career path he or she desires.

SECTION THREE FEEDBACK >

1. A variety of Web sites post jobs and publish information about careers in biotechnology. Find three sites that provide biotechnology career information. (Examples of such sites are Access Excellence, Biospace, Biofind, Science Jobs, and Bio-Link.) Prepare a table that lists and describes the three sites you have selected. For each, find an actual job title or career description and describe the educational qualifications each one requires.

2. Imagine that you have just isolated a human hair growth factor that you think will reverse the balding process in men. What are the next steps your company will need to take before this product can be marketed to the public?

3. This section describes a process for examining ethical issues. Use this process to answer the following question: Is it ethical to use genetically modified corn that contains the pesticide gene BT?

CAREERS IN TECHNOLOGY

Matching Your Interests and Abilities with Career Opportunities: Biotechnology Technicians

Significant Points

* Biotechnology technicians in production jobs can be employed on day, evening, or night shifts; other technicians work outdoors, sometimes in remote locations.
* Most technicians need an associate degree or a certificate in applied science or science-related technology; forensic science technicians usually need a bachelor's degree.
* Job opportunities are expected to be best for graduates of applied science technology programs who are well trained on equipment used in laboratories or production facilities.
* Biotechnological research and development should continue to drive employment growth for both technicians and researchers with advance degrees.

* A Ph.D. degree usually is required for independent research, but a master's degree is sufficient for some jobs in applied research or product development; temporary postdoctoral research positions are common.
* A biotechnologist researcher or technician uses knowledge of life to make products and processes benefiting people's lives while a biological researcher (also commonly termed a scientist) or technician studies organisms and their relationship to the environment.

Nature of the Industry

As indicated by the Department of Labor, Bureau of Statistics, biological researchers and technicians may specialize in one area of biology, such as zoology. Many of these individuals conduct basic research to advance our knowledge of living organisms, such as infectious bacteria and fungi or marine organisms such as sharks and whales. Industrial researchers and technicians conduct applied research or product development using knowledge gained by basic research to develop products for market, such as drugs, medical tests, improved crops, pesticides, biofuels, and plastics. They usually have less freedom than do basic researchers to choose the focus of their research, and they spend more time working on marketable items to meet the business goals of their employers.

Biotechnology researchers and technicians use a wide variety of equipment depending upon their focus. Some work only in the laboratory. Others, such as ecologists, who study organisms in the environment, work mainly outside or work equal amounts in the laboratory and outside. A researcher who is studying what moose eat may watch moose eat outside, collect some of the plants the moose are eating, and then bring them into the laboratory to identify it, using molecular biology (genetic) techniques. In today's world, a biotechnologist needs to know or partner with several different specialists to be able to apply knowledge from many different fields, even within biology, to answer both basic and applied research questions.

Working Conditions

As part of their jobs, biotechnologists, whether they are a researcher with a Ph.D. or a biotechnician with a two-year degree, have to wear the appropriate laboratory attire; at minimum, close-toed shoes and eye protection. Some laboratories work with dangerous solutions and solids that must be handled carefully and appropriately, while other laboratories do not, but all laboratories will require safety training and knowledge of what to do in an emergency. Workers have to safely assemble the correct equipment and glassware, safely prepare the necessary reagents, safely run the experiment measuring and combining the correct amounts of reagents, properly use the equipment, collect and document the data appropriately, and finally, safely clean up and properly store reagents, solutions, and experimental materials.

The majority of jobs have daytime hours, but there are jobs and locations that conduct night shifts and/or require the scientist or technician to return to the laboratory during off-duty time to collect data or check on equipment.

Training and Advancement

A solid background in biology, chemistry, and physics is helpful, along with the ability to solve problems and multitask. At a minimum, the biotechnician must have a high school education that includes a biotechnology training course or a community college biotechnology certificate or degree. Many technicians have a bachelor's degree, but that does not guarantee that you will have a higher paying job than a coworker who has graduated from a two-year community college biotechnology program. If you aspire to higher levels, a graduate-level education can place you in a supervisory role.

Outlook

As reported by the Occupational Outlook Handbook, the continued growth of scientific and medical research—particularly research related to biotechnology—as well as the development and production of technical products should stimulate the demand for science technicians in many industries. Stronger competition among pharmaceutical companies and an aging population are expected to contribute to the need for innovative and improved drugs, further spurring demand for biotechnicians. The fastest employer growth of biotechnicians should occur in the pharmaceutical and medicine manufacturing industry and education services.

Earnings

The U.S. Department of Labor's Occupational Outlook Handbook reports that in 2007, the average annual salary in nonsupervisory, supervisory, and managerial positions in the federal government for biological and biotechnologists was on average $50,000. As with all occupations, geographical location and experience can determine salaries. For instance, the median expected salary for a typical biomedical engineering technician in Alexandria, Virginia is $49,700 while in San Francisco, California, it is $55,256 annually according to Yahoo's Hot Jobs Web site (http://salary.hotjobs.com). Individuals with advanced degrees typically make much more money in industry.

[Bureau of Labor Statistics, U.S. Department of Labor, Occupational Outlook Handbook, 2008–09 Edition, 2/2009 http://www.bls.gov/oco/]

Summary >

The term biotechnology refers to the use of living organisms or their products to improve people's lives. Biotechnology probably began as an accident when microorganisms contaminated food products and ancient people realized that it sometimes made the food taste better, provided a means to store food, and increased the variety of ways food could be prepared.

Fermentation was the primary tool for thousands of years until large-scale historical events or major changes in technology occurred. These events, such as the World Wars, the advent of genetic engineering, and the Human Genome Project, with backing from society and funding from governments, have promoted basic and applied research in biotechnology with far-reaching results.

Applications of biotechnology can be found in medicine, bioremediation, forensics, and the production of fuels. The expansion of biotechnology continues with new applications in other fields and will only continue to increase. However, will it continue to enjoy support from society? Already, many countries have banned certain biotechnology products and processes. What will happen next?

FEEDBACK

1. Produce a timeline of five events mentioned in this chapter that revolutionized biotechnology.

2. Using the Internet, make a list of biotechnology companies, including one example for each of the seven different types of companies found in the biotechnology industry. Indicate at least one product or service that the company makes or does for its customers. Organize this information in a table.

3. Look up Herceptin on the Internet and find out exactly how it works against breast cancer. Also find out when the patent on this drug will expire. Once the patent expires, other companies will be able to make a generic form of this drug without having to do the expensive clinical trials required for the original drug. The cost savings also may mean that the drug would be available to third-world countries and the people who could not originally afford the drug. Companies who do the original research and testing want to extend the life of drug patents because they say they cannot make enough money to justify developing new drugs. What type of information do you need to know to decide if this is fair or not? Taking either the side of the company who holds the patent, or the company that is going to produce the generic form, and write a short paragraph arguing for your side, and include future potential benefits to patients.

4. If you wanted to genetically engineer the human growth hormone, what body organ do you think you would have to start with?

5. One of the major tools in genetic engineering is restriction enzymes, otherwise known as the "molecular scissors." They are used to cut the engineered DNA for insulin and the plasmid used to move the DNA into a bacterium. Look up restriction enzymes on the Internet and write a brief description of the job these enzymes do in nature, as well as the job they perform in the biotechnology field.

6. Many commercialized proteins used in molecular biology techniques were discovered and obtained from organisms isolated from national parks (e.g., Yellowstone National Park). Do you think these companies should pay a certain percentage of their profits to these parks (i.e., the public)? Should these companies have limited access to sources of these bacteria? Should the general public have access to soil/water/plant samples and the like from park lands even if their motives are noncommercial? Write a short paragraph supporting your position.

7. In the introduction to this chapter, you learned that there are two main points of view regarding the use of GMO crops. Using the Internet, learn about a strain of corn known as Bt corn (*Bacillus thuringiensis*). Find out if the proper tests have been done, as determined by USDA, EPA, and FDA, to ensure that Bt corn does not harm beneficial insects and that the public can consume it safely without having to worry about side effects.

8. Companies have attempted to isolate from soil a bacterial-protein–degrading enzyme that could survive a complete cycle in a clothes washer. The companies had to learn how to grow these organisms under conditions that simulated a washing machine full of water and detergent. Why? Scientists reasoned that if the organisms could survive and grow under these conditions then the enzymes inside them could survive these conditions. List all of the conditions that this enzyme would have to be able to survive to work in a washing machine. Then design a procedure for how this organism might be isolated.

DESIGN CHALLENGE 1:
Extracting DNA from Living Organisms

• Problem Situation

A complete copy of DNA is found in the nucleus of every cell in an organism. In order to release the DNA to analyze it or isolate genes for further cloning experiments, biotechnologists must break open the cells and remove structural proteins and enzymes that interfere with the DNA structure. This simplified procedure releases a great deal of DNA so that you can easily see it and observe its physical and chemical properties.

• Your Challenge

You and your team members will isolate DNA from fruits and vegetables obtained from your grocery market.

• Safety Considerations:

1. Wear eye protection at all times.
2. Wear gloves at all times.
3. Disinfect your work surfaces with ethanol to decontaminate.
4. Thoroughly wash and disinfect all glassware and measuring spoons.
5. Never touch the rapidly moving blade of the blender.
6. Ethanol is highly flammable and should be kept away from open flames at all times.
7. Do not pour any unused chemicals back into storage containers where it may contaminate the rest of the reagent. Dispose of unused chemicals in proper waste containers.

• Materials Needed

1. ½ cup of a fruit or vegetable of your choice
2. ½ cup ice cold water
3. ⅛ teaspoon salt
4. Blender or mortar and pestle
5. Strainer
6. Dish detergent
7. Small glass containers (test tubes)
8. Meat tenderizer
9. Ice cold ethanol
10. Paper clip
11. Balance
12. Weigh boat or paper towel

• Clarify the Design Specifications and Constraints

Your DNA must be isolated from a fruit or vegetable that can be found in your local grocer's market. The cells from these living organisms will be physically broken opened in a household blender and then chemically broken open (lysed) with a detergent. Enzymes will be used to stabilize the DNA. The DNA will be visible as a stringy substance that has the consistency of mucus when precipitated in ethanol. You should be able to spool the DNA onto a paperclip.

• Research and Investigate

To complete the design challenge, you need to first gather information to help you build a knowledge base.

1. In your guide, complete the Knowledge and Skill Builder I: Safety considerations.
2. In your guide, complete the Knowledge and Skill Builder II: Choosing a fruit or vegetable.
3. In your guide, complete the Knowledge and Skill Builder III: Lysing cells.
4. In your guide, complete the Knowledge and Skill Builder IV: Stabilizing DNA.
5. In your guide, complete the Knowledge and Skill Builder V: Precipitating DNA.
6. In your guide, complete the Knowledge and Skill Builder VI: Analyzing DNA yield.

• Generate Alternative Designs

Describe two of your possible alternative approaches in extracting DNA from a living organism. Discuss the decisions you made in (a) choosing a fruit or vegetable, (b) lysing the cells, and (c) stabilizing the DNA. Attach drawings if helpful and use additional sheets of paper if necessary.

• Choose and Justify the Optimal Solution

What decisions did you reach about the optimal conditions for extraction of DNA from a living organism?

• Display Your Prototypes

Spool and store the DNA you extracted. Include descriptions, photographs, or drawings of the process in your guide.

• Test and Evaluate

Explain whether your designs met the specifications and constraints. What tests did you conduct to verify this?

• Redesign the Solution

What problems did you encounter that would cause you to select a different organism or redesign the lysis or stabilization of the DNA? What modifications would you recommend for future experiments? What additional tradeoffs would you have to make?

• Communicate Your Achievements

Describe the plan you will use to present your results to your class and show what handouts you will use. (Include a media based presentation.)

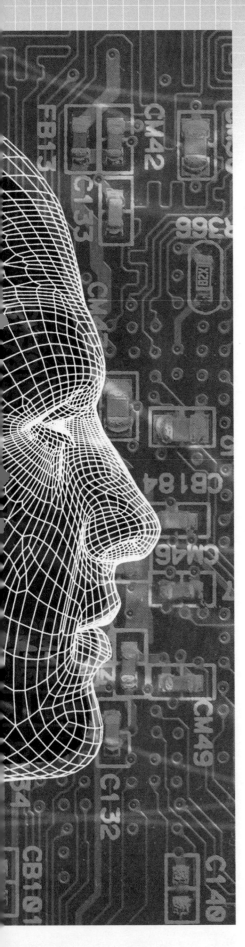

DESIGN CHALLENGE 2:
Bioremediation of Oil Spills

• Problem Situation

Oil spills in the Gulf of Mexico are a problem, resulting in oil washing up on beaches and negatively affecting both wildlife and the tourist industry along the coasts of Texas and Louisiana. In addition, spills, if they persist on the surface of the water, kill life that concentrates at the surface of the ocean. Several technologies are employed to clean up the oil, for example, dispersant technology, which is used to break up the oil so it sinks down away from the surface. Dispersants can be composed of surfactants, such as detergents, which break up the oil. Their action allows the oil to sink, and then natural processes eventually degrade the oil. For small spills, boom technology is used to sequester and concentrate surface oil slicks. Sometimes sorbents, such as sawdust, are added to help sequester and concentrate the oil. Then the oil, including the sorbent, is either skimmed or swept off the surface. These technologies work fairly well with small spills, but large spills are more difficult to rapidly disperse or are impossible to completely sequester before they cause damage. People are seeking additional technology to augment or replace these existing technologies.

Your company, Remediate, has won the bid on a contract to develop technologies to accelerate the process of large oil spill degradation. Possible choices for speeding up the process are bioaugmentation (the addition of organisms that specifically and rapidly degrade oil) and/or biostimulation (the addition of food sources to speed up the degradation of the oil).

• Your Challenge

Based on the conditions found at the site, you and your coworkers will develop a model for the purpose of determining the best method for speeding up remediation, either by bioaugmentation and/or biostimulation. Included in this challenge is how to measure the degradation of oil over a period of time so that the effectiveness of the method can be properly evaluated.

• Safety Considerations

1. Wear personal protection equipment (PPE) while working with the laboratory materials and equipment; this includes gloves and eyewear.
2. Wear closed-toed shoes, as this is a requirement of the industry when working in the laboratory.

• Materials Needed

1. Jars or large tubes the same size
2. Mineral oil
3. Powdered cocoa or oil red
4. Sodium chloride
5. Distilled water
6. Soil from the garden
7. Fertilizer
8. Surfactants, for example, dish detergent, liquid soap, or washing machine detergent
9. Devices for measuring liquids (preferably graduated cylinders)
10. Device and equipment for measuring solids (preferably a balance)
11. Device for measuring water temperature such as a thermometer

• Clarify the Design Specifications and Constraints

Your design must model the conditions found in the Gulf of Mexico the majority of time, including salinity concentration, solar radiation, and water temperature. Since crude oil is difficult to obtain, it is suggested that you use mineral oil. It is easily purchased, nontoxic, and produces approximately the same oil/water interfacial properties that crude oil produces when mixed with saltwater.

• Research and Investigate

To complete the design challenge, you need to first gather information to help you build a knowledge base.

1. In your guide, complete the Knowledge and Skill Builder I: Safety considerations.
2. In your guide, complete the Knowledge and Skill Builder II: Pick your model: Baking dish or aquarium.
3. In your guide, complete the Knowledge and Skill Builder III: Oil spill behavior and boom technology.
4. In your guide, complete the Knowledge and Skill Builder IV: Bioremediation.
5. In your guide, complete the Knowledge and Skill Builder V: Bioaugmentation.
6. In your guide, complete the Knowledge and Skill Builder VI: Measuring oil degradation.

• Generate Alternative Designs

This project requires your team to determine the best method for accelerating bioremediation under a specific set of parameters, as well as to determine the best method for evaluating the degradation of oil over a period of time. Discuss the decisions you made in (a) determining the oceanic conditions and how they were to be duplicated in the model, (b) how biostimulation and bioaugmentation can be done using the model, and (c) how oil degradation is to be measured over a period of time. Attach drawings if helpful and use additional sheets of paper if necessary.

• Choose and Justify the Optimal Solution

What decisions did you reach about the model, the conditions that needed to be duplicated, and how oil degradation was to be evaluated over a period of time?

• Display Your Prototypes

Produce your model and evaluate oil degradation. Include any necessary documentation, data measurements, tables, descriptions, photographs, or drawings of these in your guide.

• Test and Evaluate

Explain whether your designs met the specifications and constraints. What tests did you conduct to verify this?

• Redesign the Solution

What problems did you face that would cause you to redesign the (a) model, including the oceanic conditions and the conditions for bioremediation, and (b) the evaluation techniques? What changes would you recommend in your new designs? What additional tradeoffs would you have to make?

• Communicate Your Achievements

Describe the plan you will use to present your solution to your class. (Include a media-based presentation.)

CHEMISTRY HAS MADE life easier and better, beginning with the first people who figured out that stomach fluid from a goat could be used to make cheese or that chewing on willow bark would decrease pain. This information was collected, handed down from generation to generation, and shared to some degree between people and civilizations. Possessing this information made individuals powerful and invaluable to a tribe or village, so it was not

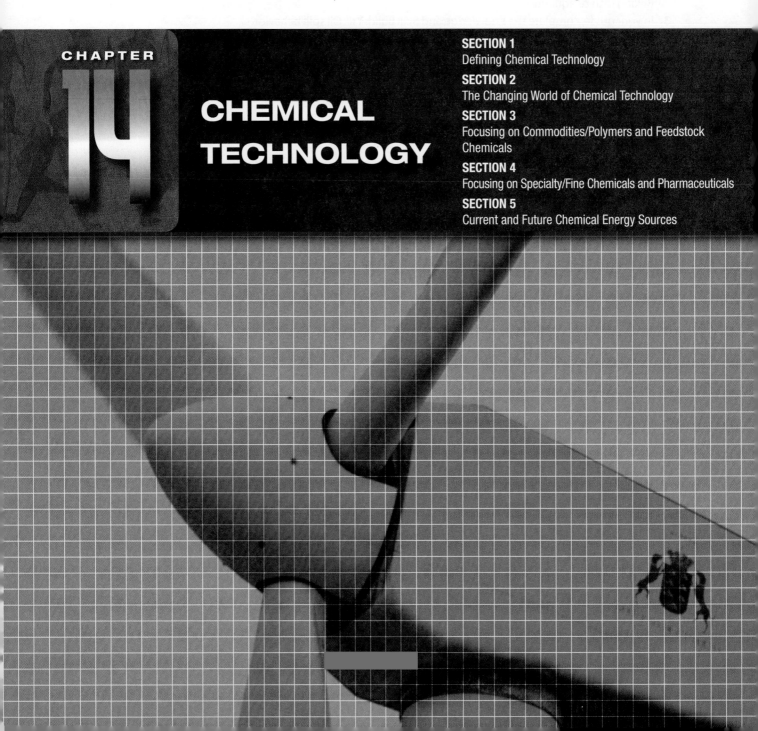

CHAPTER
14

CHEMICAL TECHNOLOGY

always openly shared, but rather kept a secret by a family, a society, or a culture. However, a lot of chemically based information was distributed because it was easier for a tribe or village to specialize in tasks instead of everyone trying to do it all for themselves. Tanners, bakers, and apothecaries are all examples of people with special chemical knowledge that benefited their cultural group.

The first apothecary shops were founded during the Middle Ages. The individuals who ran these shops became known as **apothecaries** and their study of herbal and chemical ingredients is considered the precursor of the modern sciences of chemistry and pharmacology. By the fifteenth century, the apothecary had gained the status of a skilled practitioner of medicine (see Figure 14.1). By the end of the nineteenth century, however, the role of the apothecary was more narrowly viewed as that of dispensing pharmacist and not one who could prescribe medicines.

Figure 14.1 | How much power can be wielded by a group that understands how chemicals affect the body? In England, the apothecaries merited their own trade association, founded in 1617. Today, this group continues to license doctors as a member of the United Examining Board, the only nonuniversity medical licensing body in the United Kingdom.

This chapter will review the following: the nature of chemical technology, changes in the chemical industry, and the three categories that make up the chemical industry.

KEY IDEAS >

- Chemical technologies provide a means for humans to alter or modify materials and to produce chemical products. The chemical industry is divided into three categories; commodities, specialty/fine chemicals, and energy/fuel.

- Like other technologies, chemical technology is regulated to protect the worker, the environment, and the customer who purchases the product.

Figure 14.2 | **Examples of chemical products include the components for iPods; DVDs; all types of plastics; athletic equipment; cosmetics; food preservatives and additives; fuels for transportation, heating, and powering industries; houses; airplanes; reagents for medical tests; drugs; textiles; synthetic materials for furniture; and cleaning reagents.**

Chemical technologies provide a means for humans to alter or modify materials and to produce chemical products. In today's world, the chemical industry is a multibillion dollar industry and an important contributor to society, lifestyle, and world economics. Part of the reason the United States was able to establish itself as an economic world power was due to its chemical industry, and this is still true today. The chemical industry in the United States converts raw materials, (e.g., oil, natural gas, air, water, metals, and minerals) into more than 70,000 different products like those shown in Figure 14.2. Few goods are manufactured without some input from the chemical industry; just look around your home and school for materials that are made with chemical products, either directly or indirectly. Even goods such as wood furniture or bikes are manufactured with some input from the chemical industry, in that the wood furniture is covered with varnish and the bike is covered with paint.

The process parameters for producing chemical products are the factors that govern their production. These process parameters are similar to the production of products in any industry:

- The initial materials, known as raw materials, or feedstocks, should be as inexpensive as possible and readily available.
- The process to convert the feedstocks into finished products should be cost-effective and, nowadays, energy efficient.
- The process should abide by governmental guidelines and regulations, many of which are the same or similar to the ones found in the biotechnology industry.

Review the product development process for the chemical industry shown in Figure 14.3, and compare it to the graphic that visualizes the steps in the development of recombinant insulin (rhinsulin). Are they similar? The titles for the steps are different, but many of the activities are the same.

The chemical industry is divided into three categories, shown in Figure 14.4: commodities, specialty and fine chemicals, and energy/fuel.

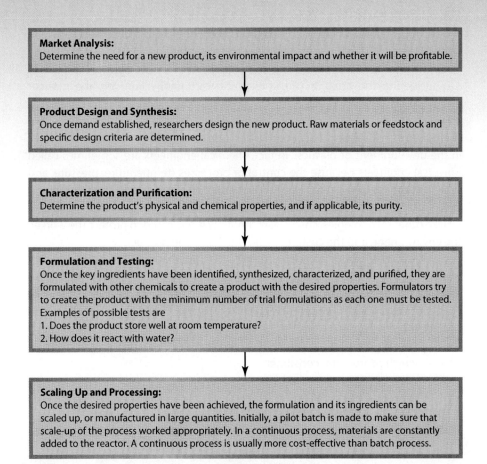

Market Analysis:
Determine the need for a new product, its environmental impact and whether it will be profitable.

Product Design and Synthesis:
Once demand established, researchers design the new product. Raw materials or feedstock and specific design criteria are determined.

Characterization and Purification:
Determine the product's physical and chemical properties, and if applicable, its purity.

Formulation and Testing:
Once the key ingredients have been identified, synthesized, characterized, and purified, they are formulated with other chemicals to create a product with the desired properties. Formulators try to create the product with the minimum number of trial formulations as each one must be tested. Examples of possible tests are
1. Does the product store well at room temperature?
2. How does it react with water?

Scaling Up and Processing:
Once the desired properties have been achieved, the formulation and its ingredients can be scaled up, or manufactured in large quantities. Initially, a pilot batch is made to make sure that scale-up of the process worked appropriately. In a continuous process, materials are constantly added to the reactor. A continuous process is usually more cost-effective than batch process.

Figure 14.3 | This figure reviews the steps for development of a chemical product. Compare it to the graphic that visualizes the steps in the development of the pharmaceutical recombinant insulin (rhinsulin) (see Figure 13.23).

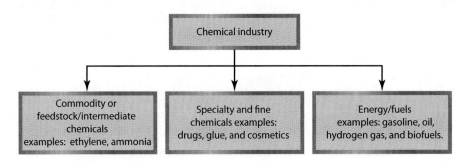

Chemical industry

Commodity or feedstock/intermediate chemicals examples: ethylene, ammonia

Specialty and fine chemicals examples: drugs, glue, and cosmetics

Energy/fuels examples: gasoline, oil, hydrogen gas, and biofuels.

Figure 14.4 | The chemical industry can be divided into three categories: commodities/intermediates such as ethylene, or ammonia, specialty/fine chemicals such as drugs, glue, cosmetics, and energy and fuels such as gasoline, oil, hydrogen gas, and biofuels.

Commodity products are the feedstocks for other industries. Examples include ethylene for plastics or surfactants (chemicals that break up or dissolve grease and similar compounds) for detergents. This is the largest category of the chemical market. Commodity products are produced in bulk and are inexpensive to make. The second category is specialty and fine chemicals. Pharmaceuticals and cosmetics are good examples of fine chemicals. The third category is energy/fuel products, such as gasoline, oil, and hydrogen gas. Many scientists, technologists, economists, business people, and consumers find this category important yet somewhat confusing. Why are our fuels so expensive?

Section 1 △ Defining Chemical Technology

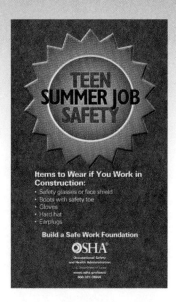

Figure 14.5 | One of several agencies that regulate industries, the Occupational Safety and Health Administration (OSHA) strives to promote the safety and health of U.S. workers by setting and enforcing standards and by providing training, outreach, and education. OSHA employs over 2,000 inspectors, investigators, engineers, and other personnel in more than 200 offices across the country to assist employers and employees.

Figure 14.6 | The Environmental Protection Agency (EPA) develops and enforces regulations, performs and funds environmental research, and publishes educational materials to protect human health and the environment. The agency monitors a wide range of areas, including air quality, wetland management, human health, pollution prevention, and handling of hazardous wastes. Its largest facility is this Research Park Triangle campus in North Carolina, which employs over 2,000 workers.

Why are we dependent on oil? Why aren't we using more "green" fuels instead of oil? These fuels that are discussed represent some of the prominent fuels of the present and future.

Considering the breadth of the chemical technology industry, the number of categories appears to be limited. Yet, the process techniques and research approaches are similar across the entire spectrum of chemical products. For example, process development approaches in drug design are similar to approaches in the development of plastics. In fact, chemical engineers are sometimes called universal engineers because the control of processes (temperature, pressure, and time) is common to the synthesis (production) of all chemicals, whether it is drugs, paints, plastics, cosmetics, rubber, or gasoline.

Like other technologies, chemical technology is regulated to protect the worker, the environment, and the customer who purchases the product.

Some of these agencies are as follows: the Occupational Safety and Health Administration (OSHA), which protects the worker (see Figure 14.5); the Environmental Protection Agency (EPA) (see Figure 14.6); the Department of Transportation (DOT); and the Nuclear Regulatory Commission (NRC), which protect the environment including people; and, depending upon the product, the Food and Drug Administration (FDA) and the U.S. Department of Agriculture (USDA), which protect the consumer.

SECTION ONE FEEDBACK >

1. What are the main functions of chemical technologies?
2. What are the three categories that make up the chemical industry? Give an example for each category.
3. Why are chemical engineers sometimes called universal engineers?
4. Name three agencies that regulate the chemical industry. What are such agencies designed to protect?

SECTION 2: The Changing World of Chemical Technology

KEY IDEAS >

● Originally, society was agriculture based, but the availability of resources, different economics, and technology, led to a chemical-based society.

● Products made from plants and animals are considered natural, whereas products made using specific chemical steps, possibly involving high temperature and pressure and using chemically-derived feedstocks, are generally considered synthetic.

● The goals for green chemistry include decreasing waste and the need for energy; more economical processes; and increased safety for the plant technician, the environment, and the consumer.

● Genetically engineered chemical processes are also considered green chemistry since they occur at lower temperatures and pressures and produce chemicals that natural processes can degrade.

- Many synthetic products are not degradable by natural processes and therefore become pollutants as they persist in the environment.

From a Bio-based Society to a Chemical-based Society

Originally, society was agriculture based, but the availability of resources, different economics, and technology led to a chemical-based society.

One hundred years ago, a major proportion of human clothing, fuel, dyes, medicines, construction materials, and industrial chemicals were still derived from plants and animals. These types of products are considered to be "natural" since they appear in nature or they are made from natural ingredients. This situation changed dramatically over the past fifty to one hundred years as synthetic products became more common, as shown in Figure 14.7.

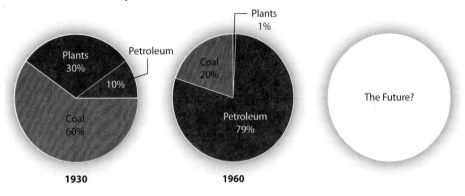

Figure 14.7 | In the 1930s, the primary source of industrial chemicals in the United States was plants. In the 1960s and early 2000s, it was petroleum. What will it be in the future?

Products made from plants and animals are considered natural, whereas products made using specific chemical steps, possibly involving high temperature and pressure and using chemically derived feedstocks, are generally considered synthetic. Synthetic rubber and synthetic fibers for clothing (e.g., nylon, polyester) are just two examples of chemically derived products that were developed to replace products once obtained exclusively from agriculture. In fact, both of these synthetic products are derived from oil or natural gas and are known as petroleum-based products. Many synthetic products are not degradable by natural processes and therefore become pollutants as they persist in the environment.

As of the year 2000, over 95 percent of organic chemicals that society used were derived from oil and gas and not from plants. Why? The trend is due to three major factors: (1) development of the automobile, which generated an enormous demand for inexpensive liquid fuel; (2) the discovery and development of large oil deposits; and (3) the chemical industry's invention of new processes that converted the cheap petroleum (oil and natural gas) into usable, inexpensive products such as gasoline. In the 1940s and 1950s, a kilogram of corn cost four to five times more than a kilogram of oil. However, times have changed, and in the past fifty years agricultural costs have decreased whereas oil production costs have increased (see Figure 14.8). A barrel of crude oil is now much more expensive than a barrel of corn.

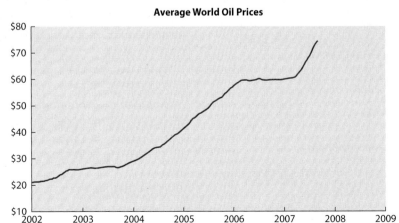

Figure 14.8 | World oil prices have more than doubled over the last several years. Some analysts say that prices may double again over the next five to ten years, due to production levels that are not rising to meet increased demand.

Section 2 △ The Changing World of Chemical Technology

Additional factors that could cause a switch of fuel sources from petroleum to other technologies are the environmental, political, and social benefits and costs associated with the cleaning up of waste products. Governmental regulations, starting thirty-five years ago, mandate that industries clean up the wastes they produce or face large fines or even lawsuits from cities or individuals affected by the waste. All of these factors have made the old processes more costly in the long run and have opened the door to green chemistry.

Green chemistry is the term that has been given to a variety of environmentally friendly products and processes, and it is a major goal of the twenty-first century chemical industry. The goals for green chemistry include decreasing waste and the need for energy, more economical processes, and increased safety for the plant technician, the environment, and the consumer.

The U.S. Environmental Protection Agency (EPA) has developed a list of the principles of green chemistry, some of which are presented below:[1]

1. Design chemical syntheses to prevent waste, leaving no waste to treat or remove. Synthetic chemical reactions where the majority of starting material atoms are found in the products are said to be atom economic, a phrase invented as part of the green-chemistry philosophy.

2. Design and use safer chemicals and run the chemical reactions under safer and more economical conditions (e.g., milder temperature and pressure).

3. Use renewable feedstocks for both starting materials (reactants) and as an energy source. Renewable feedstocks are often made from agricultural products while depleting feedstocks are from fossil fuels (oil, gas, or coal). Fossil fuels cannot be renewed.

4. Use catalysts (materials that cause a reaction but are not used up in the reaction) for speeding up processes instead of high temperatures and pressure, decreasing energy usage.

5. Analyze the process in real time (i.e., monitor while the reactions are occurring rather than simply at the end) to monitor the accumulation of undesirable by-products and to allow for immediate changes in the process.

Upon comparing these guidelines with what happens in nature, you will notice that synthetic "green" chemical production resembles bio-processing in nature. The lower temperature, the use of catalysts, the real-time analysis/feedback, and fewer by-products are all traits of natural processes. This means that "green" synthetic chemical production often will mimic natural chemical production, if it is designed appropriately. However, the definition and guiding principles of green chemical engineering do not specify that the process must be based on a natural process.

A good example of an old wasteful process revamped into a new, environmentally friendly, energy-efficient process is the synthesis of ibuprofen (see Figure 14.9).

Ibuprofen is a common anti-inflammatory agent found in a variety of pain killers (e.g., Motrin, Advil, and Medipren). The original reaction developed by Boots Company of England in the 1960s yielded only 40 percent atom economy, which means that 60 percent of the atoms from the starting materials were wasted:

Starting Materials (100% of the atoms) → (yields) Ibuprofen (40% of the original number of atoms) + Waste (60% of the number of atoms)

Figure 14.9 | Ibuprofen is a nonprescription anti-inflammatory and pain reliever marketed under many trade names worldwide, including Advil and Motrin in the United States. In 1992, the production of ibuprofen became a more efficient "green" process that eliminated the production of millions of pounds of landfill waste every year.

---

[1] Originally published by Paul Anastas and John Warner in Green Chemistry: Theory and Practice (Oxford University Press: New York, 1998.

What is an atom? It is the smallest particle defining a chemical element. Ibuprofen is a compound composed of three different elements, carbon (C), hydrogen (H), and oxygen (O). The chemical formula is thirteen carbons, eighteen hydrogens and two oxygens as shown in Table 14.1. The weight of the ibuprofen and the other compounds is determined by adding up the weight of all of the atoms found in that compound; so for ibuprofen, the weight would be 13 carbons + 18 hydrogens + 2 oxygens. You can learn the identity and weight of the elements by looking them up in the periodic table, which groups all of the elements based on their physical and chemical properties.

Table 14.1 | Comparing the Atom Economy of the Two Reactions

TOTAL NUMBER OF ATOMS IN STARTING MATERIALS	WEIGHT OF STARTING MATERIALS	TOTAL ATOMS USED	WEIGHT OF UTILIZED ATOMS	WASTE PRODUCTS	WEIGHT OF WASTE PRODUCTS
Boots process $C_{20}H_{42}NO_{10}ClNa$	514.5	Ibuprofen C_{13},H_{18},O_2	206	$C_7H_{24}NO_8ClNa$	308.5
BHC process $C_{15}H_{22}O_4$	266	Ibuprofen C_{13},H_{18},O_2	206	can be recycled $C_2H_3O_2$ (acetic acid)	60

The "wasted" atoms were therefore a source of pollution and required proper disposal. The green, BHC, reaction has an atom economy of 77 percent or actually closer to 100 percent, since its only byproduct, acetic acid, could be used over again.

Starting Material (100% of the atoms) → Ibuprofen (77% of the atoms) + (23% of the atoms—and it is recycled back to starting materials)

The green BHC process was a result of a joint venture between the Boots Company and the Celanese Corporation. The Celanese Corporation came up with the process and the Boots Company was responsible for marketing. In addition to producing less waste, the green process uses a faster, three-step catalyzed synthesis, while the older process uses six steps and no catalysis. Thus, the green synthesis is a win-win situation in terms of both the environment and the cost of the reaction. On October 15, 1992, the green synthesis was put into practice on an industrial scale by Celanese Corporation at one of the largest ibuprofen manufacturing facilities in the world in Bishop, Texas. Currently, it produces 20 to 25 percent (more than 7 million lbs.) of the world's supply of ibuprofen. This city might be considered the painkiller capital of the world, since the company also makes 20 million pounds of acetaminophen (the active ingredient in Tylenol). With the advent of new chemical technologies, it is anticipated that many other old, wasteful chemical processes will be converted to green processes.

Genetically engineered chemical processes are also considered green chemistry, since they occur at lower temperatures and pressures and produce chemicals that natural processes can degrade. See the following text box, Technology in the Real World: Green Chemical Manufacturing Utilizing Gene Technology.

TECHNOLOGY IN THE REAL WORLD: Green Chemical Manufacturing Utilizing Gene Technology

The New Zealand–based life science company, HortResearch, has discovered the gene responsible for producing the compound alpha-farnesenes. This compound gives green apples their characteristic scent (see Figure 14.10a). Up to this point, the only way to get a "green apple" smell was to manufacture the

Figure 14.10a | What is the value of manufactured smells, especially "fresh" smells? Fresh odors have been shown to have positive effects on our mood. For example, the presence of lemon fragrance in the central air-conditioning system of an office block has been known to significantly reduce the stress levels of the occupants and hence operator errors. The scent of green apple has been shown to reduce the severity of migraine headaches.

compound synthetically using an energy-inefficient process that also produced a lot of "atom-waste." Using gene technology coupled with biofermentation, similar to the process shown in Figure 14.10b, HortResearch can synthesize large amounts of this compound efficiently, at low cost and with little waste.

Industrial Systems Biology: Search chemical, gene, protein, enzyme databases for possible targets. For example, chemicals that have similar smells have similar structures, so what else smells like apple? Also, concentrate on genes and enzymes found in apples or similar fruits that might be responsible for producing the chemical(s) that make that smell. If a nature pathway for making this smell is not found, then ask, "Is it possible to change DNA so that it codes for a chemical that will make the smell?"

↓

Metabolic Engineering: Clone target DNA(s) into the appropriate cell for mass production of the necessary enzyme or enzymes to make the smell.

↓

Manufacturing Process: Grow up a huge number of cells that produce the enzymes necessary for manufacturing the chemical or chemicals that are responsible for the smell. At this point, some of the processing could become chemical processing as described for the production of biodegradable plastics as described in the next section.

Figure 14.10b | Flowchart describing how gene technology and biofermentation is applied to the production of synthetic production of "natural smells"; an example of the "new" chemical technology.

SECTION TWO FEEDBACK >

1. Give two reasons why society went from bio-based products to chemically based products.
2. Define green chemistry.
3. Name two factors that have caused a switch away from petroleum fuels.
4. What is a catalyst?
5. Is the product alpha-farnesene, made by HortResearch, natural or synthetic? Or is it both? Explain.

SECTION 3: Focusing on Commodities/ Polymers and Feedstock Chemicals

KEY IDEAS >

- Commodity chemicals are produced in bulk at low cost, are interchangeable, and are not branded.
- The largest amount of feedstock or intermediate produced in the world is ethylene.
- The majority of feedstock chemicals are produced from oil and gas, which means their cost increases with increases in energy costs.
- Polymers are large molecules made up of long chains of molecules (monomers) bonded (polymerized) together.

- Examples of natural polymers are collagen, cellulose, DNA, and starch.

- Examples of synthetic polymers are the plastics, polyvinyl chloride (PVC), and polypropylene.

- Green chemistry initiatives push for the development of biodegradable or recyclable plastics.

Commodity chemicals are produced in bulk at low cost, are interchangeable, and are not branded. Examples are ethanol, ethylene, acetic acid, and even aspirin, which was originally a specialty, branded chemical when solely made by the Bayer Corporation. Now it is cheaply produced in bulk by several manufacturers worldwide. Many of these chemicals serve as feedstock for other more expensive products. The largest amount of feedstock or intermediate produced in the world is ethylene.

Ethylene is produced by a variety of processes. The principal method uses thermal cracking (breaking apart) of hydrocarbons (compounds containing hydrogen and carbon such as natural gas or oil) in the presence of steam, and recovery from oil refinery-cracked gas. The majority of feedstock chemicals are produced from oil and gas, which means their cost increases with increases in energy costs.

If ethylene is produced from natural gas, the methane is removed, and the ethane portion of the gas is purified. The purified ethane is heated with steam (steam "cracked"), breaking down the ethane into ethylene, hydrogen, and other byproducts, as shown in Figure 14.11. Sudden cooling stops the reaction, and the subsequent mixture of gasses is compressed, changing the components from a gas to a liquid. The mixture is chilled and separated in a series of distillation towers. Since the production of ethylene is energy intensive, chemical engineers figured out how to recover the heat from the cracked gas to power the turbines that power the compression of the cracked gas. This made the process more economical. In addition, petrochemical plants that use ethylene as a feedstock are commonly placed near plants that produce ethylene. Then it can be directly piped to those plants (see Figure 14.12), decreasing transportation costs.

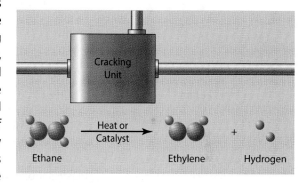

Figure 14.11 | Ethylene can be produced by several methods. One is method is thermal, using, for example, high-temperature steam (816°C) to break larger-chain hydrocarbons into smaller-chain hydrocarbons. A second method uses an inorganic catalyst at a lower temperature.

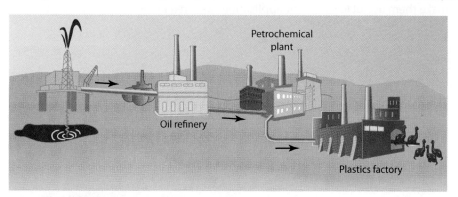

Figure 14.12 | One cost-saving approach is to pipe the oil from the field directly to the oil refinery, which distills the oil into different fractions; the fractions are piped directly into the petrochemical plant, which produces the plastic resins. The resins are then shipped directly to the plastics plant that makes the plastic products, such as in this case, plastic toy dinosaurs.

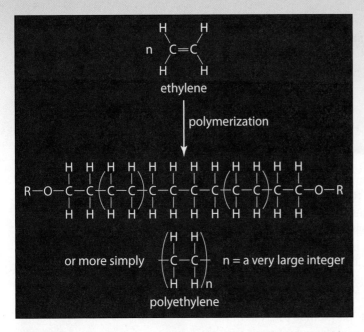

Figure 14.13 | Ethylene is joined to another ethylene in a chemical reaction known as polymerization. This reaction occurs many times to form the long-chain molecule known as polyethylene. Polyethylene is identified as the polymer made from the polymerization of the monomer ethylene.

Cellulose
Glycogen
Starch
DNA

Figure 14.14 | There are many examples of polymers in nature formed from the polymerization of monomers. The sugar monomer, glucose, is joined to another glucose to make the polymer cellulose, which is found in plants. Glycogen and starch are also polymers formed by the joining of the sugar monomer, glucose. The difference between these sugar polymers is how the glucose monomers are joined together. The polymer DNA is formed from the joining of 4 different nucleic acid monomers.

In a related process used in oil refineries, high-molecular-weight hydrocarbons found in oil are cracked (broken into smaller hydrocarbons) over catalysts. Because of the catalyst, the reaction temperature does not need to be as high as in steam cracking.

The plants that use ethylene as a feedstock produce a number of products, including several types of synthetic plastics. For example, the plastic polyethylene is formed by the joining of many ethylene subunits. The ethylene subunits are known as monomers and the resultant polyethylene product is known as a polymer. The formulae for ethylene, the ethylene monomer, and the polymer are given in Figure 14.13.

Polymers

Polymers are large molecules made up of long chains of molecules (monomers) that are bonded (polymerized) together, much like beads on a string. Polymer synthesis was not invented by researchers; many polymers occur in the natural world. Examples of natural polymers are collagen, cellulose, DNA, and starch (see Figure 14.14).

The polymer DNA is synthesized from nucleic acid monomers, and the polymer cellulose is produced by the joining of sugar monomers.

However, at first, researchers did not realize that these natural materials were polymers. Then, in 1922, the renowned German chemist Dr. Hermann Staudinger published his theories on polymers stating that natural rubbers were made up of long, repetitive chains of monomers. Researchers then started to realize how many things in the natural world could be made of polymers and saw the many possibilities for chemically synthesizing polymers such as plastics.

In today's world, there are lots of different synthetic polymers. Examples of synthetic polymers are plastics, polyvinyl chloride (PVC), and polypropylene. Plastics dominate the market. Several are made starting with the monomer ethylene, while others are synthesized from totally different monomers and chemicals. For example, there are two types of polyethylene plastics, depending upon the structure of the monomer that is polymerized. As a result, the different polymerized plastics have a wide variety of chemical and physical properties that make them useful for different applications as listed in Table 14.2.

Table 14.2 | Plastic Polymers: Properties and Uses

	MAJOR PLASTIC RESINS AND THEIR USES			
RESIN CODE	RESIN NAME	CHARACTERISTICS	USES	RECYCLED PRODUCTS
♻1	Polyethylene Terephthalate (PET or PETE)	Transparent, high impact strength, impervious to acid and atmospheric gasses, not subject to stretching	Plastic soft drink bottles, mouthwash bottles, peanut butter, and salad dressing containers	Liquid soap bottles, strapping, fiberfill for winter coats, surfboards, paint brushes, fuzz on tennis balls, soft drink bottles, film, egg cartons, skis, carpets, boats
♻2	High Density Polyethylene (HDPE)	Similar to LDPE, more opaque, denser and rigid	Milk, water, and juice containers, grocery bags, toys, liquid detergent bottles	Flower pots, drain pipes, signs, stadium seats, trash cans, recycling bins, traffic-barrier cones, golf bag liners, detergent bottles, toys

(Continued)

Table 14.2 | Continued

RESIN CODE	RESIN NAME	CHARACTERISTICS	USES	RECYCLED PRODUCTS
3	Polyvinyl Chloride (V)	Rigid, thermoplastic, impervious to oils and many organic materials, high impact strength	Clear food packaging, shampoo bottles	Floor mats, pipes, hose, mud flaps
4	Low Density Polyethylene (LDPE)	Opaque, white, soft, flexible, melts at 100 to 125°C, oxidizes on exposure to sunlight	Bread bags, frozen food bags, grocery bags	Garbage can liners, grocery bags, multipurpose bags
5	Polypropylene (PP)	Opaque, high melting point, lowest density commercial plastic, impermeable to liquid and gasses, smooth surface with high luster	Ketchup bottles, yogurt containers and margarine tubs, medicine bottles	Manhole steps, paint buckets, videocassette storage cases, ice scrapers, fast-food trays, lawn mower wheels, automobile battery parts
6	Polystyrene (PS)	Glassy, sparkling clarity, rigid, brittle, upper temperature use 90°C, soluble in many organic materials	Videocassette cases, compact disc jackets, coffee cups, knives, spoons, forks, cafeteria trays, grocery store meat trays, and fast-food sandwich containers	License plate holders, golf course and septic tank drainage systems, desk top accessories, hanging files, food service trays, flower pots, trash cans, videocassettes

For the big six, once the monomers have been polymerized to form the polymer, they are extruded (pushed out through a mold or nozzle), pelletized, or flaked, making a product known as a resin. These resins are sold to plastic plants; re-extruded, and made into containers, films, and other products. See Figure 14.15.

The Big Six Plastic Polymers

Since 1976, the United States has manufactured more synthetic plastic polymers than the volume of steel, copper, and aluminum combined. Sixty-six percent of all plastics in the United States are low- and high-density polyethylene (LDPE and HDPE), polypropylene (PP), polystyrene (PS), polyvinyl chloride (PVC), and polyethylene terephthalate (PET or PETE). These polymers have the resin recycling designation numbers from 1 to 6 (see Figure 14.16)

Figure 14.15 | Once a polymer is formed, it is palletized, at which point it becomes known as resin. Companies can then sell the resin to other companies that make a variety of plastic products.

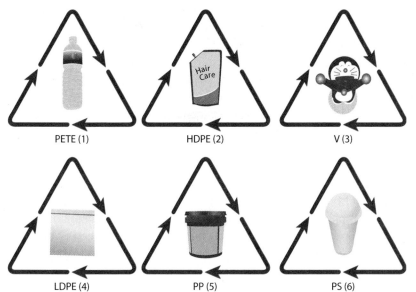

PETE (1) HDPE (2) V (3)

LDPE (4) PP (5) PS (6)

Figure 14.16 | The six major groups of plastics are labeled with numbers so that they can be properly grouped for recycling purposes.

and are often called "the big six." They are summarized in Table 14.2. All other plastic resins are designated with a 7. You can find these designations stamped on the plastic items.

As you can see from the table, many everyday items are made out of plastics. However, these items represent a large problem for disposal, especially since many of them are not biodegradable, that is, they cannot easily degraded by natural processes, such as exposure to sun and water and degradation by the action of bacteria or fungus. In addition, their production costs fluctuate as the cost of the feedstocks and the source of energy for the reactions fluctuates (usually upward). Some of the different resin products can be recycled and made into other items, as shown in the table, but even this has its problems. Recycling itself costs money in that recycled plastic trash has be sorted, and then transported to the appropriate facilities for recycling. Plus, several of the recycled items, for example, polyester clothing made from PET products, cannot be recycled into other plastic products. Green chemistry initiatives push for the development of biodegradable or recyclable plastics, plastics that are better suited for recycling over and over again, and plastics that are made from renewable feedstocks such as plants.

Green Chemistry and Plastics

In the emerging field of green chemistry, researchers area finding new ways to make plastics out of organic materials such as plants and bacteria.

Plastics Can Be Made out of Plants Significant efforts have been made to use ingredients and processes that follow green chemistry principles. Plastics can be made from biodegradable plant materials such as hemp (see Figure 14.17); non-biodegradable polymers like polyethylene can be processed with starch to make biodegradable products such as grocery bags. In addition, using plants to manufacture plastics decreases the use of petroleum feedstocks and provides a high value crop for the farmer. Balanced against these benefits is the reality that food will be diverted from populations that need it, and that the trend will likely promote the agricultural development of wildlife habitat (e.g., rainforest) or recreational land.

Bacteria Can Make Plastics Bacteria can also be used to make plastics (see Figure 14.18). Large-scale production facilities have recently been established for two molecules that can be polymerized to make plastics. A plastic with the trade name NatureWorks® is produced by CargillDow Polymers. The NatureWorks® process is based on feeding certain bacteria with glucose derived from starch to produce large amounts of lactic acid. After recovery from the fermentor, technicians chemically polymerize the lactic acid to polylactic acid (PLA), a biodegradable plastic polymer with many attractive properties for products such as carpet fibers and thin films (see Figure 14.19). By using a biologically derived feedstock (glucose), PLA uses 30 to 50 percent less fossil fuel than is required to produce conventional plastic resins, and the polymer is highly biodegradable.

Another example is the bacterial fermentation of glucose to produce the organic chemical 1,3-propanediol, which, after combining with terephthalic acid, can be polymerized to produce a plastic useful for fiber production. DuPont will market this bio-based plastic under the name Sorona®. Genetic engineering of bacterial strains to achieve high yields of the polymer feedstock has been key to economic production of both of these bio-plastics. In the case of PLA, almost 70 percent of the carbon from the glucose feedstock ends up in the final plastic.

In a third example, genetically modified bacteria are used to produce a plastic known as PHB. The bacteria actually make the final product. Therefore, chemical synthesis is not required to make the final polymer product, as in the first

Figure 14.17 | Plastics can be made from biodegradable plant materials.

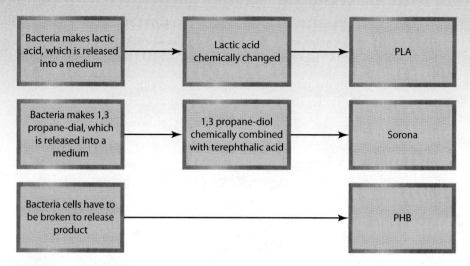

Figure 14.18 | There are several ways bacterial products can be used to make plastics. The first and second processes use bacterial-synthesized feedstocks to make plastics and in the third process, the bacteria make the final plastic product (PHB).

Figure 14.19 | NatureWorks™ is a bacterial plastic product made from the polymer PLA.

two examples. However, the bacteria must be broken open, making this process more expensive. The production of PLA only costs $1–2/kg dry weight of bacteria while the production of PHB costs $4–5/kg dry weight of bacteria. Therefore, although all steps of PHB synthesis occur biologically, the extraction/recovery costs raise the final cost of the product above that of PLA.

Future of Polymer Chemistry

In 2007, the National Science Foundation, in conjunction with several other organizations, held a workshop on polymer chemistry. Based on presentations by workshop participants, a paper was published on the importance of polymer chemistry and its future. It was found that polymer chemistry, already important, was going to be even more important in areas such as health care, communication, production of lightweight but strong structural materials, water purification, and development of new energy sources.

1. What is a polymer?
2. Go online and find information about the commodity company BASF and write a brief paragraph summarizing some of the commodities produced and what industries or companies use these products.
3. What are the "big six" plastic polymers?
4. Go online and find the following information: Prepare a table of the "big six" polymers and indicate which ones are biodegradable and which ones are not. If they are biodegradable, how long does it take for them to degrade in the environment?
5. Go online and find three products for new applications of polymers (e.g., wound care, conductivity of light or electricity). Write a brief paragraph on how these products were developed and how they will be used.

SECTION 4: Focusing on Specialty/Fine Chemicals and Pharmaceuticals

KEY IDEAS >

- While commodity chemicals are made in large batches at low cost, specialty chemicals cost more to make and are made in small amounts.
- Two tools used to visualize proteins are X-ray crystallography and nuclear magnetic resonance (NMR).

While commodity chemicals are made in large batches at low cost, specialty chemicals are made in small batches and cost more to make. Some examples of specialty chemicals are adhesives, additives, antioxidants, corrosion inhibitors, cutting fluids, dyes, lubricants, pigments, cosmetics, and pharmaceuticals. As in any industry, specialty chemical product development can occur in many ways; all of them are based on the needs of the consumer, whether that consumer is another industry or the public. A good example is the development of pharmaceutical products. In the pharmaceutical industry, product development can be the result of researchers investigating the basis for ancient remedies (e.g., discovery of aspirin), the result of a "prepared mind" viewing an accidental occurrence (e.g., discovery of penicillin), or, as it is often done in today's chemical laboratory, the methodical designing of a chemical, in this case a drug otherwise known as a "targeted drug."

Aspirin: An Ancient Remedy

The use of willow bark as a pain medication was recommended by the Greek physician Hippocrates over 2,000 years ago. He used it to alleviate the pain of childbirth and to treat eye infections. Realizing that willow bark must contain an ingredient that alleviates pain, chemists analyzed the chemicals in it and discovered the compound salicin (see Figure 14.20). Aspirin, otherwise known as acetylsalicylic acid, was isolated from this compound and became the first drug to be chemically synthesized.

Figure 14.20 | The original source for aspirin was the bark of the willow tree.

Many drugs on the market today have their origins in nature. In fact, scientists argue this is another good reason for maintaining wild and uncharacterized natural habitats, on the grounds that there may be compounds with medicinal value just waiting to be discovered (Figure 14.21).

Figure 14.21 | Rain forests and other wild environments contain many compounds with medicinal value that are just waiting to be discovered.

 ENGINEERING QUICK TAKE

Penicillin: A Discovery made by "Prepared Minds"

Dr. Alexander Fleming accidentally discovered an important drug, the antibiotic penicillin. In 1928, he was working at St. Mary's Hospital in London on the bacteria *Staphylococcus aureus*, important pathogenic bacteria. Dr. Fleming had left petri dishes of the bacteria by the sink before leaving the laboratory for vacation. After returning to the laboratory, he noticed that the plates were contaminated with mold growth. He was throwing away the plates when he noticed something odd about the plates containing both the mold and the bacteria. There was no bacterial growth near the mold. (See Figure 14.22.) He realized that the mold must be producing chemicals that killed the bacteria. He isolated the mold and later identified it as from the genus *Penicillium* (see Figure 14.23). It was later identified as *Penicillim notatum*. Most likely it came from a laboratory one floor below, where it was being studied.

Fleming presented his findings and published them in the *British Journal of Experimental Pathology* in 1929, but the presentation and article raised little interest. Fleming worked with the mold for some time, but isolating the chemical penicillin was a difficult process for him, as he was not trained as a chemist.

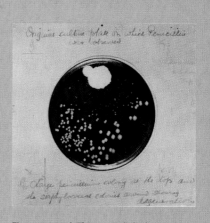

Figure 14.22 | This petri dish shows the antibacterial properties of penicillin; the large white area at the top of the dish is the mold *Penicillium notatum;* the smaller spots are the bacteria. Notice the lack of bacterial growth near the mold; this demonstrates how the mold carves out space for itself in nature by producing penicillin.

Figure 14.23 | Dr. Alexander Fleming accidentally discovered the world's first antibiotic, penicillin.

In 1938, Dr. Ernst Chain, who was part of Dr. Howard Florey's Oxford chemistry team, happened across Fleming's paper on penicillin, and realized the implications of the report.

The team began experimenting with penicillin mold, finding that it cured mice with bacterial infections. They then experimented on a few human subjects who had bacterial infections and saw positive results. However, they could not produce it in large amounts. By this time, it was 1941. England was at war. Florey realized that penicillin was needed to save lives, but the big problem was producing enough penicillin for the number of wounded soldiers that the war was generating. He knew of connections in the United States that could both solve the problem and fund the large-scale production of penicillin. The scientists met with U.S. chemical manufacturers who were producing compounds in large-scale fermentation facilities and using one such facility at an agricultural research center in Peoria, Illinois. They determined how to grow the fungus on a large scale using corn, which was not commonly grown in Britain. The yield of penicillin increased 500-fold. In searching for more productive strains of the mold, the team eventually found a vigorous and productive strain on a rotting cantaloupe from a Peoria, Illinois, market in 1943. When the United States entered World War II, the American government, knowing the benefits of penicillin, funded and enlisted twenty-one chemical companies to produce it (see Figure 14.24). By the end of the war, U.S. companies were making 650 billion units of penicillin per month.

Fleming later said that he felt the chance of that particular mold of growing on that plate at that spot on a Petri dish containing that particular bacterial strain, as well as his noticing how it inhibited the bacterial growth before he threw away the plate, was as slim as him winning the Irish Sweepstakes. (See Figure 14.25.)

Figure 14.24 | Large-scale production of penicillin soared around the time of World War II.

Figure 14.25 | In 1945, Alexander Fleming, along with Howard Walter Florey and Ernst Boris Chain, received the Nobel Prize in Medicine for their work on penicillin.

This story illustrates several traits about successful scientists and technologists: (1) They possess minds that are open and observant of the world around them at all times. (2) They persevere and learn to be patient. It was a long time before another scientist realized the value of Fleming's observations, and, luckily, Fleming was not discouraged at the lack of interest by other scientists and continued to work on the project. (3) They understand the value of collaboration and are willing to search out others to find expertise that they lack.

After the discovery of penicillin, scientists realized that there were probably many different types of potential antibiotics being produced by fungi and bacteria as they competed for survival. They isolated and purified hundreds of natural compounds and tested them to see if they would work, which took a lot of time. In addition, every time they made small chemical additions to these compounds to change the solubility of the original compound or make them work more effectively against their targets, they would have to retest their effectiveness. Sometimes these compounds worked better than the original, and sometimes they did not work at all. It was not until the penicillin was purified, crystallized, and its 3-D structure determined did researchers start thinking about designing drugs based on their structure and how they interact with their target molecules. Two tools used to visualize proteins are X-ray crystallography and nuclear magnetic resonance (NMR). These are both described in the following sections.

TECHNOLOGY AND PEOPLE:
Dorothy Crowfoot Hodgkin

Dorothy Crowfoot Hodgkin (1910–1994) was born in Cairo, Egypt. Both of her English parents were well-known archaeologists and, as a result, she developed an interest in archaeology herself. She became interested in chemistry and in crystals at about the age of ten, an interest encouraged by a friend of her parents, who gave her chemicals and helped her analyze minerals. In fact, she performed some of her early work on archaeological samples. She also lived in the Sudan and in England, where she attended Sir John Leman School and was allowed to join the boys doing chemistry. By the end of her school career, she had decided to study chemistry and possibly biochemistry at university. At Somerville College, Oxford, for her fourth-year research project, she decided to do a project based on X-ray crystallography. A new technique, X-ray crystallography involved viewing chemical structure using X-rays beamed into crystals to determine where the various atoms are located in molecules in three dimensions. The technique required a lot of math to interpret the data generated, and there were no computers to do the number crunching. Because the technique was so new, there were still many challenges to overcome, so not many people were willing to use it. Hence, there were many opportunities for research using this technique. After graduating, she went to Cambridge so she could work with one of the leaders in this new field, John Bernal. They were able to determine the 3-D structure of several organic compounds: something that had not been done before (see Figure 14.26). After receiving her Ph.D. in 1937, she joined the faculty at Oxford. That year, she married Thomas L. Hodgkin, a member of the Oxford history faculty. Dorothy Hodgkin's most significant discoveries were the determination of the organic structures of vitamin B_{12}, insulin, and penicillin using X-ray crystallography. She and her group were the first to determine the structure of penicillin, which took several years because of the difficulties of getting enough crystalline material to use this new technique. Hodgkin's structural information on penicillin proved most useful after the war when semisynthetic antibiotics like ampicillin were developed based on the structure she determined.

In 1964, Hodgkin won the Nobel Prize in Chemistry "for her determinations by X-ray techniques of the structures of important biochemical substances." She was the third woman ever to win the prize in chemistry.

Figure 14.26 | Dorothy Crowfoot Hodgkin determined the 3-D structures of several important organic molecules, including penicillin. Her work helped develop the basis for the "targeted design" approach that is used today by many pharmaceutical chemists.

[Adapted from Mary Ellen Bowden, *Chemical Achievers* (Philadelphia: Chemical Heritage Foundation, 1997).]

Designing Drugs

In order to design drugs successfully, researchers have developed a variety of sophisticated instruments. Today, the main tools are still X-ray crystallography and a technique developed later, nuclear magnetic resonance spectroscopy (NMR). Combined with computer modeling, they are powerful techniques for imaging chemical structures. Today, these tools are used to determine the chemical structure of drugs and how they bind to target sites to cause their therapeutic effects.

For example, scientists have used them to visualize the binding of penicillin to its target, an enzyme on the bacterial cell wall. When penicillin binds, it prevents the enzyme from forming cross-links. If the bacteria continue to grow, eventually the cell wall breaks and the bacteria break open, releasing their contents as shown in Figure 14.27.

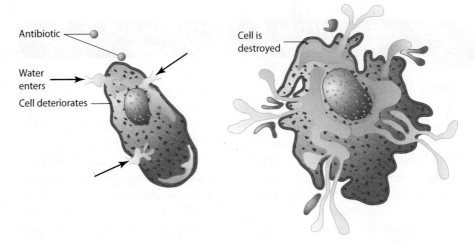

Figure 14.27 | Penicillin binds to a site on the cell wall of certain types of bacteria, preventing the synthesis of the cell wall. The bacteria keep growing, and, eventually, the wall breaks. This allows water to get inside, and the cell breaks open.

Knowing the structure of penicillin and how it binds to this enzyme provides information that researchers can use to create new antibiotics. How do these techniques work, and what do researchers have to do to prepare their samples?

Tools for Visualizing Chemical Compounds and Interactions To perform X-ray crystallography, crystallographers (researchers who specialize in this technique) have to grow solid crystals because crystal bonds are fixed and not moving around, as they would be in a chemical solution. They aim high-power X-rays at a tiny crystal containing trillions of identical molecules (see Figure 14.28). The molecules, based on their structure, scatter the rays onto an electronic detector, similar to how a

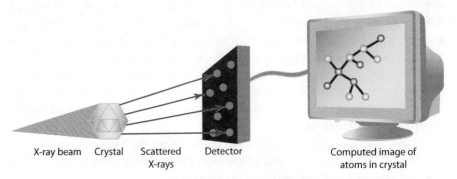

X-ray beam   Crystal   Scattered   Detector   Computed image of
                         X-rays   atoms in crystal

Figure 14.28 | X-rays are directed through the crystal, and the scattered X-rays are collected on a digital screen, similar to a digital-camera screen.

digital camera captures images. After this blast, the researchers rotate the crystal and blast it again. They do this repeatedly until the crystal has been completely rotated and enough information has been gathered to form a computerized 3-D digital image of an individual molecule. (Remember, there are trillions of these molecules in one crystal!)

A major drawback with this technique is that the crystalline form most likely is not the form found in solution. For example, the structure of DNA in a cell is also determined by the water and proteins that are bound to it; crystalline DNA does not have anything bound to it and forms a different structure. NMR examines the structure of proteins in solution, and it can even examine drugs binding to a protein target (see Figure 14.29).

To use NMR, researchers have to know a protein's amino acid sequence (the names and order of all of its amino acid building blocks). Through a series of experiments, the researcher will learn how particular atoms (e.g., carbon-13 and hydrogen) of each amino acid in the protein interact together. This information is processed by a computer program that translates it into a 3-D model of the protein. As you can imagine, this is a lot of data and it usually takes about six months to determine the structure of a small protein and up to a year for a large protein.

How does NMR work? Only the nuclei of hydrogen (H), fluorine (F), phosphorus (P), and certain carbon isotopes (C^{13}) can act as tiny magnets. When you put a compound containing these nuclei into a large magnet, they all orient themselves to align with the magnetic field. Then the molecules are "blasted" with a series of split-second radio-wave pulses that disrupt this magnetic equilibrium in the nuclei of the selected atoms. (See Figure 14.30.)

By observing how these nuclei react to the radio waves, researchers can assess their chemical nature. They measure a property of the atoms called chemical shift, which is how an element's chemical profile changes, or shifts, from one environment to another. Every type of NMR-active atom (hydrogen, fluorine, phosphorus, and carbon isotopes) has a characteristic chemical shift. NMR spectroscopists have discovered characteristic chemical shift values for different atoms (e.g., a hydrogen bonded to an oxygen has a different shift value than a hydrogen bonded to a carbon). To determine protein structures, researchers use a technique called multi-dimensional NMR, which, combining several sets of experiments, even using computer modeling, can take from six months to one year to complete.

By using these two techniques, scientists have developed a large number of molecular models that can be viewed using different software programs. Scientists and technologists use these programs to model drug interactions with protein targets such as enzymes and cell-surface receptors. The next section discusses how this type of modeling was done to design drugs to combat the viral infection HIV.

Designing Drugs for Combating HIV HIV (human immunodeficiency virus) is the virus responsible for the disease AIDS (acquired immunodeficiency

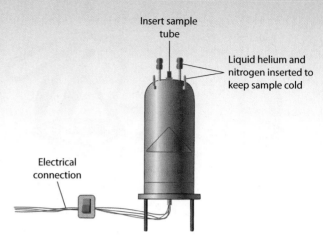

Figure 14.29 | An NMR (nuclear magnetic resonance) machine can provide 3-D information about proteins and chemical molecules in solution; even protein to protein interactions, as well as protein to chemical interactions.

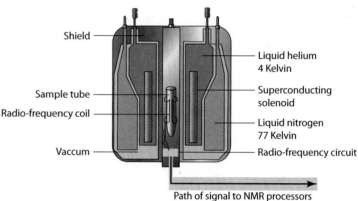

Figure 14.30 | There is a very powerful magnet inside an NMR machine. It can align particular atoms in an element or compound that are tiny magnets themselves, producing a characteristic signal that is detected by the machine, and recorded on a computer. Hydrogen is one example of an element that acts as a magnet and depending upon what element it is bonded to and what elements it is close to, it produces a different signal. For example, a hydrogen bonded to a carbon which is nearby an oxygen will produce a different signal than a hydrogen bonded to a carbon near another carbon.

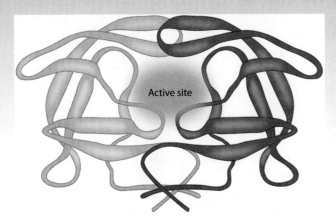

Active site

Figure 14.31 | This computer-generated drawing shows the human immunodeficiency virus (HIV) protease enzyme, which is essential in order for HIV to infect humans. This makes it a good target for drug therapy as knocking it out stops the virus. The enzyme is a symmetrical molecule with two equal but separate halves. The computer-generated model shows how each half, identified as a different color, folds over on itself, creating a site where the target protein can sit. This site is also known as the active site of the enzyme. The target proteins for this enzyme are viral products that are modified to make the mature form of the virus before it buds off the infected cell. If this enzyme does not work, the virus does not mature—only immature, noninfectious viral proteins are released.

syndrome). Presently there is no cure for this virus, but as a result of targeted drug design, infected individuals are living longer. In their efforts to determine how to combat this virus, researchers first determined the life cycle of the virus (e.g., how individuals are infected with the virus, what cells the virus infects, and how it reproduces), and the biochemistry behind its life cycle (for example, what proteins it uses in the cell and what proteins are encoded by its genetic material). From this point, researchers can determine specific HIV proteins to target, yet avoid destroying human proteins. Researchers determined that the HIV protease enzyme, vital to the virus's life cycle, was an appropriate target. This protease enzyme is responsible for generating mature viral particles right before they bud off from the infected cell; without this enzyme, only immature, noninfectious viral particles are released. This enzyme was purified and crystallized. In 1969, X-ray crystallography was used to determine the protease enzyme's 3-D structure. Figure 14.31 shows a computer-generated model using the X-ray crystallography information.

Once they knew the structure, researchers could determine which drug molecules could fit into the active site (the site where the substrate, another material acted upon in a chemical reaction, binds) and inhibit the action of the enzyme. Since both the protein structure and drug structures were known, they could use computational chemistry to select chemicals that would fit in the site. Researchers no longer had to perform expensive, time-consuming and hit-or miss-drug studies using cells or animals to figure out if a drug could work. Essentially, by being able to generate the 3-D structures, the initial screening could be done using computers. Plus researchers can take a drug that doesn't quite fit, change its 3-D structure using a computer model, and then see if it fits better.

What the computer modeling program revealed about the protease enzyme was that it was made up of two identical halves, with the active site in the middle. Pharmaceutical researchers at Abbott realized that the drug molecule needed to possess the same two-fold symmetry of the enzyme. Using a computer modeling program, they took the enzyme's natural substrate, divided it in half, rotated both halves 180 degrees, and "glued" the two halves back together. (See Figure 14.32.) They synthesized this compound, and it fitted perfectly into the active site of the enzyme, inhibiting its activity. However, it was not water soluble so it would not make an effective drug. The Abbott researchers added some water soluble structures to it and eventually ended up with a nonsymmetrical molecule they called Norvir® (Ritonavir). This entire process is known as structure-based design or the "targeted drug approach" to developing drugs.

However, this is not the end of the story for HIV drug development. Like many disease-causing bacterial or viral biological agents, or even cancers, the disease-causing agent can develop resistance to the drug. Why? Because in a biological world, genetic material changes with each generation through mutation or the reproductive process (for example, cell division, viral replication). HIV has an especially high mutation rate, which means that each time the virus replicates, some of the copies are genetically different from the original. For example, DNA mutations may result in a protease enzyme that is active even though the drug Norvir is present. This virus may be present in low amounts, but because it is the only copy that can replicate

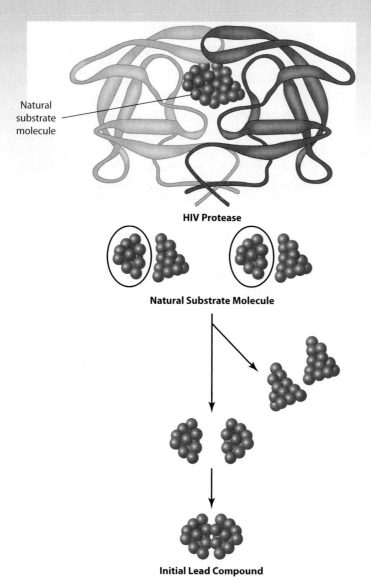

Natural
substrate
molecule

HIV Protease

Natural Substrate Molecule

Initial Lead Compound

Figure 14.32 | Knowing that HIV protease has two symmetrical halves, pharmaceutical researchers initially attempted to block the enzyme with symmetrical small molecules. They made these by chopping in half molecules of the natural substrate, then making a new molecule by fusing together two identical halves of the natural substrate.

in the presence of the drug, it becomes, over time, the predominant strain in that individual. To combat this problem, researchers mapped all of the changes (as shown in red on the enzyme in Figure 14.33) in the active site that prevented the drug from binding but that still allowed the enzyme to work. From this information, they determined what sites were necessary for enzyme activity. Then they designed a drug that bound to some of these sites, inactivating the enzyme. If the virus mutated so that this drug would not bind or act at the site, it did not matter because that mutation made the virus also inactive. Either way, virus replication was stopped.

Of course, this does not eliminate the possibility that a mutation elsewhere in the enzyme could alter the structure of the protease, preventing the binding of the drug. The current approach only decreases the number of drug resistant mutations occurring at the enzyme's active site.

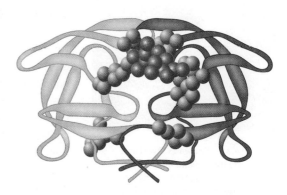

Figure 14.33 | The red balls indicate the sites where mutations in the protease enzyme resulted in drug resistance. Some of these sites are not in the active site, so these mutations must change the structure of the enzyme so that the drug still could not bind to the enzyme.

1. How is a specialty chemical different from a commodity chemical?
2. What did you learn about the discovery of penicillin?
3. Do an Internet search and find another drug, besides aspirin and penicillin, that has been isolated from an organism. Provide information on the organism, how it was discovered, and how it is used in the healthcare industry. Also provide information about how this chemical is used by the organism in nature. (Hint: Try searching with "natural medicinal chemicals" or "natural medical botanical products.")
4. Do an Internet search on the "life cycle of HIV" and also "drug therapy for HIV." Write a paragraph on the life cycle of HIV and how drug therapies target this life cycle to prevent the replication of HIV in the human body.

SECTION 5: Current and Future Chemical Energy Sources

KEY IDEAS >

- Energy production is greatly influenced by the chemical industry.

- Energy fuels are renewable or nonrenewable.

- Examples of possible energy fuels for the future include hydrogen and biofuels.

- Future fuels will be dependent on the cost of feedstocks, the costs of converting process technology, and the costs for developing and implementing the technology.

Figure 14.34 | Known as "The Land of Fire and Ice," Iceland not only has abundant water, but it also has numerous steam vents due to its active volcanoes. This geothermal energy supplies about 10 percent of the country's energy needs, and hydroelectric supplies most of the remainder. Thus, most of Iceland's energy supply comes from clean sources.

Energy production is greatly influenced by the chemical industry. Listed below are different types of energy, their definitions, and how the chemical industry affects them.

Electrical energy is the movement of electrical charges. The charges move through wires that are made in the chemical industry. All of the wires that conduct electricity made by other sources of energy such as radiant, thermal, water, and wind are made in the chemical industry. Structures such as buildings and dams are made of cement and other construction materials that are made in the chemical industry. (See Figure 14.34.)

Radiant energy is electromagnetic energy and includes visible light, X-rays, gamma rays and radio waves. Solar energy is an example of radiant energy, and the newer solar panels are made from plastic.

Thermal energy, or heat, is the internal energy in substances, and geothermal (hot water produced by volcanic activity under the ground) is an example of thermal energy.

Chemical energy is energy stored in the bonds of atoms and molecules. It is the energy that holds these particles together. Biomass (plant and animal materials), petroleum, natural gas, and propane are examples of stored chemical energy.

Nuclear energy is energy stored in the nucleus of an atom—the energy that holds the nucleus together. The energy can be released when the nuclei are combined or split apart. Nuclear energy generation is considered to be part of the chemical industry.

The use of each form of energy is affected by its availability (for example, geothermal energy in Iceland); the cost and development of the energy production technology or devices for extracting it (for example, retrieving oil from below the ocean floor); the development of processes that transform the source into useable fuels (for example, oil refineries); and, finally, the costs of implementing the transformed source (for example, gas for car engines). The chemical industry plays a major role in all of the steps that lead from harvesting the energy source to its final implementation.

Figure 14.35 shows the global demand for energy, which is governed by all of these steps, from availability to implementation.

Countries focus on the energy source that is most economical for them. For example, coal (Figure 14.36) is heavily used in Africa because it is readily available. Geothermal is used in Iceland because of the abundance of thermal vents. As oil reserves dwindle and the cost of oil refining increases, energy companies are experimenting with new technologies, such as hydrogen fuel cells, wind and solar power, as well as fuels derived from vegetation and other organic matter. Nuclear energy is also making a comeback, even though many experts doubt that it will ever be a major source because of its media-poor relationship with the public. However, because of the major investment in oil technology (from refineries to cars) for many decades to come, fossil fuels (petroleum, natural gas, and coal) will continue to supply the bulk of our energy needs; in response, the oil industry will do its utmost to extract every drop of oil from this earth. It may be up to the public and governments to encourage a switch to other sources of fuel.

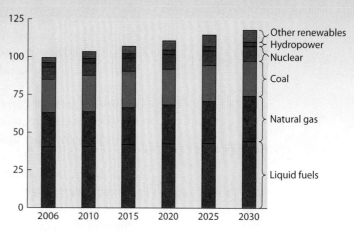

Figure 14.35 | This chart shows global energy demand, starting from the year 2006 and projecting to 2030. Source: *Annual Energy Outlook 2008*, Energy Information Administration, *www.eia.doe.gov/oiaf/aeo/*.

Figure 14.36 | Coal is a fossil fuel made from plant remains protected from biodegradation by water and mud and subjected to geological action over a long time.

Chemical Energy Sources

What makes substances, such as coal, oil, and gas, biofuels, and even hydrogen, usable as fuels? To answer this question, one must consider the chemical and physical properties of the substance; the availability of technology for harvesting, processing, and distributing energy; and, essentially, all of the costs associated with this technology. Energy fuels are renewable or nonrenewable. It is reasonable to assume that, as the demand for energy increases with the population, there will be a switch from nonrenewable, cheap sources to more expensive, renewable sources (although with technological advances, the renewable sources could become cheaper in the future).

In the past and presently, the most common energy-generating chemical reaction is combustion. Combustion is the combination of fuel with oxygen to form product compounds. In these chemical transformations, the potential energy of initial compounds (reactants) is greater than that of the products. It was the combustion of fossil fuels in the steam engine that started the Industrial Revolution about two centuries ago.

Section 5 △ Current and Future Chemical Energy Sources

People used either wood or coal, but they found that coal produced more energy. Coal is a better fuel than wood because it contains a higher percentage of carbon and a lower percentage of oxygen and water. Coal, formed from buried plant material that is subjected to elevated temperature and pressure over a long period of time, is a complex mixture of compounds that naturally occurs in varying grades—carbon, hydrogen, oxygen, and nitrogen atoms come from the original plant material. The lowest grade is soft lignite or brown coal (it has undergone the least amount of change) and the highest grade is bituminous and anthracite, which have been exposed to higher pressure, temperatures, and durations in the subsurface. Generally speaking, the less oxygen a compound contains, then the more energy per gram it will release on combustion because it is higher up on the potential energy scale. There are many detrimental side effects of using coal for energy, such as environmental deterioration (e.g., strip mining) and the production of harmful products (i.e., greenhouse gasses and acid-rain–producing sulfur oxides). However, as readily available oil reserves are depleted, it is likely that the world's dependence on coal will increase. We need coal for energy, but prefer not to have the environmental problems that accompany its use. How do we reconcile this dilemma? Most likely, coal will not be burned in its familiar solid form but rather converted to cleaner-burning and more convenient liquid and gaseous fuels, which will increase the cost of energy.

The Switch from Coal to Petroleum Globally, somewhere around 1950, petroleum surpassed coal as the major energy source for the industrialized world. The reasons are relatively easy to understand. Like coal, petroleum is transformed organic matter, formed from single cell organisms such as diatoms buried in the sediments of oceans and seas, and subjected to high temperature and pressure over a long period of time, but it is a liquid, which means it can be pumped to the surface, rather than mined, and it can be transported as a liquid via pipelines to points of use. Moreover, it is a more concentrated energy source than coal, yielding approximately 40 to 60 percent more energy per gram than coal. There is one property that hindered its acceptance; unlike coal, crude oil is not ready for immediate use when it is extracted from the ground. It must first be refined. In the refining process, the crude oil is separated into compounds (fractions) with similar properties. This fractionation (see Figure 14.37) is accomplished by distillation; a separation process in which a solution is heated to its boiling point and the vapors are condensed and collected.

The refining of a barrel (forty-two gallons) of crude oil provides a variety of products, and the percentages of the refining fractions can be varied. In general, the lowest amount goes to chemical products, including plastics, and the highest amount goes to gasoline and diesel.

Natural Gas No discussion concerning energy compounds is complete without mentioning natural gas, which is mostly methane. A distinct advantage of natural gas is that it burns much more completely and cleanly than do other fossil fuels. Because of its purity, it releases hardly any sulfur dioxide, and only low levels of unburned volatile hydrocarbons, carbon monoxide, and nitrogen oxides. In addition, it leaves no residue of ash or heavy metals. Moreover, natural gas produces 30 percent less carbon dioxide than oil and 43 percent less than coal. Its disadvantages as a transportation fuel are the short driving range it provides—about 100 miles—the heavy and awkward fuel tanks it requires, and the complexity of refueling. It is more suitable to fleet vehicles than to private automobiles.

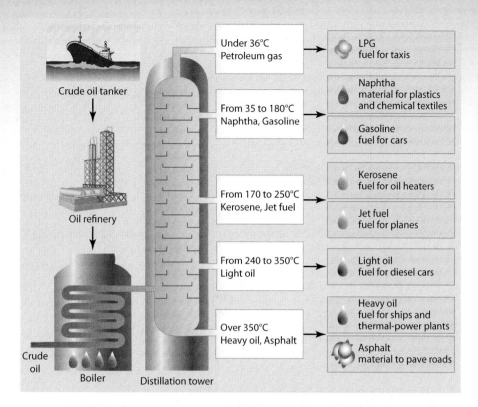

Figure 14.37 | This diagram of a fractional distillation column lists some of the fractions obtained when oil is distilled. The petroleum is pumped into the bottom of the industrial-sized still, and the mixture is heated. As the temperature increases, the components with the lowest boiling points are the first to vaporize. The gaseous molecules move up the still, where they are cooled back to a liquid state, but this time in a purer condition. By varying the temperature of the still and the fractionating column, the chemical engineer can regulate the boiling point range of the fractions distilled and condensed and therefore regulate the amount of each fraction. Notice that the compounds with more carbons atoms, and hence electrons and higher molecular weight, require more energy to volatilize (i.e., a higher boiling point). At the highest still temperature, some compounds (the asphalt-containing fraction) still do not volatilize and are siphoned off the bottom. The largest fraction is the one containing gasoline. In fact, in another type of tower, known as a cracking tower, large carbon-containing compounds (sixteen to eighteen carbons) are converted to smaller carbon compounds to maximize the amount of gasoline produced.

Chemical Fuels for the Twenty-First Century

If cheap fuel products defined the last century, what about the twenty-first century? There are more fossil fuels, such as kerogen, an organic compound found in sedimentary rocks, and tar sands, mixtures of clay or sand, water, and a dense form of petroleum, but their extraction and processing will increase the cost of fuel tremendously, and they are still dirty, nonrenewable energy sources. Examples of possible energy fuels for the future include hydrogen and biofuels. Biofuels (for example, biodiesel and ethanol from fermentation of grains), water, geothermal, and solar will all play a part in increased renewable energy sources.

Hydrogen as a Fuel Source Hydrogen is first on the list in many respects. It is the first element in the periodic table of the elements, containing only one proton and one electron. It is first in energy content of any common fuel and lowest by weight. But since it is the lowest by volume and is a gas at room temperature and normal pressure, it occupies a very large space, which makes it difficult to store and transport. Found in all living things, it is the most abundant gas in the universe, and the source of all the energy we receive from the sun. It is the tenth most abundant element in the earth's crust. Hydrogen as a gas (H_2) does not exist naturally on earth

Hydrogen Fuel Cell

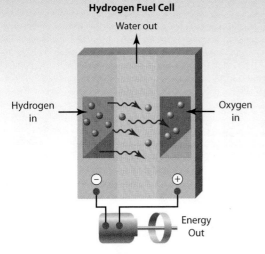

Water out

Hydrogen in

Oxygen in

⊖ ⊕

Energy Out

Figure 14.38 | The space shuttle uses hydrogen fuel cells (batteries) to run its computer systems. The fuel cells basically reverse electrolysis—hydrogen and oxygen are combined to produce electricity. Hydrogen fuel cells are very efficient and produce only water as a by-product, but they are expensive to build. Source: Energy Kids Page, Energy Information Administration.

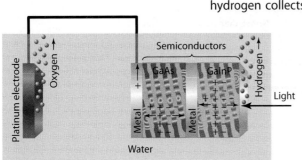

Figure 14.39 | A solar semiconductor device splits water into hydrogen and oxygen.

Figure 14.40 | General Motors displayed this fuel-cell vehicle concept at the 2007 Shanghai Auto Show. However, the high cost of fuel cells may prevent such vehicles from becoming viable in the near future.

in large concentrations but rather in combination with other elements to form compounds. Combined with oxygen, it is water (H_2O). Combined with carbon it forms compounds like methane (CH_4), coal, and petroleum.

However, as an energy source, hydrogen is not without its problems. In addition to taking up a lot of room in its gas form, to cool down and compress it requires a great deal of energy. This means that, storing liquid hydrogen in a small area demands a lot of energy. Plus, since molecular hydrogen (H_2) does not occur naturally except in combination with other elements, before we can utilize its energy, it must be converted to H_2, which also takes a lot of energy.

There are a variety of ways to utilize hydrogen as energy; one way is the fuel cell, in which it is combined with oxygen to form water and release energy in the form of electricity. Figure 14.38 shows the basic structure of a fuel cell.

As you have learned, it takes a great deal of energy to produce hydrogen gas. So how is it being made now, what are the relative costs, and what are the possibilities for the future?

Industry produces the hydrogen it needs by a process called steam reforming, in which high-temperature steam separates hydrogen from the carbon atoms in methane (CH_4). The hydrogen produced by this method isn't used as a fuel but rather as an industrial chemical. This is the most cost-effective way to produce hydrogen today, but it uses fossil fuels both in the manufacturing process and as the heat source, which means it is not a long-term solution.

Another way to make hydrogen is by electrolysis—splitting water into its basic elements—hydrogen and oxygen. Electrolysis involves passing an electric current through water to separate the atoms ($2H_2O + \text{electricity} = 2H_2 + O_2$). Molecular hydrogen collects at the negatively charged cathode and oxygen at the positive anode. Hydrogen produced by electrolysis is extremely pure, and electricity from renewable energy sources can be used to power the process, but it is very expensive at this time. On the other hand, water is abundant and renewable, and technological advances in renewable electricity could make electrolysis a more attractive way to produce hydrogen in the future.

There are also several experimental methods of producing hydrogen. Photoelectrolysis uses sunlight to split water molecules into its components. A semiconductor absorbs the energy from the sun and acts as an electrode to separate the water molecules (see Figure 14.39).

In biomass gasification, wood chips and agricultural wastes are super-heated until they turn into hydrogen and other gasses. Biomass can also be used to provide the heat.

Scientists have also discovered that some algae (e.g. *Chorella*) and bacteria (e.g., *Clostridium*) produce hydrogen under certain conditions. Experiments are underway to find ways to induce these microbes to produce hydrogen efficiently.

The first possible widespread use of hydrogen could be as an additive to transportation fuels.

Hydrogen can be combined with gasoline, ethanol, methanol, and natural gas to increase performance and reduce pollution. Adding just 5 percent hydrogen to gasoline can reduce nitrogen oxide emissions by 30 to 40 percent in today's engines. Because of the cost, hydrogen will not produce electricity on a wide scale in the near future. It may, however, be added to natural gas to reduce emissions from existing power plants. As the production of electricity from renewable sources increases, so will the need for energy storage and transportation. Many of these sources—especially solar and wind—are located far from population centers and produce electricity only part of the time. Hydrogen may be the perfect carrier for this energy. It can store the energy and distribute it to wherever it is needed. It is estimated that transmitting electricity long distances is four times more expensive than shipping hydrogen by pipeline.

Fermentation Products Unlike fuels obtained from petroleum, ethanol (ethyl alcohol) comes from the fermentation of biomass such as grains, corn, and potatoes. All of these sources are renewable. Ethanol combustion does produce carbon dioxide, but it is more carbon neutral than petroleum products since plants use carbon dioxide to make the fermentable products that are turned into ethanol. Also, ethanol has comparatively low toxicity and burns cleanly. Among its disadvantages are its relatively high cost compared to other fuel costs (its production is subsidized indirectly by the government), low energy content, and current lack of abundance. Ethanol has been widely used in gasoline. However, this increases the volatility of the mixture and thereby increases the tendency of the hydrocarbon components to vaporize and potentially escape into the atmosphere.

Biodiesel also comes from biomass, mostly from the oils of plant crops, but also from recycled restaurant greases, sewage, algae, and animal fats. It is easily produced by a transesterification reaction using methanol and triglycerides (plant and animal fats) and that only 20 percent biodiesel/ 80 percent diesel blends can be used in most unmodified diesel engines. Engines do need to be modified for higher percentages of biodiesel. It is safe, biodegradable, and can be distributed using today's infrastructure. Fuel stations are beginning to make biodiesel available to consumers, and a growing number of transport fleets use it to supplement their large fuel consumption. In the United States, a famous singer/songwriter has given it his stamp of approval and in fact is a shareholder in a company that produces a biodiesel product named BioWillie in his honor (Figure 14.41).

Figure 14.41 | Biowillie is one example of a commercially available biodiesel. Would you be more likely to buy it because it is endorsed by the famous singer/song writer Willie Nelson?

Some say, however, that use of biodiesel could increase nitrous oxide emissions and that biodiesel can't be produced in enough quantity without turning more habitable land into farmland.

Methanol costs about as much as gasoline, mile for mile, and produces fewer pollutants. It can be manufactured from a variety of sources, including wood, coal, natural gas, and even garbage. Unfortunately, it corrodes common varieties of steel, has relatively low energy content, and is toxic. As a result, fuel tanks would have to be large, made of strong polymers or expensive, corrosion-resistant stainless steel, and well sealed. One of the most serious, but minor, side-products in methanol is formaldehyde. In addition to being the active ingredient of embalming fluid, formaldehyde is also a carcinogen (it can cause cancer) and an irritant.

Completely changing from gasoline to another fuel source will be a daunting task. Just think of the thousands of fueling stations that would have to be converted, the production and distribution systems that serve them, and the engineering of the engines that would use this fuel. Cost is, in fact, a major factor in converting from any major nonrenewable fuel to a renewable one. Future fuels will be dependent on the cost of feedstocks, the costs of converting process technology, and the costs for developing and implementing the technology. Change will come slowly and will depend on many environmental, societal, and economic factors.

Section 5 △ Current and Future Chemical Energy Sources

1. Why was there a switch from coal as an energy source to oil as an energy source in the 1950s?
2. Go online and research "Vermont gasifier" at the National Renewable Energy Laboratory (NREL) Web site. Write a brief paragraph on how it was developed and who developed it. Include in your explanation the significance of its development and why it is an R&D 100 Award-winning gasification system.
3. At the National Renewable Energy Laboratory (NREL) Web site, find information on the six biomass platforms and write a brief description of each platform.
4. How can hydrogen be used as an energy source? Describe two ways that it can be produced on a large scale. How should it be stored and made available for consumer use?
5. Do an Internet search with "feedstocks for biodiesel" and find out at least three different feedstocks for biodiesel production.

Using Block Flow Diagrams

▷ Chemical engineers have much to consider when designing plants. First, chemists research how elements and compounds interact, and often, on a small scale in a "batch-type" experiment, determine optimal reaction conditions. Chemical engineers take that information, along with their own expertise, and convert it to an industrial process that is either batch or continuous synthesis. Often, a small chemical plant called a **pilot plant** is built to provide design and operating information before construction of a large plant. As part of this design process, the chemical engineer can draw a simplified drawing known as a **block flow diagram** or a more detailed diagram showing pipes and valves known as a **process flow diagram**. In the diagram, each block or rectangle represents a unit operation. The blocks are connected by straight lines to represent the process flow streams from one unit to another. These process flow streams may be mixtures of liquids, gasses and solids flowing in pipes or ducts, or solids being carried on a conveyor belt.

In order to prepare clear, easy-to-understand, and unambiguous block flow diagrams, a number of rules should be followed:

- Unit operations such as mixers, separators, reactors, distillation columns, and heat exchangers are usually denoted by a simple block or rectangle.
- Groups of unit operations may be noted by a single block or rectangle.
- Process flow streams flowing into and out of the blocks are represented by straight lines. These lines should either be horizontal or vertical.
- The flow direction of each process flow stream must be clearly indicated by arrows.
- Flow streams should be numbered sequentially in a logical order.
- Unit operations (i.e., blocks) should be labeled.
- Where possible, the diagram should be arranged so that the process material flows from left to right, with upstream units on the left and downstream units on the right.

In Figure 14.42, the simple block flow diagram shows the production of heat water for a house, with the final delivery point being a bathtub.

A more elaborate block flow diagram is shown in Figure 14 43. It is a typical block flow diagram for an oil refinery.

Figure 14.42 | A simple block diagram of a process that occurs in your house—the production of hot water for use in a bathtub.

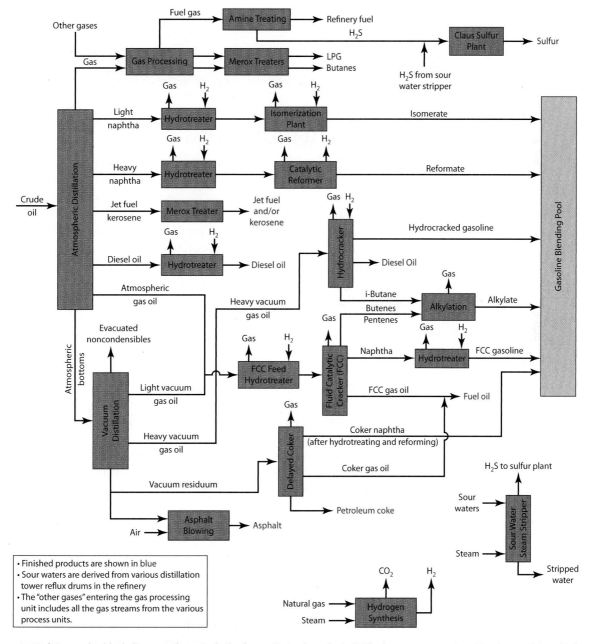

Figure 14.43 | A complex block diagram of a typical oil refinery. Notice how the individual processes are placed in a box and that the flow between the boxes is identified.

Chemical engineers also have to consider plant location, transport of feedstocks and product, employee safety, and cost of materials to build the plant. They also have to weigh the cost of different processes with the cost of feedstock, the amount of energy required, cost of waste disposal, and the cost of disposing of the product (biodegradability).

Besides the heating of water for a tub, there are several processes that occur in your house that could be represented by a block flow diagram; for example, the air conditioning of your room or house. You could find similar processes at school or your church. Pick one process and diagram the flow from one unit, to the next, to the final delivery point using a block flow diagram.

CAREERS IN TECHNOLOGY

Matching Your Interests and Abilities with Career Opportunities: Chemists and Materials Scientists

There are many possible careers in the field of chemistry: Researchers study how atoms, elements, and compounds react. Chemical technicians work in research laboratories, where they set up, operate, and maintain laboratory instruments, monitor experiments, make observations, calculate and record results, and often develop conclusions. Biological scientists study living organisms and their relationship to the environment. They perform research to develop new products or processes. They may specialize in one area of biology, such as zoology or microbiology. Chemists and materials scientists search for and use new knowledge about chemicals and develop processes such as improved oil refining and petrochemical processing that save energy and reduce pollution. This career profile focuses on chemists and materials scientists.

Significant Points

For chemists and materials scientists:
* A bachelor's degree in chemistry or a related discipline is the minimum educational requirement; however, many research jobs require a master's degree or, more often, a Ph.D.
* Job growth will occur in professional, scientific, and technical services firms as manufacturing companies continue to outsource their research and development (R&D) and testing operations to these smaller, specialized firms.
* New chemists at all levels may experience competition for jobs, particularly in declining chemical manufacturing industries. Graduates with a master's degree, and particularly those with a Ph.D., will enjoy better opportunities at larger pharmaceutical and biotechnology firms.

Nature of the Industry

Chemists and materials scientists search for and use new knowledge about chemicals, leading to new and improved synthetic fibers, paints, adhesives, drugs, cosmetics, electronic components, lubricants, and thousands of other products. They also develop processes such as improved oil refining and petrochemical processing that save energy and reduce pollution.

Many chemists and materials scientists work in R&D, where they investigate the properties, composition, and structure of matter and the laws that govern the combination of elements and reactions of substances. They create new products and processes or improve existing ones.

Chemists also work in production and quality control in chemical manufacturing plants. They prepare instructions for and monitor processes to ensure proper product yield. They test products to ensure that they meet industry and government standards and then report and analyze results. Materials scientists study the structures and chemical properties of materials to develop new products and determine ways to strengthen or combine materials or develop new materials.

Working Conditions

Chemists and materials scientists usually work regular hours in offices and laboratories. R&D chemists and materials scientists spend much time in laboratories but also work in offices when they do theoretical research or plan, record, and report on their lab research. Although some laboratories are small, others are large enough to incorporate prototype chemical manufacturing facilities as well as advanced testing equipment. Materials scientists also work with engineers and processing specialists in industrial manufacturing facilities. Chemists do some of their work in a chemical plant or outdoors—gathering water samples to test for pollutants, for example. Workers are required to have protective gear and extensive knowledge of the dangers associated with the job. Body suits with breathing devices designed to filter out any harmful fumes are mandatory for work in dangerous environments.

Chemists and materials scientists typically work regular hours. A forty-hour workweek is usual, but longer hours are not uncommon. Researchers may be required to work odd hours in laboratories or other locations, depending on the nature of their research. Manufacturing chemicals usually is a continuous process, which means that workers must be present twenty-four hours a day, seven days a week. For this reason, split, weekend, and night shifts are common.

Training and Advancement

A bachelor's degree in chemistry or a related discipline is the minimum educational requirement; however, many research jobs require a master's degree or, more often, a Ph.D. Individuals with Ph.D.s can apply for jobs in R&D at companies, state laboratories and faculty positions at four-year universities and colleges. While some materials scientists hold a degree in materials science, degrees in chemistry, physics, or electrical engineering are also common.

In government or industry, beginning chemists with a bachelor's degree work in quality control, perform analytical testing, or assist senior chemists in R&D laboratories. Many employers prefer chemists and materials scientists with a Ph.D., or at least a master's degree, to lead basic and applied research. Within materials science, a broad background in various sciences is preferred.

Advancement among chemists and materials scientists usually takes the form of greater independence in their work or larger budgets. Others choose to move into managerial positions and become natural sciences managers. People who pursue management careers spend more time preparing budgets and schedules and setting research strategy.

Outlook

New chemists at all levels may experience competition for jobs, particularly in declining chemical manufacturing industries. Graduates with a master's degree or a Ph.D. will enjoy better opportunities, especially at larger pharmaceutical and biotechnology firms.

Employment of chemists and materials scientists is expected to grow 9 percent over the 2006–16 decade. Job growth is expected to occur in professional, scientific, and technical services firms as manufacturing companies continue to outsource their R&D and testing operations to these smaller, specialized firms.

Chemists should experience employment growth in pharmaceutical and biotechnology research as recent advances in genetics open new avenues of treatment for diseases. Employment of chemists in the non-pharmaceutical chemical manufacturing industries is expected to decline, along with overall declining employment in these industries.

Employment of materials scientists should continue to grow as manufacturers of diverse products seek to improve their quality by using new materials and manufacturing processes.

Graduates with a bachelor's degree in chemistry may find science-related jobs in sales, marketing, and middle management. Some become chemical technicians, chemical technologists, or high school chemistry teachers. Community college graduates can become chemical technicians, working under the supervision of a person with a master's degree or Ph.D. in a research laboratory. In addition, bachelor's degree holders are increasingly finding assistant research positions at smaller research organizations.

Earnings

Median annual earnings of chemists in 2006 were $59,870. The middle 50 percent earned between $44,780 and $82,610. The lowest 10 percent earned less than $35,480, and the highest 10 percent earned more than $106,310. Median annual earnings of materials scientists in 2006 were $74,610. The middle 50 percent earned between $55,170 and $96,800. The lowest 10 percent earned less than $41,810, and the highest 10 percent earned more than $118,670. Median annual earnings in the industries employing the largest numbers of chemists in 2006 are shown below:

Federal executive branch	$88,930
Scientific research and development services	68,760
Basic chemical manufacturing	62,340
Pharmaceutical and medicine manufacturing	57,210
Testing laboratories	45,730

Beginning salary offers in July 2007 for graduates with bachelor's degrees in chemistry averaged $41,506 a year. In 2007, annual earnings of chemists in non-supervisory, supervisory, and managerial positions in the Federal Government averaged $89,954.

[Bureau of Labor Statistics, U.S. Department of Labor, Occupational Outlook Handbook, 2008–09 Edition, visited *May 2008*, http://www.bls.gov/oco/]

Summary >

Historically, human civilization was biologically based; most clothing, medications, and materials came from plants and animals. Today, most of these items are manufactured from oil and gas using chemical technologies. Chemical technologies provide a means for humans to alter or modify materials and to produce chemical products that make life easier, as in the production of gasoline from oil or the production of penicillin from mold.

Based on the commercialization of chemical products, the chemical industry can be divided into three categories: commodities, specialty/fine chemicals, and energy/fuel. Commodity chemicals are produced in bulk and are relatively inexpensive to make. They are the feedstocks for other industries, such as ethylene for plastics or surfactants (i.e., chemicals that break up or dissolve grease and similar compounds) for detergents. Presently, the "big six plastics" are made from the commodity product, ethylene, the largest petrochemical feedstock made worldwide. However, ethylene is not a renewable feedstock, like plants. With the push for green chemistry (products that are easily recycled or made from renewable feedstocks such as plants or bacteria), it is hoped that the majority of plastics made in the future will be from those feedstocks instead of ethylene or oil.

The second category is specialty and fine chemicals, such as lubricants, adhesives, pharmaceuticals, and cosmetics. Pharmaceutical development and production is an excellent example of the different approaches that are taken to develop and produce new chemicals. For example, there are drug compounds synthetically made but originally known as ancient remedies found in the natural world such as the active ingredient in aspirin. Drugs have been discovered accidentally as a result of keen observation of the natural world such as the inhibition of bacterial growth by the fungal-produced compound, penicillin. Thirdly, in today's world, it is common for chemists to design the drug or chemical based on the required application using techniques like X-ray crystallography or nuclear magnetic resonance spectroscopy (NMR). Designed products can range from a marine paint that must be resistant to salt water, wave action, and the growth of barnacles to a drug that must target a specific site on an enzyme or protein that a pathogenic microorganism needs to infect an individual.

The third category of the chemical industry is energy/fuel. The chemical industry plays a large part in energy production, including electrical, radiant, thermal, and nuclear sources. As global energy demand grows, countries will use the most practical energy sources available to them, which, for many countries, will continue to be oil technology. But as technological advances increase opportunities for and bring down the cost of using hydrogen, biomass fermentation, wind, and solar energy, a long, slow transition to renewable energy sources may be possible.

FEEDBACK

1. What are the three categories that compose the chemical industry? Which one is the largest category?

2. Search on the Internet for the production of ethylene and write a short paragraph on how it is made and determine some of the final products it serves as a feedstock.

3. Briefly define the differences and similarities between a chemist and a materials scientist.

4. As head of human resources, you write job descriptions for your company. Write a brief job description for two people, one person who will determine how to synthesize ibuprofen using "green chemistry" and one who will design the plant and scaled up production process to make the drug.

5. What does a biological scientist do?

6. Why is a chemical engineer known as the universal engineer?

7. Prepare a table listing sources of energy that may replace today's gasoline for our cars of tomorrow. List several of the advantages and disadvantages of each.

8. Go online and find out what biodiesel is and how it is made. Write a brief paragraph explaining your findings.

9. In a table, compare the production of the plastics PLA, PHB, and polyethylene, including the different feedstocks and processes (synthetic chemistry or biological), the amount of energy is required, and whether or not they are biodegradable.

10. The chemical industry provides numerous products. For each category in Table 14.3, list at least five different products. For example, for food, you might list fertilizers and animal feed (but not cat food and dog food as they are the same type of items); do not list brand names. Two categories already have entries to help you get started.

Category	Product
Agriculture and Food	Food Additives
Intermediate and Specialty Chemicals	Biocides (used to slow the growth of bacteria)
Electronics	
Energy and Fuel	
Environmental Technologies	
Health and Medicine	
Housing and Construction	
Industrial Coatings and Colorants	
Paper and Printing	
Polymers, Materials, and Textiles	
Personal-Care Products	

DESIGN CHALLENGE 1:
Manufacturing Glue

• Problem Situation

Your research and development group at the chemical company Solutions has been charged with developing a nontoxic glue for use by young children for school projects.

• Your Challenge

Develop a nontoxic glue that works with paper and wood. Preferably the glue will not discolor with age and the bond will last for at least two weeks.

• Safety Considerations

All chemical companies must adhere to safety regulations that are both for the safety of the employees and the customers who use their products.
1. Always wear gloves and eye protection.
2. Use special care when working with the hot plate or oven.
3. Wipe up all spills.

• Materials Needed

1. Distilled water
2. Nonfat dry milk
3. White vinegar or acetic acid
4. Cheesecloth
5. Baking soda or sodium bicarbonate
6. Metric measuring tools
7. Beakers or bowls
8. Hot plate or oven
9. Thermometer

• Clarify the Design Specifications and Constraints

The glue needs to be nontoxic and develop a bond as soon as possible, with twelve hours, or overnight, being the maximum amount of time. It should bond paper to paper, paper to wood, and wood to wood. The bond should last at least two weeks. Both tensile and shear strength need to be determined. Toxicity to biological life needs to be determined.

• Research and Investigate

To better complete the design challenge, you need to first gather information to help you build a knowledge base.
1. In your guide, complete the Knowledge and Skill Builder I: Safety considerations.
2. In your guide, complete the Knowledge and Skill Builder II: Preliminary glue recipe and testing.
3. In your guide, complete the Knowledge and Skill Builder III: Tensile and shear testing.
4. In your guide, complete the Knowledge and Skill Builder IV: Toxicity testing.
5. In your guide, complete the Knowledge and Skill Builder V: Factor analysis.

• Generate Alternative Designs

Describe two of your possible changes to the glue recipe; remember to consider the specifications and constraints.

571

- ## Choose and Justify the Optimal Solution

Refer to your guide; explain why you selected your final glue recipe and why it is the best choice.

- ## Develop Your Prototypes

Produce your glue prototype. Include any necessary documentation, data measurements, tables, descriptions, photographs, or drawings of these in your guide.

- ## Test and Evaluate

Explain if your designs met the specifications and constraints. What tests did you conduct to verify this?

- ## Redesign the Solution

If you had to redesign your glue formulation, what problems did you encounter? Did you modify your original design concept? Why? If you had to redesign your formulation, what changes would you recommend? Explain your reasoning. What additional trade-offs, if any, would you have to make?

- ## Communicate Your Achievements

Describe the plan you will use to present your glue formulation to your class. (Include a media-based presentation.)

DESIGN CHALLENGE 2:
Photobioreactor

- ### Problem Situation

The power plant company has hired your company, Solar Solutions, to design an algal photobioreactor that will remove much of the polluting nitric oxide (NO), nitrogen dioxide (NO_2), and carbon dioxide gasses being emitted by their plant and then convert it to biomass.

- ### Your Challenge

Given the necessary raw materials, you will design and build the algae photobioreactor, optimizing its operation for sequestration of carbon dioxide and nitric oxide, nitrogen dioxide, and production of biomass.

- ### Safety Considerations

1. Operating the photobioreactor with nitric oxide or nitrogen dioxide feed gas (representing power plant flue gas) is not recommended. It is preferable to use sodium nitrate in the growth medium instead of these gaseous nitrogen oxides as the source of nitrogen for algal growth.
2. Always exercise caution when working with electricity.
3. Assume that circuits are energized.
4. Follow your instructor's safety guidelines for your lab.

- ### Materials Needed

1. At least three inexpensive 2 ft double-bulb fluorescent fixtures
2. 16 gauge two-wire, water resistant electrical cord (12 ft)
3. Small aquarium air pump from local pet store (this air pump injects air into the bottom of the PVC coil such that the medium is "air-lifted" up the tubing coil to the reservoir on top and then is fed back into the coil via a small-diameter plastic tube from the reservoir to the bottom of the coil)
4. 8 ft airline tubing for the pump
5. At least 3 two-prong plugs
6. ¾ inch inner diameter, PVC (vinyl) clear, flexible tubing, 40 ft total
7. Plastic barbed connector(s) for tubing, if unable to obtain continuous 40 ft of tubing
8. ½ mesh hardware cloth to mount the PVC coil
9. Black rubber ¾ inch diameter stopper to plug bottom end of PVC coil. Stopper must be cored to accept 2 metal tubes for attaching air supply tubing and medium feed to the coiled PVC from the medium reservoir on top.
10. Plastic vessel (~1 gal.) to hold medium. It should have leak-proof barbed fittings to attach the 3/4 inch PVC tubing at the top of the coil and the air pump tubing attached to the bottom of the PVC coil.
11. 6 cool, daylight 20W lamps
12. 2 ft × ~13 ft piece of galvanized flashing sheet metal to act as a reflector surrounding the photobioreactor.
13. Cooling fan. The photobioreactor has a 5-inch square fan mounted in the base of the wood support to pull out the hot air generated by the fluorescent fixtures.
14. On/Off wall-socket timer to provide power to the fluorescent fixtures on a light/dark daily cycle of 16 hrs light and 8 hrs dark.

15. *Chorella* culture (ordered from WARDS Natural Science or Carolina Biological Supply). Bristol culture medium composition is included.
16. Chemicals for *Chorella* growth medium

● Clarify the Design Specifications and Constraints

Your design must meet the following specifications and constraints: The photobioreactor should occupy no more than 30 cubic ft and have a footprint no more than 10 square ft. It must be built following safety considerations

1. All wiring must meet code.
2. Maximize algal biomass within the budget constraints.
3. Consider ease of construction, maintenance, and repair.

● Research and Investigate

To complete the design challenge, you need to first gather information to help you build a knowledge base.

1. In your guide, complete the Knowledge and Skill Builder I: Safety considerations.
2. In your guide, complete the Knowledge and Skill Builder II: Algal basics: Choose your algae and growth medium.
3. In your guide, complete the Knowledge and Skill Builder III: Pollution generated by power plants.
4. In your guide, complete the Knowledge and Skill Builder IV: Designing a photobioreactor.
5. In your guide, complete the Knowledge and Skill Builder V: Building the photobioreactor.
6. In your guide, complete the Knowledge and Skill Builder VI: Operating the photobioreactor.
7. In your guide, complete the Knowledge and Skill Builder VII: Measurements for the photobioreactor.

● Generate Alternative Designs

There is more than one solution. Remember that form follows function. The purpose of the 40-ft coil of PVC tubing is to remove gaseous nitrogen oxides. What are alternative designs?

Describe two of your possible alternative approaches to making a photobioreactor. Discuss the decisions you made in (a) choosing the algae culture, (b) designing the photobioreactor, (c) building the photobioreactor, and (d) determining methods for measurement of algal growth, optimization of culture conditions, and sequestration of carbon dioxide and nitrogen oxides. Attach drawings if helpful and use additional sheets of paper if necessary.

● Choose and Justify the Optimal Solution

What decisions did you reach about the design of the photobioreactor?

● Display Your Prototypes

Produce your photobioreactor. Include any necessary documentation, data measurements, tables, descriptions, photographs, or drawings of these in your guide.

● Test and Evaluate

Explain if your designs met the specifications and constraints. What tests did you conduct to verify this?

• Redesign the Solution

What problems did you face that would cause you to redesign the (a) photobioreactor, (b) the algal choice, and (c) the measurement techniques? What changes would you recommend in your new designs? What additional trade offs would you have to make?

• Communicate Your Achievements

Describe the plan you will use to present your solution to an audience. (Include a media-based presentation.)

**BOUT 10,000 YEARS** ago, an agricultural revolution transformed society by allowing families and communities to produce, for the first time, more food than they needed. People accomplished this through the development of tools such as the plow and practices such as fertilization, irrigation (Figure 15.1), and selective breeding.

CHAPTER

15

AGRICULTURAL TECHNOLOGY

Figure 15.1 | Organized agricultural practices such as irrigation, maintaining animal herds for the community, and sharing communal fields were practiced by many ancient civilizations, including the Egyptians.

These advances made it possible for just a few individuals to feed large numbers of people, thus freeing people to pursue other activities than food growing.

As in biotechnology and chemical technology, worldwide events led to rapid advances in agricultural technology. The Industrial Revolution led to the mechanization of agriculture, starting with steam engines and leading to the highly specialized fuel-powered agricultural machines of today (see Figure 15.2).

World War I saw the development and large-scale production of ammonia for explosives, which led to the development and production of nitrogen-based fertilizers. World War II stimulated the development of the chemical warfare agent nerve gas, which provided the chemical basis for the production of pesticides that increased agricultural productivity by decreasing crop loss due to insect damage. These advances in fertilizer and pesticides allowed growers to produce more crops using less land and became known

Figure 15.2 | Improvements in the development of the steam engine gradually gave rise to the Industrial Revolution in the 19th century. Horse-drawn machinery was gradually replaced by steam-powered farm equipment for reaping, threshing, and plowing.

Introduction

as intensive agriculture. Intensive agricultural techniques produced large food surpluses, and the trend became known as the Green Revolution (1955–1985).

It was hoped that these surpluses would end worldwide famine. However, even though food was distributed to countries that could not feed their masses, famine, as a result of poor farming practices and war, continued, and it still persists today. Negative side effects of the Green Revolution's intensive farming and ranching also became apparent. For example, in her famous book *Silent Spring*, Rachel Carson documented the dangers of using the pesticide DDT (dichloro-diphenyl-trichloroethane) (Figure 15.3). As with many pesticides developed during the Green Revolution, it was considered a "miracle chemical," as it decreased the incidence of insect-borne diseases such as malaria and wiped out crop pests.

Figure 15.3 | The Green Revolution spread the use of DDT to destroy insects.

However, DDT use also caused the rapid decrease in populations of predatory birds such as the bald eagle. Researchers found that when animals, such as mice, ingest DDT-contaminated crops or plants near the sprayed fields, the pesticide accumulates in the animals' fat tissue. Predatory birds, such as the eagle, eat these animals, resulting in DDT accumulating in their tissues, which weakens their egg shells (Figure 15.4). The eggs are then inadvertently broken by the parents, decreasing the number of viable offspring. Eventually, DDT use was banned and predatory bird populations have increased. However, now that the incidence of malaria is on the rise in third world countries, people are considering using DDT again.

Figure 15.4 | Bald eagle populations declined sharply because of thinning eggshells due to impaired calcium absorption. This harmful effect was the unintended result of the widespread use of DDT to kill insects, such as mosquitoes, that kill crops and carry diseases.

No one disputes that advances in agricultural technology have allowed the farmer to produce more food than the overall world demand. In fact, many believe that intensive farming and ranching are necessary so that less land needs to be devoted to agriculture to support the ever-growing human population. However, people are now asking, "What will be the final price for the alteration of land, plants, and animals to meet the goal of feeding the masses?" As humankind moves from the Green Revolution to the Agricultural Gene Revolution, otherwise known as the biotechnology era, these questions and others will need to be answered.

SECTION 1: From Green Revolution to the Agricultural Gene Revolution

KEY IDEAS >

- Agriculture includes a combination of businesses that use a wide array of products and systems to produce, process, and distribute food, fiber, fuel, chemicals, and other useful products.

- Biotechnology has applications in such areas as agriculture, pharmaceuticals, food and beverages, medicine, energy, the environment, and genetic engineering.

Agriculture includes a combination of businesses that use a wide array of products and systems to produce, process, and distribute food, fiber, fuel, chemicals, and other useful products. It refers not only to growing plants and trees for food, but also to growing trees for construction materials, cork, rubber, and fuel; plants for their medicinal chemicals; and fiber such as cotton. Agriculture also includes large-scale processing of animals and crops such as pork for bacon and tomatoes for tomato sauce. Finally, agriculture includes storage and transportation of these products—for example, giant grain silos and the packaging and transportation of bananas from South America via cargo ship. (See Figure 15.5.)

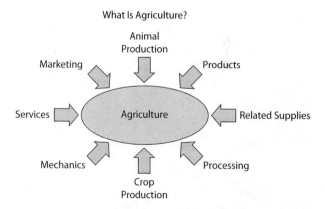

Figure 15.5 | Agriculture encompasses the steps involved in producing a plant or animal produce; starting in the field; harvesting it; storing it and then transporting to a processing plant such as a canning plant; and transporting that product to a point, such as a grocery store, for distribution to the public.

Plants and Trees: Cheap Production of Food

In order to provide food for us and other animals, agriculture relies on the processes plants use to live. To live, plants use the sun's energy (light) to convert carbon dioxide and water into sugar and energy-storage compounds. This process is known as photosynthesis (see Figure 15.6). Then, in a process known as respiration, plants use these sugars to produce energy and make products such as proteins and fats. Animals produce proteins and fats by eating plants containing these sugars. Because plants are the initial source for food, they

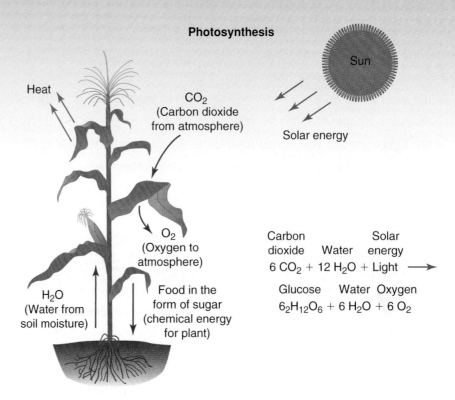

Photosynthesis

Heat

CO_2
(Carbon dioxide
from atmosphere)

Sun

Solar energy

O_2
(Oxygen to
atmosphere)

H_2O
(Water from
soil moisture)

Food in the
form of sugar
(chemical energy
for plant)

Carbon   Solar
dioxide   Water energy
$6 CO_2 + 12 H_2O + Light \longrightarrow$

Glucose   Water Oxygen
$6_2H_{12}O_6 + 6 H_2O + 6 O_2$

Figure 15.6 | During photosynthesis, carbon dioxide and water combine, using light energy, to form sugars. The plant uses the sugars for energy, as do any organisms that consume the plant.

are known as the primary producers. Animals are known as the secondary producers since they eat plants for food. Plant processes to make food are generally inexpensive compared to animal processes since food must be grown and purchased to feed animals. This is why plant protein, such as beans, is commonly cheaper to purchase than animal protein, such as steaks and hamburger.

Agriculture Takes Center Stage Since food is a necessity, agriculture takes center stage in society. It is not surprising that people historically have taken a keen and emotional interest in questions such as "Is the food safe and nutritious?" or "Can my family and I get enough food?" Because we can now get food products from all over the world, another question is becoming important: "Where does this food come from?" Less critical questions reflect how food is integrated into everyday life and special events: "Should we prepare food for the party or buy it premade?" or "Where can I get my favorite coffee?" Another question has affected people for generations: "What food dishes are we going to make for the holiday?"

Figure 15.7 | Food, a necessity, is also a part of our celebrations, such as Thanksgiving, and our daily rituals, such as meeting people for a meal at a special restaurant.

People everywhere ask these questions, reflecting how food is both necessary and integrated into everyday life and special events around the world (see Figure 15.7). If you are a goat herder in the Ahaggar region of the Sahara (Figure 15.8), an important question could be, "How long can I store fermented milk in my animal-skin sack before it goes bad?" (The answer: You can store it for ten days in winter, but only for five to six days in the summer. The cold weather of winter suppresses the bacterial growth that causes the milk to sour).

Figure 15.8 | Many people in the world, such as this goat herder, do not have access to refrigeration for the purpose of storing food products, such as milk, that eventually "sour" as a result of bacterial processes. Therefore they have to figure out other ways and how long they can store food that is sensitive to bacterial action.

Inputs: Moving from Green to Gene

Due to the central role food plays in people's lives, it is not surprising that the Green Revolution and now the Agricultural Gene Revolution both focus on the mass production of food. Both of these revolutions require purchased inputs to succeed. An input is an item that is required for or improves the nature of the agricultural process, for example, a tractor or fertilizer.

Originally, inputs on the farm were derived from the farm itself. All of the grain that the cows on a particular farm ate was produced by that farm, as was all of the fertilizer it used; the farmer did not have to buy any of these items. With the Green Revolution and the development of large-scale, intensive farming and ranching, purchased inputs such as fertilizer and grain became necessary to increase yields.

But agricultural inputs are not limited to materials dug into the soil or sprayed on crops. Other inputs come from the larger economic, political, and scientific environment. These include government policy and new, more efficient farming techniques, as well as plant and animal breeding technology.

Governmental Inputs

Not all farm inputs are purchased. When it distributes land to people, government supplies inputs that promote agriculture, too. In the United States, the Homestead Act of 1862 gave 160 acres of land, free of charge, to any adult citizen who would live on the land for five years and develop it (see Figure 15.9). The key words are "free" and "develop." People could not just subsist on the land; they were required to develop it—and they did so, producing livestock, wheat, and a variety of cash crops.

Interestingly, the U.S. government did not just distribute land for growing crops and ranching; another purpose was to encourage education in agriculture and the mechanical arts. Another important act that

Figure 15.9 | The Homestead Act of 1862 required claimants to improve the land they were granted by cultivating a certain percentage of it for crops and building a residence on it of a certain size. Claimants who fulfilled these obligations within five years received title to the land.

Figure 15.10 | A university building in Kansas, one of the first built under the Morrill Act of 1862.

promoted both agricultural research and a college education for the public was The Morrill Act, also known as the Land Grant Act. It was passed by Congress and signed into law in 1862 by then-President Abraham Lincoln. Before this act, college was reserved for a select and scholarly few, who usually went to school to study Latin, logic, and other classical topics. This act allowed the federal government to give land to each state with the stipulation that it was to be sold and the funds used by the states to establish colleges of agriculture and mechanical arts (see Figure 15.10). Examples of land-grant colleges that have traditionally focused on agriculture and mechanical arts that were established as a result are Kansas State University, Purdue, Rutgers, Cornell, Texas A&M, and Iowa State.

Both of these acts stimulated agriculture in the United States and helped to make it a leader in the Green Revolution and now the Agricultural Gene Revolution.

Governments still support agriculture. As described in the Technology in the Real World box, governments worldwide subsidize (grant money for) the production of crops and livestock. Many countries believe that it is important for their national security to maintain farming and ranching within their own borders so as not to develop a dependence on foreign countries for basic foods such as meat and bread. Without such resources, these countries would face the same problems as many countries that now rely on foreign oil. However, there are major problems as a result of this attitude.

TECHNOLOGY IN THE REAL WORLD: Farm Subsidies and Homeland Security: How Are They Related?

The United States government subsidizes farmers and ranchers. Why? The government believes that the nation cannot afford to depend totally on other countries for the production of necessary foods, even if other countries can make them more cheaply. As a result, grain farmers in the United States are subsidized by the government so that they will continue to grow grain, even though farmers in other countries can grow it more cheaply. Moreover, since these farmers are being paid to grow grain, they can sell it more cheaply, than other countries, so European animal industries prefer buying American grain for feed, even though the European Union produces a grain surplus. This allows European companies to sell their animal products for a cheaper price, meaning cheaper steak and eggs for the consumer. What happens to the grain surplus in Europe? Because transporting it costs too much, it is simply dumped.

Types of Farming and Ranching

As indicated in the introduction, world events and scientific discoveries led to advances in agriculture, such as the purchased inputs fertilizer and oil. However, not all of the agriculture that is practiced in today's world requires purchased inputs, Before reviewing some of the purchased inputs that led to the green and agricultural gene revolutions, consider some of the different types of farming and ranching that exist in today's world (see Figure 15.11). The list in Table 15.1 starts with the least intensive type of farming or ranching, known as "subsistence," and moves towards to the most complex, known as "pharming."

Table 15.1 | Types of Farming and Ranching

TYPES	DESCRIPTION / EXAMPLES
Subsistence	Mainly practiced in developing countries. The farmer only produces what the family needs in order to eat. These farmers commonly do not use a lot of "purchased inputs" to grow plants.
Community and Family	Practiced in many cities worldwide, where people plant vegetables and other plants for food and enjoyment. In a community garden, people rent space for individual plots.
Crop	Fruits, such as apples; vegetables, such as corn; grains, such as rye.
Seed	Production of seed, for example, wheat, soybeans, and corn.
Livestock	Dairy, meat.
Organic	Standards are established by governmental regulation agencies, such as USDA, FDA (e.g., no genetically modified plants or nonnatural pesticides).
Ornamental and Turfgrass	Production of both indoor and outdoor flowering and non-flowering plants and the commercially important group, turfgrasses.
Aquaculture	Production of shrimp, fish, including genetically modified fish, and shellfish.
Hydroponics	The growing of plants in nutrient solutions without soil. Plants can be grown in areas where the soil or climate is not suitable for the crop (e.g., tomatoes for winter use) and in a controlled environment, decreasing pest destruction.
Pharming	Gene-cloned drugs are produced in plants such as corn, tobacco, or bananas. The gene responsible for spider silk is placed in a goat's DNA (cloned) such that it can be isolated from goat's milk.

(a)

(b)

(c)

(d)

Figure 15.11 | **(a) Aquaculture is the science of farming fish and other sea animals and plants. (b) Hydroponics is a way of growing plants using a nutrient-rich solution instead of soil. (c) Various types of turf grasses are used as ground covers for athletic fields, golf courses, and similar venues. (d) A community garden is a shared piece of land in an urban or rural area in which individual residents cultivate their own plots, growing food and plants for personal use.**

Tilling	Planting
Moldboard, disc, and chisel plows	Grain drill Combination fertilizer/planter
 (a)	 (b)

Growing
Fertilizer, Insecticide or Herbicide sprayer Irrigation equipment
 (c)

Harvesting	Storing
Combine, specialized pickers for crops, such as cotton, peanut, potato	Grain elevator, cold-storage bins
 (d)	 (e)

Figure 15.12 | This figure shows modern and antique examples of the variety of equipment used during the various stages of crop production. The chisel plow shown in part (a) is an antique example; the machines pictured in parts (b) through (e) can be seen on working farms today.

As you have learned, agricultural mechanization was an important factor leading to the Green Revolution. It allowed for the development of large farms, automated equipment, specialized farming such as hydroponics, and, through selective breeding, pest resistant plants and improved cows, turkeys, chickens, and other animal species. The next several sections review many of the purchased inputs that were and are part of the Green Revolution and now the Agricultural Gene Revolution.

Farm Machinery

First seen in the Industrial Revolution, farm machinery played a major role in the Green Revolution, and continues to play a significant part in farming today. Machines were developed and optimized based on the particular phase of farming or ranching (for example, tilling, or harvesting) and the farming of specific crops (such as tomatoes or corn) or ranching of specific animals (cows or sheep). In farming the four major processes for crop production, in order, are tilling (preparing the soil), planting, which could including fertilizing at the same time, growing, which includes pest and weed control, and irrigation, harvesting and storing, which includes transportation to market. Examples of this equipment are shown in Figure 15.12.

Facilities for Animals

The industrialization of livestock production allowed the farmer and rancher to efficiently raise huge numbers of animals. Figure 15.13 shows how special facilities for producing chickens and eggs can produce high yields. Many animals are kept in close quarters, fed to maximize size, and harvested as soon as they are market ready.

(a)   (b)

Figure 15.13 | (a) In this intensive system for egg laying, each cage contains about four birds. A combination of natural and artificial light is provided so that birds will lay throughout the winter. Large buildings, some as big as football fields, hold up to 200,000 birds. A sloping wire mesh below the birds permits the eggs to roll down to a collection belt. (b) Chickens bred for meat production are known as broilers. They are bred to convert food into body flesh instead of eggs, so they are allowed more room to move and grow.

Many animals are raised using intensive techniques, including fish and other aquatic life. Harvesting the ocean's bounty started in natural fisheries that occur in nature without human intervention. Fishing in these areas was so successful, however, that some fish populations were reduced beyond recovery. Aquaculture is one answer to this problem. Aquaculture creates and manages controlled water environments to harvest usable plants and animals (see Figure 15.14). Aquaculture farms raise several aquatic species, including catfish, salmon, steelhead trout, clams, crabs, shrimp, prawn, and crayfish, in addition to algae and seaweed.

Figure 15.14 | Aquaculture bins permit the raising of sea life in controlled environments.

Figure 15.15 | In a VMS, cows are automatically fed after they are automatically milked. A cow enters an area where a sensor identifies her ID tag and determines if enough time has lapsed since her last milking. If enough time has passed, the cow is allowed to move into the next area where a multipurpose robotic arm extends underneath the cow to clean and dry each teat before attaching vacuum milking cups. After milking, the cow's underside is sprayed with a disinfectant, then the gate opens and releases the cow to the feeding area.

Other examples of specialized facilities and equipment for animals include automated milking machines, for example, the Voluntary Milking System (VMS). Developed in the 1980, the VMS milks cows without any direct human intervention (Figure 15.15), freeing the farmer's time for other tasks.

Precision Farming

Another technological advance that has increased yields and reduced costs is precision farming. Precision farming uses the global positioning system (GPS) for the purpose of navigation. GPS, originally established by the U.S. Department of Defense, consists of twenty-four satellites orbiting the earth. GPS can determine positions anywhere on Earth twenty-four hours a day, and it collects precise data to guide field operations. For example, it can precisely map information for a particular field, such as nitrogen levels or insect concentrations. This information can be used to precisely and automatically apply fertilizer or pesticide to correct the problem, saving time and costs (see Figures 15.16–15.19).

Laser technology is also used to survey fields and to level land to improve drainage and increase productivity. In laser land-leveling, laser beams transmitted from a base located in a field transmit signals to a receiver connected to earth-moving equipment. Based on the information received, the equipment digs into or adds soil to equalize elevation differences.

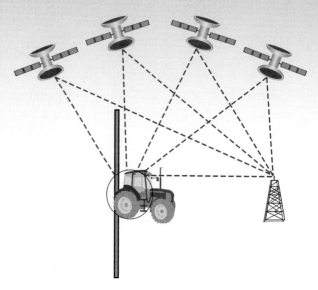

Figure 15.16 | In precision farming using GPS, satellites broadcast signals to a GPS receiver on the ground. After automatic correction for any inaccuracies, the signal is used to control a tractor to an accuracy of five to ten feet.

Herbicide (oz/ac)
- Undefined
- 0
- 16
- 24
- 32

(a)   (b)

Figure 15.17 | (a) A diagram like this is used to automate herbicide spraying. The amount applied varies depending upon plant density; the light blue areas indicate the greatest plant density. (b) An ultra-low–volume sprayer like this one delivers herbicide precisely where needed, decreasing excess run off.

Information Technology and Precision Farming

On large farms, farmers use remote sensing of land and crops by satellite or airplanes in order to adjust irrigation water, adjust fertilizer application, and to determine whether, and where, to spray for pests. This technology is based on how plants reflect and absorb light differently depending upon on leaf color. Leaf color can change due to many factors, such as nitrogen concentration, hydration, and time of year. For example, a lack of nitrogen (i.e., fertilizer) can make leaves turn yellow. A lack of hydration (water) can turn leaves a duller green. Remote sensors can record these changes in light absorption. To use this technology effectively, a farmer first establishes a baseline signature for healthy plants, and then compares it to later signatures to determine if fertilizer is necessary or if the plants need to be watered.

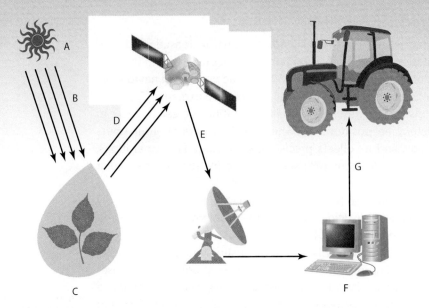

Figure 15.18 | Satellite remote sensing can be used to monitor the health and hydration of plants. The sun (A) emits electromagnetic energy (B) to plants (C). A portion of the electromagnetic energy is transmitted through the leaves. Leaves of healthy or hydrated plants are a different color than leaves of unhealthy or dry plants and therefore reflect light differently. The sensor on the satellite detects this reflected energy (D). The data are then transmitted to the ground station (E). The data is analyzed (F) and displayed on field maps (G).

Figure 15.19 | In this aerial photo, healthy plants are green. Unhealthy crops are yellow green. Light reflected by the two differently colored sections is picked up by the satellite and transmitted to the ground station. The data are used by the farmer to determine the proper action needed to remedy the problem, such as applying fertilizer.

Fertilizers

Traditionally, farmers relied only on manure and crop rotation to keep their land fertile. Later they found that ground bones and rock phosphate enhanced crop production. These traditional methods were replaced by nitrogen-based fertilizers developed after World War I, which became a staple of the Green Revolution. However, as new technologies, such as gene cloning and precision farming, are introduced in developed countries, fertilizer applications decrease. Some genetically modified crops require less nitrogen. Precision farming pinpoints the location and amount of fertilizer needed, therefore decreasing overall usage. However, in developing countries that do not practice these new technologies, fertilizer use is increasing.

Treating Plant Diseases with Herbicides, Pesticides, and Chemicals Agricultural pests include weeds, insects, and plant diseases. The Green Revolution saw the development of herbicides that could kill weeds or prevent them from germinating; the development of chemicals to treat plant diseases; and the development of pesticides, such as DDT, that killed crop-eating and disease-transmitting insects. As a result of good pest management, crop yields increased dramatically. In 1850, a single farmer in the United States could only support himself and four people; but currently a farmer can make enough food for 140 people.

To control pests, farmers and ranchers began practicing what is known today as integrated pest management. Integrated pest management (IPM) is a pest-control strategy that relies on multiple control practices instead of just one. It establishes the amount of damage that can be tolerated before control actions are taken. For example, the corn borer is a major pest (see Figure 15.20). Suppose a farmer has planted 100 acres of corn and finds ten borers in a thirty corn-cob sample. The farmer knows that since the plants are six feet high, harvesting will begin fairly soon. Should the farmer treat the field with insecticide? The farmer has to balance the amount of presumed damage this concentration of borer will do before harvest—and therefore how much money will be lost as a result of the damage—with how much money it will cost to treat the field with insecticide.

From 1940 to 1972 farmers using this type of management relied primarily on chemicals because they produced such great results at low cost. However, the negative effects of these chemicals on the ecosystem were eventually realized, and now farmers try to use other options in the integrated pest management approach. These options include natural biological agents (such as ladybugs, see Figure 15.21), genetically modified plants that are more pest resistant (Bt-corn), and other agricultural practices such as cultural control (i.e., soil tillage, crop rotation, adjustment of harvest or planting dates, and irrigation schemes). Even though IPM most is most commonly applied against pests, it is also used against weeds and plant diseases.

Figure 15.20 | The European corn borer larva invades the ears and stalks of both sweet corn and grain corn. It also attacks other vegetables. In addition to burrowing inside various parts of the plant, it can also introduce fungus and make the plants more susceptible to other infectious diseases.

Figure 15.21 | Some types of ladybugs are among the insect predators of the corn borer larva.

Section 1 △ From Green Revolution to the Agricultural Gene Revolution

Weeds are plants that are considered to be growing out of place and are undesirable because they compete with desirable plants for water and nutrients. Roundup® is an herbicide that is taken up by plant and tree leaves, killing them by inhibiting the production of important amino acids. This herbicide, developed in 1974 by Monsanto, is used worldwide to kill weeds. However, it also kills crops, so it cannot be sprayed aerially; it has to be applied row by row only to the weeds. While this process is labor-intensive, it is still better than manually pulling weeds or letting them grow. However, with the development of genetically altered Roundup Ready® crops (crops that tolerate Roundup®), the herbicide can be aerially sprayed over these altered crops, killing the surrounding plants that cannot tolerate it.

As with other pests, eventually this practice selects for a weed that can naturally resist the herbicide and the value of the Roundup Ready® crops is diminished. Already, there are pests, insects, plant diseases, and weeds that are resistant to the chemicals made by technology, forcing researchers, farmers, and ranchers to try other approaches as part of integrated pest management.

Irrigation Technology

Irrigation has become an important agricultural practice during the past forty years and accounts for much increased food production. Although only 18 percent of the world's arable land is irrigated, this land produces 40 percent of our food. Irrigated land is highly productive but is limited to areas where there is enough available water (see Figure 15.22). Moreover, irrigation in some parts of the world leads to increasing deposits of salts that were dissolved in the irrigation water, and the land gradually becomes less productive.

Figure 15.22 | In modern spray irrigation techniques, water is carried through tubes, then dispersed over crops with attached spray guns. Newer drip-irrigation techniques aim to decrease water lost through evaporation and run off. About 40 percent of the water used in the U.S. goes toward crop irrigation.

New Varieties of Animals and Plants

Technology in the form of selective breeding and genetic engineering has produced new animals and plants.

Selective Breeding Farmers and ranchers have been practicing selective breeding for a long time (see Figure 15.23). Selective breeding refers to the process of choosing animals or plants that have desirable characteristics, such as size, color, or behavior, and breeding them to produce a next generation that will have those characteristics. Over the centuries, it has been used to develop improved varieties of crops and animals, whether it is a sweeter ear of corn or a larger cow for beef production. For example, if a rancher wanted a bigger cow for meat production, he would select the biggest female and male in his herd and breed them to produce a larger breed of cattle.

Genetic Engineering Biotechnology has applications in such areas as agriculture, pharmaceuticals, food and beverages, medicine, energy, the environment, and genetic engineering. In genetic engineering, biotechnology has dramatically increased the speed of selective breeding. For example, BT-corn is genetically engineered corn that contains a pesticide made originally in a bacterium. There are numerous types of RoundUp Ready crops available to the farmer (for example, soy and canola). Other examples are genetically engineered cows that produce more milk, as well as genetically modified salmon (GM) that grow ten times faster than regular salmon.

Transgenic animals are produced by taking the gene of choice from one animal and putting it into an egg that has been removed from a different animal. The egg is then implanted into a surrogate mother. For example, Figure 15.24 shows how transgenic spider silk protein is produced in goats. When these transgenic goats grow up, the spider silk protein is produced in their milk. The milk is collected, and the silk protein is obtained by biochemical separation techniques. This transgenic animal was patented by Dr. Randy Lewis of the University of Wyoming.

Figure 15.23 | Selective breeding has been practiced throughout history to improve animal and crop quality and has significantly improved the world's food supply.

Creating a Transgenic Animal

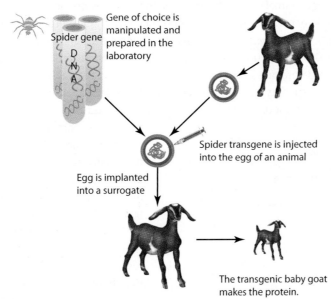

Spider gene

Gene of choice is manipulated and prepared in the laboratory

Spider transgene is injected into the egg of an animal

Egg is implanted into a surrogate

The transgenic baby goat makes the protein.

Figure 15.24 | Transgenic animals are produced by fusing a genetically modified cell (such as the cell carrying the foreign DNA that codes for spider protein) to an isolated egg that has had its nucleus removed. The egg is then implanted into a surrogate mother.

Pharming: Using Genetically Engineered Organisms to Make Products Besides becoming better food sources, animals and plants are being genetically modified to produce other products. This process is known as pharming. In the early 1900s, a song titled "Old McDonald Had a Farm" described the typical animals you might find on a farm—cow, goat, duck, horse, and so on. Old McDonald's farm will never be the same thanks to the Agricultural Gene Revolution. The farm of today and tomorrow could include genetically engineered transgenic cows that possess the genetic codes for a human blood clotting factor, or, as described previously, genetically engineered goats that possess the gene for spider silk.

The production of transgenic plants and animals has countless possibilities: cloned vaccines in transgenic chicken eggs or bananas; cloned drugs in genetically engineered tobacco plants; and many others. Since plants require only the fundamental inputs of water, minerals, sunlight, and carbon dioxide to grow, they make inexpensive chemical factories. Plants are produced for biofuels and as feedstock for making plastics. In addition, like bacteria or mammalian cells, plants can be genetically modified to make protein products.

Besides being grown to obtain a single protein, animals are also being grown for the purpose of harvesting their organs to be transplanted in human beings. For example, pig heart valves are already used routinely in human patients. Researchers are working on genetically altering pigs so that their hearts can be transplanted into humans without being rejected (Figure 15.25). Transplantation of organs between different species is defined as xenotransplantation.

Figure 15.25 | Transplantation of pig hearts into humans may soon be a reality. Their hearts are approximately the right size and have a similar construction. Scientists are working to breed pigs with hearts suitable for human transplantation.

Cloning Animals Selective breeding to produce offspring with desirable characteristics is often accomplished through artificial insemination, in which the semen of an animal with desired traits is transferred to a female. Artificial insemination also prevents animals from being injured during mating. However, breeding prized animals using this process does not ensure the reliable duplication of desired characteristics. This uncertainty became the impetus for cloning (producing duplicate) animals. In this case, the DNA of a desirable animal is used to replace the DNA of an unfertilized egg. The egg is then fertilized in a petri dish and implanted into a surrogate female animal. The first animals to be cloned were livestock, such as Dolly the sheep; prize bulls; and quarter horses. Now even house pets are cloned. Interestingly, owners of cloned animals have discovered that these animals do not behave exactly the same way as their original pets. However, this result is not unexpected, as identical twins do not necessarily act exactly the same even though they share identical DNA.

At first, researchers did not realize that using an older animal's DNA results in a cloned animal that has the characteristics of "old" DNA. This was the case for Dolly the sheep. The DNA used in cloning Dolly was from a six-year-old sheep. Researchers feel that her death at six years of age, half the average life span for a sheep, was a result of using "old DNA."

How is the cloning process done? Review Figure 15.26 to find out.

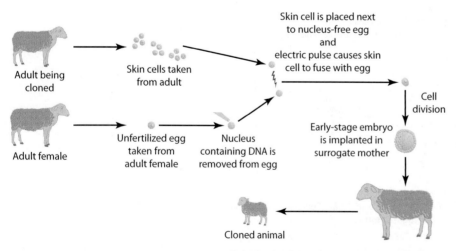

Skin cell is placed next to nucleus-free egg and electric pulse causes skin cell to fuse with egg

Adult being cloned

Skin cells taken from adult

Adult female

Unfertilized egg taken from adult female

Nucleus containing DNA is removed from egg

Cell division

Early-stage embryo is implanted in surrogate mother

Cloned animal

Figure 15.26 | Researchers clone animals by placing a harvested animal cell next to an egg (nucleus removed) of an adult from the same species. They apply an electrical pulse, which bonds the cell with the egg. The fused cell forms an embryo, which they implant into the uterus of another female. This process produces an animal called a "clone," a genetic copy of the animal that produced the cells.

New varieties of plants and animals produced by genetic techniques hold great promise for supplying pharmaceuticals, food, and materials in the next century. However, many practical and ethical barriers need to be overcome before (and if) these products become part of our everyday lives.

ENGINEERING QUICK TAKE

Patents and Genetically Engineered Crops

A landmark Canadian legal case in 2000, Monsanto Canada vs. Schmeiser, raised the issue of patent rights for biotechnology. A Canadian farmer, Percy Schmeiser, was sued by Monsanto Canada for infringing upon their patent rights.

Monsanto, an agricultural chemicals business, develops genetically modified seed lines with pesticide and herbicide resistance. Its genetically modified seeds are patented. Based on a time-honored tradition, farmers like Schmeiser typically collect seeds from good crops and save them as stocks for future plantings. By doing this, farmers and ranchers have selected for the best plants to grow in their area. The practice saves money, as well, as new seed need not be purchased each year. Over several years, Schmeiser built up a large stock pile of seed from which he took just enough to use annually. He did not purchase or use a genetically modified seed.

On the other hand, Schmeiser's neighbors purchased Monsanto's patented Roundup Ready® rapeseed (canola), shown in Figure 15.27. Some of this seed was accidentally wind dispersed from trucks as it was being transported; the seeds ended up on Schmeiser's property. He noticed that Roundup-resistant rapeseed plants were growing on his property and among his own plants when he sprayed for weeds; he noticed that these plants were not affected when he sprayed Roundup on them. He did not destroy those plants. As usual, he harvested crops and collected seed, which he added to his seed stock. Monsanto found out about this, went onto his land without his knowledge, and collected these plants as evidence for their suit against him. The Canadian Supreme Court found in favor of Monsanto and said that he must destroy all of his seed stock and pay Monsanto. Monsanto also sued other farmers but eventually dropped many of these lawsuits, especially when these events were repeated elsewhere in the world, and as these farmers started to purposely collect the seeds.

Monsanto has now developed "terminator gene technology" which makes genetically-modified seeds sterile after that first planting. This was done in response to environmental concerns that genetically modified plants, such as Bt-corn, could harm beneficial insects and accidentally spread the pesticide gene into the wild. Bt is a naturally occurring pesticide made by a bacterium. Bt crops have been genetically modified to produce this pesticide so that particular pests, such as corn borers, will die when they eat the plant. In addition, this technology also prevents spread of the herbicide gene (i.e., resistant to RoundUp). This means that farmers can only grow one crop of genetically modified seed. If they want more of these plants, they have to buy more seed.

Figure 15.27 | Monsanto's Roundup Ready® Canola incorporates a trait that tolerates the herbicide Roundup.

SECTION ONE FEEDBACK >

1. List three agricultural advances resulting from the Industrial Revolution and the World Wars.

2. Two of the new types of farming and ranching in the twenty-first century are aquaculture and hydroponics. Explain what they are and find one product for each type in markets in your city. Research these two products and explain how they were produced.

3. Give two examples of intensive agriculture that result in food products that are an essential part of the diet of many of the world's people.

4. Give an example of how information technology is used in agriculture to better manage production.

5. This chapter mentions the development of BT-corn as an example of a genetically engineered crop. On the Internet, find three other crops that have been developed using genetic engineering techniques.

SECTION 2: Agricultural Engineering Today: Problems and Solutions

KEY IDEAS >

- Conservation is the process of controlling soil erosion, reducing sediment in waterways, conserving water, and improving water quality.

- The engineering design and management of agricultural systems require knowledge of artificial ecosystems and the effects of technological development on flora and fauna.

While today's agricultural technologies and practices have untold potential for improving human lives, they come with problems—some obvious and some subtle—to which many individuals and governments are striving to find solutions.

Negative Side-Effects of Intensive Agriculture

As you have learned, intensive agriculture can have unintended negative consequences, such as species extinction resulting from DDT use, and water loss resulting from widespread irrigation. This section discusses some additional negative consequences of intensive agriculture.

A Decrease in Plant and Animal Varieties Up until the green revolution, there was a high level of biodiversity, meaning that there were thousands of different varieties of plants in cultivation, each with evolved properties that had unique desirable traits that ensured their survival. However, intensive agriculture has led to a decrease in the number of plant and livestock varieties. Instead, there is a tendency towards monocultures, or raising one type of plant or animal. Monocultures can produce an ecological vacuum in which, if a unique variety is susceptible to a particular pest, the entire population of that variety could be wiped out. For example, live oaks and red oaks are popular trees for ornamental cultivation; however oak wilt, a deadly fungal disease, kills those two species in particular, while leaving other types of oaks unharmed. Because these two species alone, rather than a mixture of oak types, were primarily planted in many areas, their destruction can leave huge gaps in ornamental landscapes. As another case in point, if a monoculture of wheat or corn is totally destroyed by unforeseen conditions or pests (Figure 15.28), widespread famine could result in large numbers in underdeveloped countries where people are already living hand to mouth.

Pesticide and Fertilizer Treadmills Monocultures also lead to the pesticide and fertilizer treadmills, in which, as insects and diseases develop resistance to existing chemicals, more chemicals must be developed to prevent destruction of these few plant and animal varieties. In addition, intensive agriculture means more plants and animals are packed into an area, without adequate time for the ground to lie fallow and restore itself. This is an important practice in agronomics, the science of land management and crop production. As a result of decreased fallow time, farmers must use an increasing amount of fertilizer to maintain soil productivity. This leads in turn to more pollution and greater demand on water resources. These practices also contribute to increased soil salinization (see Figure 15.29) and other types of soil degradation, such as loss of nutrients, minerals, organic matter, and living organisms, as well as changes in pH, which affect availability of minerals. Some minerals, such as iron, precipitate (fall out of solution) with a change in pH making them unavailable for plants to absorb.

Figure 15.28 | Monocultures can lead to crop destruction by making them susceptible to attack by a pest or disease that targets that specific plant type.

Figure 15.29 | The overuse of pesticides and fertilizer that characterizes intensive agriculture can increase soil salinization, which eventually leads to land unusable for farming or ranching.

Waste Materials and Inhumane Treatment of Animals The intensive raising of livestock produces a huge amount of waste material that must be treated or recycled as energy and fertilizer. Cows are now so numerous that they have been described as major producers of the greenhouse gas methane, which is four times worse than carbon dioxide in holding in heat when released into the atmosphere. In addition, animal rights groups feel that these huge industrialized animal farms treat the animals poorly. For example, animals are penned up in small spaces to conserve space and, in some cases, to prevent the animal from moving too much and producing a tougher meat product (see Figure 15.30).

Figure 15.30 | Veal stock has traditionally been raised in "growing stalls," which curtailed the spread of disease and increased yield. Largely due to protests by animal welfare activists, these stalls have been banned in England and throughout the European Union. In 2007, the American Veal Association announced that it was phasing out the use of crates for raising veal in favor of housing the stock together, so they can move around freely.

The Greenhouse Effect

▷ When sunlight passes through the glass of a greenhouse, it heats the air inside the greenhouse. If the warmed air is not cooled or blown out, heat builds up under the glass. People exploit this effect for the purpose of growing plants in areas of the world that are too cold for the plant, or for growing plants earlier in a season when it is still too cold outside to begin. This same greenhouse effect, known as **global warming**, is occurring in our Earth's atmosphere. The gasses that surround the earth act like glass, in that sunlight passes through them and heats the air around the earth. Some gasses keep in more heat than other gasses. These gasses are known as **greenhouse gasses**. For example, methane and carbon dioxide keep in more heat than nitrogen and oxygen. Scientists are worried that all of the methane and carbon dioxide being released in the atmosphere as a result of industry is going to heat up the atmosphere too much and negatively impact life on earth. (See Figure 15.31.)

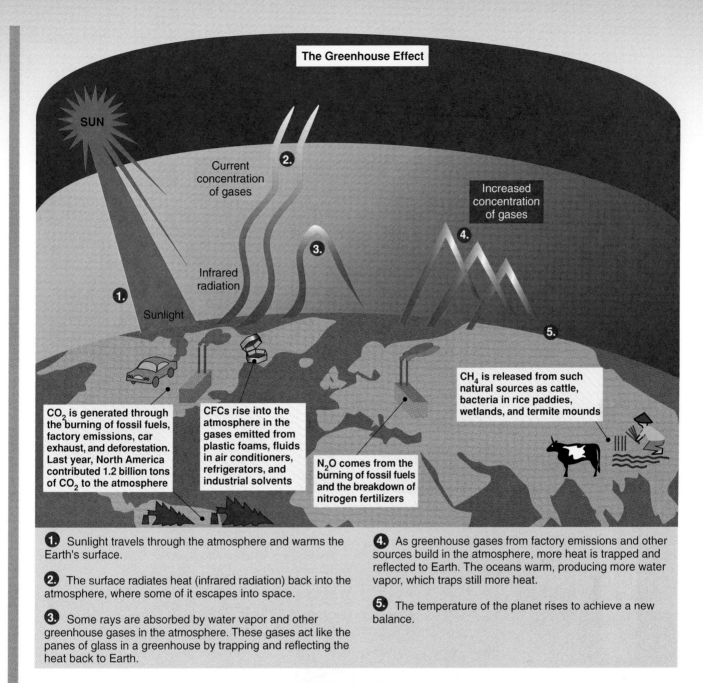

The Greenhouse Effect

SUN

Current concentration of gases

2.

Increased concentration of gases

4.

3.

Infrared radiation

5.

1.

Sunlight

CO₂ is generated through the burning of fossil fuels, factory emissions, car exhaust, and deforestation. Last year, North America contributed 1.2 billion tons of CO₂ to the atmosphere

CFCs rise into the atmosphere in the gases emitted from plastic foams, fluids in air conditioners, refrigerators, and industrial solvents

N₂O comes from the burning of fossil fuels and the breakdown of nitrogen fertilizers

CH₄ is released from such natural sources as cattle, bacteria in rice paddies, wetlands, and termite mounds

1. Sunlight travels through the atmosphere and warms the Earth's surface.

2. The surface radiates heat (infrared radiation) back into the atmosphere, where some of it escapes into space.

3. Some rays are absorbed by water vapor and other greenhouse gases in the atmosphere. These gases act like the panes of glass in a greenhouse by trapping and reflecting the heat back to Earth.

4. As greenhouse gases from factory emissions and other sources build in the atmosphere, more heat is trapped and reflected to Earth. The oceans warm, producing more water vapor, which traps still more heat.

5. The temperature of the planet rises to achieve a new balance.

Figure 15.31 | Global warming is caused by greenhouse gasses that reflect heat back toward the Earth.

Solutions: Sustainable Agriculture and Conservation Practices

Sustainable farming and ranching, also known as agroecology, is defined by the World Commission on Environment and Development as agriculture that "meets the needs of the present without compromising the ability of future generations to meet their own needs." The goals of sustainable farming are to produce crops and livestock in a way that is profitable, gives social benefits to farm families and communities, and conserves the environment. The management of soil, water, plants, and animals must be integrated to produce a healthy ecosystem that promotes diversity of plants and animals.

How should agriculture be practiced to ensure sustainability? Today's and tomorrow's farmer and rancher need to practice conservation to ensure sustainability of resources for agriculture and otherwise. Conservation is the process of controlling soil erosion, reducing sediment in waterways, conserving water, and improving water quality. Farmers can influence all of these factors through the choices they make in managing their farm assets.

Air Quality Air quality is also important to farmers and ranchers but intensive agricultural practices can lead to excessive carbon dioxide production as a result of burning vegetation to clear forestland or to encourage grass growth (see Figure 15.32). Plus even though some nutrients, such as potassium, phosphate, and calcium, are returned to the soil when forests and grasslands are burned, nitrogen, sulfur, and carbon in the plants disappear into the atmosphere as gasses such as sulfure dioxide and nitrogen oxide. Besides contributing to global warming, the resultant gasses cause acidification of lakes and rivers via acid rain. This type of pollution has destroyed forests in Canada and the northeastern United States.

(a)

(b)

Figure 15.32 | (a) Some scientists estimate that two-thirds of the carbon dioxide released into the atmosphere by humans is caused by burning fossil fuels, and one-third is caused by the burning of forest, grassland, and crops. (b) The term "acid rain" refers to rain, snow, fog, or dust mixing with sulfur dioxide and nitrogen oxide from fossil-fuel combustion, creating mild solutions of nitric and sulfuric acid that seep through the ground. This phenomenon increases the acidity of lakes and streams, affecting wildlife and damaging trees.

Figure 15.33 | Since 1980, glaciers in many areas have retreated substantially, and some have disappeared. In Asia, North and South America, Greenland, and West Antarctica, retreating glaciers will, many fear, increase sea levels around the world, with disastrous effects on coastal areas and populations.

Many believe that global warming is a natural phenomenon and that the increases in greenhouse gasses such as methane and carbon dioxide released by man do not contribute to this warming. However, the overwhelming majority of scientists believe that humankind is accelerating this process as evidenced by the loss and the decrease in size of glaciers (Figure 15.33) and the greater variability in worldwide weather patterns (e.g., droughts, typhoons, frosts, heat waves, and floods).

Water and Soil Quality We live on a world that is mainly covered in water, but most of that water is not suitable for humans to drink. This means we need to conserve the water we can drink and not allow it to become polluted. Good water quality is achieved by proper land management, careful water storage and handling, and appropriate use of water. In agriculture, some of the practices that help maintain soil and water quality are:

- Care for gardens and farmlands, improve soil by adding organic matter, mulch plants, and only use the proper amounts and types of lime and fertilizer (precision farming helps with this goal). Only till soil (prepare it for planting) until it will not erode excessively; eroded land loses water quickly. Cover exposed soil with a new crop as soon as possible to prevent soil loss. Typically, farmers and ranchers rotate crops every year or several years to prevent nutrient depletion of the soil (see Figure 15.34). For example, they grow corn for several years in one spot, and then switch to growing crops such as alfalfa in that spot for the next several years , in order to add nitrogen back into the soil (such crops are known as nitrogen fixers). They also plant economically valuable crops together, simultaneously. These combined crops benefit each other's growth. For example, planting alfalfa and clover with oats is mutually beneficial, because the oats protect the young alfalfa and clover plants. The oats are cut for silage (fermented animal food) when the alfalfa or clover plants are larger; at this point, getting rid of the oat plants stimulates the growth of the other plants. These are later harvested as a hay crop (dry, nonfermented animal food). In addition, alfalfa and clover are nitrogen fixers.
- Control soil erosion caused by wind by planting a tree barrier.
- Control damage by hard rains and water runoff by farming on the contour (following the level of the land around a hill) or by building terraces when farming on steep hillsides (as shown in Figure 15.35).
- Control soil erosion and water loss by mulching (adding a top layer, such as pine needles; or even recycled nondegradable rubber pieces) as shown in Figure 15.36.
- Water soil only when it is excessively dry, and then only to a depth of four to six inches.
- Practice rotational grazing to avoid overgrazing, which results in soil damage; animals eat too many plants at one time, killing the plants and leaving soil exposed.
- Practice sensible pest control by using insecticides sparingly to prevent water contamination.

Figure 15.34 | Prolonged use of land for agriculture can deplete soil-nitrogen levels over time. By alternating the planting of crops that use up the nitrogen in the soil, such as corn, with crops that add nitrogen to the soil, such as alfalfa (shown here), beans, and clover, farmers do not need to use excessive amounts of fertilizers to feed nitrogen-requiring crops. Plants that add nitrogen to the soil are known as nitrogen-fixers.

Figure 15.35 | These terraced farms in Vietnam enable farming on steep hillsides.

Figure 15.36 | Mulching conserves water, prevents erosion, fights weeds, and reduces soil compaction caused by heavy rain.

Besides understanding how agricultural technology can impact air, water, and soil quality, farmers and ranchers need to understand how this affects plants and animals in the different environments found on earth. For example, plants and animals, whether domestic or wild, that live in the desert have different requirements to survive in that particular environment than plants and animals that live in the jungle. A particular habitat where specific animals, plants, and microorganisms interact for the purpose of surviving to reproduce is known as an ecosystem. An ecosystem can be as complex as a large forest or as simple as a tree in that forest. All parts of an ecosystem depend on each other. If even one living species is destroyed, it affects all other living organisms, and even the health of resources such as water and soil. People, including farmers and ranchers, who understand what makes an ecosystem sustainable are better stewards of natural resources found in these different ecosystems, including the agricultural ecosystem otherwise known as the agroecosystem. They understand how an agroecosystem like a farm can maintain and not destroy natural resources.

The Importance of Understanding Ecosystems

The engineering design and management of agricultural systems require knowledge of artificial ecosystems and the effects of technological development on flora and fauna. An artificial ecosystem is an ecosystem that is established by humans, such as an agroecosystem. There are many reasons for establishing an artificial ecosystem, such as for research, recreational opportunities, and even for treatment of polluted waters. For example, farmers and ranchers can establish an artificial wetland as part of their agroecosystems to treat polluted water runoff from intensive dairy operations and intensive ranching operations. They can also use the excess manure from livestock to fertilize their crops instead of purchasing huge amounts of fertilizer.

Constructing an Artificial Wetland

▷ Constructed wetlands, just like natural ones, can be used to treat water that has been polluted with pesticides, animal waste, or excess fertilizers. They provide a habitat for microorganisms that can break down materials that cause water pollution. Since the majority of water is not moving or is moving very slowly, the microorganisms have the time to completely break down these materials. Some water plants also take up these pollutants. Wetlands are constructed in a series of rectangular plots. They are filled with gravel or porous soil on a plastic liner so that pollutants cannot leach out. Plants normally found in marshes are added to the plots, and, with time, an important microorganism population is developed. (See Figure 15.37.)

Figure 15.37 | This constructed wetland shows that such areas can be both hospitable to wildlife and attractive to humans.

Finding Appropriate Agricultural Solutions Historically, developed countries have tried to help third world countries develop agriculture and meet the food needs of their populations. Sometimes this assistance was successful and sometimes it was not. One of the ways in which it was not successful is that developed countries tried to get these countries to grow the same crops that they, themselves, could successfully grow, such as wheat and corn. Instead, it would have been more appropriate if they had supported these countries in large-scale farming and ranching of the plants and animals that were already adapted to those environments. For example, in sub-Saharan Africa, farmers now focus on the indigenous plants—cassava (Figure 15.38), sweet potato, yams, and cocoyams—not the grains grown in North America.

Using native plants, farmers could develop pest resistant, high-yield strains. For example, in sub-Saharan Africa, a cassava plant strain that is more resistant to one of its major pests, the green mite, has been developed. The locals have also developed a biologically friendly, oil-based biopesticide derived from an African fungal pathogen of locusts. The biopesticide, nicknamed Green Muscle, has no adverse effect on mammals, and it controls both grasshoppers and the migratory desert locust. It is produced and marketed by the companies named Biological Control Products in South Africa and National Plant Protection in France. These are good examples of what people can do when they understand not only technology but also flora and fauna native to their own countries.

Figure 15.38 | Indigenous to South America, cassava (also called "manioc") is also grown in Africa, India, and Indonesia. Both its greens and its roots are edible, but they must be specially prepared to remove the toxin cyanide in its flesh. Its cooked leaves are a source of protein, and the roots are a source of iron, vitamin C, and vitamin B1. The leaves and the roots are fed to livestock. The cyanide does not seem to bother pigs; perhaps the high pH of the pig stomach destroys the plant enzyme that releases it.

ENGINEERING QUICK TAKE

Calculating the Cassava Crop

In many countries, people must grow all of the vegetables their family is going to eat each year. They either cannot afford to buy them or they cannot buy enough at the market as we can, and if they grow extra, they most likely trade it for other essentials. To be able to do this, each family must figure out how much land to cultivate; which vegetable strains to grow; how much fertilizer they are going to have to add (most likely this will come from their animals rather than being purchased as in developed countries); and also how much water they are going to have to use. As an example, our family in Africa is composed of six people, two parents, two teenagers, and two grandparents. Each one of them will eat about 100 kilograms per year of this tuber, and each tuber weighs on average 0.8 kg. There are about seven to ten tubers per root.

Figure 15.39 | Cassava roots, or tubers, are one of the largest sources of carbohydrates consumed in the world.

They want to be able to grow at least twice as much as they need so that they can use some of it for bartering. How much land should they cultivate for each of these vegetable crops, given the following information?

Each cassava plant should be planted 1 meter from another cassava plant.

If the plants are healthy, the family should expect twenty-eight tons per hectare (ha), if they are unhealthy, ten tons per ha. Each ha is equivalent to 1000m².

Determine the size of the plot in meters squared this family will need to plant.

1. Write a short paragraph explaining how intensive farming and ranching could both help and lead to the destruction of natural ecosystems.
2. Search the Internet for information on the Irish Potato Famine. What was it? When and why did it occur?
3. Search the Internet and find three ways in which the U.S. Environmental Protection Agency says we can reduce acid rain.
4. Search the Internet for information concerning UNESCO's Man and the Biosphere Programme (MAB). Explain the goals for this project and give one example of a biosphere that has been established for research purposes.

SECTION 3: Preserving Food: From Plot to Table

KEY IDEA >

- The development of refrigeration, freezing, dehydration, preservation, and irradiation provide long-term storage of food and reduce the health risks caused by tainted food.

Once food is harvested, it must be stored, processed, prepared, preserved, and distributed before it arrives at our grocery stores. This is not always an easy task because, especially in developed countries, food may have to travel a long way before it reaches its destination. Yet, when consumers decide which apple, potato, or fish to purchase, they expect foods to be not only safe and nutritious, but also the "right" color and texture. The food industry uses a variety of technologies and techniques to preserve fresh food so that it looks appealing and is also tasty and safe to eat.

Preserving Foods in the Food Industry

The problem with fresh food is that it eventually spoils, whether it is stored at room temperature (25°C) or even in the refrigerator (4°C). Fresh food spoils because it becomes contaminated with bacterial and fungal growth. Even if microorganisms did not spoil food, fruits and vegetables contain enzymes that start breaking them down as soon as they are harvested. The reason is because when fruit and vegetables ripen (and eventually fall to the ground) the fleshy part starts to degrade, in order to release the seeds inside them, which then use the nutrients in the pulp to grow. Or, as the fleshy part degrades, it releases an odor in order to attract animals to eat it. The development of refrigeration, freezing, dehydration, preservation, and irradiation provide long-term storage of food and reduce the health risks caused by tainted food.

Preservation technologies include:

- Heating (for example, boiling) to kill organisms or denature (inactivate) proteins, such as enzymes, responsible for decomposition
- Oxidation (such as the use of sulphur dioxide)

Table 15.2 | Food Preservation Techniques

PROCESS	HOW IT PROTECTS FOOD	EXAMPLES
Boiling (sometimes done with high salt or in oil)	Heat destroys most but not all bacteria	All types of foods
Canning	Heat food to kill bacteria and bacterial spores	Fruits and vegetables, beans, soups
Pasteurization	Liquid is heated to 160°F for 20 seconds, then quickly cooled to about 38°F (only kills targeted organisms)	Milk that is to be used in the time indicated by the expiration date
Ultra Heat Treatment (UHT)	Liquid is heated to about 175°F and then to 300°F	Liquid that needs to be stored for several months
Refrigeration	Technically defined as 4° C. Slows down the growth of bacteria and fungi; short-term storage	Milk, fruits, vegetables, meats
Freezing	Uncooked food is steamed first to decrease bacterial concentration. Most foods are frozen by dipping the packaged food into a tank of freezing salt water.	Vegetables, fish, poultry, canned juices, red meat
Drying	Water is removed, causing the concentrations of salt and sugar to increase. The higher concentrations kill or inhibit growth of bacteria living on the food	Powdered milk, soups, potatoes, orange juice, all types of dried fruits, cereal grains
Addition of preservatives such as nitrates, sulfate ions, and acids to decrease the pH < 4.6 (e.g. vinegar)	Inhibition of specific groups of microorganisms	Meats, fruits, catsup
Addition of high salt or sugars is known as pickling; smoke the food product to add growth-inhibitory products	Inhibition of specific groups of microorganisms	Meats, vegetables, fruit
Addition of dyes, vitamins	Make foods more appealing or nutritious	Ruby red grapefruit juice, orange skins, cereal, rice
Irradiation	X-ray and gamma-ray (short wave) radiation kills microorganisms in food. Foods can be irradiated after they are packaged.	All types of packaged foods
Vacuum and oxygen-free atmosphere (also protects foods containing oils that become rancid [oxidized] in the presence of oxygen)	Low oxygen tension inhibits strict aerobes (organism that require oxygen to live) and delays growth of facultative anaerobes (organisms that prefer using oxygen but can live also in its absence).	Nuts, meats, potato products such as artificial (formed) potato chips

- Toxic inhibition to kill micoorganisms (for example, smoking, use of carbon dioxide, vinegar, alcohol, etc.)
- Dehydration (drying)
- Osmotic inhibition (for example, use of sugar syrups)
- Low temperature inactivation (freezing)
- Many combinations of these methods

Well-publicized incidents of unsafe food reaching the population, such as the bacterial contamination of fresh spinach in September 2006 from which several people died, cause people to ask, "How are our foods protected against contamination?" Refer to Table 15.2 for a review of some of the processes, how they work, and examples of foods that are prepared using that particular process.

Processing Food in the Food Industry

Before freshly harvested foods can be made into finished products and sold to consumers, they must be processed. Primary food processing includes transporting the product from the fields to a facility where they undergo preliminary storage and treatment. In secondary food processing, the product is converted into saleable food products and shipped to stores.

For a company that processes potatoes, it is cost-effective to make multiple potato-based food products, such as potato chips, French fries, potato flakes, mashed potatoes, and formed potato products like potato patties. Such a company would have multiple food-processing technologies that convert raw potatoes into finished products. Each product is made on a particular manufacturing process line that performs various processing steps, such as washing, cutting, cooking, and the like.

One process line that interests people in many countries is the processing of potatoes to make French fries. French fries are a popular food, but, depending on how they are made, they are not necessarily very healthy because they may contain a high percentage of fat. Is there any way French fries could be made that is both tasty and healthy? Let's explore all of the different ways that French fries are processed and discover if this is a possibility.

Primary Food Processing In primary food processing for potatoes, the potatoes are first harvested and inspected. When harvesting and storing foods, it is important to harvest at the proper time and to keep them at the proper temperature to avoid damage and spoilage. For example, in Oregon, potatoes are usually harvested from July 1 through the end of November. If they are harvested when the temperature is too low (below 45 degrees F), they may be cut, broken, or bruised by the mechanical harvester. If the temperature is too high (over 65 degrees F), they are too difficult to cool in storage piles. The soil should be moist but not wet or muddy.

A mechanical potato harvester called a combine usually digs several rows at once and moves the potatoes over a series of chains to eliminate soil and vines. Workers on the combine remove decaying potatoes and other debris. The clean potatoes are then moved on a belt and dumped into a truck moving along beside the harvester. The equipment and workers must handle the potatoes carefully so they are not damaged in the harvesting process (Figure 15.40): Only healthy potatoes without mud or field debris can be stored. Depending upon weather conditions, potatoes must be moved directly from field to storage in order to prevent overheating or freezing.

Figure 15.40 | Automatic potato harvesters need to handle potatoes, particularly young potatoes, gently in order to avoid damage that would affect their storage lives and grades.

Before entering final storage, the potatoes must be carefully inspected to eliminate diseased product and sorted into various piles based on grading criteria such as size, weight, color, shape, and amount of damage, if any. This grading and inspection must follow the guidelines and standards established by the U.S. Department of Agriculture (USDA), a federal agency. If products are shipped across state lines or between countries, additional state and federal agencies can become involved. The Food and Drug Administration (FDA) also plays a role, but usually further along in the process, for instance, during secondary food processing and after the product is placed on supermarket shelves.

Secondary Food Processing From storage, the potatoes are transported to food companies for secondary food processing. Secondary food processing involves the different process lines that use the raw potatoes as starting material. In general, the process line for French fries is as follows: (1) The potatoes are mechanically peeled in a hot, steam-pressurized tank that makes the skins literally fly off. (2) They are visually inspected and a pump rapidly propels them to stationary blades for slicing. (3) The correctly sized strips are allowed to continue on to the blanching step (which boils them briefly), removing excess sugars and giving all of them a consistent, pale color after cooking. This step is done for aesthetic purposes only because consumers like French fries that are a uniform color. (4) The strips are partially dried using hot air blasts, partially cooked in hot oil such as peanut oil, and then quickly frozen to –40°F. (5) After freezing, the product is bagged or boxed (see Figure 15.41), then shipped to restaurants and stores. An automated French fry machine can process 1,200 kg/hr. That is a lot of French fries!

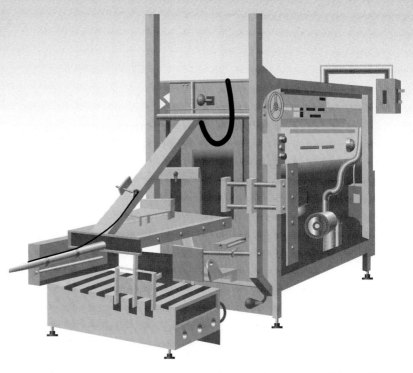

Figure 15.41 | This packing machine is used in the potato industry to pack bags of frozen French fries. It can pack up to 80 bags of French fries per minute.

Checking for Food Contaminants

▷ As part of final inspection, many food products are checked for insect and rodent filth. The Food and Drug Administration (FDA) booklet *Food Defect Action Levels* lists the amount of such filth that is permitted in the foods we buy. For example, 100 grams of peanut butter can contain thirty insect fragments or one rodent hair. The FDA calls these "natural contaminants," which also include fruit-fly eggs and maggots. One university publication estimates that we eat over a pound of insect parts each year without knowing it. The FDA sees these as "natural or unavoidable defects" that "present no health hazards for humans." It also says most food products average much lower levels than their limits; the maximum levels listed in their booklet are those they use to determine if a food is "adulterated" and therefore subject to legal action. (See Figure 15.42.)

Figure 15.42 | Foods that come from agricultural sources may contain contaminants. The FDA regulates acceptable contaminant levels in most foods.

The automated French fry machine uses oil, such as peanut oil, to fry the raw potato strips. This is the step where the fat is added. To make French fries with less oil, companies would have to bake them instead of fry them in oil. Even if the taste of the baked fries is the same as the deep-fried, the texture is different, and texture is a very important part of their appeal. If you search on the Internet, you will discover that many people have tried to duplicate the taste, texture, and smell of a deep-fried French fry, using McDonald's French fries as the "gold standard." The food-processing technology behind a low-calorie French fry is just waiting to be invented.

Country-of-Origin Labeling (COOL)

▷ The Food and Drug Administration (FDA) is responsible for assuring that foods sold in the United States are safe, wholesome, and properly labeled. This applies to foods produced domestically, as well as to foods from foreign countries. The federal Food, Drug, and Cosmetic Act (FD&C Act) and the Fair Packaging and Labeling Act are the federal laws governing food products under the FDA's jurisdiction.

We all recognize country-of-origin labeling. It is on clothes, electronic goods, and some food products. But it is the definition of "some food products" that has people and Congress worried. With recent food poisoning scares, people want to know where all of their food comes from. Right now, loose bananas displayed in a supermarket bin don't have to have a country of origin label, but strawberries packaged in a container must have such a label. Another important exception is available for goods that undergo "substantial transformation." It is recognized that if an imported article undergoes a substantial transformation in the United States. (for example, imported grapes made into jelly), the transformed article does not need to have a country-of-origin label. In this case, the jelly manufacturer is regarded as the "ultimate purchaser" of the imported grapes. Due to these exceptions, many food products do not have a country-of-origin label, even though the food may have been imported or may contain imported ingredients. Voluntary country-of-origin labels and "Made in the USA" claims are a separate matter. Forcing food manufacturers to add this information to their labels or adding more labels to food will cost them money, which, in the end, will cost the consumer more money. Do you think the consumer wants to know this information even though it means raising food costs?

Ensuring that food progresses safely from field to table is a complex task. As foods go through shipping, preservation, storage, processing, and packaging, the combination of technology, engineering, and human judgment in most cases manages to deliver safe and appealing foods to our tables.

TECHNOLOGY AND PEOPLE:
Kristen Hughes

Kristen Hughes (Figure 15.43), who grew up swimming in the Chesapeake Bay, is now a staff scientist with the Chesapeake Bay Foundation. She works with representatives from agriculture, industry, government agencies, universities, and nonprofits to identify ways to prevent agricultural air and water pollution.

While she was getting her degree in Natural Resources Management, she learned how difficult it is to control pollution as a result of excess nitrogen and phosphorous runoff from farms, golf courses, and subdivisions. She works to create new projects and markets for excess nutrients and organic materials. For example, with an industry partner, they are evaluating manure as a possible energy source.

One of Kristen's mentors at the University of Maryland was Professor Pat Kangas, who taught students about the emerging field of ecological engineering and how the combination of ecology and engineering principles could be combined to develop sustainable treatment systems.

After obtaining bachelor's degrees in Natural Resources Management and Biological Resources Engineering, as well as a master's degree in Marine and Estuarine

Figure 15.43 | Kristen Hughes, a natural resource scientist with the Chesapeake Bay Foundation, works with industries to find economically feasible ways to reduce agricultural pollution.

Environmental Science, Hughes decided to pursue a Master's of Engineering Science degree in agricultural engineering. Based on what she has accomplished with that knowledge, she feels it was the right decision.

Her studies in biological resources engineering enabled her to learn the mechanics of how nutrients are transported from land to water, as well as how to measure and prevent these losses. She learned how to measure biological processes in natural systems and develop equations to explain these processes that can be used to design ecologically engineered systems. She has found it especially important to foster powerful partnerships between environmental and agricultural organizations.

Does she find her work satisfying? Apparently so: "I can say that Mondays are not a problem for me—I love my work!"

[Based on information obtained from the ASABE or American Society of Agricultural and Biological Engineers Web site, June, 2008, http://www.asabe.org/membership/TakeFive/Hughes.html]

SECTION THREE FEEDBACK >

1. A targeted organism that needs to be killed in the canning process is *Clostridium botulinum*. Search online for information on this organism, for example: Why it is so hard to destroy in the canning process? What are the symptoms of the food-borne illness it causes? How should canning be properly done?

2. When grains are harvested and stored before being processed, insect and microbial contamination can occur. Search online and find out how insect parts are detected in grains, then find out the "allowable limit" of insect parts in grain.

3. Name five fresh produce products received by the United States, especially during the winter months. Include the country of origin for each.

4. What are three genetically modified crops currently being planted and what percentage are they of the total market for that type of crop (for example, RoundUp Ready® rapeseed).

5. What are some of the other career opportunities for an agricultural engineer besides the one described for Kristen Hughes?

6. Go to the Web site of The National Center for Agriscience and Technology Education (www.agrowknow.org). Examine and list how this organization categorizes careers in agriculture.

7. Go to the Web site developed both by USDA and Purdue known as *USDA, Living Science*, and find the three jobs that interest you most. Describe them and indicate what interests you about these jobs.

CAREERS IN TECHNOLOGY

Matching Your Interests and Abilities with Career Opportunities: Agricultural and Food Scientists

The Bureau of Labor Statistics lists several types of jobs in agriculture, including agricultural engineers, inspectors, workers, managers, and scientists. This section profiles the job outlook for Agricultural and Food Scientists. The work of agricultural and food scientists plays an important part in maintaining the nation's food supply by ensuring agricultural productivity and food safety.

Significant Points

* About 14 percent of agricultural and food scientists work for federal, state, or local governments.
* A bachelor's degree in agricultural science is sufficient for some jobs in product development; a master's or Ph.D. degree is required for research or teaching.
* Opportunities for agricultural and food scientists are expected to be good over the next decade, particularly for those holding a master's or Ph.D. degree.

Nature of the Industry

Agricultural scientists study farm crops and animals to find ways to improve their quality. They research ways to convert crops into attractive and healthy foods and also into fuels. Agricultural and food scientists in biotechnology work with the genetic material in plants and crops to make them more productive, as well as to find commercial applications for this material in industry. Some work with biologists and chemists to develop biofuels.

Some agricultural scientists conduct research in order to understand chemical and biological processes in plants and livestock. Others work to improve the quality and safety of agricultural products, or manage marketing or production in food or agricultural product companies.

Food scientists and technologists usually work in the food processing industry, universities, or the federal government to create and improve food products. *Plant scientists* study plants, helping producers of food, feed, and fiber crops to feed a growing population and conserve natural resources. *Soil scientists* study the chemical, physical, biological, and mineralogical composition of soils as they relate to plant growth. *Animal scientists* work to develop better, more efficient ways of producing and processing meat, poultry, eggs, and milk.

Working Conditions

Agricultural scientists involved in management or basic research tend to work regular hours in offices and laboratories. The work environment for those engaged in applied research or product development varies, depending on specialty and on type of employer. For example, food scientists in private industry may work in test kitchens while investigating new processing techniques. Animal scientists working for federal, state, or university research stations may spend part of their time at dairies, feedlots, farm-animal facilities, or outdoors conducting research. Soil and crop scientists also spend time outdoors conducting research on farms and agricultural research stations.

Training and Advancement

Most agricultural and food scientists need at least a master's degree to work in basic or applied research, whereas a bachelor's degree is sufficient for some jobs in applied research or product development, or jobs in other occupations related to agricultural science. Degrees in related sciences such as biology, chemistry, or physics or in related engineering specialties also may qualify people for many agricultural science jobs.

A bachelor's degree in agricultural science is sufficient for some jobs in product development or assisting in applied research, but a master's or doctoral degree is generally required for basic research or for jobs directing applied research. A Ph.D. in agricultural science usually is needed for college teaching and for advancement to senior research positions. Agricultural scientists who have advanced degrees usually begin in research or teaching. The American Society of Agronomy certifies agronomists and crop advisors, and the Soil Science Society of America certifies soil scientists and soil classifiers.

All states have a land-grant college that offers agricultural science degrees. Many other colleges and universities also offer agricultural science degrees or agricultural science courses. However, not every school offers all specialties.

Agricultural and food scientists should be able to work independently or as part of a team and be able to communicate clearly and concisely, both orally and in writing.

Outlook

Employment of agricultural and food scientists is expected to grow 9 percent between 2006 and 2016, about as fast as the average for all occupations. Recent, agricultural research has created higher-yielding crops, crops with better resistance to pests and plant pathogens, and more effective fertilizers and pesticides. Research is still necessary, however, particularly as insects and diseases continue to adapt to pesticides and as soil fertility and water quality continue to need improvement. This creates more jobs for agricultural scientists.

Emerging biotechnologies will play an ever larger role in agricultural research. Scientists will be needed to apply these technologies to the creation of new food products and other advances. Moreover, increasing demand is expected for biofuels and other agricultural products used in industrial processes. Agricultural scientists will be needed to find ways to increase the output of crops used in these products.

Earnings

Median annual earnings of food scientists and technologists were $53,810 in May 2006. The middle 50 percent earned between $37,740 and $76,960. The lowest 10 percent earned less than $29,620, and the highest 10 percent earned more than $97,350. Median annual earnings of soil and plant scientists were $56,080 in May 2006. The middle 50 percent earned between $42,410 and $72,020. The lowest 10 percent earned less than $33,650, and the highest 10 percent earned more than $93,460. In May 2006, median annual earnings of animal scientists were $47,800.

The average federal salary in 2007 was $91,491 in animal science and $79,051 in agronomy.

According to the National Association of Colleges and Employers, beginning salary offers in 2007 for graduates with a bachelor's degree in animal sciences averaged $35,035 a year; plant sciences, $31,291 a year; and in other agricultural sciences, $37,908 a year.

[Bureau of Labor Statistics, U.S. Department of Labor, Occupational Outlook Handbook, 2008–09 Edition, visited December, 2008, http://www.bls.gov/oco/]

Summary >

Over the course of history, society has moved from mere subsistence to the use of sophisticated tools and techniques to produce food for entire societies. These tools and techniques evolved to create the Green Revolution, which saw the introduction of intensive farming techniques. Today, agriculture consists of an array of products and systems to create food, fiber, fuel, and other products to supply world populations. However, this evolution has featured the continuous use of purchased inputs, such as chemical fertilizers and pesticides. These inputs have not only increased agricultural production, but they have had negative effects on the environment.

Humankind has now turned to the agricultural gene revolution, also called the biotechnology era, for solutions to many of these problems. Biotechnology advances in agriculture, pharmaceuticals, food, medicine, energy, the environment, and genetic engineering.

Farming and ranching in today's world ranges from the most basic food production, through genetically produced food and drugs. The main areas include:

- Subsistence
- Community and family
- Crop
- Seed
- Livestock
- Organic
- Ornamental and turfgrass
- Aquaculture
- Hydroponic
- Pharming

Farm machinery now plays a central role in tillage, planting, growing, harvesting, and storing crops. The industrialization of livestock production has led to large, intensive environments for animals and plants.

Information technology now plays a large part in agricultural production, from automatic milking machines, to the use of GPS and remote sensing to monitor and treat fields. These developments have drastically reduced the amount of human labor required and resulted in the more efficient use of inputs, such as fertilizer and water. The use of integrated pest management, natural biological agents, genetically modified plants and animals, and culture control has had similar benefits.

Intensive agriculture has had negative effects on the environment and has decreased animal and plant varieties. The new field of agroecology aims to create more sustainable farming and ranching techniques that emphasize conservation, improved air, water, and soil quality to create healthy agroecosystems worldwide.

Agricultural products, once harvested, must be stored, processed, and made into saleable products for human and animal consumption. Techniques that preserve foods and make them attractive to consumers include:

- Boiling
- Canning
- Pasteurization
- Ultra Heat Treatment
- Refrigeration
- Freezing
- Drying
- Preserving
- Pickling and smoking
- Addition of dyes and vitamins
- Irradiation
- Oxygen removal

Primary and secondary food processing moves agricultural products from the field to the consumer. Each step along the way must be carefully controlled to ensure that the final product is safe, tasty, and attractive. Many processing steps are streamlined to increase speed and reduce costs. Issues of food adulteration and country-of-origin labeling draw increased attention to the safety of foods that reach our tables.

FEEDBACK

1. Pick one country in the world and see if you can discover up to three pivotal events that influenced their current agricultural practices.

2. Find one company that makes a product using a transgenic animal or plant and summarize briefly the cloned gene(s) and the product.

3. Conduct a survey of your family and friends to discover their thoughts on why people in the world are starving even though enough food is being produced worldwide to feed every person. If they lack adequate background knowledge (and most will), inform them of what you have learned in this chapter. Ask them how they think this problem can be solved. How do they think the public can be better informed about these issues?

4. Discuss global warming, its likely causes and effects on Earth, and its proposed effect on food production capacity.

5. How should the political process treat farming? Should there be subsidies, and to whom should they go? How is farming subsidized in your geographic locale?

6. How has plant and animal breeding been modified over the years?

7. Plants can be "stressed out," just as people can. Discuss plant stresses and the responses of the plants.

8. Are foods produced by traditional plant-breeding methods safer to eat than those produced by genetic engineering? Why or why not?

9. How does traditional plant and animal breeding differ from genetic engineering?

10. Discuss the steps required to a) introduce a new gene into an animal or plant by traditional methods and b) introduce a new gene by genetic engineering.

11. In what way do seed crops differ from food crops? What are some specific procedures for seed production that might not be necessary for producing the crop? Discuss the patenting of genes. How should the people who preserved genes be compensated by companies interested in those genes?

12. Discuss the gradual loss of biodiversity resulting from agricultural practices.

13. Search the Internet for information about Quality Protein Maize. How can this product improve maize value in developing countries?

14. Plant breeders can select for improved plant response to predictable environmental stresses. Describe two environmental conditions, such as drought, that you believe breeders should consider.

15. Pest outbreaks are natural, but how have they been increased by intensive agricultural practices?

16. Are synthetic pesticides more or less dangerous to people than organic pesticides? Which agency in your country determines the safety of a pesticide?

17. What is a weed? Give a definition from different perspectives—even from the plant's perspective!

18. How does an herbicide's target site influence the likelihood that resistant weed populations will be selected?

19. What are the social and environmental ramifications of only a small percentage of the U.S. population being involved in agriculture and ranching?

20. Search online for the FDA pamphlet "Food Defect Action Level" and find the acceptable levels of insect and rodent filth for five of the food products in your kitchen. Follow up with additional research and find the test methods for insect and rodent filth.

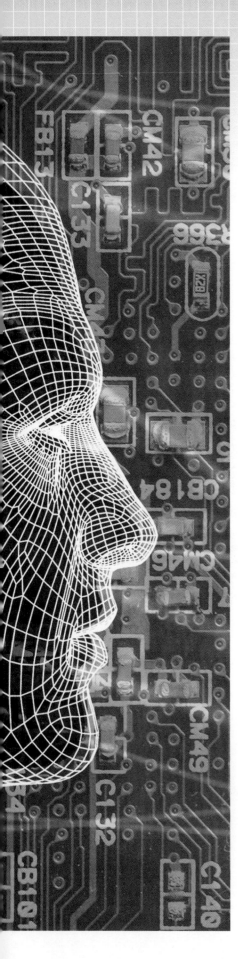

DESIGN CHALLENGE 1:
Plant Tissue Culture

• Problem Situation

Plant tissue culture is an important area of biotechnology. Many strategies for genetic engineering of plants rely on plant tissue culture. Tissue culture allows for pieces of plant tissues to grow into whole plants. Every cell within a plant has all the characteristics of the whole plant. In the proper conditions, every cell is capable of growing into a new plant. Each new plant has the same characteristics as its parent plant. This allows a grower to produce many plants in a short time from an individual piece of plant tissue. It also allows for the production of many plants in a very small area.

• Your Challenge

You and your R&D team members have been given the task of developing a plant tissue culture kit in order to produce an identical clone of a plant.

• Safety Considerations

1. Wear eye protection at all times.
2. Wear gloves at all times.
3. Disinfect your work area with ethanol to decontaminate.
4. Thoroughly wash and disinfect all glassware and instruments.
5. Ethanol is highly flammable and should be kept away from open flames at all times.
6. Do not pour any unused chemicals back into storage containers where it may contaminate the rest of the reagent. Dispose of unused chemicals in proper waste containers.
7. Bleach solutions will discolor clothing and can be harmful to the skin and eyes. In addition to gloves and goggles, you may want to wear an apron or lab coat while you are handling bleach solutions.
8. Long hair should be tied back to minimize contamination.
9. The edges of scalpels and razor blades are extremely sharp and should be handled with a great deal of caution.

• Materials Needed

1. Plant
2. Multiplication medium with agar in sterile containers
3. Mist bottle
4. 70 percent ethanol
5. 10 percent chlorine beach
6. 3 sterile wide-mouth containers
7. 2 beakers
8. Sterile forceps
9. Sterile razor blade/scalpel
10. Parafilm or florists tape
11. Cool-white fluorescent lights (optional)
12. Plastic box
13. Shoot multiplication medium
14. Soil
15. Fertilizer
16. Planting pot
17. Plastic bag

• Clarify the Design Specifications and Constraints

You will design a kit and clone a plant using plant tissue culture techniques. The first step in culturing plant tissue is to establish an explant in sterile culture. Explants are usually small pieces of leaf that have been treated to kill surface microorganisms that would contaminate the culture. The explant is grown in a medium that contains organic salts, vitamins, and hormones for shoot development. Shoots are the developing stems and leaves on the newly forming plant. Shoots formed on explants will then be transferred to shoot multiplication medium, which contains hormones that allow shoots to elongate and form roots. Finally, whole shoots are transferred to soil where mature plants will develop. You should produce a viable clone of your original plant.

• Research and Investigate

To complete the design challenge, you need to first gather information to help you build a knowledge base.

1. In your guide, complete Knowledge and Skill Builder I: Safety considerations.
2. In your guide, complete Knowledge and Skill Builder II: Building a sterile hood.
3. In your guide, complete Knowledge and Skill Builder III: Choose a plant and corresponding medium.
4. In your guide, complete Knowledge and Skill Builder IV: Transfer the explant.
5. In your guide, complete Knowledge and Skill Builder V: Root and shoot formation.
6. In your guide, complete Knowledge and Skill Builder VI: Transfer to soil.

• Generate Alternative Designs

Describe two of your possible alternative approaches in this plant tissue culture exercise. Discuss the decisions you made in (a) building a sterile hood, (b) transferring the explants, and (c) root and shoot formation. Attach drawings if helpful and use additional sheets of paper if necessary.

• Choose and Justify the Optimal Solution

What decisions did you reach about the optimal conditions for producing a clone using plant tissue culture techniques?

• Display Your Prototype

Produce your sterile hood and final cloned plant. Include descriptions, photographs, or drawings of the process in your guide.

• Test and Evaluate

Explain whether your designs met the specifications and constraints. What tests did you conduct to verify this?

• Redesign the Solution

What problems did you encounter that would cause you to redesign the sterile hood or modify the procedure for transferring the explant or the root and shoot development? What changes would you recommend in your new designs? What additional trade-offs would you have to make?

• Communicate Your Achievements

Describe the plan you will use to present your results to your class, and show what handouts you will use. (Include a media-based presentation.)

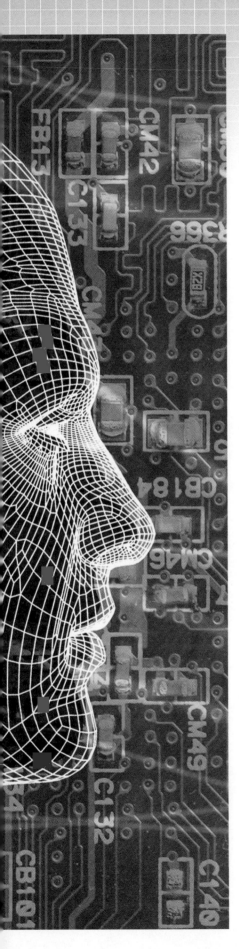

DESIGN CHALLENGE 2:
Vegetable Garden Plot

• Problem Situation

Your family would like to invest in a garden plot for the purpose of growing much of the vegetable produce they annually consume. Realizing that you have just finished reading about agricultural technology, they give you the assignment of figuring out how to establish the plot that will meet their needs.

• Your Challenge

Determine the plot size necessary for your family, its location, which vegetables need to be grown and when, how many plants of each vegetable are needed, and what soil amendments need to be added based on the soil conditions found in your area. You will also have to determine how much water needs to be added to the plot based on the vegetables being grown and the climate conditions of your area. If insect pests are anticipated, you need to determine what treatments you are willing to do to prevent crop damage.

• Safety Considerations:

1. Should you decide to construct your garden, you will have to make sure you use tools correctly.
2. Wear rubber gloves and eye protection when applying chemicals.
3. Be sure to use sunscreen and a hat if you are outside on hot, sunny days.

• Materials Needed

1. Internet access
2. Drawing tools (either for making mechanical drawings or CAD drawings)

• Clarify the Design Specifications and Constraints

The plot size is determined based on the amount of vegetables consumed annually by the family and also by the amount of vegetables that the family is willing to can or preserve for storage. Types of vegetables are determined by soil conditions, annual climate, and what soil amendments you need to make; for example, some vegetables require a great deal of nitrogen, while others do not. If your family wants to do organic gardening versus intensive gardening, then additional constraints will have to be considered when using amendments, pesticides, herbicides, and any chemicals to control plant diseases. Additional factors and items to consider are the quality of the water available (the amount of minerals such as calcium, the pH of the water); the cost of additional garden items such as hoes, rakes, and mulching; and the requirements for pesticides or herbicides. Finally, if your family is going to place a limit on the amount of money spent, you will have to discuss final choices based on the cost to grow particular vegetables compared to cheaper varieties and whether you want to start from seed or from start-ups purchased at a nursery.

• Research and Investigate

To complete the design challenge, you need to first gather information to help you build a knowledge base.

1. In your guide, complete the Knowledge and Skill Builder I: The location, climate, and choosing your vegetables.
2. In your guide, complete the Knowledge and Skill Builder II: Size of the plot and orientation of the plants.

3. In your guide, complete the Knowledge and Skill Builder III: Soil amending and preparing the bed.

4. In your guide, complete the Knowledge and Skill Builder IV: Herbicides, pesticides, and combating plant diseases.

● Generate Alternative Designs

Describe two of your possible alternative approaches in preparing a vegetable garden plot. Discuss the decisions you made in (a) the size of the plot, (b) determining what vegetables to be grown, (c) what soil amendments are needed, and (d) what integrated pest plan you were willing to use to protect your investment. Attach drawings if helpful and use additional sheets of paper if necessary.

● Choose and Justify the Optimal Solution

What decisions did you reach about the vegetable garden plot?

● Display Your Prototype

Draw your garden plot to scale, including what vegetables you plan to grow. If you plan to rotate your crops based on season, include a drawing of the plot for each season. Make sure you include any constraints you have decided on, such as integrated pest management and the proposed amount of water you will need.

● Test and Evaluate

Get feedback on your ideas from other students and from your instructor. Ask your family for feedback as well. You might also visit a garden center to discuss your prototype with professionals.

● Redesign the Solution

Consider the feedback you have received. Make changes in your design that reflect good ideas offered by those you have consulted.

● Communicate Your Achievements

Describe the plan you will use to present your results to your class, and show what handouts you will use. (Include a media-based presentation.)

MEDICAL TECHNOLOGY IS the use of products, devices, and procedures to solve medical problems. Its goals are disease prevention, diagnosis, and treatment. It is medical technology that has brought us kidney dialysis for patients whose kidneys no longer function and sonography for imaging fetuses during prenatal checkups. Medical technology has also produced prosthetic

CHAPTER

16

MEDICAL TECHNOLOGY

SECTION 1
Looking at Early Medical Technology

SECTION 2
Applying Medical Technology to Diagnosis, Therapeutics, and Rehabilitation

SECTION 3
Applying Technological Advances from Other Fields to Medical Technology

SECTION 4
Applying Basic Science Research to Advances in Medical Technology

limbs—replacements for missing body parts such as arms and legs that enable people to function independently (see Figure 16.1).

Advancements in medical technology have occurred in a short period of time as a result of advances in other technologies, basic research, and as a result of the needs and desires of society. In fact, based on what people fantasize, as evident on television and at the movies (see Figure 16.2), people would like medical technology to not only maintain them for a normal life, but make them better than what is normally possible for humans; in short, "make them faster, stronger, than they were before."

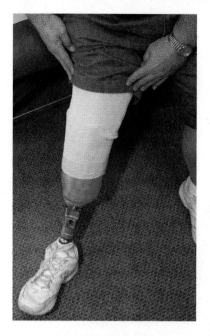

Figure 16.1 | **Artificial limbs allow people to maintain their independence.**

Figure 16.2 | **People have always been fascinated with the notion of replacement body parts that could make them stronger, faster, or even able to live forever. Many TV shows and movies have focused on this idea.**

Given the rapid development of today's medical technology, the 1970s science fiction concept of humans with medically designed body parts is not as farfetched as it seemed just twenty years ago, and the pace of research and development in this field promises to accelerate in the future.

Medical technology is often discussed alongside biomedical technology, although researchers and the medical profession define them separately. Biomedical technology attempts to prevent, diagnose, and treat disease and improve lives using living organisms like human tissue, DNA, or pharmaceutical products. In this chapter, for the sake of simplicity, both topics will be treated as one.

KEY IDEAS >

- The use of devices to diagnose or treat disease advanced in the 1800s.

- This reliance on technology has only increased over time.

- In today's world, medical technologies include prevention and rehabilitation, vaccines and pharmaceuticals, medical and surgical procedures, genetic engineering, and the systems within which health is protected and maintained.

The use of devices to diagnose or treat disease advanced in the 1800s. For example, the stethoscope was invented in 1810. As a result, medicine changed, in that, instead of relying only on the subjective evidence provided by the patient, it began to rely on the objective evidence suggested by technology. This reliance on technology has only increased over time. By the 1970s, medical technicians and data specialists were needed to run the equipment and to collect and store the data being generated for each patient. As a result, many of the people involved in the health professions today are not physicians or nurses but people who run equipment and collect and store data.

Examples of Early Technology Invented in the Nineteenth Century

Some of the early medical technological instruments were the portable six-inch clinical thermometer (invented in 1866), the stethoscope (invented in 1810), the ophthalmoscope (first invented in 1847 but not widely used until reinvented in 1851), laryngoscope (invented sometime during the 1860s), and the value of X-rays for medical purposes by Wilhelm Roentgen (who recognized its use for medical purposes in 1895). With these instruments, physicians could precisely determine a person's temperature in a short period of time and see and hear parts inside the body that they previously could only observe in cadavers. Interestingly, the thermometer had been invented much earlier, but its use as a medical diagnostic device was not widely accepted until two events occurred: (1) Several doctors made and published their observations that a patient's temperature could be correlated with a diseased state and, based on numerous readings, determined the normal temperature range for humans, and (2) the device was reengineered so that it could be used easily. Before 1866, the thermometer used with patients was a foot long and took twenty minutes to measure a patient's temperature. This technology, already in place, was not widely used until it was reengineered for efficiency and enough patient information had been gathered to show its diagnostic value.

The stethoscope is used to listen to internal sounds of the body such as heartbeat or respiration. The laryngoscope (Figure 16.3) is used to look at the vocal cords and the glottis, or the space between the vocal cords. X-rays are used to visualize bones and internal organs. The ophthalmoscope (Figure 16.4) is used to look inside the eye.

Twentieth Century

World War II (1939–1945) saw the rapid spread of X-ray technology to diagnose diseases in military troops such as pneumonia, pleurisy, and tuberculosis, as

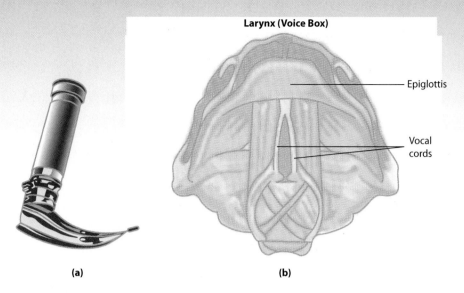

Larynx (Voice Box)

Epiglottis

Vocal
cords

(a)   (b)

Figure 16.3 | (a) Physicians use a laryngoscope to inspect a (b) person's throat. The epiglottis closes when you are eating so food does not pass into the airways.

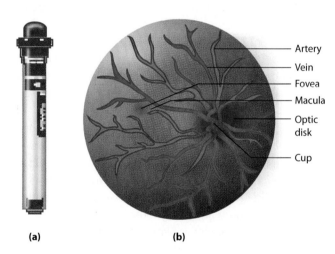

Artery

Vein

Fovea

Macula

Optic
disk

Cup

(a)   (b)

Figure 16.4 | (a) Physicians use an ophthalmoscope to look inside a person's eye. (b) This is what the ophthalmologist sees when he or she looks into your eye. The fovea has the highest concentration of cone cells which are responsible for clearest vision; some hawks have two which probably makes them better hunters.

well as to assist doctors prior to surgery. During this time, major advancements in basic science translated into medical technology advancements such as the electrocardiograph, and, in the surgery arena, the respirator, the heart-lung bypass machine, cardiac catheterization, and angiography. The electrocardiograph is an electronic test that records the electrical activity of the heart as it contracts from the beginning of the heartbeat to the end. The information recorded is known as an EKG (Figure 16.5). Today, it is a test that is often part of an annual routine checkup for people over forty. In the surgery arena, patients are routinely put on a respirator, otherwise known as a medical ventilator, when they are anesthetized because their "breathing muscles" are also asleep. The respirator breathes for the patient. The heart-lung bypass machine is used when the heart must be operated on. This machine takes over the job of pumping and oxygenating blood. Cardiac cauterization is the insertion of a catheter into a chamber or vessel of the heart, performed when a physician wants to see the heart's interior (Figure 16.6). Typically, a physician makes a small puncture in a vessel in the person's groin, inner elbow, or neck. Then a guide wire is inserted into the incision and threaded through the

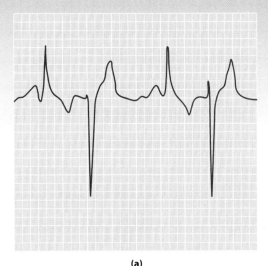

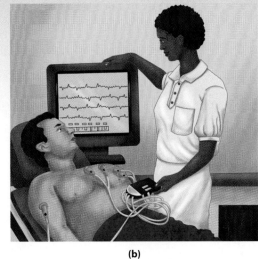

(a)   (b)

Figure 16.5 | (a) This is an example of a typical electrocardiogram recording. (b) It can be recorded while the patient, with leads attached to the chest, is moving, sitting, or in surgery to monitor the electrical output of the heart.

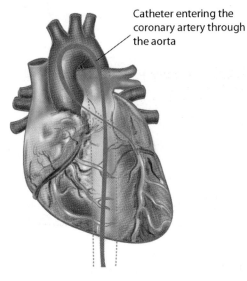

Catheter entering the coronary artery through the aorta

Figure 16.6 | In a cardiac catheterization, a small needle is threaded from the arm or the groin into the heart, usually for the purpose of doing an angiogram, where a dye is injected so that the arteries and veins in the heart can be visualized. It is minimally invasive, so the patient is awake and can watch if he or she wishes.

vessel into the area of the heart to be visualized or treated. If a coronary angiographic is to be done, a radiocontrast agent will be administered to the patient during the procedure. An angiographic is an X-ray of the interiors of organs or blood vessels; a coronary angiographic is an X-ray of the interiors of the heart or its vessels. A radiocontrast agent is a radioactive agent that can be visualized as a result of X-rays. It will reveal whether the flow of blood is blocked or is normal.

The Korean War (1950–1953) saw the development of the Mobile Army Surgical Hospital (MASH), which was designed to move expert medical assistance closer to the front, so that the wounded could be treated sooner and, therefore, with greater success. Wounded soldiers were first treated by buddy aid, then routed through a battalion aid station for emergency stabilization surgery, and finally routed to the MASH for the most extensive treatment. As a result of this plan, more than 90 percent of the soldiers who

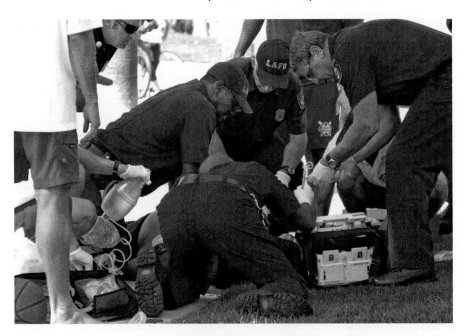

Figure 16.7 | An emergency medical technician (EMT) provides emergency medical care. These EMTs are attempting to stabilize a person who had a heart attack while playing basketball.

made it to MASH units survived. These units gave rise to a new kind of specialist, as well as an associated technology, the Emergency Medical Technician, or EMT (Figure 16.7). Today, an EMT is trained to provide prehospital emergency medical care. His or her job is to rapidly evaluate a patient's condition and provide immediate, stabilizing intervention, so that the patient can be safely transported to a hospital. Immediate interventions include providing CPR and cardiac defibrillation, controlling severe external bleeding, preventing shock, immobilization of the spine to prevent further injury, and splinting of bone fractures or breaks. A paramedic is an EMT who has received advanced training; it is the highest level an EMT can attain and is officially entitled an EMT-P.

In the 1970s, medical researchers combined computer technology with medical technology, leading to the development of a variety of visualizing, diagnostic tools, for example tomography (Figure 16.8). Tomography is an X-ray technique that creates a three-dimensional image of the body by taking X-ray pictures all around the body. A computer program takes the data provided by these pictures and creates the 3-D image.

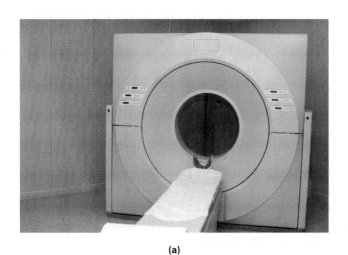

(a)

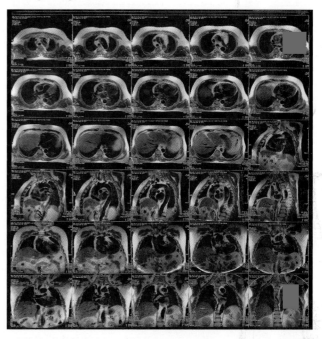

(b)

Figure 16.8 | (a) Tomagraphy, or imaging by sections, uses a software program to combine images into a three-dimensional picture. A common tomography procedure is an MRI, or magnetic resonance imaging. MRI is mainly used in medical imaging to visualize the structure or function of the body. Because it provides enhanced soft tissue contrast, it is useful in neurological, musculoskeletal, and cardiovascular imaging. (b) In this MRI image you see can see multiple views of a patient's chest.

During this time, medical technologies also developed prosthetic devices, such as artificial heart valves, artificial blood vessels, electromechanical limbs, and reconstructive skeletal joints. These devices have evolved rapidly, for example, as shown in Figure 16.9, today's electromechanical limbs are becoming more functional.

Transplantation technology had progressed to include artificial heart transplantations. Today, the technology to grow human organs in pigs and in the laboratory is being developed.

TECHNOLOGY AND PEOPLE:
Dr. Doris Taylor

Figure 16.9 | The iLimb™ was developed by the company Touch Bionics. This functional electromechanical limb can duplicate many of the activities of a real hand.

As a child, Dr. Doris Taylor woke up Saturday mornings thinking she had to invent something. She thought, "This is what adults do." As an adult, she is an inventor—in regenerative medicine. Doris Taylor, Ph.D., as the Medtronic-Bakken Chair in cardiac repair, Director of the Center for Cardiovascular Repair, in the

Figure 16.10 | Organs like this rat heart are being experimentally grown in the laboratory. This heart is only partially functional.

Stem Cell Institute at the University of Minnesota, oversees a laboratory that is working on regeneration of organs. Employing a process called whole organ decellularization, Taylor's laboratory grows living, working heart tissue by taking rat and pig hearts with all original cells removed and reseeding them with a mixture of new live cells (see Figure 16.10). The process, known as "decellularization," involves removing all of the cells from an organ using a special detergent solution, leaving only the extracellular matrix intact. This matrix is the framework between the cells. After removing all cells from a heart, the researchers inject the matrix with a mixture of cells from neonatal or newborn hearts and place the structure in a sterile container to grow. Four days after seeding, contractions are observed. Eight days later, the heart is pumping, but not at full capacity. The technique still needs further development, but this is a very auspicious beginning. Dr. Taylor has indicated that this same type of technology could be applied to the regeneration of other organs (see Figure 16.11).

Doing this type of research isn't necessarily easy; there have been a lot of ups and downs. A sign on Dr. Taylor's door reads, "Trust Your Crazy Ideas." That is her advice to students and fellow researchers. When asked what she would tell students deciding on a career, she said, "Figure out what interests you and do it, and if you dream big, that is okay, because you will be doing what you enjoy."

To watch a movie on this technique and learn more about the research in Dr. Taylor's laboratory, search the Internet with the phrase, "Dr. Taylor and Center for Cardiovascular Repair" or go to http:// www.med.umn.edu/beatingheart/home.html.

Figure 16.11 | Dr. Doris Taylor working in her laboratory at the Stem Cell Institute.

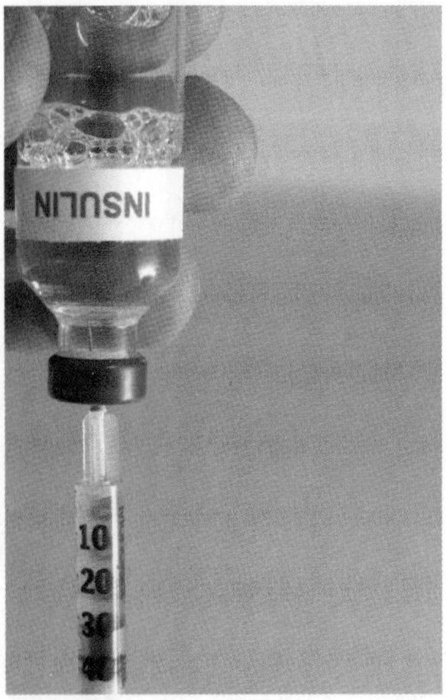

The Role of Genetic Engineering in Medical Technology

Genetic engineering is defined as the deliberate, controlled manipulation of genes in an organism, with the intent of making that organism better in some way. Recombinant DNA technology is one aspect of genetic engineering. It is defined as the combination of DNA from different species to form a hybrid DNA for the purpose of studying the DNA or the product that it represents. In the early 1970s, biochemists Paul Berg and Herb Boyer produced the first recombinant DNA molecules, heralding the beginning of recombinant DNA pharmaceutical products such as human insulin (see Figure 16.12) and human growth hormone. Insulin is a hormone that regulates the cellular uptake of sugar, and growth hormone stimulates growth and cell reproduction. The recombinant DNA version of human growth hormone is known as somatotrophin because it is produced by the somatroph cells in the pituitary gland. It is provided to children who do not produce enough of this hormone and therefore are too short for their age.

In 1986, researchers proposed the sequencing of the human genome. In the 1990s, after much debate, the Human Genome Project was launched and was

INSULIN

10
20
30
40

Figure 16.12 | Recombinant human insulin (rhInsulin) was approved by the U.S. government for public use in 1982. Before the human gene for insulin was cloned, it took seven to ten pounds of pancreas from approximately seventy pigs or fourteen cows to purify enough insulin for one year's treatment for a single diabetic.

finished fifteen years later. Examples of advances in medical technology as a result of this project are the development and implementation of recombinant DNA vaccines, cellular engineering, and DNA gene therapy.

Recombinant DNA Vaccines: An Application of Genetic Engineering You probably remember getting a vaccination, but you may not know how a vaccine is prepared. In general, vaccines are preparations of specific, killed or attenuated, whole microorganisms (e.g., bacteria or a virus) that cause disease in humans. An attenuated microorganism is alive, but weakened, and therefore considered nonpathogenic. After being vaccinated, a person develops immunity to (or protection from) the diseases caused by that particular microorganism.

Just as with other drugs, these vaccines, are tested in clinical trials before their release to the general public. However, because only a small number of people can be realistically tested compared to the huge number of people who will take these drugs or be vaccinated, these trials may not include the small number of people who will have a serious reaction. This has been true for several killed, whole cell or attenuated virus vaccines such as Hepatitis B Virus (HBV). Some of the serious side-effects of the attenuated HBV are liver damage, and sometimes cancer, especially if it reverts back to normal. So when recombinant DNA technology became available, researchers realized that vaccines, including the attenuated HBV vaccine, could comprise only the parts of the microorganism that confer immunity, and that the rest of the microorganisms that cause serious side effects could be omitted, The current HBV recombinant vaccine is composed of a protein found on the surface of the virus. The gene for this protein has been placed into yeast cells for mass production (see Figure 16.13). It confers immunity but without the possibility of serious side-effects.

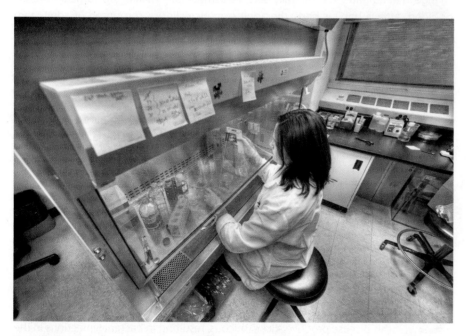

Figure 16.13 | To keep yeast and other cell cultures free from bacterial contamination, technicians have to work with them in a sterile hood wearing gloves and lab coat. The hood also protects the technician from exposure to anything that is in the hood, like cells infected with a virus, which is also important.

The Ins and Outs of Cellular Engineering: An Application of Genetic Engineering Genes, which are found on your chromosomes and, therefore, your DNA, are the functional units of heredity. Genes carry the instructions for proteins and other products. When a gene is defective, it does not carry the correct instructions, so genetic disorders can result. Since every cell in your body contains the same DNA, every cell in the body will carry the defect. Researchers proposed the idea of gene therapy, which involves replacing defective DNA with the "correct" DNA. So far, proposed methods include: (1) Cells could be removed from the person, the defective gene could be replaced with a normal gene, and the corrected cells could be replaced. (2) The defective gene could be continually "silenced" in place. (3) The defective gene could be repaired in place and returned to normal function. The three types of cells that can be repaired are somatic cells, embryonic stem cells, and germ cells. The description of these cell types and their advantages and disadvantages for repair are as follows: (1) Somatic cells are composed of both differentiated cells and undifferentiated adult stem cells. A differentiated cell is the final cell type, and it cannot become undifferentiated; examples are a red blood cell or a muscle cell. An undifferentiated adult stem cell can become several types of cells and, therefore, is considered multipotent. Its main role in the body is as a source of new cells in its tissue of origin, so it divides to produce new cells as cells die or are damaged. An example of an adult stem cell is a myeloid cell, and it can differentiate into a red blood cell, or a platelet, or a particular type of white blood cell. However, this type of stem cell cannot become a brain cell or a liver cell. Therefore, if it is engineered, it can only replace certain cells. (2) Embryonic stem cells, are pluripotent, undifferentiated cells, which means that they can become any type of cell in the body. However, they cannot be cloned to produce a human being. (3) Germ line cells are found in the egg and sperm. When the egg and sperm fuse to form a cell, this cell is totipotent, which is defined as a cell that can become any cell in the body (Figure 16.14), and theoretically, this type of cell could be cloned to form a human being. In the embryo, the totipotent cells continue to divide and eventually they eventually become pluripotent. These pluripotent cells continue to divide and change (differentiate) to eventually become somatic cells making up particular organs or blood. After a baby is born, the adult somatic cells are composed of adult stem cells which are multipotent and nondividing cells which die in time. The adult stem cells divide when the body needs to replace cells such as tissue damage or aging cells. Currently, genetic engineering of a human totipotent cell and its progeny before they become embryonic stem cells is against the law.

At this time, the only human cells that can be genetically engineered are somatic cells and embryonic stem cells; of course, the best type to repair are stem cells because they can still change (differentiate) into a variety of specialized cell types and therefore can be used to repair a variety of different cells in a person's body.

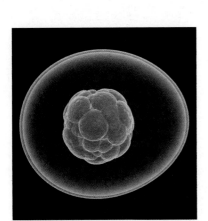

Figure 16.14 | Totipotent embryonic cells can become any type of cell. These cells in the embryo divide and become committed to certain paths, at which point they are known as pluripotent cells. Pluripotent cells continue to further divide and commit themselves to even narrower pathways eventually forming the heart, blood, and other organs and cells.

TECHNOLOGY IN THE REAL WORLD:
Genetic Engineering for Genetic Diseases

Somatic, Differentiated Cell Fix

Cystic fibrosis is a genetic disease: The protein responsible for transporting chloride across cell membranes does not work correctly. As a result, patients with this disease produce sticky mucous, resulting in bacterial infections in the lungs, food digestion problems, and, if they are male, sterility. The majority of patients die from complications arising from bacterial lung infections. In approximately 90 percent of the patients with this disease, the change in the DNA is the same, so this disease is a good candidate for gene therapy treatment. Plus, it became obvious that the lifespan and quality of life for most of these patients could be improved if only their lungs were not constantly becoming infected and scarred, which eventually resulted in their deaths. The cells targeted to be repaired are somatic cells lining the lungs. Clinical trials have proved promising as symptoms are dramatically decreased, even though the fix is temporary and only lasts as long as these cells remain alive. Once the cells are dead, the treatment must be performed again, or in this case, "inhaled" again.

Adult Stem Cell Fix

In some genetic diseases, such as those that affect the immune system, adult stem cells can be repaired and injected in the bloodstream. This was the case in 1993 for Andrew Gobea. He was born with a normally fatal genetic disease known as severe combined immunodeficiency or SCID. Genetic screening before birth showed that he had the SCID defect, which would impair his immune system. He would not be able to produce immunity against any diseases and therefore would die very young. To fix this defect, adult stem cells were removed from his umbilical cord. The normal gene was placed in these cells, and they were injected into his bloodstream. For four years, these cells produced normal white blood cells protecting Andrew from disease. After four years, the treatment had to be repeated with treated stem cells that had been stored when he was born. For the rest of his life, Andrew will need to be treated every four years or so. (See also Figure 16.15.)

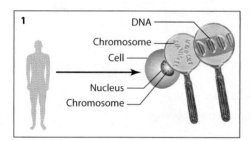

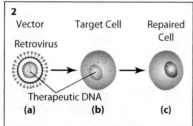

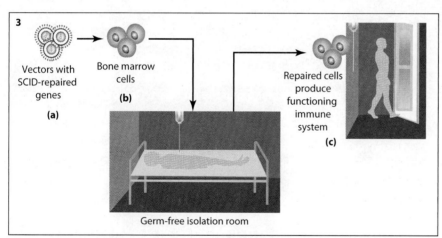

Figure 16.15 | If a gene is incorrect than it will code for an incorrect protein or not make it at all. This is true for children born with severe combined immune deficiency (SCID). Kids born with this disease have to live in a germ free environment because their immune systems do not work correctly. To reverse this disease, researchers put the correct gene in a virus and then put that virus into a bone marrow cell taken from the patient. The "corrected" bone marrow cell was put back into the patient where it colonized and increased in number in the bone marrow of the patient. The corrected cells now made the immune cell that the patient was missing, restoring the patient's immune system to normal.

Advances in Medical Technology Have Led to Ethical Situations for Society

In today's world, medical technologies include prevention and rehabilitation, vaccines and pharmaceuticals, medical and surgical procedures, genetic engineering, and the systems within which health is protected and maintained. However, these advances have not occurred without risks to patients and the emergence of ethical questions that presently do not have clearcut answers. Society's understanding has not kept up with the rapid advancement in medical technology. In addition, many of the long-term effects of these new products will not be realized for several generations. Furthermore, presently science has no way to predict what these effects will be.

As a result, misunderstandings between the public and industry lead to mistrust and ethical situations that the government, the public, and the medical care industry are ill equipped to handle. People would like limb and organ replacements that are not only effective but even better than the originals. Yet all these advances cost a tremendous amount of money to develop, test, and implement. Furthermore, they give rise to questions that are not always easy to answer: Who should pay for this technology? Should it only be available to people who can afford it? Should it all be covered by health insurance, which means that the public is going to pay for it? And finally, the biggest question of all: If technology can extend life, and possibly extend a lifetime to hundreds of years with excellent quality of life, should it?

SECTION ONE FEEDBACK >

1. Using resources found on the Internet, identify the structures in the eye that a physician can see using an ophthalmoscope. Describe one condition that a physician could diagnose using this instrument that he or she would not be able to see without it.

2. How was MASH unit treatment of the wounded different from the treatment for the wounded in World War II?

3. Search on the Internet for an animation of an EKG being generated as a heart is beating. Draw a picture of an EKG, labeling the peaks and valleys as they correspond to the different parts of the heart muscle as it contracts. Explain how you think an EKG is used to determine what part of a heart muscle has been damaged as a result of a heart attack.

4. Go to the University of Minnesota Stem Cell Institute Web site and watch the video on the making of the rat heart. Do you think this type of technology, when and if it is ready for humans, should be paid for by health insurance? Should it be available without conditions (i.e., age, reason for the failing heart)? If you believe it should be made available but with conditions, what should these conditions be?

SECTION 2: Applying Medical Technology to Diagnosis, Therapeutics, and Rehabilitation

KEY IDEAS >

- Medical technologies are an important part of diagnostic medicine.
- Advances in therapeutic medical technologies include drug delivery and surgery.
- Medical technologies are used for rehabilitation.
- Advances in medical technology have created ethical dilemmas for people in healthcare and for society in general.

Several examples of advance medical technologies now available were provided in section one. Section two focuses on advances in medical technology, specifically in diagnostic, therapeutic, and rehabilitation medicine. Both positive and negative aspects of these advances are explored.

Diagnostics

Medical technologies are an important part of diagnostic medicine, especially the noninvasive techniques. Before the development of noninvasive techniques, exploratory surgery was the only option for visualizing many conditions, such as tumors. For this reason, in order for a physician to determine what was wrong, patients had to be subjected to the trauma of surgery. The discussion that follows reviews many of the noninvasive technologies now available to the medical community.

Radiology Radiology has traditionally been defined as the branch of medicine that studies and interprets images of the human body created by X-rays on photographic film. However, since there are new, noninvasive ways of looking inside the body that do not use X-rays or ionizing radiation at all, radiological science is now often called medical imaging; one of these new methods, ultrasonography, uses ultrasound waves. Another, Magnetic Resonance Imaging (MRI), uses a magnetic field combined with radio frequency waves. These techniques and others use computers to produce and store images. Table 16.1 contrasts these noninvasive imaging technologies.

Table 16.1 | Medical Imaging

TYPE	USES	PRODUCES
Nuclear medicine	Internal ionizing radiation	Cross-sectional images (CAT and PET scans)
Ultrasonography	1–15 mhz waves	Images of soft tissues, such as a pregnant uterus
Magnetic resonance imaging	Radio waves and magnetic field	Cross-sectional anatomical images for conditions such as tendonitis
Radiology	External X-rays	X-ray photographs, such as a chest X-ray.

X-rays are produced in a vacuum tube that contains an anode (positively charged) and a cathode (source of electrons) connected to a high-voltage source. When the power supply is switched on, it heats the cathode, causing the production of electrons. Attracted by the strong positive charge, the negatively charged electrons bombard the anode, which then produces X-rays. The X-rays are then directed to the area of concern.

The X-rays penetrate the body, but their transmission through different types of tissues depends on how specific tissue attenuates, or reduces the energy of the X-rays. For example, bone attenuates X-rays more than air. Thus, a smaller amount of X-ray energy passes through bone, and its corresponding shadow on the developed radiographic film will be lighter. Conversely, the show of air will be darker. This explains why, on a chest X-ray, the ribs are relatively light, whereas the air-containing lungs are relatively dark.

Tomography Tomography is imaging by sections, or sectioning. In medicine there are many different types of tomography, depending upon the "imaging energy source" (e.g., X-rays or MRI) and how the imaging is acquired. Before computers, technicians had to take X-ray pictures from

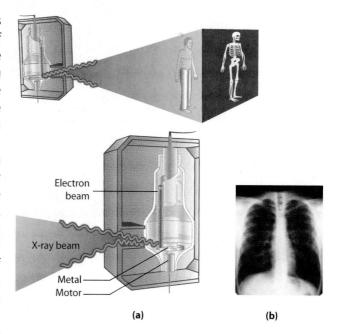

(a)   (b)

Figure 16.16 | (a) X-rays are produced when electrons in a metal target are momentarily pushed to a high energy. When these electrons drop back to their normal energy, they give off radiation, including X-rays. Some of the X-rays can be directed as beams towards specific parts of the human body, such as an arm or the chest. This is the diagram of a typical X-ray setup. (b) This is a typical X-ray image of the human chest.

different angles and then manually overlay them to get a 3-D effect. This commonly did not produce an accurate 3-D picture of the body. However, now, with computers, that can precisely record and process information, and the ability to rotate the patient or the imaging machine around the patient, accurate 3-D pictures can be obtained of the body. Case in point is Computer Axial Tomography (CAT) scan. Using a rotating X-ray device, a CAT scan takes hundreds of pictures as it moves around the patient. The image provides a cross section of the body, which is displayed on a computer screen. A CAT scan is much clearer and more complete than an ordinary X-ray image. It also uses lower radiation levels. (See Figure 16.17.)

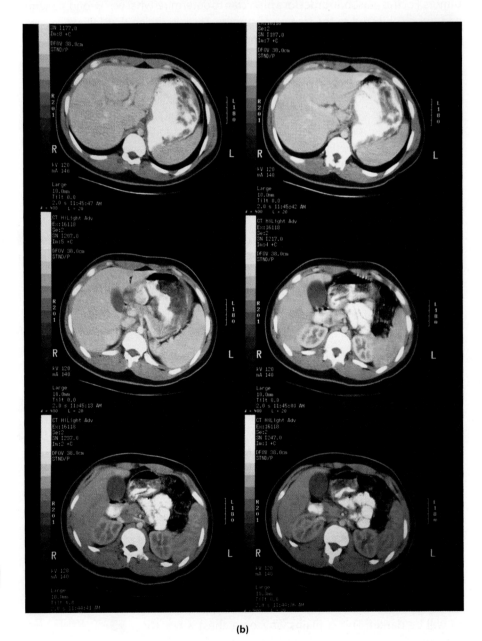

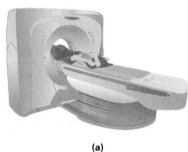

(a)

(b)

Figure 16.17 | (a) Computer Axial Tomography (CAT) scanner (b) Using a rotating X-ray device, a CAT scan takes hundreds of pictures as it moves around the patient. The image provides a cross section of the body, which is displayed on a computer screen.

The Visible Human Project®, funded by the United States National Library of Medicine, used transverse CT (computer tomography), magnetic resonance imaging, and cryosection (quick-frozen sections) images of representative male

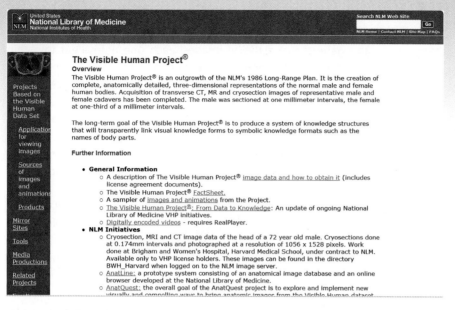

The Visible Human Project®

Overview

The Visible Human Project® is an outgrowth of the NLM's 1986 Long-Range Plan. It is the creation of complete, anatomically detailed, three-dimensional representations of the normal male and female human bodies. Acquisition of transverse CT, MR and cryosection images of representative male and female cadavers has been completed. The male was sectioned at one millimeter intervals, the female at one-third of a millimeter intervals.

The long-term goal of the Visible Human Project® is to produce a system of knowledge structures that will transparently link visual knowledge forms to symbolic knowledge formats such as the names of body parts.

Further Information

- **General Information**
  - A description of The Visible Human Project® image data and how to obtain it (includes license agreement documents).
  - The Visible Human Project® FactSheet.
  - A sampler of images and animations from the Project.
  - The Visible Human Project®: From Data to Knowledge: An update of ongoing National Library of Medicine VHP initiatives.
  - Digitally encoded videos - requires RealPlayer.
- **NLM Initiatives**
  - Cryosection, MRI and CT image data of the head of a 72 year old male. Cryosections done at 0.174mm intervals and photographed at a resolution of 1056 x 1528 pixels. Work done at Brigham and Women's Hospital, Harvard Medical School, under contract to NLM. Available only to VHP license holders. These images can be found in the directory BWH_Harvard when logged on to the NLM image server.
  - AnatLine: a prototype system consisting of an anatomical image database and an online browser developed at the National Library of Medicine.
  - AnatQuest: the overall goal of the AnatQuest project is to explore and implement new visually and compelling ways to bring anatomic images from the Visible Human dataset

Figure 16.18 | The Visible Human Project®, funded by the United States National Library of Medicine, used **transverse** CT (computer tomography), magnetic resonance imaging, and **cryosection** (quick-frozen sections) images of representative male and female cadavers to create complete, anatomically detailed, three-dimensional representations of the normal male and female human bodies.

and female cadavers to create complete, anatomically detailed, three-dimensional representations of the normal male and female human bodies. It provides a good example of what you can see with this technology (Figure 16.18).

Magnetic resonance imaging Magnetic resonance imaging (MRI) is another noninvasive imaging system that does not depend on radiation. Instead, radio waves are directed at protons, the nuclei of hydrogen atoms, in a strong magnetic field. (The field is so strong that it can pull metal objects such as mop buckets, IV poles, oxygen tanks, heart monitors, pens from pockets, etc., into the machine). All living material possesses hydrogen atoms. In most MRI units, the magnetic field is produced by passing an electric current through wire coils. Other coils located in the machine send and receive radio waves. Because the magnet surrounds the body, an MRI can image in any plane (see Figure 16.19); axial (the way in which bread is normally sliced), sagittal (slicing bread from side to side, lengthwise), or coronal (sliced as layers of a cake).

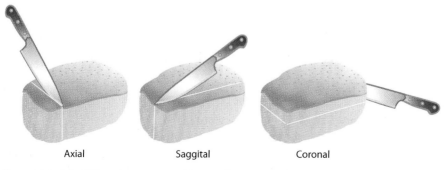

Axial   Saggital   Coronal

Figure 16.19 | An MRI can image in any plane; axial (the way in which bread is normally sliced), sagittal (slicing bread from side to side, lengthwise), or coronal (sliced as layers of a cake).

Section 2 △ Applying Medical Technology

As a result of the pull of the magnetic field, the protons change position, producing signals that are picked up by the other coils. A computer processes these signals and generates a series of images, each one equivalent to a thin slice of the body. The computer compiles the images into a three-dimensional representation of the body, which can be studied from several angles on the computer screen. Because different parts of the body contain different amounts of protons—for example, parts containing water contain more protons—this imaging technique reveals differences in water content between body tissues. As a result, MRI is especially useful in detecting disorders that increase water (e.g., as a result of inflammation, infection, or tumors) in diseased parts of the body. Some of the diseases for which MRI diagnosis is especially useful include: (1) multiple sclerosis, (2) tumors and infections in the brain, spine, or joints, (3) visualizing torn ligaments and tendonitis, and (4) strokes in their earliest stages. People whose bodies contain metal objects such as pacemakers or who are too big to fit in the machine cannot be scanned. Plus, a person who is claustrophobic or cannot hold still for very long will find it difficult to be scanned. Hopefully, in the future, small scanners will be developed so a person does not have to lie down and fit into the machine; the small scanner will just scan the part of the body that needs to be visualized.

Besides static MRI, there are already functional magnetic resonance imaging (fMRI) devices that can measure neural activity in the brain or spinal cord by measuring the blood flow through these organs. By doing this, researchers can determine what part of the brain or spine is active when a person is performing a specific activity or looking at a specific picture. This technique is known as "brain mapping" (see Figure 16.20).

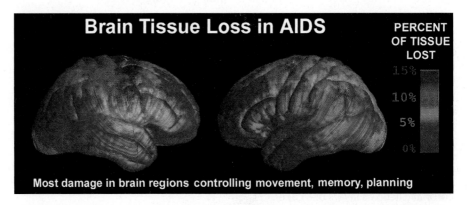

Figure 16.20 | High resolution MRI mapping of people's brains who have AIDs. There is a loss of cortical tissue.

Positron Emission Tomography Positron emission tomography (PET) uses a short-lived radioactive tracer, such as fluorine 18, as part of a compound, such as glucose, that is metabolized by the tissue to be scanned. PET works like this: (1) The compound is injected into the bloodstream of the person; (2) the body metabolizes it concentrating it in the target tissue; (3) the radioactive part decays producing positrons (the antimatter counterpart of electrons) and these positrons interact with electrons at the site, destroying both of them; (4) the destruction of the electrons produces photons (light); (5) the imaging scanner detects the light produced and a computer compiles this information, producing a three-dimensional image or map of functional body processes. (See Figure 16.21.)

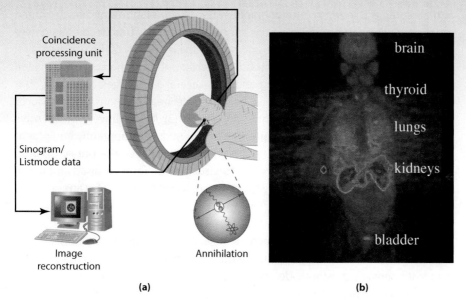

(a)   (b)

Figure 16.21 | (a) A typical PET machine showing how the signal is processed and the image reconstructed (b) This whole-body PET scan shows the concentration of the enzyme monoamine oxidase (MAO) in various body organs in a nonsmoker. This crucial enzyme breaks down neurotransmitters and dietary amines, and too much or too little MAO can adversely affect health and even personality. In a smoker, this enzyme can be reduced as much as 50% in the brain and other organs demonstrating that smoking not only affects the lungs it effects other organs as well.

Sonography Similar to the way in which whales and dolphins detect objects around them, medical ultrasonography (sonography) uses sound waves to produce images (see Figure 16.22). These sound waves are well outside human hearing range. Like X-rays, sound waves transfer energy from one point to another, but, unlike radiation, sound waves can only pass through matter and cannot pass through a vacuum. When the ultrasound beam passes through different tissues, the energy is reflected at the interfaces between the different tissues (e.g., the interface between a fetus and the fluid in which the fetus is suspended). The reflection changes at the different interfaces, and this reflection is recorded by the machine via the device that produced the sound wave. The time it takes for the echo to return is proportional to the depth

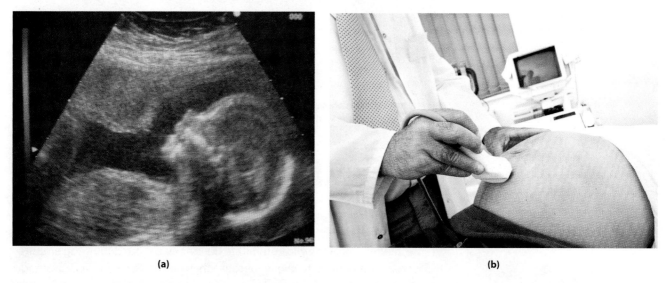

(a)   (b)

Figure 16.22 | (a) This is a sonography of a fetus. You can see its head and even its nose and eyes. (b) The transducer is the device that both produces the ultrasound waves and records the reflection of these waves as they "bounce back" from body tissues.

of the interface. The echo's intensity is dependent on the physical properties of the organs or regions on both sides of the interface, as well as on its depth. The device that both produces the sound wave and detects the reflection of the wave is called a transducer. A gel is rubbed on the transducer to enhance the contact between the transducer and the body. Sonography has several advantages, including: (1) it differentiates between solid and fluid interfaces, delineates organs, and produces live images; (2) it has no long-term side effects; (3) it is relatively inexpensive compared to other imaging techniques like CAT or MRI; (4) sonography equipment is fairly portable and widely used; (5) it is widely accepted by the public because it is commonly used; during pregnancy to visualize the fetus. A disadvantage of ultrasound is that it does not display the same anatomic detail as CT scans or MRI.

Therapeutics

Advances in therapeutic medical technologies include drug delivery and surgery. Examples of this technology are automated drug-delivery systems coupled with diagnostic sensors, prosthetic devices for amputees, kidney dialysis units (see Figure 16.23), pacemakers, stents (metal tubes inserted into blood vessels to keep them open) coupled with drug delivery to open up arteries, and artificial tissue and organ replacement, such as retinas, livers, bladders, hearts, and skin. Some of the more interesting technology involves very small sensors and devices, like those described in the section on nanotechnology, and microelectromechanical (MEM) devices.

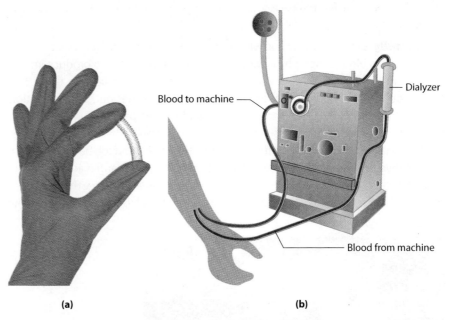

Blood to machine

Dialyzer

Blood from machine

(a)   (b)

Figure 16.23 | (a) Stents are small tubes placed in blood vessels to keep them open. (b) Kidney dialyzers do the job of the kidneys. The kidneys filter out the waste in the blood.

ENGINEERING QUICK TAKE

Artificial Limbs: An Unfair Advantage?

The International Olympic Committee does not want the double leg amputee to compete in running events because they feel he has an unfair advantage: The artificial limbs are composed out of carbon fiber. Why do you think they feel it is unfair? How are his artificial lower legs different from real lower legs? Our lower legs, including our feet, are composed of bone, muscle, arteries, veins, blood, and cartilage, and they work when muscles contract and relax. Carbon fiber lower legs act like springs, enabling the runner to push off the ground more efficiently than a person employing muscle power alone. Plus, carbon fiber lower legs weigh a lot less than real legs and feet, which means a person carries less weight in his or her lower extremities. This presents a definite advantage for the disabled runner. However, the disabled runner does not run using exactly the same "motion" as a runner with normal legs; it is unknown whether this has any negative impact on the disabled runner's performance. How would you determine if a runner with artificial lower legs has either an unfair advantage or disadvantage compared to athletes with normal legs? What tests would you do? If he does have an advantage, what would you engineer for his lower-leg replacements to get rid of these advantages but still allow him to maintain the same running abilities as the other athletes? To help you decide, research online on "artificial limbs and athletes," or similar keywords, as these questions have already been scientifically explored. Find information that will help you decide what tests should be done to show that these limbs provide a running advantage, or are neutral (no advantage) in their effect, or a disadvantage in running for the disabled individual. From your research, determine how lower-leg replacements should be weighted or changed to eliminate the advantage of being lighter than normal legs and feet. Based on what you found, write a short discussion explaining whether you think the carbon fiber replacement limbs provide an advantage or not, and if they do provide an advantage, how you would "handicap" them.

Figure 16.24 | Oscar Pistorius, an athlete with prosthetic legs, was banned from the Olympics because it was believed that he had an unfair advantaged over runners with normal legs.

Rehabilitation

Medical technologies used for rehabilitation started with Elizabeth Kenny, a nurse in the Australian Medical Corps during World War I. Kenny was responsible for introducing the use of hot packs and muscle manipulation to treat polio patients. People who received this treatment said it was painful, but that it did work. One famous person who received the treatment that she pioneered is Alan Alda.

Rehabilitation services have progressed tremendously since that time. These technologies include better wheelchair design, long-distance rehabilitation or "tele-rehabilitation," improved prosthetics, new treatments for bowel and bladder control for quadriplegics, and more accessible transportation systems so that patients can remain mobile. Consider the advances in wheelchair technology. Computer technology, combined with electric powered wheelchairs, has allowed independent mobility for people with severe impairment. A recent development is wheelchair that has been designed to climb and descend stairs and put the person in the chair at eye level when holding conversations with people who are standing. (See Figure 16.25.)

However, technology development in rehabilitation has not been free of problems; access to new treatments and devices is often limited, and electromagnetic interference has occurred between devices such as cell phones and hearing aids. Also, since primary care doctors are responsible for directing patients to resources, they need to become more knowledgeable about rehabilitation sources and about the appropriate physical medicine and rehabilitation specialists. The latter are doctors who specialize in rehabilitation and treat problems that touch upon all the major systems in the body. They are also known as "physiatrists."

People are surviving injuries and diseases from which, just a few years ago, they would have died. Nevertheless, survivors can be left with debilitating disabilities,

Figure 16.25 | The iBOT wheelchair can go up and down stairs, allowing users access without a ramp. It can also raise and lower a person vertically and handle rough terrain.

leaving them and their families to deal with the realities of the situation. To help them cope, patients and their families are being treated by a team of specialists who work together to provide the best solutions. Most likely, such a team includes a physical therapist, an occupational therapist, a doctor, and often a rehabilitation nurse and counselor. One of the most difficult situations is when the patient is left in a vegetative state; finding the best way to deal with that crisis is one of the most discussed medical dilemmas.

Medical Dilemmas

▷ Advances in medical technology have forced healthcare workers to reevaluate the traditional definitions of life and death. For example, the concept of when life begins is being revisited as fetuses can now be removed from the womb for treatment weeks prior to traditional delivery. However, in many cases, the individual survives with major medical problems. The concept of when death occurs is also being revisited as patients can now be resuscitated after prolonged heart, lung, and even brain failure. If a patient is resuscitated, but remain unconscious, the patient may develop a condition called **persistent vegetative state** (PVS), in which they have periods of sleep and wakefulness and open their eyes, but do not respond to verbal stimulation. The term was first used in 1972 by a Scottish neurosurgeon and an American neurologist. An adult with PVS has been shown to have about a 50 percent chance for recovering within the first six months. But even this time period is up for debate, as patients who have been in this state for as long as twenty years have recovered. Such cases are the exception, but even one success makes it difficult for family and healthcare workers to decide on "pulling the plug."

Karen Ann Quinlan was the first case that resulted in a debate between the right-to-die proponents versus those who favor the use of medical technology to keep a patient alive on machines after the traditional indicators of death—lung and heart failure—have occurred (see Figure 16.26). Quinlan slipped into a coma after ingesting alcohol and tranquilizers. However, despite severe brain damage and a coma, she was not considered dead because her cardiac and respiratory functions could be maintained using machines. Eventually, her family wanted to remove her from the life-support technology that kept her breathing, but her doctors disagreed. They fought a well-publicized legal battle and the family won. Quinlan was removed from the respirator. However, still in a vegetative state, Quinlan kept breathing for approximately another ten years until her ultimate death of acute pneumonia in 1985.

In 1990, Terri Shiavo, twenty-six years old, went into PVS as a result of cardiac arrest due to a potassium imbalance. After eight years, her husband sought to withdraw the feeding tube to allow her to die, but her parents and sister wanted to continue supplying her body with food and water, claiming that Terri might eventually be rehabilitated. From 1998 to 2005, her husband and family were locked in a legal struggle. Because Terri did not have a living will and because there is no conclusive test for PVS, Terri was taken off and put back on a feeding tube several times as a result of the court battle. Appeals by her family continued until March 18, 2005, at which time her feeding tube was removed for a third and final time. She died on March 31, 2005. An autopsy revealed neuropathologic changes in her brain precisely of the type seen in patients who enter a PVS following cardiac arrest.

On the other hand, some patients ultimately recover from PVS. Sarah Scantlin, severely injured by a drunk driver at the age of eighteen, had been in a coma-like state for twenty years. Sarah was only able to communicate by blinking her eyes. Friends wondered if she could comprehend the world around her. Suddenly she began to speak, her first words being "Hi, Mom." Terry Wallis, a young man injured in a traffic accident at age twenty, was paralyzed from the neck down and could not talk. Wallis received continuing care in a rehabilitation center. Then, nineteen years later, he said, "Mom," and was soon able to converse. Researchers concluded that brain-cell regrowth was responsible for his recovery. Other documented cases have patients recovering from PVS after being given a sleeping pill, a totally unexpected response. Presently, there is no general agreement in the healthcare industry as to when treatment should cease.

Figure 16.26 | People fiercely debate the issue of "the right to die," one of the major ethical dilemmas society faces as medical technology continues to advance.

SECTION TWO FEEDBACK >

1. How are X-rays produced?
2. What are the four major imaging technologies?
3. Describe PET and how it is used.

KEY IDEAS >

- Telemedicine works well for emergency situations, rural health care, forensic medicine, and monitoring chronic conditions.

- Telemedicine reflects the convergence of technological advances in a number of fields, including medicine, telecommunications, virtual presence, computer engineering, informatics, artificial intelligence, robotics, material science, and perceptual psychology.

- Nanotechnology and microelectromechanical (MEM) devices are the results of a convergence of technologies: semiconductor, computer, biology, and chemistry.

Improvements in medical technology have been accelerated by changes in other technologies. For example, improvements in information technology and communication technology have made a tremendous impact, as have advances in nanotechnology, which creates extremely small devices, as small as an atom or molecule.

Telemedicine

Telemedicine is the use of communication to consult, provide, or conduct healthcare. It has been practiced for a long time, only the method of communication has changed. African villagers used smoke signals to warn people to stay away from the village in case of serious disease. In the early 1900s, people living in remote rural areas in Australia used two-way radios, powered by bike pedaling, to communicate with the Royal Flying Doctor Service of Australia. Providing medical advice to people in rural and inaccessible areas is probably one of the first forms of telemedicine. These days, we know that telemedicine works well for emergency situations, rural health care, forensic medicine, and monitoring chronic conditions.

In today's world, telemedicine reflects the convergence of technological advances in a number of fields, including medicine, telecommunications, virtual presence, computer engineering, informatics, artificial intelligence, robotics, material science, and perceptual psychology. Telemedicine can be as simple as two healthcare professionals discussing a case over the phone or via e-mail, or as complex as doing telesurgery, when a skilled surgeon remotely conducts an operation on a patient while being assisted by operating staff, as shown in Figure 16.27. Telesurgery is done in real time, otherwise known as "synchronous" time. Telemedicine can also be done asynchronously, where information is stored and then viewed later. Asynchronous, or "store-and-forward," telemedicine involves acquiring medical data (e.g., medical images, biodata) and then transmitting this data to a healthcare professional for assessment offline. It does not require the simultaneous presence of both parties. Dermatology, radiology, and pathology are common specialties that are conducive to asynchronous telemedicine.

Figure 16.27 | Dr. Michael DeBakey participates in a telesurgery experiment in an underwater habitat, used to train astronauts, located off the Florida Keys.

The telemedicine contains video-conferencing equipment or webcam instrumentation. It also provides room for tools, including a tele-otoscope for looking inside a patient's ear, and a tele-stethoscope for monitoring heartbeats and other body sounds. An examination camera, used to visualize the patient, could also be mounted in the room or on the cart. (See Figure 16.28.)

As a result of advances and convergence with other technologies, telemedicine is being expanded to other medical specialties, such as psychiatry, internal medicine, rehabilitation, cardiology, pediatrics, obstetrics, gynecology, pharmacy, and neurology.

Figure 16.28 | Equipment used in telemedicine includes laptop cameras with webcam, a camera that is used to view the patient, and a telecart, where many of the tools are stored.

Nanotechnology and MEM Devices

Two more important fields that have made an impact on medical technology are nanotechnology and microelectromechanical (MEM) devices. They are also a result of a convergence of technologies; semiconductor, computer, biology, and chemistry.

Nanotechnology is defined as devices and materials in the nanometer range. An example of nanotechnology in medical devices is the glucose sensor in the portable diabetic device shown in Figure 16.29. The sensor is actually an enzyme that interacts with glucose and produces a product that degrades and as a result, produces a signal that is detected by the device.

MEM devices are mechanized devices that are measured in the micrometer range. If mechanized parts are in the nanometer range, they are also classified as nanotechnology. Examples of MEM devices are the pumps in the glucose monitoring device described above. More information about this device is provided later in the chapter.

The specific use of nanotechnology in medicine is known as nanomedicine. Besides the glucose sensor, we can find other examples of nanomedicine in medical imaging and drug delivery. For example, nanosilver is being used as to make a silver superlens to enhance optical imaging. The superlens is placed between the viewed object and the conventional optical microscope, dramatically improving image resolution (the ability to see two objects that are close together as two separate objects). Objects that are 60 nm apart can be resolved. The optical microscope is viewed as one of the most important imaging tools due to its noninvasive nature, low cost, and versatility. Historically, its use has been limited by an inability to resolve small objects, such as parts of cells.

A significant problem with some of these new nanomaterials is their toxicity. It seems that when some metals are converted to nanocrystals, they become toxic in that they can be directly absorbed through the skin. Nanosilver is an example of such a metal. As part of a ring or a fork, it is perfectly safe, but converted to nanocrystals, it must be handled carefully.

Figure 16.29 | Kislaya Kunjan, lead engineer with Vista Biosciences LLC, works on a glucose-detecting sensor system developed jointly with Purdue University researchers . This technology may lead to a new variety of wearable glucose monitors that would automatically determine glucose concentration even while the patient was sleeping.

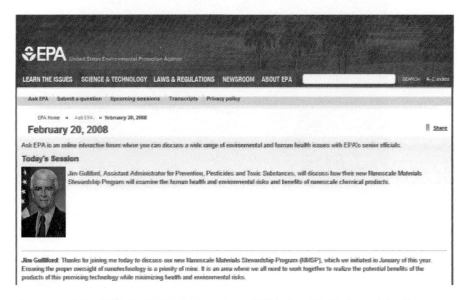

Figure 16.30 | The Environmental Protection Agency (EPA) is charged with determining if nanosilver and any other "nano" metals are toxic.

Nanotechnology has positively impacted drug delivery. For example, as shown in rats, it is possible to specifically target, image, and kill tumors using designed nanoparticles. The nanoparticle consists of an iron oxide core, an MRI contrast material on the outside of the particle making imaging of the target possible; a cancer targeting peptide (very small protein) known as F3 that binds specifically to the abnormal cancer cells; and photofrin, a compound that heats up and kills cells when exposed to red light. This nanoparticle is administered intravenously. Thus far, it has demonstrated no negative side effects nor is it absorbed or metabolized by the liver, which would decrease its concentration in the blood stream, making it an effective drug. As a result, it is assumed that 100 percent of this nanoparticle is delivered to the targeted cancer cells and kills them.

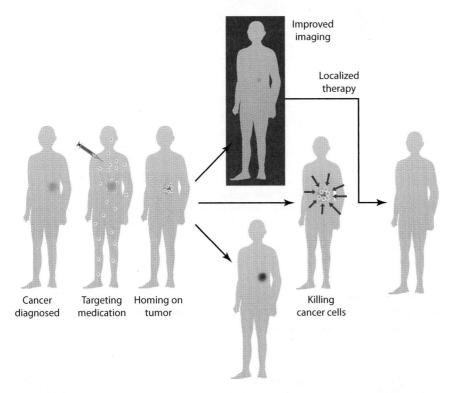

Figure 16.31 | The targeted medication consists of a nanoparticle containing an iron oxide core improving MRI visualization; a targeting peptide (very small protein) known as F3 that binds specifically to the abnormal cancer cells; and photofrin, a compound that heats up and kills cells when exposed to red light. This nanoparticle is administered intravenously; an MRI is taken to show that the particles have reached their target, and the patient is exposed to red light killing the targeted cells.

Biodegradable nanospheres 100 to 5,000 nm in diameter (small enough to pass through blood capillaries, but not so small as to be cleared by the kidneys) have been developed for either drug delivery or for removal of toxic agents found in the bloodstream. They contain an iron compound so they can be visualized by MRI, and they are "coated" to prevent the body from attacking them. Attached to their surfaces are proteins that bind to specific targets, such as tumors or toxins. They are intravenously injected so that they circulate through the patient's bloodstream. If the target is a soluble toxin or free floating cell, the nanospheres and their bound targets can be removed from the bloodstream using an external magnetic separator, where a strong magnet immobilizes the nanospheres. Clean blood flows out of the separator and back into the bloodstream. A drug payload can be added to the spheres, which can then be delivered to specific targets such as tumors or diseased

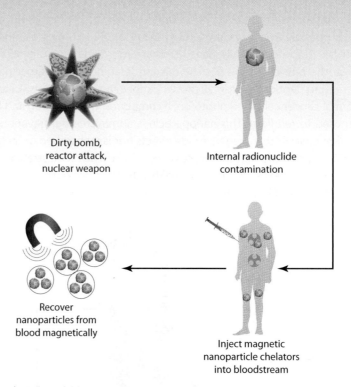

Dirty bomb,
reactor attack,
nuclear weapon

Internal radionuclide
contamination

Recover
nanoparticles from
blood magnetically

Inject magnetic
nanoparticle chelators
into bloodstream

Figure 16.32 | Biodegradable nanospheres 100 to 5,000 nm in diameter have been developed for either drug delivery or for removal of toxic agents found in the bloodstream. They contain an iron compound for MRI visualization and for removal by a large magnet. They also have antitumor or antitoxin compounds attached to their surfaces that bind specifically to the target. The patient's blood is pumped through a machine containing magnets separating the nanospheres with their bound targets out of the blood.

cells. If it is not possible to target using specific proteins. a strong magnetic, like an MRI machine, could be used to target the nanosphere into a particular location in the body. The nanospheres have already been tested for toxicity and found not toxic. (See Figure 16.32.)

MEM devices can be either a combination of *in vitro* (outside of the body) or *in vivo* (inside the body) devices. One of the most striking and useful applications of MEMs is in the treatment of diabetic patients. According to the American Diabetes Association, almost 21 million Americans (7 percent of the population) have diabetes, and this currently costs the United States more than $132 billion in direct and indirect costs. Of these 21 million people, there are more than 175,000 who have type 1, often referred to as juvenile diabetes because they developed the disease under the age of 20. Individuals with juvenile diabetes have to detect glucose levels and take insulin for all of their life.

The therapeutic technology for treating diabetes is a twice-daily finger prick to determine glucose concentration in the blood and a twice daily shot to administer the insulin. This relentless daily testing is painful and difficult, and compliance is a major problem, especially for juvenile diabetics. Even when diabetics do perform the necessary testing, they are still subject to undetected reductions in blood sugar that can occur during sleep and go undetected during the day. In a related condition, diabetic patients sometimes do not experience the usual physical warning symptoms that their blood sugar is low, known as hypoglycemia unawareness. These events can lead to seizures and loss of consciousness, and they can increase the risk of diabetic eye disease and kidney and nerve damage by 76 percent.

To help with these difficult situations, there are products on the market that continuously monitor glucose levels and/or deliver insulin. One example is the Medtronic Mini-Med Paradigm® Real-time continuous glucose monitoring/delivery device. The system shown (Figure 16.33) contains a glucose sensor implanted in the skin, consisting of a wireless radio device that connects to the external pump that delivers the insulin through a micro-needle.

As mentioned, the glucose sensor in this device is an example of nanotechnology. The sensor is composed of an enzyme that interacts with, and therefore detects, glucose. This enzyme, glucose oxidase or GOx, is immobilized in a polymer/protein composite film placed on platinum microcylinders. The enzyme interacts with the glucose to produce small hydrogen peroxide bubbles that diffuse through the coatings to produce an electrical signal. The signal is proportional to the amount of glucose that interacts with the enzyme.

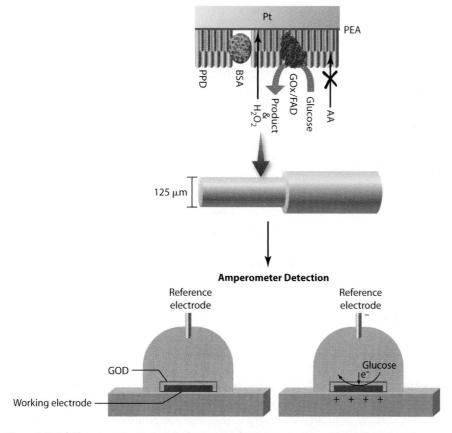

Figure 16.33 | The sensor is composed of a enzyme that interacts with, and therefore detects, glucose (bottom picture). This enzyme, glucose oxidase or GOx, is immobilized in a polymer/protein composite film placed on platinum microcylinders that are only 125 um in diameter (middle picture). The enzyme interacts with the glucose to produce small hydrogen peroxide bubbles that diffuse through the coatings to produce an electrical signal (top picture). The signal is proportional to the amount of glucose that interacts with the enzyme.

Another ground-breaking use of MEMs with tremendous implications for people who lose their eyesight is the artificial retina.

In normal vision, light enters the eye and strikes photoreceptor cells in the retina. These cells convert light signals to electric impulses that are sent to the optic nerve and brain. In retinal diseases, such as age-related macular degeneration

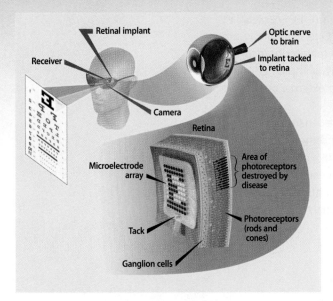

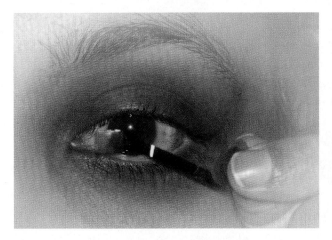

Figure 16.34 | Model 3 device is being developed and it will be constructed of flexible materials that will conform to the shape of the inner eye. Since it will be many times smaller than earlier models, it will be implantable entirely inside or around the eye-no eyeglasses will be needed.

and retinitis pigmentosa, the photoreceptor cells are destroyed. The artificial retina device bypasses these cells to transmit signals directly to the optic nerve. This device consists of a camera and microprocessor mounted in eyeglasses, a receiver implanted behind the ear, and an electrode-studded array that is tacked to the retina. A wireless battery pack is worn on the belt for power. (One laboratory is even developing a battery that can be implanted in the person's head.) The camera captures an image and sends the information to a microprocessor mounted in the eyeglasses, which converts the data to an electronic signal and transmits it to the receiver implanted behind the ear. The receiver sends the signals through a tiny cable to an electrode array, stimulating it to emit pulses. The pulses travel through the optic nerve to the brain, which perceives patterns of light and dark spots corresponding to the electrodes stimulated. Patients have to learn to interpret the visual patterns produced.

Increasing the number of electrodes results in more visual perceptions and higher-resolution vision. To date, six patients have successfully been implanted with the prototype Model 1 device, which contains 16 electrodes. With this device, they can see the equivalent of 16 pixels. The next model, model 2, provides greater resolution at 256 pixels. It is currently being tested in clinical trails. The proposed model 3 device is to be constructed of flexible materials that will conform to the shape of the inner eye—no eyeglasses will be needed (see Figure 16.34). The ultimate goal for this device is to enable facial recognition and large-print reading vision, using materials that will last for a lifetime.

The final MEM example is in MIS. Minimal Invasive Surgery (MIS) is the process of accomplishing a surgical task with the least amount of intrusion, harm, and ultimate cost to the patient. Typically, there is less postoperative pain, shorter hospital stays, quicker recoveries, and less scarring. Robots are already being used for MIS with favorable results. Surgeons find that robot-assisted MIS provides them greater range of motion over standard techniques. An example of this type of surgery is laparoscopic surgery, which is usually defined as surgery within the abdominal or pelvic cavities. A key element in laparoscopic surgery, and typically in any MIS, is the use of a telescopic rod lens system that is usually connected to a video camera. Also attached is a fiber optic cable system connected to a light source (halogen or xenon), to illuminate the operative field. The abdomen is usually inflated with carbon dioxide gas to create a working and viewing space. Carbon dioxide is used because the tissue absorbs this gas and removes it via the respiratory system, plus, it is nonflammable. The main problem with robotic-assisted surgery is that the surgeon cannot "feel" the tissue, and therefore he does not know how hard to press when cutting, nor can he feel differences in the tissue. The sense of touch in this context is known as "haptic feedback." The development of a MEM pneumatic balloon-based haptic feedback system at the Center for Advanced Surgical and Interventional Technology (CASIT) in the David Geffen Medical

School at UCLA allows the surgeon to "feel" the patient's tissue via robotic arms or a grasper with the aid of a force sensor array. The forces from the array are translated to proportional pressures that are applied to the surgeon's hands via balloon actuator arrays.

Figure 16.35 | William Peine, an assistant professor of mechanical engineering at Purdue, operates hand controls for a portable surgical robot under development. They are also working on the problem of haptic feedback.

SECTION THREE FEEDBACK >

1. Name three of the tools used in telemedicine and how are they different.
2. What are the similarities and differences between nanotechnology and MEMs?
3. Describe one nanotechnology or MEM device that has been developed for medical technology.

SECTION 4: Applying Basic Science Research to Advances in Medical Technology

KEY IDEAS >

- The sciences of biochemistry and molecular biology have made it possible to manipulate the genetic information found in living creatures.

- Advances in molecular biology, as a result of the Human Genome Project, have led to advances in molecular diagnostics, such that medical testing has moved from being descriptive to being predictive.

- The potential for misuse and misunderstanding of the genetic information being generated as a result of the Human Genome Project and everyday molecular diagnostic testing of patients has compelled society to research and establish ethical mandates for regulating the incidence of testing and the uses of test results.

The sciences of biochemistry and molecular biology have made it possible to manipulate the genetic information found in living creatures. In addition to recombinant DNA techniques, DNA sequencing, and microarray testing, the new field of molecular diagnostics signals a new era in diagnostic testing.

Molecular Diagnostics Advances in molecular biology and robotics have led to new developments in molecular diagnostics, such that medical testing has moved from being descriptive to being predictive. For example, instead of simply determining whether or not a patient has diabetes, tests are being developed that will determine whether the patient is at high risk of developing diabetes in the future (see Figure 16.36). As a result of predictive molecular diagnostic tests, genetic counseling and the consideration of ethical issues are now nearly as important as the laboratory procedure itself.

Section 4 △ Applying Basic Science Research

Figure 16.36 | For example, instead of simply determining whether or not a patient has diabetes, automated laboratory tests are being developed that will determine whether the patient is at high risk of developing diabetes in the future.

Ethical Considerations The potential for misuse and misunderstanding of the genetic information being generated as a result of the Human Genome Project and everyday molecular diagnostic testing of patients has compelled society to research and establish ethical mandates for regulating the incidence of testing and the uses of test results. Three percent of the funding for the Human Genome Project was set aside to study these issues. This represents the world's largest bioethics program.

Genetic information differs from medical information in that, where medical information only has implications for the individual patient, genetic information has relevant information that affects the extended family, grandparents, parents, children, and even cousins and uncles or aunts. (See Figure 16.37.) For example, if your uncle had a genetic test that showed he has a predisposition to heart disease, do you think he should have to tell the rest of his family, including his extended family? Also, what if the result of his tests showed that he had a genetic disease with a high chance of development later in life, such as Huntington's Disease? Do you think he should be allowed to have children? Presently the policy at the moment protects an individual's right to reproduce, no matter what their genetic makeup. It protects the right to keep that knowledge private, even from family members. Two processes that maintain that right in the United States are "informed consent" and the federal Health Insurance Portability and Accountability Act (HIPAA).

Informed consent refers to the process during which a patient agrees to surrender his/her tissue only after learning and understanding the risks and benefits of doing so. After

Figure 16.37 | Advances in molecular biology have raised a host of ethical issues as such couples, who feel they are at risk, are advised to seek professional counseling when planning families.

the healthcare provider explains this information, the patient is then asked to sign the "informed consent" form to show acceptance of these risks. The informed consent process is important for the following reasons:

- It promotes respect for patients.
- It protects an individual's right of self-determination.
- It promotes the belief that it is wrong to force a person to act against his or her will.

Once a patient has consented to genetic testing (or a waiver of consent is granted), the confidentiality of any information gathered from this testing becomes an important concern. The federal Health Insurance Portability and Accountability Act (HIPAA) is designed to do the following:

- Ensure health insurance portability (ease of movement) for workers and families when they change or lose a job
- Reduce healthcare fraud and abuse
- Guarantee security and privacy of healthcare information
- Enforce standards for health information
- Set standards for electronic data interchange transactions

Why is HIPAA needed? The following two examples illustrate why it is important that the privacy of patients be protected. The now deceased tennis star Arthur Ashe's positive HIV status was revealed by a healthcare worker to a newspaper without his permission. The medical records of the country western singer Tammy Wynette were sold to the National Enquirer by a hospital staff member.

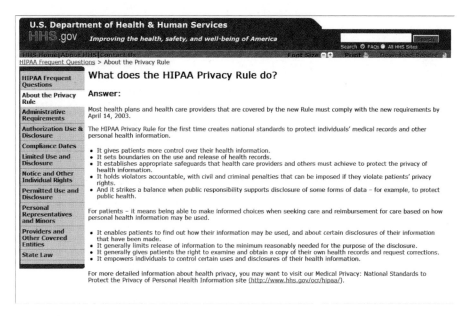

Figure 16.38 | The Web site of the U.S. Department of Health and Human Services is a major source of information about the HIPAA Privacy Rule.

All healthcare providers including hospitals, clinics, nursing homes, physicians, dentists, chiropractors must comply with HIPAA (see Figure 16.38). So must any organization that receives payment for healthcare services or that transmits electronic forms containing healthcare information. HIPAA defines protected health information (PHI) as information that can be communicated orally, in written form, or through other media and is "individually identifiable" concerning an individual's past, present, or future. This includes information about a patient's

645

physical and mental health, healthcare treatments, healthcare payments, and identification information, such as social security number, gender, or birthdate. In addition, these providers that handle and transmit such information must establish procedures that follow the "Minimum Necessary Rule." This rule states that, within an organization or institution, reasonable effort must be used to ensure that only the minimum amount of PHI is handled. For example, a patient's social security number does not need to be on every form, or record, or medical image for it be processed—only a minimal identifier, such as a number given to that person's record, is required. Employees who violate this law can be fined and spend time in jail.

SECTION FOUR FEEDBACK >

1. What are the goals of the Health Insurance Portability and Accountability Act (HIPAA)?
2. Give some examples of protected health information, according to HIPAA.
3. Go online and find out how long your immunity to the diseases whooping cough, tetanus, and diphtheria lasts when you are vaccinated with a DTaP vaccine.

CAREERS IN TECHNOLOGY

Matching Your Interests and Abilities with Career Opportunities: Medical Technicians

No matter what the economic climate, there will always be jobs in the medical field. The obvious careers are the nurse, physician, and medical researcher, but there are other important, less well-known jobs. The examples cited are biomedical engineers who design medical equipment, and the individuals who use the equipment, such as medical sonagraphers, cardiovascular technologists and technicians, and medical technologists and technicians. You can find more examples by searching the Department of Labor Occupational Outlook Handbook, which is available through the Internet.

Significant Points

* In general, employment is expected to grow much faster than average.
* Many of the technician and technology jobs are obtained after completing a two-year community college program.
* Biomedical engineering usually requires at least a four-year degree and a researcher generally has at least a master's degree.

Working Conditions

Most technologists and technicians in this industry work in hospitals, doctor's offices, and industry. Some technologist and technician positions may require lifting of heavy patients or equipment and a lot of time walking and standing. The work week is generally forty hours. Biomedical engineers generally work in academic or industry industrial laboratories and do both applied and basic research. The amount of hours per week spent on the job varies.

Occupations in the Industry

Biomedical engineering usually requires a four-year degree or higher. It requires knowledge of engineering, biology, and medicine. Its goal is to improve human health by integrating basic and applied scientific discoveries in these fields. Specialty areas include biomechanics (mechanics applied to biological or medical problems), biomaterials (includes living tissue and materials used for joints, heart valves), rehabilitation engineering (e.g., wheelchairs), and clinical engineering (technology that requires databases, software, and computers).

In medical imaging, technicians are needed for specialized equipment such as sonographs. The technician specialties are commonly named based on the type of equipment the technicians use and the part of the body that they visualize. For example, neurosonographers use sonography to visualize the nervous system.

Cardiovascular technologists and technicians assist physicians in diagnosing and treating cardiac (heart) and peripheral vascular (blood vessel) ailments. Individuals who specialize in electrocardiograms (EKGs) are known as cardiographic or EKG technicians, while individuals who specialize in an invasive procedure are known as cardiology technologists.

Radiological technologists and technicians take X-rays. Some specialize in diagnostic imaging technologies, such as CT and MRI.

Medical technologists are also known as clinical laboratory technologists, and medical technicians are also known as medical technicians. Technologists usually require four years of training, and technicians usually require two years. These people work in hospitals or company clinical laboratories and perform both nonautomated and automated tests for diagnostic purposes. As a result of the advances in genomic and proteomic science, testing is becoming more elaborate and predictive than just diagnostic. For example, besides determining the presence of heart disease, eventually, through DNA testing, tenchology will be able to determine a person's predisposition for heart disease.

Training and Advancement

The possibilities for advancements are generally excellent, as most hospital and industry sites provide several levels of employment for technologists and technicians. Since patient care is involved, all positions are governed by accrediting agencies to ensure quality. Biomedical engineers can belong to several different professional engineering societies, and these societies provide professional development, opportunities to share information, and also help disseminating information concerning employment opportunities.

Outlook

The human life span has increased and the population continues to grow. As a result, job opportunities in these fields will continue to increase. Also, because of the increasing demand for employees, wages and salaries for this type of employment are expected to grow faster than the growth projected for the entire economy. However, this increase is somewhat restricted by what the public and the health industry can afford to pay.

[Bureau of Labor Statistics, U.S. Department of Labor, Occupational Outlook Handbook, 2008–09 Edition, visited July, 2009 http://www.bls.gov/oco/]

Summary >

The use of devices to diagnose or treat disease advanced in the 1800s. For example the stethoscope was developed in 1810. This reliance on technology for diagnosis, therapy, and rehabilitation has continually increased over time. In today's world, medical technologies include predictive testing besides diagnostic testing, such as determining cholesterol levels in a person's blood, imaging technology so that organs can be visualized, eliminating the need for surgery, and the development of artificial limbs, such as the carbon fiber springs which may actually allow a person to run faster than on normal legs.

Advances in medical technology have also created ethical dilemmas for people in healthcare and for society in general. The definitions for when life starts and when life ends are being questioned since it is now possible to operate on fetuses and keep people alive whose hearts and lungs have stopped moving. These advances have not only occurred for urban society but also for people in remote rural areas. Telemedicine allows individuals to access expert medical care, even to the point that a surgeon can operate on an individual robotically when they are separated by huge distances.

Many of the advances in medical technology are the result of a convergence of technology advances occurring in separate fields. For example, nanotechnology and microelectromechanical (MEM) medical devices are a result of a convergence of technologies (i.e., semiconductor, computer, biology, and chemistry). An example of such a device is one that a diabetic patient can wear twenty-four hours a day. It detects the level of glucose, and, in response, delivers the correct amount of insulin. The detector and drug delivery devices are worn as small patches on the person, and the pump is a small device that can be worn under clothes at the waist. The sciences of biochemistry and molecular biology have made it possible to alter a person's DNA and its expression. This means people with genetic diseases such as SCIDs can have the normal gene placed strategically within their bodies to produce the correct proteins that were originally missing from their bodies. It also means that a person's predisposition to certain diseases based on their DNA can be determined. Clinically this is predictive testing at the DNA level. However, these advances have progressed so rapidly that neither society nor the healthcare industry can fully understand the ethical and economic ramifications brought about by this technology. The resulting ethical questions have compelled society to research and establish ethical mandates for regulating testing and the use of test results. However, it is likely that continued advances in medical technology will result in more questions, not fewer. These questions will likely be left for future generations to work out.

FEEDBACK

1. How is subjective evidence provided by the patient different from objective evidence as provided by diagnostic testing and interpreted by healthcare professionals? Should healthcare professionals consider subjective evidence when making a diagnosis? If so, why?

2. Search the Internet for information about drug-coated stents. What are their advantages? What are their disadvantages or possible side effects?

3. What is tomography and how is it combined with MRI technology to produce a 3-D image of a person's body?

4. Go online and search for the latest advances in prosthetic devices. Find one not mentioned in the chapter and write a short paragraph describing its advantages and disadvantages compared to previous devices.

5. Go to the Web site for the National Center for Biotechnology Information, click on OMIM™ (Online Mendelian Inheritance in Man™), go to Human Genome Resources, and then click on Genes and Disease. Find a genetic disease that interests you and write a brief paragraph starting with a one- to three-line description of the NCBI site and who is responsible for OMIM™. The rest of the paragraph should address the following questions about the disease: What are the tissues and/or organs affected? What is the incidence of the disease? What are the symptoms and long-term consequences of the disease? Are there any medical treatments? Make a list of any words that you did not understand, look them up, and supply a definition next to each word.

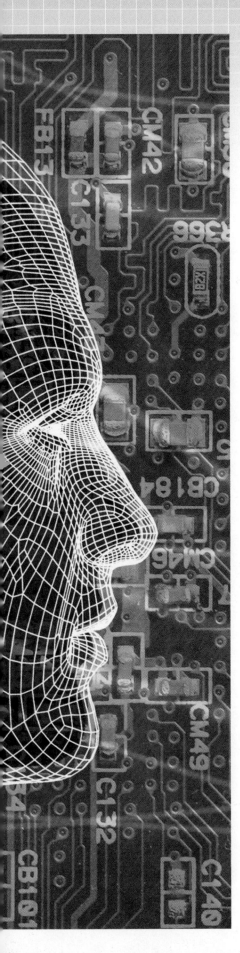

DESIGN CHALLENGE 1:
Making a Stethoscope

• Problem Situation

A doctor uses a stethoscope to listen to the human heart. As the heart beats, it causes the stethoscope to vibrate. These vibrations are transmitted to the doctor's ears as sound. A simple acoustic stethoscope consists of several parts. A chestpiece, which the doctor places against the patient's chest over the heart, vibrates with each heartbeat. The resulting acoustic waves travel through an air-filled tube that leads to a three-way junction that leads to two earpieces. The earpieces relay the acoustic waves into the doctor's ears. The doctor is then able to hear the characteristic "lub-dup" sound of the heart beating and so evaluate the patient's health.

• Your Challenge

You work in the risk management department of a large corporation. As part of the company's new safety regulations, each work area must have instructions for making basic medical equipment from everyday materials in case of dire emergency. You have been asked to come up with guidelines for creating a working stethoscope from common equipment.

• Safety Considerations

1. The information provided in this activity is not medical advice and is not intended to diagnose or treat any health conditions. Likewise, do not use the stethoscope you create to diagnose or treat any health conditions.
2. If you share a stethoscope with another person, clean the earpieces with rubbing alcohol before each use.

• Materials Needed

1. Plastic funnel or cut-off top portion of plastic water bottle
2. Balloon
3. Plastic, rubber, or silicone tubing, at least 24" long
4. 3-way connector
5. Tape
6. Other commonly found materials

• Clarify the Design Specifications and Constraints

You will make a functioning stethoscope using commonly found materials. It will consist of a chestpiece with a diaphragm and at least one earpiece. The stethoscope must allow the user to clearly hear a human heartbeat when placed on a subject's upper left chest. It must work whether the subject is sitting, standing, or lying down, and the user must be able to count the subject's heartbeats. Your stethoscope should work as well as a commercially available acoustic stethoscope.

• Research and Investigate

To complete the design challenge, you need to first gather information to help you build a knowledge base.

1. In your guide, complete the Knowledge and Skill Builder I: Acoustics.
2. In your guide, complete the Knowledge and Skill Builder II: Cardiac auscultation.
3. In your guide, complete the Knowledge and Skill Builder III: Heart rate.

Generate Alternative Designs

Describe two of your possible alternative approaches in this stethoscope exercise. Discuss the decisions you made in (a) construction of the chestpiece, (b) construction of the earpiece(s), and (c) connection of the chestpiece to the earpiece(s). Attach drawings if helpful and use additional sheets of paper if necessary.

Choose and Justify the Optimal Solution

What decisions did you reach about the construction of a stethoscope? How did you reach these decisions?

Display Your Prototypes

Produce your stethoscope. Include descriptions, photographs and/or drawings of it in your guide.

Test and Evaluate

Explain whether your designs met the specifications and constraints. What tests did you conduct to verify this?

Redesign the Solution

What problems did you encounter that would cause you to redesign the stethoscope? What changes would you recommend in your new design?

Communicate Your Achievements

Describe the plan you will use to present your results to your class, and show what handouts you will use.

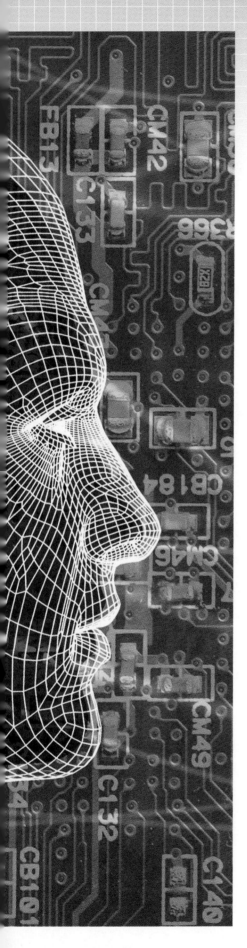

DESIGN CHALLENGE 2:
Determine What Biocompatibility Tests Need to Be Done for Food and Drug Administration (FDA) Approval

● Problem Situation

Your company, Compatibility Solutions, has been contacted by a biodevice company that manufactures stents. This company has just produced a new stent that also releases a drug for inhibiting tissue growth. Tissue growth over implanted stents has become a major problem as it clogs the stent, decreasing its lifespan. However, whenever a new drug is developed for use in humans it has to be tested to make sure it is biocompatible (i.e., does not cause negative side effects in humans). Since the initial tests cannot be performed on humans, they are conducted with other life forms, such as yeast, fish, or monkeys, to make sure the drug is at least safe in these organisms. After passing the initial tests in other organisms, the drug is finally tested in humans before it is released by the FDA to be used by the general public.

● Your Challenge

Your job is to determine what biocompatibility tests need to be done for Food and Drug Administration (FDA) approval

● Safety Considerations

Any safety issues associated with testing should be included in your plan. For example, if you decide to do laboratory testing, then safety issues concerning the laboratory need to be included.

● Materials Needed

You will need access to the Internet for information searching, word processing software, Excel if you are to do any graphing, and PowerPoint for presenting the information.

● Clarify the Design Specifications and Constraints

The presentation needs to be fifteen minutes long. The targeted audience is the CEO and president of the biodevice company that manufactures the stent. As part of the presentation, the FDA regulations that govern the testing and release of this product to the market should be reviewed as background information for the audience.

● Research and Investigate

To complete the design challenge, you need to first gather information to help you build a knowledge base.

1. In your guide, complete the Knowledge and Skill Builder I. What is a biodevice and what is a stent?
2. In your guide, complete the Knowledge and Skill Builder II: Understanding the FDA regulations that govern the testing and market release of biodevices that release drugs. Do an Internet search for the center for device and radiological health (CDRH). At this site, you will find information you need.
3. In your guide, complete the Knowledge and Skill Builder III: Biocompatibility testing for biodevices.
4. In your guide, complete the Knowledge and Skill Builder IV: Using power point effectively.

- **Generate Alternative Designs**

Describe two of your possible alternative testing approaches in this exercise. Discuss the decisions you made when selecting an organism and a test for determining biocompatibility of the drug in humans.

- **Choose and Justify the Optimal Solution**

What decisions did you reach about the test and test organism? How did you reach these decisions?

- **Display Your Prototypes**

Produce your test plan; include descriptions, photographs, and/or drawings of it in your guide.

- **Test and Evaluate**

Explain whether your plan met the specifications and constraints.

- **Redesign the Solution**

What information did you find that might cause you to redesign the plan for testing? What changes would you recommend in your new plan?

- **Communicate Your Achievements**

Present your plan, using PowerPoint, to the rest of the class.

Describe the plan you will use to present your results to your class, and show what handouts you will use.

INTRODUCTION

THE **TECHNOLOGICAL REVOLUTION** of the twentieth century is multidisciplinary and crosses all aspects of life, not just those of business and commerce. Under this umbrella you can find information, communications, and materials technology, as well as biotechnology and nanotechnology. The interaction of these technologies, together with the global nature of business and industry, changes in communication, as well as socialization

CHAPTER

17

TECHNOLOGY IN THE FUTURE

Figure 17.1 | New communication technologies enabled people around the world to take part in a global conversation about ideas and products.

and interaction brought about by the Internet (see Figure 17.1), has spawned its own global revolution.

This revolution in multiple technologies is being amplified by other significant changes taking place at the same time: the easing of trade restrictions between countries, making it easier to undertake international trade; the changing political climate in eastern Europe, India, and China, which has encouraged the development of business and industry; and finally, the impact of the Internet on the individual. Low-cost Internet access has allowed vast numbers of people around the world to add their intellectual abilities to the development of global activities. This has unleashed an enormous number of people who can participate in the development of new and existing ideas, in a climate that now actively encourages the trading in such ideas and products across the world.

The German philosopher Arthur Schopenhauer (1788–1860) observed: "Thus the task is not so much to see what no one yet has seen, but to think what nobody yet has thought, about that which everybody sees."

Now anyone, located anywhere in the world, can develop, communicate, and market new ideas by combining the technologies that surround us every day in a unique way.

Figure 17.2 | New technologies enable global participation in product development.

KEY IDEAS >

- Globalization of trade and commerce has stimulated business and industry.

- The social revolution brought about by worldwide access to computing and communications technology has changed how people work, communicate, and interact.

- The interaction of technological, educational, and social tools has made it possible for anyone to start a commercial or social transaction anywhere in the world.

Figure 17.3 | Trade is now a global phenomenon, as improved forms of communication allow companies to outsource work around the world.

Globalization of trade and commerce has stimulated business and industry. The exchange of goods and services for other goods, services, or money has been a mainstay of business and commerce. Over the latter part of the twentieth century, however, the speed, geographic spread, complexity, and volume of commerce has increased considerably as companies in the United States and Europe saw the benefits of moving production to lower-cost locations. In doing so, they were able to use rapidly improving information, communications, transportation, and business systems to achieve such efficiencies.

The corporate benefits of technological improvements in computing, communications technologies, and global interconnectivity offered by the Internet have been considerable. But they pale in significance compared to the incredible growth of interest and skill in using such tools for personal, social, educational, and business purposes by individuals.

There is little doubt that this global revolution is fueled by technological change, but its impact is most significant in the social revolution it has caused. Low-cost technology tools allow individuals, small groups, and communities to interact and communicate across the globe (see Figure 17.3).

The social revolution brought about by worldwide access to computing and communications technology has changed how people work, communicate, and interact. The impact of this social revolution cannot be underestimated, as shown in Figure 17.4: In 1995, 16 million people used the Internet; in 2007, that number had risen to over 1.2 billion (almost 20 percent of the world's population).

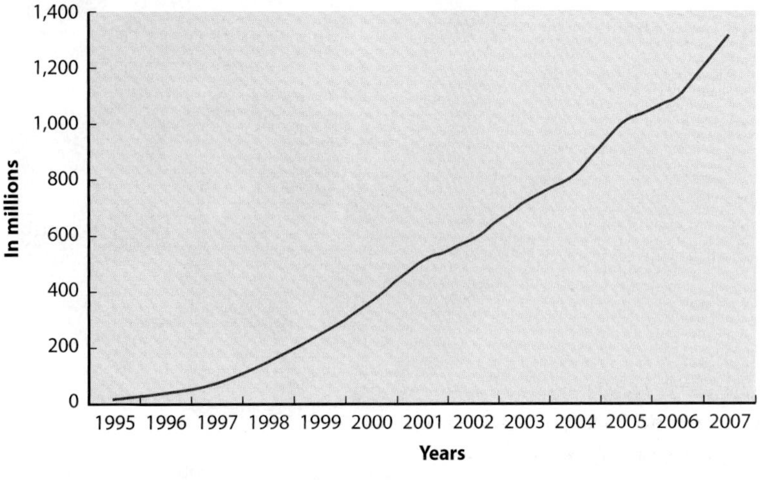

The World Is Flat

Thomas Friedman, in his book, *The World Is Flat: A Brief History of the Twenty-First Century* (which is a reference to the fact that the world market place is increasingly becoming a level or flat playing field for all players, no matter where they are located, see Figure 17.5), says that the next step in globalization is not just an expansion of worldwide corporate commerce, but something much more powerful.

Figure 17.4 | According to Internet World Stats (*www.internetworldstats.com*), Internet usage has increased more than twelvefold since 1995.

In millions (y-axis): 0, 200, 400, 600, 800, 1,000, 1,200, 1,400
Years (x-axis): 1995 1996 1997 1998 1999 2000 2001 2002 2003 2004 2005 2006 2007

Figure 17.5 | The concept of a flat world has taken on new meaning in today's world.

TECHNOLOGY AND PEOPLE:
Thomas Friedman—Journalist and Futurist

Thomas Friedman is a columnist for *The New York Times*. He is also a celebrated journalist and author who started his career as a reporter for the United Press International.

In the early 1980s, he covered the Israeli invasion of Lebanon for *The New York Times*. His coverage of that event, as well as other significant issues such as the Sabra and Shatila refugee camp massacre, won him his first Pulitzer prize for international reporting.

In his career with *The New York Times*, Friedman has served as the chief diplomatic, chief White House, and international economics correspondent. His interest in foreign and economic issues has figured highly in his writing. In 1994, he moved to his current role as the newspaper's op-ed columnist.

Friedman's first book, *From Beirut to Jerusalem*, 1989, won a number of awards, as well as becoming a basic college textbook on the subject of the Middle East. In 1999, he published his first book to address issues of globalization, *The Lexus and the Olive Tree*. After visiting the Far East and meeting with entrepreneurs in India and China, and drawing on his extensive experience in global and economic issues, he published his powerful follow up book, *The World Is Flat*.

In *The World Is Flat*, Friedman has traced the path of globalization through three milestone events: the discovery of the new world and new trade routes shrank the world from large to medium; the Industrial Revolution rendered global trade commonplace and the world became much more accessible; and finally, the advent of the "flat" world where global fiber optic networks connect all parts of the globe instantly.

Friedman is a strong believer in the need for a robust support for free trade, where the barriers to importing and exporting are reduced or removed. While acknowledging that this places low-income jobs at risk in the United States, he believes an increase in the global workforce will reduce costs and provide alternate employment opportunities from the resulting strong economic impact of growing sales. To illustrate his thinking, he looks to retail giant Wal-Mart as an example of a truly global company that maximizes its earning in all parts of the world. As it grows, it supports the employment of hundreds of thousands of workers, as well as reducing costs for customers. In becoming more successful, the company serves to increase revenue and profits that support U.S. investors and employees.

While Friedman is an advocate of globalization, he also points out the need for a country to preserve its local traditions, through a process he called glocalization, where individuals, groups, or communities think *globally* but act *locally* (see Figure 17.6).

In 1975, Friedman received a bachelor of arts in Mediterranean Studies from Brandeis University, where he first arrived as a transfer student in 1973. He then attended St. Antony's College at the University of Oxford on a Marshall scholarship, earning a master of arts in Middle Eastern studies.

Figure 17.6 | Thomas Friedman is an advocate of globalization but feels that countries must maintain their local traditions.

657

The interaction of technological, educational, and social tools has made it possible for anyone to start a commercial or social transaction anywhere in the world. In a global population with increased technological and communication skills that makes use of lower-cost information technology resources, it is now possible for a person in India to undertake tax filing for a company in Indiana without any loss of speed or accuracy.

This social and technological revolution has made it possible for anyone to participate in the global economy in ways never seen before. The unprecedented growth of business and communication systems across countries made it possible for major corporations to undertake global activities. These systems are now available to anyone with an Internet connection, a computer, and entrepreneurial skills, who can now participate in these same activities.

Freidman has identified a number of factors that are crucial to the leveling of the playing field. He lists ten significant developments that can be combined to have a leveling effect, so it is not surprising that he calls such items levelers:

1. **The fall of the Berlin Wall and the commercialization of communist countries**, combined with the relaxing of government restrictions on international trade developments in India and Pakistan, which allowed a large number of powerful and populous nations to join the world's economic mainstream;

2. **The Internet browser**, which allowed the Internet to become accessible to ordinary people, and no longer the property of only computer "geeks";

3. **Workflow software**, software tools that communicate between machines and systems to allow activities to happen without human intervention;

4. **Open communities**—including open-source software, social networking systems such as blogs, listservs, and forums—a most powerful force in expanding the intellectual power of individuals and groups;

5. **Outsourcing**, the mechanism that has allowed subsystems to be made elsewhere at a lower cost;

6. **Offshoring**, or the moving of production to a lower-cost location;

7. **Supply chaining**, where information and communication technologies are used to streamline item procurement, distribution, and sale;

8. **Insourcing**, in which one company uses another company to carry out duties it used to perform, but at a lower cost;

9. **Informing**, referring to the fact that never before have so many people on the planet had their own access to information;

10. **The steroids**, the devices that add to the above factors: digital, mobile, virtual, and personal.

Friedman suggests that the above factors have uniquely interacted to bring about a social and economic change that is so pervasive that it will alter the nature of the way that business is conducted globally.

Return to the Origins of Trade

Other commentators have noted that while Friedman might be right with regard to the importance of globalization, the basic concepts underlying this social revolution reflect the very beginning of trade, when individuals who could offer goods or services in exchange for other goods or services, or some monetary or exchangeable reward, did so locally. This all predates what we would now call commercialization. The difference is that the exchange now is not between people in the same village or in neighboring towns, but across the entire globe.

1. Research the gross domestic product value (GDP) for three major nations in 1970, 1980, 1990, and 2000.

2. Find statistics on the growth of a social networking site, such as Facebook or MySpace. Compare the growth of the site's usage in the United States and in one other country.

3. Research the development of the Internet browser from its invention through to the most popular current browsers. What were the names and manufacturers of the most popular early browsers? How has that changed up to the present time?

SECTION 2: Nanotechnology

KEY IDEAS >

- Nanotechnology allows the manipulation of atoms or molecules to create or modify nanoscale materials.

- Nanotubes offer the greatest advance in materials technology.

- Molecular manufacturing offers the possibility of creating inanimate objects by assembling atoms in the correct order and location.

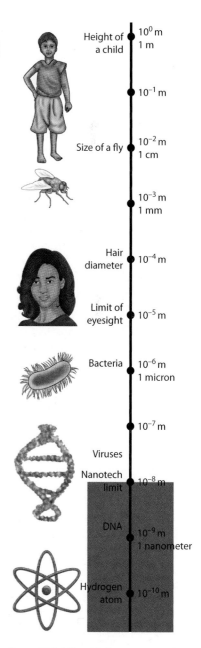

In 1959, the physicist Richard Feynman speculated that science could theoretically maneuver atoms to build new materials, almost as if it were using building blocks and incredibly small robot arms. At that time, the technology that allowed scientists and technologists to work knowledgeably at the atomic level did not exist. Even though Feynman did not coin the term, what he was describing would eventually become known as nanotechnology.

Nanotechnology allows the manipulation of atoms or molecules to create or modify nanoscale materials. At this scale, the materials are too small to be seen using the most powerful optical microscopes in use today. An atom ranges in size from 0.1 to 0.5 nanometers, or 80,000 times smaller than the diameter of a human hair, so we are looking at manipulation at the very smallest scale. (A nano, which comes from the Greek word *nanos*, meaning very small, is one billionth of a unit of measurement, so a nanometer is one billionth of a meter; see Figure 17.7.)

All materials contain atoms. Therefore, they contain nanosized particles. For centuries, scientists have worked to change the characteristics of materials, and so they have effected change at the nanoparticle level. But until recently, they have not been able to see and understand the changes at the atomic level. Nanotechnology gives scientists an unprecedented degree of control over materials at the molecular level, and, perhaps more importantly, the ability to understand atomic scale interactions.

Figure 17.7 | How big is a nanometer?

659

Consumers are already encountering aspects of nanotechnology in everyday products. The Wilson® double-core tennis ball has clay nanoparticles embedded in the polymer lining to slow the escape of air from the ball and make it last longer (Figure 17.8). Nano-care fabrics incorporate nanowhiskers (extremely small particles that are not combined to make other objects) into the fabric to make it stain resistant to water-based liquids.

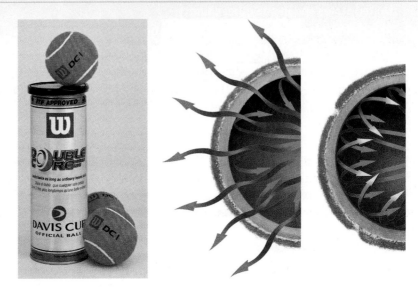

Figure 17.8 | Nanotechnology increases the life of tennis balls like the Wilson DoubleCore™ tennis ball.

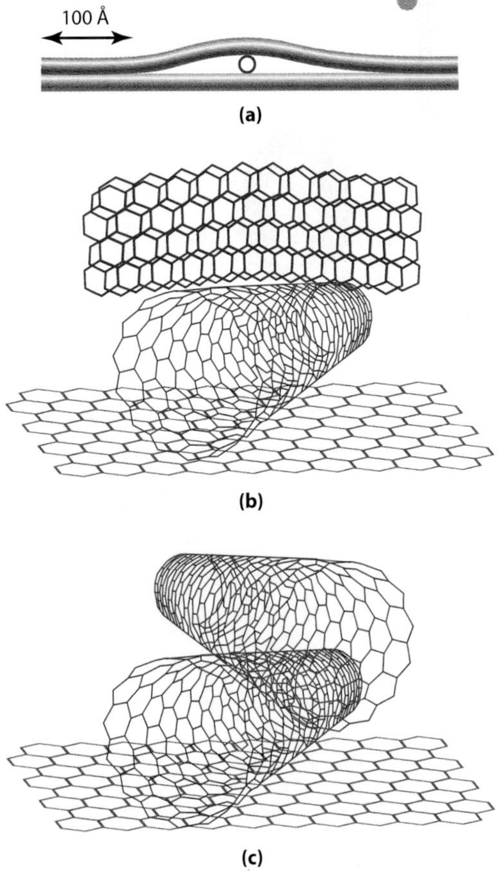

Nanotubes

Carbon nanotubes (shown in Figure 17.9) are long cylinders made from rolled up sheets of carbon that are one atom thick. Although very thin (1 nanometer), their length-to-diameter ratio is very large (in excess of 1,000,000 to 1). Like all nanomaterials, they also have a large surface area compared to other materials, and they have very low density as well.

Theoretically, carbon nanotubes are the strongest and stiffest known materials, with a tensile strength of more than 10,000,000 tons per foot, which makes them eleven times stronger than carbon fiber and thirty-seven times stronger than high carbon steel (see Figure 17.10). Not only is this material very strong, it is also very light.

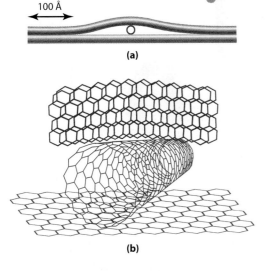

100 Å

(a)

(b)

(c)

Figure 17.9 | Carbon nanotubes are extremely long and thin, yet are exceedingly strong.

MATERIAL	TENSILE STRENGTH (GPa)	DENSITY (g/cm³)
Single wall nanotube	150	1.4
Multi wall nanotube	150	2.6
Diamond	130	3.5
Kevlar	3.6	7.8
Steel	1.0	7.8
Wood	0.008	0.6

Figure 17.10 | Nanotubes are stronger than steel yet much less dense.

Carbon nanotubes, as well as nanotubes made from other materials, offer the greatest advance in materials technology since the production of low-cost steel. This very strong but light material offers manufacturers a way to greatly reduce weight without reducing strength.

Currently, the most common method of manufacturing carbon nanotubes is chemical vapor deposition, where a gas containing carbon enters a specially prepared high-temperature reactor vessel. As the gas comes in contact with materials in the vessel, they form solids. The carbon nanotubes are "grown" from carbon particles and can be formed up to 3/4 inch in length.

Molecular Manufacturing

In the *Star Trek* television series, actors used a device called a "molecular replicator" that could create any inanimate object by assembling the atoms in the correct order to make food or whatever the user desired. This is the imaginative concept behind molecular manufacturing. Scientists today are researching processes in which the chemical properties of single molecules can be used to allow carefully controlled molecular self-assembly. They are looking at the development of nanosized fabricators that could produce their own mass in a few minutes. By using precise positioning of strongly bonded molecules, it would be possible to produce strong and durable products.

Nanofabricators alone would only produce basic products. Once a large number of nanofabricators are combined with computer control systems to form vast nanofactories, the prospect of creating an entire product, such as a computer, becomes theoretically possible.

In the future, nanoscale motors, which convert energy into rotational motion using nanoscale technology, will be thousands of times more powerful than they are now, and computers will be millions of times more compact. In almost any human-scale product, the volume occupied by the nanofabricated components will be negligible.

Alternate Energy

Nanotechnology is making it possible to develop low-cost solar power. In a normal solar cell, when a photon (a particle or a wave) of sunlight strikes the cell, a single electron in the solar cell is released as electrical energy. Solar photons have enough energy to release multiple electrons from the conventional solar cell, but current generations of solar cells do not do so. Nanocrystal solar cells can release multiple electrons when struck by a single solar photon, greatly increasing the efficiency of the solar cell (see Figure 17.11).

The Hydrogen Solar Company is using nanotechnology to enable the production of hydrogen from sunlight. The conversion cell has nanocrystalline coatings of metal oxides on the large surface area of nanotechnology materials, enabling the cell to capture the full spectrum of ultraviolet light. The cell converts the energy of sunlight directly into hydrogen gas by splitting water into its constituent elements, hydrogen and oxygen.

An engineering team at the Massachusetts Institute of Technology uses nanotubes to improve ultracapacitors, energy storage devices similar to batteries. Ultracapacitors (see Figure 7.12) are relatively large and hold energy in an

Figure 17.11 | Large-scale solar panel arrays can use nanotechnology for increased efficiency.

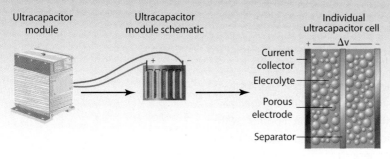

Ultracapacitor module

Ultracapacitor module schematic

Individual ultracapacitor cell

Current collector
Elecrolyte
Porous electrode
Separator

Figure 17.12 │ Ultracapacitor energy storage systems consist of cells with plates that collect voltage, reacting with an electrolyte to produce energy.

electric field, whereas batteries are generally smaller and rely on a chemical reaction to produce power. The researchers are increasing ultracapacitor storage capacity by using nanotubes to increase the amount they can store and to reduce their size.

Electric Vehicles

Nanoscale fabrication, both for hydrogen production and storage and for fuel cell manufacture, could make the dream of hydrogen-based electric vehicles a reality in the future. The nanotech research firm Cientifica, for example, reports that carbon nanotubes could enable a tenfold improvement in fuel cells performance, together with a 50 percent cost reduction for the catalyst material. As prices come down over the next several years, the firm estimates carbon nanotubes will be used in 70 percent of all fuel cells.

SECTION TWO FEEDBACK >

1. Create a graph showing the scale of objects from a baseball down to nanotubes in five steps. Use objects that differ from those shown in Figure 17.7. Make sure that each step is illustrated with a diagram or picture, and that the relative size is shown for each step of the scale.

2. Using research, identify three currently available commercial products, other than those mentioned in the chapter, that use nanotechnology. Briefly explain how each product uses this technology.

3. Research the terms "bottom up" and "top down" nanotechnology and describe how such techniques are used to develop nanoscale products.

SECTION 3: Biotechnology

KEY IDEAS >

● Biotechnology uses biological systems or living organisms to develop, make, or modify products or processes for a particular purpose.

● Genetically modified plants are being developed to increase the output of agricultural systems and to create plants resistant to damaging insects or bacteria.

Biotechnology uses biological systems or living organisms to develop, make, or modify products or processes for a particular purpose. Biotechnology has existed from the time that humans began selecting and refining plants and seeds. Over the centuries, biotechnology evolved through the processes of selection and hybridization (where different species that would not have usually existed in the same location were cross fertilized using natural techniques).

Recent developments in genetics, molecular biology, cell biology, and biochemistry have combined to form the biotechnology industry. Today's information and communication technologies have combined with biotechnology to further spread its reach.

In the medical world, biotechnology has led to manipulation of genes (the building blocks that define all characteristics of animate objects) to create medicines targeted to specific genetic conditions. Gene therapy is now considered a major area of medical research, and it offers significant opportunities for highly targeted treatments for conditions that have eluded other medical solutions.

Genetically modified plants are being developed to increase the output of agricultural systems and to create plants resistant to damaging insects or bacteria. Genetic modification is being used to develop plants that combine flavors and characteristics of two or more fruits or vegetables.

Biotechnology is playing an increasingly important role in the development of alternate energy sources. The need to move from fossil fuels has led to a great interest in development of alternate fuels developed from sources such as corn, or more promisingly, fiber-regeneration techniques.

The human genome project, and related projects looking at the genetic maps of other organisms, is mapping the genetic makeup of all living organisms (see Figures 17.13 and 17.14). From this multidimensional map of the biological world, it is hoped that a better understanding of the interaction of all the various genetic components will result. Such insights are being used to develop more appropriate and targeted medications for the future.

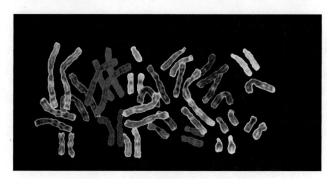

Figure 17.13 | This is how Human DNA appears in the laboratory.

Recently, the first genetic map of a single individual was published. J. Craig Venter published his entire diploid genetic sequence, which is all the DNA in both sets of chromosomes inherited from each of his parents.

Although the genomic mapping process is still ongoing, scientists have developed sufficient knowledge of the cell reproductive process to successfully undertake cloning of living organisms. A clone is a genetically identical copy of another living organism that has not been created as a result of a natural reproductive process. The cloning of Dolly (named after country western star Dolly Parton, a favorite singer of the lead scientist, Dr. Keith Campbell), a Scottish blackface sheep, by scientists at the Roslin Institute in Midlothian, a small city in Scotland, showed for the first time that it was possible to create an identical genetic copy of an animal in the laboratory.

246 million base pairs

Cataracts
Malignant transformation suppression
Ehlers-Danlos syndrome, type VI
Glaucoma, primary infantile
Hirschsprung disease, cardiac defects
Schwartz-Jampel syndrome
Hypophosphatasia, infantile, childhood
Breast cancer, ductal
Cutaneous malignant melanoma/dysplastic nevus
p53-related protein
Serotonin receptors
Schnyder crystalline corneal dystrophy
Kostmann neutropenia
Oncogene MYC, lung carcinoma-derived
Deafness, autosomal dominant
Porphyria
Epiphyseal dysplasia, multiple, type 2
Intervertebral disc disease
Lymphoma, non-Hodgkin
Breast cancer, invasive intraductal
Colon adenocarcinoma
Maple syrup urine disease, type II
Atrioventricular canal defect
Fluorouracil toxicity, sensitivity to
Zellweger syndrome
Stickler syndrome, type III
Marshall syndrome
Stargardt disease
Retinitis pigmentosa
Cone-rod dystrophy
Macular dystrophy, age-related
Fundus flavimaculatus
Hypothyroidism, nongoitrous
Exostoses, multiple
Pheochromocytoma
Psoriasis susceptibility
Limb-girdle muscular dystrophy, autosomal dominant
Pycnodysostosis
Vohwinkel syndrome with ichthyosis
Erythrokeratoderma, progressive symmetric
Anemia, hemolytic
Elliptocytosis
Pyropoikilocytosis
Spherocytosis, recessive
Schizophrenia
Lupus nephritis, susceptibility to
Migraine, familial hemiplegic
Emery-Dreifuss muscular dystrophy
Cardiomyopathy, dilated
Lipodystrophy, familial partial
Dejerine-Sottas disease, myelin P-related
Hypomyelination, congenital
Nemaline myopathy, autosomal dominant
Lupus erythematosus, systemic, susceptibility
Neutropenia, alloimmune neonatal
Viral infections, recurrent
Antithrombin III deficiency
Atherosclerosis, susceptibility to
Glaucoma
Tumor potentiating region
Nephrotic syndrome
Sjogren syndrome
Coagulation factor deficiency
Alzheimer disease
Cardiomyopathy
Factor H deficiency
Membroproliferative glomerulonephritis
Hemolytic-uremic syndrome
Nephropathy, chronic hypocomplementemic
Epidermolysis bullosa
Popliteal pterygium syndrome
Ectodermal dysplasia/skin fragility syndrome
Usher syndrome, type 2A
Kenny-Caffey syndrome
Diphenylhydantoin toxicity

Homocystinuria
Neuroblastoma (neuroblastoma suppressor)
Rhabdomyosarcoma, alveolar
Neuroblastoma, aberrant in some
Exostoses, multiple-like
Opioid receptor
Hyperprolinemia, type II
Bartter syndrome, type 3
Prostate cancer
Brain cancer
Charcot-Marie-Tooth neuropathy
Muscular dystrophy, congenital
Erythrokeratodermia variabilis
Deafness, autosomal dominant and recessive
Glucose transport defect, blood-brain barrier
Hypercholesterolemia, familial
Neuropathy, paraneoplastic sensory
Muscle-eye-brain disease
Medulloblastoma
Basal cell carcinoma
Corneal dystrophy, gelatinous drop-like
Leber congenital amaurosis
Retinal dystrophy
B-cell leukemia/lymphoma
Lymphoma, MALT and follicular
Mesothelioma
Germ cell tumor
Sezary syndrome
Colon cancer
Neuroblastoma
Glycogen storage disease
Osteopetrosis, autosomal dominant, type II
Waardenburg syndrome, type 2B
Vesicoureteral reflux
Choreoathetosis/spasticity, episodic (paroxysmal)
Hemochromatosis, type 2
Leukemia, acute
Gaucher disease
Medullary cystic kidney disease, autosomal dominant
Renal cell carcinoma, papillary
Insensitivity to pain, congenital, with anhidrosis
Medullary thyroid carcinoma
Hyperlipidemia, familial combined
Hyperparathyroidism
Lymphoma, progression of
Porphyria variegata
Hemorrhagic diathesis
Thromboembolism susceptibility
Systemic lupus erythematosus, susceptibility
Fish-odor syndrome
Prostate cancer, hereditary
Chronic granulomatous disease
Macular degeneration, age-related
Epidermolysis bullosa
Chitotriosidase deficiency
Pseudohypoaldosteronism, type II
Hypokalemic periodic paralysis
Malignant hyperthermia susceptibility
Glomerulopathy with fibronectin deposits
Metastasis suppressor
Measles, susceptibility to
van der Woude syndrome (lip pit syndrome)
Rippling muscle disease
Hypoparathyroidism-retardation-dysmorphism syndrome
Ventricular tachycardia, stress-induced polymorphic
Fumarase deficiency
Chediak-Higashi syndrome
Muckle-Wells syndrome
Zellweger syndrome
Adrenoleukodystrophy, neonatal
Endometrial bleeding-associated factor
Left-right axis malformation
Prostate cancer, hereditary
Chondrodysplasia punctata, rhizomelic, type 2

Figure 17.14 | This diagram shows selected genes, traits, and disorders identified on Chromosome 1 as part of the Human Genome Project. To see similar diagrams for other chromosomes, go to http://genomics.energy.gov/gallery/chromosomes/gallery-01.html.

Section 3 △ Biotechnology

Medicine

In the world of medicine, scientists at Harvard and the Massachusetts Institute of Technology are able to attach special ribonucleic acid (RNA) strands, measuring about ten nanometers (nm) in diameter, to nanoparticles that are filled with a chemotherapy drug. The RNA strands are attracted to cancer cells. When the RNA strand encounters a cancer cell, it adheres to it, and the nanoparticles carried by the RNA then release the drug into the cancer cell. This directed drug delivery technique has great potential for treating cancer patients while producing fewer harmful side effects than those caused by conventional chemotherapy.

SECTION THREE FEEDBACK >

1. Research the use of genetically modified corn in the United States and determine the percentage of the crops produced for the last two years that were genetically modified.
2. What is DNA? Why is it important? Name the scientists who first described the structure.
3. Find one more example of animal cloning, and briefly describe the process used.

SECTION 4: Information and Communications Technology

KEY IDEAS >

- The programmable computer is a device that can store a range of instructions and carry such instructions repeatedly.

- Moore's Law states that the number of transistors on an integrated circuit will double every two years.

- Parallel processing makes use of multiple processors to undertake different aspects of a program simultaneously.

The programmable computer is a device that can store a range of instructions and carry such instructions repeatedly. The idea that a device could be programmed to undertake a wide range of calculations stems back to the 1850s, when Charles Babbage designed his difference engine, the first programmable calculator ever developed. Although the theory was good, the machine never worked because it relied on high-precision gears that could not be successfully manufactured at that time. It wasn't until the development of the integrated circuit, which allowed thousands of transistors to be fabricated on a single chip, that the programmable computer became an affordable and practical proposition. The first personal desktop computer (comprising a processing unit, memory, keyboard, and display) was developed and put on sale by Hewlett Packard in 1972. From that time onwards, the computing power of the personal computer has grown exponentially, as has the number of personal computers and their users.

The rapid pace of development in the field of information and communication technologies continues today. Moore's Law states that the number of transistors on an integrated circuit will double every two years. This prediction was made in 1995 by the cofounder of Intel, Gordon Moore. Moore admits that he made the statement to alert people to likely growth of computing power over time, and he did not know if his prediction would ever hold true. He also admits that it is the press that has termed it a law; because the facts cannot be independently tested and verified, it is clearly not a scientific law.

However, since the time Moore made that statement, for many years, it has held true: The growth in transistor density has doubled every two years. During that time, the industry has appeared to meet a number of significant technological barriers, only to develop a solution in time to allow the growth in chip density to continue. Current technologies appear to be close to the practical limit of transistor density. To ensure that computing power continues to double every two years, the microelectronics industry has developed a range of central processing units that contain two or more central processors, called multi-core processors, working together to increase the total computing power. In the future, multi-core processors with ten or more processors will become normal.

Parallel Processing

Getting the best performance from multi-core processors has required major developments in data processing. The current generation of software products has been written for a single-processor world (see Figure 17.15), where every step of the computer program is followed in sequence. Parallel processing makes use of multiple processors to undertake different aspects of a program simultaneously (see Figure 17.16).

The best known parallel processing computer is the human brain, which can process information from multiple sources (eyes, ears, touch) and carry our actions based on the multiple inputs. Just imagine how hard it would be to catch a ball without that capability!

Spintronics

Spin-based electronics is a technology that makes use of the fact that an electron can spin in one of two ways, up or down. The momentum of the electron creates a tiny magnetic field. By using the direction of the spin, the electron can be used to encode data as a 0 or 1, depending on the direction of the spin. Unlike charge-based data storage (where the charge of an electron determines whether is stores 0 or 1), spin-based data storage retains the information when the electrical current stops (see Figure 17.17).

Spintronics is being used in the development of magnetic random access memory chips (MRAMs). MRAMs can store greater volumes of data in smaller packages and access the data more quickly, while using less power than the current range of charge-based random access memory circuits.

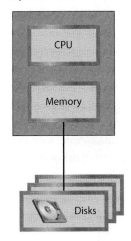

Uniprocessor Environment

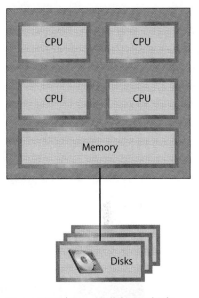

Figure 17.15 | In a single or sequential central processing unit, processing steps are sequential.

Figure 17.16 | In a parallel or multiple central processing unit, instructions can be performed simultaneously.

Figure 17.17 | The spintronic effect used in memory chips allows them to retain their charge even when electrical power is not present. Joe Orenstein, a physicist who holds a joint appointment with Berkeley Lab's Materials Sciences Division and UC Berkeley's Physics Department, is a leading researcher in this field.

Section 4 △ Information and Communications Technology

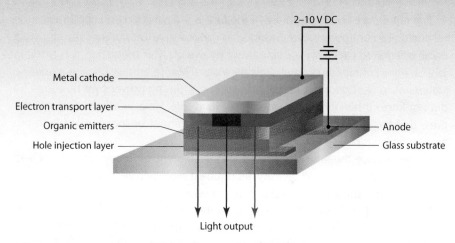

Figure 17.18 | Organic light-emitting diodes contain thin layers of organic materials that, in the presence of electricity, generate light.

Organic Light-emitting Diode Displays

Figure 17.19 | Organic light-emitting diodes are being used to create flexible LED screens. Screens like these will soon become commercially available.

The Organic Light-Emitting Diode (OLED) display, shown in Figure 17.18, uses several layers of organic materials that glow red, green, and blue, sandwiched between two layers of conductive material. The OLED display emits its own backlight, so, unlike tradition Liquid Crystal Displays (LCD), it does not need a dedicated backlight, and thus requires very little power.

Since the OLED layers can be printed on the conductive material using low-cost printers, the cost of production is low. Also, since they can be printed on any conductive substrate, flexible displays that could be rolled up when not in use are now possible, like the one shown in Figure 17.19.

Compared to LCDs, OLEDs have a much greater range of colors, brightness, and viewing angle, and they have a faster response rate. The OLED layers are very thin, making it possible to produce translucent displays that would admit daylight, and could be used as a solid-state white-light source at other times.

OLEDs can now be found in many cell phones, digital cameras, and small televisions. In the near future, given the range of uses of this very adaptable technology, it is likely to become the light bulb of the future as well.

SECTION FOUR FEEDBACK >

1. What was the transistor density per inch of a modern microprocessor chip used in desktop computers?
2. What is the number of processor cores used in a high-powered desktop gaming computer?
3. Research how a flexible OLED might be used in a modern, multimedia home environment.

HEY IDEAS >

- The interaction of powerful technological systems enables existing technology-rich systems and processes such as manufacturing, transportation, and health services to be improved and for production and service costs to be reduced.

- Rapid prototyping allows solid objects to be created using rapid fabrication techniques.

- Robotics describes any machine or device that emulates human movement, capability, or appearance.

- Automated Personal Rapid Transit allows individuals to use driverless robot vehicles to travel to self-selected destinations.

- Planned communities are those designed to ensure that all parts of the infrastructure work together to reduce energy consumption, to reduce transportation costs, and to simplify provision of services.

Few technological systems exist in isolation. For example, biotechnology employs imaging power that is dependent on sophisticated information technology systems. The interaction of powerful technological systems enables existing technology-rich systems and processes such as manufacturing, transportation, and health services to be improved and production and service costs to be reduced.

In addition, when developing technological systems are used in unique ways and in new situations, we develop the capability to develop new products. Just as you learned in the introduction, the visionary entrepreneurs (and those likely to become very wealthy) can *"think what nobody yet has thought about that which everybody sees."*

Rapid Prototyping

Rapid prototyping allows solid objects to be created using rapid fabrication techniques. From designs created using three-dimensional (3-D) software tools, the computer-aided modeling program is able to output information to a fabrication machine that creates the object by repeatedly laying down very thin, three-dimensionally precise layers.

Figure 17.20 | A rapid prototyping process creates 3-dimensional objects like this one using computer-aided design. Objects like this lawnmower are used as models for products in development.

Three-dimensional fabrication can be undertaken by a number of techniques, but all use the same basic principle, in which multiple layers of very thin, precisely dimensioned material are deposited to create a three-dimensional object, almost like layers of icing being spread on a cake. (See Figure 17.20.)

Radio Frequency Identification

In 1946, a Soviet engineer developed a spying device that could transmit audio information that it picked up as a radio wave, using only the energy from the audio waves. This is said to be one of the earliest examples of a passive radio device. It is called passive because it does not have its own source of power. Needless to say, spying devices that did not have a source of power were very difficult to detect.

667

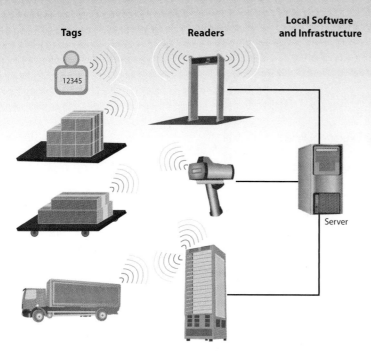

12345

Server

Figure 17.21 | A radio frequency identification (RFID) system can make product handling faster and more efficient.

This concept of a passive device that responds to a radio wave has been used to develop the next generation of product identification. We are all familiar with the product bar code that is used to help identify products at the checkout in the supermarket. Imagine a situation where the entire supermarket cart could be scanned without any human intervention: that is where Radio Frequency Identification (RFID) comes in (see Figure 17.21). This "super" bar code system can store in a very small printable circuit all the product information that would be gathered from a bar code (in fact, the RFID tag can contain much more information as well). This small RFID tag will respond to the signal from the checkout system and return the information that the checkout needs to identify each item in your cart—and it can do this for the entire cart in one scan.

RFID credit cards are now beginning to appear, and so it is possible that your cart could be scanned, the total calculated, and the cost deducted automatically from your credit card, all as you walk through the check-out at your local supermarket.

Robotics

The term robot was invented by the Czech playwright Karel Capek in his 1921 play *Rossum's Universal Robots*. The term is derived from the Czech word robota, meaning "required work." Robotics describes any machine or device that emulates human movement, capability, or appearance. Animated mechanical figures predate the term by many centuries. From the time of ancient Greece through the renaissance of the sixteenth century, android-like figures have been created using mechanical movements that mimicked the movements of humans. The term robot became used to describe mobile human service android figures by the science fiction author, Isaac Asimov.

The convergence of communication and information technologies with manufacturing technologies has created a new breed of robots that can replicate human endeavor faithfully without needing any breaks or rest, twenty-four hours a day, seven days a week.

Figure 17.22 | This production line is totally automated.

Automated factories where robotic systems carry out all of the repetitive and noncritical work already exist (see Figure 17.22). For example, Nokia manufactures its cell phones in highly automated factories with humans monitoring and checking the quality of the work and undertaking the most precise tasks. This allows more than ten phones per second to be produced with a very small but highly skilled workforce.

TECHNOLOGY IN THE REAL WORLD:
Musical Robots

Honda has used manufacturing robots for many years in its automobile production plants, and in 2000 launched a humanoid robot designed to someday assist people in their homes or offices. The human-like robot (shown in Figure 17.23) is able to walk, run, climb stairs, carry a tray, and has even conducted a symphony orchestra. Another automobile manufacturer, Toyota, has also developed musically inclined humanoid robots. One robot has artificial lips that move with precision, enabling the robot to play the trumpet. These research projects show how robots can carry out more mundane human chores without complaint or fatigue.

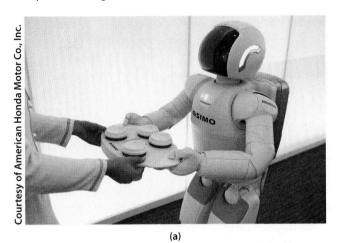

Courtesy of American Honda Motor Co., Inc.

(a)

(b)

Figure 17.23 | Honda is developing a human service robot that can perform human tasks and even conduct an orchestra.

The U.S. military is also interested in the use of robots to help soldiers become more effective. One of the most interesting developments in that field is the robotic exoskeleton, which adds a robotic framework or skeleton to the human body, and augments or adds to human capabilities. For example, if a human wanted to lift a weight, the Sarcos exoskeleton suit would make that easier by making it possible to lift 250-lb weights repeatedly while the human expends very little energy.

The Carnegie Mellon University driverless robot, Boss, based on a 2007 Chevy Tahoe SUV, successfully navigated over 200 miles of suburban and urban roads in Nevada when it won the DARPA 2007 Robotic Vehicle Challenge and a $2 million prize (Figure 17.24).

This robotic vehicle was equipped with multiple sensors that enabled it to follow the road, detect other vehicles, obey rules of the road at traffic lights and intersections, and react to other problems like broken-down vehicles or closed roads.

The technology being developed can be used in conjunction with human drivers to greatly improve accident avoidance. If the Mercedes Benz S class automobile detects an impending crash, it will warn the driver, tighten the seat belts, close the windows, and move the seats to the safest setting. Other manufacturers have undertaken research to add lane departure warnings if a vehicle departs from a lane without the use of a turn signal. Later systems even apply a robotic force to the steering system to gently steer the vehicle back into the lane.

Figure 17.24 | Carnegie-Mellon's BOSS Driverless Robotic 2007 Tahoe successfully navigated a 55-mile course at an average speed of 14 miles per hour.

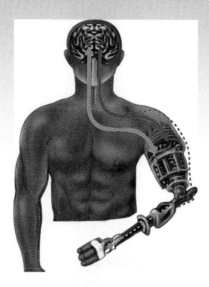

Figure 17.25 | Neuroprosthetics use signals direct from the brain to operate robotic devices.

Humans can also benefit directly from the use of robots. Those who have lost a limb can regain some functions by the use of prosthetic devices. Researchers have developed a technique known as reinnervation, in which signals from the brain are used to control muscles on the amputee's body, and from these muscle movements, microprocessors are used to tell the robot-limb motors what to do. Currently, such robot limbs provide up to four axes of control.

To move to the next level, researchers in the field of neuroprosthetics are developing techniques that will allow direct control of robotic devices by tapping into the signals from the motor control area of the brain (Figure 17.25).

Transportation

Transportation technologies are evolving rapidly to meet a variety of emerging needs: (1) to reduce the dependence on fossil fuels, the supplies of which are likely to be greatly reduced sometime in this century; (2) to reduce the environmental impact in terms of emissions and other harmful byproducts; and (3) to reduce the cost of manufacture and, at the same time, make the systems safe by using improved materials and design and automated safety and control systems.

Achieving these goals will require the use of a wide range of technologies working together (just imagine trying to produce bio-fuels without using advanced biotechnology), as well as a willingness to re-examine the other factors that increase our dependence on the current transportation infrastructure.

Automated Personal Rapid Transit Automated Personal Rapid Transit allows individuals to use driverless robot vehicles to travel to self-selected destinations. Such systems are ideal for locations such as airports where large numbers of travelers need to be able to get to a wide variety of locations within a large geographic space. An automated Personal Rapid Transport (PRT) is being designed to be used at London's Heathrow airport that will use driverless "pods" like the one shown in Figure 17.26.

A small-scale trial of the Urban Light Transport system (ULTRa) is scheduled to be operational by 2009. If the concept proves to be a technical success, it is very likely that such systems will be expanded to other airports worldwide. The ULTRa system uses electrically powered four-person pods, and are automatically routed on concrete tracks. The pods are routed according to passenger demand and traffic density. Each pod carries up to four passengers at speeds up to 25 mph. The pods use 70 percent less energy per mile than a car, and the system can ferry up to 4,800 passengers per hour.

Figure 17.26 | The ULTra Personal Rapid Transport system aims to provide driverless travel on demand.

Space Elevator

▷ Although it sounds impossible, there is a very serious attempt being made to create a **space elevator**. Such a system would be located on the equator, and a cable would extend out from the Earth's surface for some 22,000 miles. The elevator would be anchored to the ground, and electrically powered cars would rise up the elevator system until they were in orbit (see Figure 17.27).

Figure 17.27 | An Earth-to-Space elevator system could be feasible using carbon nanotubes. The LiftPort Group, which is developing such an elevator, maintains a "Countdown to Lift" clock on its Web site, at www.liftport.com.

Satellites could easily and economically be released from such a system, and if humans decide to populate the moon or exploit space, the elevator would reduce the costs considerably.

The creation of such a long, tethered cable is beyond current manufacturing capability. If the strongest construction materials available were used to construct a cable-support system, once it reached some three or four miles high, its weight would be such that it would collapse under its own weight. The production of new materials, such as carbon nanotubes, does offer the possibility of materials that are both strong and light enough to be deployed into space.

The U.S. company Liftport Inc. is so confident that this technology can be developed that it has a countdown clock on its Web site indicating when the first space elevator will rise into space.

Planned Communities Planned communities are those designed to ensure that all parts of the infrastructure work together to reduce energy consumption, to reduce transportation costs, and to simplify provision of services. If cities and transportation systems were designed to make it possible for people to travel quickly and easily using the lowest costs systems, with a small carbon impact, it would be likely to reduce the demand for cars and trucks (and therefore reduce demand for fossil fuels).

Mountain Valley in Idaho is a planned community, where the city, business and industry, and residential housing are located in convenient locations radiating out from the city center. The cities of Mountain Valley are also designed to allow for multiple urban transit systems, including walking, cycling, bus, tram, and light rail. An important goal is to reduce the dependence on using a car for local travel. Planned communities now form over 20 percent of major developments in the United States.

671

In Salt Lake City, planners are using refurbishment grants to encourage the redevelopment of the downtown to increase the city center population. At the same time, the city is using federal, state, and local dollars to build a state-of-the-art integrated transportation system using buses, trams, and light rail.

Renewable Energy Sources

The continued growth in demand for energy, combined with the fact that fossil fuel reserves will eventually become exhausted, has spurred the quest for alternate energy sources. These include alternative fuels to supplement, and perhaps eventually to replace, the dwindling reserves of fossil fuels. Secondly, much is being done to explore renewable energy sources. The Sun is the primary source of energy for the Earth. Its energy is harnessed naturally by plants and can also be harnessed by a wide range of technological devices.

ENGINEERING QUICK TAKE

Figure 17.28 | Dams are used around the world to generate hydroelectric power.

Hydroelectric Dam

Hydroelectric power is clean and efficient: All that is needed is a large quantity of running water, which can be fed to an electrical generator. (See Figure 17.28.)

The factors that determine the power generated are the distance the water falls, the quantity of water flowing through the generator, and the efficiency of the generator system.

The hydroelectric engineer needs to be able to calculate the power that can be generated from a hydroelectric project.

In our scenario, use the following figures:

Height of Dam = 10 ft
Water Flow = 500 cubic feet per second
Efficiency = 80 percent (that would appear as .80 in the formula)
Conversion Factor = Divide by 11.8 to convert power to kilowatts per hour.

All the above can be put into the following formula to calculate the power output from our dam in kilowatts of electrical power:

Power (kilowatts) = (Height of Dam − ft) × (Water Flow − cubic ft/second) × (Efficiency − %)/11.8

Since electric energy is normally measured in kilowatt-hours, we multiply the power from our dam by the number of hours in a year using the following formula:

Total Energy (kilowatt hours) = kilowatts × (24 hours) × (365 days per year)

The average annual residential energy use in the United States is about 3,000 kilowatt-hours for each person. So we can calculate how many people our dam could serve by dividing the annual energy production by 3,000.

People Served = Total Energy (kilowatt hours)/3,000 kilowatt-hours per person

In our example, list the following:

Power produced (kilowatts)
Total power (kilowatt hours)
Number of people served annually

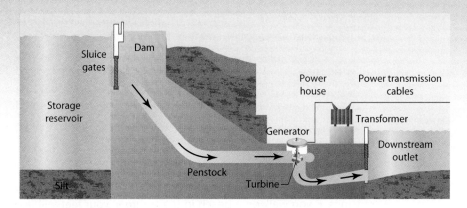

Figure 17.29 | Hydroeletric power generation uses water power to generate electricity.

Water power is one of the best-known sources of renewable energy. The potential energy stored in water flowing from the melting of snow in the mountains can be used to power water turbines to produce electrical power. In the United States, we create some 10 percent of our power using hydroelectric systems (see Figure 17.29). Although this is currently the largest single renewable energy source in the United States, other countries have done much more than we have to develop hydroelectric systems to fulfill their needs: Canada generates 60 percent of its electrical power need using hydroelectric systems, and Brazil generates more than 90 percent of its electrical power in this way.

While hydroelectric systems are well established, wave energy is just beginning to be exploited. As waves cascade on the seashore, the potential energy they contain is translated into familiar sounds, movement, and heat. Wave powered generators are now being used to harness the potential energy of waves in seas and oceans where large wave movements occur year round.

Open Sea Wave Systems A successful three megawatt open sea wave generation test system has been developed and is in use off the coast of Scotland. The Pelamis system (shown in Figure 17.30) comprises a series of hinged tubes that flex as the waves roll by. The flexing drives hydraulic systems that convert the energy into electric power, which is then transmitted to shore by an underwater power cable laid on the sea floor.

Off the coast of Atlantic City, the Powerbuoy system also harnesses power from the rolling waves. This buoy-based system exploits the fact that a buoyant vessel will move up and down as the waves roll by. The Powerbuoy (see Figure 17.31) has a lower section that is much less buoyant, and so doesn't move with the waves. This lower section is combined with a more buoyant top section that moves easily with the waves. The difference in the motion between the two parts is used to drive electric generators. The power is transmitted ashore using an underwater power cable.

Figure 17.30 | In a flexible-tube-based open sea wave generation system, wave action is converted to electricity.

Figure 17.31 | Buoy-based offshore power generation extracts energy from ocean waves.

673

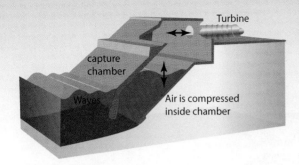

Figure 17.32 | Near-shore power generation systems like this one use a column of water to drive a turbine.

Figure 17.33 | Solar energy collector systems harness the sun's energy to drive turbines and generate electricity.

Figure 17.34 | This new micro-generator developed at Georgia Tech can produce enough power to a cell phone. It may eventually be able to power a laptop.

Near Shore Systems The effect of waves crashing on the shore during a hurricane clearly demonstrates their power. There are a number of developments that exploit the power of waves as they strike the shore. The Wavegen Near Shore Wave Generator, located in a remote part of Scotland, uses an innovative approach that harnesses the massive forces of the sea. Waves hit a carefully-designed breakwater that directs their kinetic energy into movement, which is then used to generate power (see Figure 17.32).

Solar Concentrators When the Sun's energy falls on an object, that object will absorb some of the heat and light energy the Sun transmits. If you walk on a beach in the summer, your feet will feel the heating effect of the Sun's energy as it is absorbed by the sand. Technology is now employed in a number of innovative ways to capture that energy and transmit it to homes and businesses across the United States.

It is possible to focus, or concentrate, solar power by using large reflective dishes. Concentrating Solar Power (CSP) does just that by using large parabolic mirrors that focus the power of an energy-collecting source (see Figure 17.33). Some large installations use water as the energy-collecting source, and the steam generated by concentrating the solar energy is used to drive a high-efficiency steam turbine generator. The system relies on a solar-tracking system that ensures that the mirrors are aimed at the Sun as it moves in relation to the Earth's rotation. This system could also be used to remove salt from seawater if the solar facility is close enough to a body of saltwater such as areas of North Africa.

Micro-Generators Any form of motion is capable of providing potential energy to produce power. Micro-generators are about the size of a small stack of dimes, but these diminutive electro-mechanical devices can generate enough electricity to power a pacemaker or small sensors buried deep inside the Golden Gate Bridge without requiring a change of batteries. (See Figure 17.34.) The movement of the body, or the vibrations from a machine, moves a series of magnets over a copper coil and can generate up to 50 microwatts of power. This is not enough to run a computer, but it is enough to run small medical implants. When constructed to fit inside a standard battery case, these micro-generators can be retrofitted to give indefinite power to devices that use standard cells for similar low-power applications.

Sonofusion The Sun creates energy by fusion of hydrogen into helium. Nuclear fusion occurs when atomic particles are forced to combine to create a heavier nucleus. The process can release great quantities of energy. The biggest problem to date has been controlling the fusion process so that the energy can be gathered.

Scientists at Rensselaer, Troy, NY, have developed a process that bombards a liquid with neutrons inside a pressure vessel. The resulting reaction of the liquid is such that very high pressure and temperatures are reached and that fusion has been observed in a controlled environment.

The quantity of energy released by Sonofusion is small. If the process could be scaled up, then it would offer the energy potential of nuclear power without the radioactive waste byproducts.

SECTION 6: Technological Impact

KEY IDEA >

- All technological actions and developments have intended and unintended consequences.

All technological actions and developments have intended and unintended consequences. Developers hope that the consequences are beneficial, but it is possible for such developments to have unintended impacts. For example, the development of chlorinated fluorocarbons (CFCs), developed as a safe replacement for fluids used as refrigerants in air conditioning and refrigeration systems, was applauded as a major increase in safety. This seemingly nonreactive, stable gas appeared to be an ideal replacement for unstable and dangerous chemicals used previously.

However, it was discovered in the 1970s that CFCs contributed greatly to the depletion of the ozone layer (the layer of stratospheric ozone gas that absorbs 97 to 99 percent of ultraviolet radiation), as a result of the gas breaking down to release chlorine when it ascends into the upper atmosphere (see Figure 17.35).

Through nanotechnology, we can achieve wondrous things. Nanotechnology can change how we practice medicine, how we develop new materials, and how we develop alternate energy resources, to name a few. But there are also issues and concerns regarding its use.

Intended and Unintended Consequences: Nanofactories

The nanofactory that produces nanorobots loaded with drugs to target cancerous growths in a body can be easily reprogrammed to produce much less benevolent nanorobots that could be used to harm others. A number of nanotechnologists have expressed concern about the risk of uncontrolled nanofactories producing materials or products that will harm humankind. This is not to

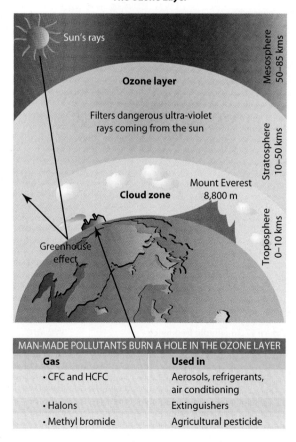

The Ozone Layer

MAN-MADE POLLUTANTS BURN A HOLE IN THE OZONE LAYER	
Gas	**Used in**
• CFC and HCFC	Aerosols, refrigerants, air conditioning
• Halons	Extinguishers
• Methyl bromide	Agricultural pesticide

Figure 17.35 | Chlorinated fluorocarbons (CFCs) deplete the ozone layer.

imply that such nanofactories would deliberately develop harmful substances autonomously. But the technologists believe we should consider the risk of a nanofactory that inadvertently produces something so toxic that its operators would be unable to shut down the process, resulting in a runaway production of dangerous materials.

Eric Drexler, a nanotechnology pioneer, in his book *Engines of Creation,* gives a worrisome analysis of nanoreplication runaway:

"Imagine such a replicator floating in a bottle of chemicals, making copies of itself. . . . the first replicator assembles a copy in one thousand seconds, the two replicators then build two more in the next thousand seconds, the four build another four, and the eight build another eight. At the end of ten hours, there are not thirty-six new replicators, but over 68 billion. In less than a day, they would weigh a ton; in less than two days, they would outweigh the Earth; in another four hours, they would exceed the mass of the sun and all the planets combined—if the bottle of chemicals hadn't run dry long before."

It is highly likely that governmental and industry regulation and oversight will be developed to control the implementation of nanofabrication.

Intended and Unintended Consequences: Robotics

In the 1960s, it was thought that robots would release humans from the drudgery of mundane work. Advanced manufacturing systems now use robotic systems that can carry out much of the repetitive manufacturing tasks that were previously performed by humans. Such robotic systems offer the opportunity to free production-line workers so they can become responsible for programming and overseeing robot workers. Unfortunately, such changes do not always happen, and the introduction of robot systems without recognizing the need to retrain and redeploy the workers replaced by them can lead to unemployment.

The introduction of robots has usually led to reduction of workforce and higher unemployment. This trend can lead to the uncomfortable situation of financial growth and success accompanied by higher unemployment and increased social tension and poverty.

Intended and Unintended Consequences: RFID

Radio frequency identification technology can be used to track animals through the implanting of an RFID chip (see Figure 17.36) just below the skin of the animal. Recently, some health concerns have emerged, and this practice is being reexamined.

The same technology can be used to track humans. A number of high-tech companies had RFID programs for workers operating in high-security areas. The information stored in the chip was used to gain access to the secure areas via readers that scan the chip to verify an individual's identity. A number of state governments have now banned this practice because it is possible for others to use RFID Scanners to read identifying information from unsuspecting individuals without their knowledge. Proposals to use RFID technology in passports and other travel documents have been abandoned for the same reason.

Figure 17.36 | Radio frequency identification chips contain integrated circuits.

We are moving toward such a system for shopping checkouts. But security concerns remain. Because unscrupulous individuals will try to scan your debit or credit cards without your knowledge, more secure RDID card systems will need to be developed before walk-through checkouts are an everyday experience.

Intended and Unintended Consequences: Transgenic Crops

Genetically modified (GM) foods (or transgenic foods) contain genetic material that has been transferred from other organisms. The crops most commonly genetically modified are soybean, corn, cottonseed oil, and wheat. While it is acknowledged that transgenic crops can substantially improve agricultural food production, there are significant concerns about the control and effect of GM crops in many parts of the world. One such concern is the possibility of errors, such as transferring a gene from Brazil nuts into soybean crops, which led to allergic reactions from those with a nut allergy when encountering foodstuffs made from the genetically modified soybeans.

The fact that GM crops cannot be made available to those in third-world and high-poverty areas is also a concern. The biotechnology companies want to protect their investments, so they will not allow their seeds to be sold at very low prices in such regions. Also, the genetic modification of the seeds currently being manufactured was targeted towards high-yield farming in the developed world. The soil, climate, and husbandry techniques in the high-poverty regions will not be suited to the crops developed for the developed nations. There is little financial incentive for the biotechnology companies to address this issue, since the enormous cost of developing GM seeds for use in such regions could never be economically recovered.

Computer Equality

▷   At the same time that chip density is doubling, there are initiatives that are using a combination of the very oldest of technologies combined with the most economical of today's computing systems to create portable, low-cost computers for disadvantaged children in third-world countries. The MIT-initiated **One Laptop Per Child Program** (**OLPC**) has developed a low-cost, robust, durable laptop computer that can be powered by a built-in, hand-cranked, high-efficiency generator. The laptop (shown in Figure 17.37) can also use local power, even if it has a variable-voltage power supply.

This specially designed laptop uses a powerful but low-power processor, combined with an innovative, free-operating system. Combined with research undertaken by the OLPC organization and the United Nations, it is likely that this device will offer Internet connectivity to students in locations that are still awaiting connection to an electric grid.

Figure 17.37 | The One Laptop Per Child (OLPC) program gives Internet access to children in remote areas. Here, a teacher helps students in Uruguay with laptops obtained through this program.

SECTION SIX FEEDBACK >

1. Using research tools, identify another technological invention (device, chemical, process) that was developed for a beneficial purpose but is resulting in unintended and unwanted consequences.

2. What are the differences between RDIF and bar code tagging? When would you use one and not the other? What are the benefits of each technology?

3. Research genetically modified foodstuffs and find a crop that has seen a significant yield increase compared to the use of regular seeds.

Matching Your Interests and Abilities with Career Opportunities: Biological Scientists

Biological scientists study living organisms and their relationship to their environment. They research problems dealing with life processes and living organisms. Most specialize in some area of biology, such as zoology (the study of animals) or microbiology (the study of microscopic organisms).

Nature of the Industry

Many biological scientists work in research and development. Some conduct basic research to advance our knowledge of living organisms, including viruses, bacteria, and other infectious agents. Biological scientists who work in applied research or product development use knowledge provided by basic research to develop new drugs, treatments, and medical diagnostic tests; increase crop yields; and protect and clean up the environment by developing new biofuels.

Because biological scientists doing applied research and product development in private industry may be required to describe their research plans or results to nonscientists who are in a position to veto or approve their ideas, they must understand the potential cost of their work and its impact on business. Scientists often work in teams, interacting with engineers, scientists of other disciplines, business managers, and technicians. Some biological scientists also work with customers or suppliers and manage budgets.

Recent advances in biotechnology and information technology are transforming the industries in which biological scientists work. In the 1980s, swift advances in basic biological knowledge related to genetics and molecules spurred growth in the field of biotechnology. Biological scientists using this technology manipulate the genetic material of animals or plants, attempting to make organisms more productive or resistant to disease. Research using biotechnology techniques, such as recombining DNA, has led to the production of important substances, including human insulin and growth hormone. Many other substances not previously available in large quantities are starting to be produced by biotechnological means; some may prove useful in treating cancer and other diseases.

Today, many biological scientists are involved in biotechnology. Those who work on the Human Genome Project isolate genes and determine their function. This work continues to lead to the discovery of the genes associated with specific diseases and inherited traits, such as certain types of cancer or obesity. These advances in biotechnology have created research opportunities in almost all areas of biology, with commercial applications in the food industry, agriculture, and environmental remediation, as well as in other emerging areas such as DNA fingerprinting.

Working Conditions

Biological scientists usually work regular hours in offices or laboratories and usually are not exposed to unsafe or unhealthy conditions. Those who work with dangerous organisms or toxic substances in the laboratory must follow strict safety procedures to avoid contamination. Many biological scientists such as botanists, ecologists, and zoologists take field trips that involve strenuous physical activity and primitive living conditions. Biological scientists in the field may work in warm or cold climates, in all kinds of weather.

Training and Advancement

A Ph.D. degree usually is necessary for independent research, industrial research, and college teaching, as well as for advancement to administrative positions. A master's degree is sufficient for some jobs in basic research, applied research or product development, management, or inspection; it also may qualify one to work as a research technician or as a teacher in an aquarium. The bachelor's degree is adequate for some non-research jobs. For example, some graduates with a bachelor's degree start as biological scientists in testing and inspection or get jobs related to biological science, such as technical sales or service representatives. In some cases, graduates with a bachelor's degree are able to work in a laboratory environment on their own projects, but this is unusual. Some may work as research assistants, while others become biological laboratory technicians or, with courses in education, high school biology teachers.

Outlook

Biological scientists held about 87,000 jobs in 2006. Federal, state, and local governments employed slightly less than half of all biological scientists. Federal biological scientists worked mainly for the U.S. Departments of Agriculture, Interior, and Defense and for the National Institutes of Health. Most of the rest worked in scientific research and testing laboratories, the pharmaceutical and medicine manufacturing industry, or hospitals. In addition, many biological scientists held biology faculty positions in colleges and universities.

Employment of biological scientists is projected to grow about 9 percent over the 2006–16 period, as biotechnological research and development continues to drive job growth.

Opportunities are expected to be better for those with a bachelor's or master's degree in biological science. The number of science-related jobs in sales, marketing, and research management for which non-Ph.D.s usually qualify is expected to exceed the number of independent research positions. Non-Ph.D.s also may fill positions as science or engineering technicians or as medical health technologists and technicians.

Biological scientists will be needed to take this knowledge to the next stage—understanding how certain genes function within an entire organism, so that gene therapies can be developed to treat diseases. Even pharmaceutical and other firms that are not solely engaged in biotechnology use biotechnology techniques extensively, spurring employment increases for biological scientists. For example, biological scientists are continuing to help farmers increase crop yields by pinpointing genes that can help crops such as wheat grow worldwide in areas that currently are hostile to the crop.

Expected expansion of research related to health issues such as AIDS, cancer, and Alzheimer's disease also should create more jobs for these scientists. In addition, efforts to discover new and improved ways to clean up and preserve the environment will continue to add to job growth. More biological scientists will be needed to determine the environmental impact of industry and government actions and to prevent or correct environmental problems, such as the negative effects of pesticide use. New industrial applications of biotechnology, such as changing how companies make ethanol for transportation fuel, also will spur demand for biological scientists.

Earnings

* Median annual earnings of biochemists and biophysicists were $76,320 in 2006.
* The middle 50 percent earned between $53,390 and $100,060.
* The lowest 10 percent earned less than $40,820, and the highest 10 percent earned more than $129,510.
* Median annual earnings of microbiologists were $57,980 in 2006.
* The middle 50 percent earned between $43,850 and $80,550.

[Bureau of Labor Statistics, U.S. Department of Labor, Occupational Outlook Handbook, 2008–09 Edition, visited May 12, 2008, http://www.bls.gov/oco/]

Summary >

Technology is the use of tools, systems, processes, and human ingenuity to solve human problems. The solutions to such problems will employ a range of technologies, depending upon the situation and the resources available. The solutions will vary in their elegance and form, but overall, humankind employs technology to solve its practical problems.

When technologies interact, they increase the capabilities of the individual technologies. This increases the opportunity to find new solutions to existing problems, or to solve problems that had defeated engineers and scientists in the past.

The use of any technology has consequences, some intended and some unintended. Noting and dealing with these consequences relies on human control and evaluation of technological systems. If society develops sufficient requirements for the use and monitoring of these systems, unintended consequences can be revealed and corrected, so that technology will play an active and beneficial role in the development of the society of the future.

FEEDBACK

1. The global economy is dependent on the functioning of the international banking system as the clearinghouse for international trade. Describe how banking supports international trade.

2. Computer Technology is just beginning to impact the world of health care. Research the benefits of such integration, and speculate on how it might impact medical decision making, as well as how it might assist in lengthening our life span.

3. The relationship between organic and inorganic/electronic nanotechnology is becoming very blurred. Research how the two fields of nanotechnology might mingle and integrate, and speculate on the benefits and drawbacks of such integration.

4. The massive growth in production of bio-fuels is having a significant impact on the availability of low-cost foodstuffs for less developed nations. Research the issue of fuel produced from bio-sources, and the impact that the production has on the availability and cost of crops in the developing world.

5. With the massive growth in technological devices in all aspects of life, explore how technology is being used to extend the life of oil wells in the United States.

6. Developments in technology hint at the creation of a single, portable technological device that will meet all our needs: Examine the issue of a single device that would contain a still and video camera, a personal computer, and a cell phone. Explain why we do not have a single device that can replace discrete systems at this time.

7. The use of completely automated and autonomous robotic devices is increasing in the manufacturing world . There are some who would like to use autonomous robots in other settings, such as border security. What social, safety, and legal issues does the use of completely autonomous robots present when they are used in a setting that brings them into operational contact with the general public?

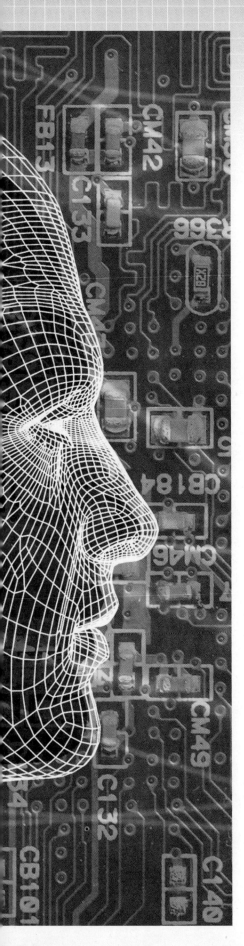

DESIGN CHALLENGE 1:
Design and Model a Water-Turbine Powered Lifting System

• Problem Situation

Because there is a real need to reduce reliance on fossil fuels as an energy source, renewable energy sources are becoming more important. Solar, wind, and wave power can be used to generate electricity. However, each of these sources of power presents a number of challenges to the technologist who is designing systems to convert energy from one form to another.

• Your Challenge

Water power was used extensively during the industrial revolution. Where the terrain is suitable, water can still be used to generate power.

Using a catch tank to provide a regular flow of water, your team is to design and make a waterwheel or water turbine device that will be powered by the flow of water. The turbine must lift as many 1-gram paperclips as possible, to a height of 0.5 meters. Record the time you take to lift the load for each test you undertake.

The time you take to lift the weight will be used to determine the horsepower and the torque (turning force) of your lifting system.

Using your *Student Activity Guide,* state the design challenge in your own words.

• Safety Considerations

1. Only use tools and machines after you have had proper instruction.
2. Wear eye protection when using tools, materials, machines, paints, and finishes.

• Materials Needed

1. Assorted wheels between 0.75"–1.5" diameter, and 0.25"–0.75" tread width, with axle hole to suit the axle rod being used, to be used as the core of the impellor
2. 2" or similar inside diameter PVC or plastic pipe
3. 1/8" diameter plastic tubing
4. 1/8" tubing connectors
5. Metal rod (such as welding rods or coat hangers) for axles
6. Assorted plastic pulleys (such as Kelvin economy pulley pack)
7. Nylon kite line or similar material
8. A 2-liter pop bottle to be used as the holding tank
9. A one-gallon milk container to be used as the catch tank
10. ¼" or similar thickness Lexan or other clear plastic sheet
11. Small, medium, and jumbo drinking straws

You can also use other construction materials as needed to construct the turbine, the impellor, and the pulley system.

• Clarify the Design Specifications and Constraints

To solve the problem, your design must meet the following specific constraints:
1. In the *Student Activity Guide,* you will see a diagram of the design of the water delivery and recovery system and the holding device that will secure your water turbine when being tested.
2. You will see that the outlet of the catch tank is located at a height of 1 meter.

3. You must use the catch tank to catch the water flowing through your waterwheel system.
4. You can combine pulleys in combination to change the mechanical advantage of the weight-lifting system.

● Research and Investigate

To better complete the design challenge, you need to gather information to help you design a solution.

In your guide, complete the Knowledge and Skill Builder I: Exploring the design of a turbine/waterwheel.

In your guide, complete the Knowledge and Skill Builder II: Using pulleys to increase mechanical efficiency.

In your guide, complete the Knowledge and Skill Builder III: Calculating the power of a turbine/waterwheel-based lifting system.

● Generate Alternative Designs

In your guide, describe two or more possible solutions to the challenge.

● Choose and Justify the Optimal Solution

Refer to the guide. Explain why you selected the solution you did, and why it is the best choice.

● Display Your Prototypes

To test your ideas and designs, try to make prototype waterwheels and test them to see how well they will lift weights. From your tests, you can modify your chosen design.

In any technological activity, you will use seven resources: people, capital, time, information, energy, materials, and tools and machines. In your guide, indicate which resources were most important in this activity and how you made trade-offs between them.

● Test and Evaluate

How will you test and evaluate your final design? In your guide, describe the testing procedures you will use. Indicate how the results will show that the design solves the problem and meets the specifications and constraints.

● Redesign the Solution

Respond to the questions in your guide about how you would redesign your solution, based upon the knowledge you gained from the testing and evaluation process.

● Communicate Your Achievements

In your guide, describe the plan you will use to present your solution to the class. Show what handouts and/or PowerPoint slides you will use.

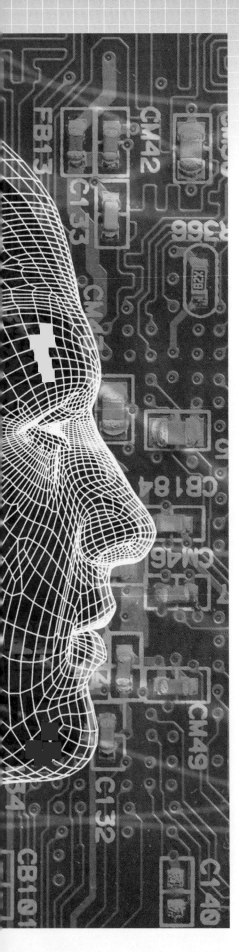

DESIGN CHALLENGE 2:
Design a Solar-Power Station

- ### Problem Situation

There is a need for low-cost, rugged solar-power stations that can be used anywhere in the world to provide local power. Twelve-volt car batteries are often used in the Third World to provide power for vital equipment. The batteries need to be charged using low-cost, reliable power sources. Solar power is one of the reliable sources of energy in such settings, so using a solar-power charging station will enable charging of 12-volt batteries, as well as providing power for other items, such as computers. For example, the One Laptop per Child project aims to provide low-cost, durable laptops to children in the Third World. There is no guarantee of reliable electrical power in such countries, so the laptop must have alternate sources of power.

- ### Your Challenge

Solar energy is readily available and so might be used to power the solar power station. Using a solar cell, design a charging unit that can be used to charge 12-volt batteries and can also be used to power other devices, such as the XO laptop. Your power station will have to be constructed to be used in harsh environments and to withstand rough handling.

Using your *Student Activity Guide,* state the design challenge in your own words.

- ### Safety Considerations

1. Ensure that a low-voltage electrical power pack or supply is used to provide the power for this activity.
2. Wear safety glasses at all times.
3. Take care when working with construction tools and materials.

- ### Materials Needed

1. One 12 V 5 W flexible solar panel
2. Two 3" alligator clips
3. One 4 mm coaxial power connector plug
4. Two feet of flexible twin core 12-gauge electric cable

You can also use other construction materials as needed to construct the power station and its case.

- ### Clarify the Design Specifications and Constraints

To solve the problem, your design must meet the following constraints:

- The charging system must be self-contained and not require any assembly in the field.
- The unit must be capable of being packed into a robust case, which must serve as part of the support structure for the charging unit.
- The unit must be constructed from suitable materials for the environment it will be used in.
- The unit must have simple graphics showing how it is set up and connected. The graphics must be permanently affixed to the charging system in some way.

• Research and Investigate

To better complete the design challenge, you need to gather information to help you design a solution.

1. In your guide, complete the Knowledge and Skill Builder I: Electrical output testing.
2. In your guide, complete the Knowledge and Skill Builder II: Enclosure design.
3. In your guide, complete the Knowledge and Skill Builder III: Positioning device design.
4. In your guide, complete the Knowledge and Skill Builder IV: "How to Use" graphics.

• Generate Alternative Designs

In your guide, describe two or more possible solutions to the challenge.

• Choose and Justify the Optimal Solution

Refer to the guide. Explain why you selected the solution you did, and why it is the best choice.

• Display Your Prototypes

To test your ideas and designs, make a prototype using cardboard to simulate the solar panel. Based on your test results, modify your chosen design.

In any technological activity, you will use seven resources: people, capital, time, information, energy, materials, tools, and machines. In your guide, indicate which resources were most important in this activity and how you made trade-offs between them.

• Test and Evaluate

How will you test and evaluate your final design? In your guide, describe the testing procedures you will use. Indicate how the results will show that the design solves the problem and meets the specifications and constraints.

• Redesign the Solution

Respond to the questions in your guide about how you would redesign your solution, based upon the knowledge gained from the testing and evaluation process.

• Communicate Your Achievements

In your guide, describe the plan you will use to present your solution to the class. Show what handouts and/or PowerPoint slides you will use.

GLOSSARY

accelerated construction techniques to reduce time of renovation or construction while maintaining quality and safety, including prefabrication and overtime work

access time the time it takes for data to appear during a read operation or to store data during a write operation

accumulator in computers, a general purpose register in which immediate results are stored; also called *A-register*

adaptive control use of a computer to optimize production process

address bus bus consisting of many wires (generally 16 to 36) on which the CPU provides addressing information necessary to access any specific piece of information

addresses the numerical way to identify and locate information in a memory chip

addressing circuitry circuitry in a memory chip that selects row locations to access

adenine one of the four amino acids that comprise DNA; denoted as A

aggregate asphalt filler, which may be sand, gravel, stone, waste slag, glass, or recycled concrete

agile manufacturing refers to an organization that has developed the processes, tools, and training to enable it to respond quickly to customer needs and market changes while still controlling costs and quality

agrarian farming

agriculture the combination of businesses that produce, process, and distribute food, fiber, fuel, chemicals, and other useful products from the land

agroecology sustainable farming and ranching

agroecosystem the agricultural ecosystem, including the ways all plants, animals, and microorganisms within that ecosystem interact

agronomics the science of land management and crop production

alloy a combination of metals

alternating current (AC) electrical current characterized by changes in level and polarity of signal over a period of time

alternative fuels fuels to eventually replace dwindling supplies of fossil fuels, such as biodiesel and hydrogen

amide a group of chemical compounds containing carbon-oxygen bonds and nitrogen

amperes (amps) measure of the amount of current flowing in a circuit; denoted as A

amphibious assault ship warship designed to allow troops to land using small amphibious craft launched from the larger vessel

amplifier device to increase signal strength (amplitude)

amplitude the peak value of a signal; on a graph of a sine wave, the maximum y value

amplitude modulation (AM) modification of a carrier wave by means of varying the height or amplitude of the wave

analog-to-digital converter (ADC) a device that converts an analog signal to a digital signal

analog controller system controller that operates according to continuously variable measures such as voltage and current

analog signal any continuous electrical signal that varies continuously rather than in discrete increments

AND gate digital device that produces a one-level output signal when all the gate's inputs are at one level and zero output when any gate input is zero

AND-OR network common type of combinatorial circuit comprised of AND gates connected to OR gates

annealing process of heating steel above the austenitic point (1375° F or 746° C) and then cooling it slowly, causing the material to become soft and ductile

annual rings growth rings visible in the cross section of a tree, each ring representing one year of growth

anthropologist scientist who studies human beings and their ancestors

Application Layer in OSI, the seventh layer, responsible for executing any program that uses the services of the network

apprentice one who works as an assistant to an artisan in preparation to become an artisan

aquaculture process of creating and managing controlled water environments to harvest usable plants and animals

aramids aromatic polyamide fibers, a type of synthetic material

arch bridge transmits load from the deck of the bridge outward along an arch to supports (abutments) on both sides

architecture the structure of logic circuitry in a CPU

arithmetic and logic unit (ALU) digital circuitry that performs mathematical operations

aromatic polyamide aramid, composed of many amide molecules connected together

artifact any object created by humans for practical purposes

artificial ecosystem any ecosystem established by humans, such as an agroecosystem

artificial intelligence (AI) expert systems that may one day be able to "think" and "learn" in order to adapt and control processes

asphalt pavement using a mixture of aggregate and bitumen

assembly line typical mass production system, best suited to large, homogeneous production runs

Asynchronous Transfer Mode (ATM) the dominant transmission protocol for communication across the Public Switched Telephone Network

atom economic synthetic chemical reactions in which the majority of starting material atoms are found in the finished products

attenuation the dying out of a signal as the destination gets farther from the source; *decay*

augmentive propulsion system potential vessel propulsion system using other sources, such as wind power, to supplement standard diesel propulsion

automated guided vehicle (AGV) mobile robot used to move materials around a manufacturing plant or warehouse

automation generic term describing use of control systems, usually computers, to control industrial machinery and processes

back end of the line (BEOL) in an IC fabrication process, the stage in which transistors are interconnected to form working circuitry

bar graph pictorial representation of a relationship between variables using columns whose heights represent values of the data being displayed

barrette column or pile of concrete with steel reinforcing rods used in the substructure of a large building

battery-powered electric vehicle (BEV) electric vehicle that needs to be plugged in to an external power source to recharge, using energy stored in rechargeable battery packs to power efficient electric traction motors

beam bridge used to span short distances, normally constructed of reinforced concrete and steel, sometimes made of wood

bearing wall superstructure used to enclose a space using walls, typically built of brick, concrete, or stone

benchmarking process of establishing what is best for a given criterion, quantifying it, and comparing a given product or process to that standard

binary number system digital system using the base 2, using only zeros and ones to correspond to voltage states

biodegradable items that can be easily degraded by natural processes such as exposure to sun and water or the action of bacteria or fungi

biodiversity numerous species present in a given area

bioethics ethical standards as applied to the biotechnology industry

biofuels fuels derived from renewable biological resources, such as plant material and organic waste products

bioinformatics the collection, organization, and analysis of large amounts of biological or genetic data, using networks of computers and databases

biological scientists those who study living organisms and their relationship to the environment

biomass accumulated vegetable and animal matter used as a source of energy

biomass gasification process of super-heating wood chips and agricultural wastes until they turn into hydrogen and other gasses

bioremediation the use of biotechnology to clean up the environment, as in the use of microorganisms to clean up oil spills

biotechnology technology using biological systems or living organisms to develop, make or modify products or processes

biotechnology industry industry using genetics, molecular biology, cell biology, and biochemistry to produce products (including some as old as cheese-making and fermentation)

bipolar junction transistor (BJT) common type of transistor, used as an amplifier

bit a single binary digit, zero or one

bitumen viscous black, sticky material obtained from distillation of crude oil, primarily aromatic hydrocarbons

block flow diagram simplified drawing illustrating the function of a chemical plant

blow molding a molding process in which a gob of molten plastic (a parison) is enclosed in a mold and inflated by air pressure so that it takes the shape of the mold

boot the action of a computer when power is first applied, reading instructions from a memory chip to prepare the system for use

bottom-up manufacturing (BUM) Building larger and more complex objects by integration of smaller building blocks or components. An alternative to top-down manufacturing

brainstorming group activity during which each person in the group can suggest any and all ideas on a given topic without criticism; intended to promote creative thinking

branch any one of the possible paths in a parallel circuit

bridge a structure that spans a valley, body of water, roadway, railroad tracks, or other obstruction to continuous travel

brittleness a measure of how easily a material under stress will fracture without significant deformation

broadcast communications communication media using radio waves, including radio and television

bronze an alloy of tin and copper

Bronze Age period from 3000 to 500 B.C.E., during which humans developed smelting and an agrarian culture arose in the Fertile Crescent

bus in computers, any common group of wires that transfers data between components

byte eight bits (binary digits) grouped together

cable-stayed bridge similar in appearance to suspension bridge, but cables are attached directly to towers, which bear the main load

cache memory the name given to high-speed SRAM fabricated right on the microprocessor chip

canal manmade waterway linking two or more points

cancellation of units mathematical technique for performing conversions, as from meters to millimeters

cantilever bridge constructed using two beams facing each other, anchored only at the originating end; the other ends of the beams are connected by a third element supported by a column (or pier)

capacitive reactance the impedance of a capacitor, denoted as Xc

capacitor electrical component constructed with alternating layers of insulating and conducting materials, used to store and release electrical energy

capital investment machinery and robotics used in manufacturing or other production systems

carbon nanotubes molecular-scale tubes of carbon having a diameter of one to three atoms

carcinogen any substance that can cause cancer

Carnot cycle a thermodynamic cycle modeled on the hypothetical Carnot heat engine, with the highest theoretical efficiency of any power-producing cycle

carrier sense in Ethernet networks, describes the fact that a transmitter listens for a carrier wave before trying to send

carrier wave analog signal used transmit a message, encoded by means of modulation

carry the most significant bit in the output of a half adder

casting the use of a mold to shape liquid materials into products

cell electrical circuit in a memory chip that stores one bit of information

cellular manufacturing a manufacturing system using an arrangement of machinery allowing processes to be grouped according to the sequence of operations required to make a product; also called *linked-cell*

cellulosic ethanol alcohol produced from agricultural plant wastes, sawdust, paper pulp, switchgrass, and other biomass materials

central processing unit (CPU) the basic software execution device in any computer

ceramic materials materials made from clay, glass, or other nonmetallic inorganic materials

ceramic matrix composites matrix composites used in advanced engines, allowing them to operate at high temperatures

charge-coupled device (CCD) semiconductor image sensor, generally a 1-cm panel with hundreds of thousands of photosites, used in digital cameras and camcorders

charging the storing of energy in a capacitor or battery.

chemical energy a type of internal energy associated with the molecular and atomic structure of matter

chemical potential a measure of chemical energy

chemical shift property of atoms describing how an element's chemical profile changes (shifts) from one environment to another

chemical technicians those who work in laboratories, setting up, operating, and maintaining laboratory instruments, monitoring experiments, making observations, calculating and recording results, and developing conclusions

chemical vapor deposition process that vaporizes a substance into a gas and deposits a thin film of high purity material onto a surface

chemists and materials scientists people who work in research and development (R&D). In basic research, they investigate the properties, composition, and structure of matter and the laws that govern the combination of elements and reactions of substances to each other.

chip fab a chip fabrication plant, the special manufacturing environment in which silicon wafers are processed into integrated circuits

chip set a group of integrated circuits, or chips, that are designed to work together

chromosomes collections of packets of DNA within the nucleus of a cell

circuit devices and connections forming paths along which electricity flows; in telecommunication systems, an electrical connection between two callers

civil engineering contractors in construction industry, those who build sewers, roads, highways, bridges, tunnels, and other projects

cladding the bonding of different metals by pressing them together under high pressure

clinical trial final series of tests, involving humans, in development of pharmaceutical products

coating processes involving paints, finishes, and chemical and electrochemical applications of layers of materials

codes and standards rules developed by government agencies and industries to ensure that products, structures, and environments meet safety and operational requirements

coefficient of thermal expansion a measure of the ability of a material to expand when heated

collision detection in Ethernet, nodes listen to the shared medium for collisions, in which case all nodes stop transmitting and are assigned random wait times which determine when they can transmit again

colonial expansion period beginning about 1500 C.E., during which Western European countries expanded their empires to control resources overseas

combinatorial circuit digital circuitry in which the output is directly dependent on circuit input signals

combine a harvesting machine that may perform several functions at once including harvesting, threshing, and cleaning crops.

combined diesel and gas turbine (CODAG) system found in many warships, in which a diesel engine is used for cruising and gas turbine engines are used for higher speeds or rapid acceleration

commercial buildings any structure such as offices, retail and wholesale outlets, shopping malls, houses of worship, hotels, libraries, or stadiums

commodity product a product that is the same no matter who makes it; usually a basic resource product, such as petroleum, ethanol, paper, coffee, or milk. In the chemical industry a commodity product, such as ethanol or petroleum, is produced in large amounts and is commonly the feedstock for other products such as plastics or gasoline. Goods can become commodity products if they are globally made the same and have a global demand such as generic aspirin or even silicon chips. A commodity's worth is determined by its global supply and demand and fluctuates daily

commons a resource, usually land, considered available for the use of all members of a community

communication channel network of phone lines or copper or fiber-optic cables, over which messages are transmitted; in broadcast communication, a specific range of frequencies in which a transmitter broadcasts

communication medium generic name for the different broad types of communication, such as graphic, broadcast, or electronic

complementary metal oxide semiconductor (CMOS) a circuit formed by connecting N-channel devices to P-channel devices; the foundation of all circuitry used in computer chips

composite materials materials made from a combination of two or more materials with different properties

compressed natural gas gas that is pressurized to about 3,600 pounds per square inch and stored in large tanks

compression algorithm mathematical formula used to reduce the size of a file, as in compressing a CD file to an MP3 file

compression-ignition a type of internal combustion engine (diesel) in which fuel combustion is induced by the heat produced by compression

compressive strength a measure of the ability of a material to withstand the stress of a load placed upon it

computer-aided design (CAD) computer software used to design products, systems, and structures

computer-aided manufacturing (CAM) the process of using computer software used to control CNC machines

computer architecture the design or organization of computer hardware

computer axial tomography (CAT) using a rotating X-ray device to take hundreds of pictures as it moves around a patient, providing an image of the cross section of the body

computer hardware the digital logic circuits combined into computer systems to execute instructions or software

computer numerical controlled (CNC) machines computer controller that reads G-code instructions and controls a machine tool to fabricate components by selective removal of material

concentrating solar power (CSP) use of large parabolic mirrors to focus, or concentrate, solar energy in an energy-collecting installation

concrete a mixture of gravel, sand, cement, and water

concurrent engineering systematic approach to integrated product design and development using a multidisciplinary team all working together during the entire course of a product life cycle

conduction heat transfer within a material caused by a temperature difference between parts of the material

conductivity a measure of how freely a particular material allows electrons to flow through it; the inverse of resistivity

conductor a material that allows electricity to flow

conservation the process of controlling resources, for example limiting soil erosion, reducing sediment in waterways, conserving water, and improving water quality

consolidation processes processes in which materials are combined, as by fastening or joining

constraints limits imposed upon a solution to a design problem

containerization the process of placing goods in a secure, movable container that does not have to be unloaded and repacked at transfer points

continuous improvement the Total Quality Management process of constantly isolating failures, identifying their causes, and fixing or improving the process

continuous process a type of flow process in which the product is one that actually flows, such as liquids, gasses, or powders

Contour Crafting layered fabrication technology projected to be able to automate construction of whole structures or subcomponents

control functions in framed superstructures, refers to how the framework controls the separation of air, moisture, heat, and sound between outside and inside

convection transfer of heat between a surface and a fluid, which may be liquid or gas

convergence the gathering of voice, video, and data onto a single network

corn-based ethanol an alcohol-based fuel produced by fermenting and distilling grain crops

corporate culture values and norms that are shared by people and groups in an organization

cottage industry an industry that relies on family units working at home, using their own equipment

cross impact analysis forecasting technique of identifying several possible futures and predicting their impact if they occurred in combination

cryosection use of quick-frozen sections from cadavers to create anatomically detailed, three-dimensional representations of bodies

culture the customary beliefs, social forms, and material traits of human groups

current conventional term for the flow of electrons; denoted as I

current good manufacturing practices (cGMPs) regulations developed and established by companies to ensure a quality product

cytosine one of the four amino acids (the others are adenine, guanine, and thymine) that comprise DNA; cytosine is denoted as C

data raw facts and figures that is processed into meaningful information

data bus common group of wires used to transfer data between computer components inside a computer or between computers

Data-Link Layer second layer of OSI, which takes the bits of the message and builds frames from them

decay dying out of a signal as the destination gets farther from the source

deciduous trees that shed their leaves each year

decoder function of a modem that translates analog signals from the communication channel to binary signals

deformation processes processes that change the shape of a material without changing its mass, including forging, rolling, machine pressing, and drawing

demultiplex to unbundle or decouple communication channels at the receiving end of a medium

denature to inactivate something, such as proteins or enzymes, as by heating

density a measure of how tightly the atoms of a material are packed together; in integrated circuits, the number of cells per chip

depleting feedstocks raw materials that cannot be renewed, such as those from fossil fuels

design constraints limitations imposed on a designer and the design, often related to resources

design specifications specific goals or performance requirements that a project must achieve in order to function as desired

destructive testing quality-control testing by stressing a part or product until it fails; *life test*

device driver the software program that specifically controls hardware, allowing it to communicate with the CPU through interrupt and I/O circuits

device programmers special equipment used to store or change data on an EEPROM chip

difference engine the first programmable calculator, developed in the 1850s by Charles Babbage

differential signaling a method of transmitting information electrically by means of two complementary signals sent on two separate wires, as in USB cabling

diffusion one of the steps in research and development; adoption of new inventions by users other than the innovators

digital controller system controller using signals that have only a finite set of states (normally, on or off, or high or low voltage)

digital signal a continuous electrical signal that varies in discrete increments, such as high/low, rather than continuous variation

Digital Signal Processor (DSP) specialized computer chip designed around a logic architecture specifically created to enhance mathematical capabilities of the chip; used in products requiring continuous processing, such as audio processing

digital-to-analog converter (DAC) device that converts a digital signal to an analog signal

direct current (DC) electrical current characterized by a constant level of voltage and current, with current flowing in one direction only

discharging the release of energy from a capacitor or battery

discrete parts transistors, resistors, and other parts soldered together on circuit boards to form an electronic system

distribution functions in framed superstructures, refers to how utilities and services are distributed through the building

division of labor the breaking down of work into its component tasks, which are distributed among different persons or groups, thereby increasing efficiency, production, and profits

DNA deoxyribonucleic acid; a nucleic acid that contains the genetic instructions used in the development and functioning of all known living organisms and some viruses

DNA sequencing the act of determining the linear sequence of DNA

domestication breeding animals for a life in close association with and to the advantage of humans

doping the process of adding materials to silicon in semiconductor manufacturing to produce N-type and P-type semiconductors

down counters digital circuits that react to input signals by producing a binary counting sequence from some maximum value down to zero

drain in an N-channel MOSFET, an N-type area embedded into a larger P-type block and thus isolated from the source

drawing process of pulling a material through a small opening (a die)

drilling the use of a rotating bit or chisel (hand drill, electric drill, drill press, or jackhammer, for example) to penetrate or break up a material

dropped packets error in data transmission across a network; with e-mail and other data, but not with voice, dropped packets can be resent to maintain the integrity of a message

Dual In-line Memory Module (DIMM) small circuit card in a computer system allowing the user to control the amount of memory

ductility a measure of how much a material will yield to stretching before breaking

durable goods goods that are designed to operate for a long period of time, generally more than three years

dynamic RAM (DRAM) type of semiconductor memory using a storage cell constructed from one transistor and one capacitor, and needs to be periodically refreshed; used when large amounts of memory are needed

earthwork in road construction, establishment of roadway's base layers and foundation, leveling the earth, watering and compaction, installation of sewers and drains, and application and compaction of gravel in preparation for paving

ecological design design approach that seeks effective adaptation to and integration with nature's processes

economic incentive the promise of financial reward that helps drive innovation by making it easier for companies to take risks and for investors to back a project

economics field of study that involves the description and analysis of the production, distribution, and consumption of goods and services.

ecosystem the sum total of animals, plants, and microorganisms within a given habitat, and the ways they interact

elasticity a measure of how much a material can be stretched without being permanently deformed

elastic limit the maximum stress to which a material can be subjected and still be able to return to its original shape when the load is released

electrical energy caused by the flow of electrical charge through an electrical field

Electrically Erasable Programmable Read Only Memory (EEPROM) a ROM technology that can be reprogrammed; used to store small amounts of data that must be saved when power is removed

electrical properties properties of materials including conductivity and resistivity

electrolysis process of splitting water into hydrogen and oxygen by means of an electric current

electromagnetism generation of a magnetic field by electricity; also, the generation of electricity when wire is moved through a magnetic field

electronic communications the use of electrical signals, pulses of light, or radio waves to transmit messages

electroplating an electrochemical process that deposits or adds a thin layer of metal onto a substrate

elliptical refers to the shape of a satellite orbit in which the altitude above Earth varies

embedded computer small computer built into the device it controls

embedded processors computer chips built into a product that control the operation of the device

encoder function of a modem that translates a computer's binary messages to an analog signal for transmission over communication channel

energy/fuel products chemical products such as gasoline, oil, and hydrogen gas that are used as energy sources in other contexts

engineer one who transforms knowledge of mathematics and sciences, such as physics, chemistry, or biology, into practical applications through the process of engineering design

engineered woods wood products made by combining particles and fibers of wood with adhesives to meet specific requirements

engineering the practical application of the information acquired by science, including a body of knowledge about the design and construction of products and a set of processes and techniques for designing and creating the products

engine management system electronic automotive control system that calculates the amount of fuel to be injected into cylinders and monitors exhaust gasses to increase fuel economy and reduce greenhouse gas emissions

environmental impact assessment government-required study of how a construction or other project would affect the natural environment

epitaxial (epi) layer of semiconductor material, a few hundred atoms deep, laid over a substrate in a diffusion furnace; the channel area for many FETs

ergonomics design and manufacture of products that are comfortable to use and conform to customers' physical constraints

execute to process computer instructions or software

expansion slots electrical and mechanical connectors that allow computer owners to add to their systems by plugging in new printed circuit cards

explants small pieces of leaf that have been treated to kill surface microorganisms that would contaminate the culture, it is grown in a medium that contains organic salts, vitamins, and hormones for shoot development

expressed word used to describe the end result of DNA transcription

external combustion type of engine in which fuel is burned outside the engine, the energy being transferred to another liquid or gas, which in turn converts thermal energy into mechanical energy

extrusion a process in which a continuous stream of a product is made by squeezing softened material through a small opening (a die)

fabrication word generally used instead of "manufacturing" to describe how materials are processed into products; in electronics, it refers to how integrated circuits are built

factor of safety a quality of any good design, accounting for the possibility of unexpected load conditions or variations in the properties of materials

facultative anaerobes organisms that prefer using oxygen but can live in its absence

fastening techniques techniques involving joining materials by means of nails, screws, nuts and bolts, rivets, and staples

feedback information about an action, process, or communication that is transmitted back to the source or controller of that action, process, or communication for purposes of evaluation and correction

feedstocks raw materials that go into any product

ferrous metals containing iron

Fertile Crescent area of the Middle East stretching from present-day Pakistan to Syria, centered on the Tigris and Euphrates rivers in present-day Iraq

fetch the first step in any CPU operation, involving retrieval of an instruction from program memory

Feynman, Richard physicist who speculated that science could theoretically maneuver individual atoms to build new materials

field effect transistor (FET) common type of transistor that uses an electric field to control the conductivity of semiconductor material; see *metal oxide semiconductor FET*

finish functions in framed superstructures, refers to the visual and aesthetic effect of the framework

fixed position a manufacturing system in which the product remains stationary and people and machinery come to it; also called *project shop*

fixture in manufacturing, a device used to hold objects in place so that the object can be processed or assembled

flash memory technology similar to EEPROM, which may be erased and reprogrammed; used for general storage and transfer of data between computers and other digital products

flexible manufacturing systems (FMS) an approach to manufacturing a product that focuses on integrating technology to enable rapid changes to production processes

flexural strength a measure of the amount of stress it takes to bend a material to the point of failure

flip-flop a digital circuit that stores a single bit in response to timing or triggering signals

floating gate special cell structure in an EEPROM that may be charged or uncharged and is nonvolatile

flowchart diagram depicting the major steps in a process, used to show how one step leads to another

flow shop a manufacturing system designed to produce only a specific product or family of products, in which the product flows from station to station during fabrication; also called *flow line*

forecasting techniques tools designed and used to predict and evaluate the impact of technologies, especially in terms of the environment

forensic scientist one who uses fingerprints, hair, footprints, blood, DNA samples, and other evidence to identify criminals or victims

forging changing the shape of metal by using hammers or presses

form mold into which concrete is placed

fossil fuel any substance formed below the earth's surface long ago from pressure on decaying plant and animal remains

foundation the substructure of a structure; the part that is in the ground

four-stroke cycle the typical gasoline or diesel engine, whose cylinders cycle through intake, compression, power, and exhaust strokes

frame in Ethernet, one of a number of smaller pieces into which a message is broken up for transmission, which are received and reassembled at the destination

framed superstructure structures enclosed and supported by a framework, typically of lumber or reinforced concrete and steel

frequency in an AC signal, the rate of signal change, measured in hertz

frequency division multiplexing multiplexing technique in which each "conversation" is assigned its own frequency

frequency modulation (FM) modification of a carrier wave by means of varying the frequency of the wave

Friedman, Thomas author of *The World Is Flat*, which discusses the state and future of globalization

front end of the line (FEOL) in an IC fabrication process, the stage in which transistors are created

functional genomics the study of how genes interact, what proteins they make, when they make them, and in what cells

fusion process of forcing atoms to combine to form new, heavier atoms, releasing great quantities of energy

futures wheel a diagram with one scenario at the center that might lead to other scenarios which, in turn, spawn other possible outcomes, and so on

futurists people who study and predict the future based on current trends

gadgeteering a trial-and-error approach to designing not based on salient knowledge

gamma-ray very short-wave radiation, sometimes used to kill microorganisms in food

Gantt chart diagram focusing on a sequence of tasks, showing them along the horizontal axis of a graph

garbled data data that are corrupted and rendered unusable as the data travel across a network

gate in an N-channel MOSFET, a semiconductor material constructed over the source and drain but insulated from them by a layer of silicon dioxide

G-code instructions computer code for programming CNC machines

general contractors in the construction industry, those who build residential, industrial, commercial, and other buildings

general purpose registers groups of flip-flops used by a CPU to temporarily store data

genes discrete packets of DNA, each coding for a particular protein

genetically modified (GM) foods food products containing genetic material transferred from other organisms

genetically modified organism (GMO) organism whose genetic material has been altered by recombinant DNA technology to produce something not found in nature

genetic anthropology emerging discipline that combines DNA and physical evidence to reveal the history of ancient human migrations

genome the entire collection of DNA in an organism

Genomics Age building on the success of recombinant DNA and the Human Genome Project, an effort to sequence the genomes of many species

geostationary a geosynchronous satellite orbit rotating at the same speed as the Earth directly above the Earth's equator

Gigabyte (GB) approximately one billion bytes, a measure of computer RAM size

Global Positioning System (GPS) a constellation of medium Earth orbit satellites enabling a GPS user to accurately determine location, speed, direction, and time

global warming process of climate change resulting from increasing amounts of greenhouse gasses in Earth's atmosphere

glocalization term coined by Thomas Friedman to describe the need for a country to preserve local traditions by thinking globally but acting locally

government subsidy funds or tax incentives set aside, usually by Congress, some of which go directly to universities or private companies to undertake important R&D projects

grain the pattern in which wood fibers grow

graphic communications the use of words or pictures to convey messages

gravitational potential energy energy associated with an object's position relative to the ground

green chemistry term given to a variety of environmentally friendly products and processes; a major goal of the twenty-first-century chemical industry

green construction another name for *sustainable development*

greenhouse gasses gasses such as carbon dioxide and methane that allow solar energy to pass through to Earth's surface, but absorb and trap heat in the atmosphere

grinding the use of abrasive materials to remove small particles of a material being processed

growing in agriculture, includes activities such as pest and weed control, as well as irrigation

guanosine one of the four amino acids that comprise DNA; denoted as G

half adder the simplest addition circuit in arithmetic and logic unit, adding two individual bits together to produce a two-bit answer

hard automation cams, stops, slides, and hard-wired circuits programmed with a controller or handheld control box to control functioning of machines; also called *fixed-position*

hardening a process in which a material's surface or internal structure is made physically harder, as by heating and then rapidly cooling steel

hardness the ability of a material to withstand scratching or penetration

hardwood wood from deciduous trees

harvesting and storing agricultural process culminating in transportation to market

Health Insurance Portability and Accountability Act (HIPAA) federal law meant to ensure health insurance portability, reduce fraud and abuse, guarantee security and privacy, and set and enforce standards for health information and its transmittal

heat thermal energy that is transferred between two objects because of a temperature difference between them

heavy engineering contractors see *civil engineering contractors*

hertz (Hz) cycles per second, the unit of measure of electrical signal frequency

high-bypass turbofans latest generation of turbojet engines, which use a large bypass fan to force air out of the engine nozzle

high-speed train (HST) any train designed to travel at speeds in excess of 90 mph

hole in semiconductors, any vacant energy level, positively charged because holes attract electrons

Homestead Act of 1862 act of Congress that offered 160 acres of land, free of charge, to any adult citizen who would live on the land for five years and develop it

HortResearch Horticulture and Food Research Institute of New Zealand, Ltd.; life science company that discovered the gene responsible for producing the compound alpha-farnesenes, which gives green apples their characteristic scent, and synthesizes it using gene technology combined with biofermentation

host in Ethernet, a computer connected to a segment and contending for the shared media

host controller the computer end of a USB connection

hovercraft versatile air-cushion vehicle (ACV), using engine-driven fans to generate lift by directing air under the craft, providing a cushion on which it rides

human-factors engineering another name for *ergonomics*

Human Genome Project (HGP) the project undertaken by the United States and other countries to sequence the entire human genome, begun in 1990

humanoid robot human-like robot able to walk, run, climb stairs, carry a tray, and perform other normal human activities

hunter-gatherers people who obtain food by hunting, fishing, and foraging, relying primarily on muscle power

hybrid electric vehicle vehicle in which an internal combustion engine drives a generator, and the power from the generator is used to drive traction motors

hydroelectric systems power-generating systems using water power, the largest single renewable energy source in the United States

hypoglycemia unawareness in some diabetic patients, an inability to experience the usual physical warning symptoms that their blood sugar is low

impedance opposition to current flow in an AC circuit, measured in ohms, denoted as Z

inductive reactance opposition to AC signals in an inductor; denoted as XL

inductor an electrical component made by coiling wire; generates a magnetic field as current flows through the coils

industrial feedstocks raw materials for industsrial production

industrial fermentation fermentation on a large scale

industrial plants structures such as factories that manufacture products, power plants, petroleum refineries, and wastewater treatment plants

Industrial Revolution period from 1750 to 1900 C.E., which saw the development of continuous manufacturing, sophisticated transportation and communication systems, advanced construction practices, and improved education and leisure time activities

industry a distinct group of humans or enterprises that provides systemic labor for the purpose of creating something of value

informed consent the requirement that patients undergo medical procedures only after learning and understanding the risks and benefits of doing so

initialization program software that tests the machine's components on startup, readying the machine for use

injection molding use of a hydraulic plunger or mechanical screw to force softened plastic into a mold under pressure

innovation the process by which preexisting products or ideas are transformed into new techniques and more useful or economic products

inorganic materials materials that occur in nature but are not carbon-based, such as stone, clay, and metals

inputs the wires leading to a circuit; in communication systems, any message entered into the system; in agriculture, any item required for the agricultural process, such as a tractor or fertilizer; in systems theory, the desired results and resources

institutional buildings schools, colleges, universities, hospitals, correctional facilities, and the like

Instruction Decoder circuitry in a CPU that examines the bit pattern of an instruction and activates specific parts of the CPU to carry out that instruction

Instruction Register a set of flip-flops used by a CPU to store each bit in the bit pattern of an instruction

instructions a software program governing the operation of the central processing unit (CPU)

instruction set the total of all commands in a microprocessor

insulator a material that does not readily conduct electricity

insulin a hormone that lowers the level of glucose in the blood; produced in the pancreas

integrated circuit (IC) computer chips comprised of thousands, even millions, of transistors built on a single piece of silicon

integrated pest management (IPM) pest-control strategy that relies on multiple control practices rather than just one, establishing the amount of damage that can be tolerated before control is undertaken

interchangeability standardization of design so that all instances of a given part are mutually substitutable

intermodal shipping transport of goods in containers by more than one shipping mode, as by sea, rail, and road

internal combustion engine an engine powered by fuel burned within the confines of the engine itself, including gasoline and diesel engines

internal energy energy associated with the molecular structure of a substance, including thermal, chemical, and nuclear energy

International Space Station (ISS) research facility built as a collaborative venture by space agencies of the United States, Russia, Japan, Canada, and the European Space Agency

internetworking the connection of LANs into a WAN

interrupt logic that permits the CPU to shift attention from one program to another; also, the signal that starts that process; *interrupt request*

inverter a NOT gate

I/O devices input/output devices, peripheral devices that send information to the CPU or receive information from it

ion implantation in IC fabrication, after unexposed photoresist is dissolved, ions of dopant are blasted into the wafer's uncovered areas, creating sources and drains of each FET

IP gateways used, for example, by VoIP to transition circuit-switched calls onto a packet-switched network and then back again near the destination

Iron Age period from 500 B.C.E. to 500 C.E., during which iron and steel came to be the primary materials for tools

jig a device used to hold a workpiece and control the location and motion of a cutting tool

jitter problem in data transmission across a network, similar to choppiness in online video

job shop a small manufacturing business using general-purpose production equipment and skilled workers capable of producing a wide variety of products

joining techniques of combining materials by means of adhesives, soldering, brazing, and welding

Just-In-Time (JIT) a manufacturing strategy in which parts and materials are produced or delivered only as needed

Karnaugh Mapping (K-map) technique in designing and simplifying combinatorial circuitry, graphically rearranging truth table combinations to make circuit maximization obvious to the trained eye

kinetic energy the energy an object possesses because of its motion

Kirchhoff's Current Law mathematical relationship describing the fact that the total current flowing into parallel branches of a circuit is equal to the sum of the current in each parallel branch

Kirchhoff's Voltage Law mathematical relationship stating that the size or magnitude of each voltage drop depends on the resistance of any individual component relative to all the resistances in the circuit; the larger the resistance, the larger the voltage drop

Knowledge and Skill Builders (KSBs) guided research and investigation activities for students that inform one's knowledge base before design solutions are proposed.

Land Grant Act the Morrill Act of 1862, which allowed the federal government to give land that could be sold to each state with the funds used to establish colleges of agriculture and mechanical arts

large-scale manufacturing mass production manufacturing process, often not undertaken until small-scale manufacturing has worked out details of the process

laser land-leveling use of laser beams to guide earth-moving equipment to equalize elevation differences

latency gaps in conversations caused by delays as data travels across a network

lean manufacturing organizational strategy involving minimal input in terms of inventory, workforce, and other resources to keep costs low and flexibility high

life test destructive testing

ligases enzymes used to join pieces of DNA together

light rail electric-powered urban trams or streetcars used as public transportation systems

line graph pictorial representation of a relationship between variables, drawn by connecting points, representing data, with a line

liner passenger ship designed to make ocean crossings at speed in all weather

linked-cell a manufacturing system using an arrangement of machinery allowing processes to be grouped according to the sequence of operations required to make a product

liquid petroleum gas (LPG) gas produced by refining crude oil or processing natural gas and then liquefying it for storage in tanks

local area network (LAN) a group of computers connected together by any media: copper, fiber optic, or wireless

logic functions the building blocks of computer hardware design, determining how digital signals are processed

logic families a group of digital circuits with similar electrical properties, enabling system designers to work at the logic level rather than designing basic electrical circuits

low Earth orbit (LEO) circular satellite orbit about 250 miles above Earth's surface, generally requiring around 90 minutes per orbit

Luddites group of people during the Industrial Revolution who smashed knitting machines, fearing they would threaten workers' jobs; today, the term disparagingly refers to people who are reluctant to adopt new technologies.

lumber wood cut into standard, usable dimensions (timber)

macadam road pavement technique using layers of stone coated with a mixture of water and stone dust, then compacted by a heavy roller

machine-level instructions binary instructions built into the computer hardware

machine pressing use of hydraulic presses to stretch a material into a desired shape

magnetic properties properties of materials related to their tendency to create or interact with magnetic fields

magnetic random access memory (MRAM) a promising nonvolatile memory technology that stores information magnetically; still in development

magnetic resonance imaging (MRI) medical imaging technique using radio waves in a strong magnetic field to compile a three-dimensional representation of the body

malleability a measure of how well a material can be hammered and pressed into a shape

management the process of leading and directing all parts of an enterprise by organizing human and financial resources and time in the most profitable and beneficial way

manufacturing the process of making raw materials into products, either by hand or by machinery

manufacturing system the entire set of objectives, people, resources, boundaries, and constraints that define a manufacturing enterprise

market research studies to determine whether potential customers will like a new product

masking technique that defines exactly where transistor features go on an IC, using a special glass covered with lines and open areas

mass production the production of large amounts of standardized products on production lines

mass superstructure made from large masses of materials, having little or no space inside, such as dams or monuments

materials the substances of which something is composed or can be made

mathematical model equations representing the operation of a system, taking into account all specifications and constraints developed during problem definition

matrix material binder that envelops a stronger reinforcement material in a composite material

mechanical energy energy that a person or object expends in moving another object; also called *mechanical work*

mechanical properties properties of materials that influence their ability to endure and withstand applied forces

mechanization use of machines to supplement human functions

media in a communication system, the means through which a message is transmitted, such as telephone wires or the atmosphere in the case of radio waves

medical imaging any of various radiological techniques for looking inside the body, such as X-rays, ultrasound, or MRI

Megabytes (MB) approximately one million bytes, a measure of computer RAM size

metal fatigue weakening of metal components, as in a bridge, under heavy stress

metallic matrix composites matrix composites with very high temperature limits, used on the skin of hypersonic aircraft, for example

metallization layers IC layers formed from aluminum or copper sandwiched with insulation, designed for interconnections

metal oxide semiconductor FET (MOSFET) type of field effect transistor used as a switch

metric numerical statement of what is to be measured in quality control systems

microarray technology technique that allows quick analysis, in one test, of which genes are expressed by given cells

microcontrollers another name for *embedded processors*, control chips built into various devices

microfactories small, desktop factories that match the size of the production system to the size of the parts being made

micro-generators small electromechanical devices capturing the energy of a body in motion or vibrations from a machine to generate up to 50 microwatts of power

microlam similar to plywood, but using many layers of thin wood glued together

microprocessor in general-purpose computers, the digital hardware doing the main computing

mixed materials a combination of natural and synthetic materials

modulation modification of the carrier wave to encode a message

modulator/demodulator (modem) device used to connect a computer to a transmission channel, coding message sent and decoding messages received

molecular manufacturing processes under research in which chemical properties of single molecules can be used to allow carefully controlled molecular self-assembly

monoculture raising only one type of plant or animal over a wide area

monomers single chemical compound that can later be combined into polymers

monopoly rights exclusive right of control over a process or commodity that enables the holder to increase the price for use of an innovation

Moore's Law the prediction that the number of transistors on an integrated circuit will double every two years

Morrill Act the Land Grant Act of 1862, which allowed the federal government to give land that could be sold to each state, with the funds used to establish colleges of agriculture and mechanical arts

motherboard in a home computer system, the main printed circuit board

multicore processors a combination of two or more CPUs working together to increase total computing power

multi-dimensional NMR nuclear magnetic resonance imaging technique used to determine protein structures

multipath effects a phenomenon in which wireless signals bounce off obstructions between source and destination, resulting in the receiver getting multiple versions of the same message

multiple access in Ethernet networks, a consequence of a segment being a shared medium requiring that parties must take turns communicating

multiplex bundling communication channels together for transmission over a medium

Multiprotocol Label Switching (MPLS) newer transmission protocol for communication across the Public Switched Telephone Network, to which telecommunications companies are transitioning from ATM

NAND gate a combination of AND and NOT gates (Not AND) that produces an inverted AND response

nanoreplication runaway potential problem in which a self-replicating nanoscale replicator makes copies of itself in an unstoppable geometrical progression

nanotechnology techniques allowing the manipulation of individual atoms or molecules to create or modify materials on a nanometer scale

natural fisheries fisheries that occur naturally, without human intervention

natural materials materials that occur in nature

N-channel MOSFET a transistor composed of N- and P-type materials, consisting of a source, a drain, and a gate

net shaping the processing of materials into products that require very little finishing or post processing

Network Layer in OSI, the third layer, responsible for addressing the data so that the data can get from network to network

nitrogen fixers crops such as alfalfa and soybeans that add nitrogen back into the soil rather than depleting it

node in Ethernet, a connection point, either a redistribution point or a communication endpoint (some terminal equipment)

nomadic life traveling in small groups, often in seasonal patterns, surviving by gathering food through hunting, fishing, and foraging, often with herds of domestic livestock

nondestructive testing quality-control testing that does not damage the part or product, such as X-rays or ultrasound

nondurable goods goods that are not designed or expected to last more than three years

nonferrous metals that do not contain iron

nonrenewable energy source an energy source that cannot be replenished in a short period of time

nonvolatile used to describe memory chips that retain their data whether the machine is on or off

NOR gate a combination of OR and NOT gates (Not OR) that produces an inverted OR function

norms those actions considered necessary and binding upon members of a group

Northbridge in some two-IC chip sets, handles the microprocessor's connections to system memory and graphics memory

NOT gate a digital device that has a single input and a single output, whose sole function is to flip the logic level

N-type semiconductor type formed by combining silicon atoms with small amounts of another element such that the resulting material has many free electrons, and hence is more conductive than silicon alone

nuclear energy a type of energy associated with the atomic structure of matter

nuclear magnetic resonance spectroscopy (NMR) family of scientific methods that exploit nuclear magnetic resonance to study chemical structures

off shoring the movement of operations of manufacturing firms to other, nations where costs are lower

off state in digital devices, used to describe devices that are off, or have no voltage; also called *low state* or *zero state*

ohm the unit of measure of electrical resistance; denoted as Ω (omega)

Ohm's Law a statement of the relationship of voltage, current, and resistance; $V = I * R$

One Laptop Per Child Program (OLPC) MIT-initiated program to develop and distribute worldwide a low-cost, robust, durable laptop computer powered by a hand-cranked, high-efficiency generator

on-off control a switch that enables a person to make or break an electrical circuit by hand

on state in digital devices, used to describe devices that are on, or have voltage present; also called *high state* or *one state*

open sea wave generation system to capture wave energy in the open ocean

Open Systems Interconnection (OSI) Model standardized networking architecture/structure established by the International Standards Organization

opportunity costs combination of direct outlay of funds or resources plus indirect costs, such as lost opportunities for financial benefit

optimization the process of improving each alternative or each part of a design

organic light-emitting diode (OLED) diode that uses layers of organic materials that glow red, green, and blue, sandwiched between two layers of conductive material

organic materials materials with carbon-based structures that originate from or relate to living organisms

OR gate digital device whose output is one if even one of its inputs is one, zero only when all inputs are zero

output the wires coming from a circuit; in communications systems, any received message; in systems theory, the actual results

oxidized refers to anything that has been chemically combined with oxygen; in foods, oils become rancid in the presence of oxygen

paradigm shift a change in the way of thinking; may be applied to individuals or whole societies

parallel circuit an electrical circuit in which there are two or more paths for electrons to follow

parallel processing the use of multiple processors to undertake different aspects of a program simultaneously

parametric programming combination of G-code with logical commands in advanced CNC controllers

passive radio device device that can transmit information picked up as a radio wave, using only the energy from the radio wave

peak value in the graph of a sine wave, the maximum value of y, indicating the *amplitude* of the signal

period the time it takes an analog waveform to repeat itself

periodic maintenance work that is done only every few years

peripheral components interface (PCI) a common expansion slot that connects devices such as sound cards and disk drives to the motherboard

peripheral device devices external to a computer, such as keyboards, monitors, and network interface cards

persistent vegetative state a prolonged, usually permanent state of semi consciousness in which patients may have periods of sleep and wakefulness and open their eyes, but do not respond to verbal stimulation or perform any normal functions

personal desktop computer complete computer system composed of a CPU, memory, keyboard, and display, first developed by Hewlett Packard in 1972

Personal Rapid Transport (PRT) automated system using driverless "pods"

personal watercraft small vessels on which the rider sits or stands, typically for one or two persons

PERT chart diagram showing how one task flows into another; used to schedule, organize, and coordinate project-related tasks

pesticide and fertilizer treadmill a result of monocultures, in which pests and diseases develop resistance to existing chemicals, requiring ever more chemicals in ever-increasing amounts

petroleum a fossil fuel in liquid form; oil

pharming the use of genetically engineered organisms to make products other than food, especially pharmaceutical products

phase change a change in phase from solid to liquid or from liquid to gas

photoelectrolysis process of using sunlight to split water molecules into hydrogen and oxygen

photoresist light-sensitive liquid chemical used in IC fabrication; hardens under ultraviolet light where mask permits the light to penetrate; unhardened chemical is dissolved by chemical solvents

photosites light-sensitive diodes on a charge-coupled device

photosynthesis the process by which plants use the Sun's energy to convert carbon dioxide and water into oxygen, sugar and energy-storage compounds

Physical Layer first level of OSI, whose job is to transmit and receive data across the medium

physical model graphical representation of a circuit or system using standardized symbols

physical properties properties of materials including density and freezing, melting, and boiling points

physical system actual circuit or system, which can be represented by a mathematical model or physical model

pie chart pictorial representation of a relationship between variables, comparing categories within the data set to the whole

piezoelectric ceramic materials that generate an electrical voltage when subjected to mechanical pressure

pilot plant small-scale chemical plant built to provide design and operating information before construction of a large plant

planned community development in which city, business, industry, and residential housing are conveniently located in patterns radiating out from the city center

planning in road construction, includes land acquisition, identifying environmental concerns, finding funding, and identifying contractors and engineers

planting application of seed to the soil, often including fertilizer application at the same time

plasmid bits of DNA that can be engineered to shuttle DNA or genes from other species into bacterial cells

plastic one of the world's most common synthetic materials, made up of long chains of molecules (polymers)

plasticity the property of a material which permits permanent deformation to occur before it ruptures

pluripotent not fixed as to developmental potentialities; used to describe undifferentiated stem cells, which are capable of becoming many different kinds of body cells

plywood an engineered wood created by cutting thin sheets of wood (plies) and laminating them with glue at alternating 90° angles

polyethylene plastic made of many ethylene molecules (C_2H_4)

Polymerase Chain Reaction (PCR) technology allowing the detection of small amounts of specific DNA sequences and the duplication of them many times in order to increase their concentration

polymers large molecules made up of long chains of molecules (monomers) that are bonded (polymerized) together

polymer matrix composites matrix material consisting of strong fibers embedded in a resilient plastic that holds them in place

positive control system computer-controlled rail system using track sensors and computers to control all the trains using a track system to ensure that each train runs safely on its own track

potentiometer a variable resistor used, for example, in volume controls on radios, TVs, and the like

power the rate at which work is done; measured in watts

power supply the source of necessary voltage and current for any system requiring electrical power

precision farming agricultural practice using GPS to precisely and automatically apply fertilizer or pesticide

Presentation Layer in OSI, the sixth layer, responsible for taking information and making it usable or presentable to the Application Layer

pressing similar to casting, except that after molten material is poured into a mold, a plunger with the shape of the inside of the object to be produced is lowered into the mold

primary fabrication conversion of raw materials from their mineral or organic origins into industrial materials that can be processed into products

primary food processing includes transporting products from fields to facilities where they are stored and undergo preliminary treatment

printed circuit board used to connect the pins of many ICs and electrical components, constructed of nonconductive materials

process in communication systems, the medium of transmission between input and output; in systems theory, the part of a system that combines resources to produce an output.

process flow diagram detailed diagram showing the process of a chemical plant, including pipes and valves

process line in secondary food processing, an individual product system including washing, cutting, cooking, and other steps necessary to produce the final product

process parameters factors that govern the production of products, including feedstocks, cost-effectiveness, energy efficiency, and adherence to government guidelines and regulations

product pipeline the process organizations use to create and eventually market a drug

product safety consideration of paramount importance in design of new products or processes; manufacturers are held responsible for safe product performance even if the product is used in a manner unintended by the manufacturer, and even if it is sold by an intermediary

product synthesis the actual making of a product, as specified by production protocols

project implementation the process of carrying out plans in ways that maximize success and preserve core values of the organization or business

project organization setting up procedures that make the best use of resources so that the plans that have been developed can be implemented successfully

project planning making decisions about what should happen to ensure that a project succeeds, including identification of goals, strategic plans, time lines, budget, and management plan

project shop a manufacturing system in which the product remains stationary and people and machinery come to it; also called *fixed position* fabrication

proportional controller controller that adjusts a system incrementally rather than simply by turning it on or off

propulsion pods system using high-efficiency electric motors to turn a propeller

protease enzyme enzyme responsible for generating mature viral particles before they bud off from an infected cell

protected health information (PHI) information defined by HIPAA as information in any format that is individually identifiable to a specific patient regarding that patient's past, present, or future

protocols the series of rules explaining exactly how to manufacture a product

prototype a working model used to test a design

P-type semiconductor material formed by combining silicon atoms with small amounts of another element such that the resulting material is more positively charged than silicon alone

Public Switched Telephone Network (PSTN) the network of the world's public circuit-switched telephone networks

qualitative refers to any information that is not stated in numerical terms

quality in general, superiority of design and function

quality control a planned process that ensures that a product, service, or system meets established criteria

quantitative refers to information that is stated in numerical terms

radiant energy electromagnetic energy, including visible light, X-rays, gamma rays, and radio waves; *radiation*

radiation energy emitted from an object in the electromagnetic spectrum, from very long radio waves to very short gamma rays

radiation heat transfer the flow of thermal energy between two bodies separated by a distance

radio frequency identification (RFID) a "super" bar code system using a small, printable circuit to store product information and transmit it instantaneously to a checkout scanner or security system

random access memory (RAM) the main memory in a computer system, in which read and write operations occur with equal speed; a volatile memory

rapid manufacturing fabrication technique for manufacturing solid objects by sequential delivery of energy and/or material to specified points in space to produce that part

rapid prototyping modeling process used in product design in which a CAD drawing of a part is processed to create a file of the part in slices, and then a part is built by depositing layer upon layer of material

rapid transit system rail-based urban public transportation system

raw materials crude or processed materials that can be converted, by manufacture, further processing, or combination, into useful products

reactants initial compounds, consumed during a chemical reaction

reading the act of retrieving information from a computer

read only memory (ROM) a nonvolatile memory in a computer

realistic constraints limitations imposed by economic, environmental, social, political, ethical, health and safety, manufacturability, and sustainability considerations

receiver in communication systems, the recipient of a message, or the device that receives the message

recombinant bacteria genetically modified bacteria used to produce large amounts of proteins such as human insulin

recombinant DNA technology the science of extracting DNA from different organisms and combining it at the molecular level to produce proteins or new, genetically modified organisms

recombinant protein a protein produced as a result of recombinant DNA, such as human insulin

refresh circuits circuitry that periodically reads and rewrites information in DRAM memory systems to counteract leakage problems in the capacitors

regenerative braking an automotive braking system that captures the kinetic energy of the vehicle to drive the electric motor to recharge the batteries while slowing the vehicle

register a combination of a number of flip-flops, used as computer storage areas, typically storing between 8 and 64 bits

reinforcement material in a composite material, the stronger material that is enveloped by the matrix

reinnervation technique in which signals from the brain are used to control muscles in an amputee's body

Renaissance period from 1400 to 1750 C.E., a time of rebirth in the arts and humanities, with major developments in technology

renewable energy source an energy source that can be replenished naturally in a short period of time, including wind, solar radiation, flowing water, geothermal sources, and biomass

renewable feedstocks raw materials that can be replenished naturally, such as those made from agricultural products

repeater device that takes a signal that may be decaying and regenerates it, allowing it to continue toward its destination

research and development (R&D) problem-solving approach used intensively in business and industry to prepare devices and systems for the marketplace

researchers in the chemical industry, those who study how atoms, elements, and compounds react

residential building any structure designed to house people, whether single- or multiple-family, apartment buildings, condominiums, or cooperatives

resin polymers extruded, pelletized, or flaked and, in turn, re-extruded and made into various plastic products

resistance opposition of a material to the passage of electrical current; denoted as R, measured in ohms

resistivity a measure of how a material restricts the flow of electrons; the inverse of conductivity

resistor component of electrical circuit designed to have a specific value of electrical resistance

respiration in plants, the process by which plants use sugars from photosynthesis to produce energy and make products such as proteins and fats

restriction enzyme the molecular scissors used to cut DNA in recombinant DNA technology

retrofitting fitting new equipment into or onto equipment already in service

ribosome a cell's protein manufacturing machinery, which decodes mRNA information to synthesize proteins

risk assessment science and art of determining the probability of the occurrence of some negative event in the future

risk aversive refers to cultures that tend to decline risks that are statistically even

RNA ribonucleic acid; a nucleic acid similar to DNA; central to the synthesis of proteins

road design includes surveying, identification of properties affected, drainage characteristics of soil, expected traffic volume, and environmental effects

robot any machine or device that emulates human movement, capability, or appearance; from the Czech word *robota*

robotic cell in linked-cell manufacturing, an unmanned cell

robotic vehicle robot equipped with sensors enabling it to follow a road, detect other vehicles, obey rules of the road, and react to other problems

rolling a process in which a material is passed through a set of rollers to either make it thinner or to shape it

root-mean-square (RMS) values used to describe AC signal magnitude, calculated by multiplying peak amplitude by 0.707

rotational molding method in which molten material is placed in a mold, and centrifugal force from rapid spinning causes the material to be distributed uniformly on the inside of the mold

routine maintenance work that is done periodically and consistently

safety in manufacturing, the imperative that products must not be dangerous to the user either in ordinary use or when they fail

sampling precision a representation of how closely each sample can be measured

sampling rate number of samples per second taken from a continuous signal, especially by an analog-to-digital converter; the greater the rate, the greater the fidelity to the original signal

satellite in astronomy, any body orbiting a planet; in communications, a manmade object positioned in Earth orbit to facilitate communication on Earth

satellite constellation a group of communication satellites working together

sawing the use of a blade with sharp teeth to separate materials into parts or shapes

schematic diagram serving as a blueprint for electric and electronic circuits

science the systematic study of the natural world, including a body of knowledge about the natural world and a process of inquiry that generates such knowledge

scrap out-of-spec parts that cannot be used

screeding process of finishing concrete by scraping off excess

screw propulsion propulsion system using screw propeller with angled blades that rotate and push against the water to propel a ship

secondary fabrication the processing of industrial materials (from primary fabrication) into products

secondary food processing process of converting raw agricultural products into saleable food products and shipping them to stores

Second Law of Thermodynamics whenever an energy transformation occurs from one form to another, there is a net decrease in energy value

segment in Ethernet, a shared piece of network media; devices on a segment contend for that shared media, but do not share the media with other segments

selective breeding process of choosing animals or plants that have desirable characteristics and breeding them to produce a next generation having those characteristics

self-adjusting machinery machines having a feedback control loop that provides information on performance and operation used to control and adjust operation

semiconductor a material that is neither a good conductor of electricity nor a good insulator; the basis of transistors and integrated circuits

semisynchronous orbit satellite orbit that requires twelve hours to circle the Earth; used for all GPS satellites

sensors energy conversion devices distributed throughout machinery to take readings on performance and operation which are then fed back to the controller

sequential circuit digital circuitry in which the output level is dependent on existing input signals as well as on data previously stored in flip-flops or memory

series circuit an electrical circuit in which there is only one path for electrons to follow

Session Layer in OSI, the fifth layer, responsible for managing conversations, or sessions

setting the combination of circumstances in which developments occur

shape memory alloys an alloy that can have one shape at a lower temperature and another shape at a higher temperature

shear strength a measure of the stress (measured in psi or MPa) required to shear a material so that the sheared parts are totally separated

signal any changing and measurable electrical quantity such as voltage, current, or electric field

sintering a process in which powdered materials are combined with a binding material and pressed into shapes in a mold under heat and high pressure

small-scale manufacturing a small-scale research and development project to design and test a manufacturing process

smelting the extraction of metal is extracted from its source, or ore

society a broad grouping of people having common traditions, institutions, and collective activities and interests

soft automation automatic control, chiefly through the use of computer processing, with relatively little reliance on computer hardware

software any computer program

softwood primarily wood from coniferous trees (evergreens) or cypress trees

solar energy energy derived directly from the Sun

solid-state phase change a phase change in which the material remains solid, but its atomic structure is rearranged

solute the smaller volume of liquid in a solution, dissolved in the solvent

solvent the larger volume of liquid in a solution, in which the solute is dissolved

source in an N-channel MOSFET, an N-type area embedded into a larger P-type block and thus isolated from the drain; in communication systems, the originator of a message

Southbridge in some two-IC chip sets, controls the I/O devices supporting serial communications, USB, and other I/O operations

space elevator hypothetical system using a cable anchored to the ground, extending up to 22,000 miles above the Earth's surface, on which electrically powered vehicles would rise until they were in orbit

spark-ignition a type of internal combustion engine in which fuel is ignited by an electrically induced spark

special maintenance work that cannot be planned in advance, such as road repairs necessitated by accident or storm damage

specialty and fine chemicals chemical products such as pharmaceuticals and cosmetics

specialty trade contractors in construction industry, those who perform specialized activities related to construction trades, such as carpentry, painting, plumbing, and electrical work

specifications performance requirements that must be addressed by the solution to a design problem

speed with reference to computer processors, how quickly a read or write operation occurs in a memory chip

spot footing a single square pad used to support a pier or a post

spread footing most common type of enlargement at the base of a wall or column to spread the load, used on hard ground and normally built around the entire perimeter of a structure

spreadsheet software program used to computerize record keeping and budgeting

static RAM (SRAM) a type of semiconductor memory in which cell structures are flip-flops and memory does not need to be periodically refreshed

statistical process control (SPC) quality control system using random sampling and testing of a fraction of production output

steady state with respect to capacitors, the point at which a capacitor is either fully charged or fully discharged; with respect to current, a current that does not change over time.

steam engine an external combustion engine using high-pressure steam to power reciprocating pistons for mechanical energy

steam locomotive in a railroad, a source of power using a steam engine

steam reforming process in which high-temperature steam separates hydrogen from carbon atoms in methane (CH_4)

steam turbine engine in which hundreds of small blades in the path of jets of high-pressure steam rotate the propeller shaft

step up/step down the function of transformers, increasing or decreasing (respectively) voltage or current

stereolithography process of building a three-dimensional model from a 3-D graphic stored in a computer, using a liquid resin that hardens when exposed to an ultraviolet laser

stewardship careful and responsible management of the environment and resources

strength a measure of the amount of stress a material can withstand

structure-change processes material processing methods that affect the atomic structure of a material, including hardening, tempering, and annealing

subassembly line distinct part of an assembly line

subsidize to grant money for some purpose; generally used in reference to government funding, including support for agriculture

substrate silicon wafer acting as the foundation for an integrated circuit

substructure the part of a structure that is in the ground; *foundation*

sum the least significant bit in the output of a half adder

supercharging use of an air pump to force greater amounts of air into the cylinders of an engine, allowing greater amounts of fuel to be burned, and thus increasing the power potential of the engine; also called *turbocharging*

superstructure the part of a structure that is visible, built upon the substructure

support functions in framed superstructures, functions related to how the enclosure helps to add structural support to the building

surface hardening a process in which steel or iron is heated while its surface is in contact with a carbon-based material so that the carbon diffuses onto the surface of the steel (also called case hardening)

surface-mount components on some printed circuit boards, small devices soldered directly to the traces

surfactants chemicals that break up or dissolve grease and similar compounds

suspension bridge used to span long distances; the bridge deck is suspended from a main suspension cable made of many strands of steel wire; the load is transmitted through suspender roads or cables to towers and anchor blocks

sustainable refers to the use of resources in ways that prevent their being depleted or permanently damaged

sustainable design design approach that seeks to address environmental effects through conservation, regeneration, and stewardship

sustainable development construction projects that limit environmental damage, reduce energy use, and use recycled and recyclable construction materials

switches electromechanical devices used to turn current flow to a device or circuit on or off

synthetic materials materials that are human-made

technical communication use of graphic communication in engineering and technology, using graphs, graphics, and other visual tools

technology the means by which humans modify the world to address their needs and wants

telecommunications broad term referring to any form of transmission of messages at a distance

tempering process of heating hardened steel and soaking it at that temperature for a short time, then cooling it quickly

tensile strength a measure of the amount of tension (stretching) a material can withstand

test equipment devices used to test circuit performance and to troubleshoot, including power supply, digital multimeter (DMM), and oscilloscope

thermal conductivity the ability of a material to conduct heat

thermal energy energy created by the motion of atoms and molecules, occurring within all matter; also called *internal energy*

thermoplastic plastic that can be softened by heat, and then hardened again by cooling, which makes them recyclable

thermoset plastic material that cannot revert to its original form after heating

thymine one of the four amino acids that comprise DNA; denoted as T

tilling preparing the soil for planting

timber wood cut into standard, usable dimensions (*lumber*)

time division multiplexing multiplexing technique in which time is divided into slots and each conversation is assigned its own time slot

toggle switch a switch that varies between two states or values, such as on or off

tomography medical imaging by sections, as in MRI

tooling shaping, forming, or finishing with a tool

top-down manufacturing (TDM) typical manufacturing process in which materials are obtained by manufacturer and then modified by humans and machine into a working product

torsional strength a measure of the stress a material can withstand when twisted

Total Quality Management (TQM) an approach toward establishing quality in an organization based on the cooperation of all workers at all levels; developed by Dr. William Edwards Deming in Japan in the 1950s

toughness a measure of how well a material can survive a sudden impact without fracturing

traces thin copper lines on a printed circuit board that connect the ICs and devices on the board

trade-offs giving up one thing in return for another

transceiver a device that can act as both a transmitter and a receiver

transcription the process of translating DNA code into messenger RNA (mRNA)

transducer in ultrasound imaging, the device that both produces the sound wave and detects the reflection of the wave

transesterification reaction using methanol and triglycerides to produce biodiesel

transformer devices that use coils to step voltage or current levels up or down

transgenic animals animals produced by recombinant DNA techniques, inserting a gene from one animal into the egg of a different animal, which is then implanted into a surrogate mother

transgenic foods genetically modified food products, containing genetic material transferred from other organisms

transient state the time it takes for a given capacitor to undergo one cycle of charging and discharging

transistor the most common semiconductor device, used to control the flow of electricity in electronic devices

transistor-transistor logic (TTL) a logic family in which both the logic gating function (e.g., AND) and the amplifying function are performed by transistors

translation the process of decoding or expressing the information carried by mRNA

transmitter device in a communication system that sends a message

transportation system the various means of moving goods or people from place to place, including the supporting infrastructure such as roads, docks, railroads, and air corridors

Transport Layer in OSI, the fourth layer, responsible for taking messages, chopping them into segments, numbering them, and preparing them for transmission; the opposite occurs at the destination

transverse CT computer tomography technique providing a cross section of a body part

trend assessment evaluative technique based on iterative procedures that involve analyzing trade-offs, estimating risks, and choosing a best course of action

truss bridge constructed using a framework of struts connected together in a triangular grid, resulting in a rigid structure

triglycerides plant and animal fats formed from a single molecule of glycerol, combined with three fatty acids; constitute most of the fats digested by humans

truth table table showing every binary combination possible for the number of inputs in a combinatorial circuit

turbojet aircraft engine mixing fuel with compressed air inside a combustion, providing forward thrust by the exit of exhaust gasses through the exhaust nozzle

turning use of a lathe to make or finish a product

ultracapacitors energy storage devices that hold energy in an electric field rather than using chemical reactions as in batteries

universal engineers term applied to chemical engineers because the control of temperature, pressure, and time is common to the synthesis of all chemicals

Universal Serial Bus (USB) a standardized bus designed to allow many peripherals to be connected using a single standardized interface socket

up counters digital circuits that react to input signals by producing a binary counting sequence from zero to some maximum value

vacuum forming a method of forming a sheet of plastic in which the sheet is heated until it softens, draped over a model, and pulled tightly down over the model by vacuum

variable resistor a type of resistor with continually adjustable resistance values

variables factors that affect the performance of a design

Venn diagram a graphical representation of relationships using overlapping circles

visual inspection a primary quality-control measure, either under a microscope or by naked eye

visual thinking an enabling intelligence empowering one to see in different ways and thus make connections among varied perspectives

Voice over Internet Protocol (VoIP) general term for a family of transmission technologies for delivery of voice communications over networks such as the Internet

volatility used to describe a memory chip's ability to retain information once it is stored; *volatile* chips lose information as soon as the power is cut

voltage electrical pressure that pushes electrons through an electrical circuit; denoted as V

voltage drop the voltage developed across a component in an electrical circuit

voltage source the source of the energy to move electrons through an electrical circuit; often denoted as E (for Electromotive Force)

waferboard engineered wood made from wood ground into thin strands, mixed with wax and glue, layered, and pressed in a hot press (also called oriented strand board, OSB)

wafers thin disks of doped silicon sawed from cylinders of silicon crystal, which will be processed into integrated circuits or chips

water energy energy derived from flowing water by taking advantage of its gravitational potential energy, as by water wheels or enclosed hydraulic turbines

water jet propulsion system using an engine to drive a water pump, which forces water through a nozzle to propel a vessel

wave energy the capturable kinetic energy of coastal wave action

wave powered generators energy-generating system under development to capture the energy of coastal wave action

Wide Area Network (WAN) a network made up of interconnected LANs; may span multiple cities or states

wind energy energy derived from wind by means of windmills or wind turbines

word size the number of bits of information available at one time, and thus the amount of data used at any one time

work energy is used to create a force that causes an object to be moved over a distance; measured in joules in the International System of Units; in foot-pounds in the English system

write operation the placing of information into a memory chip

writing the act of storing information in a computer

xenotransplantation transplantation of organs from an animal of one species into an animal of a different species

X-ray crystallography technique for viewing chemical structure using X-rays beamed into crystals to determine where atoms are located in three dimensions

yield point the amount of stress applied to a material when it starts to deform plastically

Young's modulus a measure of elastic strength (*E*) found by dividing the stress applied to a material by the amount the material moves (the strain) when subjected to that stress

Figure 1.1 Image copyright ZTS, 2008. Used under license from Shutterstock.com

Figure 1.2 Image copyright Serg64, 2008. Used under license from Shutterstock.com

Figure 1.3 Image copyright Mark Kuipers, 2008. Used under license from Shutterstock.com

Figure 1.4 Image copyright Laurence Gough, 2008. Used under license from Shutterstock.com

Figure 1.5 Image copyright Christian Lagerek, 2008. Used under license from Shutterstock.com

Figure 1.7 NASA Headquarters—Greatest Images of NASA

Figure 1.10 Image copyright silver-john, 2008. Used under license from Shutterstock.com

Figure 1.12 Courtesy NASA/JPL-Caltech

Figure 1.13 Image copyright Norma Cornes, 2008. Used under license from Shutterstock.com

Figure 1.16 Image copyright Margaret Smeaton, 2008. Used under license from Shutterstock.com

Figure 1.20 Janet Browning/Hands Around the World

Figure 1.26 Image copyright Muriel Lasure, 2008. Used under license from Shutterstock.com

Figure 1.27 Image copyright Joe Gough, 2008. Used under license from Shutterstock.com

Figure 1.28 Image copyright Terry Underwood Evans, 2008. Used under license from Shutterstock.com

Figure 1.29 French School/The Bridgeman Art Library/Getty Images

Figure 1.30 Image copyright Sally Wallis, 2008. Used under license from Shutterstock.com

Figure 1.31 Science Museum Pictorial/Science & Society Picture Library

Figure 1.32 The Granger Collection, New York

Figure 1.35 Maciej Noskowski/iStockphoto.com

Figure 1.37 Hirz/Hulton Archive/Getty Images

Figure 1.38 Image copyright Thomas M. Perkins, 2008. Used under license from Shutterstock.com

Figure 1.39 Image copyright Thor Jorgen Udvang, 2008. Used under license from Shutterstock.com

Figure 1.40 Image copyright Tobias Machhaus, 2008. Used under license from Shutterstock.com

Figure 1.43 Image copyright kristian sekulic, 2008. Used under license from Shutterstock.com

Figure 1.44 Image copyright topal, 2008. Used under license from Shutterstock.com

Figure 1.45 Image copyright SPLAV, 2008. Used under license from Shutterstock.com

Figure 1.47 Vermont Castings

Figure 1.49 United States Patent and Trademark Office

Figure 1.50 Image copyright Scott Rothstein, 2008. Used under license from Shutterstock.com

Figure 1.51 Image copyright Jean Frooms, 2008. Used under license from Shutterstock.com

Figure 1.52 Image copyright imageshunter, 2008. Used under license from Shutterstock.com

Figure 1.53 NaMCATE National Science Foundation Project, Dr. David Shaw, Principal Investigator. Contact: dshaw@buffalo.edu

Figure 2.4 Library of Congress, Prints and Photographs Division, FSA/OWI Collection. LC-USF34-024352-D

Figure 2.5 Copyright, iofoto, 2008. Used under license from Shutterstock.com

Figure 2.5 Courtesy of Cold Springs Harbor Laboratory Archives

Figure 2.6 Courtesy Facebook

Figure 2.10 Copyright, Ingvald Kaldhaussater, 2008. Used under license from Shutterstock.com

Figure 2.15 Image copyright Monkey Business Images, 2008. Used under license from Shutterstock.com

Figure 2.16 Korhan Isik/iStockphoto.com

Figure 2.18 Image copyright Kutlayev Dmitry, 2008. Used under license from Shutterstock.com

Figure 2.20 NASA Langley Research Center (NASA-LaRC)

Figure 2.21 David Paul Morris/Getty Images News

Figure 2.25 Copyright, David H. Seymour, 2008. Used under license from Shutterstock.com

Figure 2.30 NASA Stennis Space Center (NASA-SSC)

Figure 2.32 Living Technologies Ltd., UK

Figure 2.34 Image copyright GagarinARTs, 2008. Used under license from Shutterstock.com

Figure 2.37 Courtesy of Bryn Mawr College

Figure 2.38 Copyright, MW Productions, 2008. Used under license from Shutterstock.com

Figure 2.39 Image copyright Dmitry Nikolaev, 2008. Used under license from Shutterstock.com

Figure 2.40 Habitat for Humanity International

Figure 2.41 Courtesy, Outward Bound

Figure 2.43 Courtesy of the Tournament of Roses Archives

Figure 2.44 Copyright, iofot, 2008. Used under license from Shutterstock.com

Figure 2.45 American Honda Motor Co., Inc.

Figure 2.46 Copyright, Steve Lovegrove, 2008. Used under license from Shutterstock.com

Figure 2.47 Copyright, Marc Dietrich, 2008. Used under license from Shutterstock.com

Figure 2.48 Copyright, Galina Barskaya, 2008. Used under license from Shutterstock.com

Figure 2.49 Wisconsin Historical Society, Image # 30423

Figure 3.1 Texas Transportation Institute

Figure 3.2 Courtesy Lockheed Martin Corporation

Figure 3.4 Special Thanks to LandRoller, Inc.

Figures 3.5–3.9 Special Thanks to LandRoller, Inc.

Figure 3.10 Image copyright Kasia, 2008. Used under license from Shutterstock.com

Figure 3.11 Narvikk/iStockphoto.com

Figure 3.12 Courtesy Ford Motor Company

Figure 3.14 3D Systems Corporation

Figure 3.15 Rapid Processing Solutions, Inc.

Figure 3.17 The camera model was created using SolidWorks® software; © 2008 SolidWorks Coporation

Figure 3.18 Esteban Cruz García

Figure 3.19 Render Courtesy of Tom van Ryn

Figure 3.20 Fernando Luis Curutchet—University of Buenos Aires Student

Figure 3.21 Courtesy of International Business Machines Corporation. Unauthorized use not permitted

Figure 3.42 Copyright: Andrey Kozachenko 2008. Used under license from Shutterstock.com

Figure 3.43 Honeywell, UDC Universal Digital Controller

Figure 3.44 CBS Photo Archive/Hulton Archive/Getty Images

Figure 3.46 4loops/iStockphoto.com

Figure 3.47 WestfaliaSurge—Autorotor Magnum 90 Parlor

Figure 3.52 Copyright Google Inc., 2008. Reprinted with permission.

Figure 3.53 Peter Kramer/Getty Images Entertainment

Figure 4.1 Courtesy, John Bonevich, Ph.D., National Institute of Standards and Technology

Figure 4.2 Young Endeavour Youth Scheme

Figure 4.10 © Egan Visual Inc. George Gooderham 2008

Figure 4.12 Courtesy, Musical Forests Inc. www.tonewood.ca

Figure 4.13 Copyright Sabri Deniz Kizil. Used under license from Shutterstock.com.

Figure 4.16 Courtesy of Zimmer, Inc.

Figure 4.18 © DoITPoMS Micrograph Library, University of Cambridge—http://www.doitpoms.ac.uk/miclib/micrograph.php?id=712

Figure 4.22 Courtesy, Gina L. Fiore

Figure 4.23 Courtesy of S&R Photo Acquisitions, LLC

Figure 4.24 Courtesy of S&R Photo Acquisitions, LLC

Figure 4.25 Courtesy of S&R Photo Acquisitions, LLC

Figure 4.26 Courtesy of S&R Photo Acquisitions, LLC

Figure 4.27 Copyright Smit. Used under license from Shutterstock.com.

Figure 4.30 Copyright Claudio Zaccherini. Used under license from Shutterstock.com.

Figure 4.31 Courtesy of KYOCERA Corporation

Figure 4.32 Copyright EML. Used under license from Shutterstock.com.

Figure 4.33 Copyright stocksnapp. Used under license from Shutterstock.com.

Figure 4.34 Copyright, Mechanik, 2008,. Used under license from Shutterstock.com

Figure 4.36 Courtesy, Kolcums

Figure 4.37 Copyright Tatiana Sayig. Used under license from Shutterstock.com.

Figure 4.43 United States Steel Corporation

Figure 4.44 Courtesy of Instron®

Figure 4.45 Courtesy of the Council for Scientific and Industrial Research—CSIR—in South Africa

Figure 4.46 Courtesy of TRUMPF INC.

Figure 4.47 Courtesy of GF AgieCharmilles

Figure 4.48 Copyright Tihis. Used under license from Shutterstock.com.

Figure 4.49 Courtesy of JET ®

Figure 4.50 Courtesy of JET ®

Figure 4.51 Courtesy of POWERMATIC ®

Figure 4.52 Courtesy of JET ®

Figure 4.53 © Economy Industrial Corp. Used by permission.

Figure 4.54 Courtesy of Owens Illinois, Inc.

Figure 4.55 Courtesy of Ilika Technologies Ltd.

Figure 4.56 Courtesy of the Fab@Home Project (http://www.fabathome.org); photo by Floris van Breugel

Figure 4.57 The Fab@Home Project, Cornell University

Figure 4.60 Copyright ALXR. Used under license from Shutterstock.com.

Figure 4.62 Copyright Sagasan. Used under license from Shutterstock.com.

Figure 4.65 Copyright Pr2is. Used under license from Shutterstock.com.

Figure 4.67 Courtesy of Greenerd Press & Machine Company, Inc.

Figure 4.71 Courtesy of Ford Motor Company

Figure 4.73 Courtesy of the Smithsonian Institution, NMAH/Transportation

Figure 5.1 Courtesy Shaffer Technical Editing, LLC

Figure 5.3 Image copyright Glen Jones, 2008. Used under license from Shutterstock.com

Figure 5.4 Image copyright Péter Gudella, 2008. Used under license from Shutterstock.com

Figure 5.6 Image copyright Smith&Smith, 2008. Used under license from Shutterstock.com

Figure 5.11 © Fotolia

Figure 5.16 Library of Congress/Brady-Handy Photograph Collection

Figure 5.18 ASI Consulting Group, LLC, www.asiusa.com, 248-530-1395

Figure 5.20 Image copyright Mark William Richardson, 2008. Used under license from Shutterstock.com

Figure 5.21 David H. Lewis/iStockphoto.com

Figure 5.22 National Archives, Records of the U.S. Information Agency

Figure 5.23 Image copyright Baloncici, 2008. Used under license from Shutterstock.com

Figure 5.25 Haas Automation, Inc.

Figure 5.30 Jervis B. Webb Company

Figure 5.31 Center for Rapid Automated Fabrication Technologies (CRAFT), University of Southern California

Figure 5.32 Center for Rapid Automated Fabrication Technologies (CRAFT), University of Southern California

Figure 5.35 Stratasys, Inc.

Figure 5.38 Paul Gardner/iStockphoto.com

Figure 5.39 Image copyright Mechanik, 2008. Used under license from Shutterstock.com

Figure 5.44 Getty Images News

Figure 5.45 Courtesy of the William J. Clinton Presidential Library

Figure 5.46 Emrah Turudu/iStockphoto.com

Figure 6.1 Silverstein Properties, Inc.

Figure 6.2 Copyright, Ben Hays, 2008. Used under license from Shutterstock.com

Figure 6.3 Image copyright Philip Lange, 2008. Used under license from Shutterstock.com

Figure 6.4 USAID

Figure 6.5 Image copyright Elena Elisseeva, 2008. Used under license from Shutterstock.com

Figure 6.6 Foster + Partners

Figure 6.8 This photo is provided by Jeff Christian, director of the Buildings Technology Center, a US Department of Energy User Facility, housed at the Oak Ridge National Laboratory.

Figure 6.11 Courtesy Brenda Hacker

Figure 6.12 Copyright, TAOLMOR, 2008. Used under license from Shutterstock.com

Figure 6.15 Energy Information Administration

Figure 6.16 Image copyright Wade H. Massie, 2008. Used under license from Shutterstock.com

Figure 6.17 Courtesy Shai Hacker

Figure 6.18 Courtesy U.S. Department of Transportation Federal Highway Administration

Figure 6.19 Image copyright Christian Lagerek, 2008. Used under license from Shutterstock.com

Figure 6.20 Image copyright prism_68, 2008. Used under license from Shutterstock.com

Figure 6.22 Image copyright Jan van der Hoeven, 2008. Used under license from Shutterstock.com

Figure 6.24 Courtesy Brenda Hacker

Figure 6.25 Photo courtesy of Shai Hacker

Figure 6.26 Photo courtesy of Shai Hacker

Figure 6.28 Image copyright Jan van der Hoeven, 2008. Used under license from Shutterstock.com

Figure 6.30 Courtesy Shai Hacker

Figure 6.32 Courtesy, Michael Moy, photographer, and, Roads and Traffic Authority

Figure 6.33 Alberto Pomares/iStockphoto.com

Figure 6.35 Copyright, Claus Mikosch, 2008. Used under license from Shutterstock.com

Figure 6.36 Collection of the Archives, New York City Department of Environmental Protection

Figure 6.37 Courtesy Brenda Hacker

Figure 6.39 ChinaFotoPress/Getty Images News

Figure 6.43 Courtesy Shai Hacker

Figure 6.44 Courtesy Shai Hacker

Figure 6.45 Courtesy Shai Hacker

Figure 6.46 Courtesy Shai Hacker

Figure 6.47 Courtesy Brenda Hacker

Figure 6.49 Courtesy Shai Hacker

Figure 6.50 Copyright, Kristin Smith, 2008. Used under license from Shutterstock.com

Figure 6.52 Photo courtesy of Minnesota Department of Transportation

Figure 6.53 Copyright 2005, California Department of Transportation

Figure 7.10 Image copyright V.J. Matthew, 2008. Used under license from Shutterstock.com

Figure 7.14 Image copyright hfng, 2008. Used under license from Shutterstock.com

Figure 7.16 James Pharaon/iStockphoto.com

Figure 7.21 Courtesy Ivo Coelho

Figure 7.22 Courtesy M. David Burghardt

Figure 7.23 Image copyright Zacarias Pereira da Mata, 2008. Used under license from Shutterstock.com

Figure 7.26 Malcolm Romain/iStockphoto.com

Figure 11.10 Courtesy of Dr. Martin Eppler and Visual-Literacy.org/University of Lugano (USI)

Figure 11.11 Courtesy of http://newsmap.marumushi.com

Figure 11.12 Courtesy of Panopticon Software (www.panopticon.com)

Figure 11.13 Courtesy of Panopticon Software (www.panopticon.com)

Figure 11.15 Associated Press

Figure 11.16 Courtesy of www.gapminder.org

Figure 11.24 Eric and Edith Matson Photograph Collection Library of Congress Prints and Photographs Division Washington, D.C

Figure 11.29 Olivier Le Queinec/Shutterstock

Figure 11.37 Courtesy of NASA

Figure 11.38 Courtesy of NASA

Figure 11.39 Courtesy of Space Systems/Loral

Figure 11.40 Courtesy of AT&T Archives and History Center

Figure 11.44 Courtesy of Thuraya Satellite Telecommunications

Figure 11.48 © Stephen Jones/iStockphoto

Figure 11.50A © Murat Koc

Figure 11.50B © Jakub Semeniuk

Figure 12.2 Image copyright Igor Karon, 2009. Used under license from Shutterstock.com

Figure 12.3 IBM Archives

Figure 12.4a Courtesy of Nokia

Figure 12.4b PR Newswire

Figure 12.4c PRNewsFoto/Verizon Wireless

Figure 12.5 Courtesy of Dave Dunfield

Figure 12.6 IBM Archives

Figure 12.7 Courtesy of BBN Technologies

Figure 12.8 PR Newswire

Figure 12.10 Courtesy of Jesse James Garrett

Figure 12.13 Courtesy of Robert Metcalfe and Polaris Ventures

Figure 12.15 © Randy Plett/iStockphoto

Figure 12.19 Courtesy of 3Com Corporation

Figure 12.27 Image copyright KzlKurt , 2009. Used under license from Shutterstock.com

Figure 12.34a Image copyright Pablo Eder, 2009. Used under license from Shutterstock.com

Figure 12.34b © Dragan Stankovic/iStockphoto

Figure 12.39 Courtesy of 3Com Corporation

Figure 12.41 Courtesy of 3Com Corporation

Figure 12.49 Image copyright Lim Yong Hian, 2009. Used under license from Shutterstock.com

Figure 13.1 Image copyright Marie C. Fields, 2008. Used under license from Shutterstock.com

Figure 13.2A Barry Crossley/iStockphoto.com

Figure 13.2B Benoit Rousseau/iStockphoto.com

Figure 13.2C Susaro/iStockphoto.com

Figure 13.2D Yvonne Chamberlain/iStockphoto.com

Figure 13.3 Anna Kukekova

Figure 13.7 Fritz Goro/Time & Life Pictures—Getty Images

Figure 13.14 Source Unknown: Asilomar Conference 1975

Figure 13.17 Courtesy of ARS/USDA #K5764-16

Figure 13.26 Image copyright Istvan Csak, 2008. Used under license from Shutterstock.com

Figure 13.27 Courtesy of the Exxon Valdez Oil Spill Trustee Council

Figure 13.28 Courtesy of DOE/NREL—Warren Gretz

Figure 13.29 Courtesy of the U.S. Department of Energy

Figure 13.31 Image copyright Brett Atkins, 2008. Used under license from Shutterstock.com

Figure 13.32A Ken Wilson-Max/Alamy

Figure 13.32B Image copyright iCEO, 2008. Used under license from Shutterstock.com

Figure 13.33 Image copyright Niels Quist, 2008. Used under license from Shutterstock.com

Figure 13.34 Image copyright Kenetic Imagery, 2008. Used under license from Shutterstock.com

Figure 13.35 www.glofish.com

Figure 13.39 Courtesy Linnea Fletcher

Figure 14.5 Courtesy United States Department of Labor, Occupational Safety and Health Administration

Figure 14.6 Courtesy US EPA-RTP

Figure 14.9 Image copyright Rick's Photography, 2008. Used under license from Shutterstock.com

Figure 14.10a Image copyright androfroll, 2008. Used under license from Shutterstock.com

Figure 14.15 Image copyright Freerk Brouwer, 2008. Used under license from Shutterstock.com

Figure 14.17 ©creatingmore/iStockphoto.com

Figure 14.19 Courtesy of www.gogreenstarfish.com

Figure 14.22 © Mary Evans Picture Library/Alamy

Figure 14.24 Time & Life Pictures/Getty Images

Figure 14.25 Time & Life Pictures/Getty Images

Figure 14.26 Keystone/Hulton Archive/Getty Images

Figure 14.34 Image copyright Hugo de Wolf, 2008. Used under license from Shutterstock.com

Figure 14.35 Energy Information Administration

Figure 14.36 Image copyright Sally Wallis, 2008. Used under license from Shutterstock.com

Figure 14.38 Energy Kid's Page, Energy Information Aministration

Figure 14.40 Gene Chutka/iStockphoto.com

Figure 14.41 Image copyright Jim Parkin, 2008. Used under license from Shutterstock.com

Figure 15.2 Sasha Radosavljevic/iStockphoto.com

Figure 15.3 Loomis Dean//Time Life Pictures/Getty Images

Figure 15.4 Image copyright Floridastock, 2008. Used under license from Shutterstock.com

Figure 15.6 From COOPER. Agriscience Fundamentals and Applications, 4e. © 2007 Delmar Learning, a part of Cengage Learning, Inc. Reproduced with permission. www.cengage.com/permissions.

Figure 15.7 Lisa Thornberg/iStockphoto.com

Figure 15.8 Image copyright Orientaly, 2008. Used under license from Shutterstock.com

Figure 15.9 MPI/Stringer/Hulton Archive/Getty Images

Figure 15.11A Image copyright TJUK, 2008. Used under license from Shutterstock.com

Figure 15.11B From COOPER. Agriscience Fundamentals and Applications, 4e. © 2007 Delmar Learning, a part of Cengage Learning, Inc. Reproduced with permission. www.cengage.com/permissions.

Figure 15.11C Image copyright Mark Gabrenya, 2008. Used under license from Shutterstock.com

Figure 15.11D Christoopher Steer/iStockphoto.com

Figure 15.12A Image copyright Lagui, 2008. Used under license from Shutterstock.com

Figure 15.12B Image copyright Basil Tippette, 2008. Used under license from Shutterstock.com

Figure 15.12C Image copyright Roger Dale Pleis, 2008. Used under license from Shutterstock.com

Figure 15.12D Image copyright Mike Donenfeld, 2008. Used under license from Shutterstock.com

Figure 15.12E Image copyright Jack Cronkhite, 2008. Used under license from Shutterstock.com

Figure 15.13A Courtesy of Bill Muir, Purdue University

Figure 15.13B Courtesy USDA

Figure 15.14 From COOPER. Agriscience Fundamentals and Applications, 4e. © 2007 Delmar Learning, a part of Cengage Learning, Inc. Reproduced with permission. www.cengage.com/permissions.

Figure 15.15 Image copyright Beth Van Trees, 2008. Used under license from Shutterstock.com

Figure 15.17B USDA/ARS #K-5287-4

Figure 15.19B © Maciej Noskowski/iStockphoto.com

Figure 15.21 Klemens Wolf/iStockphoto.com

Figure 15.22 Image copyright Marco Alegria, 2008. Used under license from Shutterstock.com

Figure 15.23 Willi Schmitz/iStockphoto.com

Figure 15.25 Image copyright Tony Wear, 2008. Used under license from Shutterstock.com

Figure 15.27 Copyright 2008 Monsanto Company

Figure 15.28 Image copyright Igor Burchenkov, 2008. Used under license from Shutterstock.com

Figure 15.29 Photo courtesy of Montana State University Extension Water Quality Program

Figure 15.30 Courtesy of Farm Sanctuary

Figure 15.31 From COOPER. Agriscience Fundamentals and Applications, 4e. © 2007 Delmar Learning, a part of Cengage Learning, Inc. Reproduced with permission. www.cengage.com/permissions. Adapted from material provided by Electric Power Research Institute.

Figure 15.32A Image copyright Samuel Acosta, 2008. Used under license from Shutterstock.com

Figure 15.32B Image copyright Karol Kozlowski, 2008. Used under license from Shutterstock.com

Figure 15.33 Image copyright David Pruter, 2008. Used under license from Shutterstock.com

Figure 15.34 Image copyright Florin C, 2008. Used under license from Shutterstock.com

Figure 15.35 Image copyright Trinh Le Nguyen. Used under license from Shutterstock.com

Figure 15.36 Image copyright Sharon Kingston, 2008. Used under license from Shutterstock.com

Figure 15.37 City of Phoenix Water Services Department

Figure 15.38 Mark Linnard/iStockphoto.com

Figure 15.39 Image copyright Vinicius Tupinamba, 2008. Used under license from Shutterstock.com

Figure 15.40 Image copyright Francesco Ridolfi, 2008. Used under license from Shutterstock.com

Figure 15.42 Matthew Cole/iStockphoto.com

Figure 15.43 Courtesy Kristen Hughes

Figure 16.1 Image copyright Lilac Mountain, 2008. Used under license from Shutterstock.com

Figure 16.2 Image copyright Antonis Papantoniou, 2008. Used under license from Shutterstock.com

Figure 16.7 Copyright Michael Ledray, 2008. Used under license from Shutterstock.com

Figure 16.8 Image copyright Konstantin Shevtsov, 2008. Used under license from Shutterstock.com

Figure 16.8 Image copyright Beerkoff, 2008. Used under license from Shutterstock.com

Figure 16.9 Copyright Touch Bionics

Figure 16.10 Courtesy of the University of Minnesota

Figure 16.11 Courtesy of the University of Minnesota

Figure 16.12 Image copyright Rob Byron, 2008. Used under license from Shutterstock.com

Figure 16.13 Image copyright emin kuliyev, 2008. Used under license from Shutterstock.com

Figure 16.14 Image copyright sgame, 2008. Used under license from Shutterstock.com

Figure 16.16B Image copyright Damian Palus, 2008. Used under license from Shutterstock.com

Figure 16.17 Image copyright Steve Reed, 2008. Used under license from Shutterstock.com

Figure 16.20 Courtesy of Paul Thompson, Laboratory of Neuro Imaging, UCLA School of Medicine

Figure 16.21B Courtesy of Brookhaven National Laboratory

Figure 16.22A Image copyright Simon Pedersen, 2008. Used under license from Shutterstock.com

Figure 16.22B Image copyright Zsolt Nyulaszi, 2008. Used under license from Shutterstock.com

Figure 16.23 Image copyright Floris Slooff, 2008. Used under license from Shutterstock.com

Figure 16.24 PETER PARKS/AFP/Getty Images

Figure 16.25 A, B, & C Courtesy Independence Technology

Figure 16.26 © Slobo Ritic/iStockphoto.com

Figure 16.27 Courtesy of NASA

Figure 16.28 Teledoc is trademarked by Texas Tech University Health Sciences Center

Figure 16.29 Purdue News, Purdue University

Figure 16.34 A and B Courtesy of U.S. Department of Energy Artificial Retina Project. http://artificialretina.energy.gov/.

Figure 16.35 Courtesy of Purdue University

Figure 16.36 © Rich Legg/iStockphoto.com

Figure 17.1 John Woodworth/iStockphoto.com

Figure 17.3 Robert Churchill/iStockphoto.com

Figure 17.6 Copyright Nancy Ostertag/Getty Images

Figure 17.8 Courtesy Wilson Sporting Goods

Figure 17.11 Image copyright Nobor, 2008. Used under license from Shutterstock.com

Figure 17.13 Courtesy of the Cancer Genome Atlas

Figure 17.14 Genome Management Information System, Oak Ridge National Laboratory

Figure 17.17 Courtesy of Lawrence Berkeley National Laboratory. Roy Kaltschmidt, photographer

Figure 17.20 Courtesy of Stratasys, Inc.

Figure 17.22 Image copyright Tonis Valing, 2008. Used under license from Shutterstock.com

Figure 17.23 Courtesy of American Honda Motor Co., Inc.

Figure 17.24 Courtesy of GM and Carnegie Mellon University/Tartan Racing

Figure 17.26 VWU Photographic Services

Figure 17.28 Image copyright Gelpi, 2008. Used under license from Shutterstock.com

Figure 17.31 Picture provided by Ocean Power Technologies, Inc.

Figure 17.33 Courtesy DOE/NREL

Figure 17.34 Georgia Institute of Technology

Figure 17.36 Courtesy of Destron Fearing

Figure 17.37 MIGUEL ROJO/AFP/Getty Images

INDEX

B

Babbage, Charles, 664
Back end of the line (BEOL), 367
Bacteria
  plastics, 548–549
  recombinant, 505
Bang-bang controls, 109
Bar graphs, 106–107
Barrettes, 246
Batteryless flashlights, 359
Battery-powered electric vehicle (BEV),
  329, 336–337
Beam bridges, 239
Bearing wall superstructures, 246
Bell, Alexander Graham, 432
Belyaev, Dmitri, 499–500
Benchmarking criteria decision
  analysis, design, 70–71
Berg, Paul, 505
Bicycles, 325
Binary numbers, 379–381
Biocompatibility tests design
  challenge, 652–653
Biodiesel, 328
Biodiversity, 24, 594–595
Bioethics, 526
Biofuels, 327–328
Bioinformatics, 508
Biological scientists, careers, 678–680
Biomass, 284, 327–328, 562
Biomimicry, 518–520
Bioplastics, 174
Bioremediation, 516–517, 534–535
Biotechnology
  bioethics, 526
  biological scientists, careers, 678–680
  biomimicry, 518–520
  bioremediation, 516–517, 534–535
  companies, 521–528
  current good manufacturing
    practices (cGMPs), 523–524
  cystic fibrosis (CF), 514–516
  DNA technology, 502–510, 511–516,
    517–518
  domestication, plant and animal,
    499–500
  energy applications, 517
  fermentation, 500–502

  future of, 662–664
  metric conversions, 524–525
  overview, 496–498
  product pipeline, 523
  regulation, 523–524
  solutions, preparing, 525
  technician, careers, 528–530
Bipolar junction transistors (BJT),
  362–363
Bits, 379–381
Bitumen, 237
BJT (bipolar junction transistors),
  362–363
Block diagrams, 348, 564–566
Blow molding, 167
Boeing 787, 320
Boiling point, 149
*The Book of Knowledge of Ingenious
  Mechanical Devices*, 34
Booting, defined, 389
Bottom-up manufacturing (BUM), 208
Boyer, Herb, 504–505
Brainstorming design solutions, 54, 58
Breeding, selective, 591
Bridges, 238–241, 253–254, 255, 260–261
Bridges, networks, 479
Brin, Sergey, 118
Brittleness, 154
Britton, Dr. Larry, 527
Bronze Age technology, 13–15
Buckyballs, 37–38
Buildings. *See also* Construction;
  Structures
  codes, 234–235
  types, 230–234
Bulk carriers, 308
Bus
  address, 403
  data, 398, 401–402
  peripheral components interface
    (PCI), 405
Buses, transportation, 316–317

C

Cable-stayed bridges, 240–241
Cache memory, 391
CAD (computer-aided design),
  98–99, 199

Camcorders, digital, 446
Cameras, digital, 446
Canals, transportation, 301
Cantilever bridges, 241
Capacitors, 356–358
Capacity, memory, 393–394
Capek, Karel, 668
Carbon nanotubes, 131, 660–661
Carcinogen, defined, 563
Careers
  aerospace manufacturing, 214–216
  agricultural scientist, 607–609
  airline pilot, 332–334
  automotive service technician,
    79–80
  biological scientist, 678–680
  biotechnology technician, 528–530
  chemists, 566–568
  computer engineer, 408–410
  computer professional, 122–123
  computer support specialist,
    489–490
  computer systems administrator,
    489–490
  construction management, 255–257
  electrical engineer, 408–410
  engineering technician, 369–370
  environmental scientist, 40–41
  food scientist, 607–609
  materials engineering, 175–176
  materials scientist, 566–568
  medical technician, 646–647
  power-plant operator, 291
  telecommunications, 447–450
Carnot Cycle, 277, 288
Carrier Sense, Ethernet, 465
Carrier waves, 436
Carson, Rachel, 578
Case hardening, 168
Cassava, 601
Casting, 163–167
CAT (Computer Axial Tomography)
  scan, 628
Cells, memory, 387
Cellular manufacturing, 188–189
Cellulosic ethanol, 327
Central processing unit (CPU), 396–399
Ceramics, 144–145, 146–147

Systems administrators, computers, careers, 489–490

T

Tall buildings, 231–232
Tankers, 308
Taylor, Dr. Doris, 621–622
TCP/IP, 460, 471
Teamwork, design, 71–75
Technical communication, 421
Technical decision analysis, design, 68–71
Technology
　Al-Jazari, Ibn Ismail al-Razzaz, 33–34
　benefits, 7–8
　biotechnology, future of, 662–664
　Bronze Age, 13–15
　communications, future of, 664–666
　consequences, 111–117
　defined, 3, 5–6
　economics, 28–31
　environment, 34–41
　globalization, 656–659
　history, 13–24
　Industrial Revolution, 18–21
　information quality, 112
　Iron Age, 15–17
　Middle Ages, 17, 33–34
　nanotechnology, 36–39, 67
　nanotechnology, future of, 659–662
　negative consequences, 7–8
　patents, 31–33
　prehistory, 9–12
　problem solving, 7
　relationship with science, 6–7
　Renaissance, 17–18
　society, 26–34
　Stone Age, 9–12
　Twentieth-Century, 21–24
Telecommunications, 429–437, 447–450, 472–475
Telemedicine, 636–637
Telephones, 432–433, 472–475
Temperature. *See also* Heat
　calculating heat, 269–271, 281
　materials, 149, 156–157

Tempering, 168
Tensile strength, materials, 150–152
Test equipment, electricity, 355–356
Testing
　designs, 61–62
　materials, design challenge, 180–181
Therapeutics, medical technology, 632–633
Thermal
　conductivity, 156–157, 270
　energy, defined, 267
　properties, materials, 156–157
Thermodynamics
　First Law, 274–276
　Second Law, 277
Thermoplastic, 143
Thermoset, 143
Thymine, 502
Timber, 135
Time division multiplexing, 436
Time lines, 107
Tissue culture design challenge, 612–613
Tomography, 621, 627–629, 630–631
Tooling, 17
Top-down manufacturing (TDM), 208
Topology, networks, 476–477
Torsional strength, materials, 152
Total Quality Management (TQM), 100, 192–194
Toughness, 154
TQM (Total Quality Management), 100, 192–194
Traffic, opening roads to, 237
Trains, 311–313
Tranceivers, 434
Transcription, 503
Transesterification, 563
Transformers, 359
Transient state, 357–358
Transistors, 362–364, 374–375
Translation, 503
Transmittance, light, 158
Transplantations, 621, 641–642
Transportation
　air, 317–321
　air cushion vehicles, 324

　airline pilot career, 332–334
　electric vehicles, 328–330, 662
　environmental impact, 305–306
　freight, 303–305
　global trade, 302–303
　history, 300–302
　local movement systems, 324
　oil dependency, 326
　overview, 298–300
　passenger travel, 303
　Personal Rapid Transport (PRT), 670
　pipelines, 323–324
　planned communities, 671–672
　rail, 311–313
　roads, 314–317
　space, 321–323, 670–671
　steam engine, 302
　water, 307–311
Transport layer, OSI model, 468–469
Treadmills, pesticide and fertilizer, 595
Treemaps, 423–424
Trend analysis, 115–117
Trendanalyzer, 428
Trends, manufacturing, 189–190
Trevithick, Richard, 302
Trucks, freight, 316
Truss bridges, 241
Truth tables, 383
TTL (transistor-transistor logic), 378–379
Tufte, Edward, 425
Tunnels, 241–243
Turbocharging engines, 287
Turning, 163
Twentieth-Century technology, 21–24
Typewriters, 27–28

U

Ultrasound, 631–632
Underwater tunnels, 242–243
Universal Building Design symbols, 103
Universal Serial Bus (USB), 405–406
Up counters, 383–384
Uranium, 282
USB (Universal Serial Bus), 405–406